Selected Chapters from Fundamentals of Logic Design

Charles H. Roth, Jr. / Larry L. Kinney

CENGAGE Learning®

Australia • Brazil • Japan • Korea • Mexico • Singapore • Spain • United Kingdom • United States

CENGAGE Learning

Selected Chapters From Fundamentals of Logic Design

Senior Project Development Manager:
Linda deStefano

Market Development Manager:
Heather Kramer

Senior Production/
Manufacturing Manager:
Donna M. Brown

Production Editorial Manager:
Kim Fry

Sr. Rights Acquisition Account Manager:
Todd Osborne

Printed in the United States of America

ISBN-13: 978-1-285-90324-8

ISBN-10: 1-285-90324-2

Cengage Learning
5191 Natorp Boulevard
Mason, Ohio 45040
USA

Cengage Learning is a leading provider of customized learning solutions with office locations around the globe, including Singapore, the United Kingdom, Australia, Mexico, Brazil, and Japan. Locate your local office at: **international.cengage.com/region.**
Cengage Learning products are represented in Canada by Nelson Education, Ltd.
For your lifelong learning solutions, visit **www.cengage.com /custom.**
Visit our corporate website at **www.cengage.com.**

Custom Contents

(* Note – this is a "custom" textbook that has been designed specifically for this course in a joint effort between you instructor and the publisher. Please note that some chapters have been removed intentionally.)

From Fundamentals of Logic Design

Preface

Purpose of the Text

This text is written for a first course in the logic design of digital systems. It is written on the premise that the student should understand and learn thoroughly certain fundamental concepts in a first course. Examples of such fundamental concepts are the use of Boolean algebra to describe the signals and interconnections in a logic circuit, use of systematic techniques for simplification of a logic circuit, interconnection of simple components to perform a more complex logic function, analysis of a sequential logic circuit in terms of timing charts or state graphs, and use of a control circuit to control the sequence of events in a digital system.

The text attempts to achieve a balance between theory and application. For this reason, the text does not overemphasize the mathematics of switching theory; however, it does present the theory that is necessary for understanding the fundamental concepts of logic design. After completing this text, the student should be prepared for a more advanced digital systems design course that stresses more intuitive concepts like the development of algorithms for digital processes, partitioning of digital systems into subsystems, and implementation of digital systems using currently available hardware. Alternatively, the student should be prepared to go on to a more advanced course in switching theory that further develops the theoretical concepts that have been introduced here.

Contents of the Text

After studying this text, students should be able to apply switching theory to the solution of logic design problems. They will learn both the basic theory of switching circuits and how to apply it. After a brief introduction to number systems, they will study switching algebra, a special case of Boolean algebra, which is the basic mathematical tool needed to analyze and synthesize an important class of switching circuits. Starting from a problem statement, they will learn to design circuits of logic gates that have a specified relationship between signals at the input and output terminals. Then they will study the logical properties of flip-flops, which serve as memory devices in sequential switching circuits. By combining flip-flops with circuits of logic gates, they will learn to design counters, adders, sequence detectors, and similar circuits. They will also study the VHDL hardware description language and its application to the design of combinational logic, sequential logic, and simple digital systems.

As integrated circuit technology continues to improve to allow more components on a chip, digital systems continue to grow in complexity. Design of such complex systems is facilitated by the use of a hardware description language such as VHDL. This text introduces the use of VHDL in logic design and emphasizes the relationship between VHDL statements and the corresponding digital hardware. VHDL allows digital hardware to be described and simulated at a higher level before it is implemented with logic components. Computer programs for synthesis can convert a VHDL description of a digital system to a corresponding set of logic components and their interconnections. Even though use of such computer-aided design tools helps to automate the logic design process, we believe that it is important to understand the underlying logic components and their timing before writing VHDL code. By first implementing the digital logic manually, students can more fully appreciate the power and limitations of VHDL.

Although the technology used to implement digital systems has changed significantly since the first edition of this text was published, the fundamental principles of logic design have not. Truth tables and state tables still are used to specify the behavior of logic circuits, and Boolean algebra is still a basic mathematical tool for logic design. Even when programmable logic devices (PLDs) are used instead of individual gates and flip-flops, reduction of logic equations is still desirable in order to fit the equations into smaller PLDs. Making a good state assignment is still desirable, because without a good assignment, the logic equations may require larger PLDs.

Strengths of the Text

Although many texts are available in the areas of switching theory and logic design, this text is designed so that it can be used in either a standard lecture course or in a self-paced course. In addition to the standard reading material and problems, study guides and other aids for self-study are included in the text. The content of the text is divided into 20 study units. These units form a logical sequence so that mastery of the material in one unit is generally a prerequisite to the study of succeeding units. Each unit consists of four parts. First, a list of objectives states precisely what you are expected to learn by studying the unit. Next, the study guide contains reading assignments and study questions. As you work through the unit, you should write out the answers to these study questions. The text material and problem set that follow are similar to a conventional textbook. When you complete a unit, you should review the objectives and make sure that you have met them. Each of the units has undergone extensive class testing in a self-paced environment and has been revised based on student feedback.

The study units are divided into three main groups. The first 9 units treat Boolean algebra and the design of combinational logic circuits. Units 11 through 16, 18 and 19 are mainly concerned with the analysis and design of clocked sequential logic circuits, including circuits for arithmetic operations. Units 10, 17, and 20 introduce the VHDL hardware description language and its application to logic design.

The text is suitable for both computer science and engineering students. Material relating to circuit aspects of logic gates is contained in Appendix A so that this material can conveniently be omitted by computer science students or other students with no background in electronic circuits. The text is organized so that Unit 6 on the Quine-McCluskey procedure may be omitted without loss of continuity. The three units on VHDL can be studied in the normal sequence, studied together after the other units, or omitted entirely.

Supplements and Resources

This book comes with support materials for both the instructor and the student. The supplements are housed on the book's companion website. To access the additional course materials, please visit www.cengagebrain.com. At the cengagebrain.com home page, search for the ISBN of your title (from the back cover of your book) using the search box at the top of the page. This will take you to the product page where these resources can be found.

Instructor Resources

An instructor's solution manual (ISM) is available that includes suggestions for using the text in a standard or self-paced course, quizzes on each of the units, and suggestions for laboratory equipment and procedures. The instructor's manual also contains solutions to problems, to unit quizzes, and to lab exercises.

The ISM is available in both print and digital formats. The digital version is available to registered instructors at the publisher's website. This website also includes both a full set of PowerPoint slides containing all graphical images and tables in the text, and a set of Lecture Builder PowerPoint slides of all equations and example problems.

Student Resources

Since the computer plays an important role in the logic design process, integration of computer usage into the first logic design course is very important. A computer-aided logic design program, called *LogicAid,* is included on the CD that accompanies this text. *LogicAid* allows the student to easily derive simplified logic equations from minterms, truth tables, and state tables. This relieves the student of some of the more tedious computations and permits the solution of more complex design problems in a shorter time. *LogicAid* also provides tutorial help for Karnaugh maps and derivation of state graphs.

Several of the units include simulation or laboratory exercises. These exercises provide an opportunity to design a logic circuit and then test its operation. The *SimUaid* logic simulator, also available on the book's accompanying CD, may be used to verify the logic designs. The lab equipment required for testing either can be a breadboard with integrated circuit flip-flops and logic gates or a circuit board with a programmable logic device. If such equipment is not available, the lab exercises can be simulated with *SimUaid* or just assigned as design problems. This is especially important for Units 8, 16, and 20 because the comprehensive design problems in these units help to review and tie together the material in several of the preceding units.

The DirectVHDL software on the CD provides a quick way to check and simulate VHDL descriptions of hardware. This software checks the syntax of the VHDL code as it is typed in so that most syntax errors can be corrected before the simulation phase.

Changes from Previous Editions

The text has evolved considerably since the fifth edition. Programmable logic and the VHDL hardware description language were added, and an emphasis was placed on the role of simulation and computer-aided design of logic circuits. The discussion of VHDL, hazards, latches and one-hot state assignments was expanded. Numerous problems were added. Several additional changes have been made for the seventh edition. The discussion of number systems was reorganized so that one's complement number systems can be easily omitted. In the unit on Boolean algebra, the laws of switching algebra are first derived using switch networks and truth tables; these are used to define Boolean algebra and, then, further theorems of Boolean algebra are derived that are useful in simplifying switching algebra expressions. The discussion of adders is expanded to include carry-lookahead adders. Alternative implementations of multiplexers are included and also a discussion of active high and active low signals. Other types of gated latches are discussed, and a brief introduction to asynchronous sequential circuits is included. There is more discussion of incompletely specified state tables and how they may occur, and reducing incompletely specified state tables is briefly discussed. Problems have been added throughout the book with an emphasis on more challenging problems than the typical exercises. In addition, the logic design and simulation software that accompanies the text has been updated and improved.

Acknowledgments

To be effective, a book designed for self-study cannot simply be written. It must be tested and revised many times to achieve its goals. We wish to express our appreciation to the many professors, proctors, and students who participated in this process. Special thanks go to Dr. David Brown, who helped teach the self-paced course, and who made many helpful suggestions for improving the fifth edition. Special thanks to graduate teaching assistant, Mark Story, who developed many new problems and solutions for the fifth edition and who offered many suggestions for improving the consistency and clarity of the presentation.

The authors especially thank the most recent reviewers of the text. Among others, they are

Clark Guest, University of California, San Diego
Jayantha Herath, St Cloud State University
Nagarajan Kandasamy, Drexel University
Avinash Karanth Kodi, Ohio University
Jacob Savir, Newark College of Engineering
Melissa C. Smith, Clemson University
Larry M. Stephens, University of South Carolina

Feedback from the readers, both critical and appreciative, is welcome. Please send your comments, concerns, and suggestions to globalengineering@cengage.com.

Charles H. Roth, Jr. *Larry L. Kinney*

How to Use This Book for Self-Study

If you wish to learn all of the material in this text to mastery level, the following study procedures are recommended for each unit:

1. Read the *Objectives* of the unit. These objectives provide a concise summary of what you should be able to do when you complete studying the unit.
2. Work through the *Study Guide*. After reading each section of the text, write out the answers to the corresponding study guide questions. In many cases, blank spaces are left in the study guide so that you can write your answers directly in this book. By doing this, you will have the answers conveniently available for later review. The study guide questions generally will help emphasize some of the important points in each section or will guide you to a better understanding of some of the more difficult points. If you cannot answer some of the study guide questions, this indicates that you need to study the corresponding section in the text more before proceeding. The answers to selected study guide questions are given in the back of this book; answers to the remaining questions generally can be found within the text.
3. Several of the units (Units 3, 5, 6, 11, 13, 14, and 18) contain one or more programmed exercises. Each programmed exercise will guide you step-by-step through the solution of one of the more difficult types of problems encountered in this text. When working through a programmed exercise, be sure to write down your answer for each part in the space provided before looking at the answer and continuing with the next part of the exercise.
4. Work the assigned *Problems* at the end of the unit. Check your answers against those at the end of the book and rework any problems that you missed.
5. Reread the *Objectives* of the unit to make sure that you can meet all of them. If in doubt, review the appropriate sections of the text.
6. If you are using this text in a self-paced course, you will need to pass a readiness test on each unit before proceeding with the next unit. The purpose of the readiness test is to make sure that you have mastered the objectives of one unit before moving on to the next unit. The questions on the test will relate directly to the objectives of the unit, so that if you have worked through the study guide and written out answers to all of the study guide questions and to the problems assigned in the study guide, you should have no difficulty passing the test.

About the Authors

Charles H. Roth, Jr. is Professor Emeritus of Electrical and Computer Engineering at the University of Texas at Austin. He has been on the UT faculty since 1961. He received his BSEE degree from the University of Minnesota, his MSEE and EE degrees from the Massachusetts Institute of Technology, and his PhD degree in EE from Stanford University. His teaching and research interests included logic design, digital systems design, switching theory, microprocessor systems, and computer-aided design. He developed a self-paced course in logic design which formed the basis of his textbook, *Fundamentals of Logic Design*. He is also the author of *Digital Systems Design Using VHDL*, two other textbooks, and several software packages. He is the author or co-author of more than 50 technical papers and reports. Six PhD students and 80 MS students have received their degrees under his supervision. He received several teaching awards including the 1974 General Dynamics Award for Outstanding Engineering Teaching.

Larry L. Kinney is Professor Emeritus in Electrical and Computer Engineering at the University of Minnesota Twin Cities. He received the BS, MS, and PhD in Electrical Engineering from the University of Iowa in 1964, 1965, and 1968, respectively, and joined the University of Minnesota in 1968. He has taught a wide variety of courses including logic design, microprocessor/microcomputer systems, computer design, switching theory, communication systems and error-correcting codes. His major areas of research interest are testing of digital systems, built-in self-test, computer design, microprocessor-based systems, and error-correcting codes.

UNIT 1

Introduction Number Systems and Conversion

Objectives

1. Introduction
   The first part of this unit introduces the material to be studied later. In addition to getting an overview of the material in the first part of the course, you should be able to explain
   a. The difference between analog and digital systems and why digital systems are capable of greater accuracy
   b. The difference between combinational and sequential circuits
   c. Why two-valued signals and binary numbers are commonly used in digital systems

2. Number systems and conversion
   When you complete this unit, you should be able to solve the following types of problems:
   a. Given a positive integer, fraction, or mixed number in any base (2 through 16); convert to any other base. Justify the procedure used by using a power series expansion for the number.
   b. Add, subtract, multiply, and divide positive binary numbers. Explain the addition and subtraction process in terms of carries and borrows.
   c. Write negative binary numbers in sign and magnitude, 1's complement, and 2's complement forms. Add signed binary numbers using 1's complement and 2's complement arithmetic. Justify the methods used. State when an overflow occurs.
   d. Represent a decimal number in binary-coded-decimal (BCD), 6-3-1-1 code, excess-3 code, etc. Given a set of weights, construct a weighted code.

Study Guide

1. Study Section 1.1, *Digital Systems and Switching Circuits,* and answer the following study questions:

   (a) What is the basic difference between analog and digital systems?

   (b) Why are digital systems capable of greater accuracy than analog systems?

   (c) Explain the difference between combinational and sequential switching circuits.

   (d) What common characteristic do most switching devices used in digital systems have?

   (e) Why are binary numbers used in digital systems?

2. Study Section 1.2, *Number Systems and Conversion.* Answer the following study questions as you go along:

   (a) Is the first remainder obtained in the division method for base conversion the most or least significant digit?

   (b) Work through all of the examples in the text as you encounter them and make sure that you understand all of the steps.

   (c) An easy method for conversion between binary and hexadecimal is illustrated in Equation (1-1). Why should you start forming the groups of four bits at the binary point instead of the left end of the number?

   (d) Why is it impossible to convert a decimal number to binary on a digit-by-digit basis as can be done for hexadecimal?

(e) Complete the following conversion table.

Binary (base 2)	Octal (base 8)	Decimal (base 10)	Hexadecimal (base 16)
0	0	0	0
1			
10			
11			
100			
101			
110			
111			
1000			
1001			
1010			
1011			
1100			
1101			
1110			
1111			
10000	20	16	10

(f) Work Problems 1.1, 1.2, 1.3, and 1.4.

3. Study Section 1.3, *Binary Arithmetic.*

(a) Make sure that you can follow all of the examples, especially the propagation of borrows in the subtraction process.

(b) To make sure that you understand the borrowing process, work out a detailed analysis in terms of powers of 2 for the following example:

$$\begin{array}{r} 1100 \\ -\ 101 \\ \hline 111 \end{array}$$

4. Work Problems 1.5, 1.6, and 1.17(a).

5. Study Section 1.4, *Representation of Negative Numbers.*

(a) In digital systems, why are 1's complement and 2's complement commonly used to represent negative numbers instead of sign and magnitude?

(b) State two different ways of forming the 1's complement of an n-bit binary number.

(c) State three different ways of forming the 2's complement of an n-bit binary number.

(d) If the word length is $n = 4$ bits (including sign), what decimal number does 1000_2 represent in sign and magnitude?
In 2's complement?
In 1's complement?

(e) Given a negative number represented in 2's complement, how do you find its magnitude?

Given a negative number represented in 1's complement, how do you find its magnitude?

(f) If the word length is 6 bits (including sign), what decimal number does 100000_2 represent in sign and magnitude?

In 2's complement?

In 1's complement?

(g) What is meant by an overflow? How can you tell that an overflow has occurred when performing 1's or 2's complement addition?

Does a carry out of the last bit position indicate that an overflow has occurred?

(h) Work out some examples of 1's and 2's complement addition for various combinations of positive and negative numbers.

(i) What is the justification for using the end-around carry in 1's complement addition?

(j) The one thing that causes the most trouble with 2's complement numbers is the special case of the negative number which consists of a 1 followed by all 0's (1000. . .000). If this number is *n* bits long, what number does it represent and why? (It is not negative zero.)

(k) Work Problems 1.7 and 1.8.

6. Study Section 1.5, *Binary Codes.*

(a) Represent 187 in BCD code, excess-3 code, 6-3-1-1 code, and 2-out-of-5 code.

(b) Verify that the 6-3-1-1 code is a weighted code. Note that for some decimal digits, two different code combinations could have been used. For example, either 0101 or 0110 could represent 4. In each case the combination with the smaller binary value has been used.

(c) How is the excess-3 code obtained?

(d) How are the ASCII codes for the decimal digits obtained? What is the relation between the ASCII codes for the capital letters and lowercase letters?

(e) Work Problem 1.9.

7. If you are taking this course on a self-paced basis, you will need to pass a readiness test on this unit before going on to the next unit. The purpose of the readiness test is to determine if you have mastered the material in this unit and are ready to go on to the next unit. Before you take the readiness test:

(a) Check your answers to the problems against those provided at the end of this book. If you missed any of the problems, make sure that you understand why your answer is wrong and correct your solution.

(b) Make sure that you can meet all of the objectives listed at the beginning of this unit.

Introduction
Number Systems and Conversion

1.1 Digital Systems and Switching Circuits

Digital systems are used extensively in computation and data processing, control systems, communications, and measurement. Because digital systems are capable of greater accuracy and reliability than analog systems, many tasks formerly done by analog systems are now being performed digitally.

In a digital system, the physical quantities or signals can assume only discrete values, while in analog systems the physical quantities or signals may vary continuously over a specified range. For example, the output voltage of a digital system might be constrained to take on only two values such as 0 volts and 5 volts, while the output voltage from an analog system might be allowed to assume any value in the range −10 volts to +10 volts.

Because digital systems work with discrete quantities, in many cases they can be designed so that for a given input, the output is exactly correct. For example, if we multiply two 5-digit numbers using a digital multiplier, the 10-digit product will be correct in all 10 digits. On the other hand, the output of an analog multiplier might have an error ranging from a fraction of one percent to a few percent depending on the accuracy of the components used in construction of the multiplier. Furthermore, if we need a product which is correct to 20 digits rather than 10, we can redesign the digital multiplier to process more digits and add more digits to its input. A similar improvement in the accuracy of an analog multiplier would not be possible because of limitations on the accuracy of the components.

The design of digital systems may be divided roughly into three parts—system design, logic design, and circuit design. System design involves breaking the overall system into subsystems and specifying the characteristics of each subsystem. For example, the system design of a digital computer could involve specifying the number and type of memory units, arithmetic units, and input-output devices as well as the interconnection and control of these subsystems. Logic design involves determining how to interconnect basic logic building blocks to perform a specific function. An example of logic design is determining the interconnection of logic gates and flip-flops required to perform binary addition. Circuit design involves specifying the interconnection of specific components such as resistors, diodes, and transistors to form a gate, flip-flop, or other logic building block. Most contemporary circuit design is done in integrated circuit form using appropriate computer-aided design tools to lay out and interconnect the components on a chip of silicon. This book is largely devoted to a study of logic design and the theory necessary for understanding the logic design process. Some aspects of system design are treated in Units 18 and 20. Circuit design of logic gates is discussed briefly in Appendix A.

Many of a digital system's subsystems take the form of a switching circuit (Figure 1-1). A switching circuit has one or more inputs and one or more outputs which take on discrete values. In this text, we will study two types of switching circuits—combinational and sequential. In a combinational circuit, the output values depend only on the present value of the inputs and not on past values. In a sequential circuit, the outputs depend on both the present and past input values. In other words, in order to determine the output of a sequential circuit, a sequence of input values must be specified. The sequential circuit is said to have memory because it must "remember" something about the past sequence of inputs, while a combinational circuit has no memory. In general, a sequential circuit is composed of a combinational circuit with added memory elements. Combinational circuits are easier to design than sequential circuits and will be studied first.

FIGURE 1-1
Switching Circuit

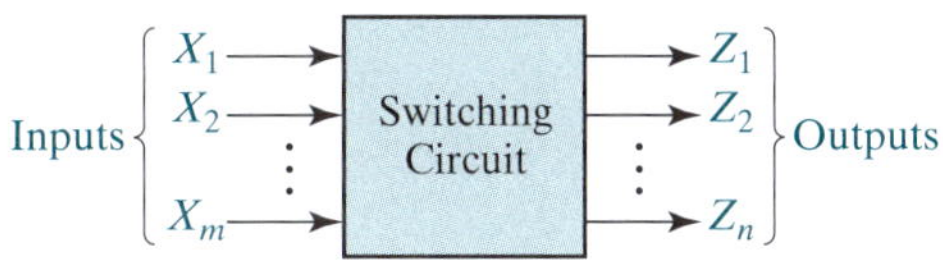

The basic building blocks used to construct combinational circuits are logic gates. The logic designer must determine how to interconnect these gates in order to convert the circuit input signals into the desired output signals. The relationship between these input and output signals can be described mathematically using Boolean algebra. Units 2 and 3 of this text introduce the basic laws and theorems of Boolean algebra and show how they can be used to describe the behavior of circuits of logic gates.

Starting from a given problem statement, the first step in designing a combinational logic circuit is to derive a table or the algebraic logic equations which describe the circuit outputs as a function of the circuit inputs (Unit 4). In order to design an economical circuit to realize these output functions, the logic equations which describe the circuit outputs generally must be simplified. Algebraic methods for this simplification are described in Unit 3, and other simplification methods (Karnaugh map and Quine-McCluskey procedure) are introduced in Units 5 and 6. Implementation of the simplified logic equations using several types of gates is described in Unit 7, and alternative design procedures using programmable logic devices are developed in Unit 9.

The basic memory elements used in the design of sequential circuits are called flip-flops (Unit 11). These flip-flops can be interconnected with gates to form counters and registers (Unit 12). Analysis of more general sequential circuits using timing diagrams, state tables, and graphs is presented in Unit 13. The first step in designing a sequential switching circuit is to construct a state table or graph which describes the relationship between the input and output sequences (Unit 14). Methods for going from a state table or graph to a circuit of gates and flip-flops are developed in Unit 15. Methods of implementing sequential circuits using programmable logic are discussed in Unit 16. In Unit 18, combinational and sequential design techniques are applied to the realization of systems for performing binary addition, multiplication, and division. The sequential circuits designed in this text are called synchronous sequential circuits because they use a common timing signal, called a clock, to synchronize the operation of the memory elements.

Use of a hardware description language, VHDL, in the design of combinational logic, sequential logic, and digital systems is introduced in Units 10, 17, and 20. VHDL is used to describe, simulate, and synthesize digital hardware. After writing VHDL code, the designer can use computer-aided design software to compile the hardware description and complete the design of the digital logic. This allows the completion of complex designs without having to manually work out detailed circuit descriptions in terms of gates and flip-flops.

The switching devices used in digital systems are generally two-state devices, that is, the output can assume only two different discrete values. Examples of switching devices are relays, diodes, and transistors. A relay can assume two states—closed or open—depending on whether power is applied to the coil or not. A diode can be in a conducting state or a nonconducting state. A transistor can be in a cut-off or saturated state with a corresponding high or low output voltage. Of course, transistors can also be operated as linear amplifiers with a continuous range of output voltages, but in digital applications greater reliability is obtained by operating them as two-state devices. Because the outputs of most switching devices assume only two different values, it is natural to use binary numbers internally in digital systems. For this reason binary numbers and number systems will be discussed first before proceeding to the design of switching circuits.

1.2 Number Systems and Conversion

When we write decimal (base 10) numbers, we use a positional notation; each digit is multiplied by an appropriate power of 10 depending on its position in the number. For example,

$$953.78_{10} = 9 \times 10^2 + 5 \times 10^1 + 3 \times 10^0 + 7 \times 10^{-1} + 8 \times 10^{-2}$$

Similarly, for binary (base 2) numbers, each binary digit is multiplied by the appropriate power of 2:

$$\begin{aligned}1011.11_2 &= 1 \times 2^3 + 0 \times 2^2 + 1 \times 2^1 + 1 \times 2^0 + 1 \times 2^{-1} + 1 \times 2^{-2}\\ &= 8 + 0 + 2 + 1 + \tfrac{1}{2} + \tfrac{1}{4} = 11\tfrac{3}{4} = 11.75_{10}\end{aligned}$$

Note that the binary point separates the positive and negative powers of 2 just as the decimal point separates the positive and negative powers of 10 for decimal numbers.

Any positive integer R ($R > 1$) can be chosen as the *radix* or *base* of a number system. If the base is R, then R digits $(0, 1, \ldots, R-1)$ are used. For example, if $R = 8$, then the required digits are 0,1,2,3,4,5,6, and 7. A number written in positional notation can be expanded in a power series in R. For example,

$$\begin{aligned} N &= (a_4 a_3 a_2 a_1 a_0 . a_{-1} a_{-2} a_{-3})_R \\ &= a_4 \times R^4 + a_3 \times R^3 + a_2 \times R^2 + a_1 \times R^1 + a_0 \times R^0 \\ &\qquad + a_{-1} \times R^{-1} + a_{-2} \times R^{-2} + a_{-3} \times R^{-3} \end{aligned}$$

where a_i is the coefficient of R^i and $0 \le a_i \le R - 1$. If the arithmetic indicated in the power series expansion is done in base 10, then the result is the decimal equivalent of N. For example,

$$\begin{aligned} 147.3_8 &= 1 \times 8^2 + 4 \times 8^1 + 7 \times 8^0 + 3 \times 8^{-1} = 64 + 32 + 7 + \tfrac{3}{8} \\ &= 103.375_{10} \end{aligned}$$

The power series expansion can be used to convert to any base. For example, converting 147_{10} to base 3 would be written as

$$147_{10} = 1 \times (101)^2 + (11) \times (101)^1 + (21) \times (101)^0$$

where all the numbers on the right-hand side are base 3 numbers. (*Note:* In base 3, 10 is 101, 7 is 21, etc.) To complete the conversion, base 3 arithmetic would be used. Of course, this is not very convenient if the arithmetic is being done by hand. Similarly, if 147_{10} is being converted to binary, the calculation would be

$$147_{10} = 1 \times (1010)^2 + (100) \times (1010)^1 + (111) \times (1010)^0$$

Again this is not convenient for hand calculation but it could be done easily in a computer where the arithmetic is done in binary. For hand calculation, use the power series expansion when converting from some base *into base 10.*

For bases greater than 10, more than 10 symbols are needed to represent the digits. In this case, letters are usually used to represent digits greater than 9. For example, in hexadecimal (base 16), A represents 10_{10}, B represents 11_{10}, C represents 12_{10}, D represents 13_{10}, E represents 14_{10}, and F represents 15_{10}. Thus,

$$A2F_{16} = 10 \times 16^2 + 2 \times 16^1 + 15 \times 16^0 = 2560 + 32 + 15 = 2607_{10}$$

Next, we will discuss conversion of a decimal *integer* to base R using the division method. The base R equivalent of a decimal integer N can be represented as

$$N = (a_n a_{n-1} \cdots a_2 a_1 a_0)_R = a_n R^n + a_{n-1} R^{n-1} + \cdots + a_2 R^2 + a_1 R^1 + a_0$$

If we divide N by R, the remainder is a_0:

$$\frac{N}{R} = a_n R^{n-1} + a_{n-1} R^{n-2} + \cdots + a_2 R^1 + a_1 = Q_1, \text{ remainder } a_0$$

Then we divide the quotient Q_1 by R:

$$\frac{Q_1}{R} = a_n R^{n-2} + a_{n-1} R^{n-3} + \cdots + a_3 R^1 + a_2 = Q_2, \text{ remainder } a_1$$

Next we divide Q_2 by R:

$$\frac{Q_2}{R} = a_n R^{n-3} + a_{n-1} R^{n-4} + \cdots + a_3 = Q_3, \text{ remainder } a_2$$

This process is continued until we finally obtain a_n. Note that the remainder obtained at each division step is one of the desired digits and the least significant digit is obtained first.

Example

Convert 53_{10} to binary.

2 /53		
2 /26	rem. = 1 = a_0	
2 /13	rem. = 0 = a_1	
2 /6	rem. = 1 = a_2	$53_{10} = 110101_2$
2 /3	rem. = 0 = a_3	
2 /1	rem. = 1 = a_4	
0	rem. = 1 = a_5	

Conversion of a decimal *fraction* to base R can be done using successive *multiplications* by R. A decimal fraction F can be represented as

$$F = (.a_{-1}a_{-2}a_{-3} \cdots a_{-m})_R = a_{-1}R^{-1} + a_{-2}R^{-2} + a_{-3}R^{-3} + \cdots + a_{-m}R^{-m}$$

Multiplying by R yields

$$FR = a_{-1} + a_{-2}R^{-1} + a_{-3}R^{-2} + \cdots + a_{-m}R^{-m+1} = a_{-1} + F_1$$

where F_1 represents the fractional part of the result and a_{-1} is the integer part. Multiplying F_1 by R yields

$$F_1R = a_{-2} + a_{-3}R^{-1} + \cdots + a_{-m}R^{-m+2} = a_{-2} + F_2$$

Next, we multiply F_2 by R:

$$F_2R = a_{-3} + \cdots + a_{-m}R^{-m+3} = a_{-3} + F_3$$

This process is continued until we have obtained a sufficient number of digits. Note that the integer part obtained at each step is one of the desired digits and the most significant digit is obtained first.

Example

Convert 0.625_{10} to binary.

$$\begin{array}{r} F = .625 \\ \times\ \ 2 \\ \hline 1.250 \\ (a_{-1} = 1) \end{array} \qquad \begin{array}{r} F_1 = .250 \\ \times\ \ 2 \\ \hline 0.500 \\ (a_{-2} = 0) \end{array} \qquad \begin{array}{r} F_2 = .500 \\ \times\ \ 2 \\ \hline 1.000 \\ (a_{-3} = 1) \end{array} \qquad .625_{10} = .101_2$$

This process does not always terminate, but if it does not terminate, the result is a repeating fraction.

Example

Convert 0.7_{10} to binary.

$$\begin{array}{r} .7 \\ 2 \\ \hline (1).4 \\ 2 \\ \hline (0).8 \\ 2 \\ \hline (1).6 \\ 2 \\ \hline (1).2 \\ 2 \\ \hline (0).4 \\ 2 \\ \hline (0).8 \end{array}$$

(0).4 ⟵ process starts repeating here because 0.4 was previously obtained

$0.7_{10} = 0.1\underline{0110}\,\underline{0110}\,\underline{0110}\ \ldots_2$

Conversion between two bases other than decimal can be done directly by using the procedures given; however, the arithmetic operations would have to be carried out using a base other than 10. It is generally easier to convert to decimal first and then convert the decimal number to the new base.

Example

Convert 231.3_4 to base 7.

$$231.3_4 = 2 \times 16 + 3 \times 4 + 1 + \tfrac{3}{4} = 45.75_{10}$$

$$\begin{array}{rll} 7\,\underline{/45} & & \\ 7\,\underline{/\;\;6} & \text{rem. } 3 & \\ 0 & \text{rem. } 6 & \end{array} \qquad \begin{array}{r} .75 \\ \underline{7} \\ (5).25 \\ \underline{7} \\ (1).75 \\ \underline{7} \\ (5).25 \\ \underline{7} \\ (1).75 \end{array} \qquad 45.75_{10} = 63.5151\ldots_7$$

Conversion from binary to hexadecimal (and conversely) can be done by inspection because each hexadecimal digit corresponds to exactly four binary digits (bits). Starting at the binary point, the bits are divided into groups of four, and each group is replaced by a hexadecimal digit:

$$1001101.010111_2 = \underbrace{0100}_{4}\ \underbrace{1101}_{D}\ .\ \underbrace{0101}_{5}\ \underbrace{1100}_{C} = 4D.5C_{16} \tag{1-1}$$

As shown in Equation (1-1), extra 0's are added at each end of the bit string as needed to fill out the groups of four bits.

1.3 Binary Arithmetic

Arithmetic operations in digital systems are usually done in binary because design of logic circuits to perform binary arithmetic is much easier than for decimal. Binary arithmetic is carried out in much the same manner as decimal, except the addition and multiplication tables are much simpler.

The addition table for binary numbers is

$$\begin{aligned} 0 + 0 &= 0 \\ 0 + 1 &= 1 \\ 1 + 0 &= 1 \\ 1 + 1 &= 0 \quad \text{and carry 1 to the next column} \end{aligned}$$

Carrying 1 to a column is equivalent to adding 1 to that column.

Example

Add 13_{10} and 11_{10} in binary.

$$\begin{array}{rr} & 1111 \longleftarrow \text{carries} \\ 13_{10} = & 1101 \\ 11_{10} = & \underline{1011} \\ & 11000 = 24_{10} \end{array}$$

The subtraction table for binary numbers is

$$\begin{array}{ll} 0 - 0 = 0 & \\ 0 - 1 = 1 & \text{and borrow 1 from the next column} \\ 1 - 0 = 1 & \\ 1 - 1 = 0 & \end{array}$$

Borrowing 1 from a column is equivalent to subtracting 1 from that column.

Examples of Binary Subtraction

(a)
$$\begin{array}{rl} 1 \longleftarrow & \text{(indicates} \\ 11101 & \text{a borrrow} \\ \underline{-\ 10011} & \text{from the} \\ 1010 & \text{3rd column)} \end{array}$$

(b)
$$\begin{array}{rl} 1111 \longleftarrow & \text{borrows} \\ 10000 & \\ \underline{-\quad 11} & \\ 1101 & \end{array}$$

(c)
$$\begin{array}{rl} 111 \longleftarrow & \text{borrows} \\ 111001 & \\ \underline{-\quad 1011} & \\ 101110 & \end{array}$$

Note how the borrow propagates from column to column in the second example. In order to borrow 1 from the second column, we must in turn borrow 1 from the third column, etc. An alternative to binary subtraction is the use of 2's complement arithmetic, as discussed in Section 1.4.

Binary subtraction sometimes causes confusion, perhaps because we are so used to doing decimal subtraction that we forget the significance of the borrowing process. Before doing a detailed analysis of binary subtraction, we will review the borrowing process for decimal subtraction.

If we number the columns (digits) of a decimal integer from right to left (starting with 0), and then we borrow 1 from column n, what we mean is that we subtract 1 from column n and add 10 to column $n - 1$. Because $1 \times 10^n = 10 \times 10^{n-1}$, the value of the decimal number is unchanged, but we can proceed with the subtraction. Consider, for example, the following decimal subtraction problem:

column 2 ↘ ↙ column 1

$$\begin{array}{r} 205 \\ \underline{-\ \ 18} \\ 187 \end{array}$$

A detailed analysis of the borrowing process for this example, indicating first a borrow of 1 from column 1 and then a borrow of 1 from column 2, is as follows:

$$
\begin{aligned}
205 - 18 &= [2 \times 10^2 + 0 \times 10^1 + 5 \times 10^0] \\
&\quad - [\qquad\qquad 1 \times 10^1 + 8 \times 10^0] \\
&\qquad \text{(note borrow from column 1)} \\
&= [2 \times 10^2 + (0 - 1) \times 10^1 + (10 + 5) \times 10^0] \\
&\quad - [\qquad\qquad 1 \times 10^1 + \quad 8 \times 10^0] \\
&\qquad \text{(note borrow from column 2)} \\
&= [(2 - 1) \times 10^2 + (10 + 0 - 1) \times 10^1 + 15 \times 10^0] \\
&\quad - [\qquad\qquad\qquad 1 \times 10^1 + 8 \times 10^0] \\
&= [1 \times 10^2 \quad + \quad 8 \times 10^1 \qquad + \ 7 \times 10^0] = 187
\end{aligned}
$$

The analysis of borrowing for binary subtraction is exactly the same, except that we work with powers of 2 instead of powers of 10. Thus for a binary number, borrowing 1 from column n is equivalent to subtracting 1 from column n and adding $2(10_2)$ to column $n - 1$. The value of the binary number is unchanged because $1 \times 2^n = 2 \times 2^{n-1}$.

A detailed analysis of binary subtraction example (c) follows. Starting with the rightmost column, $1 - 1 = 0$. To subtract in the second column, we must borrow from the third column. Rather than borrow immediately, we place a 1 over the third column to indicate that a borrow is necessary, and we will actually do the borrowing when we get to the third column. (This is similar to the way borrow signals might propagate in a computer.) Now because we have borrowed 1, the second column becomes 10, and $10 - 1 = 1$. In order to borrow 1 from the third column, we must borrow 1 from the fourth column (indicated by placing a 1 over column 4). Column 3 then becomes 10, subtracting off the borrow yields 1, and $1 - 0 = 1$. Now in column 4, we subtract off the borrow leaving 0. In order to complete the subtraction, we must borrow from column 5, which gives 10 in column 4, and $10 - 1 = 1$.

The multiplication table for binary numbers is

$$
\begin{aligned}
0 \times 0 &= 0 \\
0 \times 1 &= 0 \\
1 \times 0 &= 0 \\
1 \times 1 &= 1
\end{aligned}
$$

The following example illustrates multiplication of 13_{10} by 11_{10} in binary:

$$
\begin{array}{r}
1101 \\
\underline{1011} \\
1101 \\
1101 \\
0000 \\
\underline{1101} \\
10001111
\end{array}
= 143_{10}
$$

Note that each partial product is either the multiplicand (1101) shifted over the appropriate number of places or is zero.

When adding up long columns of binary numbers, the sum of the bits in a single column can exceed 11_2, and therefore the carry to the next column can be greater than 1. For example, if a single column of bits contains five 1's, then adding up the 1's gives 101_2, which means that the sum bit for that column is 1, and the carry to the next column is 10_2. When doing binary multiplication, a common way to avoid carries greater than 1 is to add in the partial products one at a time as illustrated by the following example:

```
    1111     multiplicand
    1101     multiplier
    ----
    1111     first partial product
   0000      second partial product
   -----
  (01111)    sum of first two partial products
  1111       third partial product
 -------
(1001011)    sum after adding third partial product
 1111        fourth partial product
--------
11000011     final product (sum after adding fourth partial product)
```

The following example illustrates division of 145_{10} by 11_{10} in binary:

```
        1101
     ________
1011 | 1001001
       1011
       -----
        1110
        1011
        ----
         1101    The quotient is 1101 with a remainder
         1011    of 10.
         ----
           10
```

Binary division is similar to decimal division, except it is much easier because the only two possible quotient digits are 0 and 1. In the above example, if we start by comparing the divisor (1011) with the upper four bits of the dividend (1001), we find that we cannot subtract without a negative result, so we move the divisor one place to the right and try again. This time we can subtract 1011 from 10010 to give 111 as a result, so we put the first quotient bit of 1 above 10010. We then bring down the next dividend bit (0) to get 1110 and shift the divisor right. We then subtract 1011 from 1110 to get 11, so the second quotient bit is 1. When we bring down the next dividend bit, the result is 110, and we cannot subtract the shifted divisor, so the third quotient bit is 0. We then bring down the last dividend bit and subtract 1011 from 1101 to get a final remainder of 10, and the last quotient bit is 1.

1.4 Representation of Negative Numbers

Up to this point we have been working with unsigned positive numbers. The most common methods for representing both positive and negative numbers are sign and magnitude, 2's complement, and 1's complement. In each of these methods, the leftmost bit of a number is 0 for positive numbers and 1 for negative numbers. As discussed below, if n bits are used to represent numbers, then the sign and magnitude and 1's complement methods represent numbers in the range $-(2^{(n-1)} - 1)$ to $+(2^{(n-1)} - 1)$ and both have two representations for 0, a positive 0 and a negative 0. In 2's complement, numbers in the range $-2^{(n-1)}$ to $+(2^{(n-1)} - 1)$ are represented and there is only a positive 0. If an operation, such as addition or subtraction, is performed on two numbers and the result is outside the range of representation, then we say that an *overflow* has occurred.

Sign and Magnitude Numbers

In an n-bit sign and magnitude system, a number is represented by a sign bit, 0 for positive and 1 for negative, followed by $n - 1$ bits that represent the magnitude of the number. With $n - 1$ bits the magnitude can be 0 to $2^{(n-1)} - 1$. With the sign bit, numbers in the range $-(2^{(n-1)} - 1)$ to $+(2^{(n-1)} - 1)$ are represented including a positive and negative 0. This is illustrated in Table 1-1 for $n = 4$. For example, 0011 represents +3 and 1011 represents −3. Note that 1000 represents minus 0.

Designing logic circuits to perform arithmetic on sign and magnitude binary numbers is awkward. One method is to convert the numbers into 2's (or 1's) complement and, after performing the arithmetic operation, convert the result back to sign and magnitude.

TABLE 1-1 Signed Binary Integers (word length: $n = 4$)

			Negative Integers		
$+N$	Positive Integers (all systems)	$-N$	Sign and Magnitude	2's Complement N^*	1's Complement $\overline{N}$
+0	0000	−0	1000	——	1111
+1	0001	−1	1001	1111	1110
+2	0010	−2	1010	1110	1101
+3	0011	−3	1011	1101	1100
+4	0100	−4	1100	1100	1011
+5	0101	−5	1101	1011	1010
+6	0110	−6	1110	1010	1001
+7	0111	−7	1111	1001	1000
		−8	——	1000	——

2's Complement Numbers

In the 2's complement number system, a positive number, N, is represented by a 0 followed by the magnitude of N as in the sign and magnitude system; however, a negative number, $-N$, is represented by its 2's complement, N^*. If the word length is n bits, the 2's complement of a positive integer N is defined as

$$N^* = 2^n - N \tag{1-2}$$

(Note that in this equation all numbers N, 2^n, and N^* are treated as unsigned positive numbers; if they are expressed in binary, $n + 1$ bits are required to represent 2^n.) Table 1-1 shows the result for $n = 4$. In Table 1-1, the 2's complement representation of negative numbers -1 through -7 can be obtained by taking the 2's complement of positive numbers 1 through 7 (i.e., by subtracting from 16). For example, the 2's complement of 5 is $16 - 5 = 11$ or, using binary numbers, $(10000) - (0101) = (1011)$. After completing the subtractions, all combinations of 4-bits have been used to represent the numbers $-7, \ldots, -1, 0, 1, \ldots 7$; the only unused combination is 1000. Since the leftmost bit of 1000 is 1, it should be a negative number. To determine its magnitude, note that the magnitude of a negative number can be obtained by taking its 2's complement; that is, from Equation (1-2),

$$N = 2^n - N^* \tag{1-3}$$

Applying Equation (1-3) to 1000 produces $(10000) - (1000) = (1000)$ or, in decimal, $16 - 8 = 8$. Hence, 1000 represents -8. In general, in an n-bit 2's complement system the number 1 followed by all 0's represents $-2^{(n-1)}$.

Using Equation (1-2) directly on binary numbers requires subtraction of $n + 1$ bit numbers. This can be avoided by noting that Equation (1-2) can be written as

$$N^* = (2^n - 1 - N) + 1$$

In binary, $2^n - 1$ consists of n 1's. Subtracting a number from all 1's does not produce any borrows, and the subtraction can be done by replacing 0's with 1's and 1's with 0's (i.e., simply complement N bit-by-bit). For example, if $n = 7$ and $N = 0101100$,

$$\begin{aligned} 2^n - 1 &= 1111111 \\ &\underline{-\ 0101100} \\ &1010011 \\ &\underline{+\ 0000001} \\ N^* &= 1010100 \end{aligned}$$

N^* is obtained by complementing N bit-by-bit and then adding 1. An alternative way to form the 2's complement of N is to start at the right and leave any 0's on the right end and the first 1 unchanged, then complement all bits to the left of the first 1. In the preceding example, the 100 on the right end of N is unchanged while the 0101 on the left is complemented bit-by-bit.

Addition of 2's Complement Numbers

The addition of n-bit signed binary numbers is straightforward using the 2's complement system. The addition is carried out just as if all the numbers were positive, and any carry from the sign position is ignored. This will always yield the correct result except when an overflow occurs. When the word length is n bits, we say that an *overflow* has occurred if the correct representation of the sum (including sign) requires more than n bits. The different cases which can occur are illustrated below for $n = 4$.

1. Addition of two positive numbers, sum $< 2^{n-1}$

+3	0011	
+4	0100	
+7	0111	(correct answer)

2. Addition of two positive numbers, sum $\geq 2^{n-1}$

+5	0101	
+6	0110	
	1011	⟵ wrong answer because of overflow (+11 requires 5 bits including sign)

3. Addition of positive and negative numbers (negative number has greater magnitude)

−5	0101	
−6	1010	
−1	1111	(correct answer)

4. Same as case 3 except positive number has greater magnitude

+5	1011	
+6	0110	
+1	(1)0001	⟵ correct answer when the carry from the sign bit is ignored (this is *not* an overflow)

5. Addition of two negative numbers, $|\text{sum}| \leq 2^{n-1}$

−3	1101	
−4	1100	
−7	(1)1001	⟵ correct answer when the last carry is ignored (this is *not* an overflow)

6. Addition of two negative numbers, $|\text{sum}| > 2^{n-1}$

−5	1011	
−6	1010	
	(1)0101	⟵ wrong answer because of overflow (−11 requires 5 bits including sign)

Note that an overflow condition (cases 2 and 6) is easy to detect because in case 2 the addition of two positive numbers yields a negative result, and in case 6 the addition of two negative numbers yields a positive answer (for four bits).

The proof that throwing away the carry from the sign bit always gives the correct answer follows for cases 4 and 5:

$$\text{Case 4:}\quad -A + B\,(\text{where } B > A)$$
$$A^* + B = (2^n - A) + B = 2^n + (B - A) > 2^n$$

Throwing away the last carry is equivalent to subtracting 2^n, so the result is $(B - A)$, which is correct.

$$\text{Case 5:}\quad -A - B\,(\text{where } A + B \le 2^{n-1})$$
$$A^* + B^* = (2^n - A) + (2^n - B) = 2^n + 2^n - (A + B)$$

Discarding the last carry yields $2^n - (A + B) = (A + B)^*$, which is the correct representation of $-(A + B)$.

1's Complement Numbers

In the 1's complement system a negative number, $-N$, is represented by the 1's complement of N, $\overline{N}$, defined as

$$\overline{N} = (2^n - 1) - N \tag{1-4}$$

As explained above, $(2^n - 1)$ consists of all 1's, and subtracting a bit from 1 is the same as complementing the bit. Hence, the 1's complement of N can be obtained by complementing N bit-by-bit. Table 1-1 illustrates 1's complement for $n = 4$. Note that the 1's complement of 0000 is 1111, which represents minus zero. Note that 1's complement has two representations of 0, as does sign and magnitude.

Addition of 1's Complement Numbers

The addition of 1's complement numbers is similar to 2's complement except that instead of discarding the last carry, it is added to the n-bit sum in the position furthest to the right. This is referred to as an *end-around* carry. The addition of positive numbers is the same as illustrated for cases 1 and 2 under 2's complement. The remaining cases are illustrated below ($n = 4$).

3. Addition of positive and negative numbers (negative number with greater magnitude)

$$\begin{array}{rrl} +5 & 0101 & \\ \underline{-6} & \underline{1001} & \\ -1 & 1110 & \text{(correct answer)} \end{array}$$

4. Same as case 3 except positive number has greater magnitude

$$\begin{array}{rrl} -5 & 1010 & \\ \underline{+6} & \underline{0110} & \\ & (1)\ 0000 & \\ & \llcorner\!\!\longrightarrow 1 & \text{(end-around carry)} \\ & \overline{0001} & \text{(correct answer, } \textit{no} \text{ overflow)} \end{array}$$

5. Addition of two negative numbers, $|\text{sum}| < 2^{n-1}$

$$\begin{array}{rr l}
-3 & 1100 & \\
-4 & \underline{1011} & \\
 & (1)\ 0111 & \\
 & \llcorner\!\longrightarrow 1 & \text{(end-around carry)} \\
 & \overline{1000} & \text{(correct answer, } no \text{ overflow)}
\end{array}$$

6. Addition of two negative numbers, $|\text{sum}| \geq 2^{n-1}$

$$\begin{array}{rr l}
-5 & 1010 & \\
-6 & \underline{1001} & \\
 & (1)\ 0111 & \\
 & \llcorner\!\longrightarrow 1 & \text{(end-around carry)} \\
 & \overline{0100} & \text{(wrong answer because of overflow)}
\end{array}$$

Again, note that the overflow in case 6 is easy to detect because the addition of two negative numbers yields a positive result.

The proof that the end-around carry method gives the correct result follows for cases 4 and 5:

$$\begin{array}{ll}
\text{Case 4:} & -A + B \quad (\text{where } B > A) \\
 & \overline{A} + B = (2^n - 1 - A) + B = 2^n + (B - A) - 1
\end{array}$$

The end-around carry is equivalent to subtracting 2^n and adding 1, so the result is $(B - A)$, which is correct.

$$\begin{array}{ll}
\text{Case 5:} & -A - B \quad (A + B < 2^{n-1}) \\
 & \overline{A} + \overline{B} = (2^n - 1 - A) + (2^n - 1 - B) = 2^n + [2^n - 1 - (A + B)] - 1
\end{array}$$

After the end-around carry, the result is $2^n - 1 - (A + B) = (\overline{A + B})$ which is the correct representation for $-(A + B)$.

The following examples illustrate the addition of 1's and 2's complement numbers for a word length of $n = 8$:

1. Add –11 and –20 in 1's complement.

+11 = 00001011 +20 = 00010100

taking the bit-by-bit complement,

–11 is represented by 11110100 and –20 by 11101011

$$\begin{array}{r l}
11110100 & \quad (-11) \\
\underline{11101011} & \underline{+(-20)} \\
(1)\ 11011111 & \\
\llcorner\!\longrightarrow 1 & \quad \text{(end-around carry)} \\
\overline{11100000} = -31 &
\end{array}$$

2. Add −8 and +19 in 2's complement

+8 = 00001000

complementing all bits to the left of the first 1, −8, is represented by 11111000

```
   11111000    (−8)
   00010011    +19
(1)00001011 = +11
 └── (discard last carry)
```

Note that in both cases, the addition produced a carry out of the furthest left bit position, but there is no overflow because the answer can be correctly represented by eight bits (including sign). A general rule for detecting overflow when adding two n-bit signed binary numbers (1's or 2's complement) to get an n-bit sum is:

> An overflow occurs if adding two positive numbers gives a negative answer or if adding two negative numbers gives a positive answer.

An alternative method for detecting overflow in 2's complement addition is as follows:

> An overflow occurs if and only if the carry out of the sign position is not equal to the carry into the sign position.

1.5 Binary Codes

Although most large computers work internally with binary numbers, the input-output equipment generally uses decimal numbers. Because most logic circuits only accept two-valued signals, the decimal numbers must be coded in terms of binary signals. In the simplest form of binary code, each decimal digit is replaced by its binary equivalent. For example, 937.25 is represented by

```
   9    3    7  .  2    5
1001 0011 0111 . 0010 0101
```

This representation is referred to as binary-coded-decimal (BCD) or more explicitly as 8-4-2-1 BCD. Note that the result is quite different than that obtained by converting the number as a whole into binary. Because there are only ten decimal digits, 1010 through 1111 are not valid BCD codes.

TABLE 1-2 Binary Codes for Decimal Digits

Decimal Digit	8-4-2-1 Code (BCD)	6-3-1-1 Code	Excess-3 Code	2-out-of-5 Code	Gray Code
0	0000	0000	0011	00011	0000
1	0001	0001	0100	00101	0001
2	0010	0011	0101	00110	0011
3	0011	0100	0110	01001	0010
4	0100	0101	0111	01010	0110
5	0101	0111	1000	01100	1110
6	0110	1000	1001	10001	1010
7	0111	1001	1010	10010	1011
8	1000	1011	1011	10100	1001
9	1001	1100	1100	11000	1000

Table 1-2 shows several possible sets of binary codes for the ten decimal digits. Many other possibilities exist because the only requirement for a valid code is that each decimal digit be represented by a distinct combination of binary digits. To translate a decimal number to coded form, each decimal digit is replaced by its corresponding code. Thus 937 expressed in excess-3 code is 1100 0110 1010. The 8-4-2-1 (BCD) code and the 6-3-1-1 code are examples of weighted codes. A 4-bit weighted code has the property that if the weights are w_3, w_2, w_1, and w_0, the code $a_3a_2a_1a_0$ represents a decimal number *N*, where

$$N = w_3a_3 + w_2a_2 + w_1a_1 + w_0a_0$$

For example, the weights for the 6-3-1-1 code are $w_3 = 6$, $w_2 = 3$, $w_1 = 1$, and $w_0 = 1$. The binary code 1011 thus represents the decimal digit

$$N = 6 \cdot 1 + 3 \cdot 0 + 1 \cdot 1 + 1 \cdot 1 = 8$$

The excess-3 code is obtained from the 8-4-2-1 code by adding 3 (0011) to each of the codes. The 2-out-of-5 code has the property that exactly 2 out of the 5 bits are 1 for every valid code combination. This code has useful error-checking properties because if any one of the bits in a code combination is changed due to a malfunction of the logic circuitry, the number of 1 bits is no longer exactly two. The table shows one example of a Gray code. A Gray code has the property that the codes for successive decimal digits differ in exactly one bit. For example, the codes for 6 and 7 differ only in the fourth bit, and the codes for 9 and 0 differ only in the first bit. A Gray code is often used when translating an analog quantity, such as a shaft position, into digital form. In this case, a small change in the analog quantity will change only one bit in the code, which gives more reliable operation than if two or more bits changed at a time. The Gray and 2-out-of-5 codes are *not* weighted codes. In general, the decimal value of a coded digit *cannot* be computed by a simple formula when a non-weighted code is used.

Many applications of computers require the processing of data which contains numbers, letters, and other symbols such as punctuation marks. In order to transmit such alphanumeric data to or from a computer or store it internally in a computer, each symbol must be represented by a binary code. One common alphanumeric code is the ASCII code (American Standard Code for Information Interchange). This is a 7-bit code, so $2^7(128)$ different code combinations are available to represent letters, numbers, and other symbols. Table 1-3 shows a portion of the ASCII code; the code combinations not listed are used for special control functions such as "form feed" or "end of transmission." The word "*Start*" is represented in ASCII code as follows:

1010011	1110100	1100001	1110010	1110100
S	t	a	r	t

TABLE 1-3 ASCII Code

	ASCII Code								ASCII Code								ASCII Code						
Character	A_6	A_5	A_4	A_3	A_2	A_1	A_0	Character	A_6	A_5	A_4	A_3	A_2	A_1	A_0	Character	A_6	A_5	A_4	A_3	A_2	A_1	A_0
space	0	1	0	0	0	0	0	@	1	0	0	0	0	0	0	'	1	1	0	0	0	0	0
!	0	1	0	0	0	0	1	A	1	0	0	0	0	0	1	a	1	1	0	0	0	0	1
"	0	1	0	0	0	1	0	B	1	0	0	0	0	1	0	b	1	1	0	0	0	1	0
#	0	1	0	0	0	1	1	C	1	0	0	0	0	1	1	c	1	1	0	0	0	1	1
$	0	1	0	0	1	0	0	D	1	0	0	0	1	0	0	d	1	1	0	0	1	0	0
%	0	1	0	0	1	0	1	E	1	0	0	0	1	0	1	e	1	1	0	0	1	0	1
&	0	1	0	0	1	1	0	F	1	0	0	0	1	1	0	f	1	1	0	0	1	1	0
'	0	1	0	0	1	1	1	G	1	0	0	0	1	1	1	g	1	1	0	0	1	1	1
(	0	1	0	1	0	0	0	H	1	0	0	1	0	0	0	h	1	1	0	1	0	0	0
)	0	1	0	1	0	0	1	I	1	0	0	1	0	0	1	i	1	1	0	1	0	0	1
*	0	1	0	1	0	1	0	J	1	0	0	1	0	1	0	j	1	1	0	1	0	1	0
+	0	1	0	1	0	1	1	K	1	0	0	1	0	1	1	k	1	1	0	1	0	1	1
,	0	1	0	1	1	0	0	L	1	0	0	1	1	0	0	l	1	1	0	1	1	0	0
–	0	1	0	1	1	0	1	M	1	0	0	1	1	0	1	m	1	1	0	1	1	0	1
.	0	1	0	1	1	1	0	N	1	0	0	1	1	1	0	n	1	1	0	1	1	1	0
/	0	1	0	1	1	1	1	O	1	0	0	1	1	1	1	o	1	1	0	1	1	1	1
0	0	1	1	0	0	0	0	P	1	0	1	0	0	0	0	p	1	1	1	0	0	0	0
1	0	1	1	0	0	0	1	Q	1	0	1	0	0	0	1	q	1	1	1	0	0	0	1
2	0	1	1	0	0	1	0	R	1	0	1	0	0	1	0	r	1	1	1	0	0	1	0
3	0	1	1	0	0	1	1	S	1	0	1	0	0	1	1	s	1	1	1	0	0	1	1
4	0	1	1	0	1	0	0	T	1	0	1	0	1	0	0	t	1	1	1	0	1	0	0
5	0	1	1	0	1	0	1	U	1	0	1	0	1	0	1	u	1	1	1	0	1	0	1
6	0	1	1	0	1	1	0	V	1	0	1	0	1	1	0	v	1	1	1	0	1	1	0
7	0	1	1	0	1	1	1	W	1	0	1	0	1	1	1	w	1	1	1	0	1	1	1
8	0	1	1	1	0	0	0	X	1	0	1	1	0	0	0	x	1	1	1	1	0	0	0
9	0	1	1	1	0	0	1	Y	1	0	1	1	0	0	1	y	1	1	1	1	0	0	1
:	0	1	1	1	0	1	0	Z	1	0	1	1	0	1	0	z	1	1	1	1	0	1	0
;	0	1	1	1	0	1	1	[	1	0	1	1	0	1	1	{	1	1	1	1	0	1	1
<	0	1	1	1	1	0	0	\	1	0	1	1	1	0	0	|	1	1	1	1	1	0	0
=	0	1	1	1	1	0	1	]	1	0	1	1	1	0	1	}	1	1	1	1	1	0	1
>	0	1	1	1	1	1	0	^	1	0	1	1	1	1	0	~	1	1	1	1	1	1	0
?	0	1	1	1	1	1	1	—	1	0	1	1	1	1	1	delete	1	1	1	1	1	1	1

Problems

1.1 Convert to hexadecimal and then to binary:
(a) 757.25_{10} (b) 123.17_{10} (c) 356.89_{10} (d) 1063.5_{10}

1.2 Convert to octal. Convert to hexadecimal. Then convert both of your answers to decimal, and verify that they are the same.
(a) 111010110001.011_2 (b) 10110011101.11_2

1.3 Convert to base 6: $3BA.25_{14}$ (do all of the arithmetic in decimal).

1.4 (a) Convert to hexadecimal: 1457.11_{10}. Round to two digits past the hexadecimal point.
(b) Convert your answer to binary, and then to octal.
(c) Devise a scheme for converting hexadecimal directly to base 4 and convert your answer to base 4.
(d) Convert to decimal: $DEC.A_{16}$.

1.5 Add, subtract, and multiply in binary:
(a) 1111 and 1010 (b) 110110 and 11101 (c) 100100 and 10110

1.6 Subtract in binary. Place a 1 over each column from which it was necessary to borrow.
(a) 11110100 – 1000111 (b) 1110110 – 111101 (c) 10110010 – 111101

1.7 Add the following numbers in binary using 2's complement to represent negative numbers. Use a word length of 6 bits (including sign) and indicate if an overflow occurs.
(a) 21 + 11 (b) (−14) + (−32) (c) (−25) + 18
(d) (−12) + 13 (e) (−11) + (−21)
Repeat (a), (c), (d), and (e) using 1's complement to represent negative numbers.

1.8 A computer has a word length of 8 bits (including sign). If 2's complement is used to represent negative numbers, what range of integers can be stored in the computer? If 1's complement is used? (Express your answers in decimal.)

1.9 Construct a table for 7-3-2-1 weighted code and write 3659 using this code.

1.10 Convert to hexadecimal and then to binary.
(a) 1305.375_{10} (b) 111.33_{10} (c) 301.12_{10} (d) 1644.875_{10}

1.11 Convert to octal. Convert to hexadecimal. Then convert both of your answers to decimal, and verify that they are the same.
(a) 101111010100.101_2 (b) 100001101111.01_2

1.12 (a) Convert to base 3: 375.54_8 (do all of the arithmetic in decimal).
(b) Convert to base 4: 384.74_{10}.
(c) Convert to base 9: $A52.A4_{11}$ (do all of the arithmetic in decimal).

1.13 Convert to hexadecimal and then to binary: 544.1_9.

1.14 Convert the decimal number 97.7_{10} into a number with exactly the same value represented in the following bases. The exact value requires an infinite repeating part in the fractional part of the number. Show the steps of your derivation.
(a) binary (b) octal (c) hexadecimal (d) base 3 (e) base 5

1.15 Devise a scheme for converting base 3 numbers directly to base 9. Use your method to convert the following number to base 9: 1110212.20211_3

1.16 Convert the following decimal numbers to octal and then to binary:
(a) $2983^{63}/_{64}$ (b) 93.70 (c) $2983^{31}/_{32}$ (d) 109.30

1.17 Add, subtract, and multiply in binary:
(a) 1111 and 1001 (b) 1101001 and 110110 (c) 110010 and 11101

1.18 Subtract in binary. Place a 1 over each column from which it was necessary to borrow.
(a) 10100100 − 01110011 (b) 10010011 − 01011001
(c) 11110011 − 10011110

1.19 Divide in binary:
(a) 11101001 ÷ 101 (b) 110000001 ÷ 1110 (c) 1110010 ÷ 1001
Check your answers by multiplying out in binary and adding the remainder.

1.20 Divide in binary:
(a) 10001101 ÷ 110 (b) 110000011 ÷ 1011 (c) 1110100 ÷ 1010

1.21 Assume three digits are used to represent positive integers and also assume the following operations are correct. Determine the base of the numbers. Did any of the additions overflow?
(a) 654 + 013 = 000
(b) 024 + 043 + 013 + 033 = 223
(c) 024 + 043 + 013 + 033 = 201

1.22 What is the lowest number of bits (digits) required in the binary number approximately equal to the decimal number 0.6117_{10} so that the binary number has the same or better precision?

1.23 Convert $0.363636\ldots_{10}$ to its exact equivalent base 8 number.

1.24 (a) Verify that a number in base b can be converted to base b^3 by partitioning the digits of the base b number into groups of three consecutive digits starting at the radix point and proceeding both left and right and converting each group into a base b^3 digit. (*Hint:* Represent the base b number using the power series expansion.)
(b) Verify that a number in base b^3 can be converted to base b by expanding each digit of the base b^3 number into three consecutive digits starting at the radix point and proceeding both left and right.

1.25 (a) Show how to represent each of the numbers $(5 - 1)$, $(5^2 - 1)$, and $(5^3 - 1)$ as base 5 numbers.
(b) Generalize your answers to part (a) and show how to represent $(b^n - 1)$ as a base b number, where b can be any integer larger than 1 and n any integer larger than 0. Give a mathematical derivation of your result.

1.26 (a) Show that the number 121_b, where b is any base greater than 2, is a perfect square (i.e., it is equal to the square of some number).
(b) Repeat part (a) for the number 12321_b, where $b > 3$.
(c) Repeat part (a) for the number 14641_b, where $b > 6$.
(d) Repeat part (a) for the number 1234321_b, where $b > 4$.

1.27 (a) Convert $(0.12)_3$ to a base 6 fraction.
(b) Convert $(0.375)_{10}$ to a base 8 fraction.
(c) Let $N = (0.a_{-1}a_{-2} \cdots a_{-m})_R$ be an *any* base R fraction with at most m nonzero digits. Determine the necessary and sufficient conditions for N to be representable as a base S fraction with a finite number of nonzero digits; say $N = (0.b_{-1}b_{-2} \cdots b_{-n})_S$. (*Hint:* Part (a) gives an example. Note that

$$(a_{-1}R^{-1} + a_{-2}R^{-2} + \cdots a_{-m}R^{-m})S^n$$

must be an integer.)
(d) Generalize part (a) to determine necessary and sufficient conditions for a *specific*, but not every, base R fraction, $N = (0.a_{-1}a_{-2} \cdots a_{-m})_R$, to be representable as a base S fraction with a finite number of nonzero digits.

1.28 Construct a table for 4-3-2-1 weighted code and write 9154 using this code.

1.29 Is it possible to construct a 5-3-1-1 weighted code? A 6-4-1-1 weighted code? Justify your answers.

1.30 Is it possible to construct a 5-4-1-1 weighted code? A 6-3-2-1 weighted code? Justify your answers.

1.31 Construct a 6-2-2-1 weighted code for decimal digits. What number does 1100 0011 represent in this code?

1.32 Construct a 5-2-2-1 weighted code for decimal digits. What numbers does 1110 0110 represent in this code?

1.33 Construct a 7-3-2-1 code for base 12 digits. Write B4A9 using this code.

1.34 (a) It is possible to have negative weights in a weighted code for the decimal digits, e.g., 8, 4, −2, and −1 can be used. Construct a table for this weighted code.
(b) If d is a decimal digit in this code, how can the code for $9 - d$ be obtained?

1.35 Convert to hexadecimal, and then give the ASCII code for the resulting hexadecimal number (including the code for the hexadecimal point):
(a) 222.22_{10} (b) 183.81_{10}

1.36 Repeat 1.7 for the following numbers:
(a) $(-10) + (-11)$ (b) $(-10) + (-6)$ (c) $(-8) + (-11)$
(d) $11 + 9$ (e) $(-11) + (-4)$

1.37 Because $A - B = A + (-B)$, the subtraction of signed numbers can be accomplished by adding the complement. Subtract each of the following pairs of 5-bit binary numbers by adding the complement of the subtrahend to the minuend. Indicate when an overflow occurs. Assume that negative numbers are represented in 1's complement. Then repeat using 2's complement.

(a)	(b)	(c)	(d)	(e)
01001	11010	10110	11011	11100
−11010	−11001	−01101	−00111	−10101

1.38 Work Problem 1.37 for the following pairs of numbers:

(a)	(b)	(c)	(d)
11010	01011	10001	10101
−10100	−11000	−01010	−11010

1.39 (a) $A = 101010$ and $B = 011101$ are 1's complement numbers. Perform the following operations and indicate whether overflow occurs.
(i) $A + B$ (ii) $A - B$
(b) Repeat part (a) assuming the numbers are 2's complement numbers.

1.40 (a) Assume the integers below are 1's complement integers. Find the 1's complement of each number, and give the decimal values of the original number and of its complement.
(i) 0000000 (ii) 1111111 (iii) 00110011 (iv) 1000000
(b) Repeat part (a) assuming the numbers are 2's complement numbers and finding the 2's complement of them.

1.41 An alternative algorithm for converting a base 20 integer, $d_{n-1}d_{n-2} \cdots d_1d_0$, into a base 10 integer is stated as follows: Multiply d_i by 2^i and add i 0's on the right, and then add all of the results.

(a) Use this algorithm to convert $GA7_{20}$ to base 10. (G_{20} is 16_{10}.)
(b) Prove that this algorithm is valid.
(c) Consider converting a base 20 fraction, $0.d_{-1}d_{-2} \cdots d_{-n+1}d_{-n}$, into a base 10 fraction. State an algorithm analogous to the one above for doing the conversion.
(d) Apply your algorithm of part (c) to $0.FA7_{20}$.

1.42 Let A and B be positive integers, and consider the addition of A and B in an n-bit 2's complement number system.
(a) Show that the addition of A and B produces the correct representation of the sum if the magnitude of $(A + B)$ is $< 2^{n-1} - 1$ but it produces a representation of a negative number of magnitude $2^n - (A + B)$ if the magnitude of $(A + B)$ is $> 2^{n-1} - 1$.
(b) Show that the addition of A and $(-B)$ always produces the correct representation of the sum. Consider both the case where $A \geq B$ and the case $A < B$.
(c) Show that the addition of $(2^n - A) + (2^n - B)$, with the carry from the sign position ignored, produces the correct 2's complement representation of $-(A + B)$ if the magnitude of $A + B$ is less than or equal to 2^{n-1}. Also, show that it produces an incorrect sum representing the positive number $2^n - (A + B)$ if the magnitude of $(A + B) > 2^{n-1}$.

1.43 Let A and B be integers and consider the addition of A and B in an n-bit 1's complement number system. Prove that addition of A and B using the end-around carry produces the correct representation of the sum provided overflow does not occur. Consider the four cases: A and B both positive, A positive and B negative with the magnitude of A greater than the magnitude of B, A positive and B negative with the magnitude of A less than or equal to the magnitude of B, and A and B both negative.

1.44 Prove that in a 2's complement number system addition overflows if and only if the carry from the sign position does not equal the carry into the sign position. Consider the three cases: adding two positive numbers, adding two negative numbers, and adding two numbers of opposite sign.

1.45 Restate the method for detecting overflow of Problem 1.44 so that it applies to 1's complement numbers.

1.46 Let $B = b_{n-1}b_{n-2} \cdots b_1b_0$ be an n-bit 2's complement integer. Show that the decimal value of B is $-b_{n-1}2^{n-1} + b_{n-2}2^{n-2} + b_{n-3}2^{n-3} + \cdots + b_12 + b_0$. (*Hint:* Consider positive ($b_{n-1} = 0$) and negative ($b_{n-1} = 1$) numbers separately, and note that the magnitude of a negative number is obtained by subtracting each bit from 1 (i.e., complementing each bit) and adding 1 to the result.)

UNIT 2

Boolean Algebra

Objectives

A list of some of the laws of switching algebra, which is a special case of Boolean algebra, is given in Table 2-3. Additional theorems of Boolean algebra are given in Table 2-4. When you complete this unit, you should be familiar with and be able to use these laws and theorems of Boolean algebra.

1. Understand the basic operations and laws of Boolean algebra.
2. Relate these operations and laws to circuits composed of AND gates, OR gates, and INVERTERS. Also relate these operations and laws to circuits composed of switches.
3. Prove any of these laws in switching algebra using a truth table.
4. Apply these laws to the manipulation of algebraic expressions including:
    a. Multiplying out an expression to obtain a sum of products (SOP)
    b. Factoring an expression to obtain a product of sums (POS)
    c. Simplifying an expression by applying one of the laws
    d. Finding the complement of an expression

Study Guide

1. In this unit you will study Boolean algebra, the basic mathematics needed for the logic design of digital systems. Just as when you first learned ordinary algebra, you will need a fair amount of practice before you can use Boolean algebra effectively. However, by the end of the course, you should be just as comfortable with Boolean algebra as with ordinary algebra. Fortunately, many of the rules of Boolean algebra are the same as for ordinary algebra, but watch out for some surprises!

2. Study Sections 2.1 and 2.2, *Introduction* and *Basic Operations.*

   (a) How does the meaning of the symbols 0 and 1 as used in this unit differ from the meaning as used in Unit 1?

   (b) Two commonly used notations for the inverse or complement of A are $\overline{A}$ and A'. The latter has the advantage that it is much easier for typists, printers, and computers. (Have you ever tried to get a computer to print a bar over a letter?) We will use A' for the complement of A. You may use either notation in your work, but please do not mix notations in the same equation. Most engineers use + for OR and • (or no symbol) for AND, and we will follow this practice. An alternative notation, often used by mathematicians, is $\vee$ for OR and $\wedge$ for AND.

   (c) Many different symbols are used for AND, OR, and INVERTER logic blocks. Initially we will use

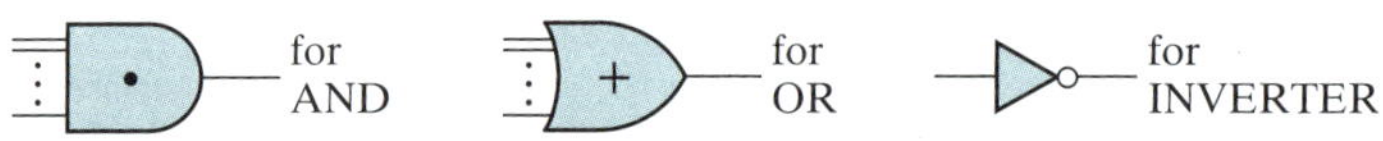

   The shapes of these symbols conform to those commonly used in industrial practice. We have added the + and • for clarity. These symbols point in the direction of signal flow. This makes it easier to read the circuit diagrams in comparison with the square or round symbols used in some books.

   (d) Determine the output of each of the following gates:

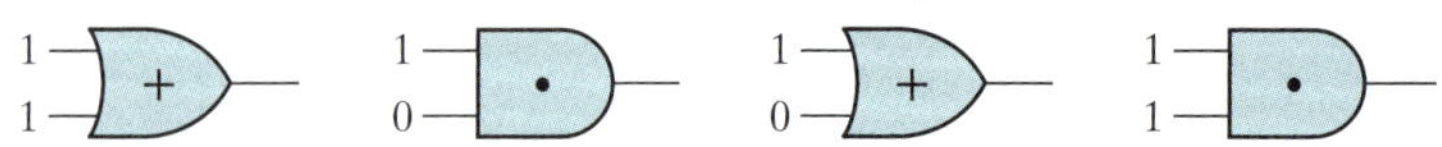

   (e) Determine the unspecified inputs to each of the following gates if the outputs are as shown:

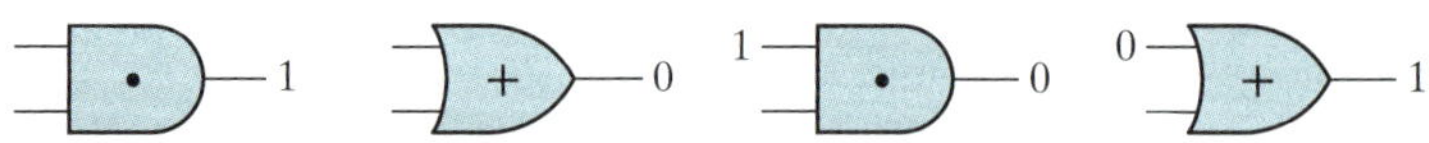

3. Study Section 2.3, *Boolean Expressions and Truth Tables.*

(a) How many *variables* does the following expression contain?
How many literals?

$$A'BC'D + AB + B'CD + D'$$

(b) For the following circuit, if $A = B = 0$ and $C = D = E = 1$, indicate the output of each gate (0 or 1) on the circuit diagram:

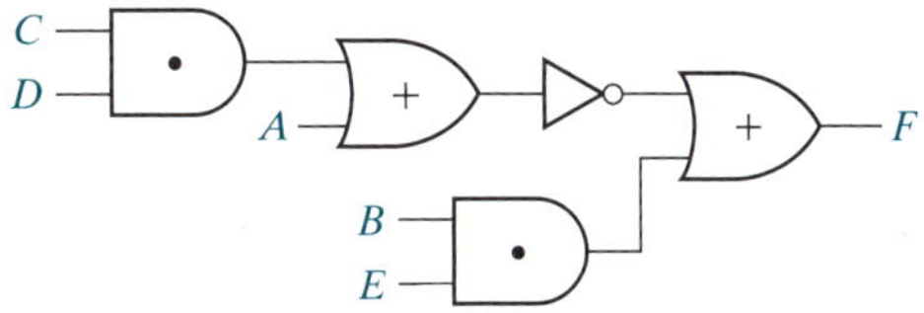

(c) Derive a Boolean expression for the circuit output. Then substitute $A = B = 0$ and $C = D = E = 1$ into your expression and verify that the value of F obtained in this way is the same as that obtained on the circuit diagram in (b).

(d) Write an expression for the output of the following circuit and complete the truth table:

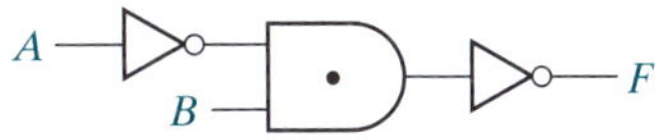

$F =$ ________

A B	A′	A′B	(A′B)′

(e) When filling in the combinations of values for the variables on the left side of a truth table, always list the combinations of 0's and 1's in binary order. For example, for a three-variable truth table, the first row should be 000, the next row 001, then 010, 011, 100, 101, 110, and 111. Write an expression for the output of the following circuit and complete the truth table:

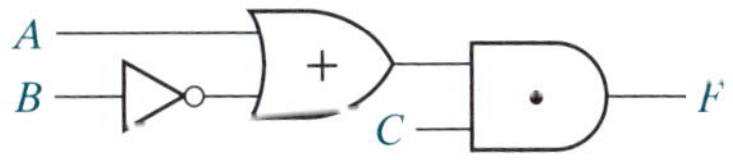

$F =$ ________

A B C	B′	A + B′	C(A + B′)

(f) Draw a gate circuit which has an output

$$Z = [BC' + F(E + AD')]'$$

(*Hint:* Start with the innermost parentheses and draw the circuit for AD' first.)

4. Study Section 2.4, *Basic Theorems.*

(a) Prove each of the Theorems (2-4) through (2-8D) by showing that it is valid for both $X = 0$ and $X = 1$.

(b) Determine the output of each of these gates:

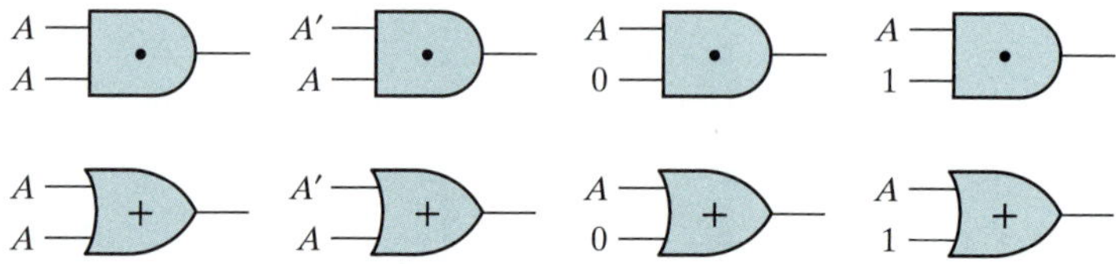

(c) State which of the basic theorems was used in simplifying each of the following expressions:

$(AB' + C) \cdot 0 = 0$ $\qquad$ $A(B + C') + 1 = 1$

$(BC' + A)(BC' + A) = BC' + A$ $\qquad$ $X(Y' + Z) + [X(Y' + Z)]' = 1$

$(X' + YZ)(X' + YZ)' = 0$ $\qquad$ $D'(E' + F) + D'(E' + F) = D'(E' + F)$

5. Study Section 2.5, *Commutative, Associative, Distributive, and DeMorgan's Laws.*

(a) State the associative law for OR.

(b) State the commutative law for AND.

(c) Simplify the following circuit by using the associative laws. Your answer should require only two gates.

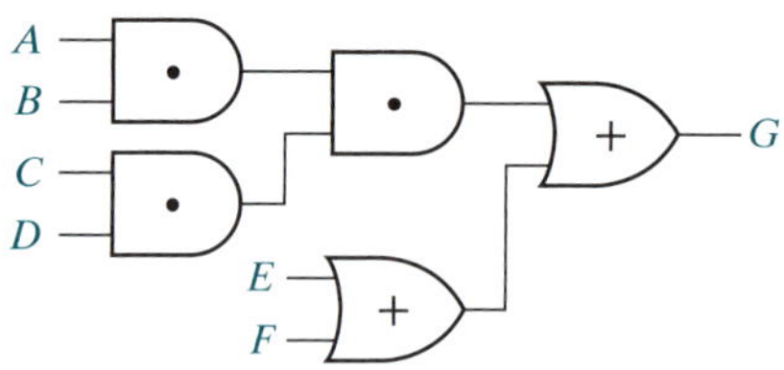

(d) For each gate determine the value of the unspecified input(s):

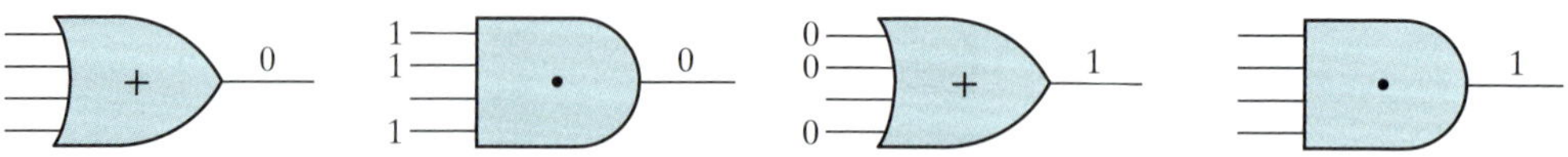

(e) Using a truth table, verify the distributive law, Equation (2-11).

(f) Illustrate the distributive laws, Equations (2-11) and (2-11D), using AND and OR gates.

(g) Verify Equation (2-3) using the second distributive law.

(h) Show how the second distributive law can be used to factor $RS + T'$.

6. Study Section 2.6, *Simplification Theorems.*

(a) By completing the truth table, prove that $XY' + Y = X + Y$.

X Y	XY′	XY′ + Y	X + Y
0 0			
0 1			
1 0			
1 1			

(b) Which one of Theorems in Table 2-4 was applied to simplify each of the following expressions? Identify X and Y in each case.

$$(A + B)(DE)' + DE = A + B + DE$$

$$AB' + AB'C'D = AB'$$

$$(A' + B)(CD + E') + (A' + B)(CD + E')' = A' + B$$

$$(A + BC' + D'E)(A + D'E) = A + D'E$$

(c) Simplify the following circuit to a single gate:

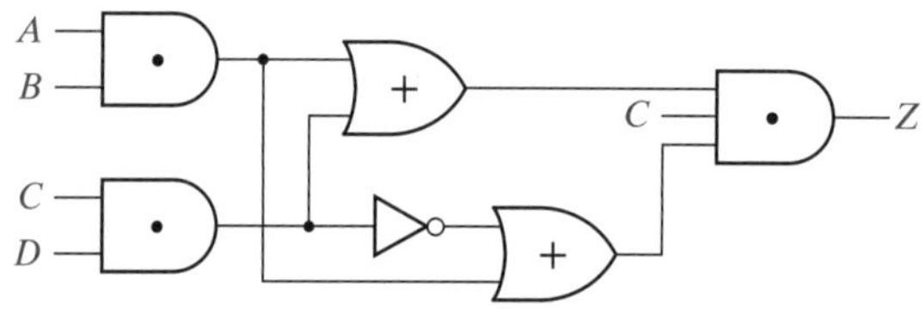

(d) Work Problems 2.1, 2.2, 2.3, and 2.4.

7. Study Section 2.7, *Multiplying Out and Factoring.*

(a) Indicate which of the following expressions are in the product-of-sums form, sum-of-products form, or neither:

$$AB' + D'EF' + G$$
$$(A + B'C')(A' + BC)$$
$$AB'(C' + D + E')(F' + G)$$
$$X'Y + WX(X' + Z) + A'B'C'$$

Your answer should indicate one expression as a product-of-sums form, one as sum-of-products form, and two as neither, not necessarily in that order.

(b) When multiplying out an expression, why should the second distributive law be applied before the ordinary distributive law when possible?

(c) Factor as much as possible using the ordinary distributive law:

$$AD + B'CD + B'DE$$

Now factor your result using the second distributive law to obtain a product of sums.

(d) Work Problems 2.5, 2.6, and 2.7.

8. Probably the most difficult part of the unit is using the second distributive law for factoring or multiplying out an expression. If you have difficulty with Problems 2.5 or 2.6, or you cannot work them *quickly*, study the examples in Section 2.7 again, and then work the following problems.

Multiply out:

(a) $(B' + D + E)(B' + D + A)(AE + C')$

(b) $(A + C')(B' + D)(C' + D')(C + D)E$

As usual, when we say multiply out, we do not mean to multiply out by brute force, but rather to use the second distributive law whenever you can to cut down on the amount of work required.

The answer to (a) should be of the following form: $XX + XX + XX$ and (b) of the form: $XXX + XXXXX$, where each X represents a single variable or its complement.

Now factor your answer to (a) to see that you can get back the original expression.

9. Study Section 2.8, *Complementing Boolean Expressions.*

10. Find the complement of each of the following expressions as indicated. In your answer, the complement operation should be applied only to single variables.

(a) $(ab'c')' =$

(b) $(a' + b + c + d')' =$

(c) $(a' + bc)' =$

(d) $(a'b' + cd)' =$

(e) $[a(b' + c'd)]' =$

11. Because $(X')' = X$, if you complement each of your answers to 10, you should get back the original expression. Verify that this is true.

(a)

(b)

(c)

(d)

(e)

12. Given that $F = a'b + b'c$, $F' =$

Complete the following truth table and verify that your answer is correct:

a b c	a′b	b′c	a′b + b′c	(a + b′)	(b + c′)	F′
0 0 0						
0 0 1						
0 1 0						
0 1 1						
1 0 0						
1 0 1						
1 1 0						
1 1 1						

13. A fully simplified expression should have nothing complemented except the individual variables. For example, $F = (X + Y)'(W + Z)$ is *not* a minimum product of sums. Find the minimum product of sums for F.

14. Work Problems 2.8 and 2.9.

15. Find the dual of $(M + N')P'$.

16. Review the laws of Table 2-3 and the first three theorems of Table 2-4. Make sure that you can recognize when to apply them even if an expression has been substituted for a variable.

17. Reread the objectives of this unit. If you are satisfied that you can meet these objectives, take the readiness test.

    [*Note:* You will be provided with a copy of Tables 2-3 and 2-4 when you take the readiness test this time. However, by the end of Unit 3, you should know all the laws and theorems by memory.]

Boolean Algebra

2.1 Introduction

The basic mathematics needed for the study of logic design of digital systems is Boolean algebra. George Boole developed Boolean algebra in 1847 and used it to solve problems in mathematical logic. Boolean algebra has many other applications, including set theory and mathematical logic; however, we primarily consider its application to switching circuits. All of the switching devices we will use are essentially two-state devices (e.g., switches which are open or closed and transistors with high or low output voltages). Consequently, we will emphasize the special case of Boolean algebra in which all of the variables assume only one of two values; this two-valued Boolean algebra is often called *switching algebra*. Claude Shannon first applied Boolean algebra to the design of switching circuits in 1939. First, we develop some of the properties of switching algebra and use these to define a general Boolean algebra.

We will use a Boolean variable, such as X or Y, to represent the input or output of a switching circuit. We will assume that each of these variables can take on only two different values. The symbols "0" and "1" are used to represent these two different values. Thus, if X is a Boolean (switching) variable, then either $X = 0$ or $X = 1$.

The symbols "0" and "1" used in Boolean algebra do not have a numeric value; instead they represent two different states in a logic circuit and are the two values of a switching variable. In a logic gate circuit, 0 (usually) represents a range of low voltages, and 1 represents a range of high voltages. In a switch circuit, 0 (usually) represents an open switch, and 1 represents a closed circuit. In general, 0 and 1 can be used to represent the two states in any binary-valued system.

2.2 Basic Operations

The basic operations of Boolean (switching) algebra are called AND, OR, and complement (or inverse). In the case of switch circuits these operations correspond to different configurations of switches. To apply switching algebra to a switch circuit, each switch contact is labeled with a variable. If contact X is open, the variable X is defined to be 0; if contact X is closed, the variable X is defined to be 1.

$X = 0 \rightarrow$ switch open
$X = 1 \rightarrow$ switch closed

The contacts in a switch can be normally open (NO) or normally closed (NC). When the switch position is changed, the NO contact closes and the NC contact opens, so the NO and NC contacts are always in opposite states. If X is the variable assigned to the NO contact, then the variable assigned to the NC contact is the complement of X, denoted as X', where the prime (′) denotes complementation.

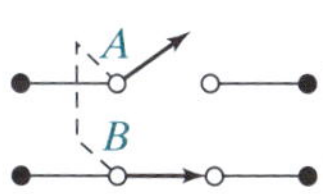

The complement of 0 is 1, and the complement of 1 is 0. Symbolically, we write

$$0' = 1 \quad \text{and} \quad 1' = 0$$

If X is a switching variable,

$$X' = 1 \text{ if } X = 0 \quad \text{and} \quad X' = 0 \text{ if } X = 1$$

An alternate name for complementation is inversion, and the electronic circuit which forms the inverse of X is referred to as an inverter. Symbolically, we represent an inverter by

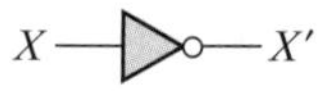

where the circle at the output indicates inversion. A low voltage at the inverter input produces a high voltage at the output and vice versa.

In a general switch circuit, the value 0 is assigned to the connection between two terminals in the circuit if there is no connection (open circuit) between the terminals, and a 1 is assigned if there is a connection (closed circuit) between the terminals. If the switch circuit only contains two switches, the switch contacts must be connected in series or in parallel. When switch contacts A and B are connected in series, there is an open circuit between the terminals if either A or B or both are open (0), and there is a closed circuit between the terminals only if both A and B are closed (1).

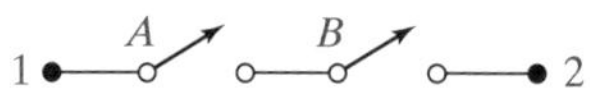

$C = 0 \rightarrow$ open circuit between terminals 1 and 2
$C = 1 \rightarrow$ closed circuit between terminals 1 and 2

This is summarized in the following truth table:

A B	C = A · B
0 0	0
0 1	0
1 0	0
1 1	1

The operation defined by the table is called AND and it is written algebraically as $C = A \cdot B$. The "·" symbol is frequently omitted in a Boolean expression, and we will usually write AB instead of $A \cdot B$. The AND operation is also referred to as logical (or Boolean) multiplication.

When switches A and B are connected in parallel, there is a closed circuit between the terminals if either A or B is closed (1), and there is an open circuit between the terminals only if both A and B are open (0).

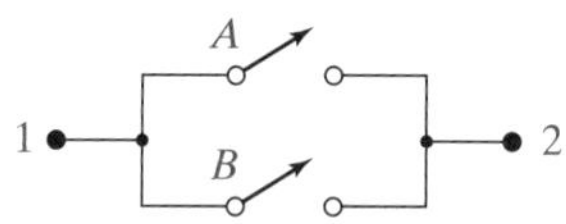

This is summarized in the following truth table:

$A\ B$	$C = A + B$
0 0	0
0 1	1
1 0	1
1 1	1

The operation defined by the table is called OR and it is written algebraically as $C = A + B$. This type of OR operation is sometimes referred to as inclusive OR as opposed to exclusive OR, which is defined later. The OR operation is also referred to as logical (or Boolean) addition.

Logic gates operate so that the voltage on inputs and outputs of a gate is either in a low voltage range or a high voltage range, except when the signals are changing. Switching algebra can be applied to logic gates by assigning 0 and 1 to the two voltage ranges. Usually, a 0 is assigned to the low voltage range and a 1 to the high voltage range.

A logic gate which performs the AND operation is represented by

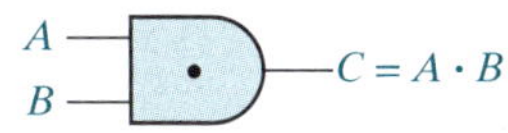

The gate output is $C = 1$ if and only if the gate inputs $A = 1$ and $B = 1$.

A logic gate which performs the OR operation is represented by

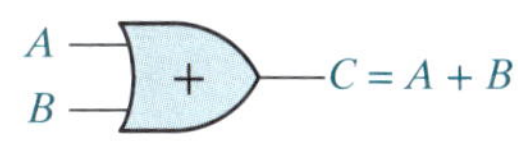

The gate output is $C = 1$ if and only if the gate inputs $A = 1$ or $B = 1$ (or both). Electronic circuits which realize inverters and AND and OR gates are described in Appendix A.

2.3 Boolean Expressions and Truth Tables

Boolean expressions are formed by application of the basic operations to one or more variables or constants. The simplest expressions consist of a single constant or variable, such as 0, X, or Y'. More complicated expressions are formed by combining two or more other expressions using AND or OR, or by complementing another expression. Examples of expressions are

$$AB' + C \tag{2-1}$$

$$[A(C + D)]' + BE \tag{2-2}$$

Parentheses are added as needed to specify the order in which the operations are performed. When parentheses are omitted, complementation is performed first followed by AND and then OR. Thus in Expression (2-1), B' is formed first, then AB', and finally $AB' + C$.

Each expression corresponds directly to a circuit of logic gates. Figure 2-1 gives the circuits for Expressions (2-1) and (2-2).

FIGURE 2-1 Circuits for Expressions (2-1) and (2-2)

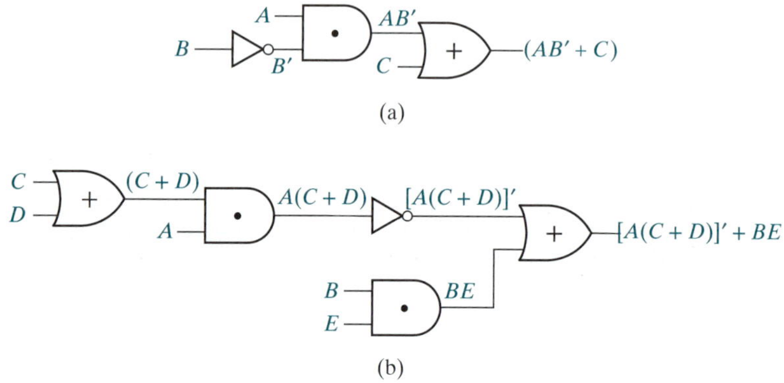

An expression is evaluated by substituting a value of 0 or 1 for each variable. If $A = B = C = 1$ and $D = E = 0$, the value of Expression (2-2) is

$$[A(C + D)]' + BE = [1(1 + 0)]' + 1 \cdot 0 = [1(1)]' + 0 = 0 + 0 = 0$$

Each appearance of a variable or its complement in an expression will be referred to as a *literal*. Thus, the following expression, which has three variables, has 10 literals:

$$ab'c + a'b + a'bc' + b'c'$$

When an expression is realized using logic gates, each literal in the expression corresponds to a gate input.

A *truth table* (also called a table of combinations) specifies the values of a Boolean expression for every possible combination of values of the variables in the expression. The name truth table comes from a similar table which is used in symbolic logic to list the truth or falsity of a statement under all possible conditions. We can use a truth table to specify the output values for a circuit of logic gates in terms of the values of the input variables. The output of the circuit in Figure 2-2(a) is $F = A' + B$. Figure 2-2(b) shows a truth table which specifies the output of the circuit for all possible combinations of values of the inputs A and B. The first two columns list the four combinations of values of A and B, and the next column gives the corresponding values of A'. The last column, which gives the values of $A' + B$, is formed by ORing together corresponding values of A' and B in each row.

FIGURE 2-2 Two-Input Circuit and Truth Table

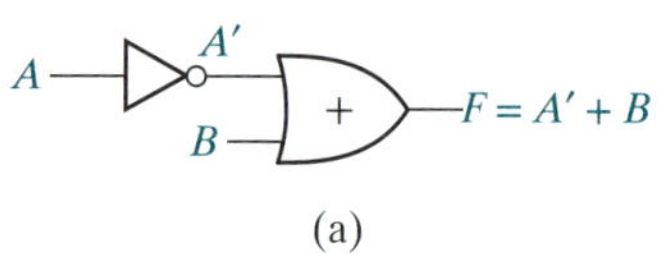

A	B	A′	F = A′ + B
0	0	1	1
0	1	1	1
1	0	0	0
1	1	0	1

(b)

Next, we will use a truth table to specify the value of Expression (2-1) for all possible combinations of values of the variables A, B, and C. On the left side of Table 2-1, we list the values of the variables A, B, and C. Because each of the three variables can assume the value 0 or 1, there are $2 \times 2 \times 2 = 8$ combinations of values of the variables. These combinations are easily obtained by listing the binary numbers 000, 001, ..., 111. In the next three columns of the truth table, we compute B', AB', and $AB' + C$, respectively.

Two expressions are equal if they have the same value for every possible combination of the variables. The expression $(A + C)(B' + C)$ is evaluated using the last three columns of Table 2-1. Because it has the same value as $AB' + C$ for all eight combinations of values of the variables A, B, and C, we conclude

TABLE 2-1

A B C	B′	AB′	AB′ + C	A + C	B′ + C	(A + C)(B′ + C)
0 0 0	1	0	0	0	1	0
0 0 1	1	0	1	1	1	1
0 1 0	0	0	0	0	0	0
0 1 1	0	0	1	1	1	1
1 0 0	1	1	1	1	1	1
1 0 1	1	1	1	1	1	1
1 1 0	0	0	0	1	0	0
1 1 1	0	0	1	1	1	1

$$AB' + C = (A + C)(B' + C) \tag{2-3}$$

If an expression has n variables, and each variable can have the value 0 or 1, the number of different combinations of values of the variables is

$$\underbrace{2 \times 2 \times 2 \times \dots}_{n \text{ times}} = 2^n$$

Therefore, a truth table for an n-variable expression will have 2^n rows.

2.4 Basic Theorems

The following basic laws and theorems of Boolean algebra involve only a single variable:

Operations with 0 and 1:

$$X + 0 = X \quad (2\text{-}4) \qquad X \cdot 1 = X \quad (2\text{-}4D)$$

$$X + 1 = 1 \quad (2\text{-}5) \qquad X \cdot 0 = 0 \quad (2\text{-}5D)$$

Idempotent laws:

$$X + X = X \qquad (2\text{-}6) \qquad\qquad X \cdot X = X \qquad (2\text{-}6D)$$

Involution law:

$$(X')' = X \qquad (2\text{-}7)$$

Laws of complementarity:

$$X + X' = 1 \qquad (2\text{-}8) \qquad\qquad X \cdot X' = 0 \qquad (2\text{-}8D)$$

Each of these theorems is easily proved by showing that it is valid for both of the possible values of X. For example, to prove $X + X' = 1$, we observe that if

$$X = 0, \quad 0 + 0' = 0 + 1 = 1, \quad \text{and if } X = 1, \quad 1 + 1' = 1 + 0 = 1$$

Any expression can be substituted for the variable X in these theorems. Thus, by Theorem (2-5),

$$(AB' + D)E + 1 = 1$$

and by Theorem (2-8D),

$$(AB' + D)(AB' + D)' = 0$$

We will illustrate some of the basic theorems with circuits of switches. As before, 0 will represent an open circuit or open switch, and 1 will represent a closed circuit or closed switch. If two switches are both labeled with the variable A, this means that both switches are open when $A = 0$ and both are closed when $A = 1$. Thus the circuit

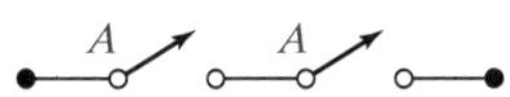

can be replaced with a single switch:

This illustrates the theorem $A \cdot A = A$. Similarly,

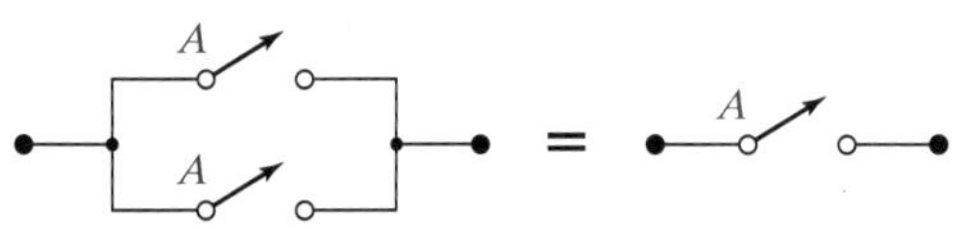

which illustrates the theorem $A + A = A$. A switch in parallel with an open circuit is equivalent to the switch alone

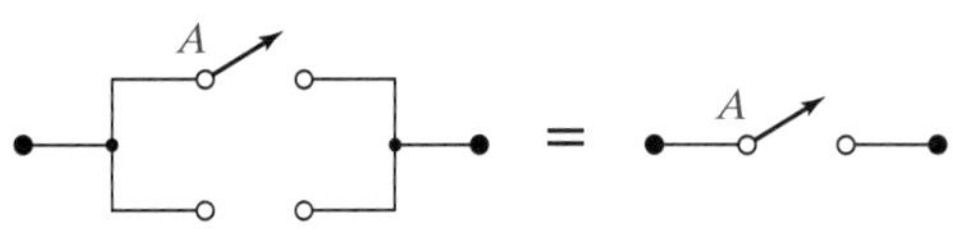

$(A + 0 = A)$

while a switch in parallel with a short circuit is equivalent to a short circuit.

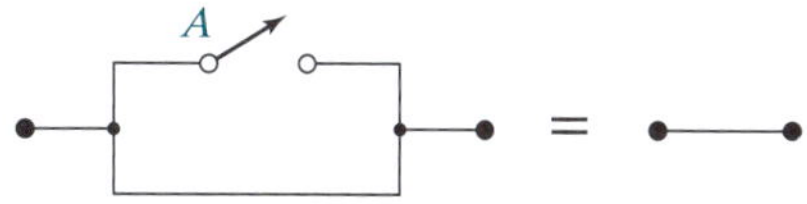

$(A + 1 = 1)$

If a switch is labeled A', then it is open when A is closed and conversely. Hence, A in parallel with A' can be replaced with a closed circuit because one or the other of the two switches is always closed.

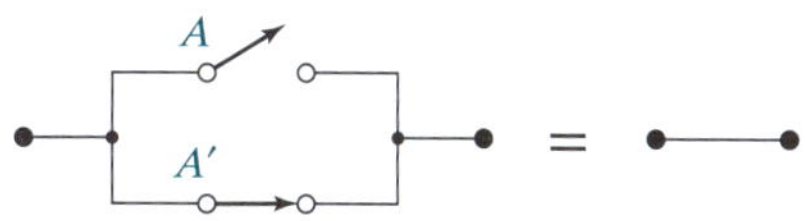

$(A + A' = 1)$

Similarly, switch A in series with A' can be replaced with an open circuit (why?).

$(A \cdot A' = 0)$

2.5 Commutative, Associative, Distributive, and DeMorgan's Laws

Many of the laws of ordinary algebra, such as the commutative and associative laws, also apply to Boolean algebra. The commutative laws for AND and OR, which follow directly from the definitions of the AND and OR operations, are

$$XY = YX \qquad (2\text{-}9) \qquad\qquad X + Y = Y + X \qquad (2\text{-}9D)$$

This means that the order in which the variables are written will not affect the result of applying the AND and OR operations.

The associative laws also apply to AND and OR:

$$(XY)Z = X(YZ) = XYZ \qquad (2\text{-}10)$$

$$(X + Y) + Z = X + (Y + Z) = X + Y + Z \qquad (2\text{-}10D)$$

When forming the AND (or OR) of three variables, the result is independent of which pair of variables we associate together first, so parentheses can be omitted as indicated in Equations (2-10) and (2-10D).

When the preceding laws are interpreted as switch circuits, they simply indicate that the order in which switch contacts are connected does not change the logic operation of the circuit.

We will prove the associative law for AND by using a truth table (Table 2-2). On the left side of the table, we list all combinations of values of the variables X, Y, and Z. In the next two columns of the truth table, we compute XY and YZ for each combination of values of X, Y, and Z. Finally, we compute $(XY)Z$ and $X(YZ)$. Because $(XY)Z$ and $X(YZ)$ are equal for all possible combinations of values of the variables, we conclude that Equation (2-10) is valid.

TABLE 2-2 Proof of Associative Law for AND

X Y Z	XY	YZ	(XY)Z	X(YZ)
0 0 0	0	0	0	0
0 0 1	0	0	0	0
0 1 0	0	0	0	0
0 1 1	0	1	0	0
1 0 0	0	0	0	0
1 0 1	0	0	0	0
1 1 0	1	0	0	0
1 1 1	1	1	1	1

Figure 2-3 illustrates the associative laws using AND and OR gates. In Figure 2-3(a) two two-input AND gates are replaced with a single three-input AND gate. Similarly, in Figure 2-3(b) two two-input OR gates are replaced with a single three-input OR gate.

FIGURE 2-3 Associative Laws for AND and OR

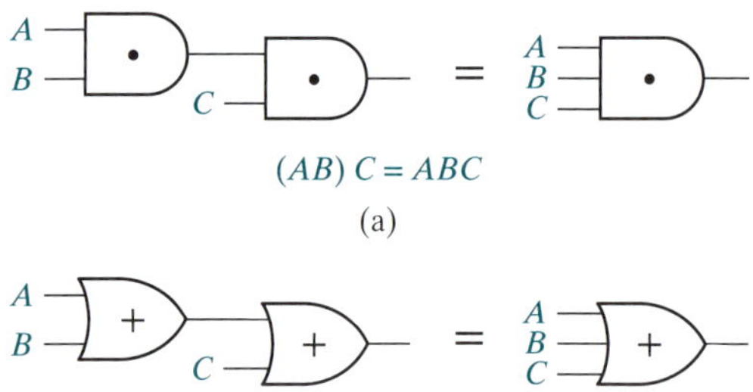

$(AB)\,C = ABC$

(a)

$(A + B) + C = A + B + C$

(b)

When two or more variables are ANDed together, the value of the result will be 1 if all of the variables have the value 1. If any of the variables have the value 0, the result of the AND operation will be 0. For example,

$$XYZ = 1 \text{ iff } X = Y = Z = 1$$

When two or more variables are ORed together, the value of the result will be 1 if any of the variables have the value 1. The result of the OR operation will be 0 iff all of the variables have the value 0. For example,

$$X + Y + Z = 0 \text{ iff } X = Y = Z = 0$$

Using a truth table, it is easy to show that the distributive law is valid:

$$X(Y + Z) = XY + XZ \tag{2-11}$$

In addition to the ordinary distributive law, a second distributive law is valid for Boolean algebra but not for ordinary algebra:

$$X + YZ = (X + Y)(X + Z) \tag{2-11D}$$

Proof of the second distributive law follows:

$$\begin{aligned}(X + Y)(X + Z) &= X(X + Z) + Y(X + Z) = XX + XZ + YX + YZ \\ &\qquad \text{(by (2-11))} \\ &= X + XZ + XY + YZ = X \cdot 1 + XZ + XY + YZ \\ &\qquad \text{(by (2-6D) and (2-4D))} \\ &= X(1 + Z + Y) + YZ = X \cdot 1 + YZ = X + YZ \\ &\qquad \text{(by (2-11), (2-5), and (2-4D))}\end{aligned}$$

The ordinary distributive law states that the AND operation distributes over OR, while the second distributive law states that OR distributes over AND. This second law is very useful in manipulating Boolean expressions. In particular, an expression like $A + BC$, which cannot be factored in ordinary algebra, is easily factored using the second distributive law:

$$A + BC = (A + B)(A + C)$$

The next laws are called DeMorgan's laws.

$$(X + Y)' = X'Y' \tag{2-12}$$

$$(XY)' = X' + Y' \tag{2-13}$$

We will verify these laws using a truth table:

X Y	X′ Y′	X + Y	(X + Y)′	X′Y′	XY	(XY)′	X′ + Y′
0 0	1 1	0	1	1	0	1	1
0 1	1 0	1	0	0	0	1	1
1 0	0 1	1	0	0	0	1	1
1 1	0 0	1	0	0	1	0	0

The laws we have derived for switching algebra are summarized in Table 2-3. One definition of Boolean algebra is a set containing at least two distinct elements with the operations of AND, OR, and complement defined on the elements that satisfy the laws in Table 2-3. This definition is not minimal (i.e., the laws are not independent since some can be derived from others). It is chosen for convenience so that other Boolean algebra theorems can be derived easily. One minimal set of laws referred to as Huntington's postulates are operations with 0 and 1, commutative laws, distributive laws, and complementation laws. The other laws can be algebraically derived from this minimal set.

TABLE 2-3 Laws of Boolean Algebra

Operations with 0 and 1:

1. $X + 0 = X$ 1D. $X \cdot 1 = X$
2. $X + 1 = 1$ 2D. $X \cdot 0 = 0$

Idempotent laws:

3. $X + X = X$ 3D. $X \cdot X = X$

Involution law:

4. $(X')' = X$

Laws of complementarity:

5. $X + X' = 1$ 5D. $X \cdot X' = 0$

Commutative laws:

6. $X + Y = Y + X$ 6D. $XY = YX$

Associative laws:

7. $(X + Y) + Z = X + (Y + Z) = X + Y + Z$ 7D. $(XY)Z = X(YZ) = XYZ$

Distributive laws:

8. $X(Y + Z) = XY + XZ$ 8D. $X + YZ = (X + Y)(X + Z)$

DeMorgan's laws:

9. $(X + Y)' = X'Y'$ 9D. $(XY)' = X' + Y'$

The Boolean algebra laws were given in pairs to show the algebra satisfies a duality. Given a Boolean algebra expression the dual of the expression is obtained by interchanging the constants 0 and 1 and interchanging the operations of AND and OR. Variables and complements are left unchanged. The laws listed in Table 2-3 show that given a Boolean algebra identity, another identity can be obtained by taking the dual of both sides of the identity. The dual of AND is OR and the dual of OR is AND:

$$(XYZ\ldots)^D = X + Y + Z + \cdots \qquad (X + Y + Z + \cdots)^D = XYZ\ldots \tag{2-14}$$

2.6 Simplification Theorems

The following theorems are useful in simplifying Boolean expressions:

Uniting: $XY + XY' = X$ (2-15) $(X + Y)(X + Y') = X$ (2-15D)

Absorption: $X + XY = X$ (2-16) $X(X + Y) = X$ (2-16D)

Elimination: $X + X'Y = X + Y$ (2-17) $X(X' + Y) = XY$ (2-17D)

Consensus: $XY + X'Z + YZ = XY + X'Z$ (2-18)

$(X + Y)(X' + Z)(Y + Z) = (X + Y)(X' + Z)$ (2-18D)

In each case, one expression can be replaced by a simpler one. Since each expression corresponds to a circuit of logic gates, simplifying an expression leads to simplifying the corresponding logic circuit.

In switching algebra, each of the above theorems can be proved by using a truth table. In a general Boolean algebra, they must be proved algebraically starting with the basic theorems.

Proof of (2-15): $XY + XY' = X(Y + Y') = X(1) = X$

Proof of (2-16): $X + XY = X \cdot 1 + XY = X(1 + Y) = X \cdot 1 = X$

Proof of (2-17): $X + X'Y = (X + X')(X + Y) = 1(X + Y) = X + Y$

Proof of (2-18): $XY + X'Z + YZ = XY + X'Z + (1)YZ =$
$XY + X'Z + (X + X')YZ = XY + XYZ + X'Z + X'YZ =$
$XY + X'Z$ (using absorption twice)

After proving one theorem in a pair of theorems, the other theorem follows by the duality property of Boolean algebra. Alternatively, the other theorem can be proved using the dual steps used to prove the first theorem. For example, (2-16D) can be proved using the dual steps of the (2-16) proof.

Proof of (2-16D): $X(X + Y) = (X + 0)(X + Y) = X + (0 \cdot Y) = X + 0 = X$

We will illustrate the elimination theorem using switches. Consider the following circuit:

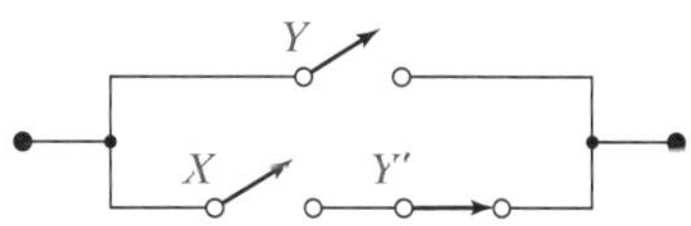

Its transmission is $T = Y + XY'$ because there is a closed circuit between the terminals if switch Y is closed *or* switch X is closed and switch Y' is closed. The following circuit is equivalent because if Y is closed $(Y = 1)$ both circuits have a transmission of 1; if Y is open $(Y' = 1)$ both circuits have a transmission of X.

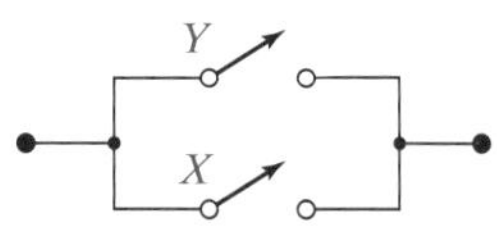

The following example illustrates simplification of a logic gate circuit using one of the theorems. In Figure 2-4, the output of circuit (a) is

$$F = A(A' + B)$$

By the elimination theorem, the expression for F simplifies to AB. Therefore, circuit (a) can be replaced with the equivalent circuit (b).

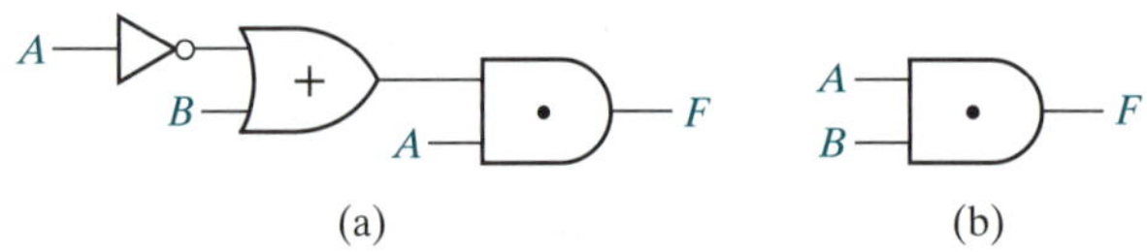

FIGURE 2-4 Equivalent Gate Circuits

Any expressions can be substituted for X and Y in the theorems.

Example 1

Simplify $Z = A'BC + A'$

This expression has the same form as absorption theorem (2-16) if we let $X = A'$ and $Y = BC$.

Therefore, the expression simplifies to $Z = X + XY = X = A'$.

Example 2

Simplify $\quad Z = [\underbrace{A + B'C} + \underbrace{D + EF}]\,[\underbrace{A + B'C} + \underbrace{(D + EF)'}]$

Substituting: $\quad Z = [\;X \;+\; Y\;]\,[\;X \;+\; Y'\;]$

Then, by the uniting theorem (2-15D), the expression reduces to

$$Z = X = A + B'C$$

Example 3

Simplify $\quad Z = \underbrace{(AB + C)}\,\underbrace{(B'D + C'E')} + \underbrace{(AB + C)'}$

Substituting: $\quad Z = \quad X' \quad\; Y \quad + \quad X$

By the elimination theorem (2-17): $Z = X + Y = B'D + C'E' + (AB + C)'$

Note that in this example we let $X = (AB + C)'$ rather than $(AB + C)$ in order to match the form of the elimination theorem (2-17).

The theorems of Boolean algebra that we have derived are summarized in Table 2-4. The theorem for multiplying out and factoring is derived in Unit 3.

TABLE 2-4 Theorems of Boolean Algebra

Uniting theorems:	
1. $XY + XY' = X$	1D. $(X + Y)(X + Y') = X$
Absorption theorems:	
2. $X + XY = X$	2D. $X(X + Y) = X$
Elimination theorems:	
3. $X + X'Y = X + Y$	3D. $X(X' + Y) = XY$
Duality:	
4. $(X + Y + Z + \cdots)^D = XYZ\ldots$	4D. $(XYZ\ldots)^D = X + Y + Z + \cdots$
Theorems for multiplying out and factoring:	
5. $(X + Y)(X' + Z) = XZ + X'Y$	5D. $XY + X'Z = (X + Z)(X' + Y)$
Consensus theorems:	
6. $XY + YZ + X'Z = XY + X'Z$	6D. $(X + Y)(Y + Z)(X' + Z) = (X + Y)(X' + Z)$

2.7 Multiplying Out and Factoring

The two distributive laws are used to multiply out an expression to obtain a sum-of-products (SOP) form. An expression is said to be in *sum-of-products* form when all products are the products of single variables. This form is the end result when an expression is fully multiplied out. It is usually easy to recognize a sum-of-products expression because it consists of a sum of product terms:

$$AB' + CD'E + AC'E' \tag{2-19}$$

However, in degenerate cases, one or more of the product terms may consist of a single variable. For example,

$$ABC' + DEFG + H \tag{2-20}$$

and

$$A + B' + C + D'E \tag{2-21}$$

are still considered to be in sum-of-products form. The expression

$$(A + B)CD + EF$$

is not in sum-of-products form because the $A + B$ term enters into a product but is not a single variable.

When multiplying out an expression, apply the second distributive law first when possible. For example, to multiply out $(A + BC)(A + D + E)$ let

$$X = A, \qquad Y = BC, \qquad Z = D + E$$

Then

$$(X + Y)(X + Z) = X + YZ = A + BC(D + E) = A + BCD + BCE$$

Of course, the same result could be obtained the hard way by multiplying out the original expression completely and then eliminating redundant terms:

$$\begin{aligned}(A + BC)(A + D + E) &= A + AD + AE + ABC + BCD + BCE \\ &= A(1 + D + E + BC) + BCD + BCE \\ &= A + BCD + BCE\end{aligned}$$

You will save yourself a lot of time if you learn to apply the second distributive law instead of doing the problem the hard way.

Both distributive laws can be used to factor an expression to obtain a product-of-sums form. An expression is in *product-of-sums* (POS) form when all sums are the sums of single variables. It is usually easy to recognize a product-of-sums expression since it consists of a product of sum terms:

$$(A + B')(C + D' + E)(A + C' + E') \tag{2-22}$$

However, in degenerate cases, one or more of the sum terms may consist of a single variable. For example,

$$(A + B)(C + D + E)F \tag{2-23}$$

and

$$AB'C(D' + E) \tag{2-24}$$

are still considered to be in product-of-sums form, but $(A + B)(C + D) + EF$ is not. An expression is fully factored iff it is in product-of-sums form. Any expression not in this form can be factored further.

The following examples illustrate how to factor using the second distributive law:

Example 1

Factor $A + B'CD$. This is of the form $X + YZ$ where $X = A, Y = B'$, and $Z = CD$, so

$$A + B'CD = (X + Y)(X + Z) = (A + B')(A + CD)$$

$A + CD$ can be factored again using the second distributive law, so

$$A + B'CD = (A + B')(A + C)(A + D)$$

Example 2

Factor $AB' + C'D$.

$AB' + C'D = (AB' + C')(AB' + D)$ ← note how $X + YZ = (X + Y)(X + Z)$ was applied here

$= (A + C')(B' + C')(A + D)(B' + D)$ ← the second distributive law was applied again to each term

Example 3

Factor $C'D + C'E' + G'H$.

$C'D + C'E' + G'H = C'(D + E') + G'H$ ← first apply the ordinary distributive law, $XY + XZ = X(Y + Z)$

$= (C' + G'H)((D + E') + G'H)$ ← then apply the second distributive law

$= (C' + G')(C' + H)(D + E' + G')(D + E' + H)$ ← now identify X, Y, and Z in each expression and complete the factoring

As in Example 3, the ordinary distributive law should be applied before the second law when factoring an expression.

A sum-of-products expression can always be realized directly by one or more AND gates feeding a single OR gate at the circuit output. Figure 2-5 shows the circuits for Equations (2-19) and (2-21). Inverters required to generate the complemented variables have been omitted.

A product-of-sums expression can always be realized directly by one or more OR gates feeding a single AND gate at the circuit output. Figure 2-6 shows the circuits for Equations (2-22) and (2-24). Inverters required to generate the complements have been omitted.

The circuits shown in Figures 2-5 and 2-6 are often referred to as two-level circuits because they have a maximum of two gates in series between an input and the circuit output.

FIGURE 2-5 Circuits for Equations (2-19) and (2-21)

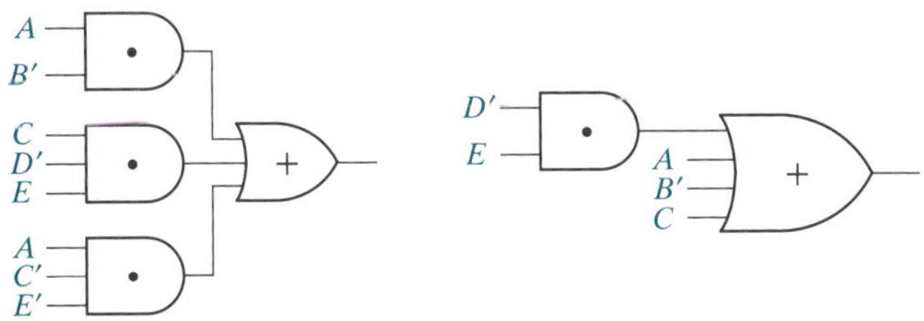

FIGURE 2-6 Circuits for Equations (2-22) and (2-24)

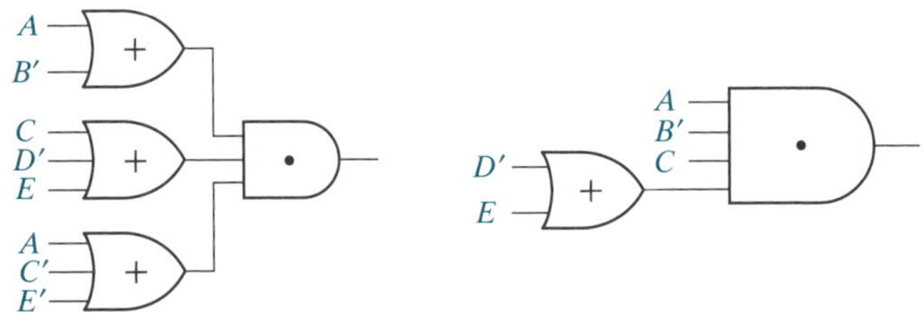

2.8 Complementing Boolean Expressions

The inverse or complement of any Boolean expression can easily be found by successively applying DeMorgan's laws.

DeMorgan's laws are easily generalized to n variables:

$$(X_1 + X_2 + X_3 + \cdots + X_n)' = X_1{}'X_2{}'X_3{}' \ldots X_n{}' \tag{2-25}$$

$$(X_1X_2X_3 \ldots X_n)' = X_1{}' + X_2{}' + X_3{}' + \cdots + X_n{}' \tag{2-26}$$

For example, for $n = 3$,

$$(X_1 + X_2 + X_3)' = (X_1 + X_2)'X'_3 = X'_1X'_2X'_3$$

Referring to the OR operation as the logical sum and the AND operation as logical product, DeMorgan's laws can be stated as

> The complement of the product is the sum of the complements.
>
> The complement of the sum is the product of the complements.

To form the complement of an expression containing both OR and AND operations, DeMorgan's laws are applied alternately.

Example 1

To find the complement of $(A' + B)C'$, first apply (2-13) and then (2-12).

$$[(A' + B)C']' = (A' + B)' + (C')' = AB' + C$$

Example 2

$$\begin{aligned}
[(AB' + C)D' + E]' &= [(AB' + C)D']'E' && \text{(by (2-12))} \\
&= [(AB' + C)' + D]E' && \text{(by (2-13))} \\
&= [(AB')'C' + D]E' && \text{(by (2-12))} \\
&= [(A' + B)C' + D]E' && \text{(by (2-13))} \quad (2\text{-}27)
\end{aligned}$$

Note that in the final expressions, the complement operation is applied only to single variables.

The inverse of $F = A'B + AB'$ is

$$F' = (A'B + AB')' = (A'B)'(AB')' = (A + B')(A' + B)$$
$$= AA' + AB + B'A' + BB' = A'B' + AB$$

We will verify that this result is correct by constructing a truth table for F and F':

A B	A′B	AB′	F = A′B + AB′	A′B′	AB	F′ = A′B′ + AB
0 0	0	0	0	1	0	1
0 1	1	0	1	0	0	0
1 0	0	1	1	0	0	0
1 1	0	0	0	0	1	1

In the table, note that for every combination of values of A and B for which $F = 0$, $F' = 1$; and whenever $F = 1$, $F' = 0$.

The dual of an expression may be found by complementing the entire expression and then complementing each individual variable. For example, to find the dual of $AB' + C$,

$$(AB' + C)' = (AB')'C' = (A' + B)'C', \quad \text{so} \quad (AB' + C)^D = (A + B')C$$

Problems

2.1 Prove the following theorems algebraically:
(a) $X(X' + Y) = XY$ (b) $X + XY = X$
(c) $XY + XY' = X$ (d) $(A + B)(A + B') = A$

2.2 Illustrate the following theorems using circuits of switches:
(a) $X + XY = X$ (b) $X + YZ = (X + Y)(X + Z)$
In each case, explain why the circuits are equivalent.

2.3 Simplify each of the following expressions by applying *one* of the theorems. State the theorem used.
(a) $X'Y'Z + (X'Y'Z)'$ (b) $(AB' + CD)(B'E + CD)$
(c) $ACF + AC'F$ (d) $A(C + D'B) + A'$
(e) $(A'B + C + D)(A'B + D)$ (f) $(A + BC) + (DE + F)(A + BC)'$

2.4 For each of the following circuits, find the output and design a simpler circuit having the same output. (*Hint:* Find the circuit output by first finding the output of each gate, going from left to right, and simplifying as you go.)

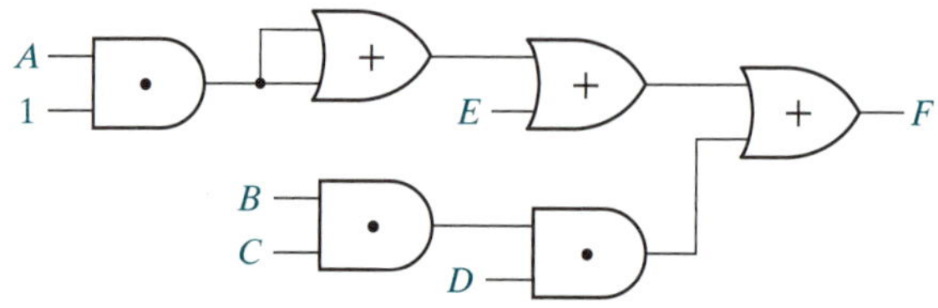

(a)

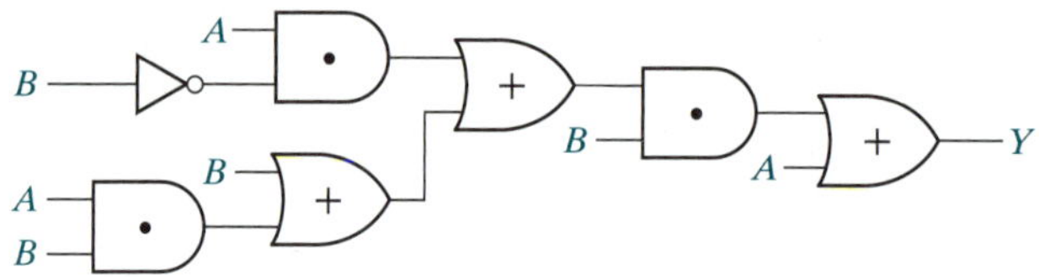

(b)

2.5 Multiply out and simplify to obtain a sum of products.
(a) $(A + B)(C + B)(D' + B)(ACD' + E)$
(b) $(A' + B + C')(A' + C' + D)(B' + D')$

2.6 Factor each of the following expressions to obtain a product of sums:
(a) $AB + C'D'$ (b) $WX + WY'X + ZYX$
(c) $A'BC + EF + DEF'$ (d) $XYZ + W'Z + XQ'Z$
(e) $ACD' + C'D' + A'C$ (f) $A + BC + DE$
(The answer to (f) should be the product of four terms, each a sum of three variables.)

2.7 Draw a circuit that uses only one AND gate and one OR gate to realize each of the following functions:
(a) $(A + B + C + D)(A + B + C + E)(A + B + C + F)$
(b) $WXYZ + VXYZ + UXYZ$

2.8 Simplify the following expressions to a minimum sum of products.
(a) $[(AB)' + C'D]'$ (b) $[A + B(C' + D)]'$ (c) $((A + B')C)'(A + B)(C + A)'$

2.9 Find F and G and simplify:

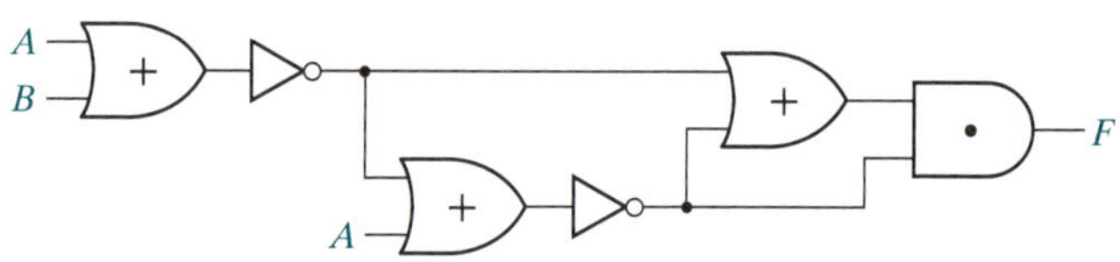

(a)

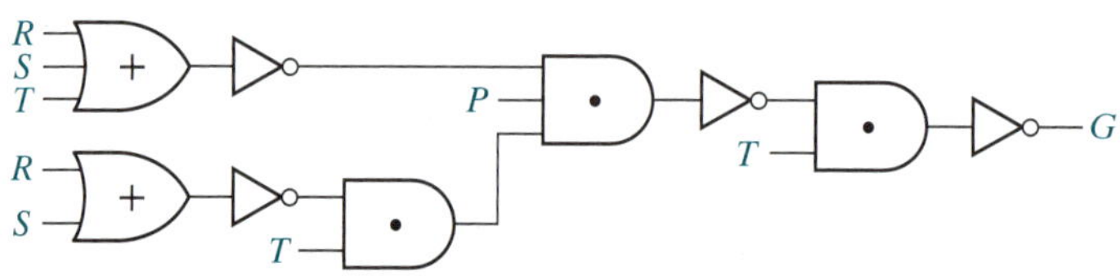

(b)

2.10 Illustrate the following equations using circuits of switches:

(a) $XY + XY' = X$ (b) $(X + Y')Y = XY$
(c) $X + X'ZY = X + YZ$ (d) $(A + B)C + (A + B)C' = A + B$
(e) $(X + Y)(X + Z) = X + YZ$ (f) $X(X + Y) = X$

2.11 Simplify each of the following expressions by applying *one* of the theorems. State the theorem used.

(a) $(A' + B' + C)(A' + B' + C)'$ (b) $AB(C' + D) + B(C' + D)$
(c) $AB + (C' + D)(AB)'$ (d) $(A'BF + CD')(A'BF + CEG)$
(e) $[AB' + (C + D)' + E'F](C + D)$ (f) $A'(B + C)(D'E + F)' + (D'E + F)$

2.12 Simplify each of the following expressions by applying *one* of the theorems. State the theorem used.

(a) $(X + Y'Z) + (X + Y'Z)'$
(b) $[W + X'(Y + Z)][W' + X'(Y + Z)]$
(c) $(V'W + UX)'(UX + Y + Z + V'W)$
(d) $(UV' + W'X)(UV' + W'X + Y'Z)$
(e) $(W' + X)(Y + Z') + (W' + X)'(Y + Z')$
(f) $(V' + U + W)[(W + X) + Y + UZ'] + [(W + X) + UZ' + Y]$

2.13 For each of the following circuits, find the output and design a simpler circuit that has the same output. (*Hint:* Find the circuit output by first finding the output of each gate, going from left to right, and simplifying as you go.)

(a)

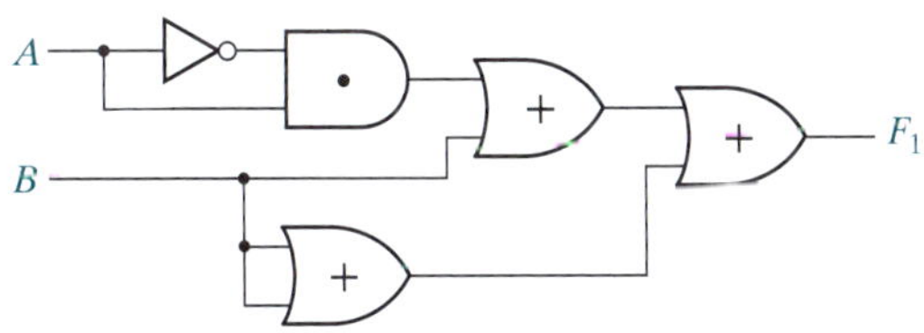

(b)

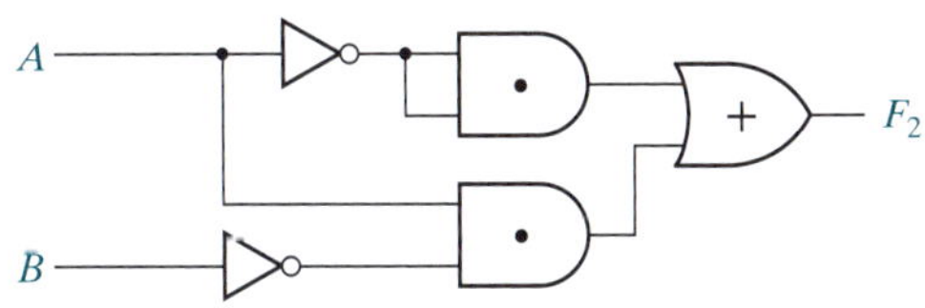

(c)

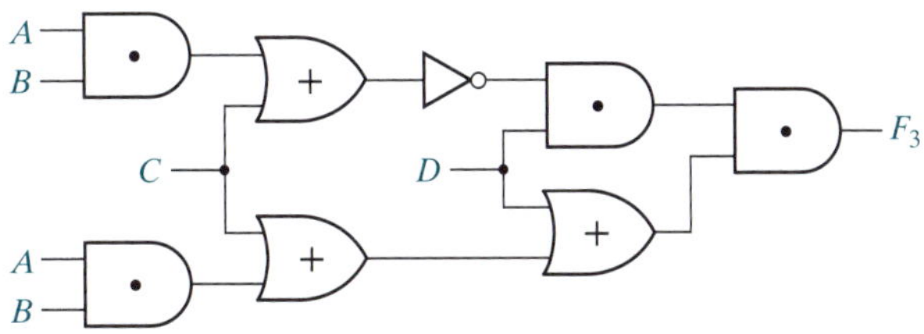

(d)

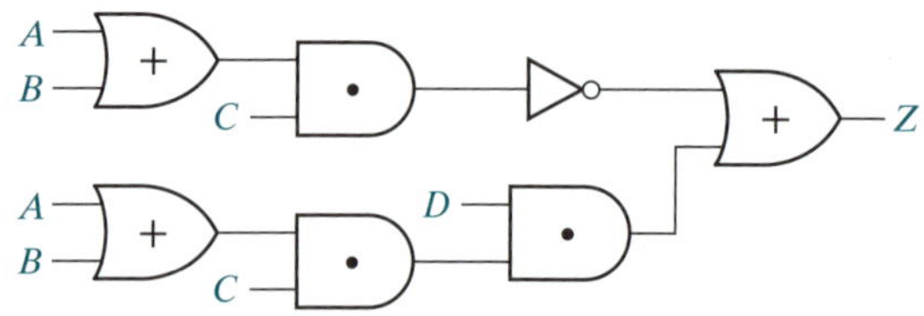

2.14 Draw a circuit that uses only one AND gate and one OR gate to realize each of the following functions:

(a) $ABCF + ACEF + ACDF$

(b) $(V + W + Y + Z)(U + W + Y + Z)(W + X + Y + Z)$

2.15 Use only DeMorgan's relationships and Involution to find the complements of the following functions:

(a) $f(A, B, C, D) = [A + (BCD)'][(AD)' + B(C' + A)]$

(b) $f(A, B, C, D) = AB'C + (A' + B + D)(ABD' + B')$

2.16 Using *just* the definition of the dual of a Boolean algebra expression, find the duals of the following expressions:

(a) $f(A, B, C, D) = [A + (BCD)'][(AD)' + B(C' + A)]$

(b) $f(A, B, C, D) = AB'C + (A' + B + D)(ABD' + B')$

2.17 For the following switching circuit, find the logic function expression describing the circuit by the three methods indicated, simplify each expression, and show they are equal.

(a) subdividing it into series and parallel connections of subcircuits until single switches are obtained

(b) finding all paths through the circuit (sometimes called *tie sets*), forming an AND term for each path and ORing the AND terms together

(c) finding all ways of breaking all paths through the circuit (sometimes called *cut sets*), forming an OR term for each cut set and ANDing the OR terms together.

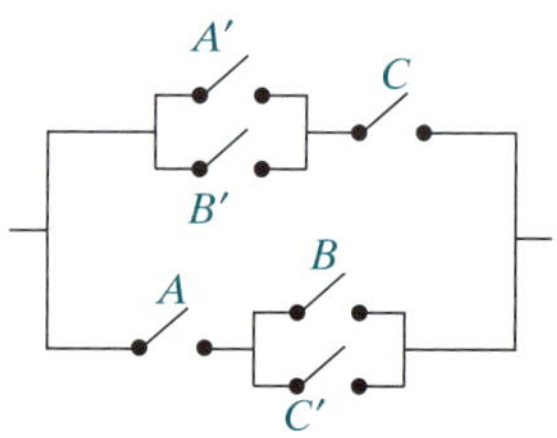

2.18 For each of the following Boolean (or switching) algebra expressions, indicate which, if any, of the following terms describe the expression: product term, sum-of-products, sum term, and product-of-sums. (More than one may apply.)

(a) $X'Y$

(b) $XY' + YZ$

(c) $(X' + Y)(WX + Z)$

(d) $X + Z$

(e) $(X' + Y)(W + Z)(X + Y' + Z')$

2.19 Construct a gate circuit using AND, OR, and NOT gates that corresponds one to one with the following switching algebra expression. Assume that inputs are available only in uncomplemented form. (Do not change the expression.)

$$(WX' + Y)[(W + Z)' + (XYZ')]$$

2.20 For the following switch circuit:
(a) derive the switching algebra expression that corresponds one to one with the switch circuit.
(b) derive an equivalent switch circuit with a structure consisting of a parallel connection of groups of switches connected in series. (Use 9 switches.)
(c) derive an equivalent switch circuit with a structure consisting of a series connection of groups of switches connected in parallel. (Use 6 switches.)

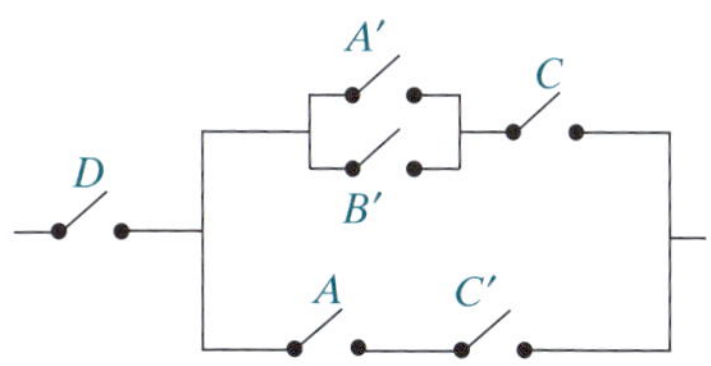

2.21 In the following circuit, $F = (A' + B)C$. Give a truth table for G so that H is as specified in its truth table. If G can be either 0 or 1 for some input combination, leave its value unspecified.

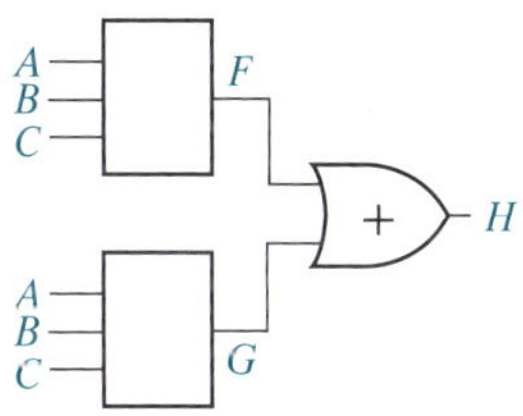

A	B	C	H
0	0	0	0
0	0	1	1
0	1	0	1
0	1	1	1
1	0	0	0
1	0	1	1
1	1	0	0
1	1	1	1

2.22 Factor each of the following expressions to obtain a product of sums:
(a) $A'B' + A'CD + A'DE'$
(b) $H'I' + JK$
(c) $A'BC + A'B'C + CD'$
(d) $A'B' + (CD' + E)$
(e) $A'B'C + B'CD' + EF'$
(f) $WX'Y + W'X' + W'Y'$

2.23 Factor each of the following expressions to obtain a product of sums:
(a) $W + U'YV$
(b) $TW + UY' + V$
(c) $A'B'C + B'CD' + B'E'$
(d) $ABC + ADE' + ABF'$

2.24 Simplify the following expressions to a minimum sum of products. Only individual variables should be complemented.
(a) $[(XY')' + (X' + Y)'Z]$
(b) $(X + (Y'(Z + W)')')'$
(c) $[(A' + B')' + (A'B'C)' + C'D]'$
(d) $(A + B)CD + (A + B)'$

2.25 For each of the following functions find a sum-of-products expression for F'.

(a) $F(P, Q, R, S) = (R' + PQ)S$

(b) $F(W, X, Y, Z) = X + YZ(W + X')$

(c) $F(A, B, C, D) = A' + B' + ACD$

2.26 Find F, G, and H, and simplify:

(a)

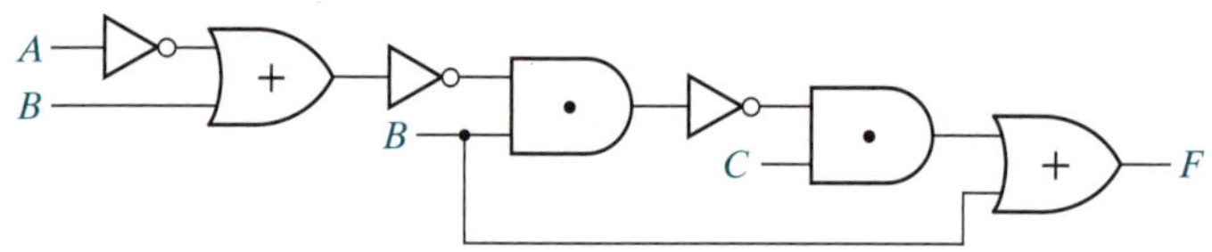

(b)

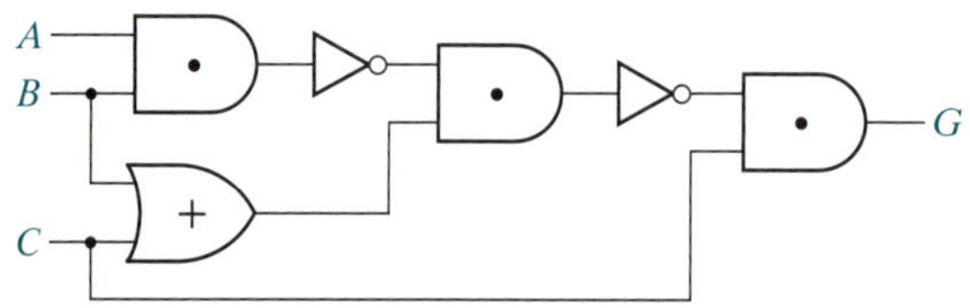

(c)

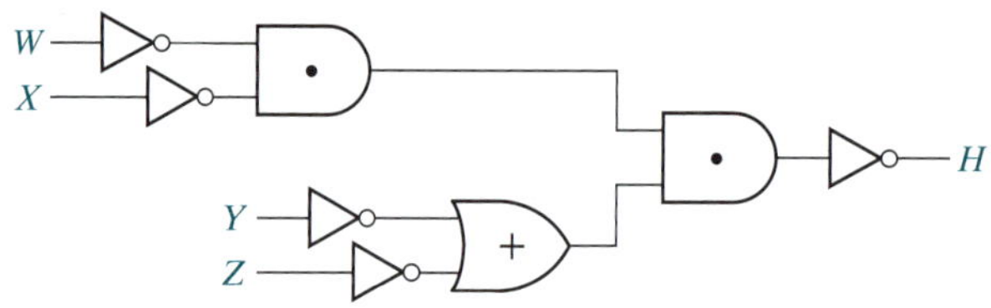

2.27 Draw a circuit that uses two OR gates and two AND gates to realize the following function:

$$F = (V + W + X)(V + X + Y)(V + Z)$$

2.28 Draw a circuit to realize the function:

$$F = ABC + A'BC + AB'C + ABC'$$

(a) using one OR gate and three AND gates. The AND gates should have two inputs.

(b) using two OR gates and two AND gates. All of the gates should have two inputs.

2.29 Prove the following equations using truth tables:

(a) $(X + Y)(X' + Z) = XZ + X'Y$

(b) $(X + Y)(Y + Z)(X' + Z) = (X + Y)(X' + Z)$

(c) $XY + YZ + X'Z = XY + X'Z$

(d) $(A + C)(AB + C') = AB + AC'$

(e) $W'XY + WZ = (W' + Z)(W + XY)$

(*Note:* Parts (a), (b), and (c) are theorems that will be introduced in Unit 3.)

2.30 Show that the following two gate circuits realize the same function.

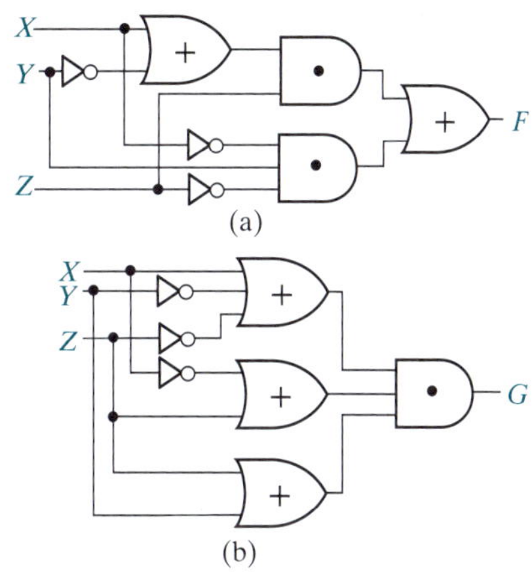

UNIT 3

Boolean Algebra (Continued)

Objectives

When you complete this unit, you should know from memory and be able to use any of the laws and theorems of Boolean algebra listed in Unit 2. Specifically, you should be able to

1. Apply these laws and theorems to the manipulation of algebraic expressions including:

   a. Simplifying an expression
   b. Finding the complement of an expression
   c. Multiplying out and factoring an expression

2. Prove any of the theorems using a truth table or give an algebraic proof.

3. Define the exclusive-OR and equivalence operations. State, prove, and use the basic theorems that concern these operations.

4. Use the consensus theorem to delete terms from and add terms to a switching algebra expression.

5. Given an equation, prove algebraically that it is valid or show that it is not valid.

Study Guide

1. Study Section 3.1, *Multiplying Out and Factoring Expressions.*

(a) List three laws or theorems which are useful when multiplying out or factoring expressions.

(b) Use Equation (3-3) to factor each of the following:

$ab'c + bd =$

$abc + (ab)'d =$

(c) In the following example, first group the terms so that (3-2) can be applied two times.

$$F_1 = (x + y' + z)(w' + x' + y)(w + x + y')(w' + y + z')$$

After applying (3-2), apply (3-3) and then finish multiplying out by using (3-1).

If we did not use (3-2) and (3-3) and used only (3-1) on the original F_1 expression, we would generate many more terms:

$$\begin{aligned} F_1 &= (w'x + w'y' + w'z + \cancel{xx'} + x'y' + x'z + xy + \cancel{yy'} + yz) \\ &\qquad (\cancel{ww'} + w'x + w'y' + wy + xy + \cancel{yy'} + wz' + xz' + y'z') \\ &= \underbrace{(w'x + w'xy' + w'xz + \cdots + yzy'z')}_{\text{49 terms in all}} \end{aligned}$$

This is obviously a *very inefficient* way to proceed! The moral to this story is to first group the terms and apply (3-2) and (3-3) where possible.

(d) Work Programmed Exercise 3.1. Then work Problem 3.6, being careful not to introduce any unnecessary terms in the process.

(e) In Unit 2 you learned how to factor a Boolean expression, using the two distributive laws. In addition, this unit introduced use of the theorem

$$XY + X'Z = (X + Z)(X' + Y)$$

in the factoring process. Careful choice of the order in which these laws and theorems are applied may cut down the amount of work required to factor an expression. When factoring, it is best to apply Equation (3-1) first, using as X the variable or variables which appear most frequently. Then Equations (3-2) and (3-3) can be applied in either order, depending on circumstances.

(f) Work Programmed Exercise 3.2. Then work Problem 3.7.

2. Checking your answers:

A good way to partially check your answers for correctness is to substitute 0's or 1's for some of the variables. For example, if we substitute $A = 1$ in the first and last expression in Equation (3-5), we get

$$1 \cdot C + 0 \cdot BD' + 0 \cdot BE + 0 \cdot C'DE = (1 + B + C')(1 + B + D) \cdot (1 + B + E)(1 + D' + E)(0 + C)$$
$$C = 1 \cdot 1 \cdot 1 \cdot 1 \cdot C \checkmark$$

Similarly, substituting $A = 0$, $B = 0$, we get

$$0 + 0 + 0 + C'DE = (0 + C')(0 + D)(0 + E)(D' + E)(1 + C)$$
$$= C'DE \checkmark$$

Verify that the result is also correct when $A = 0$ and $B = 1$.

3. The *method* which you use to get your answer is very important in this unit. If it takes you two pages of algebra and one hour of time to work a problem that can be solved in 10 minutes with three lines of work, you have not learned the material in this unit! Even if you get the correct answer, your work is not satisfactory if you worked the problem by an excessively long and time-consuming method. It is important that you learn to solve simple problems in a simple manner—otherwise, when you are asked to solve a complex problem, you will get bogged down and never get the answer. When you are given a problem to solve, do not just plunge in, but first ask yourself, "What is the easiest way to work this problem?" For example, when you are asked to multiply out an expression, do not just multiply it out by brute force, term by term. Instead, ask yourself, "How can I group the terms and which theorems should I apply first in order to reduce the amount of work?" (See Study Guide Part 1.) After you have worked out Problems 3.6 and 3.7, compare your solutions with those in the solution book. If your solution required substantially more work than the one in the solution book, rework the problem and try to get the answer in a more straightforward manner.

4. Study Section 3.2, *Exclusive-OR and Equivalence Operations.*

(a) Prove Theorems (3-8) through (3-13). You should be able to prove these both algebraically and by using a truth table.

(b) Show that $(xy' + x'y)' = xy + x'y'$. Memorize this result.

(c) Prove Theorem (3-15).

(d) Show that $(x \equiv 0) = x'$, $(x \equiv x) = 1$, and $(x \equiv y)' = (x \equiv y')$.

(e) Express $(x \equiv y)'$ in terms of exclusive OR.

(f) Work Problems 3.8 and 3.9.

5. Study Section 3.3, *The Consensus Theorem.* The consensus theorem is an important method for simplifying switching functions.

(a) In each of the following expressions, find the consensus term and eliminate it:

$$abc'd + a'be + bc'de$$
$$(a' + b + c)(a + d)(b + c + d)$$
$$ab'c + a'bd + bcd' + a'bc$$

(b) Eliminate two terms from the following expression by applying the consensus theorem:

$$A'B'C + BC'D' + A'CD + AB'D' + BCD + AC'D'$$

(*Hint:* First, compare the first term with each of the remaining terms to see if a consensus exists, then compare the second term with each of the remaining terms, etc.)

(c) Study the example given in Equations (3-22) and (3-23) carefully. Now let us start with the four-term form of the expression (Equation 3-22):

$$A'C'D + A'BD + ABC + ACD'$$

Can this be reduced directly to three terms by the application of the consensus theorem? Before we can reduce this expression, we must add another term. Which term can be added by applying the consensus theorem?

Add this term, and then reduce the expression to three terms. After this reduction, can the term which was added be removed? Why not?

(d) Eliminate two terms from the following expression by applying the dual consensus theorem:

$$(a' + c' + d)(a' + b + c)(a + b + d)(a' + b + d)(b + c' + d)$$

Use brackets to indicate how you formed the consensus terms. (*Hint:* First, find the consensus of the first two terms and eliminate it.)

(e) Derive Theorem (3-3) by using the consensus theorem.

(f) Work Programmed Exercise 3.3. Then work Problem 3.10.

6. Study Section 3.4, *Algebraic Simplification of Switching Expressions.*

(a) What theorems are used for:
Combining terms?

Eliminating terms?

Eliminating literals?

Adding redundant terms?

Factoring or multiplying out?

(b) Note that in the example of Equation (3-27), the redundant term WZ' was added and then was eliminated later after it had been used to eliminate another term. Why was it possible to eliminate WZ' in this example?

If a term has been added by the consensus theorem, it may not always be possible to eliminate the term later by the consensus theorem. Why?

(c) You will need considerable practice to develop skill in simplifying switching expressions. Work through Programmed Exercises 3.4 and 3.5.

(d) Work Problem 3.11.

(e) When simplifying an expression using Boolean algebra, two frequently asked questions are

(1) Where do I begin?

(2) How do I know when I am finished?

In answer to (1), it is generally best to try simple techniques such as combining terms or eliminating terms and literals before trying more complicated things such as using the consensus theorem or adding redundant terms. Question (2) is generally difficult to answer because it may be impossible to simplify some expressions without first adding redundant terms. We will usually tell you how many terms to expect in the minimum solution so that you will not have to waste time trying to simplify an expression which is already minimized. In Units 5 and 6, you will learn systematic techniques which will guarantee finding the minimum solution.

7. Study Section 3.5, *Proving Validity of an Equation.*

(a) When attempting to prove that an equation is valid, is it permissible to add the same expression to both sides? Explain.

(b) Work Problem 3.12.

(c) Show that (3-33) and (3-34) are true by considering both $x = 0$ and $x = 1$.

(d) Given that $a'(b + d') = a'(b + e')$, the following "proof" shows that $d = e$:

$$a'(b + d') = a'(b + e')$$
$$a + b'd = a + b'e$$
$$b'd = b'e$$
$$d = e$$

State two things that are wrong with the "proof." Give a set of values for a, b, d, and e that demonstrates that the result is incorrect.

8. Reread the objectives of this unit. When you take the readiness test, you will be expected to know from memory the laws and theorems listed in Unit 2. Where appropriate, you should know them "forward and backward"; that is, given either side of the equation, you should be able to supply the other. Test yourself to see if you can do this. When you are satisfied that you can meet the objectives, take the readiness test.

Boolean Algebra (Continued)

In this unit we continue our study of Boolean algebra to learn additional methods for manipulating Boolean expressions. We introduce another theorem for multiplying out and factoring that facilitates conversion between sum-of-products and product-of-sums expressions. These algebraic manipulations allow us to realize a switching function in a variety of forms. The exclusive-OR and equivalence operations are introduced along with examples of their use. The consensus theorem provides a useful method for simplifying an expression. Then methods for algebraic simplification are reviewed and summarized. The unit concludes with methods for proving the validity of an equation.

3.1 Multiplying Out and Factoring Expressions

Given an expression in product-of-sums form, the corresponding sum-of-products expression can be obtained by multiplying out, using the two distributive laws:

$$X(Y + Z) = XY + XZ \tag{3-1}$$

$$(X + Y)(X + Z) = X + YZ \tag{3-2}$$

In addition, the following theorem is very useful for factoring and multiplying out:

$$(X + Y)(X' + Z) = XZ + X'Y \tag{3-3}$$

Note that the variable that is paired with X on one side of the equation is paired with X' on the other side, and vice versa. In switching algebra, (3-3) can be proved by showing that both sides of the equation are the same for $X = 0$ and also for $X = 1$.

If $X = 0$, (3-3) reduces to $Y(1 + Z) = 0 + 1 \cdot Y$ or $Y = Y$.

If $X = 1$, (3-3) reduces to $(1 + Y)Z = Z + 0 \cdot Y$ or $Z = Z$.

An algebraic proof valid in any Boolean algebra is

$$(X + Y)(X' + Z) = XX' + XZ + X'Y + YZ$$

$$= 0 + XZ + X'Y + YZ = XZ + X'Y \qquad \text{(using consensus)}$$

The following example illustrates the use of Theorem (3-3) for factoring:

$$AB + A'C = (A + C)(A' + B)$$

Note that the theorem can be applied when we have two terms, one which contains a variable and another which contains its complement.

Theorem (3-3) is very useful for multiplying out expressions. In the following example, we can apply (3-3) because one factor contains the variable Q, and the other factor contains Q'.

$$(Q + AB')(C'D + Q') = QC'D + Q'AB'$$

If we simply multiplied out using the distributive law, we would get four terms instead of two:

$$(Q + AB')(C'D + Q') = QC'D + QQ' + AB'C'D + AB'Q'$$

Because the term $AB'C'D$ is difficult to eliminate, it is much better to use (3-3) instead of the distributive law.

In general, when we multiply out an expression, we should use (3-3) along with (3-1) and (3-2). To avoid generating unnecessary terms when multiplying out, (3-2) and (3-3) should generally be applied before (3-1), and terms should be grouped to expedite their application.

Example

$$\begin{aligned}
&(A + B + C')(A + B + D)(A + B + E)(A + D' + E)(A' + C) \\
&= (A + B + C'D)(A + B + E)[AC + A'(D' + E)] \\
&= (A + B + C'DE)(AC + A'D' + A'E) \\
&= AC + \cancel{ABC} + A'BD' + A'BE + A'C'DE \qquad (3\text{-}4)
\end{aligned}$$

What theorem was used to eliminate ABC? (*Hint:* let $X = AC$.)

In this example, if the ordinary distributive law (3-1) had been used to multiply out the expression by brute force, 162 terms would have been generated, and 158 of these terms would then have to be eliminated.

The same theorems that are useful for multiplying out expressions are useful for factoring. By repeatedly applying (3-1), (3-2), and (3-3), any expression can be converted to a product-of-sums form.

Example of Factoring

$$\begin{aligned}
&AC + A'BD' + A'BE + A'C'DE \\
&= \underbrace{AC}_{XZ} + \underbrace{A'}_{X'}(\underbrace{BD' + BE + C'DE}_{Y}) \\
&= (A + BD' + BE + C'DE)(A' + C) \\
&= [\underbrace{A + C'DE}_{X} + \underbrace{B}_{Y}(\underbrace{D' + E}_{Z})](A' + C)
\end{aligned}$$

$$= (A + B + C'DE)(A + \cancel{C'DE} + D' + E)(A' + C)$$
$$= (A + B + C')(A + B + D)(A + B + E)(A + D' + E)(A' + C) \qquad (3\text{-}5)$$

This is the same expression we started with in (3-4).

3.2 Exclusive-OR and Equivalence Operations

The *exclusive-OR* operation (⊕) is defined as follows:

$$0 \oplus 0 = 0 \qquad 0 \oplus 1 = 1$$
$$1 \oplus 0 = 1 \qquad 1 \oplus 1 = 0$$

The truth table for $X \oplus Y$ is

X	Y	X ⊕ Y
0	0	0
0	1	1
1	0	1
1	1	0

From this table, we can see that $X \oplus Y = 1$ iff $X = 1$ or $Y = 1$, but *not* both. The ordinary OR operation, which we have previously defined, is sometimes called inclusive OR because $X + Y = 1$ iff $X = 1$ or $Y = 1$, or both.

Exclusive OR can be expressed in terms of AND and OR. Because $X \oplus Y = 1$ iff X is 0 and Y is 1 or X is 1 and Y is 0, we can write

$$X \oplus Y = X'Y + XY' \qquad (3\text{-}6)$$

The first term in (3-6) is 1 if $X = 0$ and $Y = 1$; the second term is 1 if $X = 1$ and $Y = 0$. Alternatively, we can derive Equation (3-6) by observing that $X \oplus Y = 1$ iff $X = 1$ or $Y = 1$ *and* X and Y are not both 1. Thus,

$$X \oplus Y = (X + Y)(XY)' = (X + Y)(X' + Y') = X'Y + XY' \qquad (3\text{-}7)$$

In (3-7), note that $(XY)' = 1$ if X and Y are not both 1.

We will use the following symbol for an exclusive-OR gate:

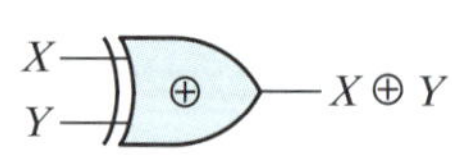

The following theorems apply to exclusive OR:

$$X \oplus 0 = X \tag{3-8}$$

$$X \oplus 1 = X' \tag{3-9}$$

$$X \oplus X = 0 \tag{3-10}$$

$$X \oplus X' = 1 \tag{3-11}$$

$$X \oplus Y = Y \oplus X \text{ (commutative law)} \tag{3-12}$$

$$(X \oplus Y) \oplus Z = X \oplus (Y \oplus Z) = X \oplus Y \oplus Z \text{ (associative law)} \tag{3-13}$$

$$X(Y \oplus Z) = XY \oplus XZ \text{ (distributive law)} \tag{3-14}$$

$$(X \oplus Y)' = X \oplus Y' = X' \oplus Y = XY + X'Y' \tag{3-15}$$

Any of these theorems can be proved by using a truth table or by replacing $X \oplus Y$ with one of the equivalent expressions from Equation (3-7). Proof of the distributive law follows:

$$\begin{aligned} XY \oplus XZ &= XY(XZ)' + (XY)'XZ = XY(X' + Z') + (X' + Y')XZ \\ &= XYZ' + XY'Z \\ &= X(YZ' + Y'Z) = X(Y \oplus Z) \end{aligned}$$

The *equivalence* operation ($\equiv$) is defined by

$$\begin{matrix} (0 \equiv 0) = 1 & (0 \equiv 1) = 0 \\ (1 \equiv 0) = 0 & (1 \equiv 1) = 1 \end{matrix} \tag{3-16}$$

The truth table for $X \equiv Y$ is

X Y	X ≡ Y
0 0	1
0 1	0
1 0	0
1 1	1

From the definition of equivalence, we see that $(X \equiv Y) = 1$ iff $X = Y$. Because $(X \equiv Y) = 1$ iff $X = Y = 1$ or $X = Y = 0$, we can write

$$(X \equiv Y) = XY + X'Y' \tag{3-17}$$

Equivalence is the complement of exclusive OR:

$$\begin{aligned} (X \oplus Y)' &= (X'Y + XY')' = (X + Y')(X' + Y) \\ &= XY + X'Y' = (X \equiv Y) \end{aligned} \tag{3-18}$$

Just as for exclusive OR, equivalence is commutative and associative.

We will use the following symbol for an equivalence gate:

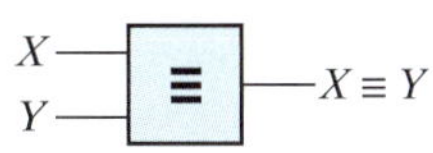

Because equivalence is the complement of exclusive OR, an alternate symbol for the equivalence gate is an exclusive-OR gate with a complemented output:

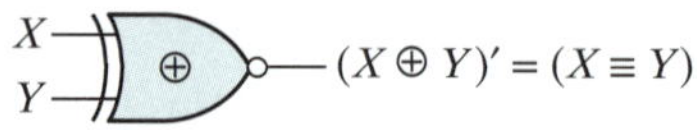

The equivalence gate is also called an exclusive-NOR gate.

In order to simplify an expression which contains AND and OR as well as exclusive OR and equivalence, it is usually desirable to first apply (3-6) and (3-17) to eliminate the $\oplus$ and $\equiv$ operations. As an example, we will simplify

$$F = (A'B \equiv C) + (B \oplus AC')$$

By (3-6) and (3-17),

$$\begin{aligned} F &= [(A'B)C + (A'B)'C'] + [B'(AC') + B(AC')'] \\ &= A'BC + (A + B')C' + AB'C' + B(A' + C) \\ &= B(A'C + A' + C) + C'(A + B' + AB') = B(A' + C) + C'(A + B') \end{aligned}$$

When manipulating an expression that contains several exclusive-OR or equivalence operations, it is useful to note that

$$(XY' + X'Y)' = XY + X'Y' \tag{3-19}$$

For example,

$$\begin{aligned} A' \oplus B \oplus C &= [A'B' + (A')'B] \oplus C \\ &= (A'B' + AB)C' + (A'B' + AB')C && \text{(by (3-6))} \\ &= (A'B' + AB)C' + (A'B + AB')C && \text{(by (3-19))} \\ &= A'B'C' + ABC' + A'BC + AB'C \end{aligned}$$

3.3 The Consensus Theorem

The consensus theorem is very useful in simplifying Boolean expressions. Given an expression of the form $XY + X'Z + YZ$, the term YZ is redundant and can be eliminated to form the equivalent expression $XY + X'Z$.

The term that was eliminated is referred to as the *consensus term*. Given a pair of terms for which a variable appears in one term and the complement of that variable in another, the consensus term is formed by multiplying the two original terms together, leaving out the selected variable and its complement. For example, the consensus of ab and $a'c$ is bc; the consensus of abd and $b'de'$ is $(ad)(de') = ade'$. The consensus of terms $ab'd$ and $a'bd'$ is 0.

The consensus theorem, given in Equation (2-18), is

$$XY + X'Z + YZ = XY + X'Z \tag{3-20}$$

The consensus theorem can be used to eliminate redundant terms from Boolean expressions. For example, in the following expression, $b'c$ is the consensus of $a'b'$ and ac, and ab is the consensus of ac and bc', so both consensus terms can be eliminated:

$$a'b' + ac + bc' + b'c + ab = a'b' + ac + bc'$$

The brackets indicate how the consensus terms are formed.

The dual form of the consensus theorem, given in Equation (2-18D), is

$$(X + Y)(X' + Z)(Y + Z) = (X + Y)(X' + Z) \tag{3-21}$$

Note again that the key to recognizing the consensus term is to first find a pair of terms, one of which contains a variable and the other its complement. In this case, the consensus is formed by adding this pair of terms together leaving out the selected variable and its complement. In the following expression, $(a + b + d')$ is a consensus term and can be eliminated by using the dual consensus theorem:

$$(a + b + c')(a + b + d')(b + c + d') = (a + b + c')(b + c + d')$$

The final result obtained by application of the consensus theorem may depend on the order in which terms are eliminated.

Example

$$A'C'D + A'BD + \cancel{BCD} + ABC + ACD' \tag{3-22}$$

First, we eliminate BCD as shown. (Why can it be eliminated?)

Now that BCD has been eliminated, it is no longer there, and it *cannot* be used to eliminate another term. Checking all pairs of terms shows that no additional terms can be eliminated by the consensus theorem.

Now we start over again:

$$A'C'D + \cancel{A'BD} + BCD + \cancel{ABC} + ACD' \tag{3-23}$$

This time, we do not eliminate BCD; instead we eliminate two other terms by the consensus theorem. After doing this, observe that BCD can no longer be eliminated. Note that the expression reduces to four terms if BCD is eliminated first, but that it can be reduced to three terms if BCD is not eliminated.

Sometimes it is impossible to directly reduce an expression to a minimum number of terms by simply eliminating terms. It may be necessary to first add a term using the consensus theorem and *then* use the added term to eliminate other terms. For example, consider the expression

$$F = ABCD + B'CDE + A'B' + BCE'$$

If we compare every pair of terms to see if a consensus term can be formed, we find that the only consensus terms are $ACDE$ (from $ABCD$ and $B'CDE$) and $A'CE'$

(from $A'B'$ and BCE'). Because neither of these consensus terms appears in the original expression, we cannot directly eliminate any terms using the consensus theorem. However, if we first add the consensus term $ACDE$ to F, we get

$$F = ABCD + B'CDE + A'B' + BCE' + ACDE$$

Then, we can eliminate $ABCD$ and $B'CDE$ using the consensus theorem, and F reduces to

$$F = A'B' + BCE' + ACDE$$

The term $ACDE$ is no longer redundant and cannot be eliminated from the final expression.

3.4 Algebraic Simplification of Switching Expressions

In this section we review and summarize methods for simplifying switching expressions, using the laws and theorems of Boolean algebra. This is important because simplifying an expression reduces the cost of realizing the expression using gates. Later, we will learn graphical methods for simplifying switching functions, but we will learn algebraic methods first. In addition to multiplying out and factoring, three basic ways of simplifying switching functions are combining terms, eliminating terms, and eliminating literals.

1. *Combining terms.* Use the theorem $XY + XY' = X$ to combine two terms. For example,

$$abc'd' + abcd' = abd' \qquad [X = abd', Y = c] \qquad (3\text{-}24)$$

When combining terms by this theorem, the two terms to be combined should contain exactly the same variables, and exactly one of the variables should appear complemented in one term and not in the other. Because $X + X = X$, a given term may be duplicated and combined with two or more other terms. For example,

$$ab'c + abc + a'bc = ab'c + abc + abc + a'bc = ac + bc$$

The theorem still can be used, of course, when X and Y are replaced with more complicated expressions. For example,

$$(a + bc)(d + e') + a'(b' + c')(d + e') = d + e'$$
$$[X = d + e', Y = a + bc, Y' = a'(b' + c')]$$

2. *Eliminating terms.* Use the theorem $X + XY = X$ to eliminate redundant terms if possible; then try to apply the consensus theorem $(XY + X'Z + YZ = XY + X'Z)$ to eliminate any consensus terms. For example,

$$a'b + a'bc = a'b \qquad [X = a'b]$$
$$a'bc' + bcd + a'bd = a'bc' + bcd \qquad [X = c, Y = bd, Z = a'b] \tag{3-25}$$

3. *Eliminating literals.* Use the theorem $X + X'Y = X + Y$ to eliminate redundant literals. Simple factoring may be necessary before the theorem is applied.

Example

$$\begin{aligned} A'B + A'B'C'D' + ABCD' &= A'(B + B'C'D') + ABCD' \\ &= A'(B + C'D') + ABCD' \\ &= B(A' + ACD') + A'C'D' \\ &= B(A' + CD') + A'C'D' \\ &= A'B + BCD' + A'C'D' \end{aligned} \tag{3-26}$$

The expression obtained after applying steps 1, 2, and 3 will not necessarily have a minimum number of terms or a minimum number of literals. If it does not and no further simplification can be made using steps 1, 2, and 3, the deliberate introduction of redundant terms may be necessary before further simplification can be made.

4. *Adding redundant terms.* Redundant terms can be introduced in several ways such as adding xx', multiplying by $(x + x')$, adding yz to $xy + x'z$, or adding xy to x. When possible, the added terms should be chosen so that they will combine with or eliminate other terms.

Example

$$\begin{aligned} &WX + XY + X'Z' + WY'Z' && \text{(add } WZ' \text{ by consensus theorem)} \\ &= WX + XY + X'Z' + WY'Z' + WZ' && \text{(eliminate } WY'Z'\text{)} \\ &= WX + XY + X'Z' + WZ' && \text{(eliminate } WZ'\text{)} \\ &= WX + XY + X'Z' && \end{aligned} \tag{3-27}$$

The following comprehensive example illustrates the use of all four methods:

Example

$$\underbrace{A'B'C'D' + A'BC'D'}_{\text{①}\ A'C'D'} + A'BD + \underbrace{\cancel{A'BC'D}}_{\text{②}} + ABCD + ACD' + B'CD'$$
$$= A'C'D' + \underbrace{BD(A' + AC)}_{\text{③}} + ACD' + B'CD'$$
$$= A'C'D' + A'BD + \underbrace{BCD + ACD'}_{+ABC\ \text{④}} + B'CD'$$

$$= A'C'D' + \underbrace{A'BD + \cancel{BCD} + \overbrace{\cancel{ACD'} + B'CD' + ABC}^{\text{consensus } ACD'}}_{\text{consensus } BCD}$$

$$= A'C'D' + A'BD + B'CD' + ABC \quad (3\text{-}28)$$

What theorems were used in steps 1, 2, 3, and 4?

If the simplified expression is to be left in a product-of-sums form instead of a sum-of-products form, the duals of the preceding theorems should be applied.

Example

$$\underbrace{(A' + B' + C')(A' + B' + C)}_{① \; (A' + B')}(B' + C)(A + C)\underset{②}{\cancel{(A + B + C)}}$$

$$= (A' + B')\underset{③}{\cancel{(B' + C)}}(A + C) = (A' + B')(A + C) \quad (3\text{-}29)$$

What theorems were used in steps 1, 2, and 3?

In general, there is no easy way of determining when a Boolean expression has a minimum number of terms or a minimum number of literals. Systematic methods for finding minimum sum-of-products and minimum product-of-sums expressions will be discussed in Units 5 and 6.

3.5 Proving Validity of an Equation

Often we will need to determine if an equation is valid for all combinations of values of the variables. Several methods can be used to determine if an equation is valid:

1. Construct a truth table and evaluate both sides of the equation for all combinations of values of the variables. (This method is rather tedious if the number of variables is large, and it certainly is not very elegant.)
2. Manipulate one side of the equation by applying various theorems until it is identical with the other side.
3. Reduce both sides of the equation independently to the same expression.
4. It is permissible to perform the same operation on both sides of the equation provided that the operation is reversible. For example, it is all right to complement both sides of the equation, but it is *not* permissible to multiply both sides of the equation by the same expression. (Multiplication is not reversible because division is not defined for Boolean algebra.) Similarly, it is *not* permissible to add the same term to both sides of the equation because subtraction is not defined for Boolean algebra.

To prove that an equation is *not* valid, it is sufficient to show one combination of values of the variables for which the two sides of the equation have different values. When using method 2 or 3 above to prove that an equation is valid, a useful strategy is to

1. First reduce both sides to a sum of products (or a product of sums).
2. Compare the two sides of the equation to see how they differ.
3. Then try to add terms to one side of the equation that are present on the other side.
4. Finally try to eliminate terms from one side that are not present on the other.

Whatever method is used, frequently compare both sides of the equation and let the difference between them serve as a guide for what steps to take next.

Example 1

Show that

$$A'BD' + BCD + ABC' + AB'D = BC'D' + AD + A'BC$$

Starting with the left side, we first add consensus terms, then combine terms, and finally eliminate terms by the consensus theorem.

$$A'BD' + BCD + ABC' + AB'D$$
$$= A'BD' + BCD + ABC' + AB'D + BC'D' + A'BC + ABD$$

(add consensus of $A'BD'$ and ABC') → $BC'D'$
(add consensus of $A'BD'$ and BCD) → $A'BC$
(add consensus of BCD and ABC') → ABD

$$= AD + A'BD' + BCD + ABC' + BC'D' + A'BC = BC'D' + AD + A'BC$$

(eliminate consensus of $BC'D'$ and AD) → ABC'
(eliminate consensus of AD and $A'BC$) → BCD
(eliminate consensus of $BC'D'$ and $A'BC$) → $A'BD'$ (3-30)

Example 2

Show that the following equation is valid:

$$A'BC'D + (A' + BC)(A + C'D') + BC'D + A'BC'$$
$$= ABCD + A'C'D' + ABD + ABCD' + BC'D$$

First, we will reduce the left side:

$$A'BC'D + (A' + BC)(A + C'D') + BC'D + A'BC'$$

(eliminate $A'BC'D$ using absorption)

$$= (A' + BC)(A + C'D') + BC'D + A'BC'$$

(multiply out using (3-3))

$$= ABC + A'C'D' + BC'D + A'BC'$$

(eliminate $A'BC'$ by consensus)

$$= ABC + A'C'D' + BC'D$$

Now we will reduce the right side:

$$= ABCD + A'C'D' + ABD + ABCD' + BC'D$$

(combine $ABCD$ and $ABCD'$)

$$= ABC + A'C'D' + ABD + BC'D$$

(eliminate ABD by consensus)

$$= ABC + A'C'D' + BC'D$$

Because both sides of the original equation were independently reduced to the same expression, the original equation is valid.

As we have previously observed, some of the theorems of Boolean algebra are not true for ordinary algebra. Similarly, some of the theorems of ordinary algebra are *not* true for Boolean algebra. Consider, for example, the cancellation law for ordinary algebra:

$$\text{If } x + y = x + z, \quad \text{then} \quad y = z \tag{3-31}$$

The cancellation law is *not* true for Boolean algebra. We will demonstrate this by constructing a counterexample in which $x + y = x + z$ but $y \neq z$. Let $x = 1, y = 0, z = 1$. Then,

$$1 + 0 = 1 + 1 \text{ but } 0 \neq 1$$

In ordinary algebra, the cancellation law for multiplication is

$$\text{If } xy = xz, \quad \text{then} \quad y = z \tag{3-32}$$

This law is valid provided $x \neq 0$.

In Boolean algebra, the cancellation law for multiplication is also *not* valid when $x = 0$. (Let $x = 0, y = 0, z = 1$; then $0 \cdot 0 = 0 \cdot 1$, but $0 \neq 1$). Because $x = 0$ about half of the time in switching algebra, the cancellation law for multiplication cannot be used.

Even though Statements (3-31) and (3-32) are generally false for Boolean algebra, the converses

$$\text{If } y = z, \quad \text{then} \quad x + y = x + z \tag{3-33}$$

$$\text{If } y = z, \quad \text{then} \quad xy = xz \tag{3-34}$$

are true. Thus, we see that although adding the same term to both sides of a Boolean equation leads to a valid equation, the reverse operation of canceling or subtracting a term from both sides generally does not lead to a valid equation. Similarly, multiplying both sides of a Boolean equation by the same term leads to a valid equation, but not conversely. When we are attempting to prove that an equation is valid, it is *not* permissible to add the same expression to both sides of the equation or to multiply both sides by the same expression, because these operations are not reversible.

Programmed Exercise 3.1

Cover the answers to this exercise with a sheet of paper and slide it down as you check your answers. Write your answer in the space provided before looking at the correct answer.

The following expression is to be multiplied out to form a sum of products:

$$(A + B + C')(A' + B' + D)(A' + C + D')(A + C' + D)$$

First, find a pair of sum terms which have two literals in common and apply the second distributive law. Also, apply the same law to the other pair of terms.

Answer $(A + C' + BD)[A' + (B' + D)(C + D')]$
(*Note:* This answer was obtained by using $(X + Y)(X + Z) = X + YZ$.)

Next, find a pair of sum terms which have a variable in one and its complement in the other. Use the appropriate theorem to multiply these sum terms together without introducing any redundant terms. Apply the same theorem a second time.

Answer $(A + C' + BD)(A' + B'D' + CD) = A(B'D' + CD) + A'(C' + BD)$ or
$A(B' + D)(C + D') + A'(C' + BD) = A(B'D' + CD) + A'(C' + BD)$
(*Note:* This answer was obtained using $(X + Y)(X' + Z) = XZ + X'Y$.)

Complete the problem by multiplying out using the ordinary distributive law.

Final Answer $AB'D' + ACD + A'C' + A'BD$

Programmed Exercise 3.2

Cover the answers to this exercise with a sheet of paper and slide it down as you check your answers. Write your answer in the space provided before looking at the correct answer.

The following expression is to be factored to form a product of sums:

$$WXY' + W'X'Z + WY'Z + W'YZ'$$

First, factor as far as you can using the ordinary distributive law.

Answer $WY'(X + Z) + W'(X'Z + YZ')$

Next, factor further by using a theorem which involves a variable and its complement. Apply this theorem twice.

Answer

$$(W + X'Z + YZ')[W' + Y'(X + Z)] = [W + (X' + Z')(Y + Z)][W' + Y'(X + Z)]$$

or

$$WY'(X + Z) + W'(X' + Z')(Y + Z) = [W + (X' + Z')(Y + Z)][W' + Y'(X + Z)]$$

[*Note:* This answer was obtained by using $AB + A'C = (A + C)(A' + B)$.]

Now, complete the factoring by using the second distributive law.

Final answer $(W + X' + Z')(W + Y + Z)(W' + Y')(W' + X + Z)$

Programmed Exercise 3.3

Cover the answers to this exercise with a sheet of paper and slide it down as you check your answers. Write your answer in the space provided before looking at the correct answer.

The following expression is to be simplified using the consensus theorem:

$$AC' + AB'D + A'B'C + A'CD' + B'C'D'$$

First, find all of the consensus terms by checking all pairs of terms.

Answer The consensus terms are indicated.

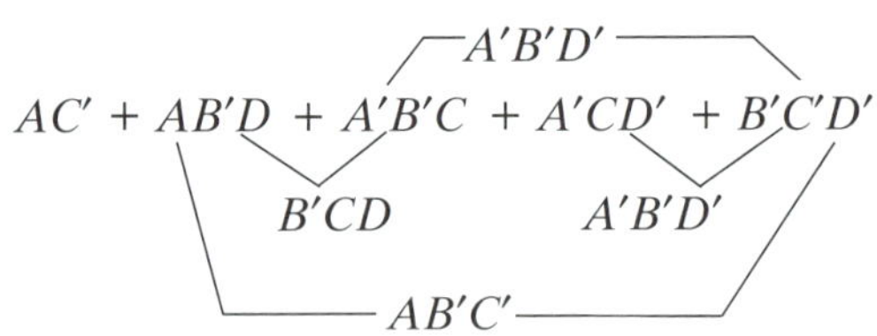

Can the original expression be simplified by the direct application of the consensus theorem?

Answer No, because none of the consensus terms appears in the original expression.

Now add the consensus term $B'CD$ to the original expression. Compare the added term with each of the original terms to see if any consensus exists. Eliminate as many of the original terms as you can.

Answer

$$\overbrace{AC' + \cancel{AB'D} + \cancel{A'B'C} + \underbrace{A'CD' + B'C'D' + B'CD}_{(A'B'C)}}^{(AB'D)}$$

Now that we have eliminated two terms, can $B'CD$ also be eliminated? What is the final reduced expression?

Answer No, because the terms used to form $B'CD$ are gone. Final answer is

$$AC' + A'CD' + B'C'D' + B'CD$$

Programmed Exercise 3.4

Keep the answers to this exercise covered with a sheet of paper and slide it down as you check your answers.

Problem: The following expression is to be simplified

$$ab'cd'e + acd + acf'gh' + abcd'e + acde' + e'h'$$

State a theorem which can be used to combine a pair of terms and apply it to combine two of the terms in the above expression.

Answer Apply $XY + XY' = X$ to the terms $ab'cd'e$ and $abcd'e$, which reduces the expression to

$$acd'e + acd + acf'gh' + acde' + e'h'$$

Now state a theorem (other than the consensus theorem) which can be used to eliminate terms and apply it to eliminate a term in this expression.

Answer Apply $X + XY = X$ to eliminate $acde'$. (What term corresponds to X?) The result is

$$acd'e + acd + acf'gh' + e'h'$$

Now state a theorem that can be used to eliminate literals and apply it to eliminate a literal from one of the terms in this expression. (*Hint:* It may be necessary to factor out some common variables from a pair of terms before the theorem can be applied.)

Answer Use $X + X'Y = X + Y$ to eliminate a literal from $acd'e$. To do this, first factor ac out of the first two terms: $acd'e + acd = ac(d + d'e)$. After eliminating d', the resulting expression is

$$ace + acd + acf'gh' + e'h'$$

(a) Can any term be eliminated from this expression by the direct application of the consensus theorem?
(b) If not, add a redundant term using the consensus theorem, and use this redundant term to eliminate one of the other terms.
(c) Finally, reduce your expression to three terms.

Answer

(a) No
(b) Add the consensus of ace and $e'h'$:
$ace + acd + acf'gh' + e'h' + ach'$
Now eliminate $acf'gh'$ (by $X + XY = X$)
$ace + acd + e'h' + ach'$
(c) Now eliminate ach' by the consensus theorem. The final answer is
$ace + acd + e'h'$

Programmed Exercise 3.5

Keep the answers to this exercise covered with a sheet of paper and slide it down as you check your answers.

$$Z = (A + C' + F' + G)(A + C' + F + G)(A + B + C' + D' + G)$$
$$(A + C + E + G)(A' + B + G)(B + C' + F + G)$$

This is to be simplified to the form

$$(X + X + X)(X + X + X)(X + X + X)$$

where each X represents a literal.

State a theorem which can be used to combine the first two sum terms of Z and apply it. (*Hint:* The two sum terms differ in only one variable.)

Answer

$$(X + Y)(X + Y') = X$$
$$Z = (A + C' + G)(A + B + C' + D' + G)(A + C + E + G)(A' + B + G)$$
$$(B + C' + F + G)$$

Now state a theorem (other than the consensus theorem) which can be used to eliminate a sum term and apply it to this expression.

Answer

$$X(X + Y) = X$$
$$Z = (A + C' + G)(A + C + E + G)(A' + B + G)(B + C' + F + G)$$

Next, eliminate one literal from the second term, leaving the expression otherwise unchanged. (*Hint:* This cannot be done by the direct application of one theorem; it will be necessary to partially multiply out the first two sum terms before eliminating the literal.)

Answer

$$(A + C' + G)(A + C + E + G) = A + G + C'(C + E) = A + G + C'E$$

Therefore,

$$Z = (A + C' + G)(A + E + G)(A' + B + G)(B + C' + F + G)$$

(a) Can any term be eliminated from this expression by the direct application of the consensus theorem?
(b) If not, add a redundant sum term using the consensus theorem, and use this redundant term to eliminate one of the other terms.
(c) Finally, reduce your expression to a product of three sum terms.

Answer (a) No
(b) Add $B + C' + G$ (consensus of $A + C' + G$ and $A' + B + G$).
Use $X(X + Y) = X$, where $X = B + C' + G$, to eliminate $B + C' + F + G$.
(c) Now eliminate $B + C' + G$ by consensus. The final answer is

$$Z = (A + C' + G)(A + E + G)(A' + B + G)$$

Problems

3.6 In each case, multiply out to obtain a sum of products: (Simplify where possible.)
(a) $(W + X' + Z')(W' + Y')(W' + X + Z')(W + X')(W + Y + Z)$
(b) $(A + B + C + D)(A' + B' + C + D')(A' + C)(A + D)(B + C + D)$

3.7 Factor to obtain a product of sums. (Simplify where possible.)
(a) $BCD + C'D' + B'C'D + CD$
(b) $A'C'D' + ABD' + A'CD + B'D$

3.8 Write an expression for F and simplify.

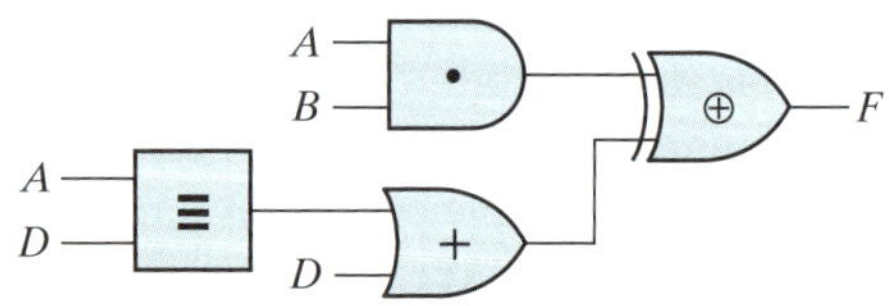

3.9 Is the following distributive law valid? $A \oplus BC = (A \oplus B)(A \oplus C)$ Prove your answer.

3.10 (a) Reduce to a minimum sum of products (three terms):
$(X + W)(Y \oplus Z) + XW'$
(b) Reduce to a minimum sum of products (four terms):
$(A \oplus BC) + BD + ACD$
(c) Reduce to a minimum product of sums (three terms):
$(A' + C' + D')(A' + B + C')(A + B + D)(A + C + D)$

3.11 Simplify algebraically to a minimum sum of products (five terms):
$(A + B' + C + E')(A + B' + D' + E)(B' + C' + D' + E')$

3.12 Prove algebraically that the following equation is valid:
$A'CD'E + A'B'D' + ABCE + ABD = A'B'D' + ABD + BCD'E$

3.13 Simplify each of the following expressions:
(a) $KLMN' + K'L'MN + MN'$
(b) $KL'M' + MN' + LM'N'$
(c) $(K + L')(K' + L' + N)(L' + M + N')$
(d) $(K' + L + M' + N)(K' + M' + N + R)(K' + M' + N + R')KM$

3.14 Factor to obtain a product of sums:
(a) $K'L'M + KM'N + KLM + LM'N'$ (four terms)
(b) $KL + K'L' + L'M'N' + LMN'$ (four terms)
(c) $KL + K'L'M + L'M'N + LM'N'$ (four terms)
(d) $K'M'N + KL'N' + K'MN' + LN$ (four terms)
(e) $WXY + WX'Y + WYZ + XYZ'$ (three terms)

3.15 Multiply out to obtain a sum of products:
(a) $(K' + M' + N)(K' + M)(L + M' + N')(K' + L + M)(M + N)$ (three terms)
(b) $(K' + L' + M')(K + M + N')(K + L)(K' + N)(K' + M + N)$
(c) $(K' + L' + M)(K + N')(K' + L + N')(K + L)(K + M + N')$
(d) $(K + L + M)(K' + L' + N')(K' + L' + M')(K + L + N)$
(e) $(K + L + M)(K + M + N)(K' + L' + M')(K' + M' + N')$

3.16 Eliminate the exclusive OR, and then factor to obtain a minimum product of sums:
(a) $(KL \oplus M) + M'N'$
(b) $M'(K \oplus N') + MN + K'N$

3.17 Algebraically prove identities involving the equivalence (exclusive-NOR) operation:
(a) $x \equiv 0 = x'$
(b) $x \equiv 1 = x$
(c) $x \equiv x = 1$
(d) $x \equiv x' = 0$
(e) $x \equiv y = y \equiv x$
(f) $(x \equiv y) \equiv z = x \equiv (y \equiv z)$
(g) $(x \equiv y)' = x' \equiv y = x \equiv y'$

3.18 Algebraically prove identities involving the exclusive-OR operation:
(a) $x \oplus 0 = x$
(b) $x \oplus 1 = x'$
(c) $x \oplus x = 0$
(d) $x \oplus x' = 1$
(e) $x \oplus y = y \oplus x$
(f) $(x \oplus y) \oplus z = x \oplus (y \oplus z)$
(g) $(x \oplus y)' = x' \oplus y = x \oplus y'$

3.19 Algebraically prove the following identities:
(a) $x + y = x \oplus y \oplus xy$
(b) $x + y = x \equiv y \equiv xy$

3.20 Algebraically prove or disprove the following distributive identities:
(a) $x(y \oplus z) = xy \oplus xz$
(b) $x + (y \oplus z) = (x + y) \oplus (x + z)$
(c) $x(y \equiv z) = xy \equiv xz$
(d) $x + (y \equiv z) = (x + y) \equiv (x + z)$

3.21 Simplify each of the following expressions using only the consensus theorem (or its dual):
(a) $BC'D' + ABC' + AC'D + AB'D + A'BD'$ (reduce to three terms)
(b) $W'Y' + WYZ + XY'Z + WX'Y$ (reduce to three terms)
(c) $(B + C + D)(A + B + C)(A' + C + D)(B' + C' + D')$
(d) $W'XY + WXZ + WY'Z + W'Z'$
(e) $A'BC' + BC'D' + A'CD + B'CD + A'BD$
(f) $(A + B + C)(B + C' + D)(A + B + D)(A' + B' + D')$

3.22 Factor $Z = ABC + DE + ACF + AD' + AB'E'$ and simplify it to the form $(X + X)(X + X)(X + X + X + X)$ (where each X represents a literal). Now express Z as a minimum sum of products in the form:

$$XX + XX + XX + XX$$

3.23 Repeat Problem 3.22 for $F = A'B + AC + BC'D' + BEF + BDF$.

3.24 Factor to obtain a product of four terms and then reduce to three terms by applying the consensus theorem: $X'Y'Z' + XYZ$

3.25 Simplify each of the following expressions:
(a) $xy + x'yz' + yz$
(b) $(xy' + z)(x + y')z$
(c) $xy' + z + (x' + y)z'$
(d) $a'd(b' + c) + a'd'(b + c') + (b' + c)(b + c')$
(e) $w'x' + x'y' + yz + w'z'$
(f) $A'BCD + A'BC'D + B'EF + CDE'G + A'DEF + A'B'EF$ (reduce to a sum of three terms)
(g) $[(a' + d' + b'c)(b + d + ac')]' + b'c'd' + a'c'd$ (reduce to three terms)

3.26 Simplify to a sum of three terms:
(a) $A'C'D' + AC' + BCD + A'CD' + A'BC + AB'C'$
(b) $A'B'C' + ABD + A'C + A'CD' + AC'D + AB'C'$

3.27 Reduce to a minimum sum of products:

$$F = WXY' + (W'Y' \equiv X) + (Y \oplus WZ)$$

3.28 Determine which of the following equations are always valid (give an algebraic proof):

(a) $a'b + b'c + c'a = ab' + bc' + ca'$

(b) $(a + b)(b + c)(c + a) = (a' + b')(b' + c')(c' + a')$

(c) $abc + ab'c' + b'cd + bc'd + ad = abc + ab'c' + b'cd + bc'd$

(d) $xy' + x'z + yz' = x'y + xz' + y'z$

(e) $(x + y)(y + z)(x + z) = (x' + y')(y' + z')(x' + z')$

(f) $abc' + ab'c + b'c'd + bcd = ab'c + abc' + ad + bcd + b'c'd$

3.29 The following circuit is implemented using two half-adder circuits. The expressions for the half-adder outputs are $S = A \oplus B$, and $C = AB$. Derive simplified sum-of-products expressions for the circuit outputs SUM and C_o. Give the truth table for the outputs.

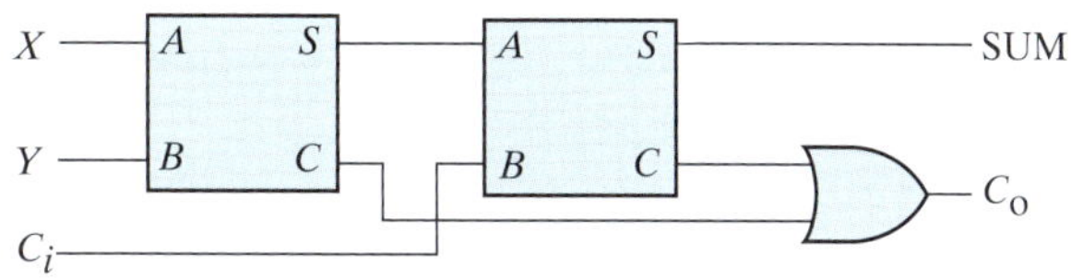

3.30 The output of a majority circuit is 1 if a majority (more than half) of its inputs are equal to 1, and the output is 0 otherwise. Construct a truth table for a three-input majority circuit and derive a simplified sum-of-products expression for its output.

3.31 Prove algebraically:

(a) $(X' + Y')(X \equiv Z) + (X + Y)(X \oplus Z) = (X \oplus Y) + Z'$

(b) $(W' + X + Y')(W + X' + Y)(W + Y' + Z) = X'Y' + WX + XYZ + W'YZ$

(c) $ABC + A'C'D' + A'BD' + ACD = (A' + C)(A + D')(B + C' + D)$

3.32 Which of the following statements are always true? Justify your answers.

(a) If $A + B = C$, then $AD' + BD' = CD'$

(b) If $A'B + A'C = A'D$, then $B + C = D$

(c) If $A + B = C$, then $A + B + D = C + D$

(d) If $A + B + C = C + D$, then $A + B = D$

3.33 Find all possible terms that could be added to each expression using the consensus theorem. Then reduce to a minimum sum of products.

(a) $A'C' + BC + AB' + A'BD + B'C'D' + ACD'$

(b) $A'C'D' + BC'D + AB'C' + A'BC$

3.34 Simplify the following expression to a sum of two terms and then factor the result to obtain a product of sums: $abd'f' + b'cegh' + abd'f + acd'e + b'ce$

3.35 Multiply out the following expression and simplify to obtain a sum-of-products expression with three terms: $(a + c)(b' + d)(a + c' + d')(b' + c' + d')$

3.36 Factor and simplify to obtain a product-of-sums expression with four terms: $abc' + d'e + ace + b'c'd'$

3.37 (a) Show that $x \oplus y = (x \equiv y)'$
(b) Realize $a'b'c' + a'bc + ab'c + abc'$ using only two-input equivalence gates.

3.38 In a Boolean algebra, which of the following statements are true? Prove your answer.
(a) If $x(y + a') = x(y + b')$, then $a = b$.
(b) If $a'b + ab' = a'c + ac'$, then $b = c$.

3.39 The definition of Boolean algebra given in Unit 2 is redundant (i.e., not all of the properties are independent). For example, show that the associative property $a + (b + c) = (a + b) + c$ can be proved using the other properties of Boolean algebra. (*Hint:* Consider expanding $[a + (b + c)][(a + b) + c]$ in two different ways. Be sure to not use the associative property.)

UNIT 4

Applications of Boolean Algebra Minterm and Maxterm Expansions

Objectives

1. Given a word description of the desired behavior of a logic circuit, write the output of the circuit as a function of the input variables. Specify this function as an algebraic expression or by means of a truth table, as is appropriate.
2. Given a truth table, write the function (or its complement) as both a minterm expansion (standard sum of products) and a maxterm expansion (standard product of sums). Be able to use both alphabetic and decimal notation.
3. Given an algebraic expression for a function, expand it algebraically to obtain the minterm or maxterm form.
4. Given one of the following: minterm expansion for F, minterm expansion for F', maxterm expansion for F, or maxterm expansion for F', find any of the other three forms.
5. Write the general form of the minterm and maxterm expansion of a function of n variables.
6. Explain why some functions contain don't-care terms.
7. Explain the operation of a full adder and a full subtracter and derive logic equations for these modules. Draw a block diagram for a parallel adder or subtracter and trace signals on the block diagram.

Study Guide

In the previous units, we placed a dot (·) inside the AND-gate symbol, a plus sign (+) inside the OR-gate symbol, and a ⊕ inside the exclusive OR. Because you are now familiar with the relationship between the shape of the gate symbol and the logic function performed, we will omit the · , + , and ⊕ and use the standard gate symbols for AND, OR, and exclusive OR in the rest of the book.

1. Study Section 4.1, *Conversion of English Sentences to Boolean Equations.*

(a) Use braces to identify the phrases in each of the following sentences:

(1) The tape reader should stop if the manual stop button is pressed,

if an error occurs, or if an end-of-tape signal is present.

(2) He eats eggs for breakfast if it is not Sunday and

he has eggs in the refrigerator.

(3) Addition should occur iff an add instruction is given and

the signs are the same, or if a subtract instruction is given and

the signs are not the same.

(b) Write a Boolean expression which represents each of the sentences in (a). Assign a variable to each phrase, and use a complemented variable to represent a phrase which contains "not".

(Your answers should be in the form $F = S'E$, $F = AB + SB'$, and $F = A + B + C$, but not necessarily in that order.)

(c) If X represents the phrase "N is greater than 3", how can you represent the phrase "N is less than or equal to 3"?

(d) Work Problems 4.1 and 4.2.

2. Study Section 4.2, *Combinational Logic Design Using a Truth Table.* Previously, you have learned how to go from an algebraic expression for a function to a truth table; in this section you will learn how to go from a truth table to an algebraic expression.

(a) Write a product term which is 1 iff $a = 0, b = 0$, and $c = 1$.

(b) Write a sum term which is 0 iff $a = 0, b = 0$, and $c = 1$.

(c) Verify that your answers to (a) and (b) are complements.

(d) Write a product term which is 1 iff $a = 1, b = 0, c = 0$, and $d = 1$.

(e) Write a sum term which is 0 iff $a = 0, b = 0, c = 1$, and $d = 1$.

(f) For the given truth table, write F as a sum of four product terms which correspond to the four 1's of F.

a b c	F
0 0 0	1
0 0 1	1
0 1 0	0
0 1 1	1
1 0 0	1
1 0 1	0
1 1 0	0
1 1 1	0

(g) From the truth table write F as a product of four sum terms which correspond to the four 0's of F.

(h) Verify that your answers to both (f) and (g) reduce to $F = b'c' + a'c$.

3. Study Section 4.3, *Minterm and Maxterm Expansions.*

(a) Define the following terms:
minterm (for n variables)

maxterm (for n variables)

(b) Study Table 4-1 and observe the relation between the values of A, B, and C and the corresponding minterms and maxterms.
If $A = 0$, then does A or A' appear in the minterm?
In the maxterm?
If $A = 1$, then does A or A' appear in the minterm?
In the maxterm?
What is the relation between minterm, m_i, and the corresponding maxterm, M_i?

(c) For the table given in Study Guide Question 2(f), write the minterm expansion for F in m-notation and in decimal notation.

For the same table, write the maxterm expansion for F in M-notation and in decimal notation.

Check your answers by converting your answer to 2(f) to m-notation and your answer to 2(g) to M-notation.

(d) Given a sum-of-products expression, how do you expand it to a standard sum of products (minterm expansion)?

(e) Given a product-of-sums expression, how do you expand it to a standard product of sums (maxterm expansion)?

(f) In Equation (4-11), what theorems were used to factor f to obtain the maxterm expansion?

(g) Why is the following expression not a maxterm expansion?

$$f(A, B, C, D) = (A + B' + C + D)(A' + B + C')(A' + B + C + D')$$

(h) Assuming that there are three variables (A, B, C), identify each of the following as a minterm expansion, maxterm expansion, or neither:

(1) $AB + B'C'$ (2) $(A' + B + C')(A + B' + C)$
(3) $A + B + C$ (4) $(A' + B)(B' + C)(A' + C)$
(5) $A'BC' + AB'C + ABC$ (6) $AB'C'$

Note that it is possible for a minterm or maxterm expansion to have only one term.

4. (a) Given a minterm in terms of its variables, the procedure for conversion to decimal notation is

(1) Replace each complemented variable with a _____ and replace each uncomplemented variable with a _____.
(2) Convert the resulting binary number to decimal.

(b) Convert the minterm $AB'C'DE$ to decimal notation.

(c) Given that m_{13} is a minterm of the variables A, B, C, D, and E, write the minterm in terms of these variables.

(d) Given a maxterm in terms of its variables, the procedure for conversion to decimal notation is

(1) Replace each complemented variable with a _____ and replace each uncomplemented variable with a _____.
(2) Group these 0's and 1's to form a binary number and convert to decimal.

(e) Convert the maxterm $A' + B + C + D' + E'$ to decimal notation.

(f) Given that M_{13} is a maxterm of the variables A, B, C, D, and E, write the maxterm in terms of these variables.

(g) Check your answers to (b), (c), (e), and (f) by using the relation $M_i = m_i'$.

(h) Given $f(a, b, c, d, e) = \Pi M(0, 10, 28)$, express f in terms of a, b, c, d, and e. (Your answer should contain only five complemented variables.)

5. Study Section 4.4, *General Minterm and Maxterm Expansions.* Make sure that you understand the notation here and can follow the algebra in all of the equations. If you have difficulty with this section, ask for help *before* you take the readiness test.

(a) How many different switching functions of four variables are possible?

(b) Explain why there are 2^{2^n} switching functions of n variables.

(c) Write the function of Figure 4-1 in the form of Equation (4-13) and show that it reduces to Equation (4-3).

(d) For Equation (4-19), write out the indicated summations in full for the case $n = 2$.

(e) Study Tables 4-3 and 4-4 carefully and make sure you understand why each table entry is valid. Use the truth table for f and f' (Figure 4-1) to verify the entries in Table 4-4. If you understand the relationship between Table 4-3 and the truth table for f and f', you should be able to perform the conversions without having to memorize the table.

(f) Given that $f(A, B, C) = \Sigma m(0, 1, 3, 4, 7)$

The maxterm expansion for f is ____________________

The minterm expansion for f' is ____________________

The maxterm expansion for f' is ____________________

(g) Work Problems 4.3 and 4.4.

6. Study Section 4.5, *Incompletely Specified Functions.*

(a) State two reasons why some functions have don't-care terms.

(b) Given the following table, write the minterm expansion for Z in decimal form.

A B C	Z
0 0 0	1
0 0 1	X
0 1 0	0
0 1 1	X
1 0 0	X
1 0 1	1
1 1 0	0
1 1 1	0

(c) Write the maxterm expansion in decimal form.

(d) Work Problems 4.5 and 4.6.

7. Study Section 4.6, *Examples of Truth Table Construction.* Finding the truth table from the problem statement is probably the most difficult part of the process of designing a switching circuit. Make sure that you understand how to do this.

8. Work Problems 4.7 through 4.10.

9. Study Section 4.7, *Design of Binary Adders.*

   (a) For the given parallel adder, show the 0's and 1's at the full adder (FA) inputs and outputs when the following unsigned numbers are added: 11 + 14 = 25. Verify that the result is correct if $C_4S_3S_2S_1S_0$ is taken as a 5-bit sum. If the sum is limited to 4 bits, explain why this is an overflow condition.

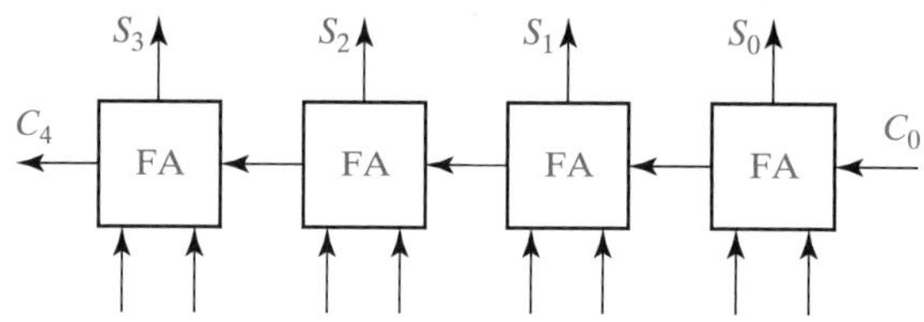

   (b) Review Section 1.4, *Representation of Negative Numbers.* If we use the 2's complement number system to add (−5) + (−2), verify that the FA inputs and outputs are exactly the same as in part (a). However, for 2's complement, the interpretation of the results is quite different. After discarding C_4, verify that the resultant 4-bit sum is correct, and therefore no overflow has occurred.

   (c) If we use the 1's complement number system to add (−5) + (−2), show the FA inputs and outputs on the diagram below before the end-around carry is added in. Assume that C_0 is initially 0. Then add the end-around carry (C_4) to the rightmost FA, add the new carry (C_1) into the next cell, and continue until no further changes occur. Verify that the resulting sum is the correct 1's complement representation of −7.

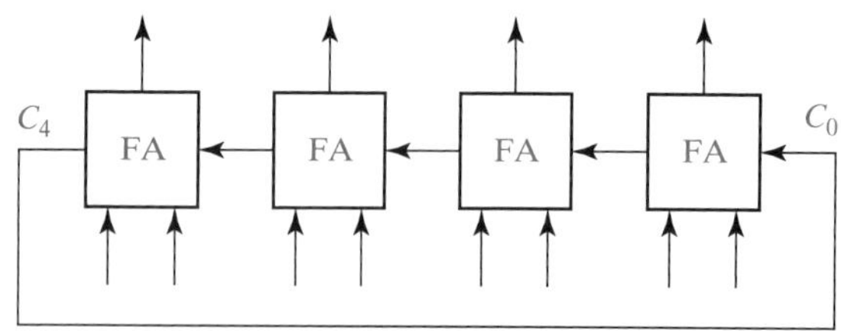

10. (a) Work the following subtraction example. As you subtract each column, place a 1 over the next column if you have to borrow, otherwise place a 0. For each column, as you compute $x_i - y_i - b_i$, fill in the corresponding values of b_{i+1} and d_i in the truth table. If you have done this correctly, the resulting table should match the full subtracter truth table (Table 4-6).

	← borrows
11000110	← X
−01011010	← Y
	← difference

x_i	y_i	b_i	b_{i+1}	d_i
0	0	0		
0	0	1		
0	1	0		
0	1	1		
1	0	0		
1	0	1		
1	1	0		
1	1	1		

(b) Work Problems 4.11 and 4.12.

11. Read the following and then work Problem 4.13 or 4.14 as assigned:
When looking at an expression to determine the required number of gates, keep in mind that the number of required gates is generally *not* equal to the number of AND and OR operations which appear in the expression. For example,

$$AB + CD + EF(G + H)$$

contains four AND operations and three OR operations, but it only requires three AND gates and two OR gates:

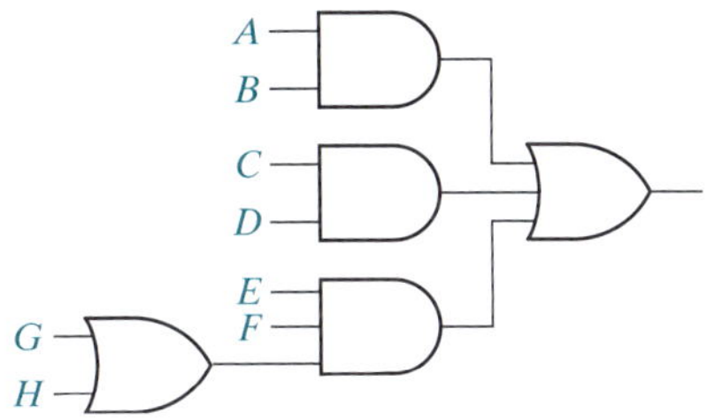

12. Reread the objectives of this unit. Make sure that you understand the difference in the procedures for converting maxterms and minterms from decimal to algebraic notation. When you are satisfied that you can meet the objectives, take the readiness test. When you come to take the readiness test, turn in a copy of your solution to assigned simulation exercise.

Applications of Boolean Algebra Minterm and Maxterm Expansions

In this unit you will learn how to design a combinational logic circuit starting with a word description of the desired circuit behavior. The first step is usually to translate the word description into a truth table or into an algebraic expression. Given the truth table for a Boolean function, two standard algebraic forms of the function can be derived—the standard sum of products (minterm expansion) and the standard product of sums (maxterm expansion). Simplification of either of these standard forms leads directly to a realization of the circuit using AND and OR gates.

4.1 Conversion of English Sentences to Boolean Equations

The three main steps in designing a single-output combinational switching circuit are

1. Find a switching function that specifies the desired behavior of the circuit.
2. Find a simplified algebraic expression for the function.
3. Realize the simplified function using available logic elements.

For simple problems, it may be possible to go directly from a word description of the desired behavior of the circuit to an algebraic expression for the output function. In other cases, it is better to first specify the function by means of a truth table and then derive an algebraic expression from the truth table.

Logic design problems are often stated in terms of one or more English sentences. The first step in designing a logic circuit is to translate these sentences into Boolean equations. In order to do this, we must break down each sentence into phrases and associate a Boolean variable with each phrase. If a phrase can have a value of true or false, then we can represent that phrase by a Boolean variable. Phrases such as "she goes to the store" or "today is Monday" can be either true or false, but a command like "go to the store" has no truth value. If a sentence has several phrases, we will mark each phrase with a brace. The following sentence has three phrases:

$$\underbrace{\text{Mary watches TV}} \text{ if } \underbrace{\text{it is Monday night}} \text{ and } \underbrace{\text{she has finished her homework}}.$$

The "if" and "and" are not included in any phrase; they show the relationships among the phrases.

We will define a two-valued variable to indicate the truth or falsity of each phrase:

$F = 1$ if "Mary watches TV" is true; otherwise, $F = 0$.
$A = 1$ if "it is Monday night" is true; otherwise, $A = 0$.
$B = 1$ if "she has finished her homework" is true; otherwise $B = 0$.

Because F is "true" if A and B are both "true", we can represent the sentence by $F = A \cdot B$

The following example illustrates how to go from a word statement of a problem directly to an algebraic expression which represents the desired circuit behavior. An alarm circuit is to be designed which operates as follows:

> The alarm will ring iff the alarm switch is turned on and the door is not closed, or it is after 6 P.M. and the window is not closed.

The first step in writing an algebraic expression which corresponds to the above sentence is to associate a Boolean variable with each phrase in the sentence. This variable will have a value of 1 when the phrase is true and 0 when it is false. We will use the following assignment of variables:

$$\underbrace{\text{The alarm will ring}}_{Z} \quad \text{iff} \quad \underbrace{\text{the alarm switch is on}}_{A} \quad \text{and}$$
$$\underbrace{\text{the door is not closed}}_{B'} \quad \textit{or} \quad \underbrace{\text{it is after 6 P.M.}}_{C} \quad \text{and}$$
$$\underbrace{\text{the window is not closed.}}_{D'}$$

This assignment implies that if $Z = 1$, the alarm will ring. If the alarm switch is turned on, $A = 1$, and if it is after 6 P.M., $C = 1$. If we use the variable B to represent the phrase "the door is closed", then B' represents "the door is not closed". Thus, $B = 1$ if the door is closed, and $B' = 1 (B = 0)$ if the door is not closed. Similarly, $D = 1$ if the window is closed, and $D' = 1$ if the window is not closed. Using this assignment of variables, the above sentence can be translated into the following Boolean equation:

$$Z = AB' + CD'$$

This equation corresponds to the following circuit:

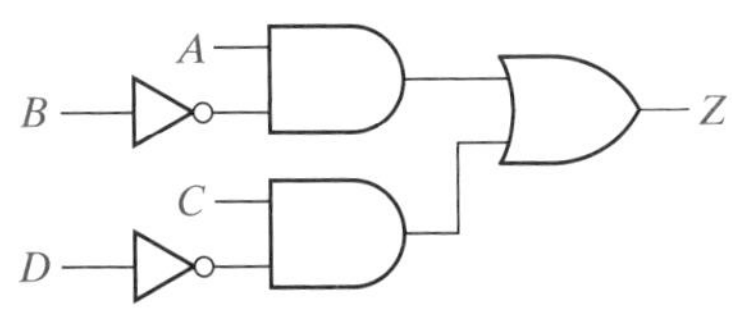

In this circuit, A is a signal which is 1 when the alarm switch is on, C is a signal from a time clock which is 1 when it is after 6 P.M., B is a signal from a switch on the door which is 1 when the door is closed, and similarly D is 1 when the window is closed. The output Z is connected to the alarm so that it will ring when $Z = 1$.

4.2 Combinational Logic Design Using a Truth Table

The next example illustrates logic design using a truth table. A switching circuit has three inputs and one output, as shown in Figure 4-1(a). The inputs A, B, and C represent the first, second, and third bits, respectively, of a binary number N. The output of the circuit is to be $f = 1$ if $N \geq 011_2$ and $f = 0$ if $N < 011_2$. The truth table for f is shown in Figure 4-1(b).

FIGURE 4-1 Combinational Circuit with Truth Table

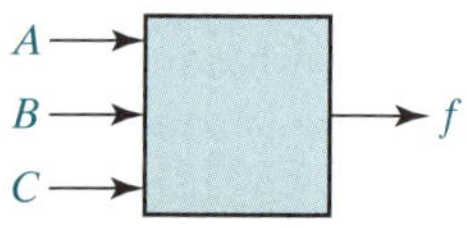

(a)

A B C	f	f′
0 0 0	0	1
0 0 1	0	1
0 1 0	0	1
0 1 1	1	0
1 0 0	1	0
1 0 1	1	0
1 1 0	1	0
1 1 1	1	0

(b)

Next, we will derive an algebraic expression for f from the truth table by using the combinations of values of A, B, and C for which $f = 1$. The term $A'BC$ is 1 only if $A = 0$, $B = 1$, and $C = 1$. Similarly, the term $AB'C'$ is 1 only for the combination 100, $AB'C$ is 1 only for 101, ABC' is 1 only for 110, and ABC is 1 only for 111. ORing these terms together yields

$$f = A'BC + AB'C' + AB'C + ABC' + ABC \tag{4-1}$$

This expression equals 1 if A, B, and C take on any of the five combinations of values 011, 100, 101, 110, or 111. If any other combination of values occurs, f is 0 because all five terms are 0.

Equation (4-1) can be simplified by first combining terms using the uniting theorem and then eliminating A' using the elimination theorem:

$$f = A'BC + AB' + AB = A'BC + A = A + BC \tag{4-2}$$

Equation (4-2) leads directly to the following circuit:

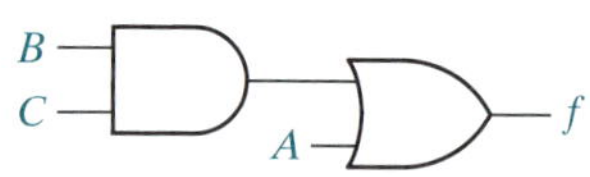

Instead of writing f in terms of the 1's of the function, we may also write f in terms of the 0's of the function. The function defined by Figure 4-1 is 0 for three combinations of input values. Observe that the term $A + B + C$ is 0 only if $A = B = C = 0$. Similarly, $A + B + C'$ is 0 only for the input combination 001, and $A + B' + C$ is 0 only for the combination 010. ANDing these terms together yields

$$f = (A + B + C)(A + B + C')(A + B' + C) \tag{4-3}$$

This expression equals 0 if A, B, and C take on any of the combinations of values 000, 001, or 010. For any other combination of values, f is 1 because all three terms are 1. Because Equation (4-3) represents the same function as Equation (4-1) they must both reduce to the same expression. Combining terms and using the second distributive law, Equation (4-3) simplifies to

$$f = (A + B)(A + B' + C) = A + B(B' + C) = A + BC \tag{4-4}$$

which is the same as Equation (4-2).

Another way to derive Equation (4-3) is to first write f' as a sum of products, and then complement the result. From Figure 4-1, f' is 1 for input combinations $ABC = 000, 001,$ and 010, so

$$f' = A'B'C' + A'B'C + A'BC'$$

Taking the complement of f' yields Equation (4-3).

4.3 Minterm and Maxterm Expansions

Each of the terms in Equation (4-1) is referred to as a minterm. In general, a *minterm* of n variables is a product of n literals in which each variable appears exactly once in either true or complemented form, but not both. (A *literal* is a variable or its complement.) Table 4-1 lists all of the minterms of the three variables A, B, and C. Each minterm has a value of 1 for exactly one combination of values of the variables A, B, and C. Thus if $A = B = C = 0$, $A'B'C' = 1$; if $A = B = 0$ and $C = 1$, $A'B'C = 1$; and so forth. Minterms are often written in abbreviated form—$A'B'C'$ is designated m_0, $A'B'C$ is designated m_1, etc. In general, the minterm which corresponds to row i of the truth table is designated m_i (i is usually written in decimal).

TABLE 4-1 Minterms and Maxterms for Three Variables

Row No.	A B C	Minterms	Maxterms
0	0 0 0	$A'B'C' = m_0$	$A + B + C = M_0$
1	0 0 1	$A'B'C = m_1$	$A + B + C' = M_1$
2	0 1 0	$A'BC' = m_2$	$A + B' + C = M_2$
3	0 1 1	$A'BC = m_3$	$A + B' + C' = M_3$
4	1 0 0	$AB'C' = m_4$	$A' + B + C = M_4$
5	1 0 1	$AB'C = m_5$	$A' + B + C' = M_5$
6	1 1 0	$ABC' = m_6$	$A' + B' + C = M_6$
7	1 1 1	$ABC = m_7$	$A' + B' + C' = M_7$

When a function f is written as a sum of minterms as in Equation (4-1), this is referred to as a *minterm expansion* or a *standard sum of products.*[1] If $f = 1$ for row i of the truth table, then m_i must be present in the minterm expansion because $m_i = 1$ only for the combination of values of the variables corresponding to row i of the table. Because the minterms present in f are in one-to-one correspondence with the 1's of f in the truth table, the minterm expansion for a function f *is* unique. Equation (4-1) can be rewritten in terms of m-notation as

$$f(A, B, C) = m_3 + m_4 + m_5 + m_6 + m_7 \tag{4-5}$$

This can be further abbreviated by listing only the decimal subscripts in the form

$$f(A, B, C) = \Sigma m(3, 4, 5, 6, 7) \tag{4-5a}$$

Each of the sum terms (or factors) in Equation (4-3) is referred to as a *maxterm.* In general, a maxterm of n variables is a sum of n literals in which each variable appears exactly once in either true or complemented form, but not both. Table 4-1 lists all of the maxterms of the three variables A, B, and C. Each maxterm has a value of 0 for exactly one combination of values for A, B, and C. Thus, if $A = B = C = 0$, $A + B + C = 0$; if $A = B = 0$ and $C = 1$, $A + B + C' = 0$; and so forth. Maxterms are often written in abbreviated form using M-notation. The maxterm which corresponds to row i of the truth table is designated M_i. Note that each maxterm is the complement of the corresponding minterm, that is, $M_i = m_i'$.

When a function f is written as a product of maxterms, as in Equation (4-3), this is referred to as a *maxterm expansion* or *standard product of sums.* If $f = 0$ for row i of the truth table, then M_i must be present in the maxterm expansion because $M_i = 0$ only for the combination of values of the variables corresponding to row i of the table. Note that the maxterms are multiplied together so that if any one of them is 0, f will be 0. Because the maxterms are in one-to-one correspondence with the 0's of f in the truth table, the maxterm expansion for a function f *is* unique. Equation (4-3) can be rewritten in M-notation as

$$f(A, B, C) = M_0 M_1 M_2 \tag{4-6}$$

This can be further abbreviated by listing only the decimal subscripts in the form

$$f(A, B, C) = \Pi M(0, 1, 2) \tag{4-6a}$$

where Π means a product.

Because if $f \neq 1$ then $f = 0$, it follows that if m_i is *not* present in the minterm expansion of f, then M_i is present in the maxterm expansion. Thus, given a minterm expansion of an n-variable function f in decimal notation, the maxterm expansion is obtained by listing those decimal integers ($0 \leq i \leq 2^n - 1$) not in the minterm list. Using this method, Equation (4-6a) can be obtained directly from Equation (4-5a).

[1]Other names used in the literature for standard sum of products are canonical sum of products and disjunctive normal form. Similarly, a standard product of sums may be called a canonical product of sums or a conjunctive normal form.

Given the minterm or maxterm expansions for f, the minterm or maxterm expansions for the complement of f are easy to obtain. Because f' is 1 when f is 0, the minterm expansion for f' contains those minterms not present in f. Thus, from Equation (4-5),

$$f' = m_0 + m_1 + m_2 = \Sigma\, m(0, 1, 2) \tag{4-7}$$

Similarly, the maxterm expansion for f' contains those maxterms not present in f. From Equation (4-6),

$$f' = \Pi\, M(3, 4, 5, 6, 7) = M_3M_4M_5M_6M_7 \tag{4-8}$$

Because the complement of a minterm is the corresponding maxterm, Equation (4-8) can be obtained by complementing Equation (4-5):

$$f' = (m_3 + m_4 + m_5 + m_6 + m_7)' = m_3'm_4'm_5'm_6'm_7' = M_3M_4M_5M_6M_7$$

Similarly, Equation (4-7) can be obtained by complementing Equation (4-6):

$$f' = (M_0M_1M_2)' = M_0' + M_1' + M_2' = m_0 + m_1 + m_2$$

A general switching expression can be converted to a minterm or maxterm expansion either using a truth table or algebraically. If a truth table is constructed by evaluating the expression for all different combinations of the values of the variables, the minterm and maxterm expansions can be obtained from the truth table by the methods just discussed. Another way to obtain the minterm expansion is to first write the expression as a sum of products and then introduce the missing variables in each term by applying the theorem $X + X' = 1$.

Example

Find the minterm expansion of $f(a, b, c, d) = a'(b' + d) + acd'$.

$$\begin{aligned} f &= a'b' + a'd + acd' \\ &= a'b'(c + c')(d + d') + a'd(b + b')(c + c') + acd'(b + b') \\ &= a'b'c'd' + a'b'c'd + a'b'cd' + a'b'cd + \cancel{a'b'c'd} + \cancel{a'b'cd} \\ &\qquad + a'bc'd + a'bcd + abcd' + ab'cd' \end{aligned} \tag{4-9}$$

Duplicate terms have been crossed out, because $X + X = X$. This expression can then be converted to decimal notation:

$$\begin{array}{llllllllll} f = & a'b'c'd' & + \; a'b'c'd & + \; a'b'cd' & + \; a'b'cd & + \; a'bc'd & + \; a'bcd & + \; abcd' & + \; ab'cd' \\ & 0\,0\,0\,0 & 0\,0\,0\,1 & 0\,0\,1\,0 & 0\,0\,1\,1 & 0\,1\,0\,1 & 0\,1\,1\,1 & 1\,1\,1\,0 & 1\,0\,1\,0 \end{array}$$

$$f = \Sigma\, m(0, 1, 2, 3, 5, 7, 10, 14) \tag{4-10}$$

The maxterm expansion for f can then be obtained by listing the decimal integers (in the range 0 to 15) which do not correspond to minterms of f:

$$f = \Pi\, M(4, 6, 8, 9, 11, 12, 13, 15)$$

An alternate way of finding the maxterm expansion is to factor f to obtain a product of sums, introduce the missing variables in each sum term by using $XX' = 0$, and then factor again to obtain the maxterms. For Equation (4-9),

$$
\begin{aligned}
f &= a'(b' + d) + acd' \\
&= (a' + cd')(a + b' + d) = (a' + c)(a' + d')(a + b' + d) \\
&= (a' + bb' + c + dd')(a' + bb' + cc' + d')(a + b' + cc' + d) \\
&= (a' + bb' + c + d)(a' + bb' + c + d')\cancel{(a' + bb' + c + d')} \\
&\qquad (a' + bb' + c' + d')(a + b' + cc' + d) \\
&= \underset{1000}{(a' + b + c + d)}\underset{1100}{(a' + b' + c + d)}\underset{1001}{(a' + b + c + d')}\underset{1101}{(a' + b' + c + d')} \\
&\qquad \underset{1011}{(a' + b + c' + d')}\underset{1111}{(a' + b' + c' + d')}\underset{0100}{(a + b' + c + d)}\underset{0110}{(a + b' + c' + d)} \\
&= \Pi\, M(4, 6, 8, 9, 11, 12, 13, 15) \qquad (4\text{-}11)
\end{aligned}
$$

Note that when translating the maxterms to decimal notation, a primed variable is first replaced with a 1 and an unprimed variable with a 0.

Because the terms in the minterm expansion of a function F correspond one-to-one with the rows of the truth table for which $F = 1$, the minterm expansion of F is unique. Thus, we can prove that an equation is valid by finding the minterm expansion of each side and showing that these expansions are the same.

Example

Show that $a'c + b'c' + ab = a'b' + bc + ac'$.

We will find the minterm expansion of each side by supplying the missing variables. For the left side,

$$
\begin{aligned}
&a'c(b + b') + b'c'(a + a') + ab(c + c') \\
&\qquad = a'bc + a'b'c + ab'c' + a'b'c' + abc + abc' \\
&\qquad = m_3 + m_1 + m_4 + m_0 + m_7 + m_6
\end{aligned}
$$

For the right side,

$$
\begin{aligned}
&a'b'(c + c') + bc(a + a') + ac'(b + b') \\
&\qquad = a'b'c + a'b'c + abc + a'bc + abc' + ab'c' \\
&\qquad = m_1 + m_0 + m_7 + m_3 + m_6 + m_4
\end{aligned}
$$

Because the two minterm expansions are the same, the equation is valid.

4.4 General Minterm and Maxterm Expansions

Table 4-2 represents a truth table for a general function of three variables. Each a_i is a constant with a value of 0 or 1. To completely specify a function, we must assign values to all of the a_i's. Because each a_i can be specified in two ways, there are 2^8

TABLE 4-2 General Truth Table for Three Variables

A	B	C	F
0	0	0	a_0
0	0	1	a_1
0	1	0	a_2
0	1	1	a_3
1	0	0	a_4
1	0	1	a_5
1	1	0	a_6
1	1	1	a_7

ways of filling the F column of the truth table; therefore, there are 256 different functions of three variables (this includes the degenerate cases, F identically equal to 0 and F identically equal to 1). For a function of n variables, there are 2^n rows in the truth table, and because the value of F can be 0 or 1 for each row, there are 2^{2^n} possible functions of n variables.

From Table 4-2, we can write the minterm expansion for a general function of three variables as follows:

$$F = a_0m_0 + a_1m_1 + a_2m_2 + \cdots + a_7m_7 = \sum_{i=0}^{7} a_im_i \tag{4-12}$$

Note that if $a_i = 1$, minterm m_i is present in the expansion; if $a_i = 0$, the corresponding minterm is not present. The maxterm expansion for a general function of three variables is

$$F = (a_0 + M_0)(a_1 + M_1)(a_2 + M_2) \cdots (a_7 + M_7) = \prod_{i=0}^{7} (a_i + M_i) \tag{4-13}$$

Note that if $a_i = 1$, $a_i + M_i = 1$, and M_i drops out of the expansion; however, M_i is present if $a_i = 0$.

From Equation (4-13), the minterm expansion of F' is

$$F' = \left[\prod_{i=0}^{7} (a_i + M_i)\right]' = \sum_{i=0}^{7} a_i'M_i' = \sum_{i=0}^{7} a_i'm_i \tag{4-14}$$

Note that all minterms which are not present in F are present in F'.

From Equation (4-12), the maxterm expansion of F' is

$$F' = \left[\sum_{i=0}^{7} a_im_i\right]' = \prod_{i=0}^{7} (a_i' + m_i') = \prod_{i=0}^{7} (a_i' + M_i) \tag{4-15}$$

Note that all maxterms which are not present in F are present in F'. Generalizing Equations (4-12), (4-13), (4-14), and (4-15) to n variables, we have

$$F = \sum_{i=0}^{2^n-1} a_im_i = \prod_{i=0}^{2^n-1} (a_i + M_i) \tag{4-16}$$

$$F' = \sum_{i=0}^{2^n-1} a_i' m_i = \prod_{i=0}^{2^n-1} (a_i' + M_i) \tag{4-17}$$

Given two different minterms of n variables, m_i and m_j, at least one variable appears complemented in one of the minterms and uncomplemented in the other. Therefore, if $i \neq j$, $m_i m_j = 0$. For example, for $n = 3$, $m_1 m_3 = (A'B'C)(A'BC) = 0$. Given minterm expansions for two functions

$$f_1 = \sum_{i=0}^{2^n-1} a_i m_i \quad f_2 = \sum_{j=0}^{2^n-1} b_j m_j \tag{4-18}$$

the product is

$$f_1 f_2 = \left(\sum_{i=0}^{2^n-1} a_i m_i\right)\left(\sum_{j=0}^{2^n-1} b_j m_j\right) = \sum_{i=0}^{2^n-1} \sum_{j=0}^{2^n-1} a_i b_j m_i m_j$$

$$= \sum_{i=0}^{2^n-1} a_i b_i m_i \quad \text{(because } m_i m_j = 0 \text{ unless } i = j\text{)} \tag{4-19}$$

Note that all of the cross-product terms $(i \neq j)$ drop out so that $f_1 f_2$ contains only those minterms which are present in both f_1 and f_2. For example, if

$$f_1 = \Sigma\, m(0, 2, 3, 5, 9, 11) \qquad \text{and} \qquad f_2 = \Sigma\, m(0, 3, 9, 11, 13, 14)$$
$$f_1 f_2 = \Sigma\, m(0, 3, 9, 11)$$

Table 4-3 summarizes the procedures for conversion between minterm and maxterm expansions of F and F', assuming that all expansions are written as lists of decimal numbers. When using this table, keep in mind that the truth table for an n-variable function has 2^n rows so that the minterm (or maxterm) numbers range from 0 to $2^n - 1$. Table 4-4 illustrates the application of Table 4-3 to the three-variable function given in Figure 4-1.

TABLE 4-3 Conversion of Forms

GIVEN FORM ↓ / DESIRED FORM →	Minterm Expansion of F	Maxterm Expansion of F	Minterm Expansion of F'	Maxterm Expansion of F'
Minterm Expansion of F	———	maxterm nos. are those nos. not on the minterm list for F	list minterms not present in F	maxterm nos. are the same as minterm nos. of F
Maxterm Expansion of F	minterm nos. are those nos. not on the maxterm list for F	———	minterm nos. are the same as maxterm nos. of F	list maxterms not present in F

TABLE 4-4
Application of Table 4.3

GIVEN FORM	DESIRED FORM: Minterm Expansion of f	Maxterm Expansion of f	Minterm Expansion of f'	Maxterm Expansion of f'
$f = \Sigma m(3, 4, 5, 6, 7)$	______	$\Pi M(0, 1, 2)$	$\Sigma m(0, 1, 2)$	$\Pi M(3, 4, 5, 6, 7)$
$f = \Pi M(0, 1, 2)$	$\Sigma m(3, 4, 5, 6, 7)$	______	$\Sigma m(0, 1, 2)$	$\Pi M(3, 4, 5, 6, 7)$

4.5 Incompletely Specified Functions

A large digital system is usually divided into many subcircuits. Consider the following example in which the output of circuit N_1 drives the input of circuit N_2.

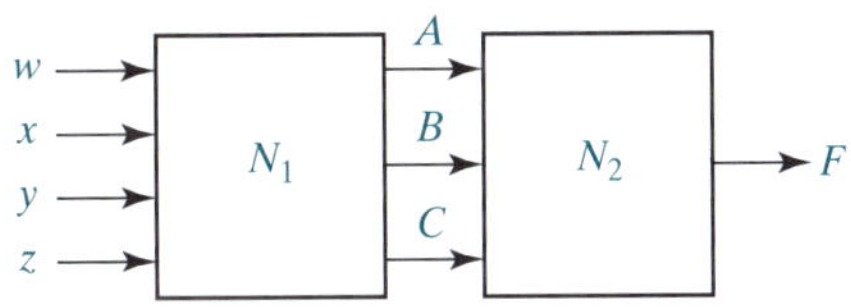

Let us assume that the output of N_1 does not generate all possible combinations of values for A, B, and C. In particular, we will assume that there are no combinations of values for w, x, y, and z which cause A, B, and C to assume values of 001 or 110. Hence, when we design N_2, it is not necessary to specify values of F for $ABC = 001$ or 110 because these combinations of values can never occur as inputs to N_2. For example, F might be specified by Table 4-5.

The X's in the table indicate that we don't care whether the value of 0 or 1 is assigned to F for the combinations $ABC = 001$ or 110. In this example, we don't care what the value of F is because these input combinations never occur anyway. The function F is then *incompletely specified*. The minterms $A'B'C$ and ABC' are referred to as don't-care minterms, since we don't care whether they are present in the function or not.

TABLE 4-5
Truth Table with Don't-Cares

A B C	F
0 0 0	1
0 0 1	X
0 1 0	0
0 1 1	1
1 0 0	0
1 0 1	0
1 1 0	X
1 1 1	1

When we realize the function, we must specify values for the don't-cares. It is desirable to choose values which will help simplify the function. If we assign the value 0 to both X's, then

$$F = A'B'C' + A'BC + ABC = A'B'C' + BC$$

If we assign 1 to the first X and 0 to the second, then

$$F = A'B'C' + A'B'C + A'BC + ABC = A'B' + BC$$

If we assign 1 to both X's, then

$$F = A'B'C' + A'B'C + A'BC + ABC' + ABC = A'B' + BC + AB$$

The second choice of values leads to the simplest solution.

We have seen one way in which incompletely specified functions can arise, and there are many other ways. In the preceding example, don't-cares were present because certain combinations of circuit inputs did not occur. In other cases, all input combinations may occur, but the circuit output is used in such a way that we do not care whether it is 0 or 1 for certain input combinations.

When writing the minterm expansion for an incompletely specified function, we will use m to denote the required minterms and d to denote the don't-care minterms. Using this notation, the minterm expansion for Table 4-5 is

$$F = \Sigma\, m(0, 3, 7) + \Sigma\, d(1, 6)$$

For each don't-care minterm there is a corresponding don't-care maxterm. For example, if $F = X$ (don't-care) for input combination 001, m_1 is a don't-care minterm and M_1 is a don't-care maxterm. We will use D to represent a don't-care maxterm, and we write the maxterm expansion of the function in Table 4-5 as

$$F = \Pi\, M(2, 4, 5) \cdot \Pi\, D(1, 6)$$

which implies that maxterms M_2, M_4, and M_5 are present in F and don't-care maxterms M_1 and M_6 are optional.

4.6 Examples of Truth Table Construction

Example 1

We will design a simple binary adder that adds two 1-bit binary numbers, a and b, to give a 2-bit sum. The numeric values for the adder inputs and output are as follows:

a	b	Sum	
0	0	00	(0 + 0 = 0)
0	1	01	(0 + 1 = 1)
1	0	01	(1 + 0 = 1)
1	1	10	(1 + 1 = 2)

We will represent inputs to the adder by the logic variables A and B and the 2-bit sum by the logic variables X and Y, and we construct a truth table:

A B	X Y
0 0	0 0
0 1	0 1
1 0	0 1
1 1	1 0

Because a numeric value of 0 is represented by a logic 0 and a numeric value of 1 by a logic l, the 0's and 1's in the truth table are exactly the same as in the previous table. From the truth table,

$$X = AB \quad \text{and} \quad Y = A'B + AB' = A \oplus B$$

Example 2

An adder is to be designed which adds two 2-bit binary numbers to give a 3-bit binary sum. Find the truth table for the circuit. The circuit has four inputs and three outputs as shown:

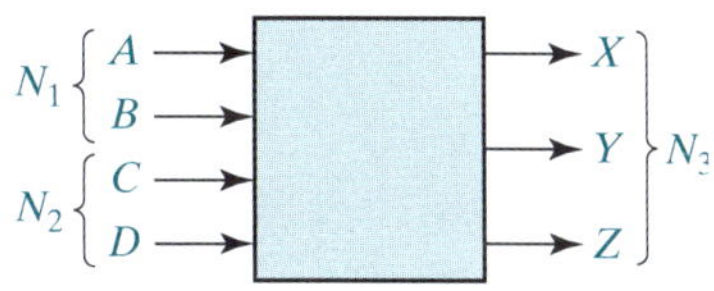

TRUTH TABLE:

N_1	N_2	N_3
A B	C D	X Y Z
0 0	0 0	0 0 0
0 0	0 1	0 0 1
0 0	1 0	0 1 0
0 0	1 1	0 1 1
0 1	0 0	0 0 1
0 1	0 1	0 1 0
0 1	1 0	0 1 1
0 1	1 1	1 0 0
1 0	0 0	0 1 0
1 0	0 1	0 1 1
1 0	1 0	1 0 0
1 0	1 1	1 0 1
1 1	0 0	0 1 1
1 1	0 1	1 0 0
1 1	1 0	1 0 1
1 1	1 1	1 1 0

Inputs A and B taken together represent a binary number N_1. Inputs C and D taken together represent a binary number N_2. Outputs X, Y, and Z taken together represent a binary number N_3, where $N_3 = N_1 + N_2$ (+ of course represents ordinary addition here).

In this example we have used A, B, C, and D to represent both numeric values and logic values, but this should not cause any confusion because the numeric and logic values are the same. In forming the truth table, the variables were treated like binary numbers having *numeric* values. Now we wish to derive the switching functions for the output variables. In doing so, we will treat A, B, C, D, X, Y, and Z as switching variables having nonnumeric values 0 and 1. (Remember that in this case the 0 and 1 may represent low and high voltages, open and closed switches, etc.)

From inspection of the table, the output functions are

$$X(A, B, C, D) = \Sigma\, m(7, 10, 11, 13, 14, 15)$$
$$Y(A, B, C, D) = \Sigma\, m(2, 3, 5, 6, 8, 9, 12, 15)$$
$$Z(A, B, C, D) = \Sigma\, m(1, 3, 4, 6, 9, 11, 12, 14)$$

Example 3

Design an error detector for 6-3-1-1 binary-coded-decimal digits. The output (F) is to be 1 iff the four inputs (A, B, C, D) represent an invalid code combination.

The valid 6-3-1-1 code combinations are listed in Table 1-2. If any other combination occurs, this is not a valid 6-3-1-1 binary-coded-decimal digit, and the circuit output should be $F = 1$ to indicate that an error has occurred. This leads to the following truth table:

A	B	C	D	F
0	0	0	0	0
0	0	0	1	0
0	0	1	0	1
0	0	1	1	0
0	1	0	0	0
0	1	0	1	0
0	1	1	0	1
0	1	1	1	0
1	0	0	0	0
1	0	0	1	0
1	0	1	0	1
1	0	1	1	0
1	1	0	0	0
1	1	0	1	1
1	1	1	0	1
1	1	1	1	1

The corresponding output function is

$$\begin{aligned} F &= \Sigma\, m(2, 6, 10, 13, 14, 15) \\ &= A'B'CD' + A'BCD' + AB'CD' + ABCD' + ABC'D + ABCD \\ &= A'CD' + ACD' + ABD = CD' + ABD \end{aligned}$$

The realization using AND and OR gates is

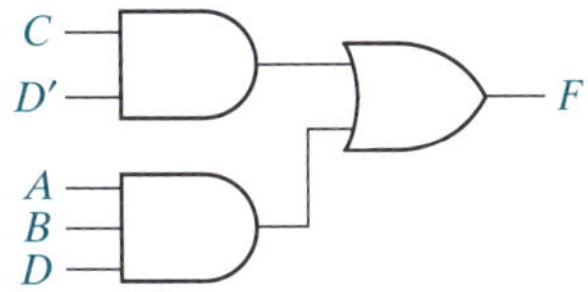

Example 4

The four inputs to a circuit (A, B, C, D) represent an 8-4-2-1 binary-coded-decimal digit. Design the circuit so that the output (Z) is 1 iff the decimal number represented by the inputs is exactly divisible by 3. Assume that only valid BCD digits occur as inputs.

The digits 0, 3, 6, and 9 are exactly divisible by 3, so $Z = 1$ for the input combinations $ABCD = 0000, 0011, 0110$, and 1001. The input combinations 1010, 1011, 1100, 1101, 1110, and 1111 do not represent valid BCD digits and will never occur, so Z is a don't-care for these combinations. This leads to the following truth table:

A	B	C	D	Z
0	0	0	0	1
0	0	0	1	0
0	0	1	0	0
0	0	1	1	1
0	1	0	0	0
0	1	0	1	0
0	1	1	0	1
0	1	1	1	0
1	0	0	0	0
1	0	0	1	1
1	0	1	0	X
1	0	1	1	X
1	1	0	0	X
1	1	0	1	X
1	1	1	0	X
1	1	1	1	X

The corresponding output function is

$$Z = \Sigma\, m(0, 3, 6, 9) + \Sigma\, d(10, 11, 12, 13, 14, 15)$$

In order to find the simplest circuit which will realize Z, we must choose some of the don't-cares (X's) to be 0 and some to be 1. The easiest way to do this is to use a Karnaugh map as described in Unit 5.

4.7 Design of Binary Adders and Subtracters

In this section, we will design a parallel adder that adds two 4-bit unsigned binary numbers and a carry input to give a 4-bit sum and a carry output (see Figure 4-2). One approach would be to construct a truth table with nine inputs and five outputs and then derive and simplify the five output equations. Because each equation would be a function of nine variables before simplification, this approach would be very difficult, and the resulting logic circuit would be very complex. A better method is to design a logic module that adds two bits and a carry, and then connect four of these modules together to form a 4-bit adder as shown in Figure 4-3. Each of the modules is called a *full adder*. The carry output from the first full adder serves as the carry input to the second full adder, etc.

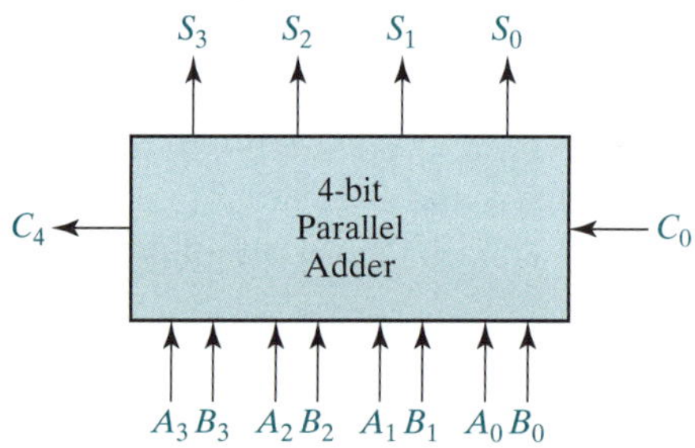

FIGURE 4-2 Parallel Adder for 4-Bit Binary Numbers

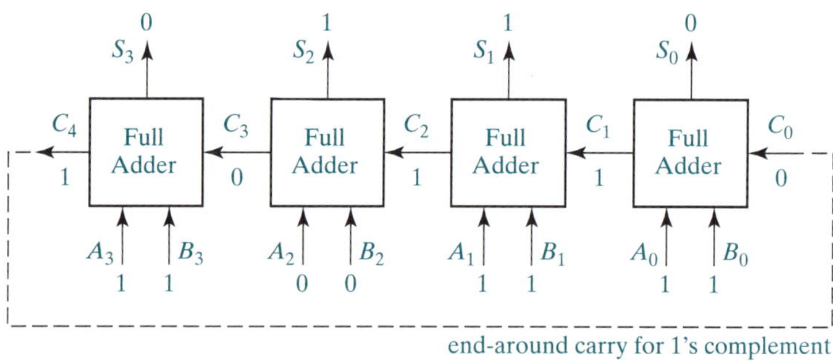

FIGURE 4-3 Parallel Adder Composed of Four Full Adders

In the example of Figure 4-3, we perform the following addition:

$$\begin{array}{rl} 10110 & \text{(carries)} \\ 1011 & \\ +\ 1011 & \\ \hline 10110 & \end{array}$$

The full adder to the far right adds $A_0 + B_0 + C_0 = 1 + 1 + 0$ to give a sum of 10_2, which gives a sum $S_0 = 0$ and a carry out of $C_1 = 1$. The next full adder adds $A_1 + B_1 + C_1 = 1 + 1 + 1 = 11_2$, which gives a sum $S_1 = 1$ and a carry $C_2 = 1$. The carry continues to propagate from right to left until the left cell produces a final carry of $C_4 = 1$.

FIGURE 4-4
Truth Table for a Full Adder

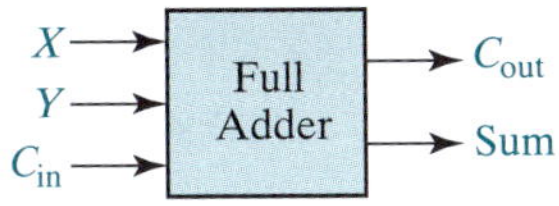

X	Y	C_{in}	C_{out}	Sum
0	0	0	0	0
0	0	1	0	1
0	1	0	0	1
0	1	1	1	0
1	0	0	0	1
1	0	1	1	0
1	1	0	1	0
1	1	1	1	1

Figure 4-4 gives the truth table for a full adder with inputs X, Y, and C_{in}. The outputs for each row of the table are found by adding up the input bits $(X + Y + C_{in})$ and splitting the result into a carry out (C_{i+1}) and a sum bit (S_i). For example, in the 101 row $1 + 0 + 1 = 10_2$, so $C_{i+1} = 1$ and $S_i = 0$. Figure 4-5 shows the implementation of the full adder using gates. The logic equations for the full adder derived from the truth table are

$$\begin{aligned} Sum &= X'Y'C_{in} + X'YC'_{in} + XY'C'_{in} + XYC_{in} \\ &= X'(Y'C_{in} + YC'_{in}) + X(Y'C'_{in} + YC_{in}) \\ &= X'(Y \oplus C_{in}) + X(Y \oplus C_{in})' = X \oplus Y \oplus C_{in} \end{aligned} \tag{4-20}$$

$$\begin{aligned} C_{out} &= X'YC_{in} + XY'C_{in} + XYC'_{in} + XYC_{in} \\ &= (X'YC_{in} + XYC_{in}) + (XY'C_{in} + XYC_{in}) + (XYC'_{in} + XYC_{in}) \\ &= YC_{in} + XC_{in} + XY \end{aligned} \tag{4-21}$$

Note that the term XYC_{in} was used three times in simplifying C_{out}. Figure 4-5 shows the logic circuit for Equations (4-20) and (4-21).

FIGURE 4-5
Implementation of Full Adder

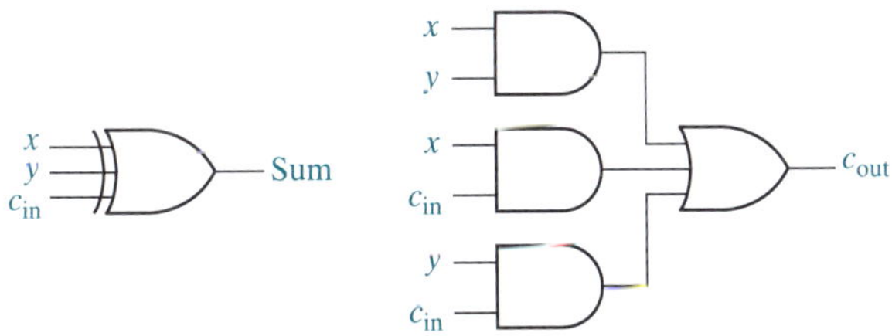

Although designed for unsigned binary numbers, the parallel adder of Figure 4-3 can also be used for signed binary numbers with negative numbers expressed in complement form. When 2's complement is used, the last carry (C_4) is discarded, and there is no carry into the first cell. Because $C_0 = 0$, the equations for the first cell may be simplified to

$$S_0 = A_0 \oplus B_0 \text{ and } C_1 = A_0 B_0$$

When 1's complement is used, the end-around carry is accomplished by connecting C_4 to the C_0 input, as shown by the dashed line in Figure 4-3.

When adding signed binary numbers with negative numbers expressed in complement form, the sign bit of the sum is wrong when an overflow occurs. That is, an overflow has occurred if adding two positive numbers gives a negative result, or adding two negative numbers gives a positive result. We will define a signal V that is 1 when an overflow occurs. For Figure 4-3, we can use the sign bits of A, B, and S (the sum) to determine the value of V:

$$V = A_3' B_3' S_3 + A_3 B_3 S_3' \tag{4-22}$$

If the number of bits is large, a parallel binary adder of the type shown in Figure 4-4 may be rather slow because the carry generated in the first cell might have to propagate all of the way to the last cell.

Subtraction of binary numbers is most easily accomplished by adding the complement of the number to be subtracted. To compute $A - B$, add the complement of B to A. This gives the correct answer because $A + (-B) = A - B$. Either 1's or 2's complement is used depending on the type of adder employed.

The circuit of Figure 4-6 may be used to form $A - B$ using the 2's complement representation for negative numbers. The 2's complement of B can be formed by first finding the 1's complement and then adding 1. The 1's complement is formed by inverting each bit of B, and the addition of 1 is effectively accomplished by putting a 1 into the carry input of the first full adder.

FIGURE 4-6
Binary Subtracter Using Full Adders

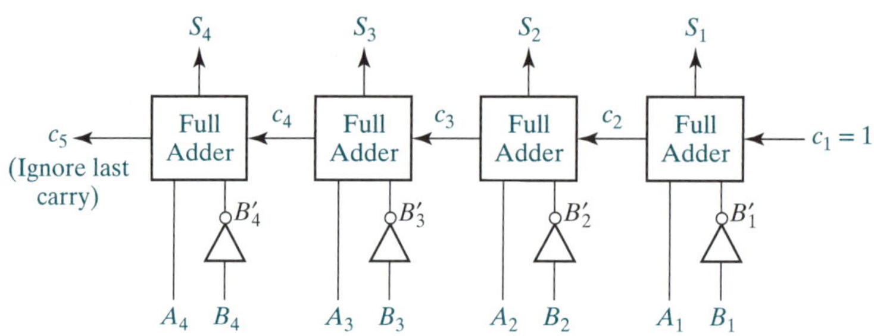

Example

$A = 0110 \quad (+6)$
$B = 0011 \quad (+3)$

The adder output is

$$\begin{array}{rrl} & 0110 & (+6) \\ & +1100 & (\text{1's complement of 3}) \\ & + \quad 1 & (\text{first carry input}) \\ \hline (1) & 0011 & = 3 = 6 - 3 \end{array}$$

Alternatively, direct subtraction can be accomplished by employing a full subtracter in a manner analogous to a full adder. A block diagram for a parallel subtracter which subtracts Y from X is shown in Figure 4-7. The first two bits are subtracted in the rightmost cell to give a difference d_1, and a borrow signal ($b_2 = 1$) is generated if it is necessary to borrow from the next column. A typical cell (cell i) has inputs x_i, y_i, and b_i, and outputs b_{i+1} and d_i. An input $b_i = 1$ indicates that we must borrow 1 from x_i in that cell, and borrowing 1 from x_i is equivalent to subtracting 1 from x_i. In cell i,

FIGURE 4-7 Parallel Subtracter

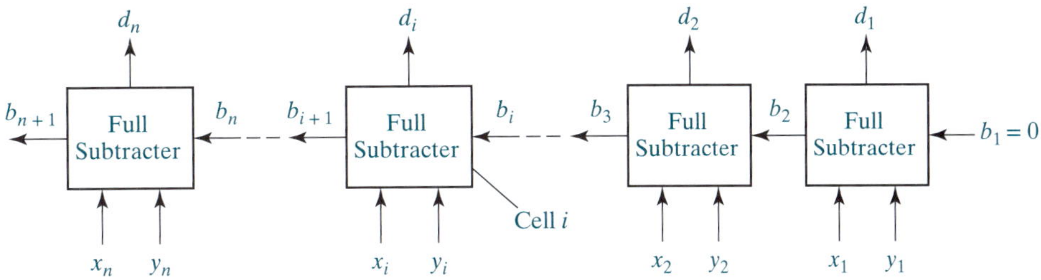

TABLE 4-6 Truth Table for Binary Full Subtracter

x_i	y_i	b_i	b_{i+1}	d_i
0	0	0	0	0
0	0	1	1	1
0	1	0	1	1
0	1	1	1	0
1	0	0	0	1
1	0	1	0	0
1	1	0	0	0
1	1	1	1	1

bits b_i and y_i are subtracted from x_i to form the difference d_i, and a borrow signal ($b_{i+1} = 1$) is generated if it is necessary to borrow from the next column.

Table 4-6 gives the truth table for a binary full subtracter. Consider the following case, where $x_i = 0$, $y_i = 1$ and $b_i = 1$:

	Column i Before Borrow	Column i After Borrow	
x_i	0	10	
$-b_i$	−1	−1	
$-y_i$	−1	−1	
d_i		0	($b_{i+1} = 1$)

Note that in column i, we cannot immediately subtract y_i and b_i from x_i. Hence, we must borrow from column $i + 1$. Borrowing 1 from column $i + 1$ is equivalent to setting b_{i+1} to 1 and adding 10 (2_{10}) to x_i. We then have $d_i = 10 - 1 - 1 = 0$. Verify that Table 4-6 is correct for the other input combinations and use it to work out several examples of binary subtraction.

The ripple carry adder is relatively slow because, in the worst case, a carry propagates through all stages of the adder, and there are two gate delays per stage. There are several techniques for reducing the carry propagation time. One is called a *carry-lookahead adder*. In the parallel adder the carry out of the ith stage can be written as

$$C_{i+1} = A_iB_i + C_i(A_i + B_i) = A_iB_i + C_i(A_i \oplus B_i) = G_i + P_iC_i$$

where $G_i = A_iB_i$ indicates the condition for the ith stage to *generate* a carry out and $P_i = A_i \oplus B_i$ (or $P_i = A_i + B_i$) indicates the condition for the ith stage to propagate a carry in to the carry out. Then C_{i+2} can be expressed in terms of C_i.

$$C_{i+2} = G_{i+1} + P_{i+1}C_{i+1} = G_{i+1} + (G_i + C_iP_i)P_{i+1} = G_{i+1} + P_{i+1}G_i + P_{i+1}P_iC_i$$

This can be continued to express C_{i+2}, C_{i+3}, etc. in terms of C_i.

$$\begin{aligned}
C_{i+1} &= G_i + P_iC_i \\
C_{i+2} &= G_{i+1} + P_{i+1}G_i + P_{i+1}P_iC_i \\
C_{i+3} &= G_{i+2} + P_{i+2}G_{i+1} + P_{i+2}P_{i+1}G_i + P_{i+2}P_{i+1}P_iC_i \\
C_{i+4} &= G_{i+3} + P_{i+3}G_{i+2} + P_{i+3}P_{i+2}G_{i+1} + P_{i+3}P_{i+2}P_{i+1}G_i + P_{i+3}P_{i+2}P_{i+1}P_iC_i
\end{aligned} \tag{4-23}$$

Assuming that the maximum fan-in of the gates is not exceeded, each of these equations can be implemented in a two-level circuit so, if a change in C_i propagates to C_j ($j = i + 1, i + 2, \cdots$), it does so with a delay of two gates. Equations (4-23) are the carry-lookahead equations. If a circuit implements, for example, four of the equations, it is a 4-bit carry-lookahead circuit.

Figure 4-8 shows a 4-bit parallel adder using a 4-bit carry-lookahead circuit. (The sum outputs are not shown.) After the generate and propagate outputs of the full adders are stable, if a change in C_0 propagates to C_i ($i = 1, 2, 3,$ or 4), it does so in two gate delays. Similarly, if a change in C_1 propagates to C_i ($i = 2, 3,$ or 4), it does so in two gate delays. In the 4-bit ripple-carry adder a change in C_0 propagating to C_4 requires 8 gate delays.

FIGURE 4-8
4-Bit Adder with Carry-Lookahead

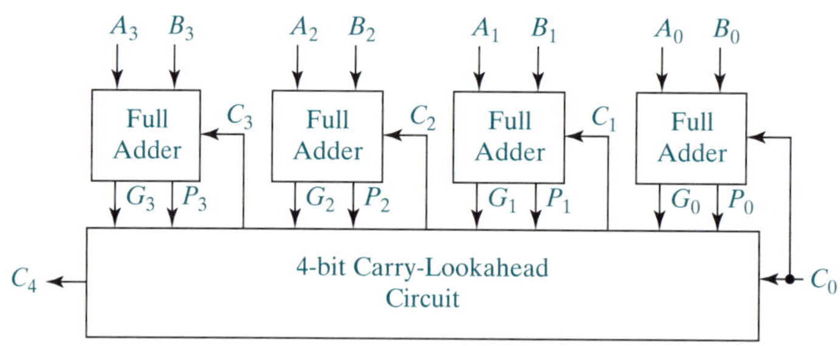

The carry-lookahead circuit can be increased in size to reduce the delay in longer parallel adders; however, the gate fan-in increases linearly with the size of the carry-lookahead circuit so the size is limited by the maximum fan-in available. For longer adders the carry-lookahead circuits can be cascaded. For example, a 16-bit parallel adder can be implemented using four 4-bit carry-lookahead circuits, as shown in Figure 4-9. Now the speed of the circuit is determined by the number of carry-lookahead circuits required. In Figure 4-9 the propagation delay from C_0 to C_{16} would be 8 gate delays; a 16-bit ripple-carry adder would have a delay of 32 gates.

To reduce the delay of the adder without increasing the size of the carry-lookahead circuit, a second level of carry-lookahead circuits can be connected to the first level carry-lookahead circuits. To illustrate this, the equations for the four carry-lookahead circuits in Figure 4-9 are written in same form as Equation (4-23).

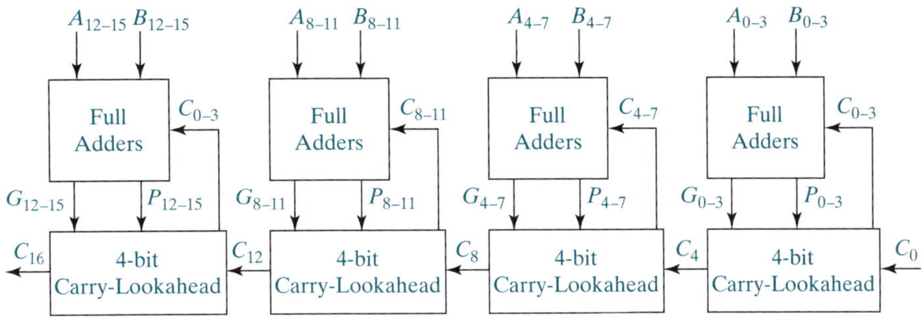

FIGURE 4-9
16-Bit Adder with Carry-Lookahead

$C_4 = \boldsymbol{G}_0 + \boldsymbol{P}_0 C_0$

where $\boldsymbol{G}_0 = G_3 + P_3G_2 + P_3P_2G_1 + P_3P_2P_1G_0$ and $\boldsymbol{P}_0 = P_3P_2P_1P_0$

$C_8 = \boldsymbol{G}_4 + \boldsymbol{P}_4 C_4$

where $\boldsymbol{G}_4 = G_7 + P_7G_6 + P_7P_6G_5 + P_7P_6P_5G_4$ and $\boldsymbol{P}_4 = P_7P_6P_5P_4$

$C_{12} = \boldsymbol{G}_8 + \boldsymbol{P}_8 C_8$

where $\boldsymbol{G}_8 = G_{11} + P_{11}G_{10} + P_{11}P_{10}G_9 + P_{11}P_{10}P_9G_8$ and $\boldsymbol{P}_8 = P_{11}P_{10}P_9P_8$

$C_{16} = \boldsymbol{G}_{12} + \boldsymbol{P}_{12} C_{12}$

where $\boldsymbol{G}_{12} = G_{15} + P_{15}G_{14} + P_{15}P_{14}G_{13} + P_{15}P_{14}P_{13}G_{12}$ and $\boldsymbol{P}_{12} = P_{15}P_{14}P_{13}P_{12}$

Now these equations for C_4, C_8, C_{12}, and C_{16} can be written in terms of C_0.

$$
\begin{aligned}
C_4 &= \boldsymbol{G}_0 + \boldsymbol{P}_0 C_0 \\
C_8 &= \boldsymbol{G}_4 + \boldsymbol{P}_4\boldsymbol{G}_0 + \boldsymbol{P}_4\boldsymbol{P}_0 C_0 \\
C_{12} &= \boldsymbol{G}_8 + \boldsymbol{P}_8\boldsymbol{G}_4 + \boldsymbol{P}_8\boldsymbol{P}_4\boldsymbol{G}_0 + \boldsymbol{P}_8\boldsymbol{P}_4\boldsymbol{P}_0 C_0 \\
C_{16} &= \boldsymbol{G}_{12} + \boldsymbol{P}_{12}\boldsymbol{G}_8 + \boldsymbol{P}_{12}\boldsymbol{P}_8\boldsymbol{G}_4 + \boldsymbol{P}_{12}\boldsymbol{P}_8\boldsymbol{P}_4\boldsymbol{G}_0 + \boldsymbol{P}_{12}\boldsymbol{P}_8\boldsymbol{P}_4\boldsymbol{P}_0 C_0
\end{aligned}
$$

These equations are the same as those for a 4-bit carry-lookahead circuit. The first level carry-lookahead circuits can be modified to produce $\boldsymbol{G}_i$ and $\boldsymbol{P}_i$ instead of C_i, $i = 0, 4, 8$, and 12. These provide inputs to a second level carry-lookahead circuit, as shown in Figure 4-10. Now the propagation delay from C_0 to C_i, $i = 4, 8, 12$, and 16, is just two gate delays.

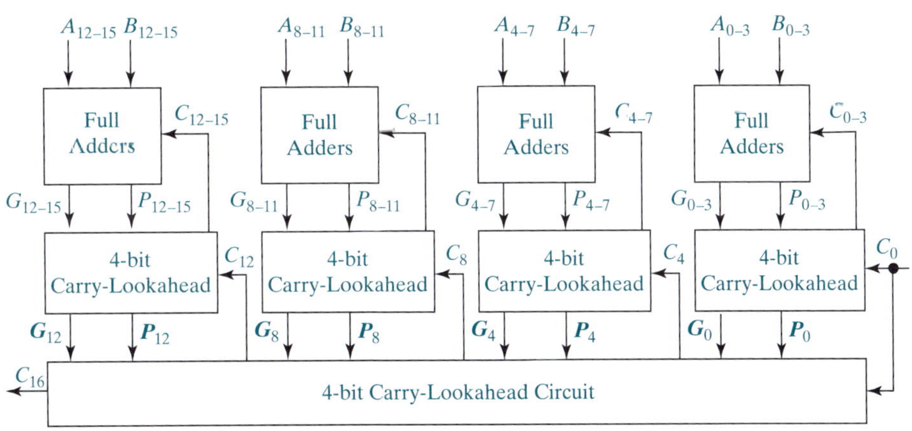

FIGURE 4-10
16-Bit Adder with Second Level Carry-Lookahead

$$C_{i+1} = G_i + P_iG_{i-1} + P_iP_{i-1}G_{i-2} + P_iP_{i-1}P_{i-2}G_{i-3} + P_iP_{i-1}P_{i-2}P_{i-3}C_{i-3}$$
$$C_4 = G_3 + P_3G_2 + P_3P_2G_1 + P_3P_2P_1G_0 + P_3P_2P_1P_0C_0 = \mathbf{G}_0 + \mathbf{P}_0C_0$$

This expression can be expanded to express C_{i+1} in terms of C_{i-1}.

$$C_{i+1} = G_i + C_iP_i + G_i = (G_{i-1} + C_{i-1}P_{i-1})P_i = G_i + P_iG_{i-1} + P_iP_{i-1}C_{i-1}$$

This procedure can be continued to obtain

$$C_{i+1} = G_i + C_iP_i$$
$$C_{i+1} = G_i + P_iG_{i-1} + P_iP_{i-1}C_{i-1}$$
$$C_{i+1} = G_i + P_iG_{i-1} + P_iP_{i-1}G_{i-2} + P_iP_{i-1}P_{i-2}C_{i-2}$$
$$C_{i+1} = G_i + P_iG_{i-1} + P_iP_{i-1}G_{i-2} + P_iP_{i-1}P_{i-2}G_{i-3} + P_iP_{i-1}P_{i-2}P_{i-3}C_{i-3}$$

and so on.

Problems

4.1 Represent each of the following sentences by a Boolean equation.

(a) The company safe should be unlocked only when Mr. Jones is in the office or Mr. Evans is in the office, and only when the company is open for business, and only when the security guard is present.

(b) You should wear your overshoes if you are outside in a heavy rain and you are wearing your new suede shoes, or if your mother tells you to.

(c) You should laugh at a joke if it is funny, it is in good taste, and it is not offensive to others, or if it is told in class by your professor (regardless of whether it is funny and in good taste) and it is not offensive to others.

(d) The elevator door should open if the elevator is stopped, it is level with the floor, and the timer has not expired, or if the elevator is stopped, it is level with the floor, and a button is pressed.

4.2 A flow rate sensing device used on a liquid transport pipeline functions as follows. The device provides a 5-bit output where all five bits are zero if the flow rate is less than 10 gallons per minute. The first bit is 1 if the flow rate is at least 10 gallons per minute; the first and second bits are 1 if the flow rate is at least 20 gallons per minute; the first, second, and third bits are 1 if the flow rate is at least 30 gallons per minute; and so on. The five bits, represented by the logical variables A, B, C, D, and E, are used as inputs to a device that provides two outputs Y and Z.

(a) Write an equation for the output Y if we want Y to be 1 iff the flow rate is less than 30 gallons per minute.

(b) Write an equation for the output Z if we want Z to be 1 iff the flow rate is at least 20 gallons per minute but less than 50 gallons per minute.

4.3 Given $F_1 = \Sigma\, m(0, 4, 5, 6)$ and $F_2 = \Sigma\, m(0, 3, 6, 7)$ find the minterm expression for $F_1 + F_2$. State a general rule for finding the expression for $F_1 + F_2$ given the minterm expansions for F_1 and F_2. Prove your answer by using the general form of the minterm expansion.

4.4 (a) How many switching functions of two variables (x and y) are there?
(b) Give each function in truth table form and in reduced algebraic form.

4.5 A combinational circuit is divided into two subcircuits N_1 and N_2 as shown. The truth table for N_1 is given. Assume that the input combinations $ABC = 110$ and $ABC = 010$ never occur. Change as many of the values of D, E, and F to don't-cares as you can without changing the value of the output Z.

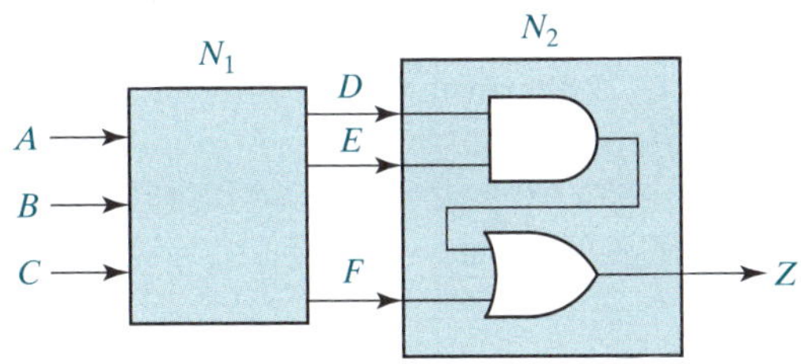

A	B	C	D	E	F
0	0	0	1	1	0
0	0	1	0	0	1
0	1	0	0	1	1
0	1	1	1	1	1
1	0	0	1	0	0
1	0	1	1	0	1
1	1	0	0	1	0
1	1	1	0	0	0

4.6 Work (a) and (b) with the following truth table:

A	B	C	F	G
0	0	0	1	0
0	0	1	X	1
0	1	0	0	X
0	1	1	0	1
1	0	0	0	0
1	0	1	X	1
1	1	0	1	X
1	1	1	1	1

(a) Find the simplest expression for F, and specify the values of the don't-cares that lead to this expression.
(b) Repeat (a) for G. (*Hint:* Can you make G the same as one of the inputs by properly choosing the values for the don't-care?)

4.7 Each of three coins has two sides, heads and tails. Represent the heads or tails status of each coin by a logical variable (A for the first coin, B for the second coin, and C for the third) where the logical variable is 1 for heads and 0 for tails. Write a logic function $F(A, B, C)$ which is 1 iff exactly one of the coins is heads after a toss of the coins. Express F
(a) as a minterm expansion.
(b) as a maxterm expansion.

4.8 A switching circuit has four inputs as shown. A and B represent the first and second bits of a binary number N_1. C and D represent the first and second bits of a binary number N_2. The output is to be 1 only if the product $N_1 \times N_2$ is less than or equal to 2.
(a) Find the minterm expansion for F.
(b) Find the maxterm expansion for F.
Express your answers in both decimal notation and algebraic form.

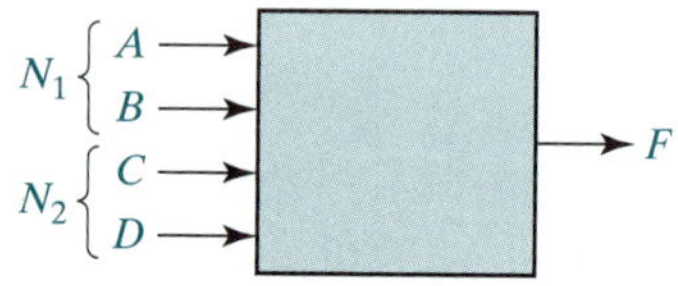

4.9 Given: $F(a, b, c) = abc' + b'$.
(a) Express F as a minterm expansion. (Use m-notation.)
(b) Express F as a maxterm expansion. (Use M-notation.)
(c) Express F' as a minterm expansion. (Use m-notation.)
(d) Express F' as a maxterm expansion. (Use M-notation.)

4.10 Work Problem 4.9 using:
$F(a, b, c, d) = (a + b + d)(a' + c)(a' + b' + c')(a + b + c' + d')$

4.11 (a) Implement a full subtracter using a minimum number of gates.
(b) Compare the logic equations for the full adder and full subtracter. What is the relation between s_i and d_i? Between c_{i+1} and b_{i+1}?

4.12 Design a circuit which will perform the following function on three 4-bit numbers:

$$(X_3X_2X_1X_0 + Y_3Y_2Y_1Y_0) - Z_3Z_2Z_1Z_0$$

It will give a result $S_3S_2S_1S_0$, a carry, and a borrow. Use eight full adders and any other type of gates. Assume that negative numbers are represented in 2's complement.

4.13 A combinational logic circuit has four inputs (A, B, C, and D) and one output Z. The output is 1 iff the input has three consecutive 0's or three consecutive 1's. For example, if $A = 1$, $B = 0$, $C = 0$, and $D = 0$, then $Z = 1$, but if $A = 0$, $B = 1$, $C = 0$, and $D = 0$, then $Z = 0$. Design the circuit using one four-input OR gate and four three-input AND gates.

4.14 Design a combinational logic circuit which has one output Z and a 4-bit input $ABCD$ representing a binary number. Z should be 1 iff the input is at least 5, but is no greater than 11. Use one OR gate (three inputs) and three AND gates (with no more than three inputs each).

4.15 A logic circuit realizing the function f has four inputs A, B, C, and D. The three inputs A, B, and C are the binary representation of the digits 0 through 7 with A being the most-significant bit. The input D is an odd-parity bit, i.e., the value of D is such that

A, B, C, and D always contain an odd number of 1's. (For example, the digit 1 is represented by $ABC = 001$ and $D = 0$, and the digit 3 is represented by $ABCD = 0111$.) The function f has value 1 if the input digit is a prime number. (A number is prime if it is divisible only by itself and 1; 1 is considered to be prime and 0 is not.)

(a) List the minterms and don't-care minterms of f in algebraic form.

(b) List the maxterms and don't-care maxterms of f in algebraic form.

4.16 A priority encoder circuit has four inputs, x_3, x_2, x_1, and x_0. The circuit has three outputs: z, y_1, and y_0. If one of the inputs is 1, z is 1 and y_1 and y_0 represent a 2-bit, binary number whose value equals the index of the highest numbered input that is 1. For example, if x_2 is 1 and x_3 is 0, then the outputs are $z = 1$ and $y_1 = 1$ and $y_0 = 0$. If all inputs are 0, $z = 0$ and y_1 and y_0 are don't-cares.

(a) List in decimal form the minterms and don't-care minterms of each output.

(b) List in decimal form the maxterms and don't-care maxterms of each output.

4.17 The 9's complement of a decimal digit d (0 to 9) is defined to be $9 - d$. A logic circuit produces the 9's complement of an input digit where the input and output digits are represented in BCD. Label the inputs A, B, C, and D, and label the outputs W, X, Y and Z.

(a) Determine the minterms and don't-care minterms for each of the outputs.

(b) Determine the maxterms and don't-care maxterms for each of the outputs.

4.18 Repeat Problem 4.17 for the case where the input and output digits are represented using the 4-2-2-1 weighted code. (If only one weight of 2 is required for decimal digits less than 5, select the rightmost 2. In addition, select the codes so that $W = A'$, $X = B'$, $Y = C'$, and $Z = D'$. (There are two possible codes with these restrictions.)

4.19 Each of the following sentences has two possible interpretations depending on whether the AND or OR is done first. Write an equation for each interpretation.

(a) The buzzer will sound if the key is in the ignition switch, and the car door is open, or the seat belts are not fastened.

(b) You will gain weight if you eat too much, or you do not exercise enough, and your metabolism rate is too low.

(c) The speaker will be damaged if the volume is set too high, and loud music is played, or the stereo is too powerful.

(d) The roads will be very slippery if it snows, or it rains, and there is oil on the road.

4.20 A bank vault has three locks with a different key for each lock. Each key is owned by a different person. To open the door, at least two people must insert their keys into the assigned locks. The signal lines A, B, and C are 1 if there is a key inserted into lock 1, 2, or 3, respectively. Write an equation for the variable Z which is 1 iff the door should open.

4.21 A paper tape reader used as an input device to a computer has five rows of holes as shown. A hole punched in the tape indicates a logic 1, and no hole indicates a logic 0. As each hole pattern passes under the photocells, the pattern is translated into logic signals on lines A, B, C, D, and E. All patterns of holes indicate a valid character with two exceptions. A pattern consisting of none of the possible holes punched is not used because it is impossible to distinguish between this pattern and the unpunched space between patterns. An incorrect pattern punched on the tape is erased by punching all five holes in that position. Therefore, a valid character punched on the tape will have at least one hole but will not have all five holes punched.

(a) Write an equation for a variable Z which is 1 iff a valid character is being read.

(b) Write an equation for a variable Y which is 1 iff the hole pattern being read has holes punched only in rows C and E.

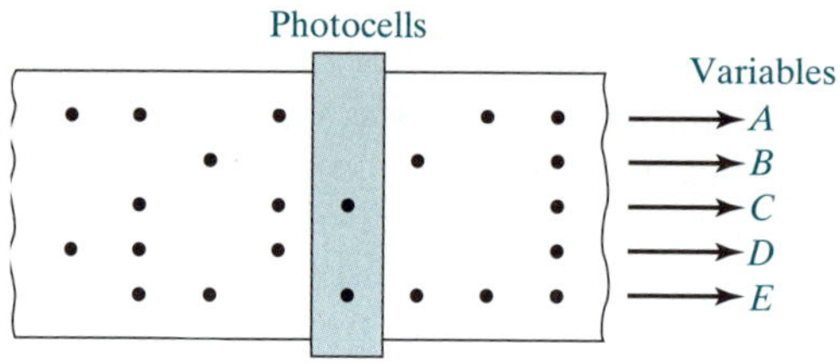

4.22 A computer interface to a line printer has seven data lines that control the movement of the paper and the print head and determine which character to print. The data lines are labeled A, B, C, D, E, F, and G, and each represents a binary 0 or 1. When the data lines are interpreted as a 7-bit binary number with line A being the most significant bit, the data lines can represent the numbers 0 to 127_{10}. The number 13_{10} is the command to return the print head to the beginning of a line, the number 10_{10} means to advance the paper by one line, and the numbers 32_{10} to 127_{10} represent printing characters.

(a) Write an equation for the variable X which is 1 iff the data lines indicate a command to return the print head to the beginning of the line.

(b) Write an equation for the variable Y which is 1 iff there is an advance paper command on the data lines.

(c) Write an equation for the variable Z which is 1 iff the data lines indicate a printable character. (*Hint:* Consider the binary representations of the numbers 0–31 and 32–127 and write the equation for Z with only two terms.)

4.23 Given $F_1 = \Pi M(0, 4, 5, 6)$ and $F_2 = \Pi M(0, 4, 7)$, find the maxterm expansion for F_1F_2. State a general rule for finding the maxterm expansion of F_1F_2 given the maxterm expansions of F_1 and F_2.
Prove your answer by using the general form of the maxterm expansion.

4.24 Given $F_1 = \Pi M(0, 4, 5, 6)$ and $F_2 = \Pi M(0, 4, 7)$, find the maxterm expansion for $F_1 + F_2$.
State a general rule for finding the maxterm expansion of $F_1 + F_2$, given the maxterm expansions of F_1 and F_2.
Prove your answer by using the general form of the maxterm expansion.

4.25 Four chairs are placed in a row:

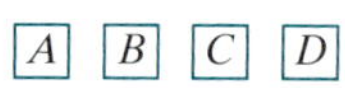

Each chair may be occupied (1) or empty (0). Give the minterm and maxterm expansion for each logic function described.

(a) $F(A, B, C, D)$ is 1 iff there are no adjacent empty chairs.
(b) $G(A, B, C, D)$ is 1 iff the chairs on the ends are both empty.
(c) $H(A, B, C, D)$ is 1 iff at least three chairs are full.
(d) $J(A, B, C, D)$ is 1 iff there are more people sitting in the left two chairs than in the right two chairs.

4.26 Four chairs (A, B, C, and D) are placed in a circle: A next to B, B next to C, C next to D, and D next to A. Each chair may be occupied (1) or empty (0). Give the minterm and maxterm expansion for each of the following logic functions:

(a) $F(A, B, C, D)$ is 1 iff there are no adjacent empty chairs.
(b) $G(A, B, C, D)$ is 1 iff there are at least three adjacent empty chairs.
(c) $H(A, B, C, D)$ is 1 iff at least three chairs are full.
(d) $J(A, B, C, D)$ is 1 iff there are more people sitting in chairs A and B than chairs C and D.

4.27 Given $f(a, b, c) = a(b + c')$.

(a) Express f as a minterm expansion (use m-notation).
(b) Express f as maxterm expansion (use M-notation).
(c) Express f' as a minterm expansion (use m-notation).
(d) Express f' as a maxterm expansion (use M-notation).

4.28 Work Problem 4.27 using $f(a, b, c, d) = acd + bd' + a'c'd + ab'cd + a'b'cd'$.

4.29 Find both the minterm expansion and maxterm expansion for the following functions, using *algebraic manipulations*:

(a) $f(A, B, C, D) = AB + A'CD$
(b) $f(A, B, C, D) = (A + B + D')(A' + C)(C + D)$

4.30 Given $F'(A, B, C, D) = \Sigma\, m(0, 1, 2, 6, 7, 13, 15)$.

(a) Find the minterm expansion for F (both decimal and algebraic form).
(b) Find the maxterm expansion for F (both decimal and algebraic form).

4.31 Repeat Problem 4.30 for $F'(A, B, C, D) = \Sigma\, m(1, 2, 5, 6, 10, 15)$.

4.32 Work parts (a) through (d) with the given truth table.

A	B	C	F_1	F_2	F_3	F_4
0	0	0	1	1	0	1
0	0	1	X	0	0	0
0	1	0	0	1	X	0
0	1	1	0	0	1	1
1	0	0	0	1	1	1
1	0	1	X	0	1	0
1	1	0	0	X	X	X
1	1	1	1	X	1	X

(a) Find the simplest expression for F_1, and specify the values for the don't-cares that lead to this expression.
(b) Repeat for F_2.
(c) Repeat for F_3.
(d) Repeat for F_4.

4.33 Work Problem 4.5 using the following circuits and truth table. Assume that the input combinations of $ABC = 011$ and $ABC = 110$ will never occur.

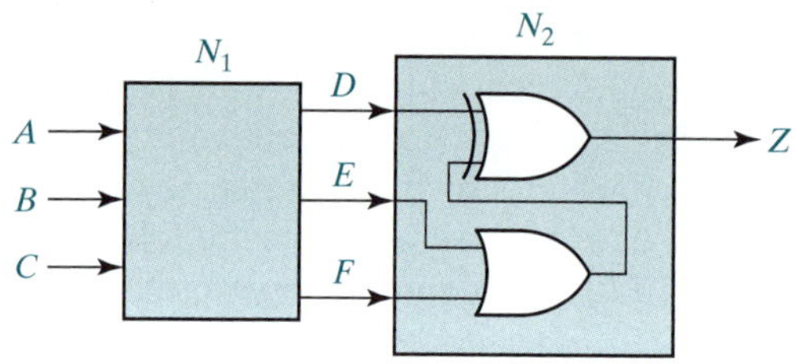

A	B	C	D	E	F
0	0	0	1	1	0
0	0	1	0	1	0
0	1	0	0	0	1
0	1	1	0	0	0
1	0	0	0	1	0
1	0	1	0	0	1
1	1	0	0	0	1
1	1	1	1	0	1

4.34 Work Problem 4.7 for the following logic functions:
(a) $G_1(A, B, C)$ is 1 iff all the coins landed on the same side (heads or tails).
(b) $G_2(A, B, C)$ is 1 iff the second coin landed on the same side as the first coin.

4.35 A combinational circuit has four inputs (A, B, C, D) and three outputs (X, Y, Z). XYZ represents a binary number whose value equals the number of 1's at the input. For example if $ABCD = 1011$, $XYZ = 011$.
(a) Find the minterm expansions for X, Y, and Z.
(b) Find the maxterm expansions for Y and Z.

4.36 A combinational circuit has four inputs (A, B, C, D) and four outputs (W, X, Y, Z). $WXYZ$ represents an excess-3 coded number whose value equals the number of 1's at the input. For example, if $ABCD = 1101$, $WXYZ = 0110$.
(a) Find the minterm expansions for X, Y, and Z.
(b) Find the maxterm expansions for Y and Z.

4.37 A combinational circuit has four inputs (A, B, C, D), which represent a binary-coded-decimal digit. The circuit has two groups of four outputs—S, T, U, V, and W, X, Y, Z. Each group represents a BCD digit. The output digits represent a decimal number which is five times the input number. For example, if $ABCD = 0111$, the outputs are 0011 0101. Assume that invalid BCD digits do not occur as inputs.
(a) Construct the truth table.
(b) Write down the minimum expressions for the outputs by inspection of the truth table. (*Hint:* Try to match output columns in the table with input columns.)

4.38 Work Problem 4.37 where the BCD outputs represent a decimal number that is 1 more than four times the input number. For example, if $ABCD = 0011$, the outputs are 0001 0011.

4.39 Design a circuit which will add a 4-bit binary number to a 5-bit binary number. Use five full adders. Assume negative numbers are represented in 2's complement. (*Hint:* How do you make a 4-bit binary number into a 5-bit binary number, without making a negative number positive or a positive number negative? Try writing down the representation for –3 as a 3-bit 2's complement number, a 4-bit 2's complement number, and a 5-bit 2's complement number. Recall that one way to find the 2's complement of a binary number is to complement *all* bits to the left of the first 1.)

4.40 A half adder is a circuit that adds two bits to give a sum and a carry. Give the truth table for a half adder, and design the circuit using only two gates. Then design a circuit which will find the 2's complement of a 4-bit binary number. Use four half adders and any additional gates. (*Hint:* Recall that one way to find the 2's complement of a binary number is to complement *all* bits, and then add 1.)

4.41 (a) Write the switching function $f(x, y) = x + y$ as a sum of minterms and as a product of maxterms.

(b) Consider the Boolean algebra of four elements $\{0, 1, a, b\}$ specified by the following operation tables and the Boolean function $f(x, y) = ax + by$ where a and b are two of the elements in the Boolean algebra. Write $f(x, y)$ in a sum-of-minterms form.

(c) Write the Boolean function of part (b) in a product-of-maxterms form.

(d) Give a table of combinations for the Boolean function of part (b). (*Note:* The table of combinations has 16 rows, not just 4.)

(e) Which four rows of the table of combinations completely specify the function of part (b)? Verify your answer.

	′
0	1
1	0
a	b
b	a

+	0	1	a	b
0	0	1	a	b
1	1	1	1	1
a	a	1	a	1
b	b	1	1	b

·	0	1	a	b
0	0	0	0	0
1	0	1	a	b
a	0	a	a	0
b	0	b	0	b

4.42 (a) If m_1 and m_2 are minterms of n variables, prove that $m_1 + m_2 = m_1 \oplus m_2$.

(b) Prove that any switching function can be written as the exclusive OR sum of products where each product does not contain a complemented literal. (*Hint:* Start with the function written as a sum of minterms and use part (a).)

4.43 (a) Show that the full adder of Figure 4-5 can be implemented using two 2-input exclusive OR gates and three 2-input NAND gates. (*Hint:* Rewrite Equation (4-2) in terms of $X \oplus Y$.)

(b) Compare the maximum addition time of the ripple-carry adder of Figure 4-3 using the full adder of part (a) versus the full adder of Figure 4-5 assuming the same gate types are used in both.

4.44 Show that a full subtractor can be implemented using two 2-input exclusive OR gates, one inverter, and three 2-input NOR gates. (*Hint:* Write the borrow out equation in product-of-sums form.)

4.45 The full adder of Figure 4-5 is modified by adding two control inputs, E_1 and E_0, and implemented as shown in the figure below.

(a) For each combination of values for E_1 and E_0, give the algebraic expression for the outputs of the full adder.

(b) Assume this modified full adder is used in the parallel adder of Figure 4-3. For each combination of values for E_1 and E_0, specify the function (Add, Exclusive OR, etc.) performed by the parallel adder.

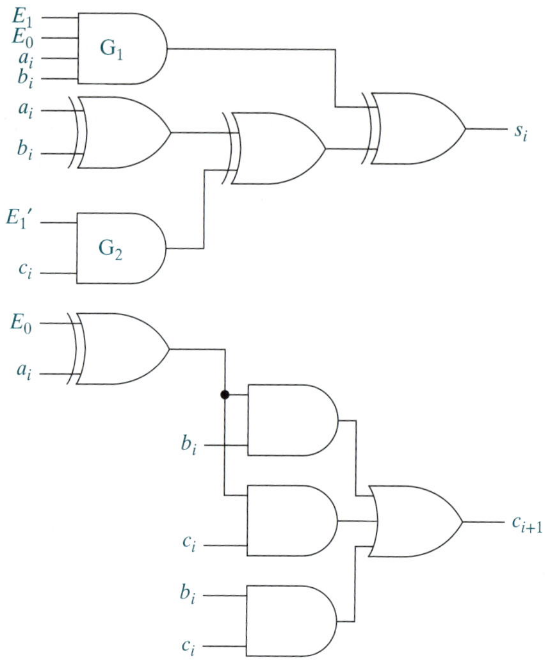

4.46 Redo Problem 4.45 if gates G_1 and G_2 are NAND gates rather than AND gates.

4.47 Redo Problem 4.45 if gates G_1 and G_2 are NOR gates rather than AND gates and an inverter is inserted in the c_i input of G_2.

4.48 Redo Problem 4.45 if gates G_1 and G_2 are OR gates rather than AND gates and an inverter is inserted in the c_i input of G_2.

UNIT 5

Karnaugh Maps

Objectives

1. Given a function (completely or incompletely specified) of three to five variables, plot it on a Karnaugh map. The function may be given in minterm, maxterm, or algebraic form.
2. Determine the essential prime implicants of a function from a map.
3. Obtain the minimum sum-of-products or minimum product-of-sums form of a function from the map.
4. Determine all of the prime implicants of a function from a map.
5. Understand the relation between operations performed using the map and the corresponding algebraic operations.

Study Guide

In this unit we will study the Karnaugh (pronounced "car-no") map. Just about any type of algebraic manipulation we have done so far can be facilitated by using the map, provided the number of variables is small.

1. Study Section 5.1, *Minimum Forms of Switching Functions.*

   (a) Define a minimum sum of products.

   (b) Define a minimum product of sums.

2. Study Section 5.2, *Two- and Three-Variable Karnaugh Maps.*

   (a) Plot the given truth table on the map. Then, loop two pairs of 1's on the map and write the simplified form of F.

PQ	F
0 0	1
0 1	1
1 0	0
1 1	1

Q \ P	0	1
0		
1		

F

$F =$ __________

   Now simplify F algebraically and verify that your answer is correct.

   (b) $F(a, b, c)$ is plotted below. Find the truth table for F.

bc \ a	0	1
00	**0**	**1**
01	**1**	**1**
11	**0**	**1**
10	**1**	**0**

F

a b c	F
0 0 0	
0 0 1	
0 1 0	
0 1 1	
1 0 0	
1 0 1	
1 1 0	
1 1 1	

(c) Plot the following functions on the given Karnaugh maps:

$$F_1(R, S, T) = \Sigma\, m(0, 1, 5, 6) \quad F_2(R, S, T) = \Pi\, M(2, 3, 4, 7)$$

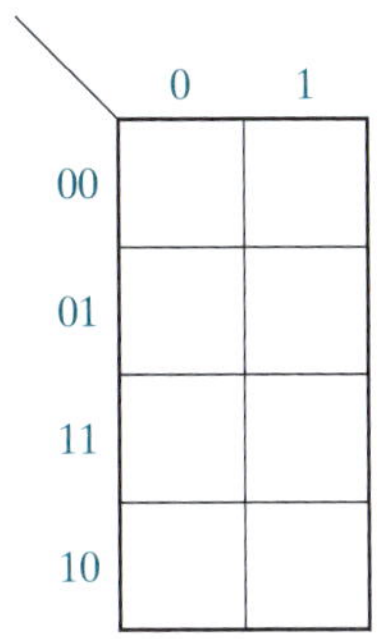

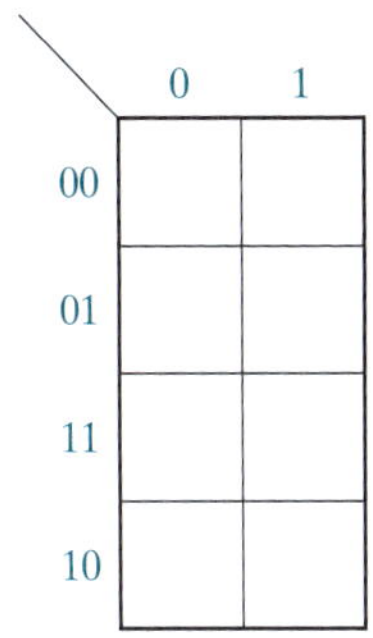

Why are the two maps the same?

(d) Plot the following function on the given map:

$$f(x, y, z) = z' + x'z + yz$$

Do *not* make a minterm expansion or a truth table before plotting.

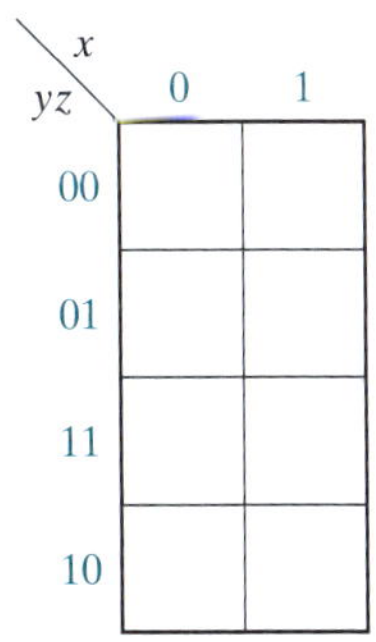

(e) For a three-variable map, which squares are "adjacent" to square 2?

____

(f) What theorem is used when two terms in adjacent squares are combined?

(g) What law of Boolean algebra justifies using a given 1 on a map in two or more loops?

(h) Each of the following solutions *is not* minimum.

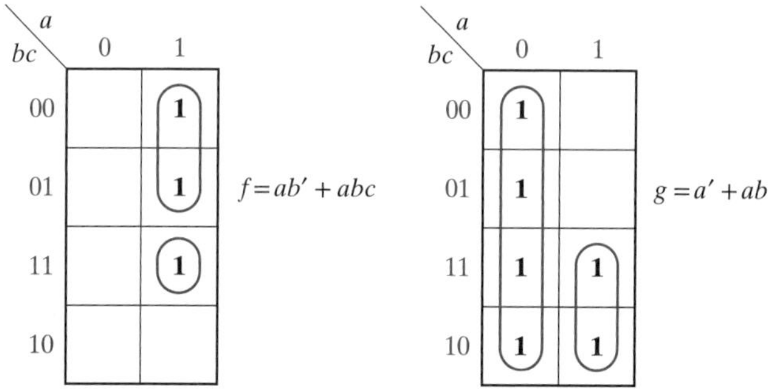

In each case, change the looping on the map so that the minimum solution is obtained.

(i) Work Problem 5.3.

(j) Find two different minimum sum-of-products expressions for the function G, which is plotted below.

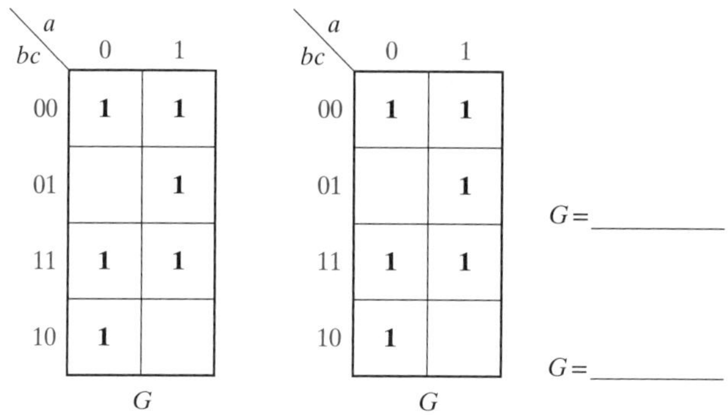

3. Study Section 5.3, *Four-Variable Karnaugh Maps*.

(a) Note the locations of the minterms on three- and four-variable maps (Figures 5-3(b) and 5-10). Memorize this ordering. This will save you a lot of time when you are plotting Karnaugh maps.

This ordering is valid only for the order of the variables given. If we label the maps as shown below, fill in the locations of the minterms:

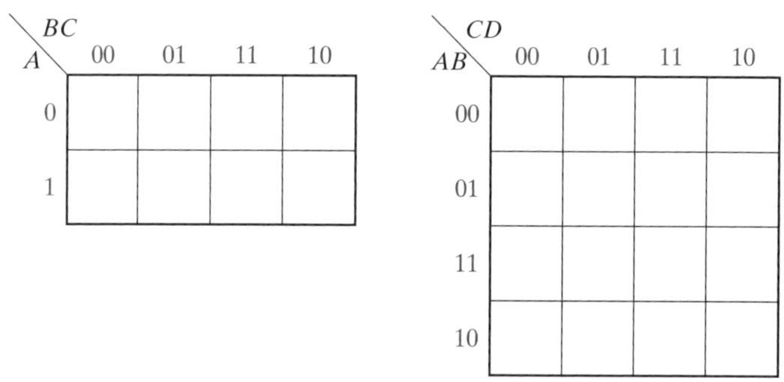

(b) Given the following map, write the minterm and maxterm expansions for F in decimal form:

$cd \backslash ab$	00	01	11	10
00			**1**	**1**
01	**1**	**1**		**1**
11		**1**	**1**	
10			**1**	

$F =$ ________________

$F =$ ________________

(c) Plot the following functions on the given maps:

(1) $f(w, x, y, z) = \Sigma\, m(0, 1, 2, 5, 7, 8, 9, 10, 13, 14)$

(2) $f(w, x, y, z) = x'z' + y'z + w'xz + wyz'$

$yz \backslash wx$	00	01	11	10
00				
01				
11				
10				

$yz \backslash wx$	00	01	11	10
00				
01				
11				
10				

Your answers to (1) and (2) should be the same.

(d) For a four-variable map, which squares are adjacent to square 14? ________

To square 8? ________

(e) When we combine two adjacent 1's on a map, this corresponds to applying the theorem $xy' + xy = x$ to eliminate the variable in which the two terms differ. Thus, looping the two 1's as indicated on the following map is equivalent to combining the corresponding minterms algebraically:

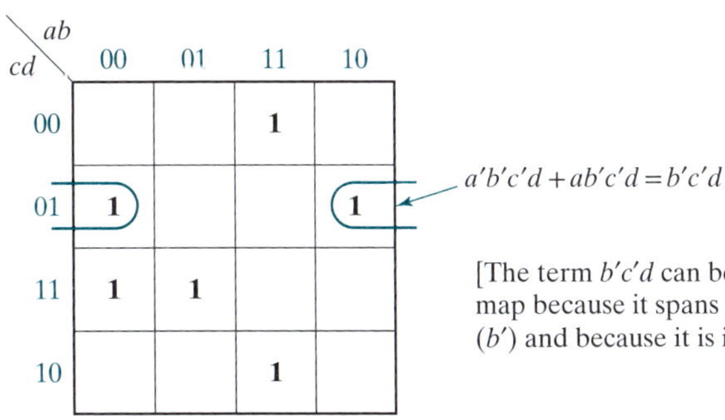

[The term $b'c'd$ can be read directly from the map because it spans the first and last columns (b') and because it is in the second row ($c'd$).]

Loop two other pairs of adjacent 1's on this map and state the algebraic equivalent of looping these terms. Now read the loops directly off the map and check your algebra.

(f) When we combine four adjacent 1's on a map (either four in a line or four in a square) this is equivalent to applying $xy + xy' = x$ three times:

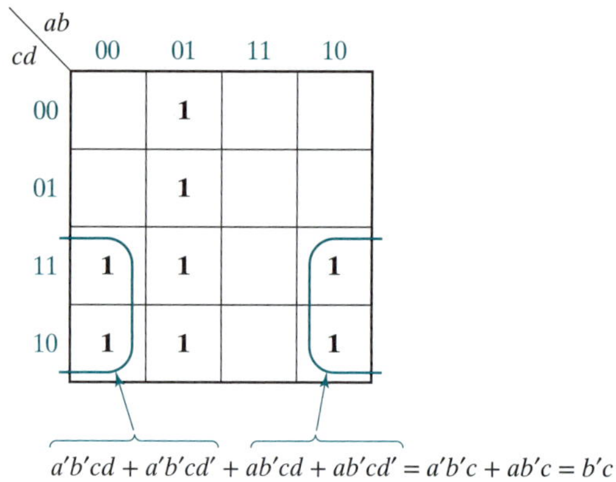

Loop the other four 1's on the map and state the algebraic equivalent.

(g) For each of the following maps, loop a minimum number of terms which will cover all of the 1's.

cd \ ab	00	01	11	10
00		1	1	
01		1	1	1
11	1	1		
10		1		

f_1

cd \ ab	00	01	11	10
00				1
01		1		
11	1	1	1	1
10	1			1

f_2

(For each part you should have looped two groups of four 1's and two groups of two 1's).

Write down the minimum sum-of-products expression for f_1 and f_2 from these maps.

$f_1 =$ ____

$f_2 =$ ____

(h) Why is it not possible to combine three or six minterms together rather than just two, four, eight, etc.?

(i) Note the procedure for deriving the minimum *product of sums* from the map. You will probably make fewer mistakes if you write down f' as a sum of products first and then complement it, as illustrated by the example in Figure 5-14.

(j) Work Problems 5.4 and 5.5.

4. Study Section 5.4, *Determination of Minimum Expressions Using Essential Prime Implicants.*

(a) For the map of Figure 5-15, list three implicants of F other than those which are labeled.

For the same map, is $ac'd'$ a prime implicant of F?
Why or why not?

(b) For the given map, are any of the circled terms prime implicants?

Why or why not?

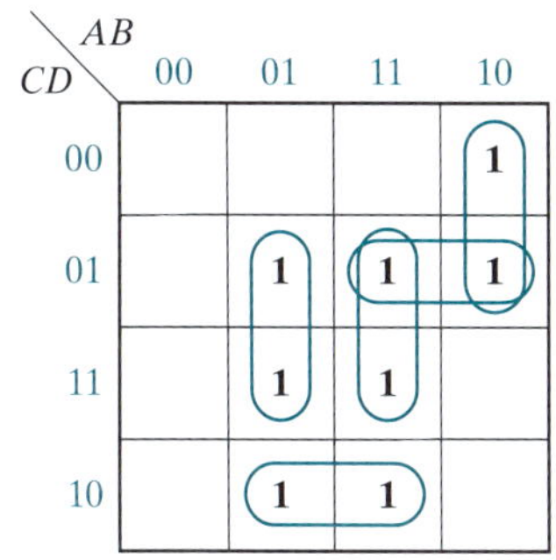

5. Study Figure 5-18 carefully and then answer the following questions for the given map:

(a) How many 1's are adjacent to m_0?

(b) Are all these 1's covered by a single prime implicant?

(c) From your answer to (b), can you determine whether $B'C'$ is essential?

(d) How many 1's are adjacent to m_9?

(e) Are all of these 1's covered by a single prime implicant?

(f) From your answer to (e), is $B'C'$ essential?

(g) How many 1's are adjacent to m_7?

(h) Why is $A'C$ essential?

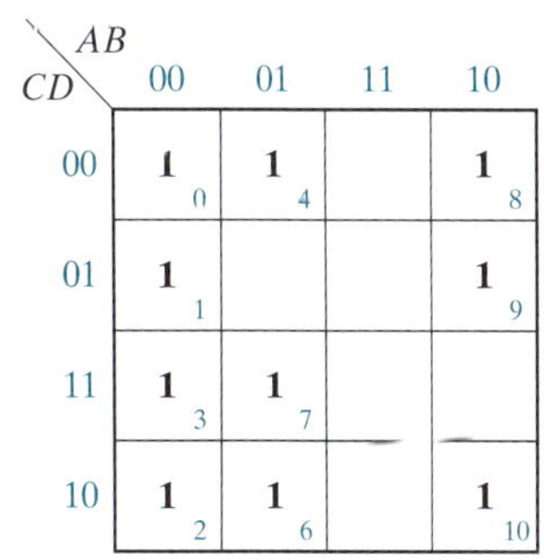

(i) Find two other essential prime implicants and tell which minterm makes them essential.

6. (a) How do you determine if a prime implicant is essential using a Karnaugh map?

(b) For the following map, why is $A'B'$ *not* essential?

Why is BD' essential?

Is $A'D'$ essential? Why?

Is BC' essential? Why?

Is $B'CD$ essential? Why?

Find the minimum sum of products.

CD \ AB	00	01	11	10
00	**1**	**1**	**1**	
01	**1**	**1**	**1**	
11	**1**			**1**
10	**1**	**1**	**1**	

(c) Work Programmed Exercise 5.1.

(d) List all 1's and X's that are adjacent to 1_0.

CD \ AB	00	01	11	10
00	1_0	1_4	1_{12}	8
01	X_1	1_5	X_{13}	9
11	3	X_7	1_{15}	1_{11}
10	2	6	X_{14}	10

Why is $A'C'$ an essential prime implicant?

List all 1's and X's adjacent to 1_{15}.

Based on this list, why can you not find an essential prime implicant that covers 1_{15}?

Does this mean that there is no essential prime implicant that covers 1_{15}?

What essential prime implicant covers 1_{11}?

Can you find an essential prime implicant that covers 1_{12}? Explain.

Find two prime implicants that cover 1_{12}.

Give two minimum expressions for F.

(e) Work Problem 5.6.

(f) If you have a copy of the *LogicAid* program available, use the Karnaugh map tutorial mode to help you learn to find minimum solutions from Karnaugh maps. This program will check your work at each step to make sure that you loop the terms in the correct order. It also will check your final answer. Work Problem 5.7 using the Karnaugh map tutor.

7. (a) In Example 4, page 107, we derived the following function:

$$Z = \Sigma\, m(0, 3, 6, 9) + \Sigma\, d(10, 11, 12, 13, 14, 15)$$

Plot Z on the given map using X's to represent don't-care terms.

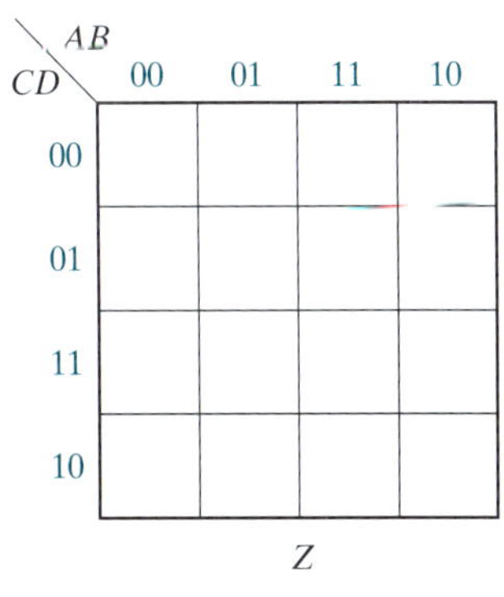

(b) Show that the minimum sum of products is

$$Z = A'B'C'D' + B'CD + AD + BCD'$$

Which *four* don't-care minterms were assigned the value 1 when forming your solution?

(c) Show that the minimum product of sums for Z is

$$Z = (B' + C)(B' + D')(A' + D)(A + C + D')(B + C' + D)$$

Which *one* don't-care term of Z was assigned the value 1 when forming your solution?

(d) Work Problem 5.8.

8. Study Section 5.5, *Five-Variable Karnaugh Maps.*

(a) The figure below shows a three-dimensional five-variable map. Plot the 1's and loops on the corresponding two-dimensional map, and give the minimum sum-of-products expression for the function.

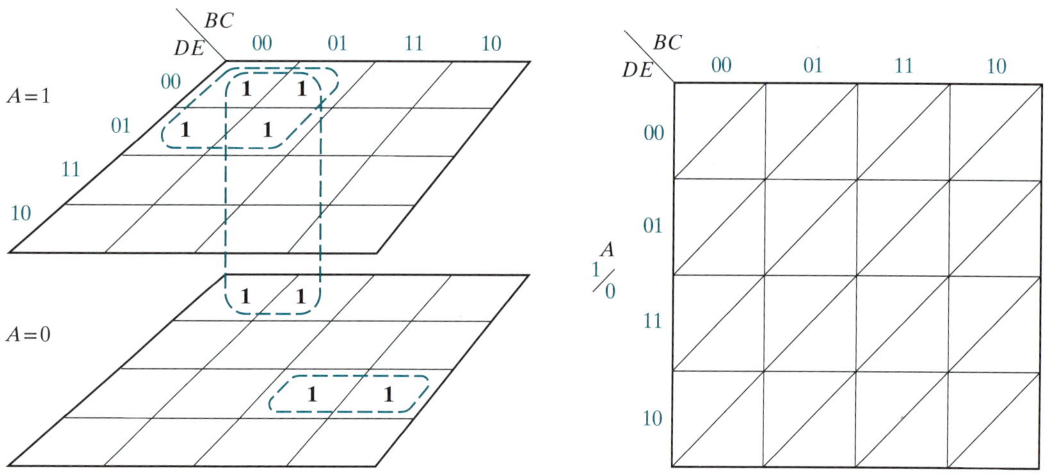

$F =$ ________________

(b) On a five-variable map (Figure 5-21), what are the five minterms adjacent to minterm 24?

(c) Work through all of the examples in this section carefully and make sure that you understand all of the steps.

(d) Two minimum solutions are given for Figure 5-24. There is a third minimum sum-of-products solution. What is it?

(e) Work Programmed Exercise 5.2.

(f)

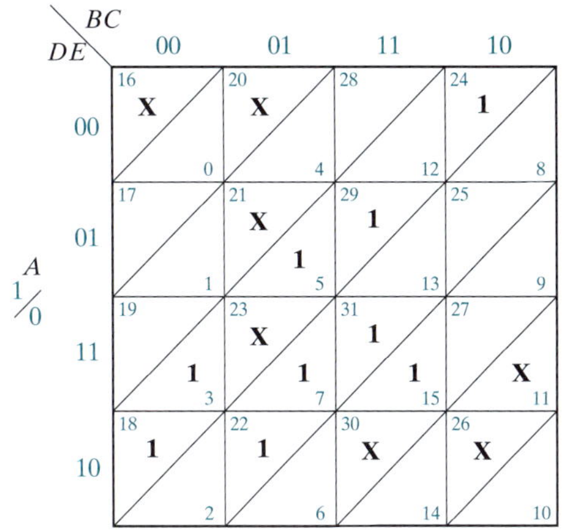

Find the three 1's and X's adjacent to 1_{18}. Can these all be looped with a single loop?
Find the 1's and X's adjacent to 1_{24}. Loop the essential prime implicant that covers 1_{24}.
Find the 1's and X's adjacent to 1_3. Loop the essential prime implicant that covers 1_3.
Can you find an essential prime implicant that covers 1_{22}? Explain.

Find and loop two more essential prime implicants.
Find three ways to cover the remaining 1 on the map and give the corresponding minimum solutions.

(g) If you have the *LogicAid* program available, work Problem 5.9, using the Karnaugh map tutor.

9. Study Section 5.6, *Other Uses of Karnaugh Maps*. Refer to Figure 5-8 and note that a consensus term exists if there are two adjacent, but nonoverlapping prime implicants. Observe how this principle is applied in Figure 5-26.

10. Work Problems 5.10, 5.11, 5.12, and 5.13. When deriving the minimum solution from the map, always write down the *essential* prime implicants first. If you do not, it is quite likely that you will not get the minimum solution. In addition, make sure you can find *all* of the prime implicants from the map (see Problem 5.10(b)).

11. Review the objectives.

Karnaugh Maps

Switching functions can generally be simplified by using the algebraic techniques described in Unit 3. However, two problems arise when algebraic procedures are used:

1. The procedures are difficult to apply in a systematic way.
2. It is difficult to tell when you have arrived at a minimum solution.

The Karnaugh map method studied in this unit and the Quine-McCluskey procedure studied in Unit 6 overcome these difficulties by providing systematic methods for simplifying switching functions. The Karnaugh map is an especially useful tool for simplifying and manipulating switching functions of three or four variables, but it can be extended to functions of five or more variables. Generally, you will find the Karnaugh map method is faster and easier to apply than other simplification methods.

5.1 Minimum Forms of Switching Functions

When a function is realized using AND and OR gates, the cost of realizing the function is directly related to the number of gates and gate inputs used. The Karnaugh map techniques developed in this unit lead directly to minimum cost two-level circuits composed of AND and OR gates. An expression consisting of a sum of product terms corresponds directly to a two-level circuit composed of a group of AND gates feeding a single OR gate (see Figure 2-5). Similarly, a product-of-sums expression corresponds to a two-level circuit composed of OR gates feeding a single AND gate (see Figure 2-6). Therefore, to find minimum cost two-level AND-OR gate circuits, we must find minimum expressions in sum-of-products or product-of-sums form.

A *minimum sum-of-products* expression for a function is defined as a sum of product terms which (a) has a minimum number of terms and (b) of all those expressions which have the same minimum number of terms, has a minimum number of literals. The minimum sum of products corresponds directly to a minimum two-level gate circuit which has (a) a minimum number of gates and (b) a minimum number of gate inputs. Unlike the minterm expansion for a function, the minimum sum of products is not necessarily unique; that is, a given function may have two different minimum sum-of-products forms, each with the same number of terms and the same number of literals. Given a minterm expansion, the minimum sum-of-products form can often be obtained by the following procedure:

1. Combine terms by using the uniting theorem $XY' + XY = X$. Do this repeatedly to eliminate as many literals as possible. A given term may be used more than once because $X + X = X$.
2. Eliminate redundant terms by using the consensus theorem or other theorems.

Unfortunately, the result of this procedure may depend on the order in which terms are combined or eliminated so that the final expression obtained is not necessarily minimum.

Example

Find a minimum sum-of-products expression for

$$F(a, b, c) = \Sigma m\,(0, 1, 2, 5, 6, 7)$$
$$F = a'b'c' + a'b'c + a'bc' + ab'c + abc' + abc$$
$$= a'b' + b'c + bc' + ab \qquad (5\text{-}1)$$

None of the terms in the above expression can be eliminated by consensus. However, combining terms in a different way leads directly to a minimum sum of products:

$$F = a'b'c' + a'b'c + a'bc' + ab'c + abc' + abc$$
$$= a'b' + bc' + ac \qquad (5\text{-}2)$$

If the uniting theorem is applied to all possible pairs of minterms, six two-literal products are obtained: $a'b'$, $a'c'$, $b'c$, bc', ac, ab. Then, the consensus theorem can be applied to obtain a second minimal solution:

$$a'c' + b'c + ab \qquad (5\text{-}3)$$

A *minimum product-of-sums* expression for a function is defined as a product of sum terms which (a) has a minimum number of terms, and (b) of all those expressions which have the same number of terms, has a minimum number of literals. Unlike the maxterm expansion, the minimum product-of-sums form of a function is not necessarily unique. Given a maxterm expansion, the minimum product of sums can often be obtained by a procedure similar to that used in the minimum sum-of-products case, except that the uniting theorem $(X + Y)(X + Y') = X$ is used to combine terms.

Example

$$(A + B' + C + D')(A + B' + C' + D')(A + B' + C' + D)(A' + B' + C' + D)(A + B + C' + D)(A' + B + C' + D)$$

$$= (A + B' + D') \quad (A + B' + C') \quad (B' + C' + D) \quad (B + C' + D)$$

$$= (A + B' + D') \quad \underbrace{(A + B' + C')} \quad (C' + D)$$

$\leftarrow$ eliminate by consensus

$$= (A + B' + D')(C' + D) \tag{5-4}$$

The uniting theorem $XY' + XY = X$ can be applied to minterms and products where the minterms and products are represented in algebraic notation or binary notation. The first four-variable example below illustrates this for minterms and the second for products containing three literals. The dash indicates a missing variable.

$$ab'cd' + ab'cd = ab'c$$
$$1010 + 1011 = 101-$$

$$ab'c + abc = ac$$
$$101- + 111- = 1\text{–}1\text{–}$$

Note that minterms only combine if they differ in one variable, and products only combine if they have dashes in the same position (same missing variables) and differ in one other variable. The examples below do not combine.

$ab'cd' + ab'c'd$ (will not combine)
$1010 + 1001$

$ab'c + abd$ (will not combine)
$101- + 11\text{–}1$

The Karnaugh maps introduced next arrange the minterms of a function so that it is easy to recognize visually when the simplification theorem applies to two minterms, two products with one missing variable, two products with two missing variables, etc.

5.2 Two- and Three-Variable Karnaugh Maps

Just like a truth table, the Karnaugh map of a function specifies the value of the function for every combination of values of the independent variables. A two-variable Karnaugh map is shown. The values of one variable are listed across the top of the map, and the values of the other variable are listed on the left side. Each square of the map corresponds to a pair of values for A and B as indicated.

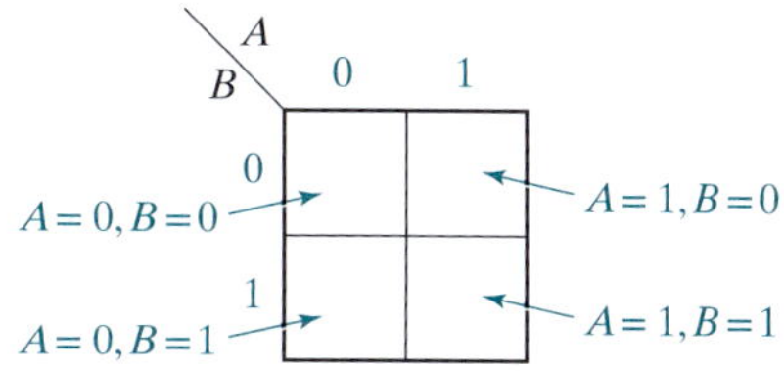

Figure 5-1 shows the truth table for a function F and the corresponding Karnaugh map. Note that the value of F for $A = B = 0$ is plotted in the upper left square, and the other map entries are plotted in a similar way in Figure 5-1(b). Each 1 on the map corresponds to a minterm of F. We can read the minterms from the map just like we can read them from the truth table. A 1 in square 00 of Figure 5-1(c) indicates that $A'B'$ is a minterm of F. Similarly, a 1 in square 01 indicates that $A'B$ is a minterm. Minterms in adjacent squares of the map can be combined since they differ in only one variable. Thus, $A'B'$ and $A'B$ combine to form A', and this is indicated by looping the corresponding 1's on the map in Figure 5-1(d).

FIGURE 5-1

A B	F
0 0	1
0 1	1
1 0	0
1 1	0

(a)

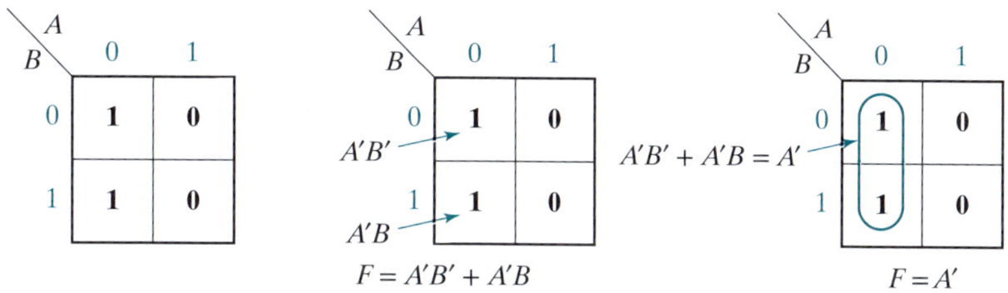

Figure 5-2 shows a three-variable truth table and the corresponding Karnaugh map (see Figure 5-27 for an alternative way of labeling maps). The value of one variable (A) is listed across the top of the map, and the values of the other two variables (B, C) are listed along the side of the map. The rows are labeled in the sequence 00, 01, 11, 10 so that values in adjacent rows differ in only one variable. For each combination of values of the variables, the value of F is read from the truth table and plotted in the appropriate map square. For example, for the input combination $ABC = 001$, the value $F = 0$ is plotted in the square for which $A = 0$ and $BC = 01$. For the combination $ABC = 110$, $F = 1$ is plotted in the $A = 1$, $BC = 10$ square.

FIGURE 5-2
Truth Table and Karnaugh Map for Three-Variable Function

A B C	F
0 0 0	0
0 0 1	0
0 1 0	1
0 1 1	1
1 0 0	1
1 0 1	0
1 1 0	1
1 1 1	0

(a)

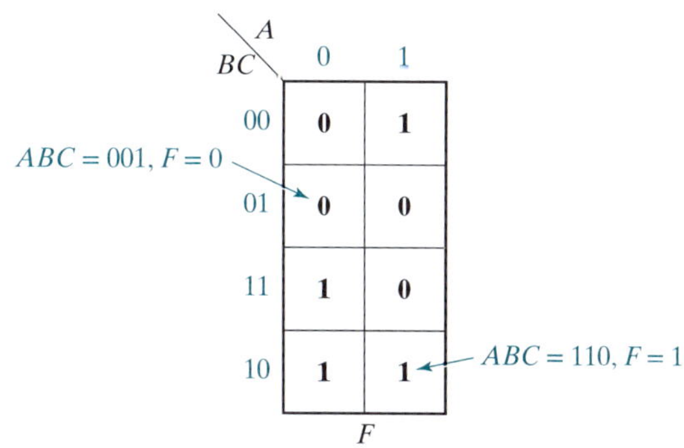

(b)

Figure 5-3 shows the location of the minterms on a three-variable map. Minterms in adjacent squares of the map differ in only one variable and therefore can be combined using the uniting theorem $XY' + XY = X$. For example, minterm 011 ($a'bc$) is adjacent to the three minterms with which it can be combined—001 ($a'b'c$), 010 ($a'bc'$), and 111 (abc). In addition to squares which are physically adjacent, the top and bottom rows of the map are defined to be adjacent because the corresponding minterms in these rows differ in only one variable. Thus 000 and 010 are adjacent, and so are 100 and 110.

FIGURE 5-3
Location of Minterms on a Three-Variable Karnaugh Map

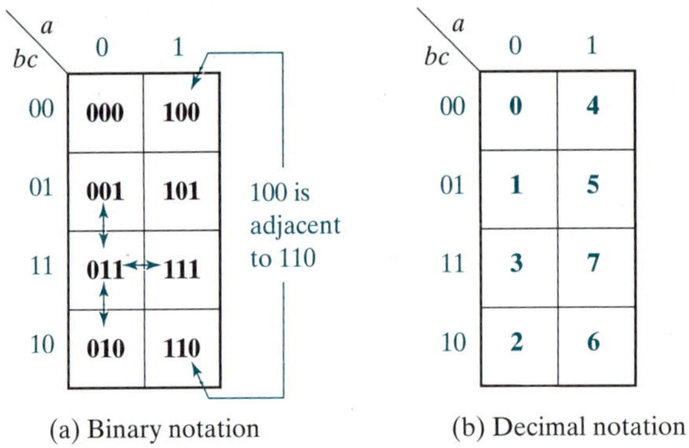

Given the minterm expansion of a function, it can be plotted on a map by placing 1's in the squares which correspond to minterms of the function and 0's in the remaining squares (the 0's may be omitted if desired). Figure 5-4 shows the plot of $F(a, b, c) = m_1 + m_3 + m_5$. If F is given as a maxterm expansion, the map is plotted by placing 0's in the squares which correspond to the maxterms and then by filling in the remaining squares with 1's. Thus, $F(a, b, c) = M_0M_2M_4M_6M_7$ gives the same map as Figure 5-4.

FIGURE 5-4
Karnaugh Map of $F(a, b, c) = \Sigma\, m(1, 3, 5) = \Pi\, M(0, 2, 4, 6, 7)$

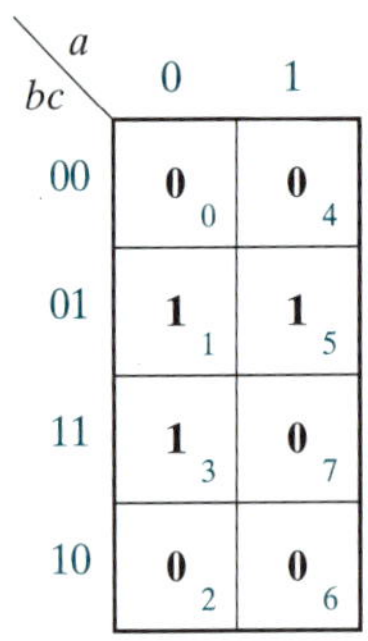

Figure 5-5 illustrates how product terms can be plotted on Karnaugh maps. To plot the term b, 1's are entered in the four squares of the map where $b = 1$. The term bc' is 1 when $b = 1$ and $c = 0$, so 1's are entered in the two squares in the $bc = 10$ row. The term ac' is 1 when $a = 1$ and $c = 0$, so 1's are entered in the $a = 1$ column in the rows where $c = 0$.

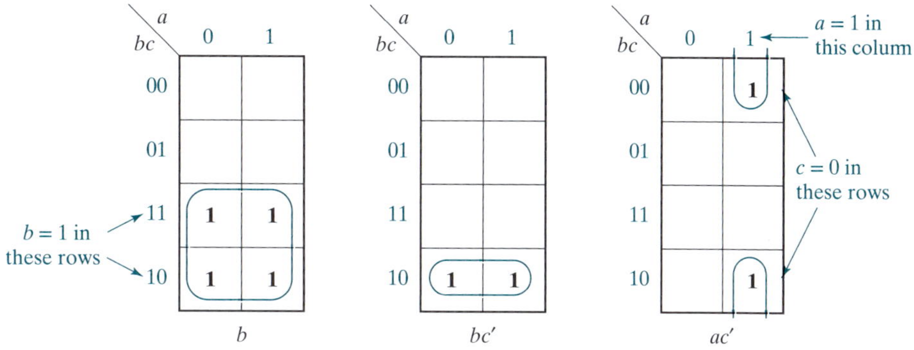

FIGURE 5-5
Karnaugh Maps for Product Terms

If a function is given in algebraic form, it is unnecessary to expand it to minterm form before plotting it on a map. If the algebraic expression is converted to sum-of-products form, then each product term can be plotted directly as a group of 1's on the map. For example, given that

$$f(a, b, c) = abc' + b'c + a'$$

we would plot the map as follows:

1. The term abc' is 1 when $a = 1$ and $bc = 10$, so we place a 1 in the square which corresponds to the $a = 1$ column and the $bc = 10$ row of the map.
2. The term $b'c$ is 1 when $bc = 01$, so we place 1's in both squares of the $bc = 01$ row of the map.
3. The term a' is 1 when $a = 0$, so we place 1's in all the squares of the $a = 0$ column of the map. (Note: Since there already is a 1 in the abc = 001 square, we do not have to place a second 1 there because $x + x = x$.)

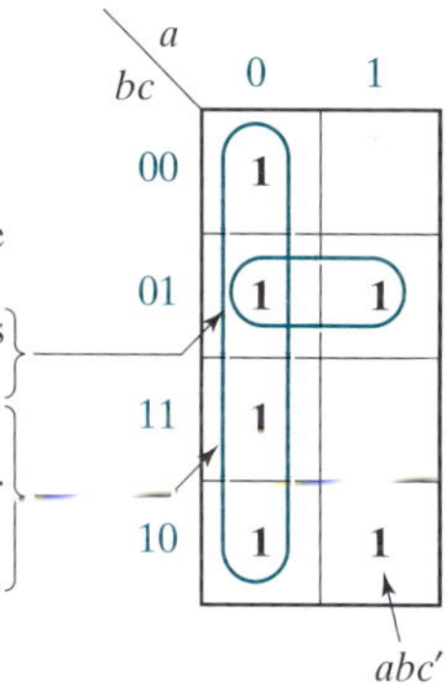

Figure 5-6 illustrates how a simplified expression for a function can be derived using a Karnaugh map. The function to be simplified is first plotted on a Karnaugh map in Figure 5-6(a). Terms in adjacent squares on the map differ in only one variable and can be combined using the uniting theorem $XY' + XY = X$. Thus $a'b'c$ and $a'bc$ combine to form $a'c$, and $a'b'c$ and $ab'c$ combine to form $b'c$, as shown in Figure 5-6(b). A loop around a group of minterms indicates that these terms have been combined. The looped terms can be read directly off the map. Thus, for Figure 5-6(b), term T_1 is in the $a = 0$ (a') column, and it spans the rows where $c = 1$, so $T_1 = a'c$. Note that b has been eliminated because the two minterms in T_1 differ in the variable b. Similarly, the term T_2 is in the $bc = 01$ row so $T_2 = b'c$, and a has been eliminated because T_2 spans the $a = 0$ and $a = 1$ columns. Thus, the minimum sum-of-products form for F is $a'c + b'c$.

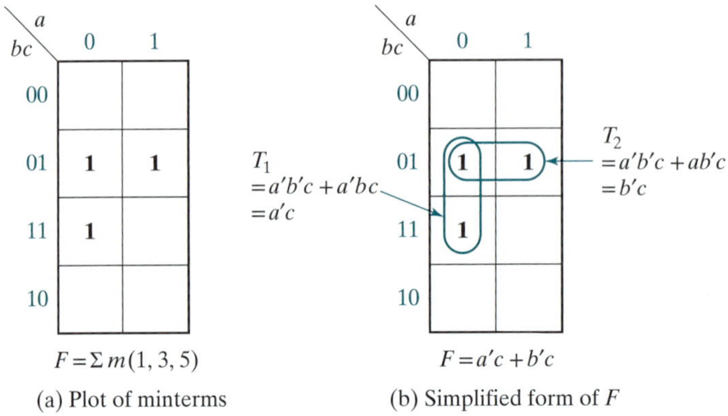

FIGURE 5-6 Simplification of a Three-Variable Function

The map for the complement of F (Figure 5-7) is formed by replacing 0's with 1's and 1's with 0's on the map of F. To simplify F', note that the terms in the top row combine to form $b'c'$, and the terms in the bottom row combine to form bc'. Because $b'c'$ and bc' differ in only one variable, the top and bottom rows can then be combined to form a group of four 1's, thus eliminating two variables and leaving $T_1 = c'$. The remaining 1 combines, as shown, to form $T_2 = ab$, so the minimum sum-of-products form for F' is $c' + ab$.

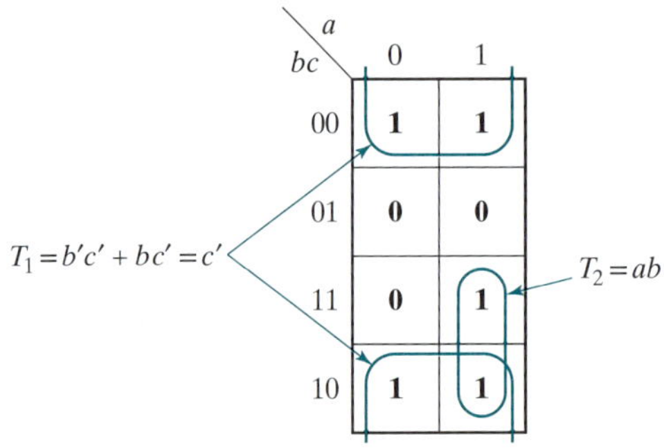

FIGURE 5-7 Complement of Map in Figure 5-6(a)

The Karnaugh map can also illustrate the basic theorems of Boolean algebra. Figure 5-8 illustrates the consensus theorem, $XY + X'Z + YZ = XY + X'Z$. Note that the consensus term (YZ) is redundant because its 1's are covered by the other two terms.

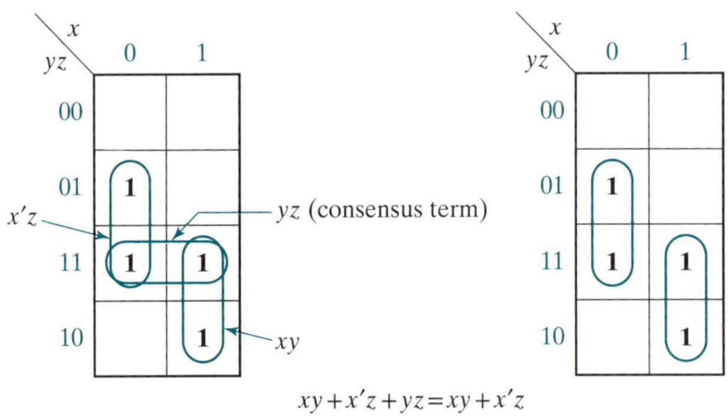

FIGURE 5-8 Karnaugh Maps that Illustrate the Consensus Theorem

If a function has two or more minimum sum-of-products forms, all of these forms can be determined from a map. Figure 5-9 shows the two minimum solutions for $F = \Sigma m(0, 1, 2, 5, 6, 7)$.

FIGURE 5-9
Function with Two Minimum Forms

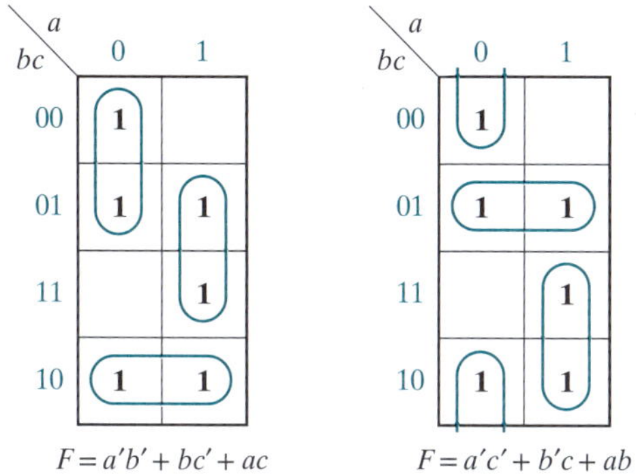

5.3 Four-Variable Karnaugh Maps

Figure 5-10 shows the location of minterms on a four-variable map. Each minterm is located adjacent to the four terms with which it can combine. For example, m_5 (0101) could combine with m_1 (0001), m_4 (0100), m_7 (0111), or m_{13} (1101) because it differs in only one variable from each of the other minterms. The definition of adjacent squares must be extended so that not only are top and bottom rows adjacent as in the three-variable map, but the first and last columns are also adjacent. This requires numbering the columns in the sequence 00, 01, 11, 10 so that minterms 0 and 8, 1 and 9, etc., are in adjacent squares.

FIGURE 5-10
Location of Minterms on Four-Variable Karnaugh Map

CD \ AB	00	01	11	10
00	0	4	12	8
01	1	5	13	9
11	3	7	15	11
10	2	6	14	10

We will now plot the following four-variable expression on a Karnaugh map (Figure 5-11):

$$f(a, b, c, d) = acd + a'b + d'$$

The first term is 1 when $a = c = d = 1$, so we place 1's in the two squares which are in the $a = 1$ column and $cd = 11$ row. The term $a'b$ is 1 when $ab = 01$, so we place four 1's in the $ab = 01$ column. Finally, d' is 1 when $d = 0$, so we place eight 1's in the two rows for which $d = 0$. (Duplicate 1's are not plotted because $1 + 1 = 1$.)

FIGURE 5-11
Plot of
acd + *a'b* + *d'*

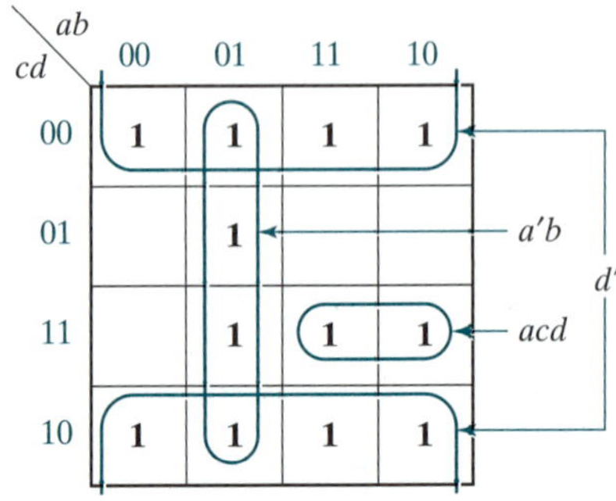

Next, we will simplify the functions f_1 and f_2 given in Figure 5-12. Because the functions are specified in minterm form, we can determine the locations of the 1's on the map by referring to Figure 5-10. After plotting the maps, we can then combine adjacent groups of 1's. Minterms can be combined in groups of two, four, or eight to eliminate one, two, or three variables, respectively. In Figure 5-12(a), the pair of 1's in the *ab* = 00 column and also in the *d* = 1 rows represents *a'b'd*. The group of four 1's in the *b* = 1 columns and *c* = 0 rows represents *bc'*.

FIGURE 5-12
Simplification
of Four-Variable
Functions

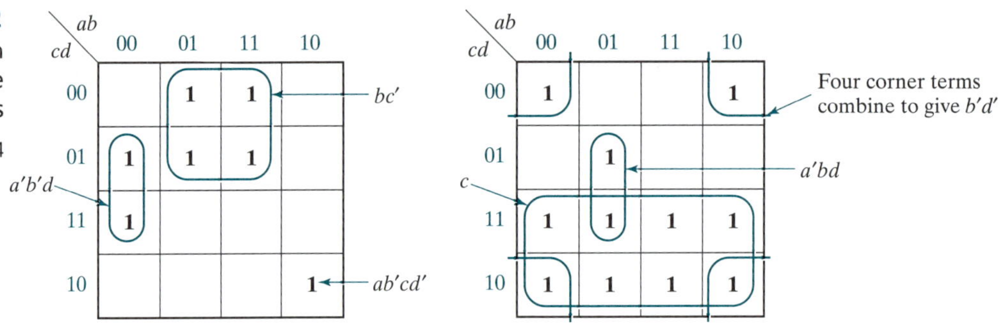

$f_1 = \Sigma\, m(1, 3, 4, 5, 10, 12, 13)$
$= bc' + a'b'd + ab'cd'$

(a)

$f_2 = \Sigma\, m(0, 2, 3, 5, 6, 7, 8, 10, 11, 14, 15)$
$= c + b'd' + a'bd$

(b)

In Figure 5-12(b), note that the four corner 1's span the *b* = 0 columns and *d* = 0 rows and, therefore, can be combined to form the term *b'd'*. The group of eight 1's covers both rows where *c* = 1 and, therefore, represents the term *c*. The pair of 1's which is looped on the map represents the term *a'bd* because it is in the *ab* = 01 column and spans the *d* = 1 rows.

The Karnaugh map method is easily extended to functions with don't-care terms. The required minterms are indicated by 1's on the map, and the don't-care minterms are indicated by X's. When choosing terms to form the minimum sum of products, all the 1's must be covered, but the X's are only used if they will simplify the resulting expression. In Figure 5-13, the only don't-care term used in forming the simplified expression is 13.

FIGURE 5-13 Simplification of an Incompletely Specified Function

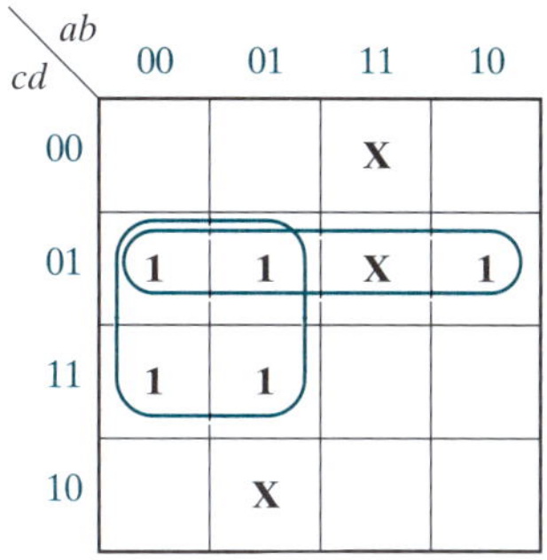

$f = \Sigma\, m(1, 3, 5, 7, 9) + \Sigma\, d(6, 12, 13)$
$= a'd + c'd$

The use of Karnaugh maps to find a minimum sum-of-products form for a function has been illustrated in Figures 5-1, 5-6, and 5-12. A minimum product of sums can also be obtained from the map. Because the 0's of f are 1's of f', the minimum sum of products for f' can be determined by looping the 0's on a map of f. The complement of the minimum sum of products for f' is then the minimum product of sums for f. The following example illustrates this procedure for

$$f = x'z' + wyz + w'y'z' + x'y$$

First, the 1's of f are plotted in Figure 5-14. Then, from the 0's,

$$f' = y'z + wxz' + w'xy$$

and the minimum product of sums for f is

$$f = (y + z')(w' + x' + z)(w + x' + y')$$

FIGURE 5-14

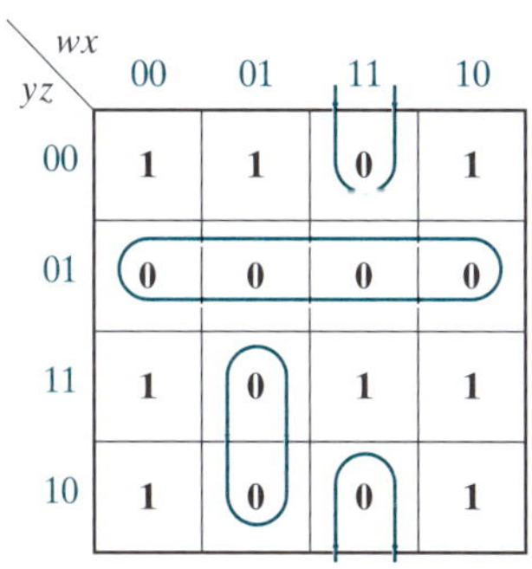

5.4 Determination of Minimum Expressions Using Essential Prime Implicants

Any single 1 or any group of 1's which can be combined together on a map of the function F represents a product term which is called an *implicant* of F (see Section 6.1 for a formal definition of implicant and prime implicant). Several implicants of F are indicated in Figure 5-15. A product term implicant is called a *prime implicant* if it cannot be combined with another term to eliminate a variable. In Figure 5-15, $a'b'c$, $a'cd'$, and ac' are prime implicants because they cannot be combined with other terms to eliminate a variable. On the other hand, $a'b'c'd'$ is not a prime implicant because it can be combined with a$'b'cd'$ or $ab'c'd'$. Neither abc', nor $ab'c'$*is* a prime implicant because these terms can be combined together to form ac'.

FIGURE 5-15

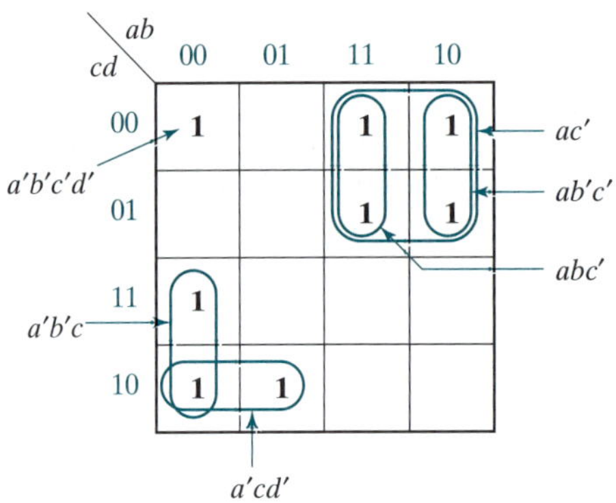

All of the prime implicants of a function can be obtained from a Karnaugh map. A single 1 on a map represents a prime implicant if it is not adjacent to any other 1's. Two adjacent 1's on a map form a prime implicant if they are not contained in a group of four 1's; four adjacent 1's form a prime implicant if they are not contained in a group of eight 1's, etc.

The minimum sum-of-products expression for a function consists of some (but not necessarily all) of the prime implicants of a function. In other words, a sum-of-products expression containing a term which is not a prime implicant cannot be minimum. This is true because if a nonprime term were present, the expression could be simplified by combining the nonprime term with additional minterms. In order to find the minimum sum of products from a map, we must find a minimum number of prime implicants which cover all of the 1's on the map. The function plotted in Figure 5-16 has six prime implicants. Three of these prime implicants cover all of the 1's on the map, and the minimum solution is the sum of these three prime implicants. The shaded loops represent prime implicants which are not part of the minimum solution.

When writing down a list of *all* of the prime implicants from the map, note that there are often prime implicants which are not included in the minimum sum of products. Even though all of the 1's in a term have already been covered by prime

FIGURE 5-16 Determination of All Prime Implicants

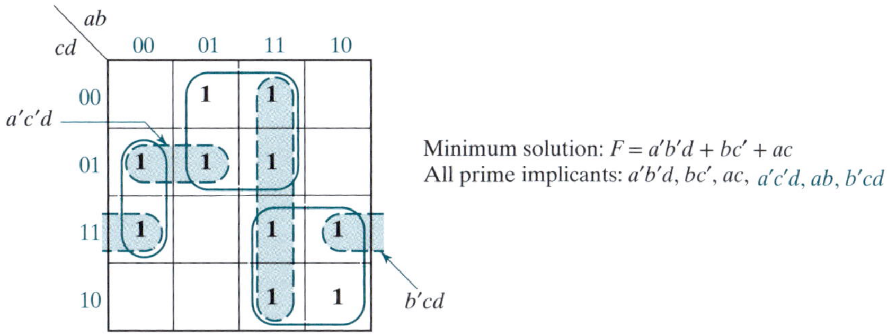

implicants, that term may still be a prime implicant provided that it is not included in a larger group of 1's. For example, in Figure 5-16, $a'c'd$ is a prime implicant because it cannot be combined with other 1's to eliminate another variable. However, abd is not a prime implicant because it can be combined with two other 1's to form ab. The term $b'cd$ is also a prime implicant even though both of its 1's are already covered by other prime implicants. In the process of finding prime implicants, don't-cares are treated just like 1's. However, a prime implicant composed entirely of don't-cares can never be part of the minimum solution.

Because all of the prime implicants of a function are generally not needed in forming the minimum sum of products, a systematic procedure for selecting prime implicants is needed. If prime implicants are selected from the map in the wrong order, a nonminimum solution may result. For example, in Figure 5-17, if CD is chosen first, then BD, $B'C$, and AC are needed to cover the remaining 1's, and the solution contains four terms. However, if the prime implicants indicated in Figure 5-17(b) are chosen first, all 1's are covered and CD is not needed.

In Section 6.2, prime implicant charts are defined. They can be used systematically to find (all) minimum solutions. The procedure described below can be used to find minimum solutions for functions that are not too complicated.

Note that some of the minterms on the map of Figure 5-17(a) can be covered by only a single prime implicant, but other minterms can be covered by two different prime implicants. For example, m_2 is covered only by $B'C$, but m_3 is covered by both

FIGURE 5-17

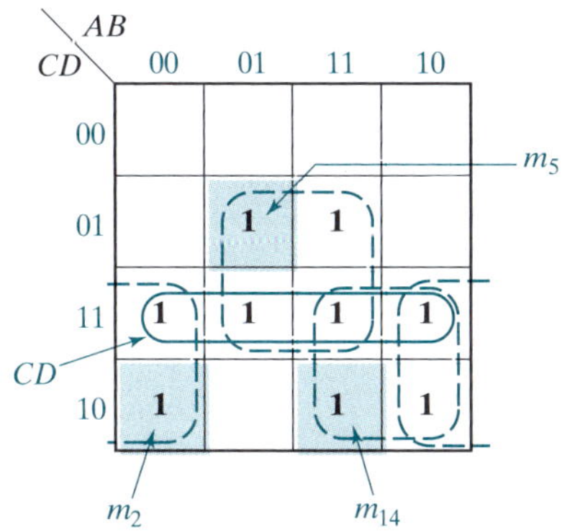

$f = CD + BD + B'C + AC$

(a)

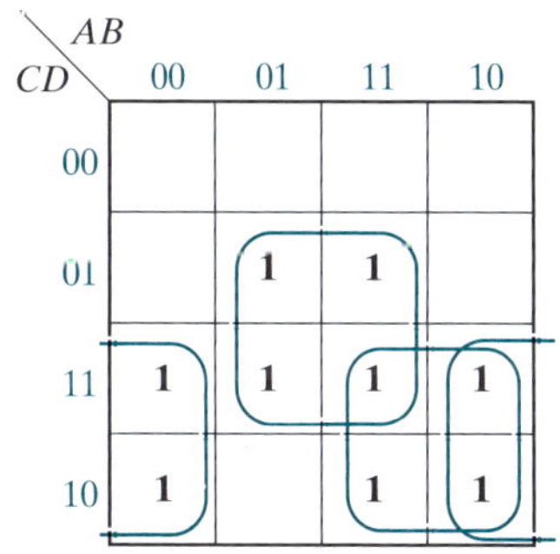

$f = BD + B'C + AC$

(b)

$B'C$ and CD. If a minterm is covered by only one prime implicant, that prime implicant is said to be *essential*, and it must be included in the minimum sum of products. Thus, $B'C$ *is* an essential prime implicant because m_2 is not covered by any other prime implicant. However, CD is *not* essential because each of the 1's in CD can be covered by another prime implicant. The only prime implicant which covers m_5 is BD, so BD is essential. Similarly, AC is essential because no other prime implicant covers m_{14}. In this example, if we choose all of the essential prime implicants, all of the 1's on the map are covered and the nonessential prime implicant CD is not needed.

In general, in order to find a minimum sum of products from a map, we should first loop all of the essential prime implicants. One way of finding essential prime implicants on a map is simply to look at each 1 on the map that has not already been covered, and check to see how many prime implicants cover that 1. If there is only one prime implicant which covers the 1, that prime implicant is essential. If there are two or more prime implicants which cover the 1, we cannot say whether these prime implicants are essential or not without checking the other minterms. For simple problems, we can locate the essential prime implicants in this way by inspection of each 1 on the map. For example, in Figure 5-16, m_4 is covered only by the prime implicant bc', and m_{10} is covered only by the prime implicant ac. All other 1's on the map are covered by two prime implicants; therefore, the only essential prime implicants are bc' and ac.

For more complicated maps, and especially for maps with five or more variables, we need a more systematic approach for finding the essential prime implicants. When checking a minterm to see if it is covered by only one prime implicant, we must look at all squares adjacent to that minterm. If the given minterm and all of the 1's adjacent to it are covered by a single term, then that term is an *essential* prime implicant.[1] If all of the 1's adjacent to a given minterm are *not* covered by a single term, then there are two or more prime implicants which cover that minterm, and we cannot say whether these prime implicants are essential or not without checking the other minterms. Figure 5-18 illustrates this principle.

FIGURE 5-18

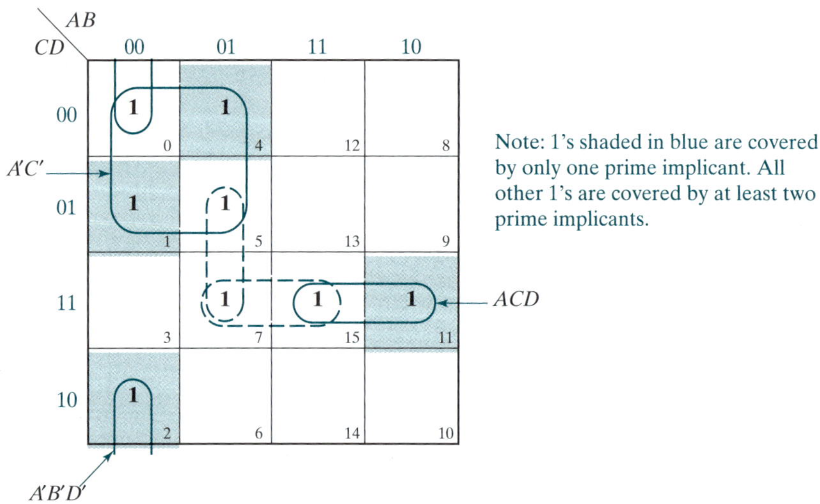

[1]This statement is proved in Appendix D.

The adjacent 1's for minterm m_0 (1_0) are 1_1, 1_2, and 1_4. Because no single term covers these four 1's, no essential prime implicant is yet apparent. The adjacent 1's for 1_1 are 1_0 and 1_5, so the term which covers these three 1's ($A'C'$) is an essential prime implicant. Because the only 1 adjacent to 1_2 is 1_0, $A'B'D'$ is also essential. Because the 1's adjacent to 1_7 (1_5 and 1_{15}) are not covered by a single term, neither $A'BD$ nor BCD is essential at this point. However, because the only 1 adjacent to 1_{11} is 1_{15}, ACD is essential. To complete the minimum solution, one of the nonessential prime implicants is needed. Either $A'BD$ or BCD may be selected. The final solution is

$$A'C' + A'B'D' + ACD + \begin{Bmatrix} A'BD \\ \text{or} \\ BCD \end{Bmatrix}$$

If a don't-care minterm is present on the map, we do not have to check it to see if it is covered by one or more prime implicants. However, when checking a 1 for adjacent 1's, we treat the adjacent don't-cares as if they were 1's because don't-cares may be combined with 1's in the process of forming prime implicants. The following procedure can then be used to obtain a minimum sum of products from a Karnaugh map:

1. Choose a minterm (a 1) which has not yet been covered.
2. Find all 1's and X's adjacent to that minterm. (Check the n adjacent squares on an n-variable map.)
3. If a single term covers the minterm and all of the adjacent 1's and X's, then that term is an essential prime implicant, so select that term. (Note that don't-care terms are treated like 1's in steps 2 and 3 but not in step 1.)
4. Repeat steps 1, 2, and 3 until all essential prime implicants have been chosen.
5. Find a minimum set of prime implicants which cover the remaining 1's on the map. (If there is more than one such set, choose a set with a minimum number of literals.)

Figure 5-19 gives a flowchart for this procedure. The following example (Figure 5-20) illustrates the procedure. Starting with 1_4, we see that the adjacent 1's and X's (X_0, 1_5, and 1_6) are *not* covered by a single term, so no essential prime implicant is apparent. However, 1_6 and its adjacent 1's and X's (1_4 and X_7) are covered by $A'B$, so $A'B$ is an essential prime implicant. Next, looking at 1_{13}, we see that its adjacent 1's and X's (1_5, 1_9, and X_{15}) are *not* covered by a single term, so no essential prime implicant is apparent. Similarly, an examination of the terms adjacent to 1_8 and 1_9 reveals no essential prime implicants. However, 1_{10} has only 1_8 adjacent to it, so $AB'D'$ *is* an essential prime implicant because it covers both 1_{10} and 1_8. Having first selected the essential prime implicants, we now choose $AC'D$ because it covers both of the remaining 1's on the map.

Judicious selection of the order in which the minterms are selected (step 1) reduces the amount of work required in applying this procedure. As will be seen in the next section, this procedure is especially helpful in obtaining minimum solutions for five- and six-variable problems.

There are two equivalent methods of obtaining minimum product-of-sum expressions for a function f. As mentioned above, one method is to find minimum a sum-of-products expression for f', and then complement f' to obtain a minimum

FIGURE 5-19 Flowchart for Determining a Minimum Sum of Products Using a Karnaugh Map

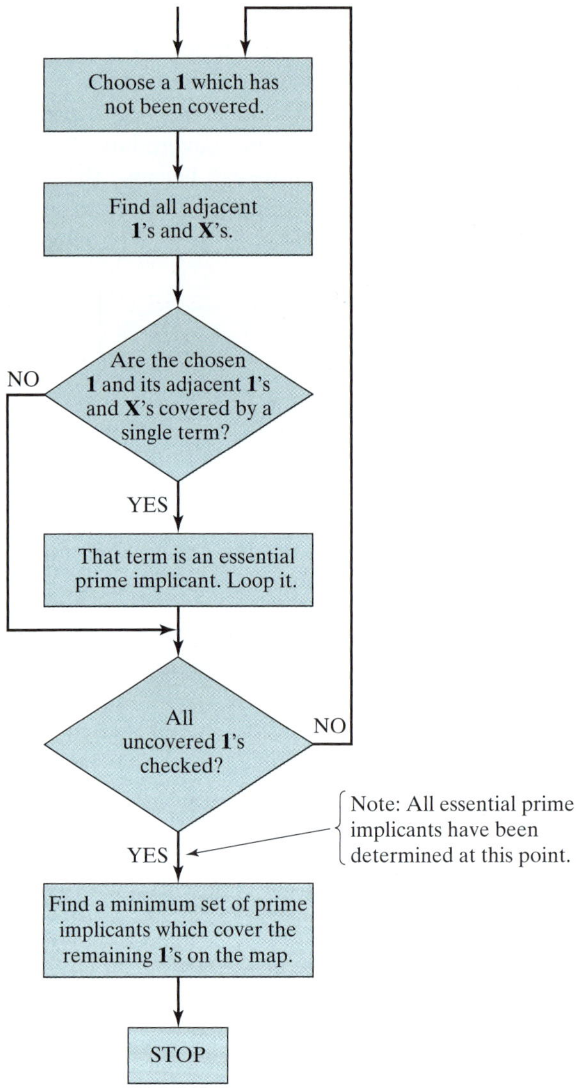

FIGURE 5-20

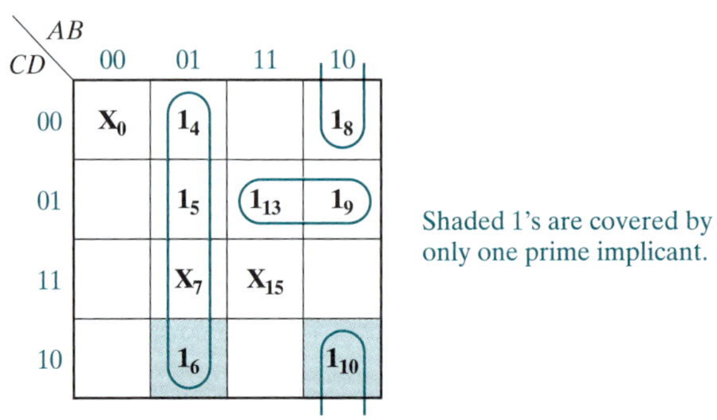

product-of-sums expression for f. Alternatively, we can perform the dual of the procedure for finding minimum sum of products. Let S be a sum term. If every input combination for which $S = 0$ f is also 0, then S can be a term in a product-of-sums expression for F. We will call such a sum term an *implicate* of f. Implicate S is a *prime implicate* if it cannot be combined with any other implicate to eliminate a literal from S. All implicates in a minimum product-of-sums expression for f must be prime implicates. The prime implicates of f can be found by looping the largest groups of adjacent zeros on the Karnaugh map for f. If a prime implicate is the only prime implicate covering a maxterm (zero) of f, then it is an *essential* prime implicate and must be included in any minimum product-of-sums expression for f.

5.5 Five-Variable Karnaugh Maps

A five-variable map can be constructed in three dimensions by placing one four-variable map on top of a second one. Terms in the bottom layer are numbered 0 through 15 and corresponding terms in the top layer are numbered 16 through 31, so that terms in the bottom layer contain A' and those in the top layer contain A. To represent the map in two dimensions, we will divide each square in a four-variable map by a diagonal line and place terms in the bottom layer below the line and terms in the top layer above the line (Figure 5-21). Terms in the top or bottom layer combine just like terms on a four-variable map. In addition, two terms in the same square which are separated by a diagonal line differ in only one variable and can be combined. However, some terms which appear to be physically adjacent are not. For example, terms 0 and 20 are not adjacent because they appear in a different column and a

FIGURE 5-21
A Five-Variable Karnaugh Map

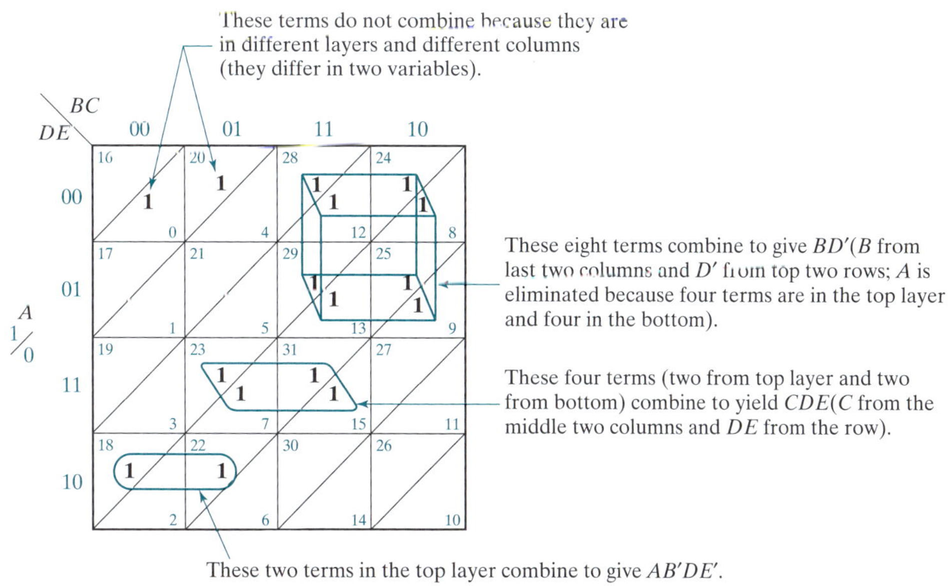

different layer. Each term can be adjacent to exactly five other terms, four in the same layer and one in the other layer (Figure 5-22). An alternate representation for five-variable maps is to draw the two layers side-by-side, as in Figure 5-28, but most individuals find adjacencies more difficult to see when this form is used.

When checking for adjacencies, each term should be checked against the five possible adjacent squares. (In general, the number of adjacent squares is equal to the number of variables.) Two examples of five-variable minimization using maps follow. Figure 5-23 is a map of

$$F(A, B, C, D, E) = \Sigma\, m(0, 1, 4, 5, 13, 15, 20, 21, 22, 23, 24, 26, 28, 30, 31)$$

FIGURE 5-22

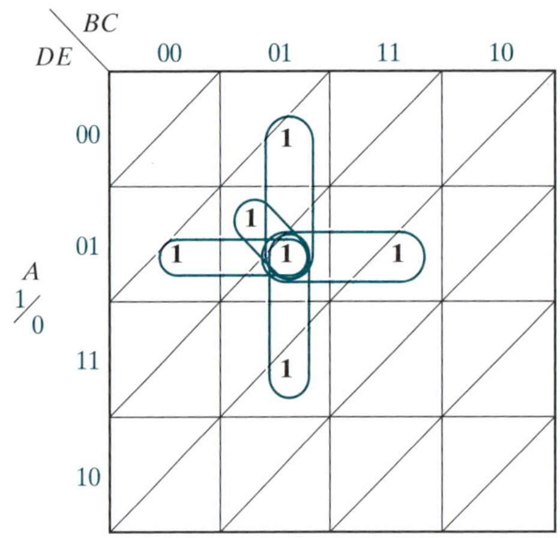

FIGURE 5-23

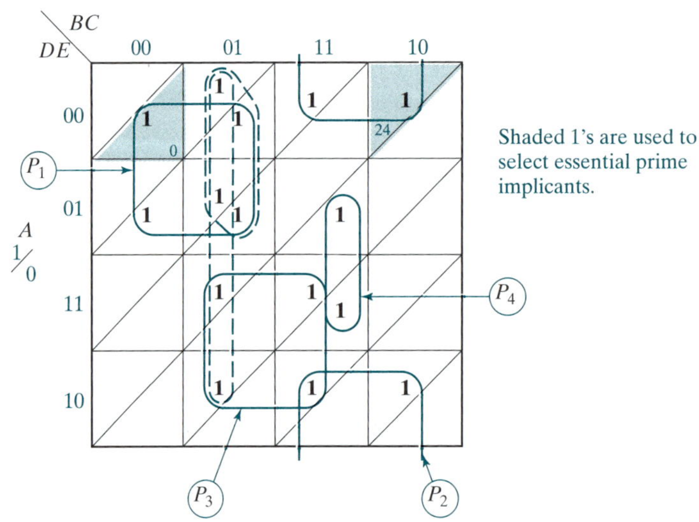

Prime implicant P_1 is chosen first because all of the 1's adjacent to minterm 0 are covered by P_1. Prime implicant P_2 is chosen next because all of the 1's adjacent to minterm 24 are covered by P_2. All of the remaining 1's on the map can be covered by at least two different prime implicants, so we proceed by trial and error. After a few tries, it becomes apparent that the remaining 1's can be covered by three prime implicants. If we choose prime implicants P_3 and P_4 next, the remaining two 1's can be covered by two different groups of four. The resulting minimum solution is

$$F = \underset{P_1}{A'B'D'} + \underset{P_2}{ABE'} + \underset{P_3}{ACD} + \underset{P_4}{A'BCE} + \begin{Bmatrix} AB'C \\ \text{or} \\ B'CD' \end{Bmatrix}$$

Figure 5-24 is a map of

$$F(A, B, C, D, E) = \Sigma\, m(0, 1, 3, 8, 9, 14, 15, 16, 17, 19, 25, 27, 31)$$

All 1's adjacent to m_{16} are covered by P_1, so choose P_1 first. All 1's adjacent to m_3 are covered by P_2, so P_2 is chosen next. All 1's adjacent to m_8 are covered by P_3, so P_3 is chosen. Because m_{14} is only adjacent to m_{15}, P_4 is also essential. There are no more essential prime implicants, and the remaining 1's can be covered by two terms, P_5 and (1-9-17-25) or (17-19-25-27). The final solution is

$$F = \underset{P_1}{B'C'D'} + \underset{P_2}{B'C'E} + \underset{P_3}{A'C'D'} + \underset{P_4}{A'BCD} + \underset{P_5}{ABDE} + \begin{Bmatrix} C'D'E \\ \text{or} \\ AC'E \end{Bmatrix}$$

FIGURE 5-24

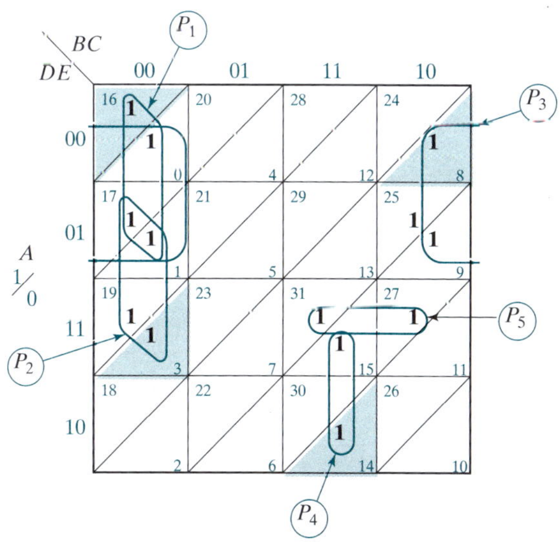

5.6 Other Uses of Karnaugh Maps

Many operations that can be performed using a truth table or algebraically can be done using a Karnaugh map. A map conveys the same information as a truth table—it is just arranged in a different format. If we plot an expression for F on a map, we can read off the minterm and maxterm expansions for F and for F'. From the map of Figure 5-14, the minterm expansion of f is

$$f = \Sigma\, m(0, 2, 3, 4, 8, 10, 11, 15)$$

and because each 0 corresponds to a maxterm, the maxterm expansion of f is

$$f = \Pi\, M(1, 5, 6, 7, 9, 12, 13, 14)$$

We can prove that two functions are equal by plotting them on maps and showing that they have the same Karnaugh map. We can perform the AND operation (or the OR operation) on two functions by ANDing (or ORing) the 1's and 0's which appear in corresponding positions on their maps. This procedure is valid because it is equivalent to doing the same operations on the truth tables for the functions.

A Karnaugh map can facilitate factoring an expression. Inspection of the map reveals terms which have one or more variables in common. For the map of Figure 5-25, the two terms in the first column have $A'B'$ in common; the two terms in the lower right corner have AC in common.

FIGURE 5-25

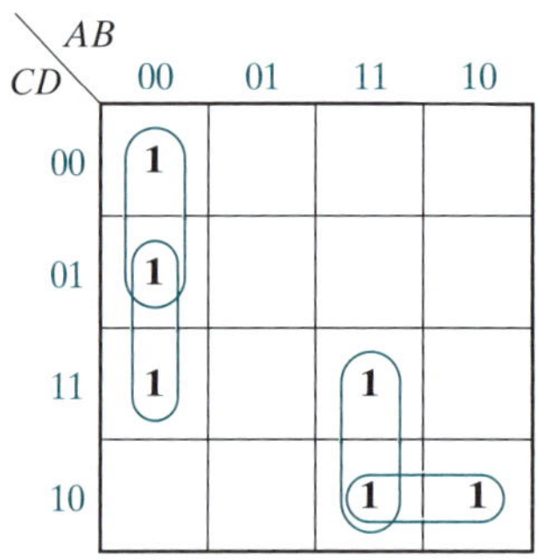

$$F = A'B'(C' + D) + AC(B + D')$$

When simplifying a function algebraically, the Karnaugh map can be used as a guide in determining what steps to take. For example, consider the function

$$F = ABCD + B'CDE + A'B' + BCE'$$

From the map (Figure 5-26), we see that in order to get the minimum solution, we must add the term $ACDE$. We can do this using the consensus theorem:

$$F = ABCD + B'CDE + A'B' + BCE' + ACDE$$

As can be seen from the map, this expression now contains two redundant terms, $ABCD$ and $B'CDE$. These can be eliminated using the consensus theorem, which gives the minimum solution:

$$F = A'B + BCE' + ACDE$$

FIGURE 5-26

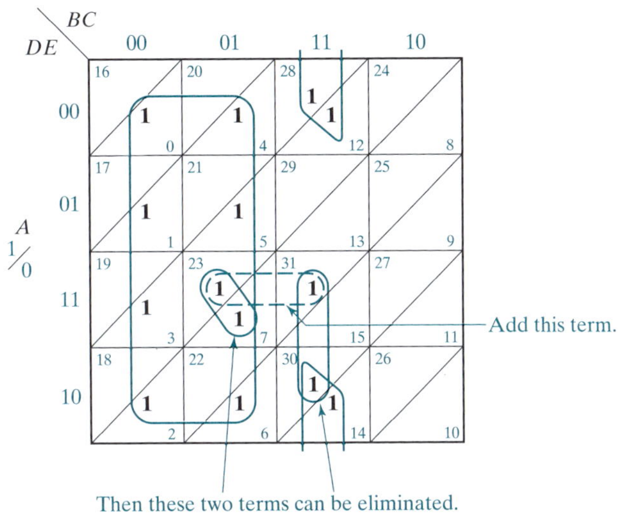

5.7 Other Forms of Karnaugh Maps

Instead of labeling the sides of a Karnaugh map with 0's and 1's, some people prefer to use the labeling shown in Figure 5-27. For the half of the map labeled A, $A = 1$; and for the other half, $A = 0$. The other variables have a similar interpretation. A map labeled this way is sometimes referred to as a Veitch diagram. It is particularly useful for plotting functions given in algebraic form rather than in minterm or maxterm form. However, when utilizing Karnaugh maps to solve sequential circuit problems (Units 12 through 16), the use of 0's and 1's to label the maps is more convenient.

FIGURE 5-27
Veitch Diagrams

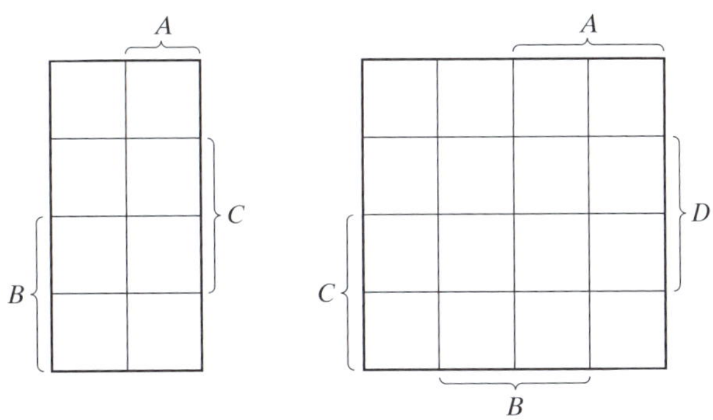

Two alternative forms for five-variable maps are used. One form simply consists of two four-variable maps side-by-side as in Figure 5-28(a). A modification of this uses a *mirror image* map as in Figure 5-28(b). In this map, first and eighth columns are "adjacent" as are second and seventh columns, third and sixth columns, and fourth and fifth columns. The same function is plotted on both these maps.

FIGURE 5-28 Other Forms of Five-Variable Karnaugh Maps

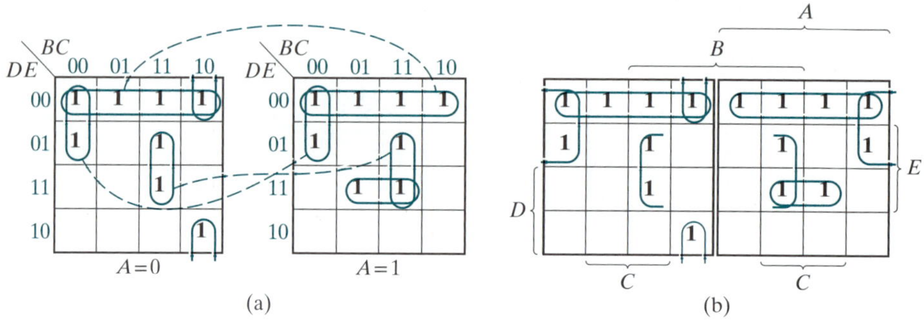

$$F = D'E' + B'C'D' + BCE + A'BC'E' + ACDE$$

Programmed Exercise 5.1

Cover the answers to this exercise with a sheet of paper and slide it down as you check your answers. Write your answers in the space provided before looking at the correct answer.

Problem Determine the minimum sum of products and minimum product of sums for

$$f = b'c'd' + bcd + acd' + a'b'c + a'bc'd$$

First, plot the map for f.

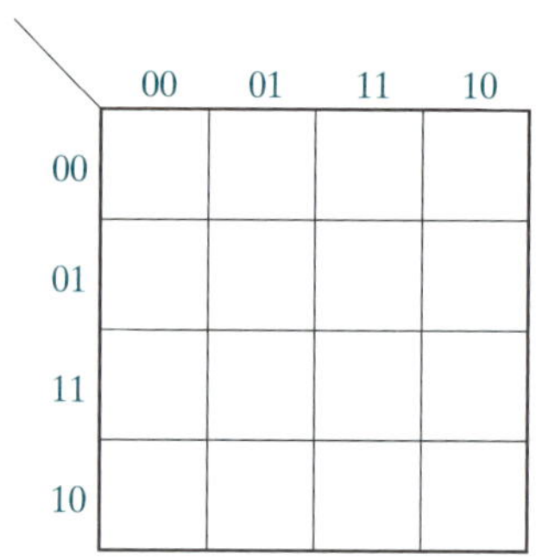

Answer:

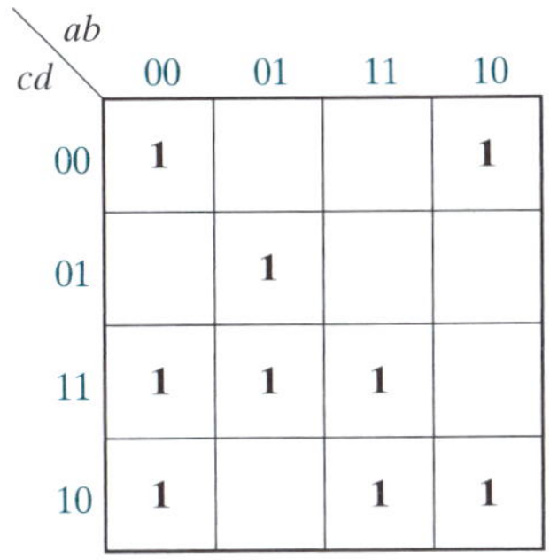

(a) The minterms adjacent to m_0 on the preceding map are ________ and ________.
(b) Find an essential prime implicant containing m_0 and loop it.

(c) The minterms adjacent to m_3 are ________ and ________.
(d) Is there an essential prime implicant which contains m_3?
(e) Find the remaining essential prime implicant(s) and loop it (them).

Answers:

(a) m_2 and m_8 (b)
(c) m_2 and m_7 (e)
(d) No

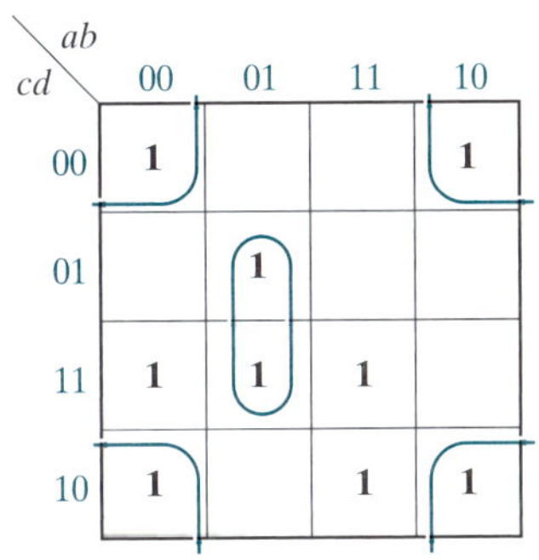

Loop the remaining 1's using a minimum number of loops.
The two possible minimum sum-of-products forms for f are
$f =$ ______________________________ and
$f =$ ______________________________

Answer:

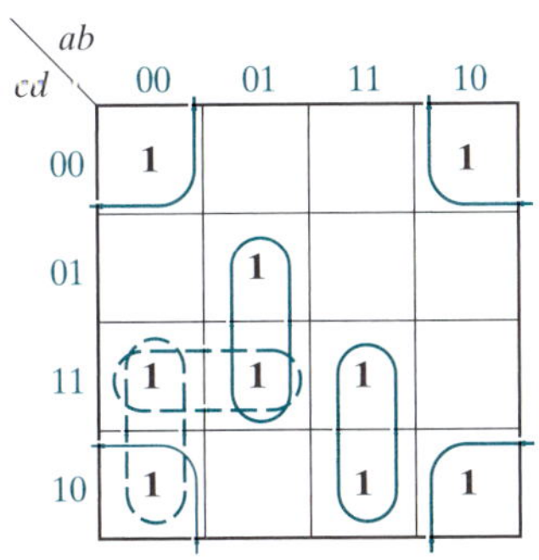

$$f = b'd' + a'bd + abc + \left\{ \begin{matrix} a'cd \\ or \\ a'b'c \end{matrix} \right\}$$

Next, we will find the minimum product of sums for f. Start by plotting the map for f'.

Loop all essential prime implicants of f' and indicate which minterm makes each one essential.

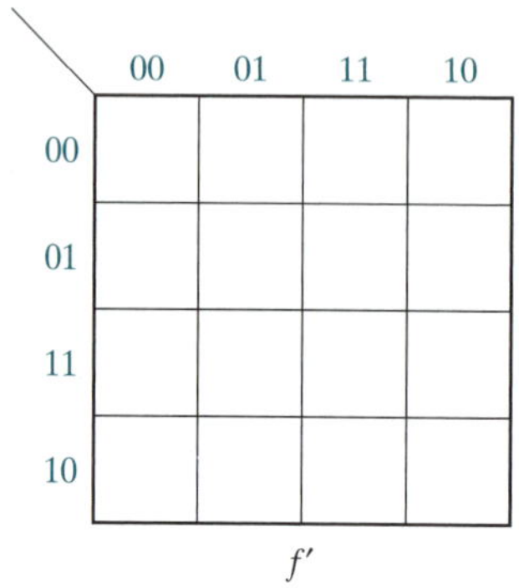

Answer:

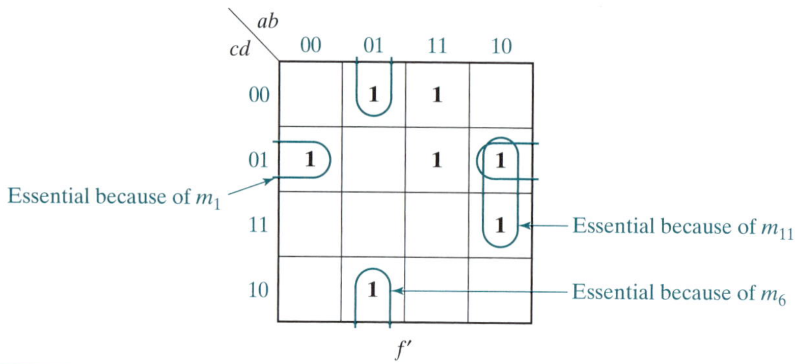

Loop the remaining 1's and write the minimum sum of products for f'.

$f' =$ ______________________

The minimum product of sums for f is therefore

$f =$ ______________________

Final Answer: $f' = b'c'd + a'bd' + ab'd + abc'$

$f = (b + c + d')(a + b' + d)(a' + b + d')(a' + b' + c)$

Programmed Exercise 5.2

Problem: Determine a minimum sum-of-products expression for

$$f(a, b, c, d, e) = (a' + c + d)(a' + b + e)(a + c' + e')(c + d + e')(b + c + d' + e)(a' + b' + c + e')$$

The first step in the solution is to plot a map for f. Because f is given in product-of-sums form, it is easier to first plot the map for f' and then complement the map. Write f' as a sum of products:

$f' =$ ______________________________

Now plot the map for f'. (Note that there are three terms in the upper layer, one term in the lower layer, and two terms which span the two layers.)

Next, convert your map for f' to a map for f.

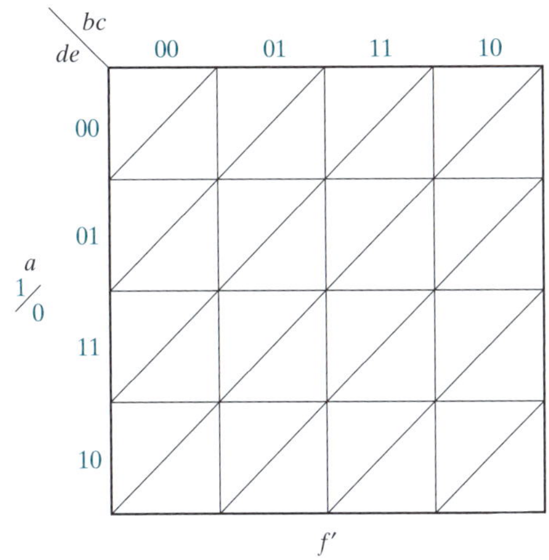

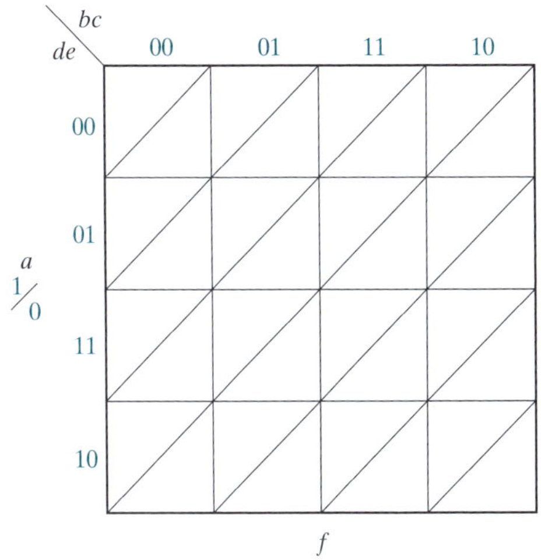

Answer:

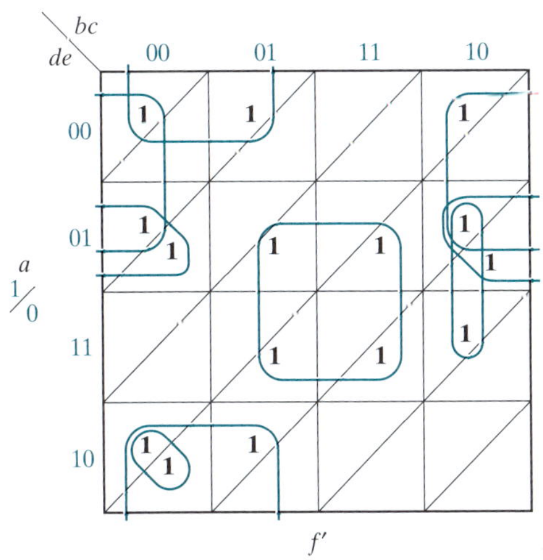

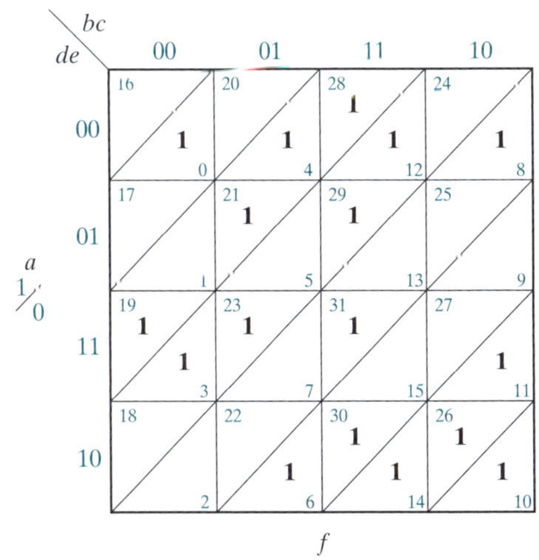

The next step is to determine the essential prime implicants of f.

(a) Why is $a'd'e'$ an essential prime implicant?

(b) Which minterms are adjacent to m_3? __________ To m_{19}? __________

(c) Is there an essential prime implicant which covers m_3 and m_{19}?

(d) Is there an essential prime implicant which covers m_{21}?

(e) Loop the essential prime implicants which you have found. Then, find two more essential prime implicants and loop them.

Answers:

(a) It covers m_0 and both adjacent minterms.
(b) m_{19} and m_{11}; m_3 and m_{23}
(c) No
(d) Yes
(e)

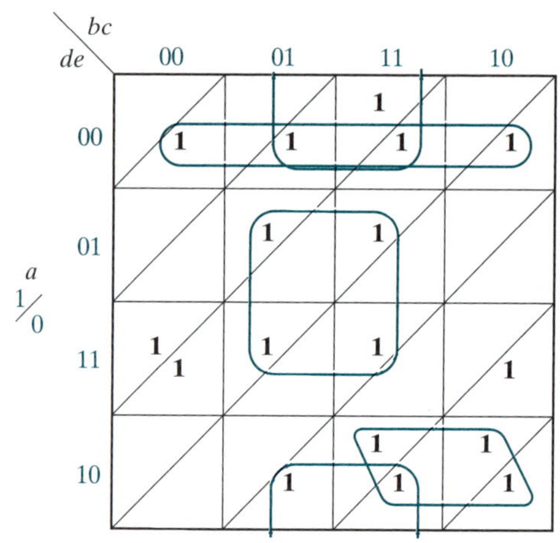

(a) Why is there no essential prime implicant which covers m_{11}?
(b) Why is there no essential prime implicant which covers m_{28}?

Because there are no more essential prime implicants, loop a minimum number of terms which cover the remaining 1's.

Answers:

(a) All adjacent 1's of $m_{11}(m_3, m_{10})$ cannot be covered by one grouping.
(b) All adjacent 1's of $m_{28}(m_{12}, m_{30}, m_{29})$ cannot be covered by one grouping.

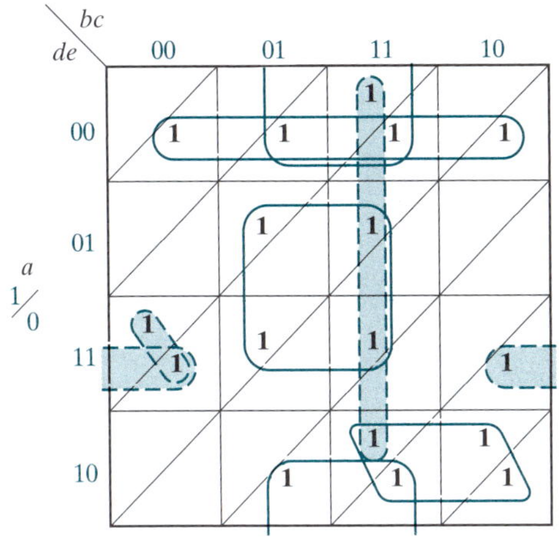

Note: There are five other possible ways to loop the four remaining 1's.

Write down two different minimum sum-of-products expressions for f.

$f =$ ________________________________

$f =$ ________________________________

Answer:

$$f = a'd'e' + ace + a'ce' + bde' + \begin{Bmatrix} abc \\ or \\ bce' \end{Bmatrix} + \begin{Bmatrix} b'c'de + a'c'de \\ b'c'de + a'bc'd \\ ab'de + a'c'de \end{Bmatrix}$$

Problems

5.3 Find the minimum sum of products for each function using a Karnaugh map.

(a) $f_1(a, b, c) = m_0 + m_2 + m_5 + m_6$ (b) $f_2(d, e, f) = \Sigma\, m(0, 1, 2, 4)$

(c) $f_3(r, s, t) = rt' + r's' + r's$ (d) $f_4(x, y, z) = M_0 \cdot M_5$

5.4 (a) Plot the following function on a Karnaugh map. (Do not expand to minterm form before plotting.)

$$F(A, B, C, D) = BD' + B'CD + ABC + ABC'D + B'D'$$

(b) Find the minimum sum of products.

(c) Find the minimum product of sums.

5.5 A switching circuit has two control inputs (C_1 and C_2), two data inputs (X_1 and X_2), and one output (Z). The circuit performs one of the logic operations AND, OR, EQU (equivalence), or XOR (exclusive OR) on the two data inputs. The function performed depends on the control inputs:

C_1	C_2	Function Performed by Circuit
0	0	OR
0	1	XOR
1	0	AND
1	1	EQU

(a) Derive a truth table for Z.
(b) Use a Karnaugh map to find a minimum AND-OR gate circuit to realize Z.

5.6 Find the minimum sum-of-products expression for each function. Underline the essential prime implicants in your answer and tell which minterm makes each one essential.
(a) $f(a, b, c, d) = \Sigma\, m(0, 1, 3, 5, 6, 7, 11, 12, 14)$
(b) $f(a, b, c, d) = \Pi\, M(1, 9, 11, 12, 14)$
(c) $f(a, b, c, d) = \Pi\, M(5, 7, 13, 14, 15) \cdot \Pi\, D(1, 2, 3, 9)$

5.7 Find the minimum sum-of-products expression for each function.
(a) $f(a, b, c, d) = \Sigma\, m(0, 2, 3, 4, 7, 8, 14)$
(b) $f(a, b, c, d) = \Sigma\, m(1, 2, 4, 15) + \Sigma\, d(0, 3, 14)$
(c) $f(a, b, c, d) = \Pi\, M(1, 2, 3, 4, 9, 15)$
(d) $f(a, b, c, d) = \Pi\, M(0, 2, 4, 6, 8) \cdot \Pi\, D(1, 12, 9, 15)$

5.8 Find the minimum sum of products and the minimum product of sums for each function:
(a) $f(a,b,c,d) = \Pi\, M(0,1,6,8,11,12) \cdot \Pi\, D(3,7,14,15)$
(b) $f(a,b,c,d) = \Sigma\, m(1,3,4,11) + \Sigma\, d(2,7,8,12,14,15)$

5.9 Find the minimum sum of products and the minimum product of sums for each function:
(a) $F(A, B, C, D, E) = \Sigma\, m(0, 1, 2, 6, 7, 9, 10, 15, 16, 18, 20, 21, 27, 30) + \Sigma\, d(3, 4, 11, 12, 19)$
(b) $F(A, B, C, D, E) = \Pi\, M(0, 3, 6, 9, 11, 19, 20, 24, 25, 26, 27, 28, 29, 30) \cdot \Pi\, D(1, 2, 12, 13)$

5.10 $F(a, b, c, d, e) = \Sigma\, m(0, 3, 4, 5, 6, 7, 8, 12, 13, 14, 16, 21, 23, 24, 29, 31)$
(a) Find the essential prime implicants using a Karnaugh map, and indicate why each one of the chosen prime implicants is essential (there are four essential prime implicants).
(b) Find all of the prime implicants by using the Karnaugh map. (There are nine in all.)

5.11 Find a minimum product-of-sums solution for f. Underline the essential prime implicates.

$$f(a, b, c, d, e) = \Sigma\, m(2, 4, 5, 6, 7, 8, 10, 12, 14, 16, 19, 27, 28, 29, 31) + \Sigma\, d(1, 30)$$

5.12 Given $F = AB'D' + A'B + A'C + CD$.

(a) Use a Karnaugh map to find the maxterm expression for F (express your answer in both decimal and algebric notation).

(b) Use a Karnaugh map to find the minimum sum-of-products form for F'.

(c) Find the minimum product of sums for F.

5.13 Find the minimum sum of products for the given expression. Then, make minterm 5 a don't-care term and verify that the minimum sum of products is unchanged. Now, start again with the original expression and find each minterm which could *individually* be made a don't-care without changing the minimum sum of products.

$$F(A, B, C, D) = A'C' + B'C + ACD' + BC'D$$

5.14 Find the minimum sum-of-products expressions for each of these functions.

(a) $f_1(A, B, C) = m_1 + m_2 + m_5 + m_7$
(b) $f_2(d, e, f) = \Sigma\, m(1, 5, 6, 7)$
(c) $f_3(r, s, t) = rs' + r's' + st'$
(d) $f_4(a, b, c) = m_0 + m_2 + m_3 + m_7$
(e) $f_5(n, p, q) = \Sigma\, m(1, 3, 4, 5)$
(f) $f_6(x, y, z) = M_1 M_7$

5.15 Find the minimum product-of-sums expression for each of the functions in Problem 5.14.

5.16 Find the minimum sum of products for each of these functions.

(a) $f_1(A, B, C) = m_1 + m_3 + m_4 + m_6$
(b) $f_2(d, e, f) = \Sigma\, m(1, 4, 5, 7)$
(c) $f_3(r, s, t) = r't' + rs' + rs$
(d) $f_1(a, b, c) = m_3 + m_4 + m_6 + m_7$
(e) $f_2(n, p, q) = \Sigma\, m(2, 3, 5, 7)$
(f) $f_4(x, y, z) = M_3 M_6$

5.17 (a) Plot the following function on a Karnaugh map. (Do not expand to minterm form before plotting.)

$$F(A, B, C, D) = A'B' + CD' + ABC + A'B'CD' + ABCD'$$

(b) Find the minimum sum of products.

(c) Find the minimum product of sums.

5.18 Work Problem 5.17 for the following:

$$f(A, B, C, D) = A'B' + A'B'C' + A'BD' + AC'D + A'BD + AB'CD'$$

5.19 A switching circuit has two control inputs (C_1 and C_2), two data inputs (X_1 and X_2), and one output (Z). The circuit performs logic operations on the two data inputs, as shown in this table:

C_1	C_2	Function Performed by Circuit
0	0	X_1X_2
0	1	$X_1 \oplus X_2$
1	0	$X_1' + X_2$
1	1	$X_1 \equiv X_2$

(a) Derive a truth table for Z.
(b) Use a Karnaugh map to find a minimum OR-AND gate circuit to realize Z.

5.20 Use Karnaugh maps to find all possible minimum sum-of-products expressions for each function.
(a) $F(a, b, c) = \Pi M(3, 4)$
(b) $g(d, e, f) = \Sigma m(1, 4, 6) + \Sigma d(0, 2, 7)$
(c) $F(p, q, r) = (p + q' + r)(p' + q + r')$
(d) $F(s, t, u) = \Sigma m(1, 2, 3) + \Sigma d(0, 5, 7)$
(e) $f(a, b, c) = \Pi M(2, 3, 4)$
(f) $G(D, E, F) = \Sigma m(1, 6) + \Sigma d(0, 3, 5)$

5.21 Simplify the following expression first by using a map and then by using Boolean algebra. Use the map as a guide to determine which theorems to apply to which terms for the algebraic simplification.

$$F = a'b'c' + a'c'd + bcd + abc + ab'$$

5.22 Find all prime implicants and all minimum sum-of-products expressions for each of the following functions.
(a) $f(A, B, C, D) = \Sigma m(4, 11, 12, 13, 14) + \Sigma d(5, 6, 7, 8, 9, 10)$
(b) $f(A, B, C, D) = \Sigma m(3, 11, 12, 13, 14) + \Sigma d(5, 6, 7, 8, 9, 10)$
(c) $f(A, B, C, D) = \Sigma m(1, 2, 4, 13, 14) + \Sigma d(5, 6, 7, 8, 9, 10)$
(d) $f(A, B, C, D) = \Sigma m(4, 15) + \Sigma d(5, 6, 7, 8, 9, 10)$
(e) $f(A, B, C, D) = \Sigma m(3, 4, 11, 15) + \Sigma d(5, 6, 7, 8, 9, 10)$
(f) $f(A, B, C, D) = \Sigma m(4) + \Sigma d(5, 6, 7, 8, 9, 10, 11, 12, 13, 14)$
(g) $f(A, B, C, D) = \Sigma m(4, 15) + \Sigma d(0, 1, 2, 5, 6, 7, 8, 9, 10)$

5.23 For each function in Problem 5.22, find all minimum product-of-sums expressions.

5.24 Find the minimum sum-of-products expression for
(a) $\Sigma m(0, 2, 3, 5, 6, 7, 11, 12, 13)$
(b) $\Sigma m(2, 4, 8) + \Sigma d(0, 3, 7)$
(c) $\Sigma m(1, 5, 6, 7, 13) + \Sigma d(4, 8)$
(d) $f(w, x, y, z) = \Sigma m(0, 3, 5, 7, 8, 9, 10, 12, 13) + \Sigma d(1, 6, 11, 14)$
(e) $\Pi M(0, 1, 2, 5, 7, 9, 11) \cdot \Pi D(4, 10, 13)$

5.25 Work Problem 5.24 for the following:

(a) $f(a, b, c, d) = \Sigma\, m(1, 3, 4, 5, 7, 9, 13, 15)$

(b) $f(a, b, c, d) = \Pi\ M(0, 3, 5, 8, 11)$

(c) $f(a, b, c, d) = \Sigma\, m(0, 2, 6, 9, 13, 14) + \Sigma\, d(3, 8, 10)$

(d) $f(a, b, c, d) = \Pi\ M(0, 2, 6, 7, 9, 12, 13) \cdot \Pi\ D(1, 3, 5)$

5.26 Find the minimum product of sums for the following. Underline the essential prime implicates in your answer.

(a) $\Pi\ M(0, 2, 4, 5, 6, 9, 14) \cdot \Pi\ D(10, 11)$

(b) $\Sigma\, m(1, 3, 8, 9, 15) + \Sigma\, d(6, 7, 12)$

5.27 Find a minimum sum-of-products and a minimum product-of-sums expression for each function:

(a) $f(A, B, C, D) = \Pi\ M(0, 2, 10, 11, 12, 14, 15) \cdot \Pi\ D(5, 7)$

(b) $f(w, x, y, z) = \Sigma\, m(0, 3, 5, 7, 8, 9, 10, 12, 13) + \Sigma\, d(1, 6, 11, 14)$

5.28 A logic *circuit* realizes the function $F(a, b, c, d) = a'b' + a'cd + ac'd + ab'd'$. Assuming that $a = c$ never occurs when $b = d = 1$, find a simplified expression for F.

5.29 Given $F = AB'D' + A'B + A'C + CD$.

(a) Use a Karnaugh map to find the maxterm expression for F (express your answer in both decimal and algebric notation).

(b) Use a Karnaugh map to find the minimum sum-of-products form for F'.

(c) Find the minimum product of sums for F.

5.30 Assuming that the inputs $ABCD = 0101$, $BCD = 1001$, $ABCD = 1011$ never occur, find a simplified expression for

$$F = A'BC'D + A'B'D + A'CD + ABD + ABC$$

5.31 Find all of the prime implicants for each of the functions plotted on page 157.

5.32 Find all of the prime implicants for each of the plotted functions:

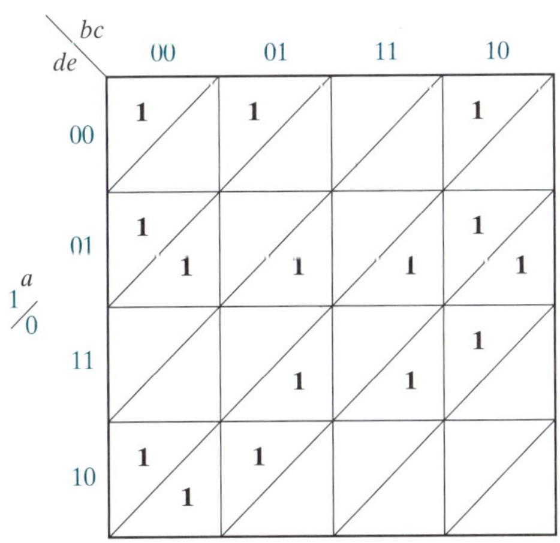

F

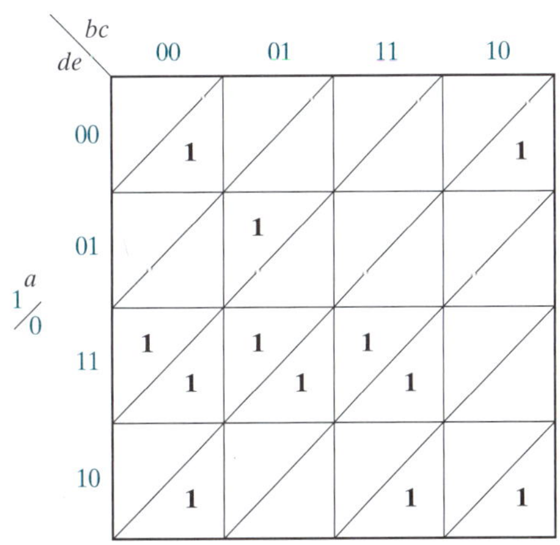

G

5.33 Given that $f(a, b, c, d, e) = \Sigma\, m(6, 7, 9, 11, 12, 13, 16, 17, 18, 20, 21, 23, 25, 28)$, using a Karnaugh map,
(a) Find the essential prime implicants (three).
(b) Find the minimum sum of products (7 terms).
(c) Find all of the prime implicants (twelve).

5.34 A logic circuit realizing the function f has four inputs a, b, c, d. The three inputs $a, b,$ and c are the binary representation of the digits 0 through 7 with a being the most significant bit. The input d is an odd-parity bit; that is, the value of d is such that $a, b, c,$ and d always contains an odd number of 1's. (For example, the digit 1 is represented by $abc = 001$ and $d = 0$, and the digit 3 is represented by $abcd = 0111$.) The function f has value 1 if the input digit is a prime number. (A number is prime if it is divisible only by itself and 1; 1 is considered to be prime, and 0 is not.)
(a) Draw a Karnaugh map for f.
(b) Find all prime implicants of f.
(c) Find all minimum sum of products for f.
(d) Find all prime implicants of f'.
(e) Find all minimum product of sums for f.

5.35 The decimal digits 0 though 9 are represented using five bits $A, B, C, D,$ and E. The bits $A, B, C,$ and D are the BCD representation of the decimal digit, and bit E is a parity bit that makes the five bits have odd parity. The function $F(A, B, C, D, E)$ has value 1 if the decimal digit represented by $A, B, C, D,$ and E is divisible by either 3 or 4. (Zero is divisible by 3 and 4.)
(a) Draw a Karnaugh map for f.
(b) Find all prime implicants of f. (Prime implicants containing only don't-cares need not be included.)
(c) Find all minimum sum of products for f.
(d) Find all prime implicants of f'.
(e) Find all minimum product of sums for f.

5.36 Rework Problem 5.35 assuming the decimal digits are represented in excess-3 rather than BCD.

5.37 The function $F(A, B, C, D, E) = \Sigma\, m(1, 7, 8, 13, 16, 19) + \Sigma\, d(0, 3, 5, 6, 9, 10, 12, 15, 17, 18, 20, 23, 24, 27, 29, 30)$.
(a) Draw a Karnaugh map for f.
(b) Find all prime implicants of f. (Prime implicants containing only don't-cares need not be included.)
(c) Find all minimum sum of products for f.
(d) Find all prime implicants of f'.
(e) Find all minimum product of sums for f.

5.38 $F(a, b, c, d, e) = \Sigma\, m(0, 1, 4, 5, 9, 10, 11, 12, 14, 18, 20, 21, 22, 25, 26, 28)$

(a) Find the essential prime implicants using a Karnaugh map, and indicate why each one of the chosen prime implicants is essential (there are four essential prime implicants).

(b) Find all of the prime implicants by using the Karnaugh map (there are 13 in all).

5.39 Find the minimum sum-of-products expression for F. Underline the essential prime implicants in this expression.

(a) $f(a, b, c, d, e) = \Sigma\, m(0, 1, 3, 4, 6, 7, 8, 10, 11, 15, 16, 18, 19, 24, 25, 28, 29, 31) + \Sigma\, d(5, 9, 30)$

(b) $f(a, b, c, d, e) = \Sigma\, m(1, 3, 5, 8, 9, 15, 16, 20, 21, 23, 27, 28, 31)$

5.40 Work Problem 5.39 with

$$F(A, B, C, D, E) = \Pi\, M(2, 3, 4, 8, 9, 10, 14, 15, 16, 18, 19, 20, 23, 24, 30, 31)$$

5.41 Find the minimum sum-of-products expression for F. Underline the essential prime implicants in your expression.

$$F(A, B, C, D, E) = \Sigma\, m(0, 2, 3, 5, 8, 11, 13, 20, 25, 26, 30) + \Sigma\, d(6, 7, 9, 24)$$

5.42 $F(V, W, X, Y, Z) = \Pi\, M(0, 3, 5, 6, 7, 8, 11, 13, 14, 15, 18, 20, 22, 24) \cdot \Pi\, D(1, 2, 16, 17)$

(a) Find a minimum sum-of-products expression for F. Underline the essential prime implicants.

(b) Find a minimum product-of-sums expression for F. Underline the essential prime implicates.

5.43 Find the minimum product of sums for

(a) $F(a, b, c, d, e) = \Sigma\, m(1, 2, 3, 4, 5, 6, 25, 26, 27, 28, 29, 30, 31)$

(b) $F(a, b, c, d, e) = \Sigma\, m(1, 5, 12, 13, 14, 16, 17, 21, 23, 24, 30, 31) + \Sigma\, d(0, 2, 3, 4)$

5.44 Find a minimum product-of-sums expression for each of the following functions:

(a) $F(v, w, x, y, z) = \Sigma\, m(4, 5, 8, 9, 12, 13, 18, 20, 21, 22, 25, 28, 30, 31)$

(b) $F(a, b, c, d, e) = \Pi\, M(2, 4, 5, 6, 8, 10, 12, 13, 16, 17, 18, 22, 23, 24) \cdot \Pi\, D(0, 11, 30, 31)$

5.45 Find the minimum sum of products for each function. Then, make the specified minterm a don't-care and verify that the minimum sum of products is unchanged. Now, start again with the original expression and find each minterm which could individually be made a don't-care, without changing the minimum sum of products.

(a) $F(A, B, C, D) = A'C' + A'B' + ACD' + BC'D$, minterm 2

(b) $F(A, B, C, D) = A'BD + AC'D + AB' + BCD + A'C'D$, minterm 7

5.46 $F(V,W,X,Y,Z) = \Pi\ M(0, 3, 6, 9, 11, 19, 20, 24, 25, 26, 27, 28, 29, 30) \cdot \Pi\ D(1,2,12,13)$

(a) Find two minimum sum-of-products expressions for F.

(b) Underline the essential prime implicants in your answer and tell why each one is essential.

5.47 Four of the minterms of the completely specified function $f(a, b, c, d)$ are $m_0, m_1, m_4,$ and m_5.

(a) Specify additional minterms for f so that f has eight prime implicants with two literals and no other prime implicants.

(b) For each prime implicant, give its algebraic representation and specify whether it is an essential prime implicant.

(c) Determine all minimum sum-of-products expressions for f.

5.48 Four of the minterms of the completely specified function $f(a, b, c, d)$ are $m_0, m_1, m_4,$ and m_5.

(a) Specify additional minterms for f so that f has one prime implicant with one literal, six prime implicants with two literals, and no other prime implicants.

(b) For each prime implicant, give its algebraic representation and specify whether it is an essential prime implicant.

(c) Determine all minimum sum-of-products expressions for f.

5.49 Four of the minterms of the completely specified function $f(a, b, c, d)$ are $m_0, m_1, m_4,$ and m_5.

(a) Specify additional minterms for f so that f has two prime implicants with one literal, two prime implicants with two literals, and no other prime implicants.

(b) For each prime implicant, give its algebraic representation and specify whether it is an essential prime implicant.

(c) Determine all minimum sum-of-products expressions for f.

5.50 Four of the minterms of an incompletely specified function $f(a, b, c, d)$ are $m_0, m_1, m_4,$ and m_5.

(a) Specify additional minterms and don't-cares for f so that f has five prime implicants with two literals and no other prime implicants and, in addition, f has one prime implicate with one literal and two prime implicates with two literals.

(b) For each prime implicant, give its algebraic representation and specify whether it is an essential prime implicant.

(c) Determine all minimum sum-of-products expressions for f.

(d) For each prime implicate, give its algebraic representation and specify whether it is an essential prime implicate.

(e) Determine all minimum product-of-sums expressions for f.

UNIT 7

Multi-Level Gate Circuits NAND and NOR Gates

Objectives

1. Design a minimal two-level or multi-level circuit of AND and OR gates to realize a given function. (Consider *both* circuits with an OR gate at the output and circuits with an AND gate at the output.)
2. Design or analyze a two-level gate circuit using any one of the eight basic forms (AND-OR, NAND-NAND, OR-NAND, NOR-OR, OR-AND, NOR-NOR, AND-NOR, and NAND-AND).
3. Design or analyze a multi-level NAND-gate or NOR-gate circuit.
4. Convert circuits of AND and OR gates to circuits of NAND gates or NOR gates, and conversely, by adding or deleting inversion bubbles.
5. Design a minimal two-level, multiple-output AND-OR, OR-AND, NAND-NAND, or NOR-NOR circuit using Karnaugh maps.

Study Guide

1. Study Section 7.1, *Multi-Level Gate Circuits*.

   (a) What are two ways of changing the number of levels in a gate circuit?

   (b) By constructing a tree diagram, determine the number of gates, gate inputs, and levels of gates required to realize Z_1 and Z_2:

   $$Z_1 = [(A + B)C + DE(F + G)]H \qquad Z_2 = A + B[C + DE(F + G)]$$

   Check your answers by drawing the corresponding gate circuits.

   (c) In order to find a minimum two-level solution, why is it necessary to consider both a sum-of-products form and a product-of-sums form for the function?

   (d) One realization of $Z = ABC(D + E) + FG$ is

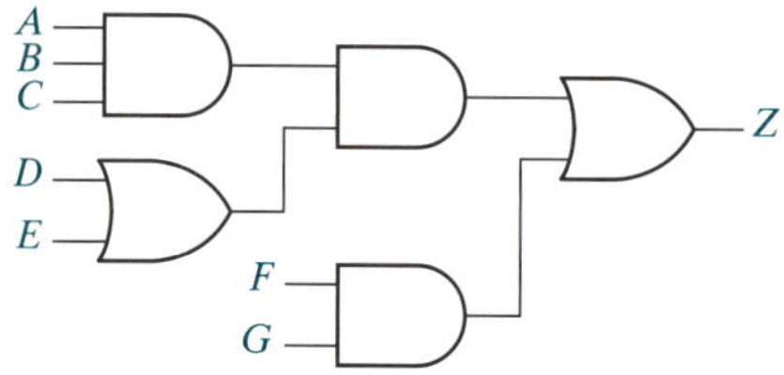

   Redraw the circuit so that it uses one less gate and so that the output of an AND gate never goes directly to the input of another AND gate.

(e) Work Problems 7.1 and 7.2. Unless otherwise specified, you may always assume that both the variables and their complements are available as circuit inputs.

2. Study Section 7.2, *NAND and NOR Gates*.

(a) For each gate, specify the missing inputs:

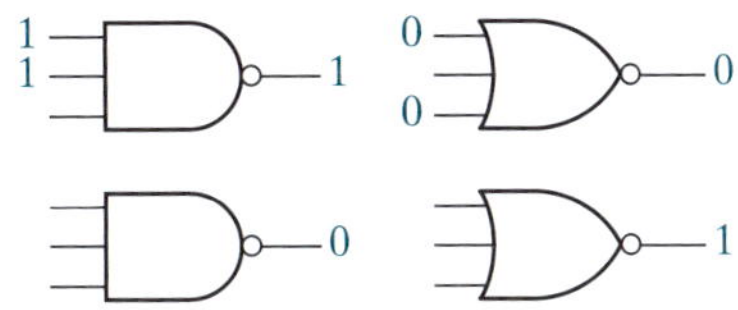

(b) What is meant by functionally complete set of logic gates?

(c) How can you show that a set of logic gates is functionally complete?

(d) Show that the NOR gate itself is functionally complete.

(e) Using NAND gates, draw a circuit for $F = (A'(BC)')'$.

(f) Using NOR gates, draw a circuit for $F = ((X + Y)' + (X' + Z)')'$.

3. Study Section 7.3, *Design of Two-Level NAND- and NOR-Gate Circuits*.

(a) Draw the circuit corresponding to Equation (7-17).

(b) Derive Equation (7-18).

(c) Make sure that you understand the relation between Equations (7-13) through (7-21) and the diagrams of Figure 7-11.

(d) Why is the NOR-NAND form degenerate?

(e) What assumption is made about the types of inputs available when the procedures for designing two-level NAND-NAND and NOR-NOR circuits are used?

(f) For these procedures the literal inputs to the output gate are complemented but not the literal inputs to the other gates. Explain why. Use an equation to illustrate.

(g) A general OR-AND circuit follows. Transform this to a NOR-NOR circuit and prove that your transformation is valid.

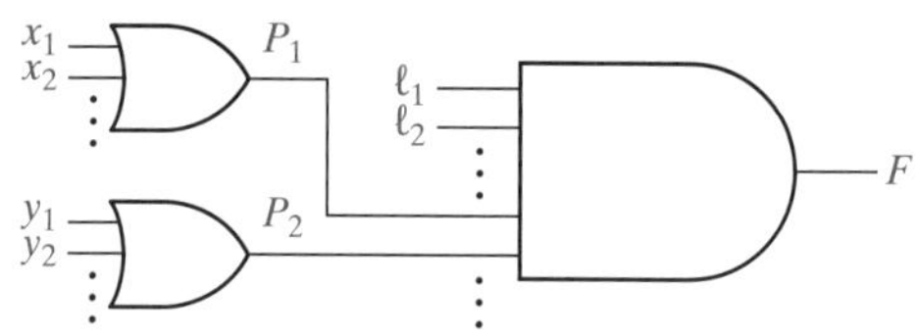

(h) Work Problem 7.3.

4. Study Section 7.4, *Design of Multi-Level NAND- and NOR-Gate Circuits.*

(a) Verify that the NAND circuit of Figure 7-13 is correct by dividing the corresponding circuit of AND and OR gates into two-level subcircuits and transforming each subcircuit.

(b) If you wish to design a two-level circuit using only NOR gates, should you start with a minimum sum of products or a minimum product of sums?

(c) Note that direct conversion of a circuit of AND and OR gates to a NAND-gate circuit requires starting with an OR gate at the output, but the direct conversion to a NOR-gate circuit requires starting with an AND gate at the output. This is easy to remember because a NAND is equivalent to an OR with the inputs inverted:

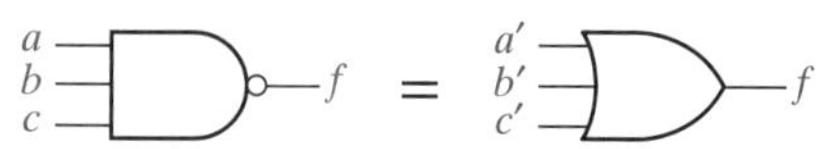

and a NOR is equivalent to an AND with the inputs inverted:

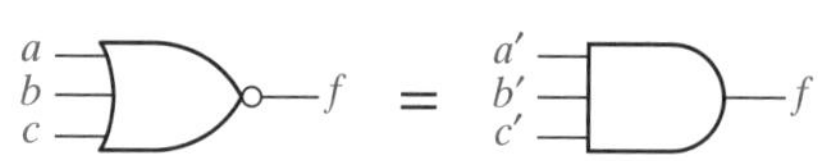

(d) Convert the circuit of Figure 7-1(b) to all NAND gates.
(e) Work Problems 7.4, 7.5, 7.6, and 7.7.

5. Study Section 7.5, *Circuit Conversion Using Alternative Gate Symbols*.

(a) Determine the logic function realized by each of the following circuits:

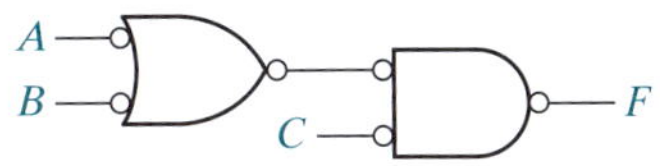

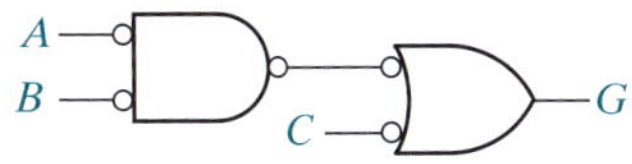

$F =$ __________ $G =$ __________

(b) Convert the circuit of Figure 7-13(a) to NAND gates by adding bubbles and complementing input variables when necessary. (You should have added 12 bubbles. Your result should be similar to Figure 7-13(b), except some of the NAND gates will use the alternative symbol.)
(c) Draw a circuit of AND and OR gates for the following equation:

$$Z = A[BC + D + E(F + GH)]$$

Then convert to NOR gates by adding bubbles and complementing inputs when necessary. (You should have added 10 bubbles and complemented six input variables.)

(d) Work Problem 7.8.

6. Study Section 7.6, *Design of Two-Level, Multiple-Output Circuits*.

(a) In which of the following cases would you replace a term xy' with $xy'z + xy'z'$?
(1) Neither $xy'z$ or $xy'z'$ is used in another function.
(2) Both $xy'z$ and $xy'z'$ are used in other functions.
(3) Term $xy'z$ is used in another function, but $xy'z'$ is not.

(b) In the second example (Figure 7-23), in f_2, c could have been replaced by $bc + b'c$ because bc and $b'c$ were available "free" from f_1 and f_3. Why was this replacement not made?

(c) In the following example, compute the cost of realizing f_1 and f_2 separately; then compute the cost using the term $a'b'c$ in common between the two functions. Use a two-level AND-OR circuit in both cases.

bc \ a	0	1
00		
01	1	
11	1	1
10		

f_1

bc \ a	0	1
00	1	1
01	1	
11		
10	1	1

f_2

(d) Find expressions which correspond to a two-level, minimum multiple-output, AND-OR realization of F_1, F_2, and F_3. Why should the term cd *not* be included in F_1?

cd \ ab	00	01	11	10
00				1
01				1
11	1	1	1	1
10				

F_1

cd \ ab	00	01	11	10
00		1		
01		1		
11	1	1	1	
10			1	1

F_2

cd \ ab	00	01	11	10
00				
01	1	1		
11		1	1	1
10				

F_3

$F_1 =$

$F_2 =$

$F_3 =$

(e) Work Problems 7.9, 7.10, and 7.11.

(f) Work Problem 7.12. (*Hint:* Work with the 0's on the maps and first find a minimum solution for f_1', f_2', and f_3'.)

7. Study Section 7.7, *Multiple-Output NAND- and NOR-Gate Circuits.*

(a) Derive expressions for the F_1 and F_2 outputs of the NOR circuits of Figure 7-26(b) by finding the equation for each gate output, and show that these expressions reduce to the original expressions for F_1 and F_2.

(b) Convert Figure 7-26(a) to 7-26(b) by using the bubble method.

(c) Work Problem 7.13.

Multi-Level Gate Circuits NAND and NOR Gates

In the first part of this unit, you will learn how to design circuits which have more than two levels of AND and OR gates. In the second part you will learn techniques for designing with NAND and NOR gates. These techniques generally consist of first designing a circuit of AND and OR gates and then converting it to the desired type of gates. These techniques are easy to apply *provided* that you start with the proper form of circuit.

7.1 Multi-Level Gate Circuits

The maximum number of gates cascaded in series between a circuit input and the output is referred to as the number of *levels* of gates (not to be confused with voltage levels). Thus, a function written in sum-of-products form or in product-of-sums form corresponds directly to a two-level gate circuit. As is usually the case in digital circuits where the gates are driven from flip-flop outputs (as discussed in Unit 11), we will assume that all variables and their complements are available as circuit inputs. For this reason, we will not normally count inverters which are connected directly to input variables when determining the number of levels in a circuit. In this unit we will use the following terminology:

1. *AND-OR circuit* means a two-level circuit composed of a level of AND gates followed by an OR gate at the output.
2. *OR-AND circuit* means a two-level circuit composed of a level of OR gates followed by an AND gate at the output.
3. *OR-AND-OR circuit* means a three-level circuit composed of a level of OR gates followed by a level of AND gates followed by an OR gate at the output.
4. Circuit of AND and OR gates implies no particular ordering of the gates; the output gate may be either AND or OR.

The number of levels in an AND-OR circuit can usually be increased by factoring the sum-of-products expression from which it was derived. Similarly, the number of levels in an OR-AND circuit can usually be increased by multiplying out some of the terms in the product-of-sums expression from which it was derived. Logic designers are concerned with the number of levels in a circuit for several reasons. Sometimes factoring (or multiplying out) to increase the number of levels of gates will reduce the required number of gates and gate inputs and, thus, reduce the cost of building the circuit, but in other cases increasing the number of levels will increase the cost. In many applications, the number of gates which can be cascaded is limited by gate delays. When the input of a gate is switched, there is a finite time before the output changes. When several gates are cascaded, the time between an input change and the corresponding change in the circuit output may become excessive and slow down the operation of the digital system.

The number of gates, gate inputs, and levels in a circuit can be determined by inspection of the corresponding expression. In the example of Figure 7-1(a), the tree diagram drawn below the expression for Z indicates that the corresponding circuit will have four levels, six gates, and 13 gate inputs, as verified in Figure 7-1(b). Each

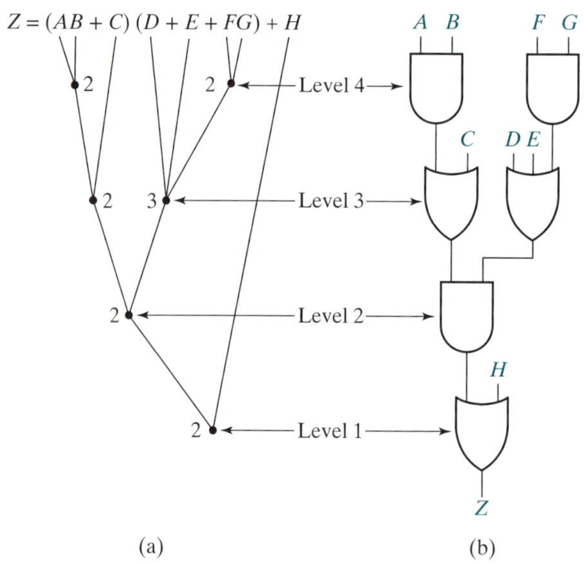

FIGURE 7-1 Four-Level Realization of Z

FIGURE 7-2
Three-Level Realization of Z

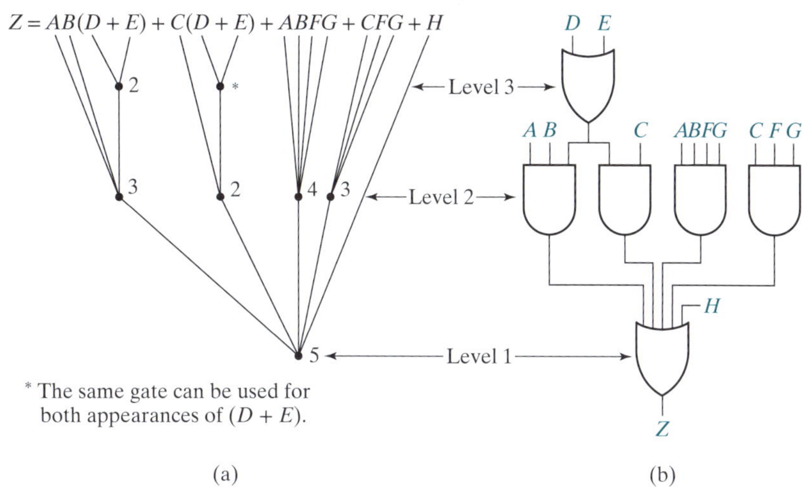

node on the tree diagram represents a gate, and the number of gate inputs is written beside each node.

We can change the expression for Z to three levels by partially multiplying it out:

$$
\begin{aligned}
Z &= (AB + C)[(D + E) + FG] + H \\
&= AB(D + E) + C(D + E) + ABFG + CFG + H
\end{aligned}
$$

As shown in Figure 7-2, the resulting circuit requires three levels, six gates, and 19 gate inputs.

Example of Multi-Level Design Using AND and OR Gates

Problem: Find a circuit of AND and OR gates to realize

$$f(a, b, c, d) = \Sigma\, m(1, 5, 6, 10, 13, 14)$$

Consider solutions with two levels of gates and three levels of gates. Try to minimize the number of gates and the total number of gate inputs. Assume that all variables and their complements are available as inputs.

Solution: First, simplify f by using a Karnaugh map (Figure 7-3):

FIGURE 7-3

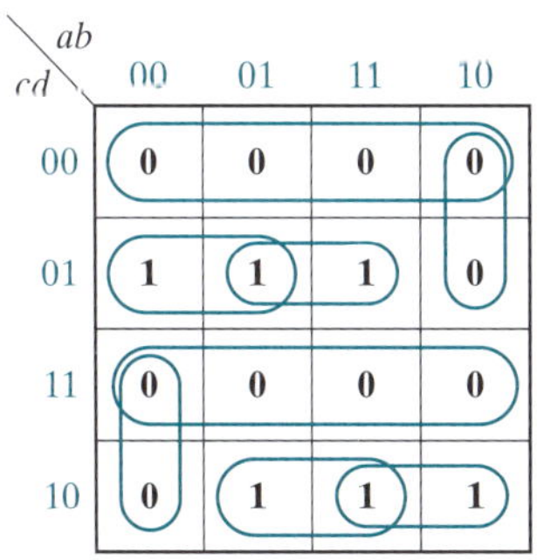

$$f = a'c'd + bc'd + bcd' + acd' \tag{7-1}$$

This leads directly to a two-level AND-OR gate circuit (Figure 7-4):

FIGURE 7-4

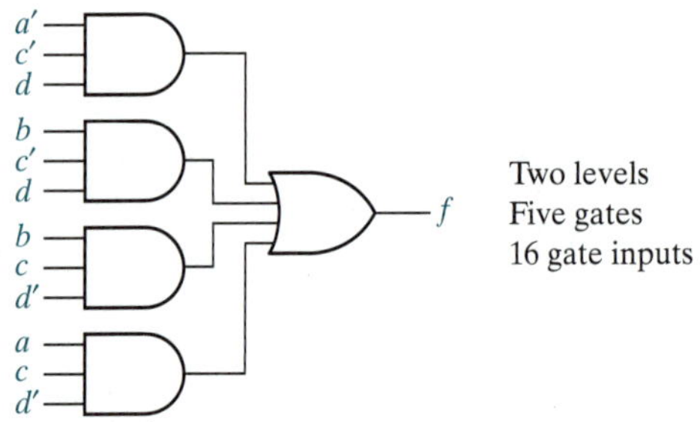

Factoring Equation (7-1) yields

$$f = c'd(a' + b) + cd'(a + b) \tag{7-2}$$

which leads to the following three-level OR-AND-OR gate circuit (Figure 7-5):

FIGURE 7-5

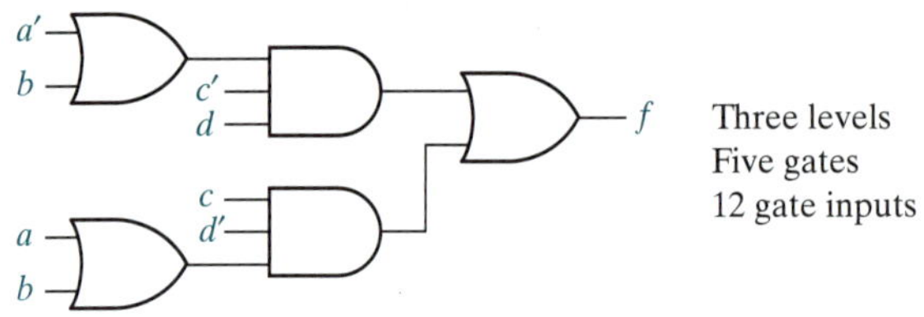

Both of these solutions have an OR gate at the output. A solution with an AND gate at the output might have fewer gates or gate inputs. A two-level OR-AND circuit corresponds to a product-of-sums expression for the function. This can be obtained from the 0's on the Karnaugh map as follows:

$$f' = c'd' + ab'c' + cd + a'b'c \tag{7-3}$$

$$f = (c + d)(a' + b + c)(c' + d')(a + b + c') \tag{7-4}$$

Equation (7-4) leads directly to a two-level OR-AND circuit (Figure 7-6):

FIGURE 7-6

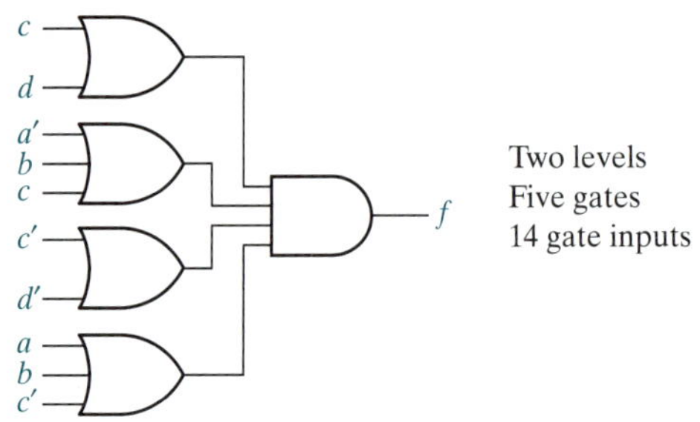

To get a three-level circuit with an AND-gate output, we partially multiply out Equation (7-4) using $(X + Y)(X + Z) = X + YZ$:

$$f = [c + d(a' + b)][c' + d'(a + b)] \tag{7-5}$$

Equation (7-5) would require four levels of gates to realize; however, if we multiply out $d'(a + b)$ and $d(a' + b)$, we get

$$f = (c + a'd + bd)(c' + ad' + bd') \tag{7-6}$$

which leads directly to a three-level AND-OR-AND circuit (Figure 7-7):

FIGURE 7-7

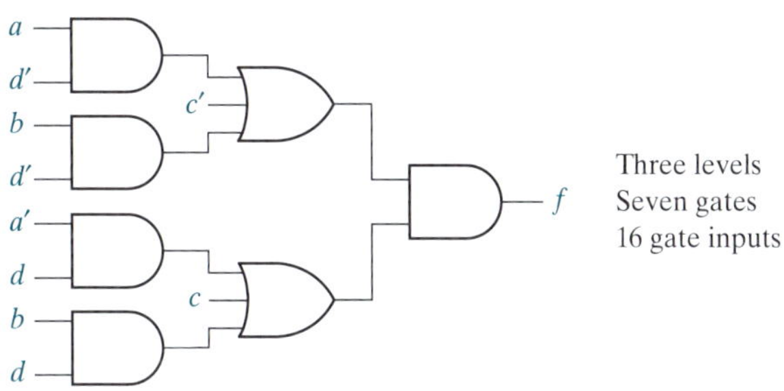

For this particular example, the best two-level solution had an AND gate at the output (Figure 7-6), and the best three-level solution had an OR gate at the output (Figure 7-5). In general, to be sure of obtaining a minimum solution, one must find *both* the circuit with the AND-gate output and the one with the OR-gate output.

If an expression for f' has n levels, the complement of that expression is an n-level expression for f. Therefore, to realize f as an n-level circuit with an AND-gate output, one procedure is first to find an n-level expression for f' with an OR operation at the output level and then complement the expression for f'. In the preceding example, factoring Equation (7-3) gives a three-level expression for f':

$$\begin{aligned} f' &= c'(d' + ab') + c(d + a'b') \\ &= c'(d' + a)(d' + b') + c(d + a')(d + b') \end{aligned} \tag{7-7}$$

Complementing Equation (7-7) gives Equation (7-6), which corresponds to the three-level AND-OR-AND circuit of Figure 7-7.

7.2 NAND and NOR Gates

Until this point we have designed logic circuits using AND gates, OR gates, and inverters. Exclusive-OR and equivalence gates have also been introduced in Unit 3. In this section we will define NAND and NOR gates. Logic designers frequently use NAND and NOR gates because they are generally faster and use fewer components than AND or OR gates. As will be shown later, any logic function can be implemented using only NAND gates or only NOR gates.

Figure 7-8(a) shows a three-input NAND gate. The small circle (or "bubble") at the gate output indicates inversion, so the NAND gate is equivalent to an AND gate followed by an inverter, as shown in Figure 7-8(b). A more appropriate name would be an AND-NOT gate, but we will follow common usage and call it a NAND gate. The gate output is

$$F = (ABC)' = A' + B' + C'$$

The output of the n-input NAND gate in Figure 7-8(c) is

$$F = (X_1X_2 \ldots X_n)' = X_1' + X_2' + \cdots + X_n' \tag{7-8}$$

The output of this gate is 1 iff one or or more of its inputs are 0.

FIGURE 7-8
NAND Gates

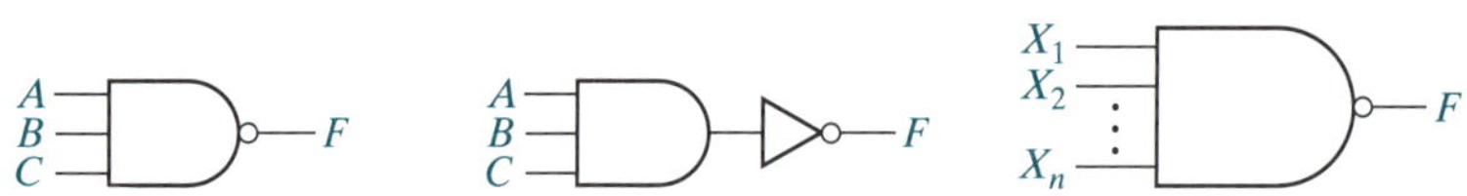

(a) Three-input NAND gate (b) NAND gate equivalent (c) n-input NAND gate

Figure 7-9(a) shows a three-input NOR gate. The small circle at the gate output indicates inversion, so the NOR gate is equivalent to an OR gate followed by an inverter. A more appropriate name would be an OR-NOT gate, but we will follow common usage and call it a NOR gate. The gate output is

$$F = (A + B + C)' = A'B'C'$$

FIGURE 7-9
NOR Gates

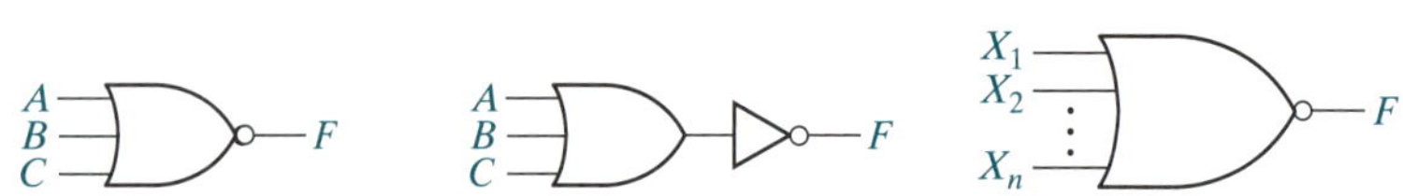

(a) Three-input NOR gate (b) NOR gate equivalent (c) n-input NOR gate

The output of an n-input NOR gate, shown in Figure 7-9(c), is

$$F = (X_1 + X_2 + \cdots + X_n)' = X_1'X_2' \ldots X_n' \tag{7-9}$$

A set of logic operations is said to be *functionally complete* if any Boolean function can be expressed in terms of this set of operations. The set AND, OR, and NOT is obviously functionally complete because any function can be expressed in sum-of-products form, and a sum-of-products expression uses only the AND, OR, and NOT operations. Similarly, a set of logic gates is functionally complete if all switching functions can be realized using this set of gates. Because the set of operations AND, OR, and NOT is functionally complete, any set of logic gates which can realize AND, OR, and NOT is also functionally complete. AND and NOT are a functionally complete set of gates because OR can also be realized using AND and NOT:

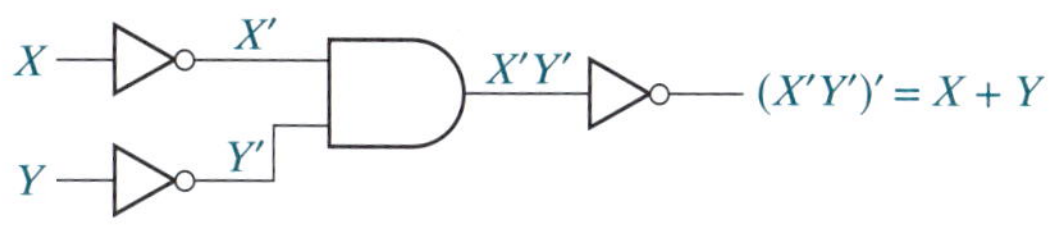

If a single gate forms a functionally complete set by itself, then any switching function can be realized using only gates of that type. The NAND gate is an example of such a gate. Because the NAND gate performs the AND operation followed by an inversion, NOT, AND, and OR can be realized using only NAND gates, as shown in Figure 7-10. Thus, any switching function can be realized using only NAND gates. An easy method for converting an AND-OR circuit to a NAND circuit is discussed in the next section. Similarly, any function can be realized using only NOR gates.

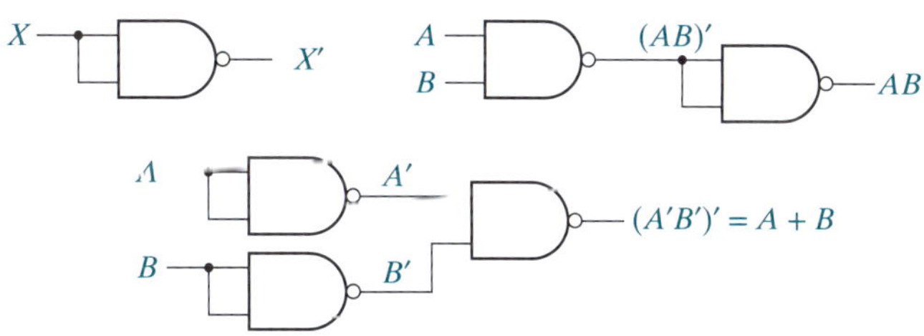

FIGURE 7-10 NAND Gate Realization of NOT, AND, and OR

The following procedure can be used to determine if a given set of gates is functionally complete. First, write out a minimum sum-of-products expression for the function realized by each gate. If no complement appears in any of these expressions, then NOT cannot be realized, and the set is not functionally complete. If a complement appears in one of the expressions, then NOT can generally be realized by an appropriate choice of inputs to the corresponding gate. (We will always assume that 0 and 1 are available as gate inputs). Next, attempt to realize AND or OR, keeping in mind that NOT is now available. Once AND or OR has been realized, the other one can always be realized using DeMorgan's laws if no more direct procedure is apparent. For example, if OR and NOT are available, AND can be realized by

$$XY = (X' + Y')' \tag{7-10}$$

7.3 Design of Two-Level NAND- and NOR-Gate Circuits

In this section two-level circuits realizing a function F using various combinations of NAND, NOR, AND, and OR gates are obtained by converting the switching algebra expression for F into the form matching the desired gate circuit. It is difficult to extend this approach to multiple-level circuits because it requires repeated complementation of parts of the expression for F. In Sections 7.4 and 7.5, an alternative method is developed which first realizes F in the desired form using AND and OR gates. The circuit with AND and OR gates is converted to one containing NAND or NOR gates by inserting inverters in pairs to convert each AND and OR gate to a NAND or NOR gate. This approach avoids manipulation of the expression for F and is less error prone.

A two-level circuit composed of AND and OR gates is easily converted to a circuit composed of NAND gates or NOR gates. This conversion is carried out by using $F = (F')'$ and then applying DeMorgan's laws:

$$(X_1 + X_2 + \cdots + X_n)' = X_1'X_2' \ldots X_n' \tag{7-11}$$

$$(X_1X_2 \ldots X_n)' = X_1' + X_2' + \cdots + X_n' \tag{7-12}$$

The following example illustrates conversion of a minimum sum-of-products form to several other two-level forms:

$$F = A + BC' + B'CD = [(A + BC' + B'CD)']' \tag{7-13}$$

$$= [A' \cdot (BC')' \cdot (B'CD)']' \qquad \text{(by 7-11)} \tag{7-14}$$

$$= [A' \cdot (B' + C) \cdot (B + C' + D')]' \qquad \text{(by 7-12)} \tag{7-15}$$

$$= A + (B' + C)' + (B + C' + D')' \qquad \text{(by 7-12)} \tag{7-16}$$

Equations (7-13), (7-14), (7-15), and (7-16) represent the AND-OR, NAND-NAND, OR-NAND, and NOR-OR forms, respectively, as shown in Figure 7-11.

Rewriting Equation (7-16) in the form

$$F = \{[A + (B' + C)' + (B + C' + D')']'\}' \tag{7-17}$$

leads to a three-level NOR-NOR-INVERT circuit. However, if we want a two-level circuit containing only NOR gates, we should start with the minimum product-of-sums form for F instead of the minimum sum of products. After obtaining the minimum product of sums from a Karnaugh map, F can be written in the following two-level forms:

$$\begin{aligned} F &= (A + B + C)(A + B' + C')(A + C' + D) \\ &= \{[(A + B + C)(A + B' + C')(A + C' + D)]'\}' \end{aligned} \tag{7-18}$$

$$= [(A + B + C)' + (A + B' + C')' + (A + C' + D)']' \qquad \text{(by 7-12)} \tag{7-19}$$

$$= (A'B'C' + A'BC + A'CD')' \qquad \text{(by 7-11)} \tag{7-20}$$

$$= (A'B'C')' \cdot (A'BC)' \cdot (A'CD')' \qquad \text{(by 7-11)} \tag{7-21}$$

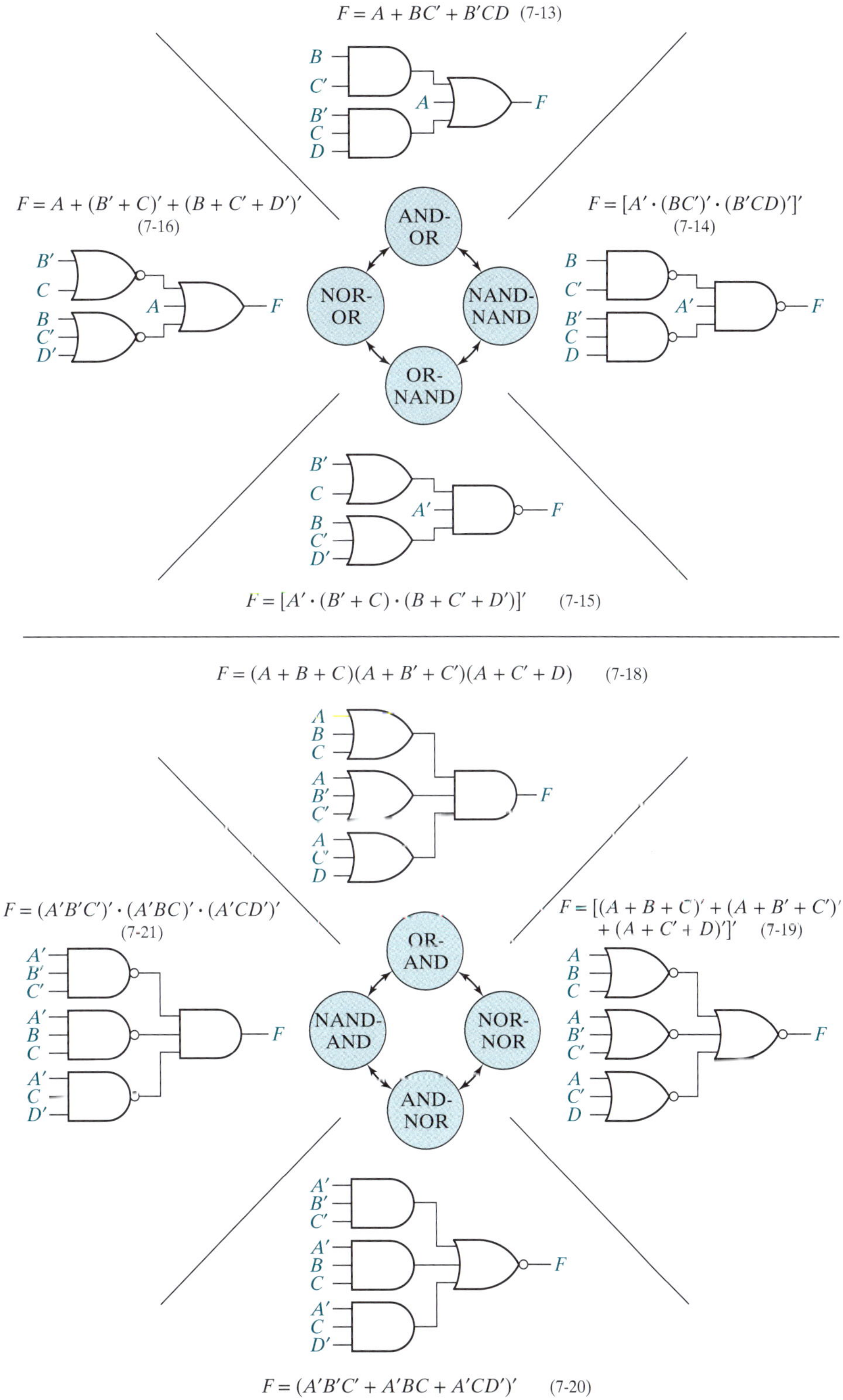

FIGURE 7-11 Eight Basic Forms for Two-Level Circuits

Equations (7-18), (7-19), (7-20), and (7-21) represent the OR-AND, NOR-NOR, AND-NOR, and NAND-AND forms, respectively, as shown in Figure 7-11. Two-level AND-NOR (AND-OR-INVERT) circuits are available in integrated-circuit form. Some types of NAND gates can also realize AND-NOR circuits when the so-called *wired OR* connection is used.

The other eight possible two-level forms (AND-AND, OR-OR, OR-NOR, AND-NAND, NAND-NOR, NOR-NAND, etc.) are degenerate in the sense that they cannot realize all switching functions. Consider, for example, the following NAND-NOR circuit:

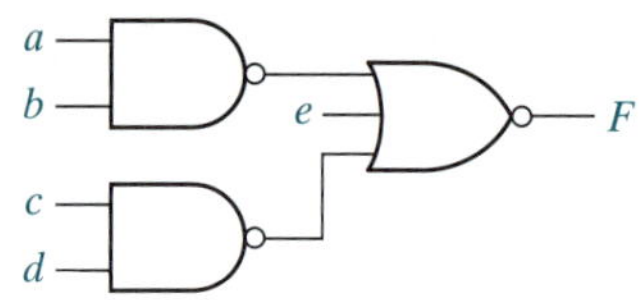

$$F = [(ab)' + (cd)' + e]' = abcde'$$

From this example, it is clear that the NAND-NOR form can realize only a product of literals and not a sum of products.

Because NAND and NOR gates are readily available in integrated circuit form, two of the most commonly used circuit forms are the NAND-NAND and the NOR-NOR. Assuming that all variables and their complements are available as inputs, the following method can be used to realize F with NAND gates:

Procedure for designing a minimum two-level NAND-NAND circuit:

1. Find a minimum *sum-of-products* expression for F.
2. Draw the corresponding two-level AND-OR circuit.
3. Replace all gates with NAND gates leaving the gate interconnections unchanged. If the output gate has any single literals as inputs, complement these literals.

Figure 7-12 illustrates the transformation of step 3. Verification that this transformation leaves the circuit output unchanged follows. In general, F is a sum of literals ($\ell_1, \ell_2, \ldots$) and product terms ($P_1, P_2, \ldots$):

$$F = \ell_1 + \ell_2 + \cdots + P_1 + P_2 + \cdots$$

After applying DeMorgan's law,

$$F = (\ell_1'\ell_2' \ldots P_1'P_2' \ldots)'$$

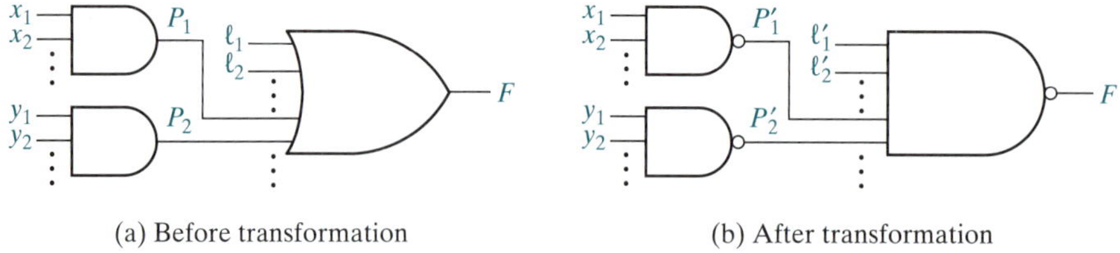

FIGURE 7-12 AND-OR to NAND-NAND Transformation

So the output OR gate is replaced with a NAND gate with inputs, ℓ_1', ℓ_2', . . ., P_1', P_2', ... Because product terms P_1, P_2, . . . are each realized with an AND gate, P_1', P_2', . . . are each realized with a NAND gate in the transformed circuit.

Assuming that all variables and their complements are available as inputs, the following method can be used to realize F with NOR gates:

Procedure for designing a minimum two-level NOR-NOR circuit:

1. Find a minimum *product-of-sums* expression for F.
2. Draw the corresponding two-level OR-AND circuit.
3. Replace all gates with NOR gates leaving the gate interconnections unchanged. If the output gate has any single literals as inputs, complement these literals.

This procedure is similar to that used for designing NAND-NAND circuits. Note, however, that for the NOR-NOR circuit, the starting point is a minimum *product of sums* rather than a sum of products.

7.4 Design of Multi-Level NAND- and NOR-Gate Circuits

The following procedure may be used to design multi-level NAND-gate circuits:

1. Simplify the switching function to be realized.
2. Design a multi-level circuit of AND and OR gates. The output gate must be OR. AND-gate outputs cannot be used as AND-gate inputs; OR-gate outputs cannot be used as OR-gate inputs.
3. Number the levels starting with the output gate as level 1. Replace all gates with NAND gates, leaving all interconnections between gates unchanged. Leave the inputs to levels 2, 4, 6, . . . unchanged. Invert any literals which appear as inputs to levels 1, 3, 5, ...

The validity of this procedure is easily proven by dividing the multi-level circuit into two-level subcircuits and applying the previous results for two-level circuits to each of the two-level subcircuits. The example of Figure 7-13 illustrates the procedure. Note that if step 2 is performed correctly, each level of the circuit will contain only AND gates or only OR gates.

The procedure for the design of multi-level NOR-gate circuits is exactly the same as for NAND-gate circuits except the output gate of the circuit of AND and OR gates must be an AND gate, and all gates are replaced with NOR gates.

Example

$F_1 = a'[b' + c(d + e') + f'g'] + hi'j + k$

Figure 7-13 shows how the AND-OR circuit for F_1 is converted to the corresponding NAND circuit.

FIGURE 7-13 Multi-Level Circuit Conversion to NAND Gates

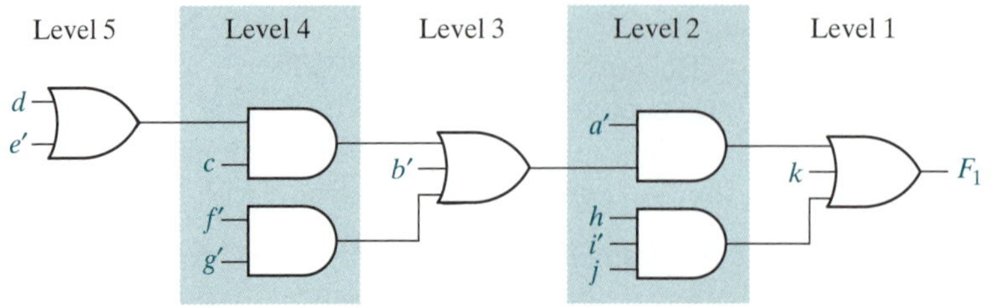

(a) AND-OR network

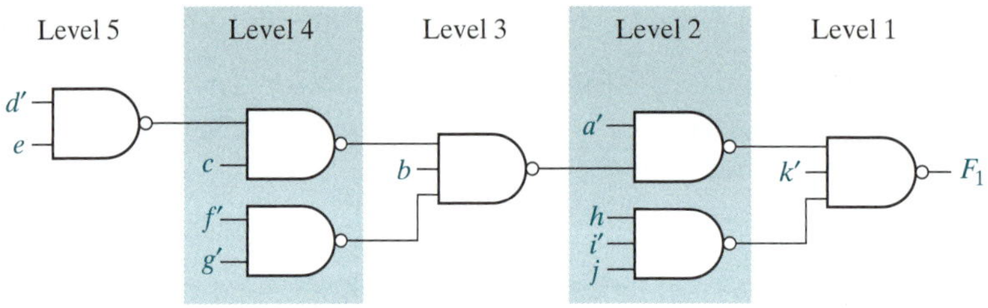

(b) NAND network

7.5 Circuit Conversion Using Alternative Gate Symbols

Logic designers who design complex digital systems often find it convenient to use more than one representation for a given type of gate. For example, an inverter can be represented by

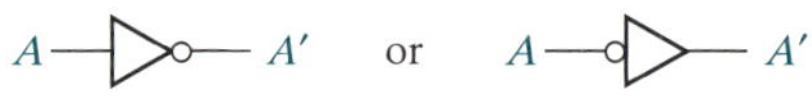

In the second case, the inversion "bubble" is at the input instead of the output. Figure 7-14 shows some alternative representations for AND, OR, NAND, and NOR gates. These equivalent gate symbols are based on DeMorgan's laws.

FIGURE 7-14 Alternative Gate Symbols

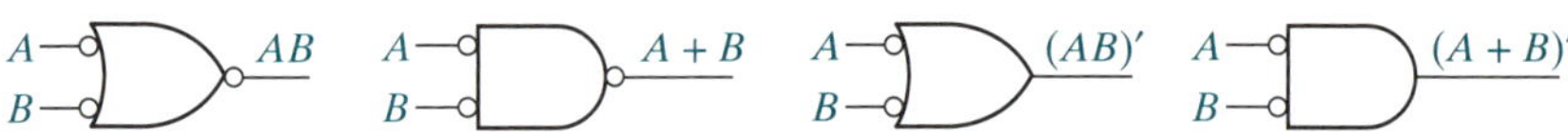

$AB = (A' + B')'$ (a) AND

$A + B = (A'B')'$ (b) OR

$(AB)' = A' + B'$ (c) NAND

$(A + B)' = A'B'$ (d) NOR

These alternative symbols can be used to facilitate the analysis and design of NAND- and NOR-gate circuits. Figure 7-15(a) shows a simple NAND-gate circuit. To analyze the circuit, we will replace the NAND gates at the first and third levels with the alternative NAND gate symbol. This eliminates the inversion bubble at the circuit output.

FIGURE 7-15
NAND Gate Circuit Conversion

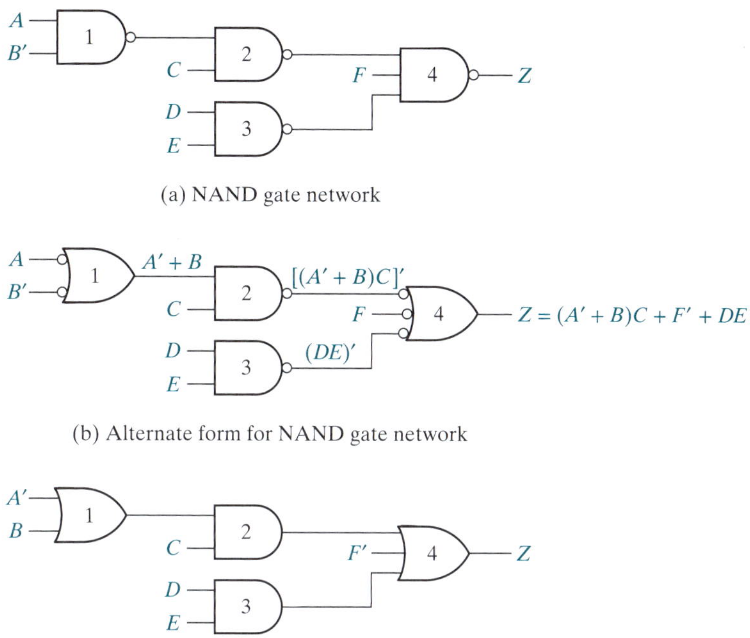

(a) NAND gate network

(b) Alternate form for NAND gate network

(c) Equivalent AND-OR network

In the resulting circuit (Figure 7-15(b)), inverted outputs (those with a bubble) are always connected to inverted inputs, and noninverted outputs are connected to noninverted inputs. Because two inversions in a row cancel each other out, we can easily analyze the circuit without algebraically applying DeMorgan's laws. Note, for example, that the output of gate 2 is $[(A' + B)C]'$, but the term $(A' + B)C$ appears in the output function. We can also convert the circuit to an AND-OR circuit by simply removing the double inversions (see Figure 7-15(c)). When a single input variable is connected to an inverted input, we must also complement that variable when we remove the inversion from the gate input. For example, A in Figure 7-15(b) becomes A' in Figure 7-15(c).

The circuit of AND and OR gates shown in Figure 7-16(a) can easily be converted to a NOR-gate circuit because the output gate is an AND-gate, and AND and OR gates alternate throughout the circuit. That is, AND-gate outputs connect only to OR-gate inputs, and OR-gate outputs connect only to AND-gate inputs. To carry out conversion to NOR-gates, we first replace all of the OR and AND gates with NOR gates, as shown in Figure 7-16(b). Because each inverted gate output drives an inverted gate input, the pairs of inversions cancel. However, when an input variable drives an inverted input, we have added a single inversion, so we must complement the variable to compensate. Therefore, we have complemented C and G. The resulting NOR-gate circuit is equivalent to the original AND-OR circuit.

Even if AND and OR gates do not alternate, we can still convert an AND-OR circuit to a NAND or NOR circuit, but it may be necessary to add extra inverters so that each added inversion is cancelled by another inversion. The following procedure may be used to convert to a NAND (or NOR) circuit:

1. Convert all AND gates to NAND gates by adding an inversion bubble at the output. Convert all OR gates to NAND gates by adding inversion bubbles at the

FIGURE 7-16
Conversion to NOR Gates

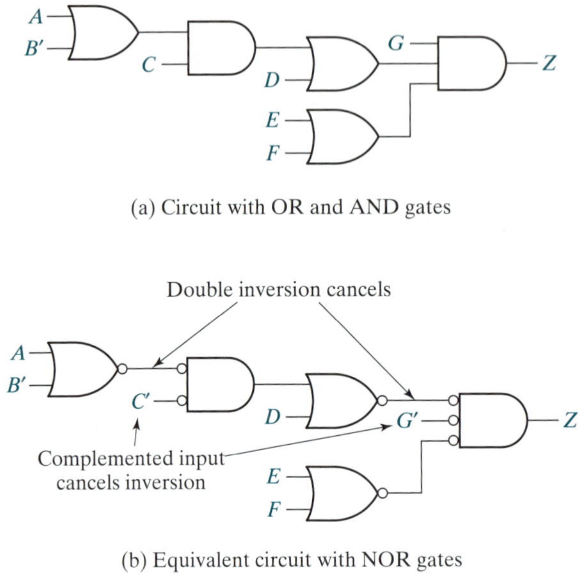

inputs. (To convert to NOR, add inversion bubbles at all OR-gate outputs and all AND-gate inputs.)

2. Whenever an inverted output drives an inverted input, no further action is needed because the two inversions cancel.
3. Whenever a noninverted gate output drives an inverted gate input or vice versa, insert an inverter so that the bubbles will cancel. (Choose an inverter with the bubble at the input or output as required.)

FIGURE 7-17
Conversion of AND-OR Circuit to NAND Gates

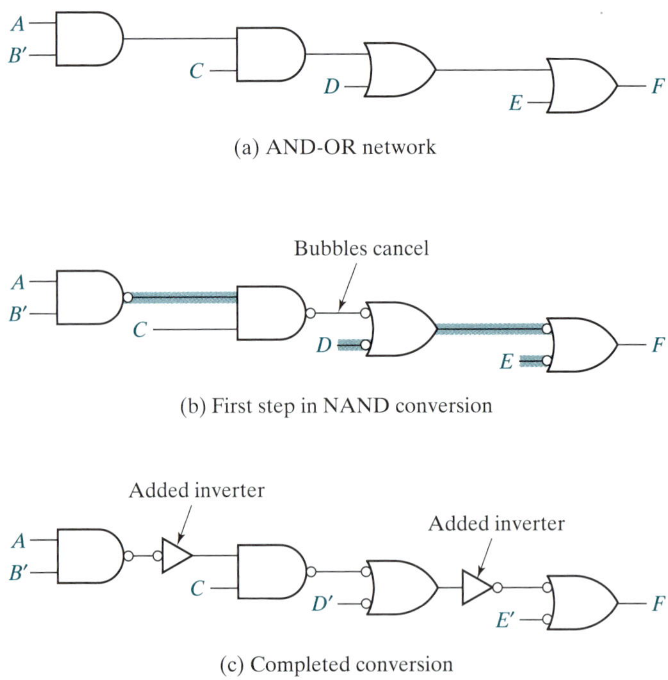

4. Whenever a variable drives an inverted input, complement the variable (or add an inverter) so the complementation cancels the inversion at the input.

In other words, if we always add bubbles (or inversions) in pairs, the function realized by the circuit will be unchanged. To illustrate the procedure we will convert Figure 7-17(a) to NANDs. First, we add bubbles to change all gates to NAND gates (Figure 7-17(b)). In four places (highlighted in blue), we have added only a single inversion. This is corrected in Figure 7-17(c) by adding two inverters and complementing two variables.

Note that when an inverter is added between two gates during the conversion procedure, the number of levels in the circuit is increased by 1. This is avoided if each path through the circuit alternately passes through AND and OR gates. Similarly, if the circuit containing AND and OR gates has an output OR (AND) gate and it is converted to a circuit with NOR (NAND) gates, then it is necessary to add an inverter at the output, which also increases the number of levels by 1. Hence, if a NAND (NOR) gate circuit is desired, it is usually best to start with a circuit containing AND and OR gates that has an output OR (AND) gate.

An advantage of multi-level circuits is that gate fan-in can be reduced. As an example, consider

$$F = D'E + BCE + AB' + AC' \tag{7-22}$$

A two-level AND-OR circuit implementing F requires one 4-input OR, one 3-input AND, and three 2-input ANDs. To reduce the fan-in, F can be factored.

$$F = A(B' + C') + E(D' + BC) \tag{7-23}$$

The resulting four-level circuit using AND and OR gates is shown in Figure 7-18.

FIGURE 7-18 Limited Fan-In Circuit

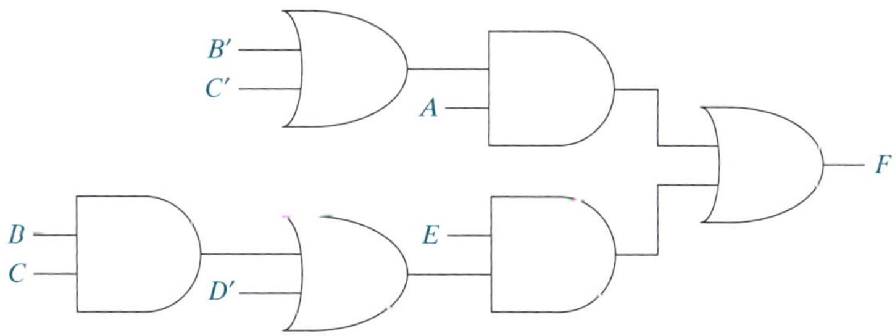

Since the output gate is an OR, the circuit can be converted to NAND gates without increasing the number of levels; Figure 7-19 is the result. Note that the three-level OR with inputs B' and C' and the four-level AND with inputs B and C both become a NAND with inputs B and C; hence, both can be replaced by the same gate.

FIGURE 7-19 NAND Gate Equivalent of Figure 7-18

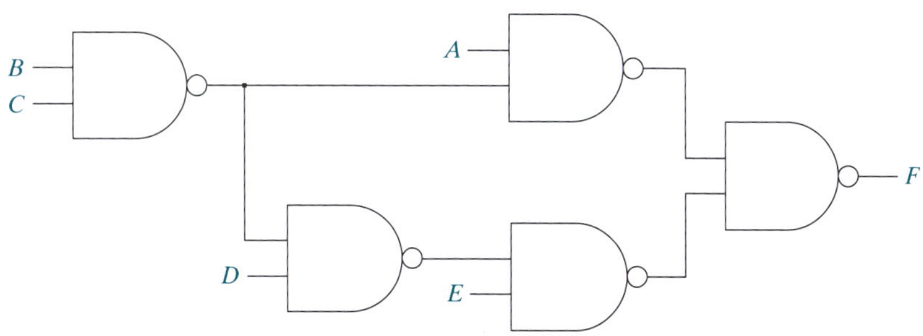

Reducing the fan-in for some functions requires inserting inverters. The fan-in for $F = ABC + D$ can be reduced to 2 by factoring F as $F = (AB)C + D$. If this is implemented using two-input NAND gates, an inverter is required and the resulting circuit has four levels.

7.6 Design of Two-Level, Multiple-Output Circuits

Solution of digital design problems often requires the realization of several functions of the same variables. Although each function could be realized separately, the use of some gates in common between two or more functions sometimes leads to a more economical realization. The following example illustrates this:

Design a circuit with four inputs and three outputs which realizes the functions

$$
\begin{aligned}
F_1(A, B, C, D) &= \Sigma\, m(11, 12, 13, 14, 15)\\
F_2(A, B, C, D) &= \Sigma\, m(3, 7, 11, 12, 13, 15)\\
F_3(A, B, C, D) &= \Sigma\, m(3, 7, 12, 13, 14, 15) \qquad (7\text{-}24)
\end{aligned}
$$

First, each function will be realized individually. The Karnaugh maps, functions, and resulting circuit are given in Figures 7-20 and 7-21. The cost of this circuit is 9 gates and 21 gate inputs.

An obvious way to simplify this circuit is to use the same gate for AB in both F_1 and F_3. This reduces the cost to eight gates and 19 gate inputs. (Another, but less obvious, way to simplify the circuit is possible.) Observing that the term ACD is necessary for the realization of F_1 and $A'CD$ is necessary for F_3, if we replace CD in F_2 by $A'CD + ACD$, the realization of CD is unnecessary and one gate is saved. Figure 7-22 shows the reduced circuit, which requires seven gates and 18 gate inputs. Note that F_2 is realized by the expression $ABC' + A'CD + ACD$, which is not a minimum sum of products, and two of the terms are not prime implicants of F_2. Thus in realizing multiple-output circuits, the use of a minimum sum of prime implicants

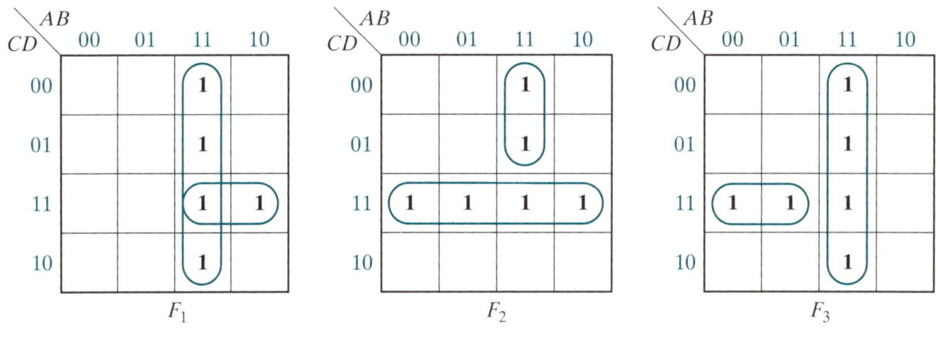

FIGURE 7-20 Karnaugh Maps for Equations (7-24)

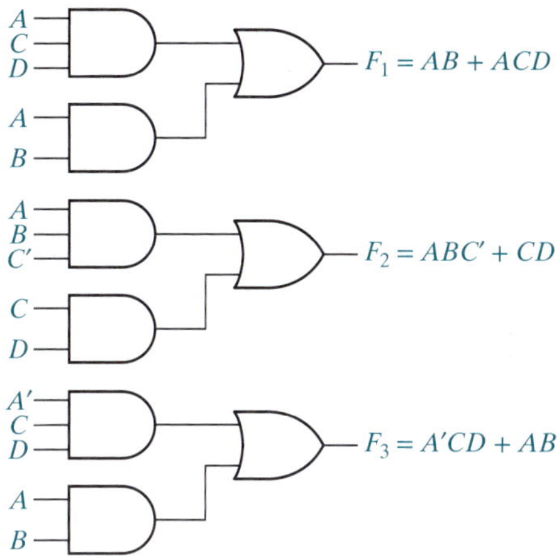

FIGURE 7-21 Realization of Equations (7-24)

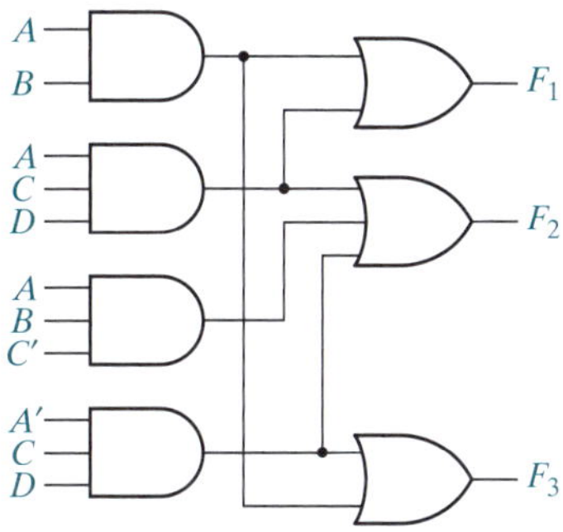

FIGURE 7-22 Multiple-Output Realization of Equations (7-24)

for each function does not necessarily lead to a minimum cost solution for the circuit as a whole.

When designing multiple-output circuits, you should try to minimize the total number of gates required. If several solutions require the same number of gates, the one with the minimum number of gate inputs should be chosen. The next example further illustrates the use of common terms to save gates. A four-input, three-output circuit is to be designed to realize

$$\begin{aligned} f_1 &= \Sigma\, m(2, 3, 5, 7, 8, 9, 10, 11, 13, 15) \\ f_2 &= \Sigma\, m(2, 3, 5, 6, 7, 10, 11, 14, 15) \\ f_3 &= \Sigma\, m(6, 7, 8, 9, 13, 14, 15) \end{aligned} \tag{7-25}$$

First, we plot maps for f_1, f_2, and f_3 (Figure 7-23). If each function is minimized separately, the result is

$$\begin{aligned} f_1 &= bd + b'c + ab' \\ f_2 &= c + a'bd \\ f_3 &= bc + ab'c' + \begin{Bmatrix} abd \\ \text{or} \\ ac'd \end{Bmatrix} \begin{matrix} \text{10 gates,} \\ \text{25 gate inputs} \end{matrix} \end{aligned} \tag{7-25(a)}$$

FIGURE 7-23

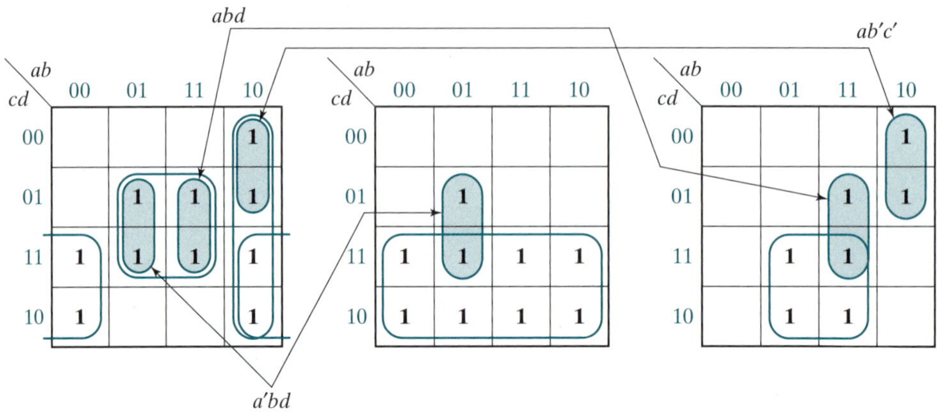

By inspecting the maps, we can see that terms $a'bd$ (from f_2), abd (from f_3), and $ab'c'$ (from f_3) can be used in f_1. If bd is replaced with $a'bd + abd$, then the gate needed to realize bd can be eliminated. Because m_{10} and m_{11} in f_1 are already covered by $b'c$, $ab'c'$ (from f_3) can be used to cover m_8 and m_9, and the gate needed to realize ab' can be eliminated. The minimal solution is therefore

$$\begin{aligned} f_1 &= \underline{a'bd} + \underline{abd} + \underline{ab'c'} + b'c \\ f_2 &= c + \underline{a'bd} && \text{eight gates} \qquad (7\text{-}25(b)) \\ f_3 &= bc + \underline{ab'c'} + \underline{abd} && \text{22 gate inputs} \end{aligned}$$

(Terms which are used in common between two functions are underlined.)

When designing multiple-output circuits, it is sometimes best not to combine a 1 with its adjacent 1's, as illustrated in the example of Figure 7-24.

The solution with the maximum number of common terms is not necessarily best, as illustrated in the example of Figure 7-25.

Determination of Essential Prime Implicants for Multiple-Output Realization

As a first step in determining a minimum two-level, multiple-output realization, it is often desirable to determine essential prime implicants. However, we must be careful because some of the prime implicants essential to an individual function may not be essential to the multiple-output realization. For example, in Figure 7-23, bd is an essential prime implicant of f_1 (only prime implicant which covers m_5), but it is not essential to the multiple-output realization. The reason that bd is *not* essential is that m_5 also appears on the f_2 map and, hence, might be covered by a term which is shared by f_1 and f_2.

We can find prime implicants which are essential to one of the functions *and* to the multiple-output realization by a modification of the procedure used for the single-output case. In particular, when we check each 1 on the map to see if it is covered by only one prime implicant, we will only check those 1's which do not appear on the other function maps. Thus, in Figure 7-24 we find that $c'd$ *is* essential to f_1 for the multiple-output realization (because of m_1), but abd is not essential because m_{15} also

FIGURE 7-24

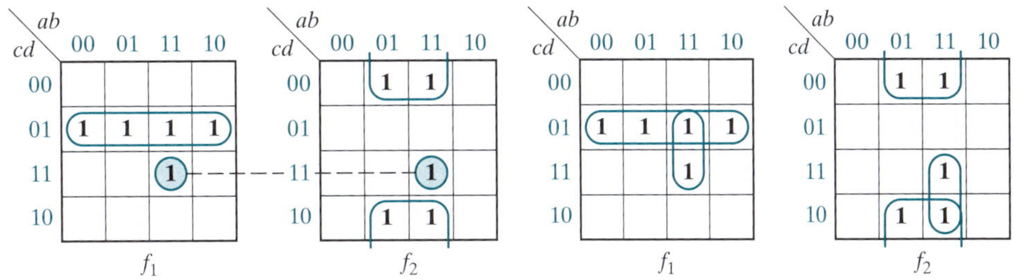

(a) Best solution

(b) Solution requires an extra gate

FIGURE 7-25

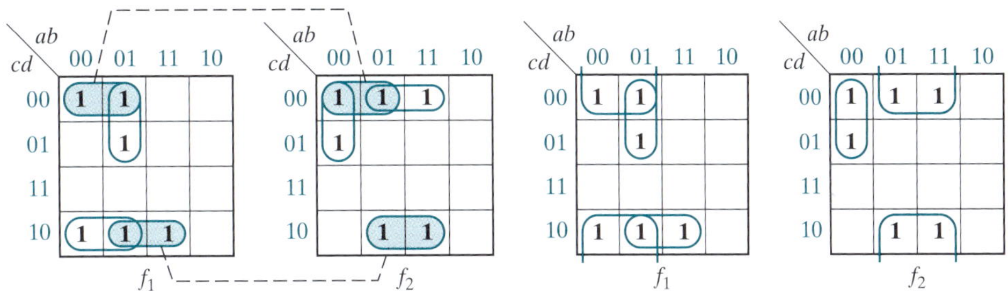

(a) Solution with maximum number of common terms requires 8 gates, 26 inputs

(b) Best solution requires 7 gates, 18 inputs and has no common terms

appears on the f_2 map. In Figure 7-25, the only minterms of f_1 which do not appear on the f_2 map are m_2 and m_5. The only prime implicant which covers m_2 is $a'd'$; hence, $a'd'$ *is* essential to f_1 in the multiple-output realization. Similarly, the only prime implicant which covers m_5 is $a'bc'$, and $a'bc'$ is essential. On the f_2 map, bd' *is* essential. Why?

Once the essential prime implicants for f_1 and f_2 have been looped, selection of the remaining terms to form the minimum solution is obvious in this example. The techniques for finding essential prime implicants outlined above cannot be applied in a problem such as Figure 7-23, where every minterm of f_1 also appears on the f_2 or f_3 map.

A general procedure for finding the minimum multiple output AND-OR circuit requires finding the prime implicants of not only each function but, also, of the product of all functions. If three functions f_1, f_2, and f_3 are being realized, then the prime implicants of $f_1, f_2, f_3, f_1f_2, f_1f_3, f_2f_3$, and $f_1f_2f_3$ are required. The optimum solution is obtained by selecting the fewest prime implicants from these prime implicants to realize f_1, f_2, and f_3. This procedure is not discussed further in this text.

7.7 Multiple-Output NAND- and NOR-Gate Circuits

The procedure given in Section 7.4 for design of single-output, multi-level NAND- and NOR-gate circuits also applies to multiple-output circuits. If all of the output gates are OR gates, direct conversion to a NAND-gate circuit is possible. If all of the output gates are AND, direct conversion to a NOR-gate circuit is possible. Figure 7-26 gives an example of converting a two-output circuit to NOR gates. Note that the inputs to the first and third levels of NOR gates are inverted.

$$F_1 = [(a + b')c + d](e' + f) \qquad F_2 = [(a + b')c + g'](e' + f)h$$

FIGURE 7-26 Multi-Level Circuit Conversion to NOR Gates

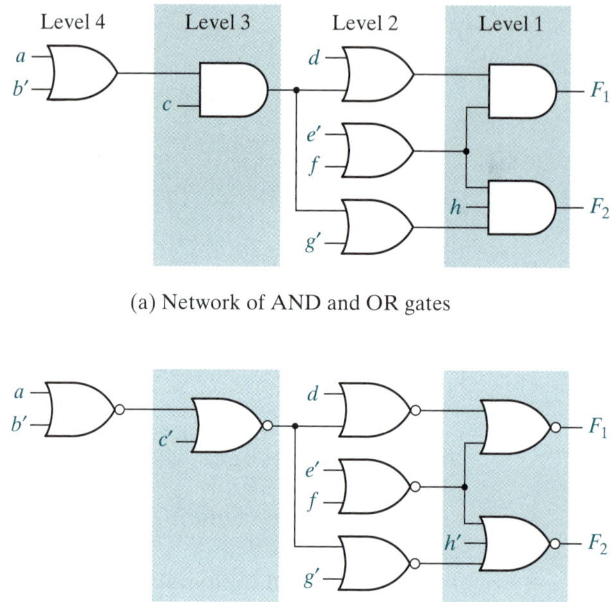

(a) Network of AND and OR gates

(b) NOR network

Problems

7.1 Using AND and OR gates, find a minimum circuit to realize

$$f(a, b, c, d) = m_4 + m_6 + m_7 + m_8 + m_9 + m_{10}$$

(a) using two-level logic
(b) using three-level logic (12 gate inputs minimum)

7.2 Realize the following functions using AND and OR gates. Assume that there are no restrictions on the number of gates which can be cascaded and minimize the number of gate inputs.

(a) $AC'D + ADE' + BE' + BC' + A'D'E'$
(b) $AE + BDE + BCE + BCFG + BDFG + AFG$

7.3 Find eight different simplified two-level gate circuits to realize

$$F(a, b, c, d) = a'bd + ac'd$$

7.4 Find a minimum three-level NAND-gate circuit to realize

$$F(A, B, C, D) = \Sigma\, m(5, 10, 11, 12, 13) \quad \text{(four gates)}$$

7.5 Realize $Z = A'D + A'C + AB'C'D'$ using four NOR gates.

7.6 Realize $Z = ABC + AD + C'D'$ using only two-input NAND gates. Use as few gates as possible.

7.7 Realize $Z = AE + BDE + BCEF$ using only two-input NOR gates. Use as few gates as possible.

7.8 (a) Convert the following circuit to all NAND gates, by adding bubbles and inverters where necessary.
(b) Convert to all NOR gates (an inverter at the output is allowed).

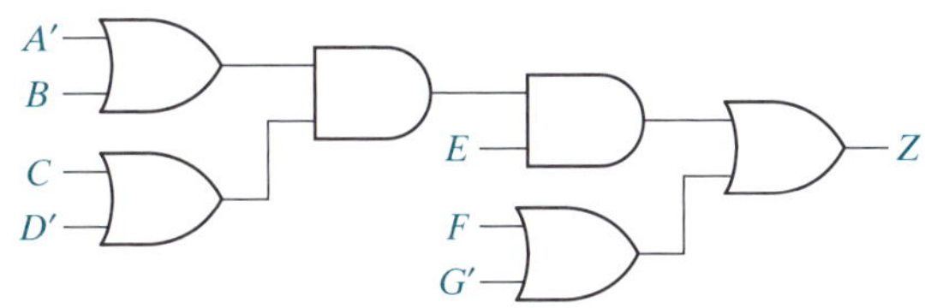

7.9 Find a two-level, multiple-output AND-OR gate circuit to realize the following functions. Minimize the required number of gates (six gates minimum).

$$f_1 = ac + ad + b'd \quad \text{and} \quad f_2 = a'b' + a'd' + cd'$$

7.10 Find a minimum two-level, multiple-output AND-OR gate circuit to realize these functions.

$$\begin{aligned} f_1(a, b, c, d) &= \Sigma\, m(3, 4, 6, 9, 11) \\ f_2(a, b, c, d) &= \Sigma\, m(2, 4, 8, 10, 11, 12) \\ f_3(a, b, c, d) &= \Sigma\, m(3, 6, 7, 10, 11) \qquad \text{(11 gates minimum)} \end{aligned}$$

7.11 Find a minimum two-level OR-AND circuit to simultaneously realize

$$\begin{aligned} F_1(a, b, c, d) &= \Sigma\, m(2, 3, 8, 9, 14, 15) \\ F_2(a, b, c, d) &- \Sigma\, m(0, 1, 5, 8, 9, 14, 15) \end{aligned}$$

(minimum solution has eight gates)

7.12 Find a minimum two-level OR-AND circuit to realize the functions given in Equations (7-25) on page 215 (nine gates minimum).

7.13 (a) Find a minimum two-level NAND-NAND circuit to realize the functions given in Equations (7-25) on page 215.
(b) Find a minimum two-level NOR-NOR circuit to realize the functions given in Equations (7-25).

7.14 Using AND and OR gates, find a minimum circuit to realize

$$f(a, b, c, d) = M_0 M_1 M_3 M_{13} M_{14} M_{15}$$

(a) using two-level logic
(b) using three-level logic (12 gate inputs minimum)

7.15 Using AND and OR gates, find a minimum two-level circuit to realize
(a) $F = a'c + bc'd + ac'd$
(b) $F = (b' + c)(a + b' + d)(a + b + c' + d)$
(c) $F = a'cd' + a'bc + ad$
(d) $F = a'b + ac + bc + bd'$

7.16 Realize the following functions using AND and OR gates. Assume that there are no restrictions on the number of gates which can be cascaded and minimize the number of gate inputs.
(a) $ABC' + ACD + A'BC + A'C'D$
(b) $ABCE + ABEF + ACD' + ABEG + ACDE$

7.17 A combinational switching circuit has four inputs (A, B, C, D) and one output (F). $F = 0$ iff three or four of the inputs are 0.
(a) Write the maxterm expansion for F.
(b) Using AND and OR gates, find a minimum three-level circuit to realize F (five gates, 12 inputs).

7.18 Find eight different simplified two-level gate circuits to realize
(a) $F(w, x, y, z) = (x + y' + z)(x' + y + z)w$
(b) $F(a, b, c, d) = \Sigma\, m(4, 5, 8, 9, 13)$

7.19 Implement $f(x, y, z) = \Sigma\, m(0, 1, 3, 4, 7)$ as a two-level gate circuit, using a minimum number of gates.
(a) Use AND gates and NAND gates.
(b) Use NAND gates only.

7.20 Implement $f(a, b, c, d) = \Sigma\, m(3, 4, 5, 6, 7, 11, 15)$ as a two-level gate circuit, using a minimum number of gates.
(a) Use OR gates and NOR gates.
(b) Use NOR gates only.

7.21 Realize each of the following functions as a minimum two-level NAND-gate circuit and as a minimum two-level NOR-gate circuit.
(a) $F(A,B,C,D) = BD' + B'CD + A'BC + A'BC'D + B'D'$
(b) $f(a, b, c, d) = \Pi\, M(0, 1, 7, 9, 10, 13) \cdot \Pi\, D(2, 6, 14, 15)$
(c) $f(a, b, c, d) = \Sigma\, m(0, 2, 5, 10) + \Sigma\, d(3, 6, 9, 13, 14, 15)$
(d) $F(A, B, C, D, E) = \Sigma\, m(0, 2, 4, 5, 11, 14, 16, 17, 18, 22, 23, 25, 26, 31)$
$+ \Sigma\, d(3, 19, 20, 27, 28)$

(e) $F(A, B, C, D, E) = \Pi\, M(3, 4, 8, 9, 10, 11, 12, 13, 14, 16, 19, 22, 25, 27) \cdot \Pi\, D(17, 18, 28, 29)$

(f) $f(a, b, c, d) = \Pi\, M(1, 3, 10, 11, 13, 14, 15) \cdot \Pi\, D(4, 6)$

(g) $f(w, x, y, z) = \Sigma\, m(1, 2, 4, 6, 8, 9, 11, 12, 13) + \Sigma\, d(0, 7, 10, 15)$

7.22 A combinational switching circuit has four inputs and one output as shown. $F = 0$ iff three or four of the inputs are 1.

(a) Write the maxterm expansion for F.

(b) Using AND and OR gates, find a minimum three-level circuit to realize F (5 gates, 12 inputs).

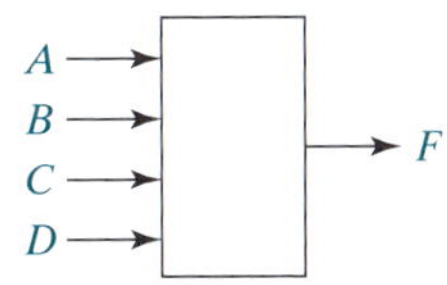

7.23 Implement $f(a, b, c, d) = \Sigma\, m(3, 4, 5, 6, 7, 11, 15)$ as a two-level gate circuit, using a minimum number of gates.

(a) Use AND gates and NAND gates.

(b) Use OR gates and NAND gates.

(c) Use NAND gates only.

7.24 (a) Use gate equivalences to convert the circuit into a four-level circuit containing only NAND gates and a minimum number of inverters. (Assume the inputs are available only in uncomplemented form.)

(b) Derive a minimum SOP expression for f.

(c) By manipulating the expression for f, find a three-level circuit containing only five NAND gates and inverters.

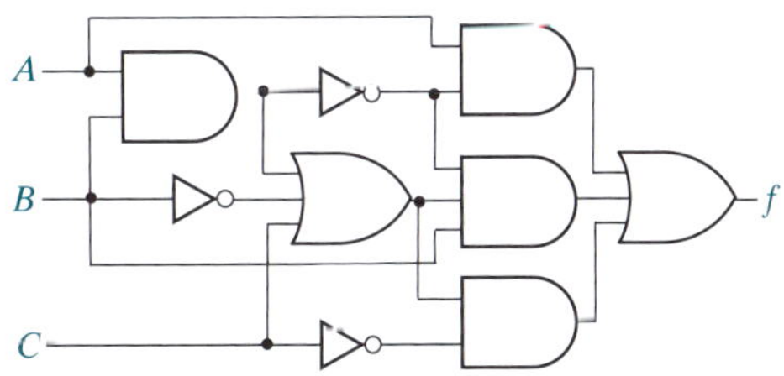

7.25 (a) Use gate equivalences to convert the circuit of Problem 7.24 into a five-level circuit containing only NOR gates and a minimum number of inverters. (Assume the inputs are available only in uncomplemented form.)

(b) Derive a minimum POS expression for f.

(c) By manipulating the expression for f, find a four-level circuit containing only six NOR gates and inverters.

7.26 In the circuit, replace each NOR gate by an AND or OR gate so that the resulting circuit contains the fewest inverters possible. Assume the inputs are available in both true and complemented form. Do *not* replace the exclusive-OR gates.

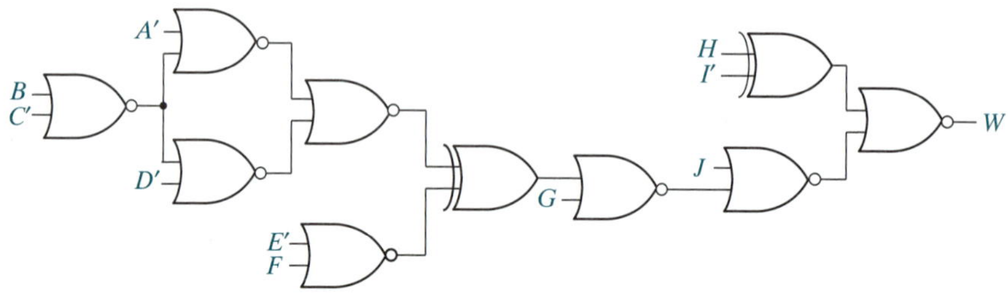

7.27 (a) Convert the circuit shown into a four-level circuit only containing AND and OR gates and a minimum number of inverters.
(b) Derive a sum-of-products expression for f.
(c) Find a circuit that realizes f' containing only NOR gates (no internal inverters). (*Hint:* Use gate conversions to convert the NAND gates in the given circuit to NOR gates.)

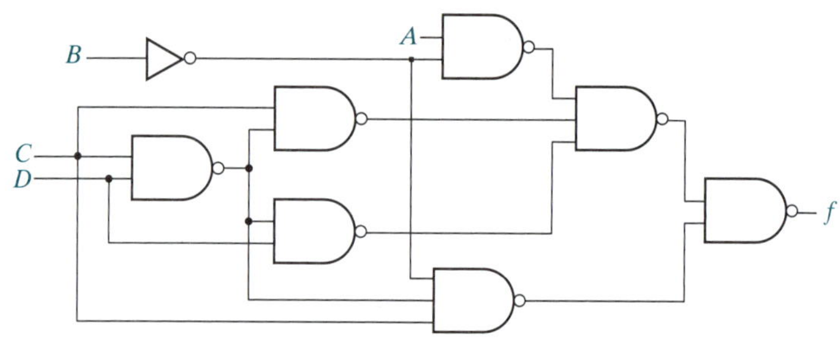

7.28 $f(a, b, c, d, e) = \Sigma\, m(2, 3, 6, 12, 13, 16, 17, 18, 19, 22, 24, 25, 27, 28, 29, 31)$
(a) Find a minimum two-level NOR-gate circuit to realize f.
(b) Find a minimum three-level NOR-gate circuit to realize f.

7.29 Design a minimum three-level NOR-gate circuit to realize

$$f = a'b' + abd + acd$$

7.30 Find a minimum four-level NAND- or NOR-gate circuit to realize
(a) $Z = abe'f + c'e'f + d'e'f + gh$
(b) $Z = (a' + b' + e + f)(c' + a' + b)(d' + a' + b)(g + h)$

7.31 Implement $abde' + a'b' + c$ using four NOR gates.

7.32 Implement $x'yz + xvy'w' + xvy'z'$ using a three-level NAND-gate circuit.

7.33 Design a logic circuit that has a 4-bit binary number as an input and one output. The output should be 1 iff the input is a prime number (greater than 1) or zero.
(a) Use a two-level NAND-gate circuit.
(b) Use a two-level NOR-gate circuit.
(c) Use only two-input NAND gates.

7.34 Work Problem 7.33 for a circuit that has an output 1 iff the input is evenly divisible by 3 (0 is divisible by 3).

7.35 Realize the following functions, using only two-input NAND gates. Repeat using only two-input NOR gates.
(a) $F = A'BC' + BD + AC + B'CD'$
(b) $F = A'CD + AB'C'D + ABD' + BC$

7.36 (a) Find a minimum circuit of two-input AND and two-input OR gates to realize $F(A, B, C, D) = \Sigma\, m(0, 1, 2, 3, 4, 5, 7, 9, 11, 13, 14, 15)$
(b) Convert your circuit to two-input NAND gates. Add inverters where necessary.
(c) Repeat (b), except convert to two-input NOR gates.

7.37 Realize $Z = A[BC' + D + E(F' + GH)]$ using NOR gates. Add inverters if necessary.

7.38 Show that the function of Equation (7-22) can be realized using four 2-input NOR gates and one 3-input NOR gate. Assume the inputs are available both complemented and uncomplemented. (No inverters are required.)

7.39 $F(A, B, C)$ equals 1 if exactly two of A, B, and C are 1.
(a) Find the minimum two-level OR-AND circuit for F.
(b) Find a four-level circuit for F that has six 2-input NOR gates.
(c) Find the minimum two level AND-OR circuit for F.
(d) Find a three-level circuit for F that has three 2-input NAND gates and two 2-input XOR gates.

7.40 (a) Use gate conversions to convert the circuit below into one containing NAND and XOR gates.
(b) Find the minimum two-level OR-AND circuit for F.
(c) Find a three-level circuit for F that has five 2-input NAND gates.

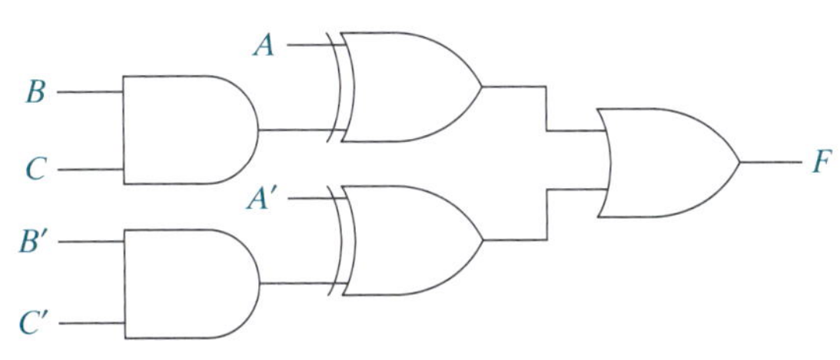

7.41 In which of the following two-level circuit forms can an arbitrary switching function be realized? Verify your answers. (Assume the inputs are available in both complemented and uncomplemented form.)

(a) NOR-AND
(b) NOR-OR
(c) NOR-NAND
(d) NOR-XOR
(e) NAND-AND
(f) NAND-OR
(g) NAND-NOR
(h) NAND-XOR

7.42 Find a minimum two-level, multiple-output AND-OR gate circuit to realize these functions (eight gates minimum).

$$f_1(a, b, c, d) = \Sigma\, m(10, 11, 12, 15) + \Sigma\, d(4, 8, 14)$$
$$f_2(a, b, c, d) = \Sigma\, m(0, 4, 8, 9) + \Sigma\, d(1, 10, 12)$$
$$f_3(a, b, c, d) = \Sigma\, m(4, 11, 13, 14, 15) + \Sigma\, d(5, 9, 12)$$

7.43 Repeat 7.42 for the following functions (six gates).

$$f_1(a, b, c, d) = \Sigma\, m(2, 3, 5, 6, 7, 8, 10)$$
$$f_2(a, b, c, d) = \Sigma\, m(0, 1, 2, 3, 5, 7, 8, 10)$$

7.44 Repeat 7.42 for the following functions (eight gates).

$$f_1(x, y, z) = \Sigma\, m(2, 3, 4, 5)$$
$$f_2(x, y, z) = \Sigma\, m(1, 3, 5, 6)$$
$$f_3(x, y, z) = \Sigma\, m(1, 2, 4, 5, 6)$$

7.45 (a) Find a minimum two-level, multiple-output OR-AND circuit to realize $f_1 = b'd + a'b' + c'd$ and $f_2 = a'd' + bc' + bd'$.
(b) Realize the same functions with a minimum two-level NAND-NAND circuit.

7.46 Repeat Problem 7.45 for $f_1 = ac' + b'd + c'd$ and $f_2 = b'c + a'd + cd'$.

7.47 (a) Find a minimum two-level, multiple-output NAND-NAND circuit to realize $f_1 = \Sigma\, m(3, 6, 7, 11, 13, 14, 15)$ and $f_2 = \Sigma\, m(3, 4, 6, 11, 12, 13, 14)$.
(b) Repeat for a minimum two-level, NOR-NOR circuit.

7.48 (a) Find a minimum two-level, multiple-output NAND-NAND circuit to realize $f_1 = \Sigma\, m(0, 2, 4, 6, 7, 10, 14)$ and $f_2 = \Sigma\, m(0, 1, 4, 5, 7, 10, 14)$.
(b) Repeat for a minimum two-level, multiple-output NOR-NOR circuit.

7.49 Draw a multi-level, multiple-output circuit equivalent to Figure 7-26(a) using:
(a) NAND and AND gates
(b) NAND gates only (a direct conversion is not possible)

UNIT 8

Combinational Circuit Design and Simulation Using Gates

Objectives

1. Draw a timing diagram for a combinational circuit with gate delays.
2. Define static 0- and 1-hazards and dynamic hazards. Given a combinational circuit, find all of the static 0- and 1-hazards. For each hazard, specify the order in which the gate outputs must switch in order for the hazard to actually produce a false output.
3. Given a switching function, realize it using a two-level circuit which is free of static and dynamic hazards (for single input variable changes).
4. Design a multiple-output NAND or NOR circuit using gates with limited fan-in.
5. Explain the operation of a logic simulator that uses four-valued logic.
6. Test and debug a logic circuit design using a simulator.

Study Guide

1. Obtain your design problem assignment from your instructor.
2. Study Section 8.1, *Review of Combinational Circuit Design.*
3. Generally, it is possible to redesign a circuit which has two AND gates cascaded or two OR gates cascaded so that AND and OR gates alternate. If this is not practical, the conversion to a NAND or NOR circuit by the techniques of Section 7.4 is still possible by introducing a dummy one-input OR (AND) gate between the two AND (OR) gates. When the conversion is carried out, the dummy gate becomes an inverter. Try this technique and convert the following circuit to all NAND gates. Alternatively, you may use the procedures given in Section 7.5 to do the conversion.

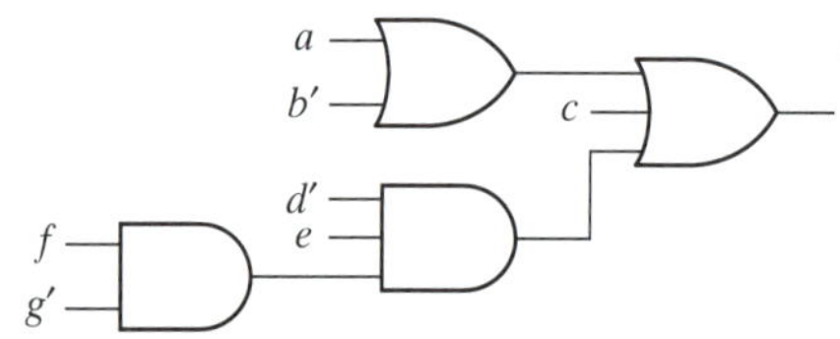

4. Study Section 8.2, *Design of Circuits with Limited Gate Fan-In.*
   (a) If a realization of a switching expression requires too many inputs on one or more gates, what should be done?

   (b) Assuming that all variables and their complements are available as inputs and that both AND and OR gates are available, does realizing the complement of an expression take the same number of gates and gate inputs as realizing the original expression?

   (c) When designing multiple-output circuits with limited gate fan-in, why is the procedure of Section 7.6 of little help?

5. (a) Study Section 8.3, *Gate Delays and Timing Diagrams.* Complete the timing diagram for the given circuit. Assume that the AND gate has a 30-nanosecond (ns) propagation delay and the inverter has a 20-ns delay.

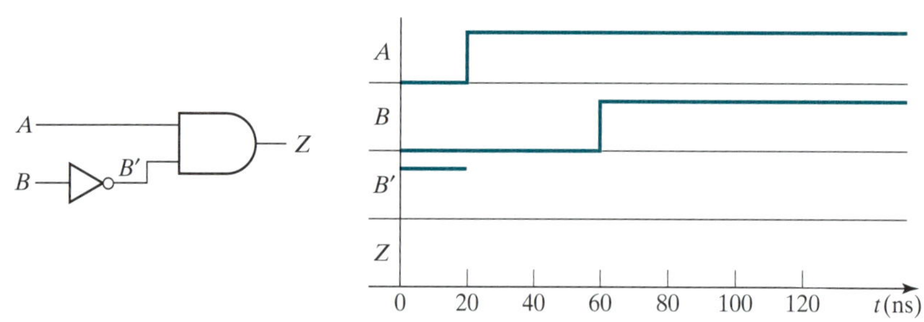

(b) Work Problem 8.1.

6. Study Section 8.4, *Hazards in Combinational Logic.*

(a) Even though all of the gates in a circuit are of the same type, each individual gate may have a different propagation delay. For example, for one type of TTL NAND gate the manufacturer specifies a minimum propagation delay of 5 ns and a maximum delay of 30 ns. Sketch the gate outputs for the following circuit when the x input changes from 1 to 0, assuming the following gate delays:

(a) gate 1–5 ns (b) gate 2–20 ns (c) gate 3–10 ns

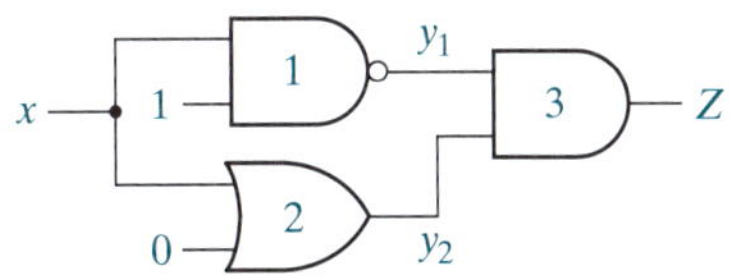

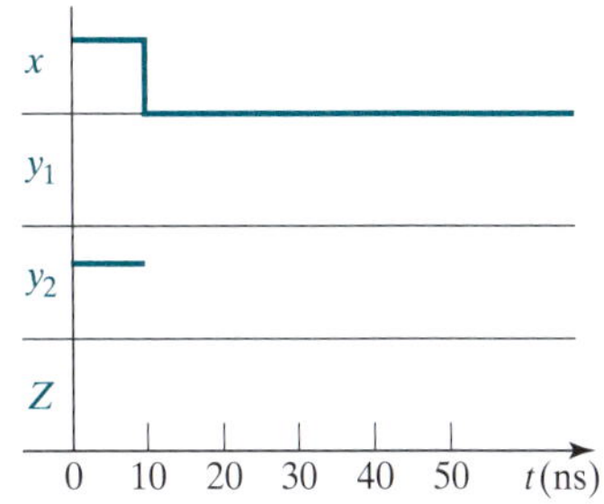

(b) Define static 0-hazard, static 1-hazard, and dynamic hazard.

(c) Using a Karnaugh map, explain why $F = a'b + ac$ has a 1-hazard for the input change abc = 011 to 111, but not for 011 to 010. Then explain it without using the map.

(d) Explain why $F = (a' + b')(b + c)$ has a 0-hazard for the input change abc = 100 to 110, but not for 100 to 000.

(e) Under what condition does a sum-of-products expression represent a hazard-free, two-level AND-OR circuit?

(f) Under what condition does a product-of-sums expression represent a hazard-free, two-level OR-AND circuit?

(g) If a hazard-free circuit of AND and OR gates is transformed to NAND or NOR gates using the procedure given in Unit 7, why will the results be hazard-free?

(h) Work Problems 8.2 and 8.3.

7. Study Section 8.5, *Simulation and Testing of Logic Circuits*.

   (a) Verify that Table 8-1 is correct. Consider both the case where the unknown value, X, is 0 and the case where it is 1.

   (b) The following circuit was designed to realize the function

$$F = [A' + B + C'D][A + B' + (C' + D')(C + D)]$$

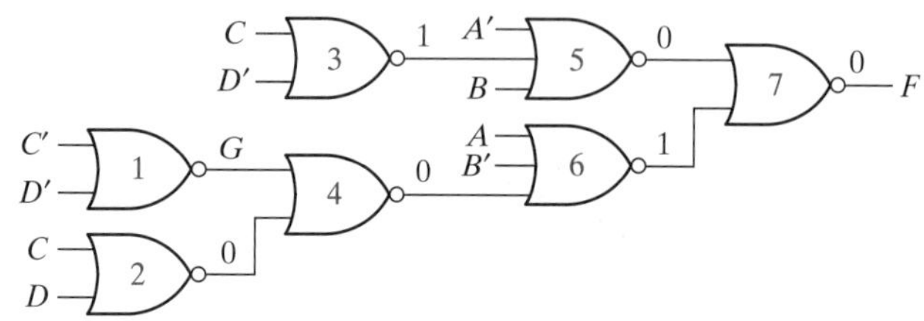

When a student builds the circuit in lab, he finds that when $A = C = 0$ and $B = D = 1$, the output F has the wrong value and that the gate outputs are as shown. Determine some possible causes of the incorrect output if $G = 0$ and if $G = 1$

   (c) Work Problems 8.4 and 8.5.

8. Study your assigned design problem and prepare a design which meets specifications. Note that only two-, three-, and four-input NAND gates (or NOR gates as specified) and inverters are available for this project; therefore, factoring some of the equations will be necessary. Try to make an economical design by using common terms; however, do not waste time trying to get an absolute minimum solution. When counting gates, count both NAND (or NOR) gates and inverters, but do not count the inverters needed for the input variables.

9. Check your design carefully before simulating it. Test it *on paper* by applying some input combinations of 0's and 1's and tracing the signals through to make sure that the outputs are correct. If you have a CAD program such as *LogicAid* available, enter the truth table for your design into the computer, derive the minimum two-level equations, and compare them with your solution.

10. In designing multi-level, multiple-output circuits of the type used in the design problems in this unit, it is very difficult and time-consuming to find a minimum solution. You are not expected to find the best possible solution to these problems. All of these solutions involve some "tricks," and it is unlikely that you could find them without trying a large number of different ways of factoring your equations. Therefore, if you already have an acceptable solution, do not waste time trying to find the minimum solution. Because integrated circuit gates are quite inexpensive, it is not good engineering practice to spend a large amount of time finding the absolute minimum solution unless a very large number of units of the same type are to be manufactured.

11. Obtain a Unit 8 supplement from your instructor and follow the instructions therein regarding simulating and testing your design.

Combinational Circuit Design and Simulation Using Gates

8.1 Review of Combinational Circuit Design

The first step in the design of a combinational switching circuit is usually to set up a truth table which specifies the output(s) as a function of the input variables. For n input variables this table will have 2^n rows. If a given combination of values for the input variables can never occur at the circuit inputs, the corresponding output values are don't-cares. The next step is to derive simplified algebraic expressions for the output functions using Karnaugh maps, the Quine-McCluskey method, or a similar procedure. In some cases, particularly if the number of variables is large and the number of terms is small, it may be desirable to go directly from the problem statement to algebraic equations, without writing down a truth table. The resulting equations can then be simplified algebraically. The simplified algebraic expressions are then manipulated into the proper form, depending on the type of gates to be used in realizing the circuit.

The number of levels in a gate circuit is equal to the maximum number of gates through which a signal must pass when going between the input and output terminals. The minimum sum of products (or product of sums) leads directly to a minimum two-level gate circuit. However, in some applications it is desirable to increase the number of levels by factoring (or multiplying out) because this may lead to a reduction in the number of gates or gate inputs.

When a circuit has two or more outputs, common terms in the output functions can often be used to reduce the total number of gates or gate inputs. If each function is minimized separately, this does not always lead to a minimum multiple-output circuit. For a two-level circuit, Karnaugh maps of the output functions can be used to find the common terms. All of the terms in the minimum multiple-output circuit will not necessarily be prime implicants of the individual functions. When designing circuits with three or more levels, looking for common terms on the Karnaugh maps may be of little value. In this case, the designer will often minimize the functions separately and, then, use ingenuity to factor the expressions in such a way to create common terms.

Minimum two-level AND-OR, NAND-NAND, OR-NAND, and NOR-OR circuits can be realized using the minimum sum of products as a starting point. Minimum two-level OR-AND, NOR-NOR, AND-NOR, and NAND-AND circuits can be realized using the minimum product of sums as a starting point. Design of multi-level, multiple-output NAND-gate circuits is most easily accomplished by first designing a circuit of AND and OR gates. Usually, the best starting point is the minimum sum-of-products expressions for the output functions. These expressions are then factored in various ways until an economical circuit of the desired form can be found. If this circuit has an OR gate at each output and is arranged so that an AND-gate (or OR-gate) output is never connected to the same type of gate, a direct conversion to a NAND-gate circuit is possible. Conversion is accomplished by replacing all of the AND and OR gates with NAND gates and then inverting any literals which appear as inputs to the first, third, fifth, . . . levels (output gates are the first level).

If the AND-OR circuit has an AND-gate (or OR-gate) output connected to the same type of gate, then extra inverters must be added in the conversion process (see Section 7.5, *Circuit Conversion Using Alternative Gate Symbols*).

Similarly, design of multi-level, multiple-output NOR-gate circuits is most easily accomplished by first designing a circuit of AND and OR gates. In this case the best starting point is usually the minimum sum-of-products expressions for the *complements* of the output functions. *After* factoring these expressions to the desired form, they are then complemented to get expressions for the output functions, and the corresponding circuit of AND and OR gates is drawn. If this circuit has an AND gate at each output, and an AND-gate (or OR-gate) output is never connected to the same type of gate, a direct conversion to a NOR-gate circuit is possible. Otherwise, extra inverters must be added in the conversion process.

8.2 Design of Circuits with Limited Gate Fan-In

In practical logic design problems, the maximum number of inputs on each gate (or the fan-in) is limited. Depending on the type of gates used, this limit may be two, three, four, eight, or some other number. If a two-level realization of a circuit requires more gate inputs than allowed, factoring the logic expression to obtain a multi-level realization is necessary.

Example

Realize $f(a, b, c, d) = \Sigma m(0, 3, 4, 5, 8, 9, 10, 14, 15)$ using three-input NOR gates.

map of f:

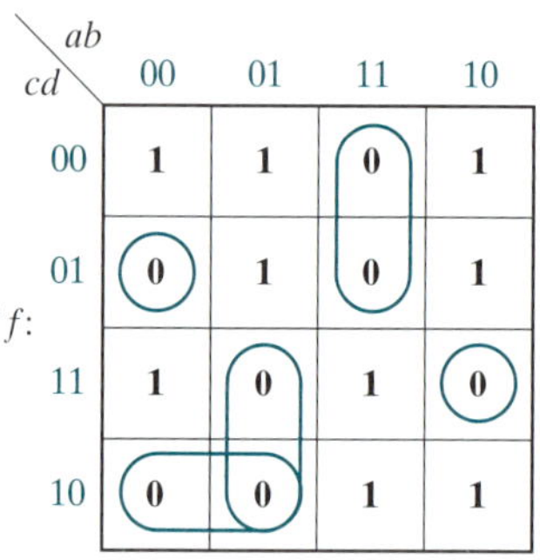

$$f' = a'b'c'd + ab'cd + abc' + a'bc + a'cd'$$

As can be seen from the preceding expression, a two-level realization requires two four-input gates and one five-input gate. The expression for f' is factored to reduce the maximum number of gate inputs to three and, then, it is complemented:

$$f' = b'd(a'c' + ac) + a'c(b + d') + abc'$$
$$f = [b + d' + (a + c)(a' + c')][a + c' + b'd][a' + b' + c]$$

The resulting NOR-gate circuit is shown in Figure 8-1.

FIGURE 8-1

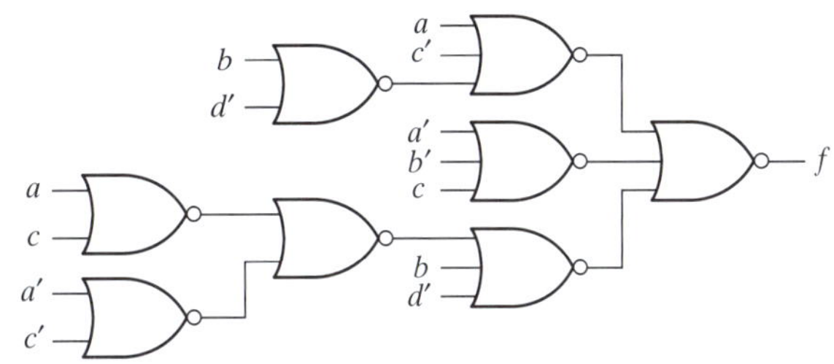

The techniques for designing two-level, multiple-output circuits given in Section 7.6 are not very effective for designing multiple-output circuits with more than two levels. Even if the two-level expressions had common terms, most of these common terms would be lost when the expressions were factored. Therefore, when designing multiple-output circuits with more than two levels, it is usually best to minimize each function separately. The resulting two-level expressions must then be factored to increase the number of levels. This factoring should be done in such a way as to introduce common terms wherever possible.

Example

Realize the functions given in Figure 8-2, using only two-input NAND gates and inverters. If we minimize each function separately, the result is

$$f_1 = b'c' + ab' + a'b$$
$$f_2 = b'c' + bc + a'b$$
$$f_3 = a'b'c + ab + bc'$$

FIGURE 8-2

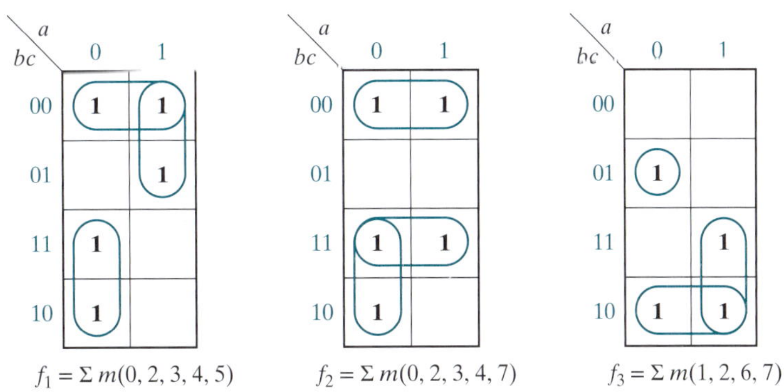

Each function requires a three-input OR gate, so we will factor to reduce the number of gate inputs:

$$f_1 = b'(\underline{a + c'}) + \underline{a'b}$$
$$f_2 = b(a' + c) + b'c' \qquad \text{or} \qquad f_2 = (b' + c)(b + c') + \underline{a'b}$$
$$f_3 = a'b'c + b(\underline{a + c'})$$

The second expression for f_2 has a term common to f_1, so we will choose the second expression. We can eliminate the remaining three-input gate from f_3 by noting that

$$a'b'c = a'(b'c) = a'(b + c')'$$

Figure 8-3(a) shows the resulting circuit, using common terms $a'b$ and $a + c'$. Because each output gate is an OR, the conversion to NAND gates, as shown in Figure 8-3(b), is straightforward.

FIGURE 8-3 Realization of Figure 8-2

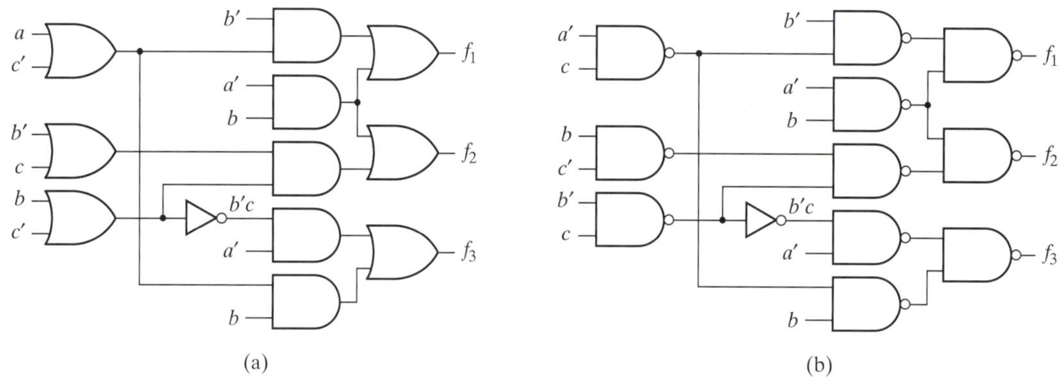

8.3 Gate Delays and Timing Diagrams

When the input to a logic gate is changed, the output will not change instantaneously. The transistors or other switching elements within the gate take a finite time to react to a change in input, so that the change in the gate output is delayed with respect to the input change. Figure 8-4 shows possible input and output waveforms for an inverter. If the change in output is delayed by time, ε, with respect to the input, we say that this gate has a propagation delay of ε. In practice, the propagation delay for a 0 to 1 output change may be different than the delay for a 1 to 0 change. Propagation delays for integrated circuit gates may be as short as a few nanoseconds (1 nanosecond $= 10^{-9}$ second), and in many cases these delays can be neglected. However, in the analysis of some types of sequential circuits, even short delays may be important.

Timing diagrams are frequently used in the analysis of sequential circuits. These diagrams show various signals in the circuit as a function of time. Several variables are usually plotted with the same time scale so that the times at which these variables change with respect to each other can easily be observed.

FIGURE 8-4 Propagation Delay in an Inverter

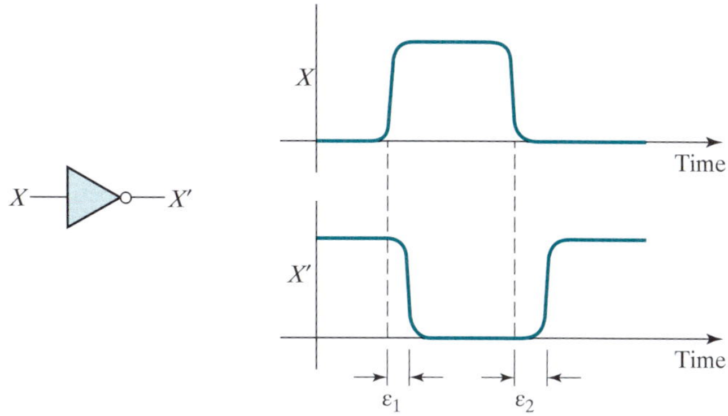

Figure 8-5 shows the timing diagram for a circuit with two gates. We will assume that each gate has a propagation delay of 20 ns (nanoseconds). This timing diagram indicates what happens when gate inputs B and C are held at constant values 1 and 0, respectively, and input A is changed to 1 at $t = 40$ ns and then changed back to 0 at $t = 100$ ns. The output of gate G_1 changes 20 ns after A changes, and the output of gate G_2 changes 20 ns after G_1 changes.

Figure 8-6 shows a timing diagram for a circuit with an added delay element. The input X consists of two pulses, the first of which is 2 microseconds (2×10^{-6} second) wide and the second is 3 microseconds wide. The delay element has an output Y

FIGURE 8-5 Timing Diagram for AND-NOR Circuit

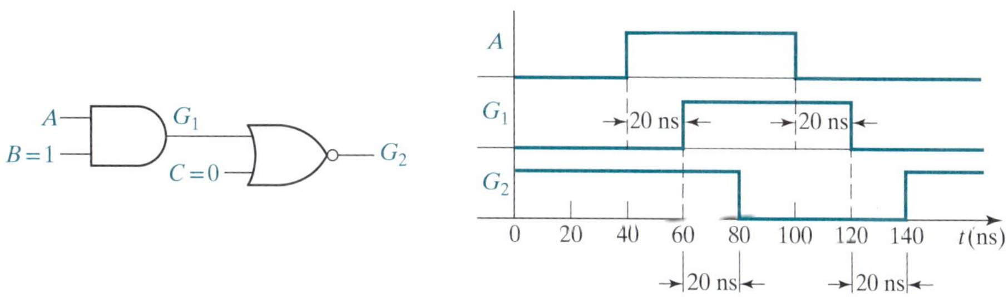

which is the same as the input except that it is delayed by 1 microsecond. That is, Y changes to a 1 value 1 microsecond after the rising edge of the X pulse and returns to 0 1 microsecond after the falling edge of the X pulse. The output (Z) of the AND gate should be 1 during the time interval in which both X and Y are 1. If we assume a small propagation delay in the AND gate (ε), then Z will be as shown in Figure 8-6.

FIGURE 8-6 Timing Diagram for Circuit with Delay

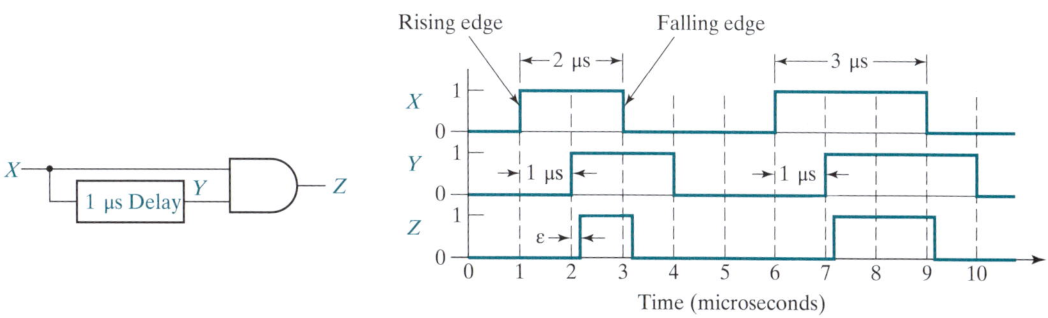

8.4 Hazards in Combinational Logic

When the input to a combinational circuit changes, unwanted switching transients may appear in the output. These transients occur when different paths from input to output have different propagation delays. If, in response to any single input change and for some combination of propagation delays, a circuit output may momentarily go to 0 when it should remain a constant 1, we say that the circuit has a static 1-hazard. Similarly, if the output may momentarily go to 1 when it should remain a 0, we say that the circuit has a static 0-hazard. If, when the output is supposed to change from 0 to 1 (or 1 to 0), the output may change three or more times, we say that the circuit has a dynamic hazard. Figure 8-7 shows possible outputs from a circuit with hazards. In each case the steady-state output of the circuit is correct, but a switching transient appears at the circuit output when the input is changed.

FIGURE 8-7 Types of Hazards

(a) Static 1-hazard (b) Static 0-hazard (c) Dynamic hazards

Figure 8-8(a) illustrates a circuit with a static 1-hazard. If $A = C = 1$, then $F = B + B' = 1$, so the F output should remain a constant 1 when B changes from 1 to 0. However, as shown in Figure 8-8(b), if each gate has a propagation delay of 10 ns, E will go to 0 before D goes to 1, resulting in a momentary 0 (a glitch caused by the 1-hazard) appearing at the output F. Note that right after B changes to 0, both the inverter input (B) and output (B') are 0 until the propagation delay has elapsed. During this period, both terms in the equation for F are 0, so F momentarily goes to 0.

Note that hazards are properties of the circuit and are independent of the delays existing in the circuit. If the circuit is free of hazards, then for any combination of delays that might exist in the circuit and for any single input change, the output will not contain a transient. On the other hand, if a circuit contains a hazard, then there is some combination of delays and some input change for which the circuit output contains a transient. The combination of delays that produces the transient may or may not be likely to occur in an implementation of the circuit; in some cases it is very unlikely that such delays would occur.

Besides depending on the delays existing in a circuit, the occurrence of transients depends on how gates respond to input changes. In some cases, if multiple input changes to a gate occur within a short time period, a gate may not respond to the input changes. For example, in Figure 8-8 assume the inverter has a delay of 2 ns rather than 10 ns. Then the D and E changes reaching the output OR gate are 2 ns apart, in which case the OR gate may not generate the 0 glitch. A gate exhibiting

FIGURE 8-8
Detection of a 1-Hazard

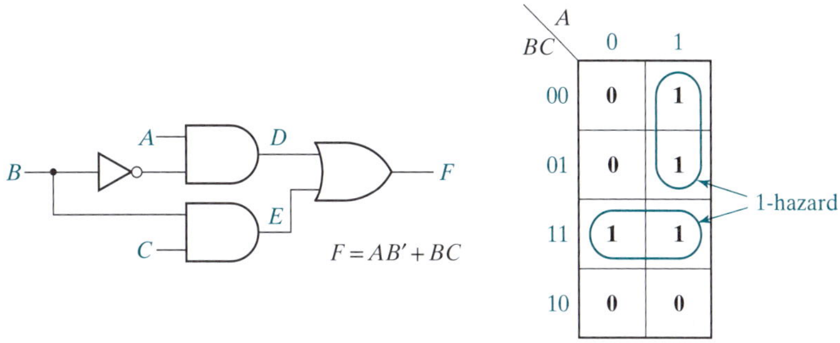

(a) Circuit with a static 1-hazard

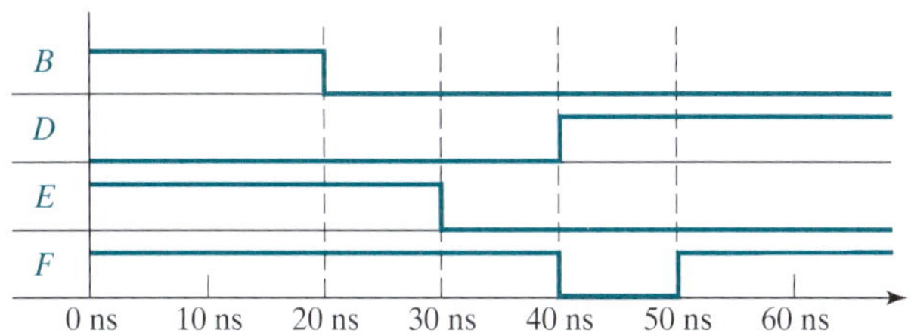

(b) Timing chart

this behavior is said to have an *inertial* delay. Quite often the inertial delay value is assumed to be the same as the propagation delay of the gate; if this is the case, the circuit of Figure 8-8 will generate the 0 glitch only for inverter delays greater than 10 ns. In contrast, if a gate always responds to input changes (with a propagation delay), no matter how closely spaced the input changes may be, the gate is said to have an *ideal* or *transport* delay. If the OR gate in Figure 8-8 has an ideal delay, then the 0 glitch would be generated for any nonzero value of the inverter delay. (Inertial and transport delay models are discussed more in Unit 10.) Unless otherwise noted, the examples and problems in this unit assume that gates have an ideal delay.

Hazards can be detected using a Karnaugh map (Figure 8-8(a)). As seen on the map, no loop covers both minterms ABC and $AB'C$. So if $A = C = 1$ and B changes, both terms can momentarily go to 0, resulting in a glitch in F. We can detect hazards in a two-level AND-OR circuit, using the following procedure:

1. Write down the sum-of-products expression for the circuit.
2. Plot each term on the map and loop it.
3. If any two adjacent 1's are not covered by the same loop, a 1-hazard exists for the transition between the two 1's. For an n-variable map, this transition occurs when one variable changes and the other $n - 1$ variables are held constant.

If we add a loop to the map of Figure 8-8(a) and, then, add the corresponding gate to the circuit (Figure 8-9), this eliminates the hazard. The term AC remains 1 while B is changing, so no glitch can appear in the output. Note that F is no longer a minimum sum of products.

FIGURE 8-9
Circuit with Hazard Removed

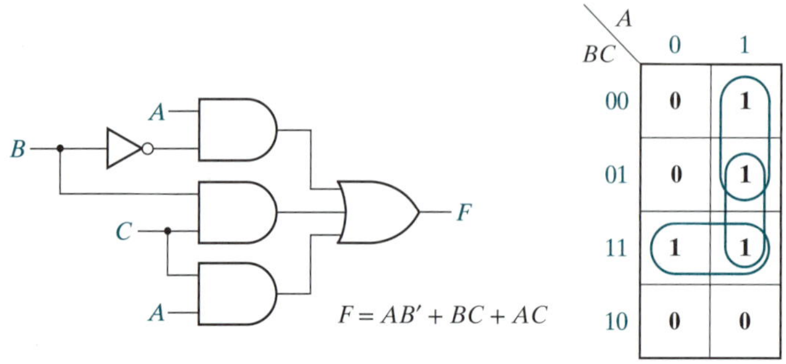

Figure 8-10(a) shows a circuit with several 0-hazards. The product-of-sums representation for the circuit output is

$$F = (A + C)(A' + D')(B' + C' + D)$$

The Karnaugh map for this function (Figure 8-10(b)) shows four pairs of adjacent 0's that are not covered by a common loop as indicated by the arrows. Each of these pairs corresponds to a 0-hazard. For example, when $A = 0$, $B = 1$, $D = 0$, and C changes from 0 to 1, a spike may appear at the Z output for some combination of gate delays. The timing diagram of Figure 8-10(c) illustrates this, assuming gate delays of 3 ns for each inverter and of 5 ns for each AND gate and each OR gate.

FIGURE 8-10
Detection of a Static 0-Hazard

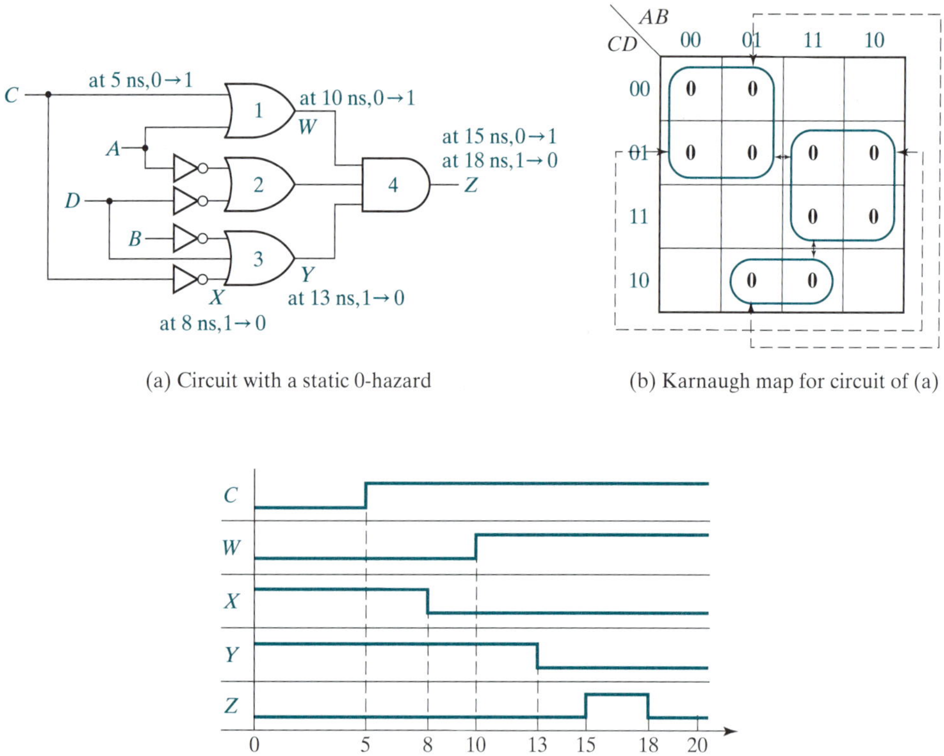

(a) Circuit with a static 0-hazard

(b) Karnaugh map for circuit of (a)

(c) Timing diagram illustrating 0-hazard of (a)

We can eliminate the 0-hazards by looping additional prime implicants that cover the adjacent 0's that are not already covered by a common loop. This requires three additional loops as shown in Figure 8-11. The resulting equation is

$$F = (A + C)(A' + D')(B' + C' + D)(C + D')(A + B' + D)(A' + B' + C')$$

and the resulting circuit requires seven gates in addition to the inverters.

FIGURE 8-11 Karnaugh Map Removing Hazards of Figure 8-10

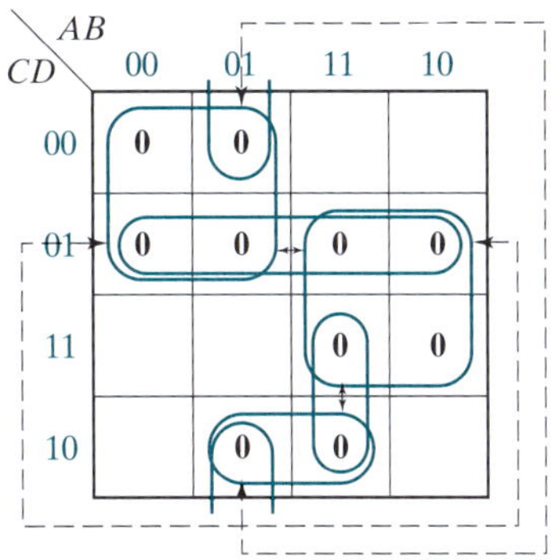

Hazards in circuits with more than two levels can be determined by deriving either a SOP or POS expression for the circuit that represents a two-level circuit containing the same hazards as the original circuit. The SOP or POS expression is derived in the normal manner except that the complementation laws are *not* used, i.e., $xx' = 0$ and $x + x' = 1$ are not used. Consequently, the resulting SOP (POS) expression may contain products (sums) of the form $xx'\alpha(x + x' + \beta)$. ($\alpha$ is a product of literals or it may be null; β is a sum of literals or it may be empty.) The complementation laws are not used because we are analyzing the circuit behavior resulting from an input change. As that input change propagates through the circuit, at a given point in time a line tending toward the value x may not have the value that is the complement of a line tending toward the value x'. In the SOP expression, a product of the form $xx'\alpha$ represents a pseudo gate that may temporarily have the output value 1 as x changes and if $\alpha = 1$.

Given the SOP expression, the circuit is analyzed for static 1-hazards the same as for a two-level AND-OR circuit, i.e., the products are mapped on a Karnaugh map and if two 1's are adjacent on the map and not included in one of the products, they correspond to a static 1-hazard. The circuit can have a static 0-hazard or a dynamic hazard only if the SOP expression contains a term of the form $xx'\alpha$. A static 0-hazard exists if there are two adjacent 0's on the Karnaugh map for which $\alpha = 1$ and the two input combinations differ just in the value of x. A dynamic hazard exists if there is a term of the form $xx'\alpha$ and two conditions are satisfied: (1) There are adjacent input combinations on the Karnaugh map differing in the value of x, with $\alpha = 1$ and with opposite function values, and (2) for these input combinations the change in x propagates over at least three paths through the circuit.

As an example consider the circuit of Figure 7-7. The expression for the circuit output is

$$\begin{aligned} f &= (c' + ad' + bd')(c + a'd + bd) \\ &= cc' + acd' + bcd' + a'c'd + aa'dd' + a'bdd' + bc'd + abdd' + bdd' \\ &= cc' + acd' + bcd' + a'c'd + aa'dd' + bc'd + bdd' \end{aligned}$$

The Karnaugh map for this function is shown as the circled 1's in Figure 7-3. It is derived in the normal way ignoring the product terms containing both a variable and its complement. The circuit does not contain any static 1-hazards because each pair of adjacent 1's are covered by one of the product terms. Potentially, the terms cc' and bdd' may cause either static 0- or dynamic hazards or both; the first for c changing and the second for d changing. (The term $aa'dd'$ cannot cause either hazard because, for example, if a changes the dd' part of the product forces it to 0.) With $a = 0, b = 0$, and $d = 0$ and c changing, the circuit output is 0 before and after the change, and because the cc' term can cause the output to temporarily become 1, this transition is a static 0-hazard. Similarly, a change in c, with $a = 1, b = 0$, and $d = 1$, is a static 0-hazard. The cc' term cannot cause a dynamic hazard because there are only two physical paths from input c to the circuit output.

The term bdd' can cause a static 0- or dynamic hazard only if $b = 1$. From the Karnaugh map, it is seen that, with $b = 1$ and d changing, the circuit output changes for any combination of a and c, so the only possibility is that of a dynamic hazard. There are four physical paths from d to the circuit output, so a dynamic hazard exists if a d change can propagate over at least three of those paths. However, this cannot happen because, with $c = 0$, propagation over the upper two paths is blocked at the upper OR gate because $c' = 1$ forces the OR gate output to be 1, and with $c = 1$ propagation over the lower two paths is blocked at the lower OR gate. The circuit does not contain a dynamic hazard.

Another approach to finding the hazards is as follows: If we factor the original expression for the circuit output (without using the complementation laws), we get

$$f = (c' + a + b)(c' + d')(c + a' + b)(c + d)$$

Plotting the 0's of f from this expression on a Karnaugh map reveals that there are 0-hazards when $a = b = d = 0$ and c changes, and also when $b = 0, a = d = 1$, and c changes. An expression of the form $x + x'$ does not appear in any sum term of f, and this indicates that there are no 1-hazards or dynamic hazards.

As another example of finding static and dynamic hazards from a SOP expression, consider the circuit of Figure 8-12(a). The SOP expression for f is

$$f = (A'C' + B'C)(C + D) = A'CC' + A'C'D + B'C$$

The Karnaugh map for f in Figure 8-12(b) shows that $f = 1$ for the input combinations $(A, B, C, D) = (0, 0, 0, 1)$ and $(0, 0, 1, 1)$ and neither product of f covers these two minterms; hence, these two input combinations imply a static 1-hazard for C changing. The product $A'CC'$ in f indicates the possibility of a static 0-hazard and a dynamic hazard for $A = 0$ and C changing. The Karnaugh map shows that when $f = 0$,

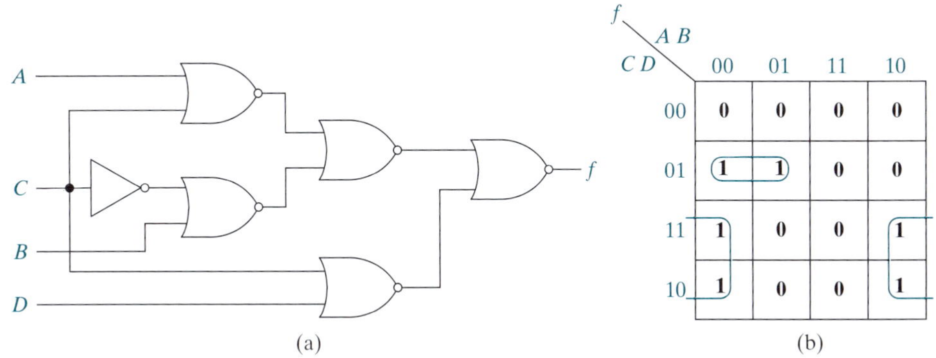

FIGURE 8-12
Hazard Example

the two input combinations (0, 1, 0, 0) and (0, 1, 1, 0) meet these conditions and, hence, they imply a static 0-hazard. The Karnaugh map shows two pairs of input combinations with f changing for $A = 0$ and C changing—namely, (0, 0, 0, 0), (0, 0, 1, 0), and (0, 1, 0, 1), (0, 1, 1, 1). In order for these to be dynamic hazards, the C change must propagate over three or more paths to the output. The circuit shows that propagation over the three paths requires $B = 0$ and $D = 0$ as well as $A = 0$; thus, a dynamic hazard only occurs for (0, 0, 0, 0) and (0, 0, 1, 0). For (0, 1, 0, 1) and (0, 1, 1, 1), the C change only propagates over one path, and f can only change once.

To design a circuit which is free of static and dynamic hazards, the following procedure may be used:

1. Find a sum-of-products expression (F^t) for the output in which every pair of adjacent 1's is covered by a 1-term. (The sum of all prime implicants will always satisfy this condition.) A two-level AND-OR circuit based on this F^t will be free of 1-, 0-, and dynamic hazards.
2. If a different form of the circuit is desired, manipulate F^t to the desired form by simple factoring, DeMorgan's laws, etc. Treat each x_i and x_i' as independent variables to prevent introduction of hazards.

Alternatively, you can start with a product-of-sums expression in which every pair of adjacent 0's is covered by a 0 term, and follow the dual procedure to design a hazard-free two-level OR-AND circuit.

It should be emphasized that the discussion of hazards and the possibility of resulting glitches in this section has assumed that only a single input can change at a time and that no other input will change until the circuit has stabilized. If more than one input can change at one time, then nearly all circuits will contain hazards, and they cannot be eliminated by modifying the circuit implementation. The circuit corresponding to the Karnaugh map of Figure 8-11 illustrates this. Consider the input change $(A, B, C, D) = (0, 1, 0, 1)$ to $(0, 1, 1, 0)$ with both C and D changing. The output is 0 before the change and will be 0 after the circuit has stabilized; however, if the C change propagates through the circuit before the D change, then the circuit will output a transient 1. Effectively, the input combination to the circuit can temporarily become $(A, B, C, D) = (0, 1, 1, 1)$, and the circuit output will temporarily become 1 no matter how it is implemented.

Glitches are of most importance in asynchronous sequential circuits. The latches and flip-flops discussed in Unit 11 are the most important examples of asynchronous sequential circuits. Although more than one input can change at the same time for some of these circuits, restrictions are placed on the changes so that it is necessary to analyze the circuits for hazards only when a single input changes. Consequently, the discussion in this section is relevant to this important class of circuits.

8.5 Simulation and Testing of Logic Circuits

An important part of the logic design process is verifying that the final design is correct and debugging the design if necessary. Logic circuits may be tested either by actually building them or by simulating them on a computer. Simulation is generally easier, faster, and more economical. As logic circuits become more and more complex, it is very important to simulate a design before actually building it. This is particularly true when the design is built in integrated circuit form, because fabricating an integrated circuit may take a long time and correcting errors may be very expensive. Simulation is done for several reasons, including (1) verification that the design is logically correct, (2) verification that the timing of the logic signals is correct, and (3) simulation of faulty components in the circuit as an aid to finding tests for the circuit.

To use a computer program for simulating logic circuits, you must first specify the circuit components and connections; then, specify the circuit inputs; and, finally, observe the circuit outputs. The circuit description may be input into a simulator in the form of a list of connections between the gates and other logic elements in the circuit, or the description may be in the form of a logic diagram drawn on a computer screen. Most modern logic simulators use the latter approach. A typical simulator which runs on a personal computer uses switches or input boxes to specify the inputs and probes to read the logic outputs. Alternatively, the inputs and outputs may be specified as sequences of 0's and 1's or in the form of timing diagrams.

A simple simulator for combinational logic works as follows:

1. The circuit inputs are applied to the first set of gates in the circuit, and the outputs of those gates are calculated.
2. The outputs of the gates which changed in the previous step are fed into the next level of gate inputs. If the input to any gate has changed, then the output of that gate is calculated.
3. Step 2 is repeated until no more changes in gate inputs occur. The circuit is then in a steady-state condition, and the outputs may be read.
4. Steps 1 through 3 are repeated every time a circuit input changes.

The two logic values, 0 and 1, are not sufficient for simulating logic circuits. At times, the value of a gate input or output may be unknown, and we will represent this unknown value by X. At other times we may have no logic signal at an input, as in the case of an open circuit when an input is not connected to any output. We use the logic value Z to represent an open circuit, or *high impedance* (hi-Z) connection. The discussion that follows assumes we are using a four-valued logic simulator with logic values 0, 1, X (unknown), and Z (hi-Z).

Figure 8-13(a) shows a typical simulation screen on a personal computer. The switches are set to 0 or 1 for each input. The probes indicate the value of each gate output. In Figure 8-13(b), one gate has no connection to one of its inputs. Because that gate has a 1 input and a hi-Z input, we do not know what the hardware will do, and the gate output is unknown. This is indicated by an X in the probe.

FIGURE 8-13

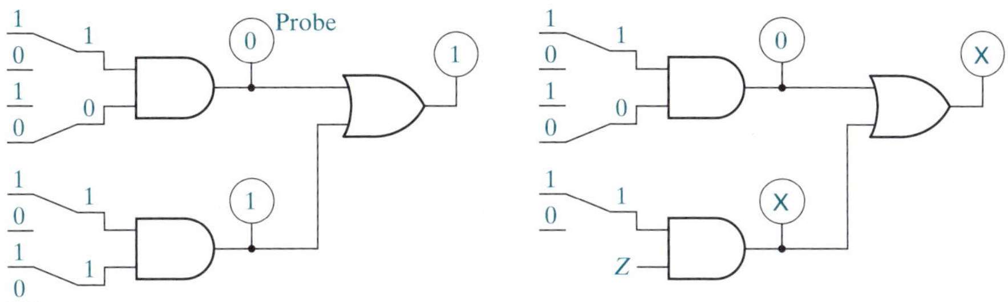

(a) Simulation screen showing switches (b) Simulation screen with missing gate input

Table 8-1 shows AND and OR functions for four-valued logic simulation. These functions are defined in a manner similar to the way real gates work. For an AND gate, if one of the inputs is 0, the output is always 0 regardless of the other input. If one input is 1 and the other input is X (we do not know what the other input is), then the output is X (we do not know what the output is). If one input is 1 and the other input is Z (it has no logic signal), then the output is X (we do not know what the hardware will do). For an OR gate, if one of the inputs is 1, the output is 1 regardless of the other input. If one input is 0 and the other input is X or Z, the output is unknown. For gates with more than two inputs, the operations may be applied several times.

TABLE 8-1 AND and OR Functions for Four-Valued Simulation

·	0	1	X	Z
0	0	0	0	0
1	0	1	X	X
X	0	X	X	X
Z	0	X	X	X

+	0	1	X	Z
0	0	1	X	X
1	1	1	1	1
X	X	1	X	X
Z	X	1	X	X

A combinational logic circuit with a small number of inputs may easily be tested with a simulator or in lab by checking the circuit outputs for all possible combinations of the input values. When the number of inputs is large, it is usually possible to find a relatively small set of input test patterns which will test for all possible faulty gates in the circuit.[1]

[1]Methods for test pattern generation are described in Alexander Miczo, *Digital Logic Testing and Simulation*, 2nd ed. (John Wiley & Sons, 2003).

If a circuit output is wrong for some set of input values, this may be due to several possible causes:

1. Incorrect design
2. Gates connected wrong
3. Wrong input signals to the circuit

If the circuit is built in lab, other possible causes include

4. Defective gates
5. Defective connecting wires

Fortunately, if the output of a combinational logic circuit is wrong, it is very easy to locate the problem systematically by starting at the output and working back through the circuit until the trouble is located. For example, if the output gate has the wrong output and its inputs are correct, this indicates that the gate is defective. On the other hand, if one of the inputs is wrong, then either the gate is connected wrong, the gate driving this input has the wrong output, or the input connection is defective.

Example

The function $F = AB(C'D + CD') + A'B'(C + D)$ is realized by the circuit of Figure 8-14:

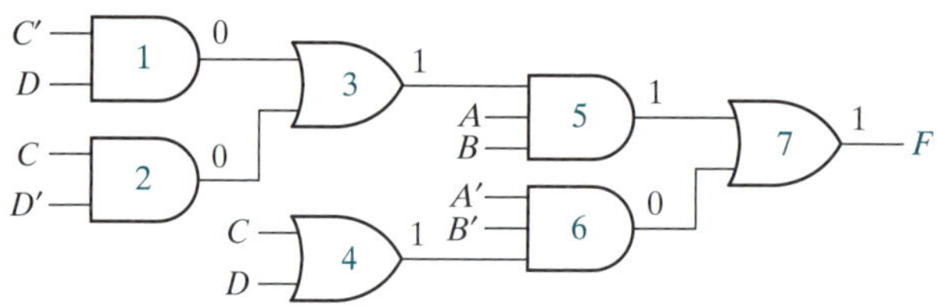

FIGURE 8-14 Logic Circuit with Incorrect Output

A student builds the circuit in a lab and finds that when $A = B = C = D = 1$, the output F has the wrong value, and that the gate outputs are as shown in Figure 8-14. The reason for the incorrect value of F can be determined as follows:

1. The output of gate 7 (F) is wrong, but this wrong output is consistent with the inputs to gate 7, that is, $1 + 0 = 1$. Therefore, one of the inputs to gate 7 must be wrong.
2. In order for gate 7 to have the correct output ($F = 0$), both inputs must be 0. Therefore, the output of gate 5 is wrong. However, the output of gate 5 is consistent with its inputs because $1 \cdot 1 \cdot 1 = 1$. Therefore, one of the inputs to gate 5 must be wrong.
3. Either the output of gate 3 is wrong, or the A or B input to gate 5 is wrong. Because $C'D + CD' = 0$, the output of gate 3 is wrong.
4. The output of gate 3 is not consistent with the outputs of gates 1 and 2 because $0 + 0 \neq 1$. Therefore, either one of the inputs to gate 3 is connected wrong, gate 3 is defective, or one of the input connections to gate 3 is defective.

This example illustrates how to troubleshoot a logic circuit by starting at the output gate and working back until the wrong connection or defective gate is located.

Problems

8.1 Complete the timing diagram for the given circuit. Assume that both gates have a propagation delay of 5 ns.

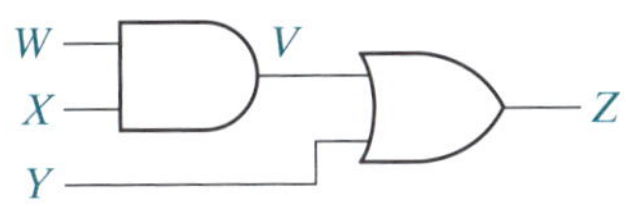

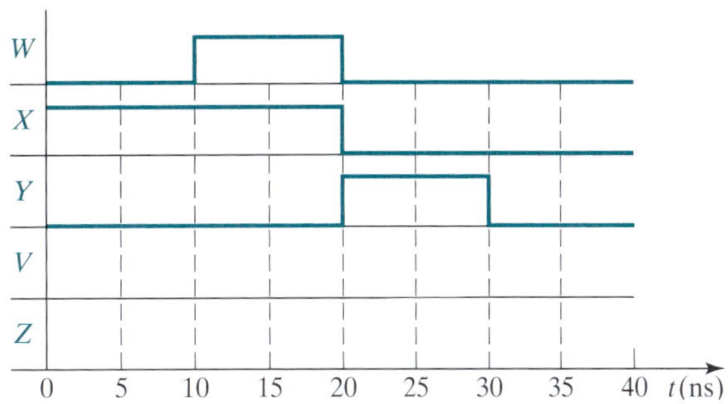

8.2 Consider the following logic function.

$$F(A, B, C, D) = \Sigma m(0, 4, 5, 10, 11, 13, 14, 15)$$

(a) Find two different minimum circuits which implement F using AND and OR gates. Identify two hazards in each circuit.
(b) Find an AND-OR circuit for F which has no hazards.
(c) Find an OR-AND circuit for F which has no hazards.

8.3 For the following circuit:

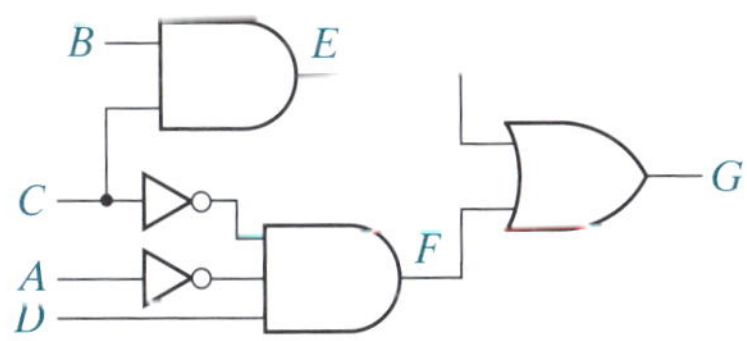

(a) Assume that the inverters have a delay of 1 ns and the other gates have a delay of 2 ns. Initially $A = 0$ and $B = C = D = 1$, and C changes to 0 at time = 2 ns. Draw a timing diagram and identify the transient that occurs.
(b) Modify the circuit to eliminate the hazard.

8.4 Using four-valued logic, find $A, B, C, D, E, F, G,$ and H.

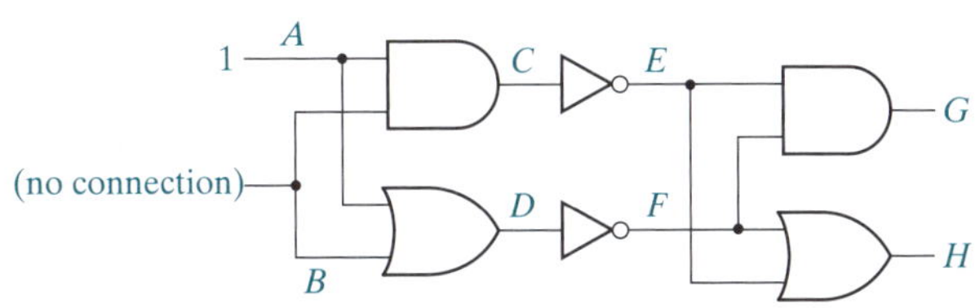

8.5 The circuit below was designed to implement the logic equation $F = AB'D + BC'D' + BCD$, but it is not working properly. The input wires to gates 1, 2, and 3 are so tightly packed, it would take you a while to trace them all back to see whether the inputs are correct. It would be nice to only have to trace whichever one is incorrectly wired. When $A = B = 0$ and $C = D = 1$, the inputs and outputs of gate 4 are as shown. Is gate 4 working properly? If so, which of the other gates either is connected incorrectly or is malfunctioning?

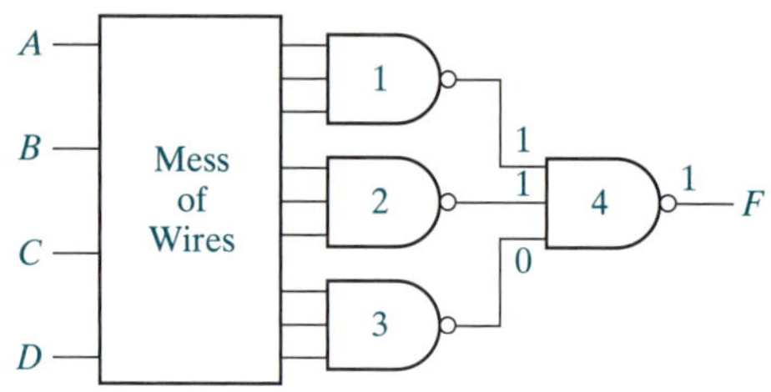

8.6 (a) Assume the inverters have a delay of 1 ns and the other gates have a delay of 2 ns. Initially $A = B = C = 0$ and $D = 1$; C changes to 1 at time 2 ns. Draw a timing diagram showing the glitch corresponding to the hazard.

(b) Modify the circuit so that it is hazard free. (Leave the circuit as a two-level, OR-AND circuit.)

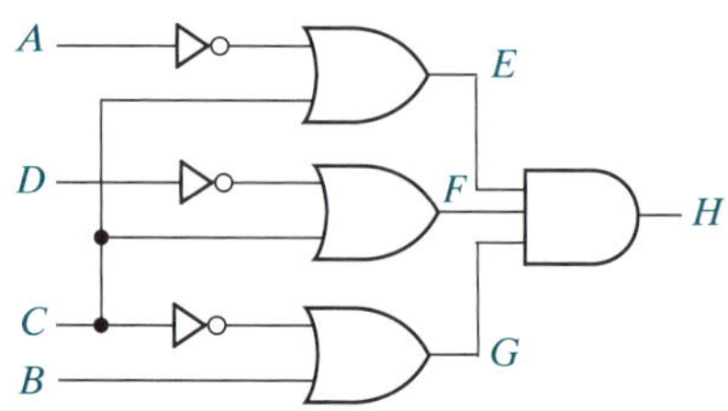

8.7 A two-level, NOR-NOR circuit implements the function
$f(a, b, c, d) = (a + d')(b' + c + d)(a' + c' + d')(b' + c' + d)$.

(a) Find all hazards in the circuit.

(b) Redesign the circuit as a two-level, NOR-NOR circuit free of all hazards and using a minimum number of gates.

8.8 $F(A, B, C, D) = \Sigma m(0, 2, 3, 5, 6, 7, 8, 9, 13, 15)$

(a) Find three different minimum AND-OR circuits that implement F. Identify two hazards in each circuit. Then find an AND-OR circuit for F that has no hazards.

(b) There are two minimum OR-AND circuits for F; each has one hazard. Identify the hazard in each circuit, and then find an OR-AND circuit for F that has no hazards.

8.9 Consider the following three-level NOR circuit:
(a) Find all hazards in this circuit.
(b) Redesign the circuit as a three-level NOR circuit that is free of all hazards.

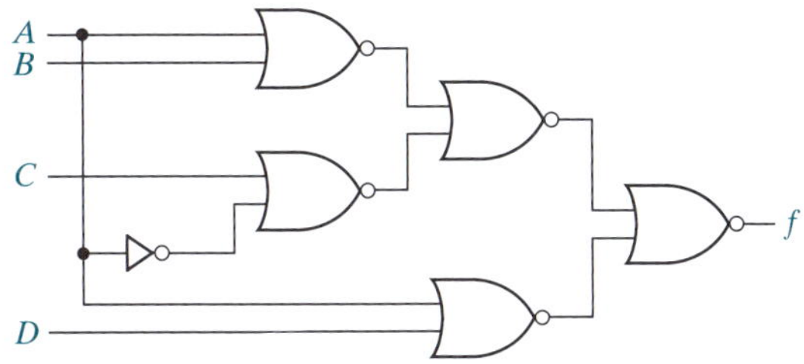

8.10 Draw the timing diagram for V and Z for the circuit. Assume that the AND gate has a delay of 10 ns and the OR gate has a delay of 5 ns.

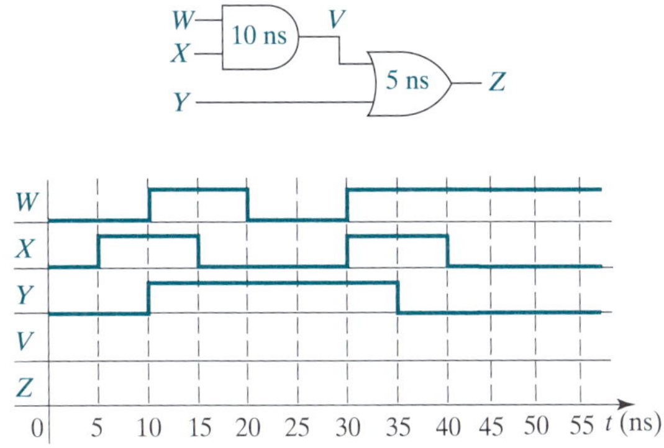

8.11 Consider the three-level circuit corresponding to the expression $f(A, B, C, D) = (A + B)(B'C' + BD')$.
(a) Find all hazards in this circuit.
(b) Redesign the circuit as a three-level NOR circuit that is free of all hazards.

8.12 Complete the timing diagram for the given circuit. Assume that both gates have a propagation delay of 5 ns.

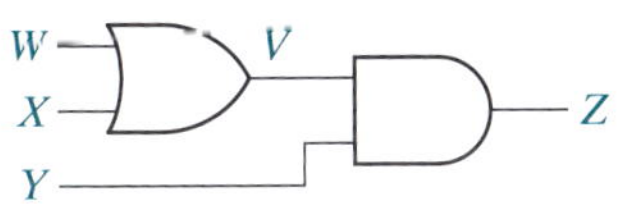

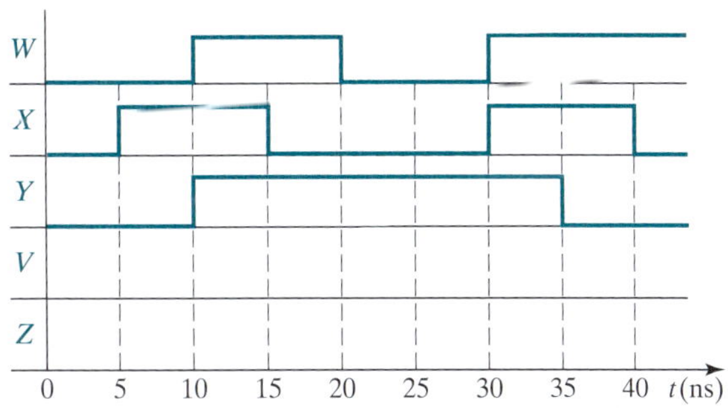

8.13 Implement the logic function from Figure 8.10(b) as a minimum sum of products. Find the static hazards and tell what minterms they are between. Implement the same logic function as a sum of products without any hazards.

8.14 Using four-valued logic, find A, B, C, D, E, F, G, and H.

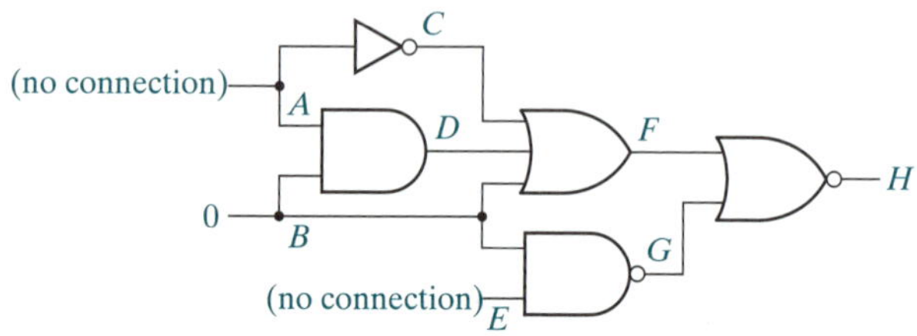

8.15 The following circuit was designed to implement the logic equation $F = (A + B' + C')(A' + B + C')(A' + B' + C)$, but it is not working properly. The input wires to gates 1, 2, and 3 are so tightly packed, it would take you a while to trace them all back to see whether the inputs are correct. It would be nice to only have to trace whichever one is incorrectly wired. When $A = B = C = 1$, the inputs and outputs of gate 4 are as shown. Is gate 4 working properly? If so, which of the other gates either is connected incorrectly or is malfunctioning?

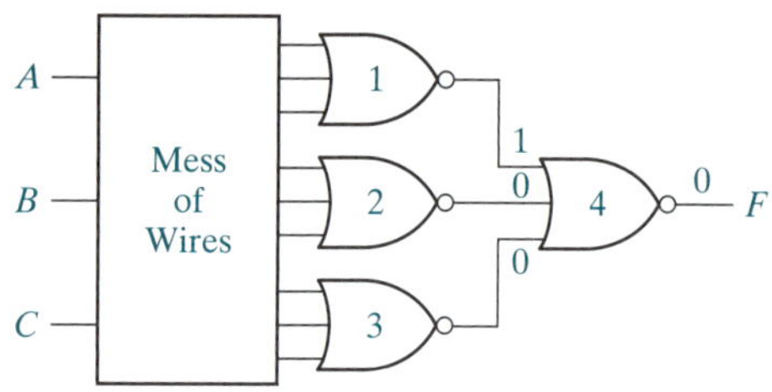

8.16 Consider the following logic function.

$$F(A, B, C, D) = \Sigma\, m(0, 2, 5, 6, 7, 8, 9, 12, 13, 15)$$

(a) Find two different minimum AND-OR circuits which implement F. Identify two hazards in each circuit. Then find an AND-OR circuit for F that has no hazards.

(b) The minimum OR-AND circuit for F has one hazard. Identify it, and then find an OR-AND circuit for F that has no hazards.

Design Problems

Seven-Segment Indicator

Several of the problems involve the design of a circuit to drive a seven-segment indicator (see Figure 8-15). The seven-segment indicator can be used to display any one of the decimal digits 0 through 9. For example, "1" is displayed by lighting segments

2 and 3, "2" by lighting segments 1, 2, 7, 5, and 4, and "8" by lighting all seven segments. A segment is lighted when a logic 1 is applied to the corresponding input on the display module.

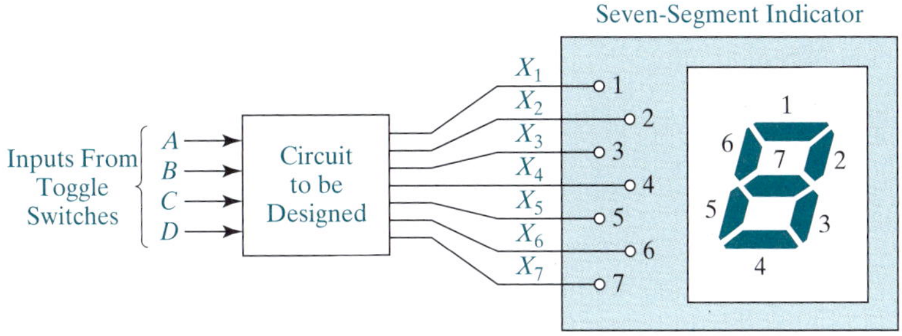

FIGURE 8-15 Circuit Driving Seven-Segment Module

8.A Design an 8-4-2-1 BCD code converter to drive a seven-segment indicator. The four inputs to the converter circuit (A, B, C, and D in Figure 8-15) represent an 8-4-2-1 binary-coded-decimal digit. Assume that only input combinations representing the digits 0 through 9 can occur as inputs, so that the combinations 1010 through 1111 are don't-cares. Design your circuit using only two-, three-, and four-input NAND gates and inverters. Try to minimize the number of gates required. The variables A, B, C, and D will be available from toggle switches.

Use  (not ) for 6. Use  (not ) for 9.

Any solution that uses 18 or fewer gates and inverters (not counting the four inverters for the inputs) is acceptable.

8.B Design an excess-3 code converter to drive a seven-segment indicator. The four inputs to the converter circuit (A, B, C, and D in Figure 8-15) represent an excess-3 coded decimal digit. Assume that only input combinations representing the digits 0 through 9 can occur as inputs, so that the six unused combinations are don't-cares. Design your circuit using only two-, three-, and four-input NAND gates and inverters. Try to minimize the number of gates and inverters required. The variables A, B, C, and D will be available from toggle switches.

Use  (not ) for 6. Use  (not ) for 9.

Any solution with 16 or fewer gates and inverters (not counting the four inverters for the inputs) is acceptable.

8.C Design a circuit which will yield the product of two binary numbers, n_2 and m_2, where $00_2 \le n_2 \le 11_2$ and $000_2 \le m_2 \le 101_2$. For example, if $n_2 = 10_2$ and $m_2 = 001_2$, then the product is $n_2 \times m_2 = 10_2 \times 001_2 = 0010_2$. Let the variables A and B represent the first and second digits of n_2, respectively (i.e., in this example $A = 1$ and $B = 0$). Let the variables C, D, and E represent the first, second, and third digits of

m_2, respectively (in this example $C = 0$, $D = 0$, and $E = 1$). Also let the variables W, X, Y, and Z represent the first, second, third, and fourth digits of the product. (In this example $W = 0$, $X = 0$, $Y = 1$, and $Z = 0$.) Assume that $m_2 > 101_2$ never occurs as a circuit input.

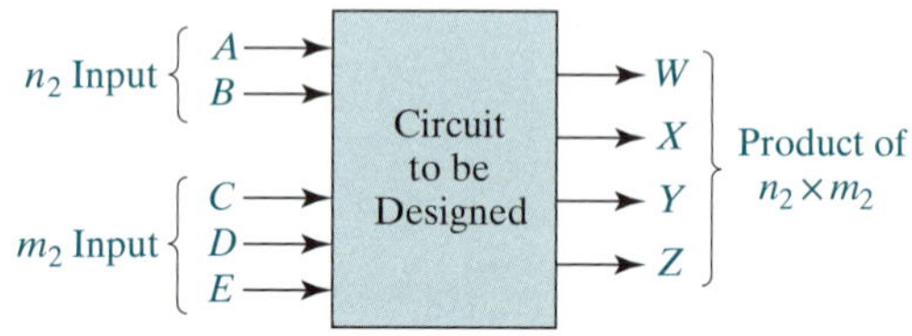

Design the circuit using only two-, three-, and four-input NOR gates and inverters. Try to minimize the total number of gates and inverters required. The variables A, B, C, D, and E will be available from toggle switches. Any solution that uses 15 or fewer gates and inverters (not counting the five inverters for the inputs) is acceptable.

8.D Work Design Problem 8.C using two-, three-, and four-input NAND gates and inverters instead of NOR gates and inverters. Any solution that uses 14 gates and inverters or less (not counting the five inverters for the inputs) is acceptable.

8.E Design a circuit which multiplies two 2-bit binary numbers and displays the answer in decimal on a seven-segment indicator. In Figure 8-15, A and B are two bits of a binary number N_1, and C and D are two bits of a binary number N_2. The product ($N_1 \times N_2$) is to be displayed in decimal by lighting appropriate segments of the seven-segment indicator. For example, if $A = 1$, $B = 0$, $C = 1$, and $D = 0$, the number "4" is displayed by lighting segments 2, 3, 6, and 7.

Use  (not ) for 6. Use  (not ) for 9.

Design your circuit using only two-, three-, and four-input NAND gates and inverters. Try to minimize the number of gates required. The variables A, B, C, and D will be available from toggle switches. Any solution that uses 18 or fewer gates and inverters (not counting the four inverters for the inputs) is acceptable.

8.F Design a Gray code converter to drive a seven-segment indicator. The four inputs to the converter circuit (A, B, C, and D in Figure 8-15) represent a decimal digit coded using the Gray code of Table 1-2. Assume that only input combinations representing the digits 0 through 9 can occur as inputs, so that the six unused combinations are don't-care terms. Design your circuit using only two-, three-, and four-input NAND gates and inverters. Try to minimize the numbers of gates and inverters required. The variables A, B, C, and D will be available from toggle switches.

Use  (not ) for 6. Use  (not ) for 9.

Any solution with 20 or fewer gates and inverters (not counting the four inverters for the inputs) is acceptable.

8.G Design a circuit that will add either 1 or 2 to a 4-bit binary number N. Let the inputs N_3, N_2, N_1, N_0 represent N. The input K is a control signal. The circuit should have outputs M_3, M_2, M_1, M_0, which represent the 4-bit binary number M. When $K = 0$, $M = N + 1$. When $K = 1$, $M = N + 2$. Assume that the inputs for which $M > 1111_2$ will never occur.

Design the circuit using only two-, three-, and four-input NAND gates and inverters. Try to minimize the total number of gates and inverters required. The input variables K, N_3, N_2, N_1, and N_0 will be available from toggle switches. Any solution that uses 13 or fewer gates and inverters (not counting the five inverters for the inputs) is acceptable.

8.H Work Problem 8.A, except use 4-2-1-8 code instead of 8-4-2-1 code. For example, in 4-2-1-8 code, 9 is represented by 0011. Also change the representations of digits 6 and 9 to the opposite form given in Problem 8.A. Any solution with 20 or fewer gates and inverters (not counting the four inverters for the inputs) is acceptable.

8.I Work Problem 8.B, except use excess-2 code instead of excess-3 code. (In excess-2 code, 0 is represented by 0010, 1 by 0011, 2 by 0100, etc.). Any solution with 17 or fewer gates and inverters (not counting the four inverters for the inputs) is acceptable.

8.J Design a circuit which will multiply a 3-bit binary number CDE by 2, 3, or 5, depending on the value of a 2-bit code AB (00, 01, or 10), to produce a 4-bit result $WXYZ$. If the result has a value greater than or equal to 15, $WXYZ$ should be 1111 to indicate an overflow. Assume that the code $AB = 11$ will never occur. Design your circuit using only two-, three-, and four-input NOR gates and inverters. Try to minimize the number of gates required. The inputs A, B, C, D, and E will be available from toggle switches. Any solution which uses 19 or fewer gates and inverters (not counting the five inverters for the inputs) is acceptable.

8.K Design a circuit which will divide a 5-bit binary number by 3 to produce a 4-bit binary quotient. Assume that the input number is in the range 0 through 27 and that numbers in the range 28 through 31 will never occur as inputs. Design your circuit using only two-, three-, and four-input NAND gates and inverters. Try to minimize the number of gates required. The inputs A, B, C, D, and E will be available from toggle switches. Any solution which uses 22 or fewer gates and inverters (not counting the five inverters for the inputs) is acceptable.

8.L Design an excess-3 code converter to drive a seven-segment indicator. The four inputs (A, B, C, D) to the converter circuit represent an excess-3 digit. Input combinations representing the numbers 0 through 9 should be displayed as decimal digits. The input combinations 0000, 0001, and 0010 should be interpreted as an error, and an "E" should be displayed. Assume that the input combinations 1101, 1110, and 1111 will never occur. Design your circuit using only two-, three-, and four-input NOR gates and inverters. Any solution with 18 or fewer gates and inverters (not counting the four inverters for the inputs) is acceptable.

Use  (not ) for 6. Use  (not ) for 9.

8.M Design a circuit which displays the letters A through J on a seven-segment indicator. The circuit has four inputs W, X, Y, Z which represent the last 4 bits of the ASCII code for the letter to be displayed. For example, if $WXYZ = 0001$, "A" will be displayed. The letters should be displayed in the following form:

Design your circuit using only two-, three-, and four-input NOR gates and inverters. Any solution with 22 or fewer gates and inverters (not counting the four inverters for the inputs) is acceptable.

8.N A simple security system for two doors consists of a card reader and a keypad.

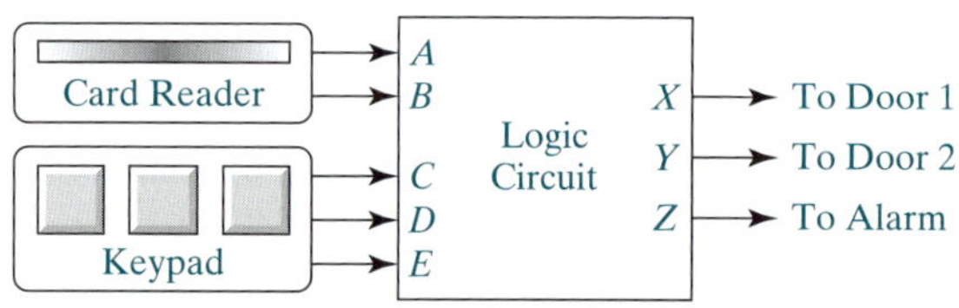

A person may open a particular door if he or she has a card containing the corresponding code and enters an authorized keypad code for that card. The outputs from the card reader are as follows:

	A	B
No card inserted	0	0
Valid code for door 1	0	1
Valid code for door 2	1	1
Invalid card code	1	0

To unlock a door, a person must hold down the proper keys on the keypad and, then, insert the card in the reader. The authorized keypad codes for door 1 are 101 and 110, and the authorized keypad codes for door 2 are 101 and 011. If the card has an invalid code or if the wrong keypad code is entered, the alarm will ring when the card is inserted. If the correct keypad code is entered, the corresponding door will be unlocked when the card is inserted.

Design the logic circuit for this simple security system. Your circuit's inputs will consist of a card code AB, and a keypad code CDE. The circuit will have three outputs XYZ (if X or $Y = 1$, door 1 or 2 will be opened; if $Z = 1$, the alarm will sound). Design your circuit using only two-, three-, and four-input NOR gates and inverters. Any solution with 19 or fewer gates and inverters (not counting the five inverters for the inputs) is acceptable. Use toggle switches for inputs A, B, C, D, and E when you test your circuit.

8.O Work Design Problem 8.A using two-, three-, and four-input NOR gates and inverters instead of NAND gates and inverters. Any solution that uses 19 gates and inverters or fewer (not counting the four inverters for the inputs) is acceptable.

8.P Work Design Problem 8.F using two-, three-, and four-input NOR gates and inverters instead of NAND gates and inverters. Any solution that uses 21 gates and inverters or fewer (not counting the four inverters for the inputs) is acceptable.

8.Q Work Design Problem 8.H using two-, three-, and four-input NOR gates and inverters instead of NAND gates and inverters. Any solution that uses 17 gates and inverters or fewer (not counting the four inverters for the inputs) is acceptable.

8.R Work Design Problem 8.I using two-, three-, and four-input NOR gates and inverters instead of NAND gates and inverters. Any solution that uses 16 gates and inverters or fewer (not counting the four inverters for the inputs) is acceptable.

8.S Design a "disk spinning" animation circuit for a CD player. The input to the circuit will be a 3-bit binary number $A_1A_2A_3$ provided by another circuit. It will count from 0 to 7 in binary, and then it will repeat. (You will learn to design such counters in Unit 12.) The animation will appear on the top four lights of the LED display of Figure 8-15, i.e., on X_1, X_2, X_7, and X_6, going clockwise. The animation should consist of a blank spot on a disk spinning around once, beginning with X_1. Then, the entire disk should blink on and off twice. The pattern is shown.

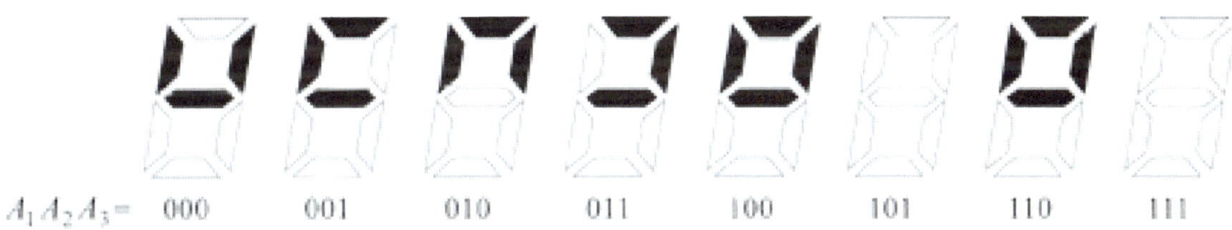

Design your circuit using only two-, three-, and four-input NOR gates and inverters. Try to minimize the number of gates required. Any solution which uses 11 or fewer gates (not counting the four inverters for the inputs) is acceptable.

UNIT 9

Multiplexers, Decoders, and Programmable Logic Devices

Objectives

1. Explain the function of a multiplexer. Implement a multiplexer using gates.
2. Explain the operation of three-state buffers. Determine the resulting output when three-state buffer outputs are connected together. Use three-state buffers to multiplex signals onto a bus.
3. Explain the operation of a decoder and encoder. Use a decoder with added gates to implement a set of logic functions. Implement a decoder or priority encoder using gates.
4. Explain the operation of a read-only memory (ROM). Use a ROM to implement a set of logic functions.
5. Explain the operation of a programmable logic array (PLA). Use a PLA to implement a set of logic functions. Given a PLA table or an internal connection diagram for a PLA, determine the logic functions realized.
6. Explain the operation of a programmable array logic device (PAL). Determine the programming pattern required to realize a set of logic functions with a PAL.
7. Explain the operation of a complex programmable logic device (CPLD) and a field-programmable gate array (FPGA).
8. Use Shannon's expansion theorem to decompose a switching function.

Study Guide

1. Read Section 9.1, *Introduction*.

2. Study Section 9.2, *Multiplexers*.

(a) Draw a logic circuit for a 2-to-1 multiplexer (MUX) using gates.

(b) Write the equation for a 4-to-1 MUX with control inputs A and C.

$Z =$ ____

(c) By tracing signals on Figure 9-3, determine what will happen to Z if $A = 1$, $B = 0$ and C changes from 0 to 1.

(d) Use three 2-to-1 MUXes to make a 4-to-1 MUX with control inputs A and B. Draw the circuit. (*Hint:* One MUX should have I_0 and I_1 inputs, and another should have I_2 and I_3 inputs.)

(e) Observe that if $A = 0$, $A \oplus B = B$, and that if $A = 1$, $A \oplus B = B'$. Using this observation, construct an exclusive-OR gate using a 2-to-1 multiplexer and one inverter.

(f) Work Problems 9.1 and 9.2.

(g) This section introduces bus notation. The bus symbol 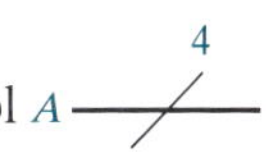

represents a group of four wires:

A_3 ________________
A_2 ________________
A_1 ________________
A_0 ________________

Draw the bus symbol for

B_2________________

B_1________________

B_0________________

(h) Represent the circuit of Figure 4-3 by one 4-bit full adder with two bus inputs, one bus output, and terminals for carry input C_0 and output C_4. Note that the carries C_3, C_2, and C_1 will not appear on your circuit diagram because they are signals internal to the 4-bit adder.

3. Study Section 9.3, *Three-State Buffers.*

(a) Determine the output of each three-state buffer:

(b) Determine the inputs for each three-state buffer (use X if an input is a don't-care).

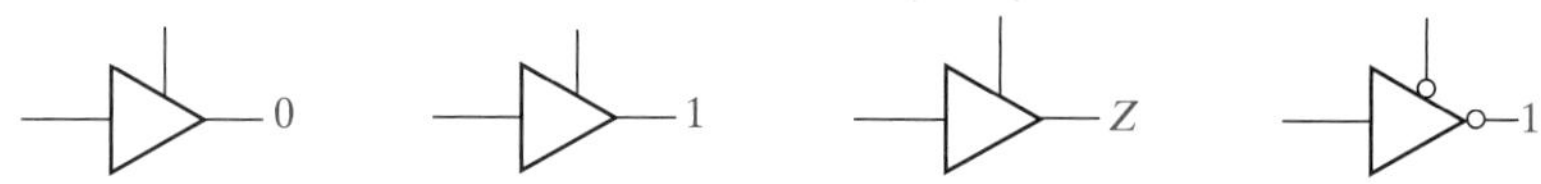

(c) Determine the output for each circuit. Use X to represent an unknown output.

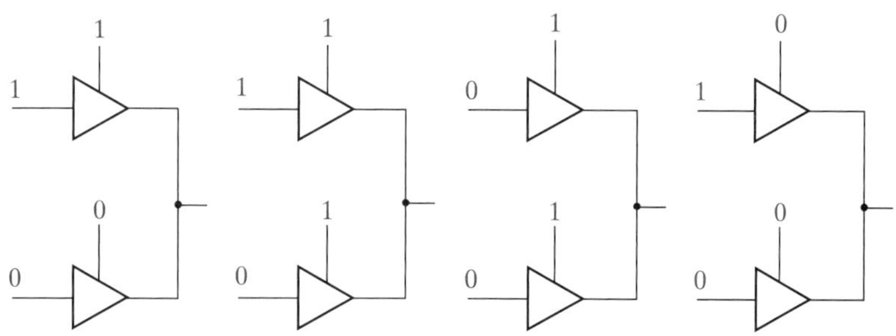

(d) The symbol 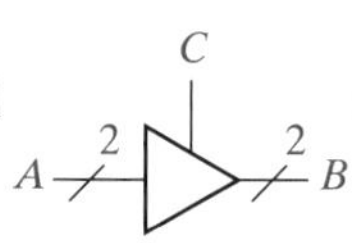

represents 2 three-state buffers with a common control input:

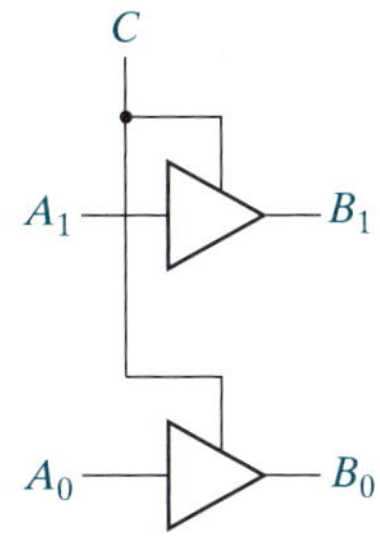

Using bus notation, draw an equivalent circuit for:

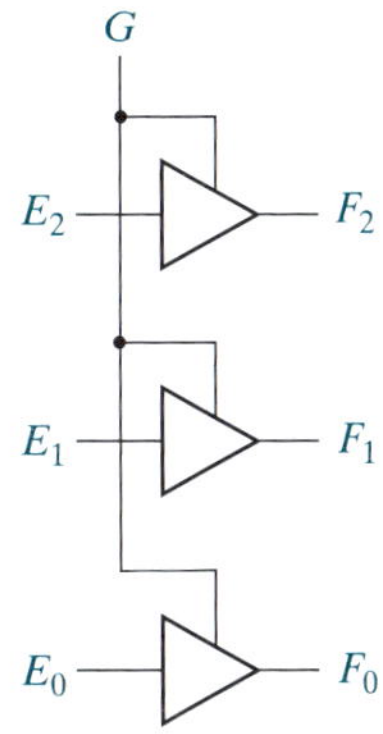

(e) For the following circuit, determine the 4-bit output (P) if $M = 0$. ____________ Repeat for $M = 1$. ____________

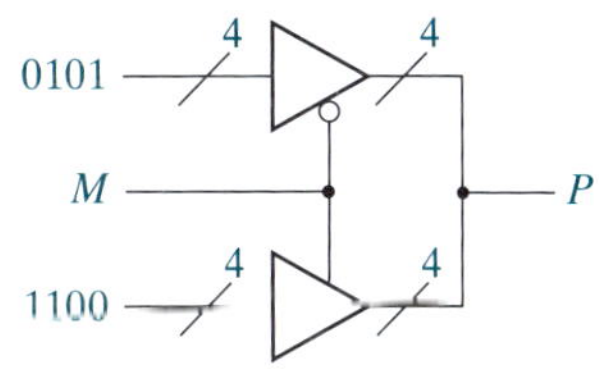

(f) Specify the AND-gate inputs so that the given circuit is equivalent to the 4-to-1 MUX in Figure 9-2. (Z in the following figure represents an output terminal, not high impedance.)

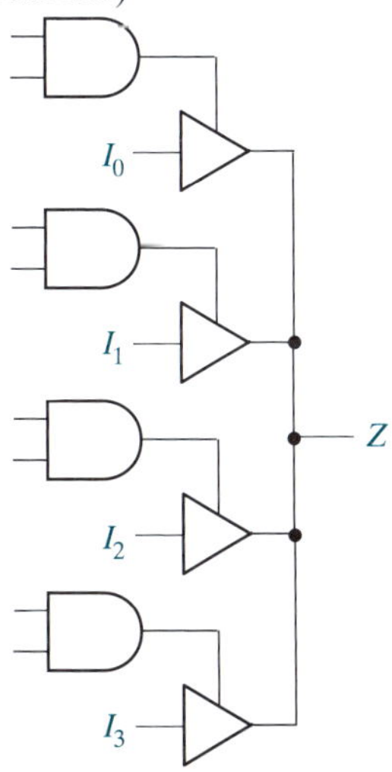

(g) Work Problem 9.3.

4. Study Section 9.4, *Decoders and Encoders.*

(a) The 7442 4-to-10 line decoder (Figure 9-18) can be used as a 3-to-8 line decoder. To do this, which three lines should be used as inputs?

____

The remaining input line should be set equal to ______________.

(b) Complete the following table for a 4-to-2 priority encoder:

y_0	y_1	y_2	y_3	a	b	c

What will a,b, and c be if $y_0\ y_1\ y_2\ y_3$ is 0101?

(c) Work Problem 9.4, 9.5, and 9.6.

5. Study Section 9.5, *Read-Only Memories.*

(a) The following diagram shows the pattern of 0's and 1's stored in a ROM with eight words and four bits per word. What will be the values of F_1, F_2, F_3, and F_4 if $A = 0$ and $B = C = 1$?

Give the minterm expansions for F_1 and F_2:

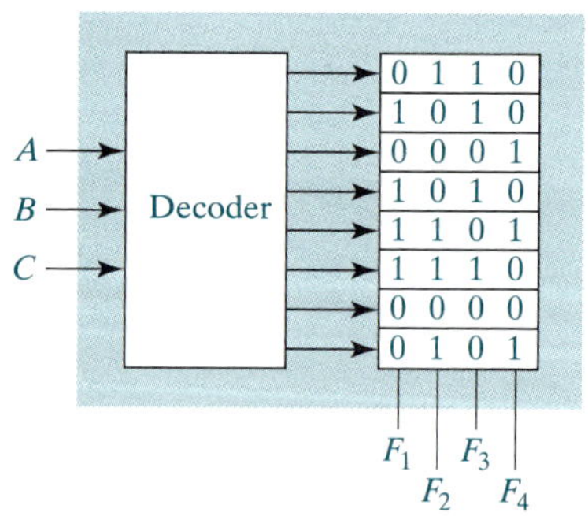

$F_1 =$

$F_2 =$

(b) When asked to specify the size of a ROM, give the number of words and the number of bits per word.
What size ROM is required to realize four functions of 5 variables?

What size ROM is required to realize eight functions of 10 variables?

(c) When specifying the size of a ROM, assume that you are specifying a standard size ROM with 2^n words. What size ROM is required to convert 8-4-2-1 BCD code to 2-out-of-5 code? (See Table 1-2, page 22.)

What size ROM would be required to realize the decoder given in Figure 9-18?

(d) Draw an internal connection diagram for a ROM which would perform the same function as the circuit of Figure 7-22. (Indicate the presence of switching elements by dots at the intersection of the word lines and output lines.)

(e) Explain the difference between a mask-programmable ROM and an EEPROM. Which would you use for a new design which had not yet been debugged?

(f) Work Problem 9.7.

6. Study Section 9.6, *Programmable Logic Devices*.

(a) When you are asked to specify the size of a PLA, give the number of inputs, the number of product terms, and the number of outputs.
What size PLA would be required to realize Equations (7-24) if no simplification of the minterm expansions were performed?

(b) If the realization of Equations (7-24) shown in Figure 7-22 were converted to a PLA realization, what size PLA would be required?

(c) Specify the contents of the PLA of question (b) in tabular form. Your table should have four rows. (You will only need seven 1's on the right side of your table. If you get eight 1's, you are probably doing more work than is necessary.)

(d) Draw an internal connection diagram for the PLA of (b). (Use X's to indicate the presence of switching elements in the AND and OR arrays.)

(e) Given the following PLA table, plot maps for Z_1, Z_2, and Z_3.

A	B	C	Z_1	Z_2	Z_3
–	0	0	1	1	0
0	1	–	1	1	0
1	0	–	1	0	0
1	1	1	0	1	1
0	–	1	1	0	1
0	0	0	0	0	1

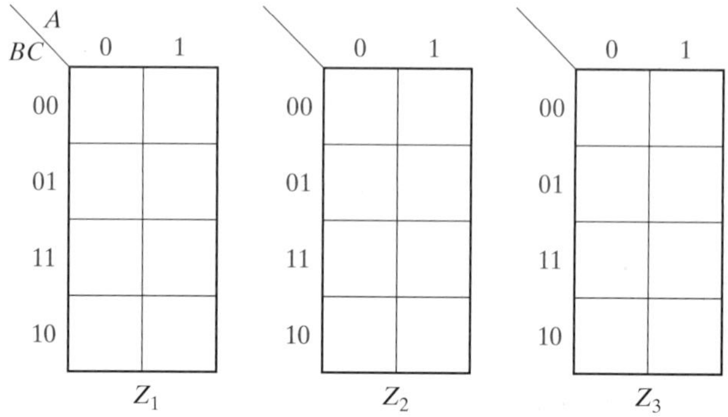

(The Z_1 map should have six 1's, Z_2 should have five, and Z_3 should have four.)

(f) For a truth table, any combination of input values will select exactly one row. Is this statement true for a PLA table?

For any combination of input values, the output values from a PLA can be determined by inspection of the PLA table. Consider Table 9-1, which represents a PLA with three inputs and four outputs. If the inputs are $ABC = 110$, which three rows in the table are selected?

In a given output column, what is the output if some of the selected rows are 1's and some are 0's? (Remember that the output bits for the selected rows are ORed together.)

When $ABC = 110$, what are the values of $F_0F_1F_2F_3$ at the PLA output?

When $ABC = 010$, which rows are selected and what are the values of $F_0F_1F_2F_3$ at the PLA output?

(g) Which interconnection points in Figure 9-32(a) must be set in order to realize the function shown in Figure 9-32(b)?

(h) What size of PAL could be used to realize the 8-to-1 MUX of Figure 9-3? The quad MUX of Figure 9-7? Give the number of inputs, the number of OR gates, and the maximum number of inputs to an OR gate.

(i) Work Problems 9.8, 9.9, and 9.10.

7. Study Section 9.7, *Complex Programmable Logic Devices.*
Work Problem 9.11.

8. Study Section 9.8, *Field-Programmable Gate Arrays.*

(a) For the CLB of Figure 9-37, write a logic equation for H in terms of F, G, and H_1.

(b) How many 4-variable function generators are required to implement a four-input OR gate? A 4-variable function with 13 minterms?

(c) Expand the function of Equation (9-9) about the variable c instead of a. Expand it algebraically and, then, expand it by using the Karnaugh map of Figure 9-39. (*Hint:* How should you split the map into two halves?)

(d) Draw a diagram showing how to implement Equation (9-12) using four function generators and a 4-to-1 MUX.

(e) In the worst case, how many 4-variable function generators are required to realize a 7-variable function (assume the necessary MUXes are available).

(f) Show how to realize $K = abcdefg$ using only two 4-variable function generators. (*Hint:* Use the output of one function generator as an input to the other.)

(g) Work Problems 9.12 and 9.13.

9. When you are satisfied that you can meet all of the objectives, take the readiness test.

Multiplexers, Decoders, and Programmable Logic Devices

9.1 Introduction

Until this point we have mainly been concerned with basic principles of logic design. We have illustrated these principles using gates as our basic building blocks. In this unit we introduce the use of more complex integrated circuits (ICs) in logic design. Integrated circuits may be classified as small-scale integration (SSI), medium-scale integration (MSI), large-scale integration (LSI), or very-large-scale integration (VLSI), depending on the number of gates in each integrated circuit package and the type of function performed. SSI functions include NAND, NOR, AND, and OR gates, inverters, and flip-flops. SSI integrated circuit packages typically contain one to four gates, six inverters, or one or two flip-flops. MSI integrated circuits, such as adders, multiplexers, decoders, registers, and counters, perform more complex functions. Such integrated circuits typically contain the equivalent of 12 to 100 gates in one package. More complex functions such as memories and microprocessors are classified as LSI or VLSI integrated circuits. An LSI integrated circuit generally contains 100 to a few thousand gates in a single package, and a VLSI integrated circuit contains several thousand gates or more.

It is generally uneconomical to design digital systems using only SSI and MSI integrated circuits. By using LSI and VLSI functions, the required number of integrated circuit packages is greatly reduced. The cost of mounting and wiring the integrated circuits as well as the cost of designing and maintaining the digital system may be significantly lower when LSI and VLSI functions are used.

This unit introduces the use of multiplexers, decoders, encoders, and three-state buffers in logic design. Then read-only memories (ROMs) are described and used to implement multiple-output combinational logic circuits. Finally, other types of programmable logic devices (PLDs), including programmable logic arrays (PLAs), programmable array logic devices (PALs), complex programmable logic devices (CPLDs), and field- programmable gate arrays (FPGAs) are introduced and used in combinational logic design.

9.2 Multiplexers

A multiplexer (or data selector, abbreviated as MUX) has a group of data inputs and a group of control inputs. The control inputs are used to select one of the data inputs and connect it to the output terminal. Figure 9-1 shows a 2-to-1 multiplexer and its switch analog. When the control input A is 0, the switch is in the upper position and the MUX output is $Z = I_0$; when A is 1, the switch is in the lower position and the MUX output is $Z = I_1$. In other words, a MUX acts like a switch that selects one of the data inputs (I_0 or I_1) and transmits it to the output. The logic equation for the 2-to-1 MUX is therefore:

$$Z = A'I_0 + AI_1$$

FIGURE 9-1
2-to-1 Multiplexer and Switch Analog

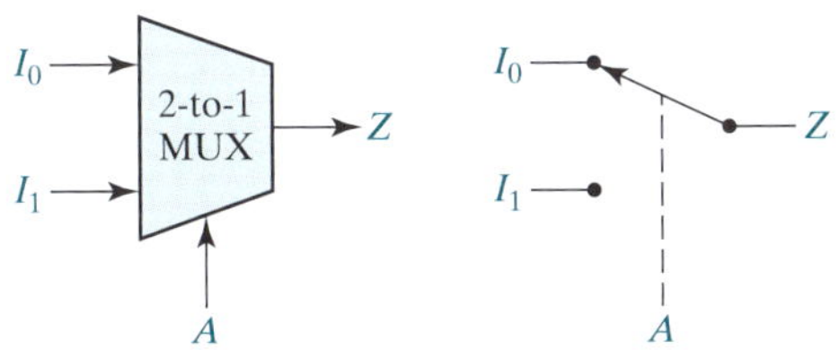

Figure 9-2 shows diagrams for a 4-to-1 multiplexer, 8-to-1 multiplexer, and 2^n-to-1 multiplexer. The 4-to-1 MUX acts like a four-position switch that transmits one of the four inputs to the output. Two control inputs (A and B) are needed to select one of the four inputs. If the control inputs are $AB = 00$, the output is I_0; similarly, the control inputs 01, 10, and 11 give outputs of I_1, I_2, and I_3, respectively. The 4-to-1 multiplexer is described by the equation

$$Z = A'B'I_0 + A'BI_1 + AB'I_2 + ABI_3 \tag{9-1}$$

FIGURE 9-2
Multiplexers

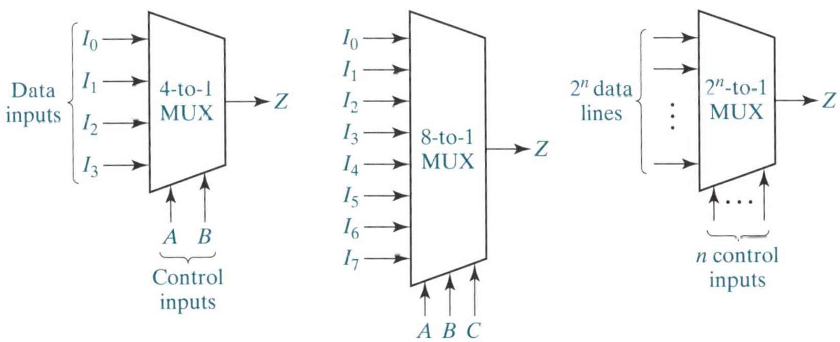

Similarly, the 8-to-1 MUX selects one of eight data inputs using three control inputs. It is described by the equation

$$Z = A'B'C'I_0 + A'B'CI_1 + A'BC'I_2 + A'BCI_3$$
$$+ AB'C'I_4 + AB'CI_5 + ABC'I_6 + ABCI_7 \tag{9-2}$$

When the control inputs are $ABC = 011$, the output is I_3, and the other outputs are selected in a similar manner. Figure 9-3 shows an internal logic diagram for the 8-to-1 MUX. In general, a multiplexer with n control inputs can be used to select any one of 2^n data inputs. The general equation for the output of a MUX with n control inputs and 2^n data inputs is

$$Z = \sum_{k=0}^{2^n-1} m_k I_k$$

where m_k is a minterm of the n control variables and I_k is the corresponding data input.

FIGURE 9-3
Logic Diagram for 8-to-1 MUX

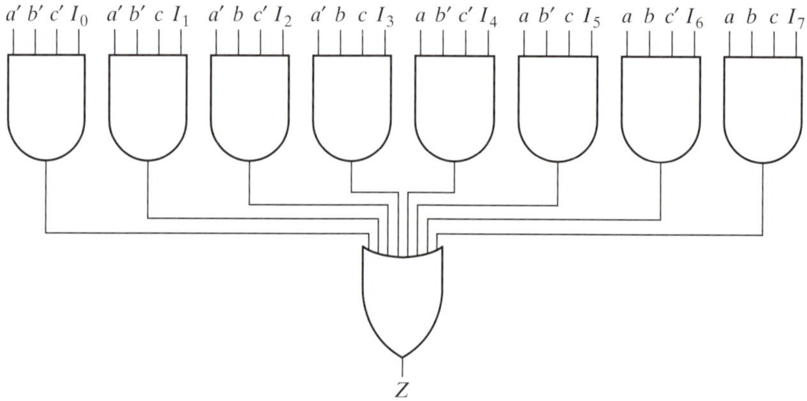

Of course, there are several other implementations of the 8-to-1 MUX. Each of the gates in Figure 9-3 can be replaced by NAND gates to obtain a NAND gate implementation. If a NOR gate implementation is wanted, the equation for Z can be written as a product of sums:

$$\begin{aligned} Z = &(A + B + C + I_0)(A + B + C' + I_1)(A + B' + C + I_2) \\ &(A + B' + C' + I_3)(A' + B + C + I_4)(A' + B + C' + I_5) \\ &(A' + B' + C + I_6)(A' + B' + C' + I_7) \end{aligned} \tag{9-3}$$

Implementations with more than two levels of gates can be obtained by factoring the equation for Z. For example, if a multiple-level NAND-gate implementation is desired, Equation (9-2) can be factored. One factorization is

$$\begin{aligned} Z = &A'B'(C'I_0 + CI_1) + A'B(C'I_2 + CI_3) + AB'(C'I_4 + CI_5) \\ &+ AB(C'I_6 + CI_7) \end{aligned} \tag{9-4}$$

The corresponding NAND-gate circuit is shown in Figure 9-4. Note that the data inputs are connected to four 2-to-1 MUXs with C as the select line, and the outputs of these 2-to-1 MUXs are connected to a 4-to-1 MUX with A and B as the select lines. Figure 9-5 shows this in block diagram form.

FIGURE 9-4 A Multi-Level Implementation of an 8-to-1 MUX

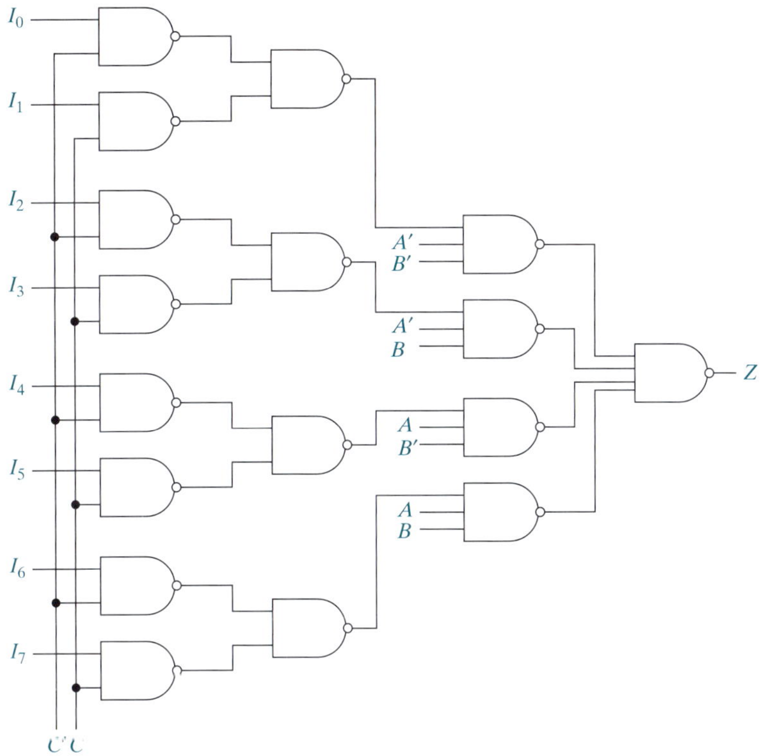

FIGURE 9-5 Component MUXs of Figure 9-4

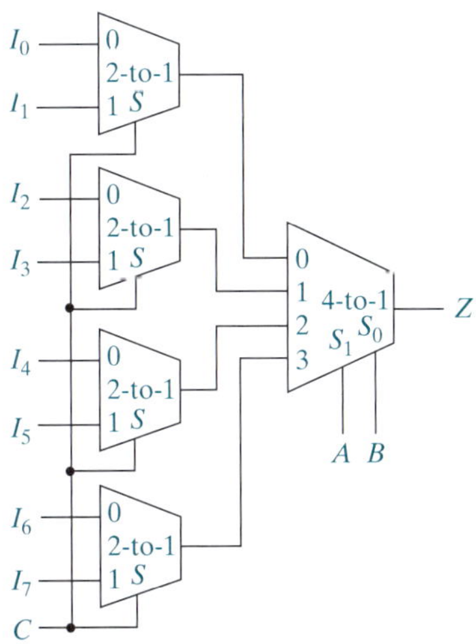

Multiplexers are frequently used in digital system design to select the data which is to be processed or stored. Figure 9-6 shows how a quadruple 2-to-1 MUX is used to select one of two 4-bit data words. If the control is $A = 0$, the values of x_0, x_1, x_2, and x_3 will appear at the z_0, z_1, z_2, and z_3 outputs; if $A = 1$, the values of y_0, y_1, y_2, and y_3 will appear at the outputs.

FIGURE 9-6 Quad Multiplexer Used to Select Data

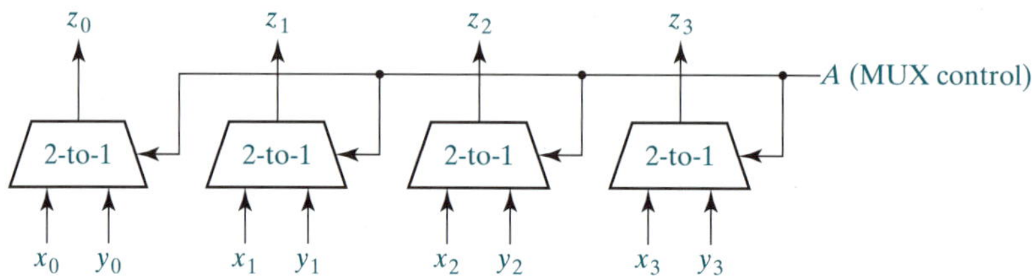

Several logic signals that perform a common function may be grouped together to form a bus. For example, the sum outputs of a 4-bit binary adder can be grouped together to form a 4-bit bus. Instead of drawing the individual wires that make up a bus, we often represent a bus by a single heavy line. The quad MUX of Figure 9-6 is redrawn in Figure 9-7, using bus inputs X and Y, and bus output Z. The X bus represents the four signals x_0, x_1, x_2, and x_3, and similarly for the Y and Z buses. When $A = 0$, the signals on bus X appear on bus Z; otherwise, the signals on bus Y appear. A diagonal slash through a bus with a number beside it specifies the number of bits in the bus.

FIGURE 9-7 Quad Multiplexer with Bus Inputs and Output

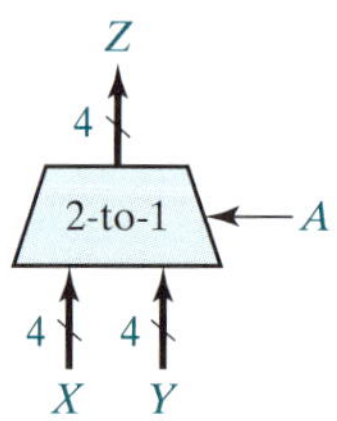

The preceding multiplexers do not invert the data inputs as they are routed to the output. Some multiplexers do invert the inputs, e.g., if the OR gate in Figure 9-3 is replaced by a NOR gate, then the 8-to-1 MUX inverts the selected input. To distinguish between these two types of multiplexers, we will say that the multiplexers without the inversion have active high outputs, and the multiplexers with the inversion have active low outputs.

Another type of multiplexer has an additional input called an *enable*. The 8-to-1 MUX in Figure 9-3 can be modified to include an enable by changing the AND gates to five-input gates. The enable signal E is connected to the fifth input of each of the AND gates. Then, if $E = 0$, $Z = 0$ independent of the gate inputs I_i and the select inputs a, b, and c. However, if $E = 1$, then the MUX functions as an ordinary 8-to-1 multiplexer. The terminology used for the MUX output, i.e., active high and active low, can be used for the enable as well. As described above, the enable is active high; E must be 1 for the MUX to function as a multiplexer. If an inverter is inserted between E and the AND gates, E must be 0 for the MUX to function as a multiplexer; the enable is active low.

Four combinations of multiplexers with an enable are possible. The output can be active high or active low, whereas the enable can be active high or active low. In a block diagram for the MUX, an active low line is indicated by inserting a bubble on the line to indicate the inclusion of an inversion. Figure 9-8 shows these combinations for a 4-to-1 MUX.

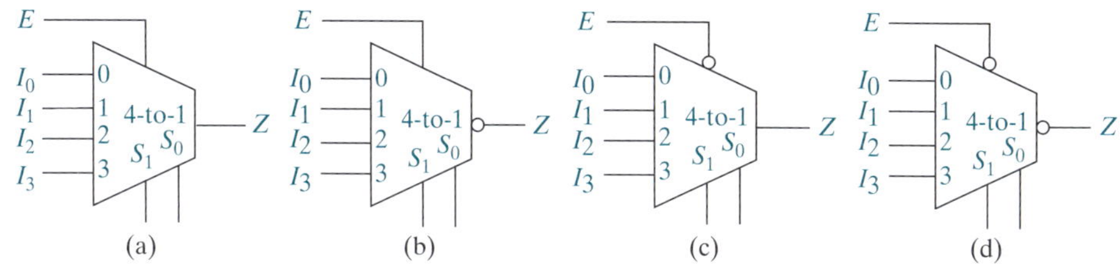

FIGURE 9-8 Active-High, Active-Low Enable and Output Combinations

In addition to acting as a data selector, a MUX can implement more general logic functions. In Figure 9-9 a 4-to-1 MUX is used to implement the function

$$\begin{aligned} Z &= C'D'(A' + B') + C'D(A') + CD'(AB' + A'B) + CD'(0) \\ &= A'C' + A'BD' + AB'D' \end{aligned}$$

Given a switching function, a MUX implementation can be obtained using Shannon's expansion of the function. (See the Subsection *Decomposition of Switching Functions* in Section 9.8.) In general, the complexity of the implementation will depend upon which function inputs are used as the MUX select inputs, so it is necessary to try different combinations to obtain the simplest solution.

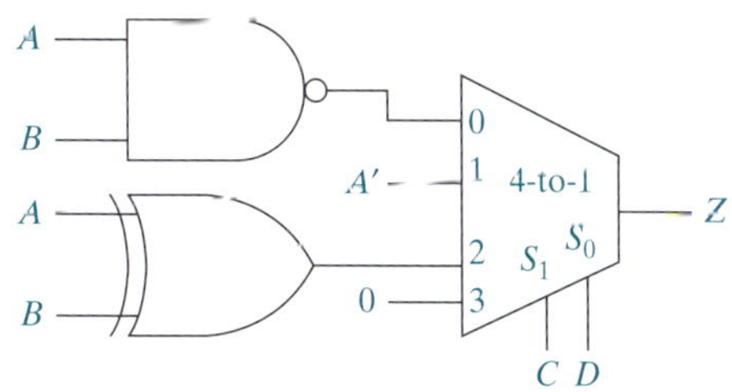

FIGURE 9-9 Four-Variable Function Implemented with a 4-to-1 MUX

9.3 Three-State Buffers

A gate output can only be connected to a limited number of other device inputs without degrading the performance of a digital system. A simple buffer may be used to increase the driving capability of a gate output. Figure 9-10 shows a buffer connected between a gate output and several gate inputs. Because no bubble is present

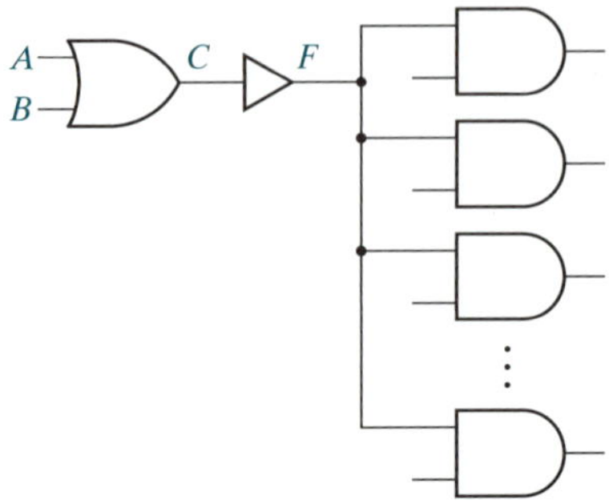

FIGURE 9-10 Gate Circuit with Added Buffer

at the buffer output, this is a noninverting buffer, and the logic values of the buffer input and output are the same, that is, $F = C$.

Normally, a logic circuit will not operate correctly if the outputs of two or more gates or other logic devices are directly connected to each other. For example, if one gate has a 0 output (a low voltage) and another has a 1 output (a high voltage), when the gate outputs are connected together the resulting output voltage may be some intermediate value that does not clearly represent either a 0 or a 1. In some cases, damage to the gates may result if the outputs are connected together.

Use of three-state logic permits the outputs of two or more gates or other logic devices to be connected together. Figure 9-11 shows a three-state buffer and its logical equivalent. When the enable input B is 1, the output C equals A; when B is 0, the output C acts like an open circuit. In other words, when B is 0, the output C is effectively disconnected from the buffer output so that no current can flow. This is often referred to as a Hi-Z (high-impedance) state of the output because the circuit offers a very high resistance or impedance to the flow of current. Three-state buffers are also called tri-state buffers.

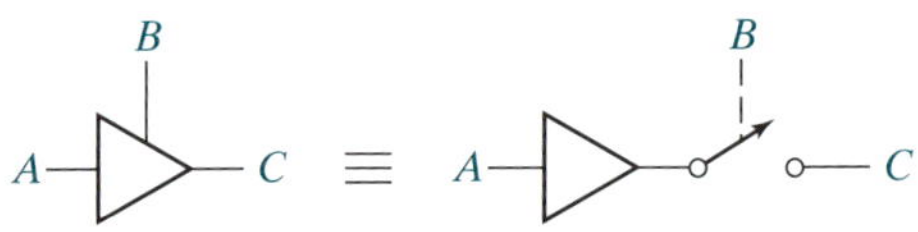

FIGURE 9-11 Three-State Buffer

Figure 9-12 shows the truth tables for four types of three-state buffers. In Figures 9-12(a) and (b), the enable input B is not inverted, so the buffer output is enabled when $B = 1$ and disabled when $B = 0$. That is, the buffer operates normally when $B = 1$, and the buffer output is effectively an open circuit when $B = 0$. We use the symbol **Z** to represent this high-impedance state. In Figure 9-12(b), the buffer output is inverted so that $C = A'$ when the buffer is enabled. The buffers in 9-12(c) and (d) operate the same as in (a) and (b) except that the enable input is inverted, so the buffer is enabled when $B = 0$.

In Figure 9-13, the outputs of two three-state buffers are tied together. When $B = 0$, the top buffer is enabled, so that $D = A$; when $B = 1$, the lower buffer is enabled, so that $D = C$. Therefore, $D = B'A + BC$. This is logically equivalent to using a 2-to-1 multiplexer to select the A input when $B = 0$ and the C input when $B = 1$.

FIGURE 9-12 Four Kinds of Three-State Buffers

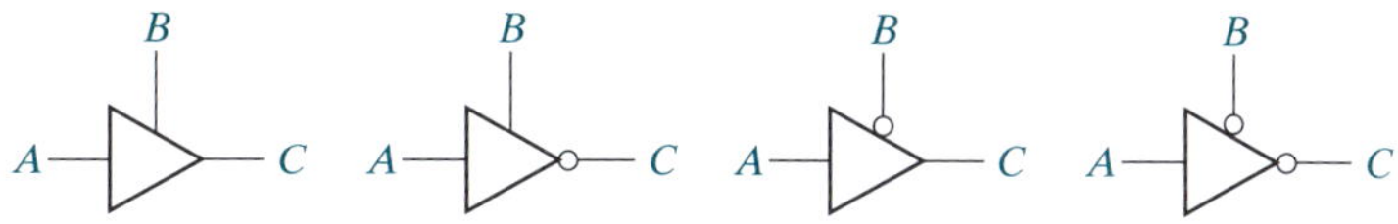

B	A	C
0	0	Z
0	1	Z
1	0	0
1	1	1

(a)

B	A	C
0	0	Z
0	1	Z
1	0	1
1	1	0

(b)

B	A	C
0	0	0
0	1	1
1	0	Z
1	1	Z

(c)

B	A	C
0	0	1
0	1	0
1	0	Z
1	1	Z

(d)

When we connect two three-state buffer outputs together, as shown in Figure 9-14, if one of the buffers is disabled (output = Z), the combined output F is the same as the other buffer output. If both buffers are disabled, the output is Z. If both buffers are enabled, a conflict can occur. If $A = 0$ and $C = 1$, we do not know what the hardware will do, so the F output is unknown (X). If one of the buffer inputs is unknown, the F output will also be unknown. The table in Figure 9-14 summarizes the operation of the circuit. S_1 and S_2 represent the outputs the two buffers would have if they were not connected together. When a bus is driven by three-state buffers, we call it a three-state bus. The signals on this bus can have values of 0, 1, Z, and perhaps X.

A multiplexer may be used to select one of several sources to drive a device input. For example, if an adder input must come from four different sources, a 4-to-1 MUX may be used to select one of the four sources. An alternative is to set up a three-state bus, using three-state buffers to select one of the sources (see Figure 9-15). In this

FIGURE 9-13 Data Selection Using Three-State Buffers

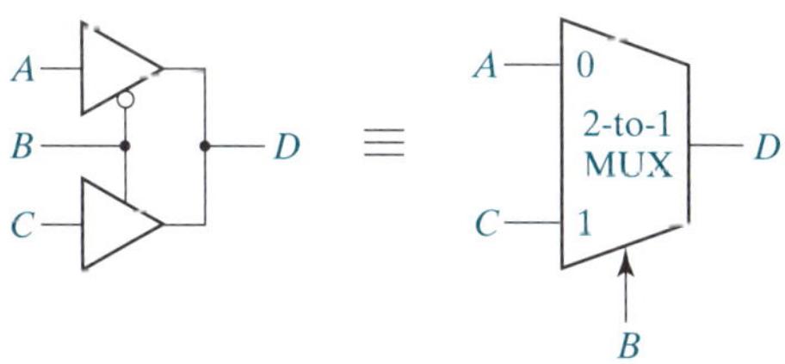

FIGURE 9-14 Circuit with Two Three-State Buffers

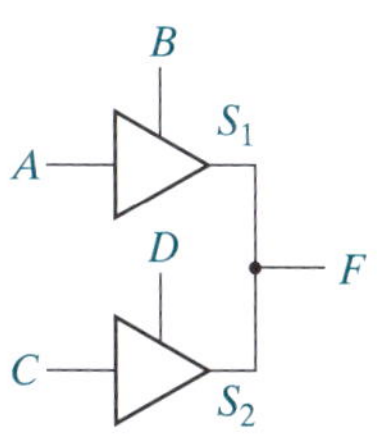

S_1 \ S_2	X	0	1	Z
X	X	X	X	X
0	X	0	X	0
1	X	X	1	1
Z	X	0	1	Z

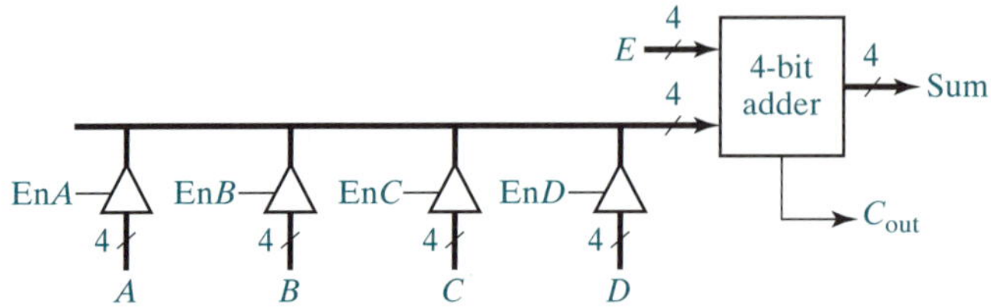

FIGURE 9-15
4-Bit Adder with Four Sources for One Operand

circuit, each buffer symbol actually represents four three-state buffers that have a common enable signal.

Integrated circuits are often designed using bi-directional pins for input and output. Bi-directional means that the same pin can be used as an input pin and as an output pin, but not both at the same time. To accomplish this, the circuit output is connected to the pin through a three-state buffer, as shown in Figure 9-16. When the buffer is enabled, the pin is driven with the output signal. When the buffer is disabled, an external source can drive the input pin.

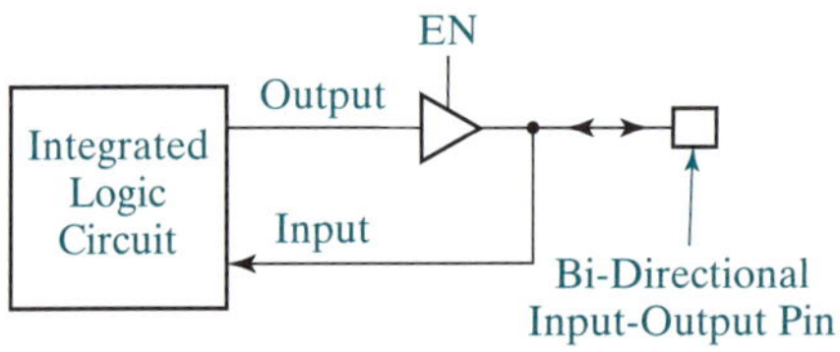

FIGURE 9-16
Integrated Circuit with Bi-Directional Input-Output Pin

9.4 Decoders and Encoders

The decoder is another commonly used type of integrated circuit. Figure 9-17 shows the diagram and truth table for a 3-to-8 line decoder. This decoder generates all of the minterms of the three input variables. Exactly one of the output lines will be 1 for each combination of the values of the input variables.

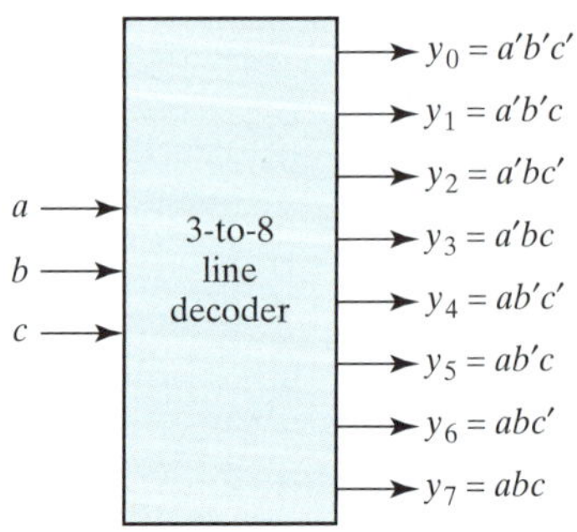

a b c	y_0	y_1	y_2	y_3	y_4	y_5	y_6	y_7
0 0 0	1	0	0	0	0	0	0	0
0 0 1	0	1	0	0	0	0	0	0
0 1 0	0	0	1	0	0	0	0	0
0 1 1	0	0	0	1	0	0	0	0
1 0 0	0	0	0	0	1	0	0	0
1 0 1	0	0	0	0	0	1	0	0
1 1 0	0	0	0	0	0	0	1	0
1 1 1	0	0	0	0	0	0	0	1

FIGURE 9-17
A 3-to-8 Line Decoder

Figure 9-18 illustrates a 4-to-10 decoder. This decoder has inverted outputs (indicated by the small circles). For each combination of the values of the inputs, exactly one of the output lines will be 0. When a binary-coded-decimal digit is used as an input to this decoder, one of the output lines will go low to indicate which of the 10 decimal digits is present.

FIGURE 9-18 A 4-to-10 Line Decoder

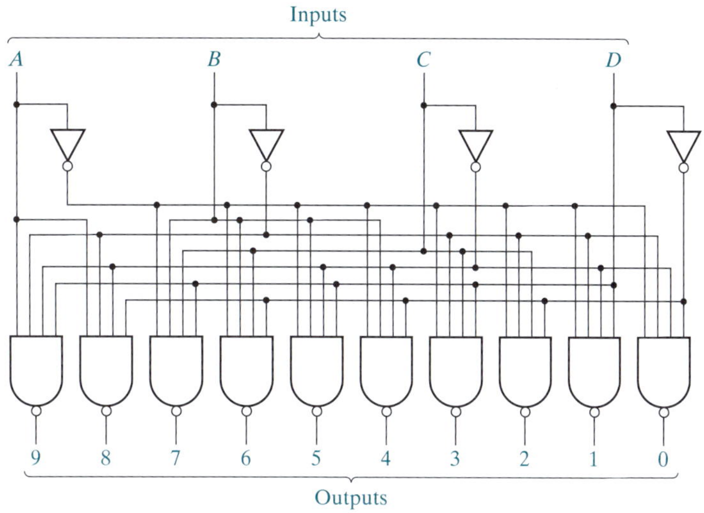

(a) Logic diagram

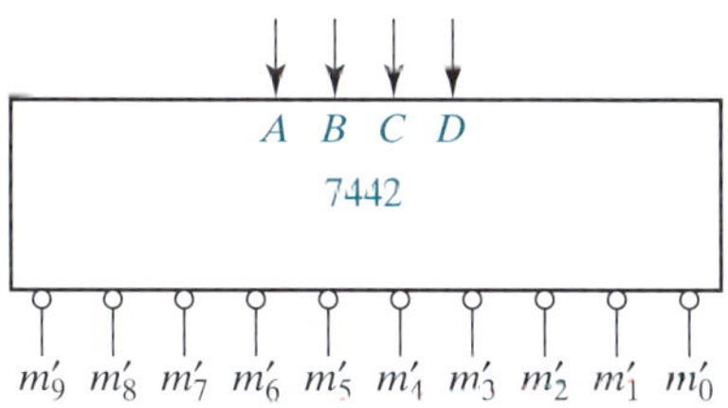

(b) Block diagram

BCD Input				Decimal Output									
A	*B*	*C*	*D*	0	1	2	3	4	5	6	7	8	9
0	0	0	0	0	1	1	1	1	1	1	1	1	1
0	0	0	1	1	0	1	1	1	1	1	1	1	1
0	0	1	0	1	1	0	1	1	1	1	1	1	1
0	0	1	1	1	1	1	0	1	1	1	1	1	1
0	1	0	0	1	1	1	1	0	1	1	1	1	1
0	1	0	1	1	1	1	1	1	0	1	1	1	1
0	1	1	0	1	1	1	1	1	1	0	1	1	1
0	1	1	1	1	1	1	1	1	1	1	0	1	1
1	0	0	0	1	1	1	1	1	1	1	1	0	1
1	0	0	1	1	1	1	1	1	1	1	1	1	0
1	0	1	0	1	1	1	1	1	1	1	1	1	1
1	0	1	1	1	1	1	1	1	1	1	1	1	1
1	1	0	0	1	1	1	1	1	1	1	1	1	1
1	1	0	1	1	1	1	1	1	1	1	1	1	1
1	1	1	0	1	1	1	1	1	1	1	1	1	1
1	1	1	1	1	1	1	1	1	1	1	1	1	1

(c) Truth Table

In general, an n-to-2^n line decoder generates all 2^n minterms (or maxterms) of the n input variables. The outputs are defined by the equations

$$y_i = m_i = M'_i, \qquad i = 0 \text{ to } 2^n - 1 \qquad \text{(noninverted outputs)} \qquad (9\text{-}5)$$

or

$$y_i = m'_i = M_i, \qquad i = 0 \text{ to } 2^n - 1 \qquad \text{(inverted outputs)} \qquad (9\text{-}6)$$

where m_i is a minterm of the n input variables and M_i is a maxterm.

Because an n-input decoder generates all of the minterms of n variables, n-variable functions can be realized by ORing together selected minterm outputs from a decoder. If the decoder outputs are inverted, then NAND gates can be used to generate the functions, as illustrated in the following example. Realize

$$f_1(a, b, c, d) = m_1 + m_2 + m_4 \text{ and } f_2(a, b, c, d) = m_4 + m_7 + m_9$$

using the decoder of Figure 9-18. Since a NAND gate ORs inverted signals, f_1 and f_2 can be generated using NAND gates, as shown in Figure 9-19.

An encoder performs the inverse function of a decoder. Figure 9-20 shows an 8-to-3 priority encoder with inputs y_0 through y_7. If input y_i is 1 and the other inputs are 0, then the abc outputs represent a binary number equal to i. For example, if

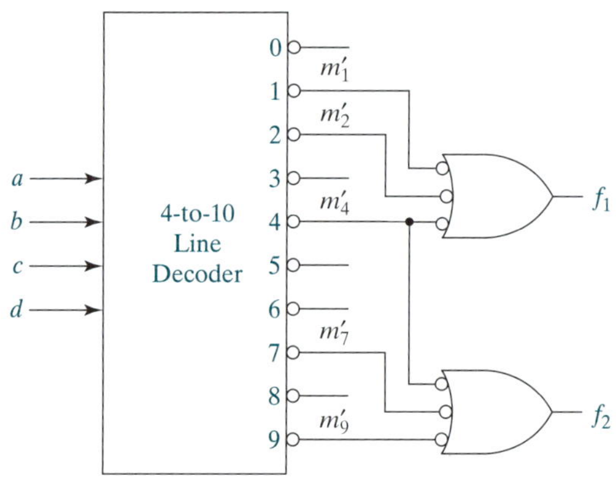

FIGURE 9-19 Realization of a Multiple-Output Circuit Using a Decoder

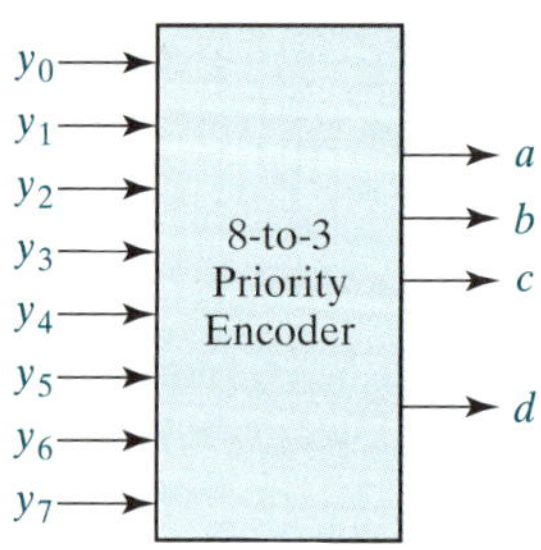

y_0	y_1	y_2	y_3	y_4	y_5	y_6	y_7	a	b	c	d
0	0	0	0	0	0	0	0	0	0	0	0
1	0	0	0	0	0	0	0	0	0	0	1
X	1	0	0	0	0	0	0	0	0	1	1
X	X	1	0	0	0	0	0	0	1	0	1
X	X	X	1	0	0	0	0	0	1	1	1
X	X	X	X	1	0	0	0	1	0	0	1
X	X	X	X	X	1	0	0	1	0	1	1
X	X	X	X	X	X	1	0	1	1	0	1
X	X	X	X	X	X	X	1	1	1	1	1

FIGURE 9-20 An 8-to-3 Priority Encoder

$y_3 = 1$, then $abc = 011$. If more than one input can be 1 at the same time, the output can be defined using a priority scheme. The truth table in Figure 9-20 uses the following scheme: If more than one input is 1, the highest numbered input determines the output. For example, if inputs y_1, y_4, and y_5 are 1, the output is $abc = 101$. The X's in the table are don't-cares; for example, if y_5 is 1, we do not care what inputs y_0 through y_4 are. Output d is 1 if any input is 1, otherwise, d is 0. This signal is needed to distinguish the case of all 0 inputs from the case where only y_0 is 1.

9.5 Read-Only Memories

A read-only memory (ROM) consists of an array of semiconductor devices that are interconnected to store an array of binary data. Once binary data is stored in the ROM, it can be read out whenever desired, but the data that is stored cannot be changed under normal operating conditions. Figure 9-21(a) shows a ROM which has three input lines and four output lines. Figure 9-21(b) shows a typical truth table which relates the ROM inputs and outputs. For each combination of input values on the three input lines, the corresponding pattern of 0's and 1's appears on the ROM output lines. For example, if the combination $ABC = 010$ is applied to the input lines, the pattern $F_0F_1F_2F_3 = 0111$ appears on the output lines. Each of the output patterns that is stored in the ROM is called a *word*. Because the ROM has three input lines, we have 2^3 = eight different combinations of input values. Each input combination serves as an *address* which can select one of the eight words stored in the memory. Because there are four output lines, each word is four bits long, and the size of this ROM is 8 words × 4 bits.

A ROM which has n input lines and m output lines (Figure 9-22) contains an array of 2^n words, and each word is m bits long. The input lines serve as an address to select one of the 2^n words. When an input combination is applied to the ROM, the pattern of 0's and 1's which is stored in the corresponding word in the memory appears at the output lines. For the example in Figure 9-22, if 00 . . . 11 is applied to the input (address lines) of the ROM, the word 110 . . . 010 will be selected and transferred to the output lines. A $2^n \times m$ ROM can realize m functions of n variables because it can store a truth table with 2^n rows and m columns. Typical sizes for commercially available ROMs range from 32 words × 4 bits to 512K words × 8 bits, or larger.

FIGURE 9-21
An 8-Word × 4-Bit ROM

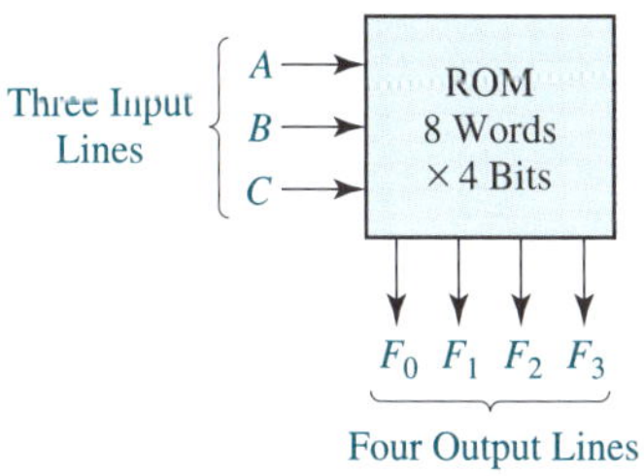

(a) Block diagram

A B C	F_0 F_1 F_2 F_3
0 0 0	1 0 1 0
0 0 1	1 0 1 0
0 1 0	0 1 1 1
0 1 1	0 1 0 1
1 0 0	1 1 0 0
1 0 1	0 0 0 1
1 1 0	1 1 1 1
1 1 1	0 1 0 1

Typical Data Stored in ROM (2^3 words of 4 bits each)

(b) Truth table for ROM

FIGURE 9-22
Read-Only Memory with n Inputs and m Outputs

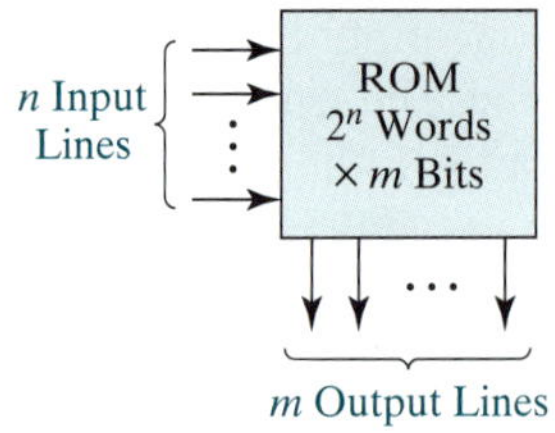

n Input Variables	m Output Variables
00 · · · 00	100 · · · 110
00 · · · 01	010 · · · 111
00 · · · 10	101 · · · 101
00 · · · 11	110 · · · 010
⋮	⋮
11 · · · 00	001 · · · 011
11 · · · 01	110 · · · 110
11 · · · 10	011 · · · 000
11 · · · 11	111 · · · 101

Typical Data Array Stored in ROM (2^n words of m bits each)

A ROM basically consists of a decoder and a memory array, as shown in Figure 9-23. When a pattern of n 0's and 1's is applied to the decoder inputs, exactly one of the 2^n decoder outputs is 1. This decoder output line selects one of the words in the memory array, and the bit pattern stored in this word is transferred to the memory output lines.

Figure 9-24 illustrates one possible internal structure of the 8-word × 4-bit ROM shown in Figure 9-21. The decoder generates the eight minterms of the three input variables. The memory array forms the four output functions by ORing together selected minterms. A switching element is placed at the intersection of a *word line* and an *output line* if the corresponding minterm is to be included in the output function; otherwise, the switching element is omitted (or not connected). If a switching element connects an output line to a word line which is 1, the output line will be 1. Otherwise, the pull-down resistors at the top of Figure 9-24 cause the output line to be 0. So the switching elements which are connected in this way in the memory array effectively form an OR gate for each of the output functions. For example, m_0, m_1, m_4, and m_6 are ORed together to form F_0. Figure 9-25 shows the equivalent OR gate.

In general, those minterms which are connected to output line F by switching elements are ORed together to form the output F_i. Thus, the ROM in Figure 9-24 generates the following functions:

$$
\begin{aligned}
F_0 &= \Sigma\, m(0, 1, 4, 6) = A'B' + AC' \\
F_1 &= \Sigma\, m(2, 3, 4, 6, 7) = B + AC' \\
F_2 &= \Sigma\, m(0, 1, 2, 6) = A'B' + BC' \\
F_3 &= \Sigma\, m(2, 3, 5, 6, 7) = AC + B
\end{aligned} \tag{9-7}
$$

FIGURE 9-23
Basic ROM Structure

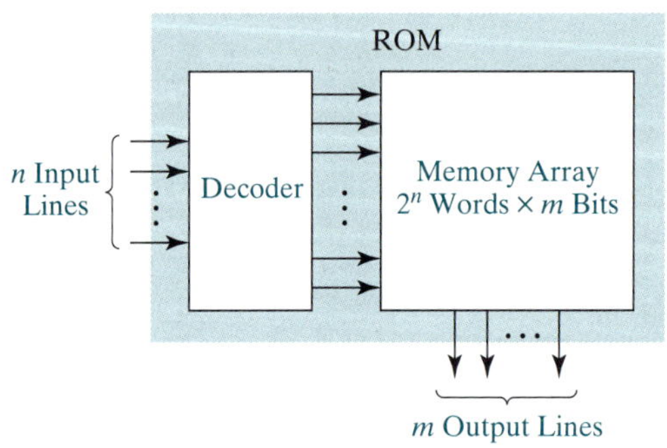

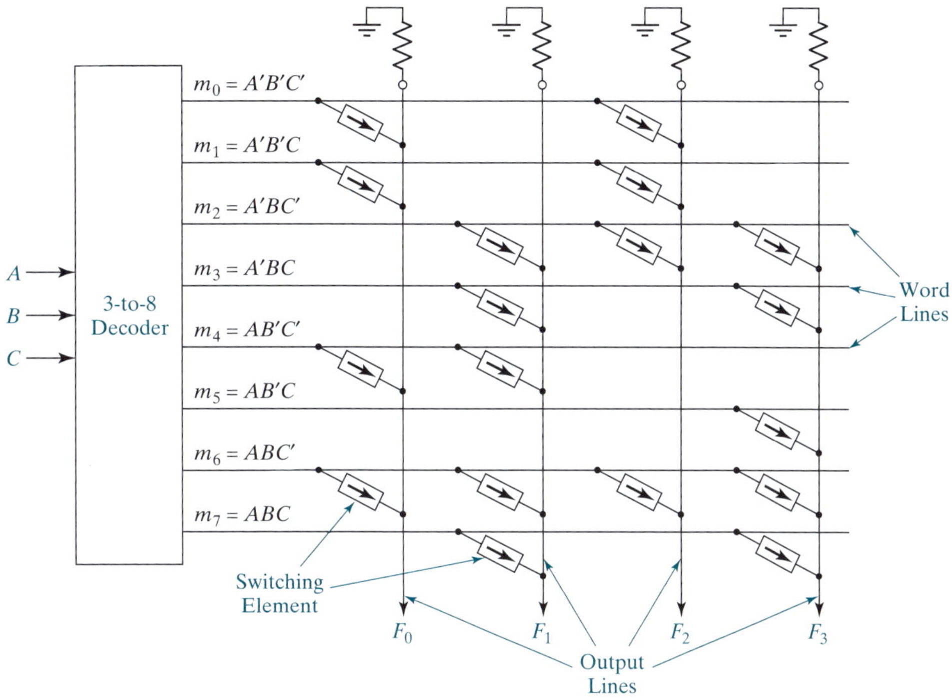

FIGURE 9-24 An 8-Word × 4-Bit ROM

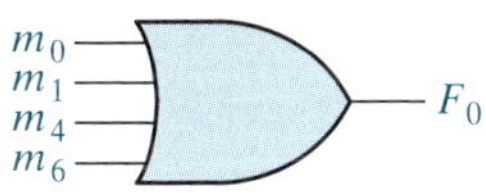

FIGURE 9-25 Equivalent OR Gate for F_0

The contents of a ROM are usually specified by a truth table. The truth table of Figure 9-21(b) specifies the ROM in Figure 9-24. Note that a 1 or 0 in the output part of the truth table corresponds to the presence or absence of a switching element in the memory array of the ROM.

Multiple-output combinational circuits can easily be realized using ROMs. As an example, we will realize a code converter that converts a 4-bit binary number to a hexadecimal digit and outputs the 7-bit ASCII code. Figure 9-26 shows the truth table and logic circuit for the converter. Because $A_5 = A_4$, and $A_6 = A_4'$ the ROM needs only five outputs. Because there are four address lines, the ROM size is 16 words by 5 bits. Columns $A_4 A_3 A_2 A_1 A_0$ of the truth table are stored in the ROM. Figure 9-27 shows an internal diagram of the ROM. The switching elements at the intersections of the rows and columns of the memory array are indicated using X's. An X indicates that the switching element is present and connected, and no X indicates that the corresponding element is absent or not connected.

Three common types of ROMs are mask-programmable ROMs, programmable ROMs (PROMs), and electrically erasable programmable ROMs (EEPROMs). At the time of manufacture, the data array is permanently stored in a mask-programmable ROM. This is accomplished by selectively including or omitting the switching elements at the row- column intersections of the memory array. This requires preparation of a

FIGURE 9-26
Hexadecimal-to-ASCII Code Converter

Input $W\ X\ Y\ Z$	Hex Digit	ASCII Code for Hex Digit $A_6\ A_5\ A_4\ A_3\ A_2\ A_1\ A_0$
0 0 0 0	0	0 1 1 0 0 0 0
0 0 0 1	1	0 1 1 0 0 0 1
0 0 1 0	2	0 1 1 0 0 1 0
0 0 1 1	3	0 1 1 0 0 1 1
0 1 0 0	4	0 1 1 0 1 0 0
0 1 0 1	5	0 1 1 0 1 0 1
0 1 1 0	6	0 1 1 0 1 1 0
0 1 1 1	7	0 1 1 0 1 1 1
1 0 0 0	8	0 1 1 1 0 0 0
1 0 0 1	9	0 1 1 1 0 0 1
1 0 1 0	A	1 0 0 0 0 0 1
1 0 1 1	B	1 0 0 0 0 1 0
1 1 0 0	C	1 0 0 0 0 1 1
1 1 0 1	D	1 0 0 0 1 0 0
1 1 1 0	E	1 0 0 0 1 0 1
1 1 1 1	F	1 0 0 0 1 1 0

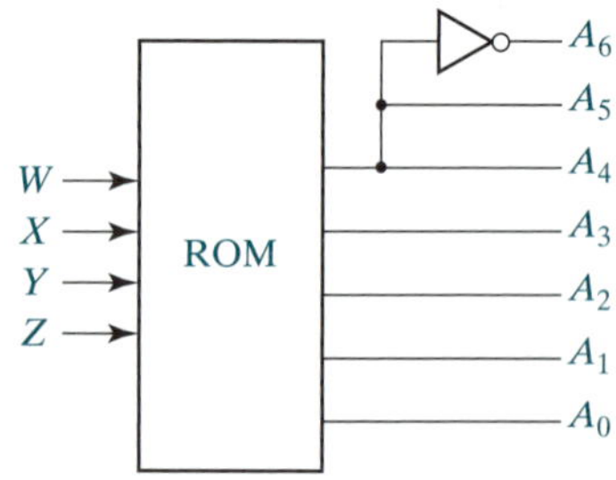

FIGURE 9-27
ROM Realization of Code Converter

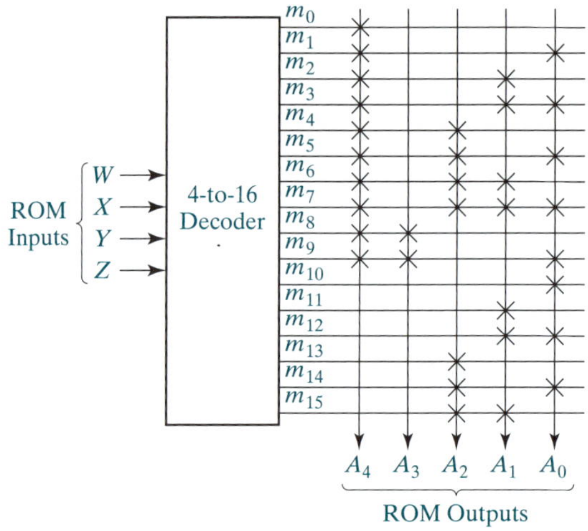

special *mask*, which is used during fabrication of the integrated circuit. Preparation of this mask is expensive, so the use of mask-programmable ROMs is economically feasible only if a large quantity (typically several thousand or more) is required with the same data array. If a small quantity of ROMs is required with a given data array, EEPROMs may be used.

Modification of the data stored in a ROM is often necessary during the developmental phases of a digital system, so EEPROMs are used instead of mask-programmable ROMs. EEPROMs use a special charge-storage mechanism to enable or disable the switching elements in the memory array. A PROM programmer is used to provide appropriate voltage pulses to store electronic charges in the memory array locations. Data stored in this manner is generally permanent until erased.

After erasure, a new set of data can be stored in the EEPROM. An EEPROM can be erased and reprogrammed only a limited number of times, typically 100 to 1000 times. Flash memories are similar to EEPROMs, except that they use a different charge-storage mechanism. They usually have built-in programming and erase capability so that data can be written to the flash memory while it is in place in a circuit without the need for a separate programmer.

9.6 Programmable Logic Devices

A programmable logic device (or PLD) is a general name for a digital integrated circuit capable of being programmed to provide a variety of different logic functions. In this section we will discuss several types of combinational PLDs, and later we will discuss sequential PLDs. Simple combinational PLDs are capable of realizing from 2 to 10 functions of 4 to 16 variables with a single integrated circuit. More complex PLDs may contain thousands of gates and flip-flops. Thus, a single PLD can replace a large number of integrated circuits, and this leads to lower cost designs. When a digital system is designed using a PLD, changes in the design can easily be made by changing the programming of the PLD without having to change the wiring in the system.

Programmable Logic Arrays

A programmable logic array (PLA) performs the same basic function as a ROM. A PLA with n inputs and m outputs (Figure 9-28) can realize m functions of n variables. The internal organization of the PLA is different from that of the ROM. The decoder is replaced with an AND array which realizes selected product terms of the input variables. The OR array ORs together the product terms needed to form the output functions, so a PLA implements a sum-of-products expression, while a ROM directly implements a truth table.

Figure 9-29 shows a PLA which realizes the same functions as the ROM of Figure 9-24. Product terms are formed in the AND array by connecting switching elements at appropriate points in the array. For example, to form $A'B'$, switching elements are used to connect the first word line with the A' and B' lines.

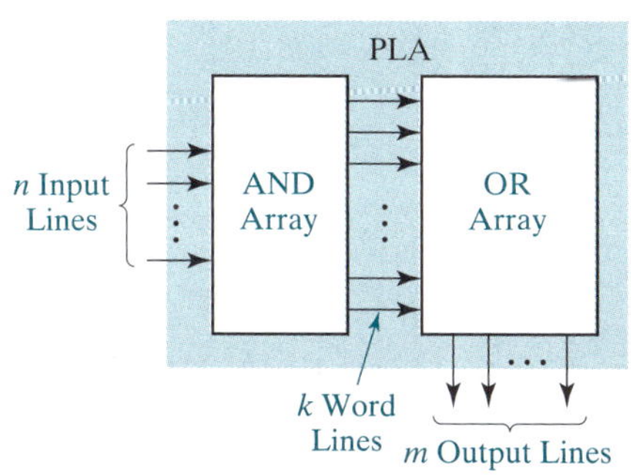

FIGURE 9-28 Programmable Logic Array Structure

FIGURE 9-29
PLA with Three Inputs, Five Product Terms, and Four Outputs

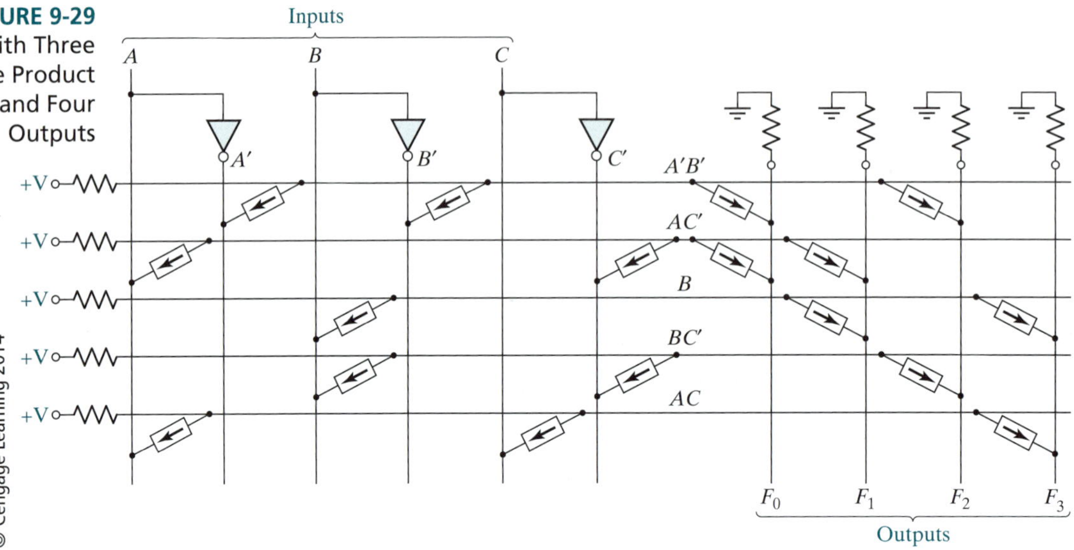

Switching elements are connected in the OR array to select the product terms needed for the output functions. For example, because $F_0 = A'B' + AC'$, switching elements are used to connect the $A'B'$ and AC' lines to the F_0 line. The connections in the AND and OR arrays of this PLA make it equivalent to the AND-OR array of Figure 9-30.

The contents of a PLA can be specified by a PLA table. Table 9-1 specifies the PLA in Figure 9-29. The input side of the table specifies the product terms. The symbols 0, 1, and – indicate whether a variable is complemented, not complemented, or

FIGURE 9-30
AND-OR Array Equivalent to Figure 9-29

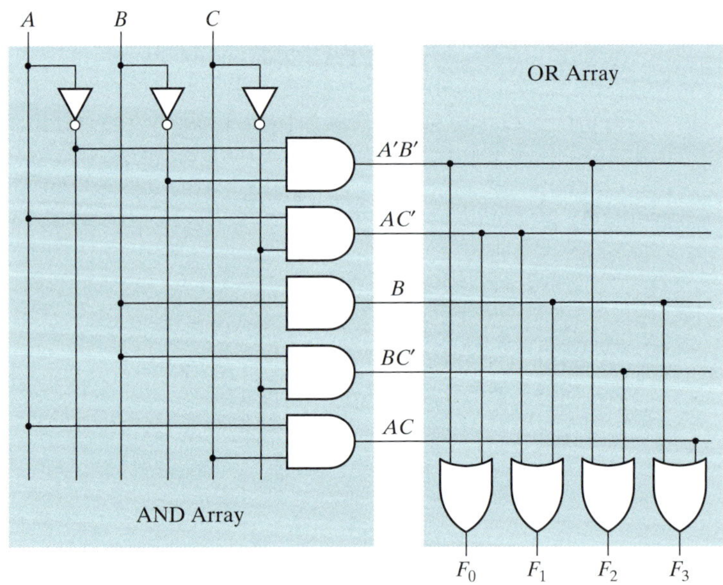

TABLE 9-1
PLA Table for Figure 9-29

Product Term	Inputs $A\ B\ C$	Outputs $F_0\ F_1\ F_2\ F_3$	
$A'B'$	0 0 –	1 0 1 0	$F_0 = A'B' + AC'$
AC'	1 – 0	1 1 0 0	$F_1 = AC' + B$
B	– 1 –	0 1 0 1	$F_2 = A'B' + BC'$
BC'	– 1 0	0 0 1 0	$F_3 = B + AC$
AC	1 – 1	0 0 0 1	

not present in the corresponding product term. The output side of the table specifies which product terms appear in each output function. A 1 or 0 indicates whether a given product term is present or not present in the corresponding output function. Thus, the first row of Table 9-1 indicates that the term $A'B'$ is present in output functions F_0 and F_2, and the second row indicates that AC' is present in F_0 and F_1.

Next, we will realize Equation (7-25) using a PLA. Using the minimum multiple-output solution given in Equation (7-25b), we can construct a PLA table, Figure 9-31(a), with one row for each distinct product term. Figure 9-31(b) shows the corresponding PLA structure, which has four inputs, six product terms, and three outputs. A dot at the intersection of a word line and an input or output line indicates the presence of a switching element in the array.

FIGURE 9-31
PLA Realization of Equation (7-25b)

$a\ b\ c\ d$	$f_1\ f_2\ f_3$
0 1 – 1	1 1 0
1 1 – 1	1 0 1
1 0 0 –	1 0 1
– 0 1 –	1 0 0
– – 1 –	0 1 0
– 1 1 –	0 0 1

(a) PLA table

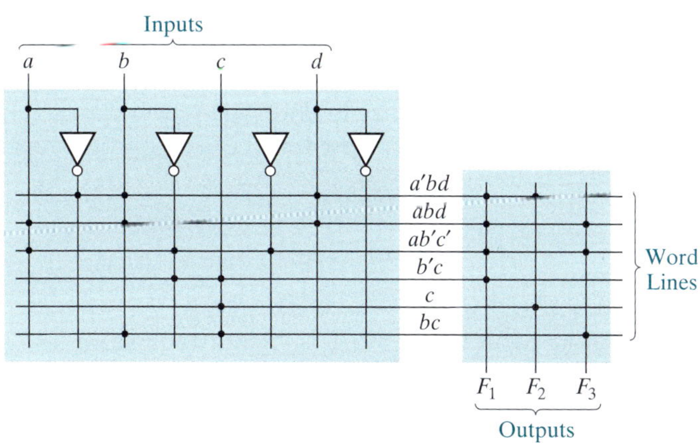

(b) PLA structure

A PLA table is significantly different than a truth table for a ROM. In a truth table each row represents a minterm; therefore, exactly one row will be selected by each combination of input values. The 0's and 1's of the output portion of the selected row determine the corresponding output values. On the other hand, each row in a PLA table represents a general product term. Therefore, zero, one, or more rows may be selected by each combination of input values. To determine the value of f_i for a given input combination, the values of f_i in the selected rows of the PLA table must be ORed together. The following examples refer to the PLA table of Figure 9-31(a). If $abcd = 0001$, no rows are selected, and all f's are 0. If $abcd = 1001$, only the third row is selected, and $f_1f_2f_3 = 101$. If $abcd = 0111$, the first, fifth, and sixth rows are selected. Therefore, $f_1 = 1 + 0 + 0 = 1$, $f_2 = 1 + 1 + 0 = 1$, and $f_3 = 0 + 0 + 1 = 1$.

Both mask-programmable and field-programmable PLAs are available. The mask-programmable type is programmed at the time of manufacture in a manner similar to mask-programmable ROMs. The field-programmable logic array (FPLA) has programmable interconnection points that use electronic charges to store a pattern in the AND and OR arrays. An FPLA with 16 inputs, 48 product terms, and eight outputs can be programmed to implement eight functions of 16 variables, provided that the total number of product terms does not exceed 48.

When the number of input variables is small, a PROM may be more economical to use than a PLA. However, when the number of input variables is large, PLAs often provide a more economical solution than PROMs. For example, to realize eight functions of 24 variables would require a PROM with over 16 million 8-bit words. Because PROMs of this size are not readily available, the functions would have to be decomposed so that they could be realized using a number of smaller PROMs. The same eight functions of 24 variables could easily be realized using a single PLA, provided that the total number of product terms is small. If more terms are required, the outputs of several PLAs can be ORed together.

Programmable Array Logic

The PAL (programmable array logic) is a special case of the programmable logic array in which the AND array is programmable and the OR array is fixed. The basic structure of the PAL is the same as the PLA shown in Figure 9-28. Because only the AND array is programmable, the PAL is less expensive than the more general PLA, and the PAL is easier to program. For this reason, logic designers frequently use PALs to replace individual logic gates when several logic functions must be realized.

Figure 9-32(a) represents a segment of an unprogrammed PAL. The symbol

represents an input buffer which is logically equivalent to

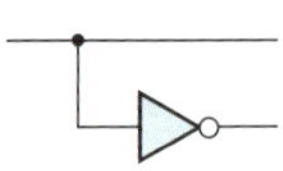

A buffer is used because each PAL input must drive many AND gate inputs. When the PAL is programmed, some of the interconnection points are programmed to make the desired connections to the AND gate inputs. Connections to the AND gate inputs in a PAL are represented by X's as shown:

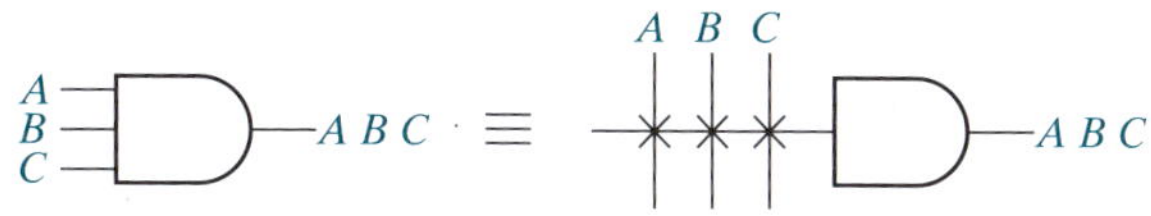

As an example, we will use the PAL segment of Figure 9-32(a) to realize the function $I_1I_2' + I_1'I_2$. The X's in Figure 9-32(b) indicate that I_1 and I_2' lines are connected to the first AND gate, and the I_1' and I_2 lines are connected to the other gate.

When designing with PALs, we must simplify our logic equations and try to fit them into one (or more) of the available PALs. Unlike the more general PLA, the AND terms cannot be shared among two or more OR gates; therefore, each function to be realized can be simplified by itself without regard to common terms. For a given type of PAL, the number of AND terms that feed each output OR gate is fixed and limited. If the number of AND terms in a simplified function is too large, we may be forced to choose a PAL with more gate inputs and fewer outputs.

FIGURE 9-32
PAL Segment

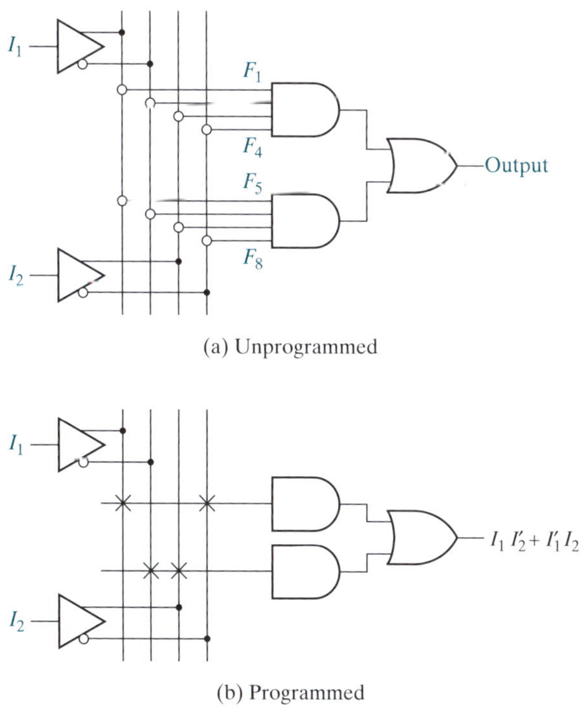

As an example of programming a PAL, we will implement a full adder. The logic equations for the full adder are

$$Sum = X'Y'C_{in} + X'YC'_{in} + XY'C'_{in} + XYC_{in}$$
$$C_{out} = XC_{in} + YC_{in} + XY$$

Figure 9-33 shows a section of a PAL where each OR gate is driven by four AND gates. The X's on the diagram show the connections that are programmed into the PAL to implement the full adder equations. For example, the first row of X's implements the product term $X'Y'C_{in}$.

FIGURE 9-33 Implementation of a Full Adder Using a PAL

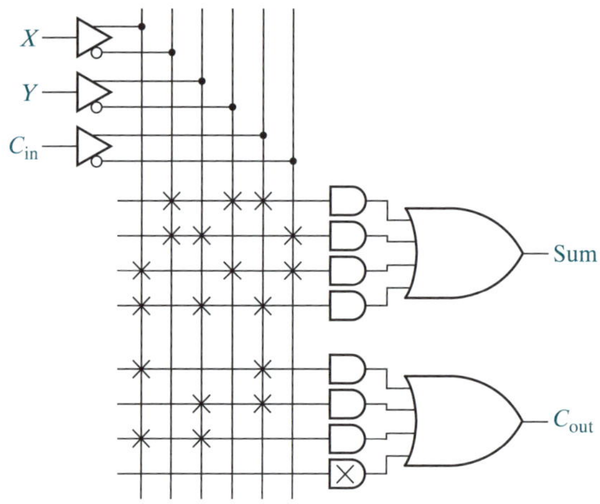

9.7 Complex Programmable Logic Devices

As integrated circuit technology continues to improve, more and more gates can be placed on a single chip. This has allowed the development of complex programmable logic devices (CPLDs). Instead of a single PAL or PLA on a chip, many PALs or PLAs can be placed on a single CPLD chip and interconnected. When storage elements such as flip-flops are also included on the same IC, a small digital system can be implemented with a single CPLD.

Figure 9-34 shows the basic architecture of a Xilinx XCR3064XL CPLD. This CPLD has four function blocks, and each block has 16 associated macrocells (MC1, MC2, . . .). Each function block is a programmable AND-OR array that is configured as a PLA. Each macrocell contains a flip-flop and multiplexers that route signals from the function block to the input-output (I/O) block or to the interconnect array (IA). The IA selects signals from the macrocell outputs or I/O blocks and connects them back to function block inputs. Thus, a signal generated in one function block can be used as an input to any other function block. The I/O blocks provide an interface between the bi-directional I/O pins on the IC and the interior of the CPLD.

FIGURE 9-34 Architecture of Xilinx XCR3064XL CPLD (Figure based on figures and text owned by Xilinx, Inc., Courtesy of Xilinx, Inc. © Xilinx, Inc. 1999–2003. All rights reserved.)

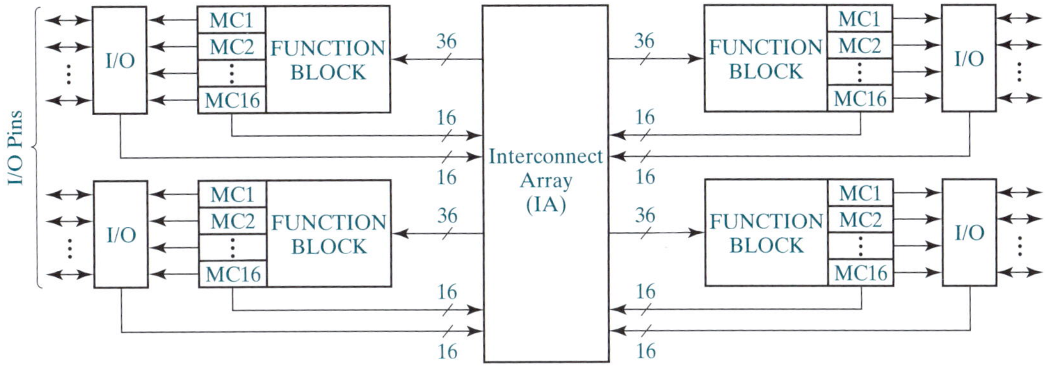

Figure 9-35 shows how a signal generated in the PLA is routed to an I/O pin through a macrocell. Any of the 36 outputs from the IA (or their complements) can be connected to any inputs of the 48 AND gates. Each OR gate can accept up to 48 product term inputs from the AND array. The macrocell logic in this diagram is a simplified version of the actual logic. The first MUX (1) can be programmed to select the OR-gate output or its complement. Details of the flip-flop operation will be discussed in Unit 11. The MUX (2) at the output of the macrocell can be programmed to select either the combinational output (G) or the flip-flop output (Q). This output goes to the interconnect array and to the output cell. The output cell includes a three-state buffer (3) to drive the I/O pin. The buffer enable input can be programmed from several sources. When the I/O pin is used as an input, the buffer must be disabled.

Sophisticated CAD software is available for fitting logic circuits into a PLD and for programming the interconnections within the PLD. The input to this software can be in several forms such as a logic circuit diagram, a set of logic equations, or code written in a hardware description language (HDL). Unit 10 discusses the use of an HDL. The CAD software processes the input, determines the logic equations to be implemented, fits these equations into the PLD, determines the required interconnections within the PLD, and generates a bit pattern for programming the PLD.

FIGURE 9-35 CPLD Function Block and Macrocell (A Simplified Version of XCR3064XL)

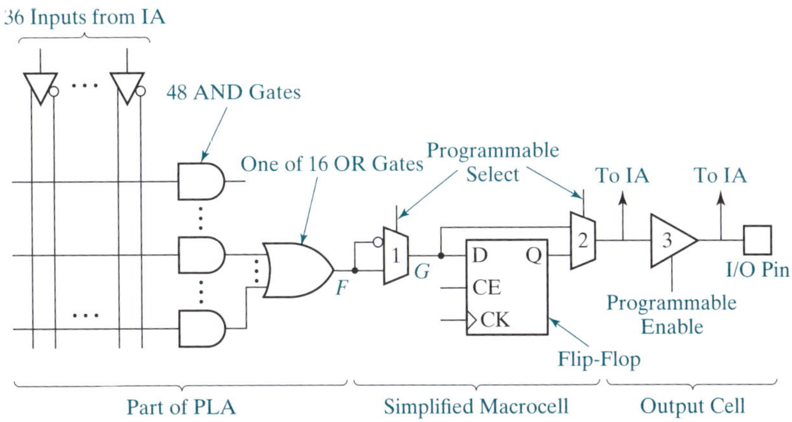

9.8 Field-Programmable Gate Arrays

In this section we introduce the use of field-programmable gate arrays (FPGAs) in combinational logic design. An FPGA is an IC that contains an array of identical logic cells with programmable interconnections. The user can program the functions realized by each logic cell and the connections between the cells. Figure 9-36 shows the layout of part of a typical FPGA. The interior of the FPGA consists of an array of logic cells, also called configurable logic blocks (CLBs). The array of CLBs is surrounded by a ring of input-output interface blocks. These I/O blocks connect the CLB signals to IC pins. The space between the CLBs is used to route connections between the CLB outputs and inputs.

Figure 9-37 shows a simplified version of a CLB. This CLB contains two function generators, two flip-flops, and various multiplexers for routing signals within the CLB. Each function generator has four inputs and can implement any function of up to four variables. The function generators are implemented as lookup tables (LUTs). A four-input LUT is essentially a reprogrammable ROM with 16 1-bit words. This ROM stores the truth table for the function being generated. The H multiplexer selects either F or G depending on the value of H_1. The CLB has two combinational outputs (X and Y) and two flip-flop outputs (XQ and YQ). The X and Y outputs and

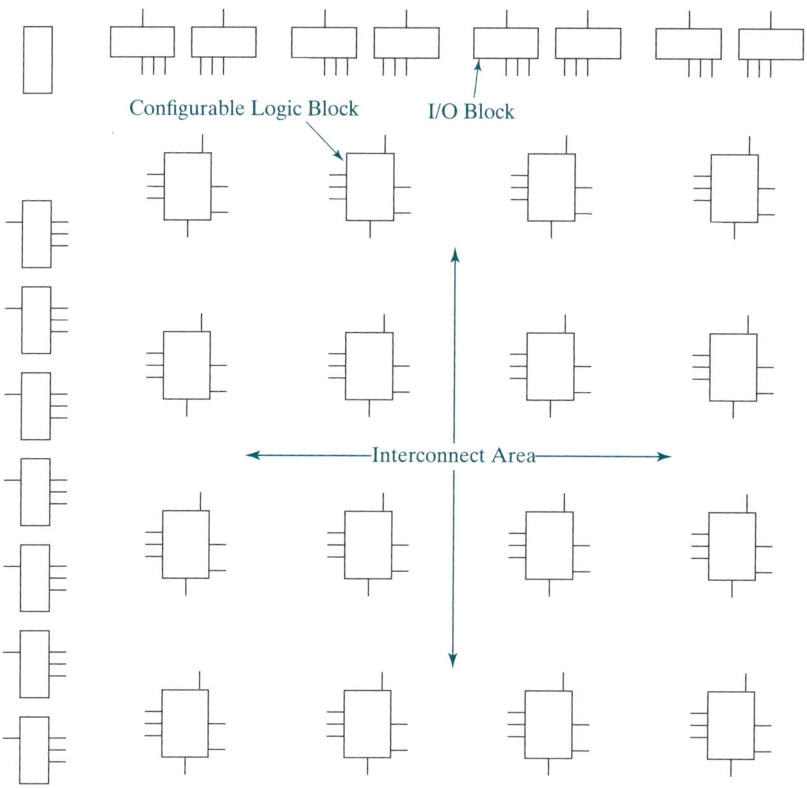

FIGURE 9-36 Layout of a Typical FPGA

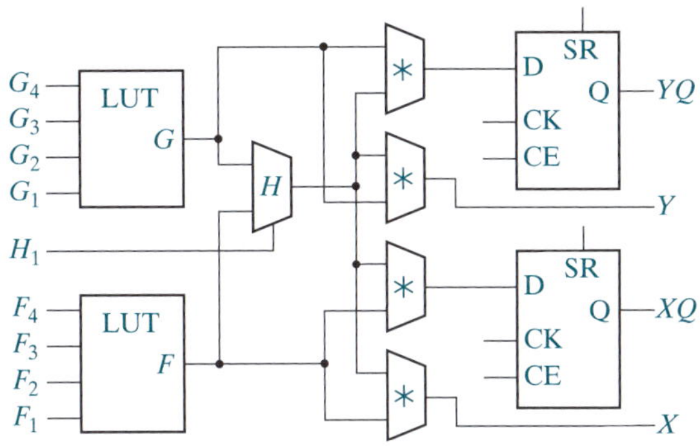

FIGURE 9-37 Simplified Configurable Logic Block (CLB)

the flip-flop inputs are selected by programmable multiplexers. The select inputs to these MUXes are programmed when the FPGA is configured. For example, the X output can come from the F function generator, and the Y output from the H multiplexer. Operation of the CLB flip-flops will be described in Unit 11.

Figure 9-38 shows one way to implement a function generator with inputs a, b, c, d. The numbers in the squares represent the bits stored in the LUT. These bits enable particular minterms. Because the function being implemented is stored as a truth table, a function with only one minterm or with as many as 15 minterms requires a single function generator. The functions

$$F = abc$$

and

$$F = a'b'c'd + a'b'cd + a'bc'd + a'bcd' + ab'c'd + ab'cd' + abc'd' + abcd$$

each require a single function generator.

FIGURE 9-38 Implementation of a Lookup Table (LUT)

a b c d	F
0 0 0 0	0
0 0 0 1	1
⋮	⋮
1 1 1 1	1

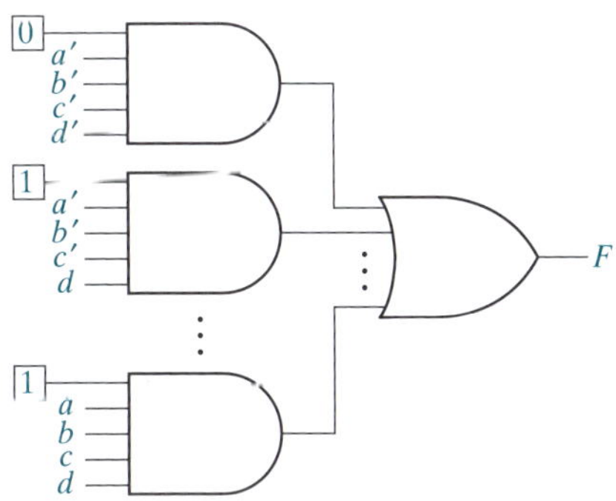

Decomposition of Switching Functions

In order to implement a switching function of more than four variables using 4-variable function generators, the function must be decomposed into subfunctions where each subfunction requires only four variables. One method of decomposition is based on Shannon's expansion theorem. We will first illustrate this theorem by expanding a function of the variables a, b, c, and d about the variable a:

$$f(a, b, c, d) = a' f(0, b, c, d) + a f(1, b, c, d) = a' f_0 + a f_1 \tag{9-8}$$

The 3-variable function $f_0 = f(0, b, c, d)$ is formed by replacing a with 0 in $f(a, b, c, d)$, and $f_1 = f(1, b, c, d)$ is formed by replacing a with 1 in $f(a, b, c, d)$. To verify that Equation (9-8) is correct, first set a to 0 on both sides, and then set a to 1 on both sides. An example of applying Equation (9-8) is as follows:

$$\begin{aligned} f(a, b, c, d) &= c'd' + a'b'c + bcd + ac' \\ &= a'(c'd' + b'c + bcd) + a(c'd' + bcd + c') \\ &= a'(c'd' + b'c + cd) + a(c' + bd) = a'f_0 + a f_1 \end{aligned} \tag{9-9}$$

Note that before simplification, the terms $c'd'$ and bcd appear in both f_0 and f_1 because neither term contains a' or a.

Expansion can also be accomplished using a truth table or a Karnaugh map. Figure 9-39 shows the map for Equation (9-9). The left half of the map where $a = 0$ is in effect a 3-variable map for $f_0(b, c, d)$. Looping terms on the left half gives $f_0 = c'd' + b'c + cd$, which is the same as the previous result. Similarly the right half where $a = 1$ is a 3-variable map for $f_1(b, c, d)$, and looping terms on the right half gives $f_1 = c' + bd$. The expressions for f_0 and f_1 obtained from the map are the same as those obtained algebraically in Equation (9-9).

The general form of Shannon's expansion theorem for expanding an n-variable function about the variable x_i is

$$\begin{aligned} &f(x_1, x_2, \ldots, x_{i-1}, x_i, x_{i+1}, \ldots, x_n) \\ &\quad = x_i' f(x_1, x_2, \ldots, x_{i-1}, 0, x_{i+1}, \ldots, x_n) + x_i f(x_1, x_2, \ldots, x_{i-1}, 1, x_{i+1}, \ldots, x_n) \\ &\quad = x_i' f_0 + x_i f_1 \end{aligned} \tag{9-10}$$

where f_0 is the $(n - 1)$-variable function obtained by setting x_i to 0 in the original function and f_1 is the $(n - 1)$-variable function obtained by setting x_i to 1 in the

FIGURE 9-39
Function Expansion Using a Karnaugh Map

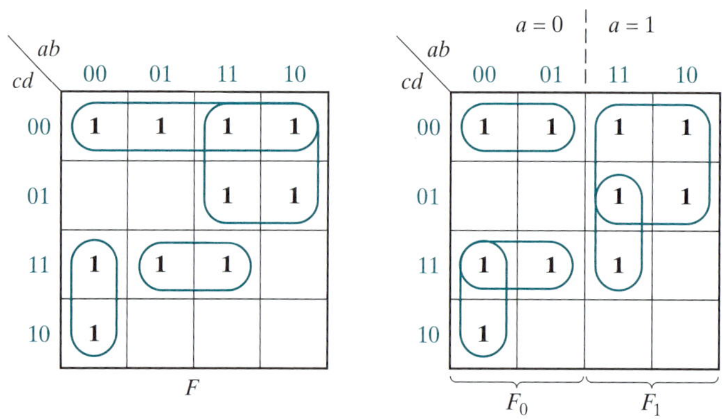

original function. The theorem is easily proved for switching algebra by first setting x_i to 0 in Equation (9-10), and, then, setting x_i to 1. Because both sides of the equation are equal for $x_i = 0$ and for $x_i = 1$, the theorem is true for switching algebra.

Applying the expansion theorem to a 5-variable function gives

$$f(a, b, c, d, e) = a' f(0, b, c, d, e) + a f(1, b, c, d, e) = a' f_0 + a f_1 \qquad (9\text{-}11)$$

This shows that any 5-variable function can be realized using two 4-variable function generators and a 2-to-1 MUX (Figure 9-40(a)). This implies that any 5-variable function can be implemented using a CLB of the type shown in Figure 9-37.

To realize a 6-variable function using 4-variable function generators, we apply the expansion theorem twice:

$$G(a, b, c, d, e, f) = a'G(0, b, c, d, e, f) + a\,G(1, b, c, d, e, f) = a'G_0 + a\,G_1$$
$$G_0 = b'G(0, 0, c, d, e, f) + b\,G(0, 1, c, d, e, f) = b'G_{00} + b\,G_{01}$$
$$G_1 = b'G(1, 0, c, d, e, f) + b\,G(1, 1, c, d, e, f) = b'G_{10} + bG_{11}$$

Because G_{00},G_{01},G_{10}, and G_{11} are all 4-variable functions, we can realize any 6-variable function using four 4-variable function generators and three 2-to-1 MUXes, as shown in Figure 9-40(b). Thus, we can realize any 6-variable function using two CLBs of the type shown in Figure 9-35. Alternatively, we can write

$$G(a, b, c, d, e, f) = a'b'G_{00} + a'b\,G_{01} + ab'G_{10} + ab\,G_{11} \qquad (9\text{-}12)$$

and realize G using four function generators and a 4-to-1 MUX. In general, we can realize any n-variable function ($n > 4$) using 2^{n-4} 4-variable function generators and one 2^{n-4}-to-1 MUX. This is a worst-case situation because many functions of n-variables can be realized with fewer function generators.

FIGURE 9-40 Realization of 5- and 6-Variable Functions with Function Generators

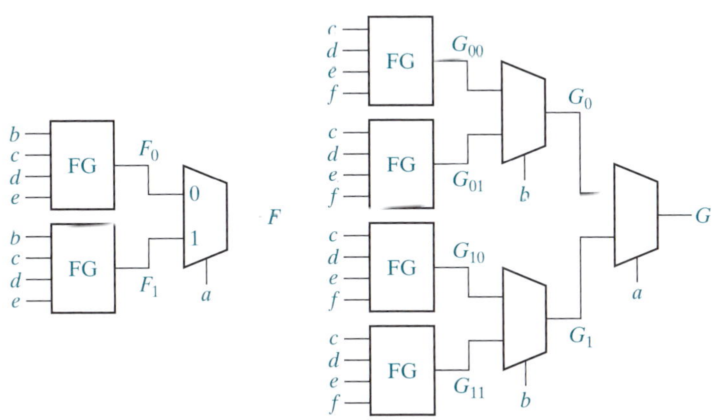

(a) 5-variable function (b) 6-variable function

Problems

9.1 (a) Show how two 2-to-1 multiplexers (with no added gates) could be connected to form a 3-to-1 MUX. Input selection should be as follows:
If $AB = 00$, select I_0
If $AB = 01$, select I_1
If $AB = 1-$ (B is a don't-care), select I_2
(b) Show how two 4-to-1 and one 2-to-1 multiplexers could be connected to form an 8-to-1 MUX with three control inputs.
(c) Show how four 2-to-1 and one 4-to-1 multiplexers could be connected to form an 8-to-1 MUX with three control inputs.

9.2 Design a circuit which will either subtract X from Y or Y from X, depending on the value of A. If $A = 1$, the output should be $X - Y$, and if $A = 0$, the output should be $Y - X$. Use a 4-bit subtracter and two 4-bit 2-to-1 multiplexers (with bus inputs and outputs as in Figure 9-7).

9.3 Repeat 9.2 using a 4-bit subtracter, four 4-bit three-state buffers (with bus inputs and outputs), and one inverter.

9.4 Realize a full adder using a 3-to-8 line decoder (as in Figure 9-17) and
(a) two OR gates.
(b) two NOR gates.

9.5 Derive the logic equations for a 4-to-2 priority encoder. Refer to your table in the Study Guide, Part 4(b).

9.6 Design a circuit equivalent to Figure 9-15 using a 4-to-1 MUX (with bus inputs as in Figure 9-7). Use a 4-to-2 line priority encoder to generate the control signals.

9.7 An adder for Gray-coded-decimal digits (see Table 1-2) is to be designed using a ROM. The adder should add two Gray-coded digits and give the Gray-coded sum and a carry. For example, $1011 + 1010 = 0010$ with a carry of $1 (7 + 6 = 13)$. Draw a block diagram showing the required ROM inputs and outputs. What size ROM is required? Indicate how the truth table for the ROM would be specified by giving some typical rows.

9.8 The following PLA will be used to implement the following equations:

$$X = AB'D + A'C' + BC + C'D'$$
$$Y = A'C' + AC + C'D'$$
$$Z = CD + A'C' + AB'D$$

(a) Indicate the connections that will be made to program the PLA to implement these equations.

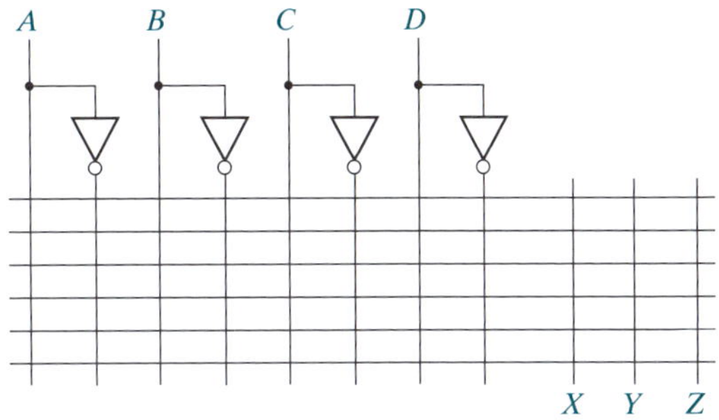

(b) Specify the truth table for a ROM which realizes these same equations.

9.9 Show how to implement a full subtracter using a PAL. See Figure 9-33.

9.10 (a) If the ROM in the hexadecimal to ASCII code converter of Figure 9-26 is replaced with a PAL, give the internal connection diagram.
(b) If the same ROM is replaced with a PLA, give the PLA table.

9.11 (a) Sometimes the programmable MUX (1) in Figure 9-35 helps us to save AND gates. Consider the case in which $F = c'd' + bc' + a'c$. If programmable MUX (1) is not set to invert F (i.e., $G = F$), how many AND gates are needed? If the MUX is set to invert F (i.e., $G = F'$), how many AND gates are needed?
(b) Repeat (a) for $F = a'b' + c'd'$.

9.12 (a) Implement a 3-variable function generator using a PAL with inputs a, b, c, and 1 (use the input inverter to get 0 also). Give the internal connection diagram. Leave the connections to 0 and 1 disconnected, so that any 3-variable function can be implemented by connecting only 0 and 1.
(b) Now connect 0 and 1 so that the function generator implements the sum function for a full adder. See Figure 9-38.

9.13 Expand the following function about the variable b.
$F = ab'cde' + bc'd'e + a'cd'e + ac'de'$

9.14 (a) Implement the following function using only 2-to-1 MUXes:
$R = ab'h' + bch' + eg'h + fgh$.
(b) Repeat using only tri-state buffers.

9.15 Show how to make a 4-to-1 MUX, using an 8-to-1 MUX.

9.16 Implement a 32-to-1 multiplexer using two 16-to-1 multiplexers and a 2-to-1 multiplexer in two ways: (a) Connect the most significant select line to the 2-to-1 multiplexer, and (b) connect the least significant select line to the 2-to-1 multiplexer.

9.17 2-to-1 multiplexers with an active high output and active high enable are to be used in the following implementations:

(a) Show how to implement a 4-to-1 multiplexer with an active high output and no enable using two of the 2-to-1 MUXes and a minimum number of additional gates.

(b) Repeat part (a) for a 4-to-1 multiplexer with an active low output.

(c) Repeat part (b) assuming the output of the 2-to-1 MUX is 1 (rather than 0) when the enable is 0.

9.18 Realize a BCD to excess-3 code converter using a 4-to-10 decoder with active low outputs and a minimum number of gates.

9.19 Use a 4-to-1 multiplexer and a minimum number of external gates to realize the function $F(w, x, y, z) = \Sigma\, m(3, 4, 5, 7, 10, 14) + \Sigma\, d(1, 6, 15)$.
The inputs are only available uncomplemented.

9.20 Realize the function $f(a, b, c, d, e) = \Sigma\, m(6, 7, 9, 11, 12, 13, 16, 17, 18, 20, 21, 23, 25, 28)$ using a 16-to-1 MUX with control inputs b, c, d, and e. Each data input should be 0, 1, a, or a'. (*Hint:* Start with a minterm expansion of F and combine minterms to eliminate a and a' where possible.)

9.21 Implement a full adder

(a) using two 8-to-1 MUXes. Connect X, Y, and C_{in} to the control inputs of the MUXes and connect 1 or 0 to each data input.

(b) using two 4-to-1 MUXes and one inverter. Connect X and Y to the control inputs of the MUXes, and connect 1's, 0's, C_{in}, or C'_{in} to each data input.

(c) again using two 4-to-1 MUXes, but this time connect C_{in} and Y to the control inputs of the MUXes, and connect 1's, 0's, X, or X' to each data input. Note that in this fashion, any N-variable logic function may be implemented using a $2^{(N-1)}$-to-1 MUX.

9.22 Repeat Problem 9.21 for a full subtracter, except use B_{in} instead of C_{in}.

9.23 Make a circuit which gives the absolute value of a 4-bit binary number. Use four full adders, four multiplexers, and four inverters. Assume negative numbers are represented in 2's complement. Recall that one way to find the 2's complement of a binary number is to invert all of the bits and then add 1.

9.24 Show how to make a 4-to-1 MUX using four three-state buffers and a decoder.

9.25 Show how to make an 8-to-1 MUX using two 4-to-1 MUXes, two three-state buffers, and one inverter.

9.26 Realize a full subtracter using a 3-to-8 line decoder with inverting outputs and

(a) two NAND gates

(b) two AND gates

9.27 Show how to make the 8-to-3 priority encoder of Figure 9-20 using two 4-to-2 priority encoders and any additional necessary gates.

9.28 Design an adder for excess-3 decimal digits (see Table 1-2) using a ROM. Add two excess-3 digits and give the excess-3 sum and a carry. For example, 1010 + 1001 = 0110 with a carry of 1 (7 + 6 = 13). Draw a block diagram showing the required ROM inputs and outputs. What size ROM is required? Indicate how the truth table for the ROM would be specified by giving some typical rows.

9.29 A circuit has four inputs $RSTU$ and four outputs $VWYZ$. $RSTU$ represents a binary-coded-decimal digit. VW represents the quotient and YZ the remainder when $RSTU$ is divided by 3 (VW and YZ represent 2-bit binary numbers). Assume that invalid inputs do not occur. Realize the circuit using

(a) a ROM

(b) a minimum two-level NAND-gate circuit

(c) a PLA (specify the PLA table)

9.30 Repeat Problem 9.29 if the inputs $RSTU$ represent a decimal digit in Gray code (see Table 1-2).

9.31 (a) Find a minimum two-level NOR-gate circuit to realize F_1 and F_2. Use as many common gates as possible.

$F_1(a, b, c, d) = \Sigma\, m(1, 2, 4, 5, 6, 8, 10, 12, 14)$

$F_2(a, b, c, d) = \Sigma\, m(2, 4, 6, 8, 10, 11, 12, 14, 15)$

(b) Realize F_1 and F_2 using a PLA. Give the PLA table and internal connection diagram for the PLA.

9.32 Braille is a system which allows a blind person to read alphanumerics by feeling a pattern of raised dots. Design a circuit that converts BCD to Braille. The table shows the correspondence between BCD and Braille.

(a) Use a multiple-output NAND-gate circuit.

A	B	C	D
0	0	0	0
0	0	0	1
0	0	1	0
0	0	1	1
0	1	0	0
0	1	0	1
0	1	1	0
0	1	1	1
1	0	0	0
1	0	0	1

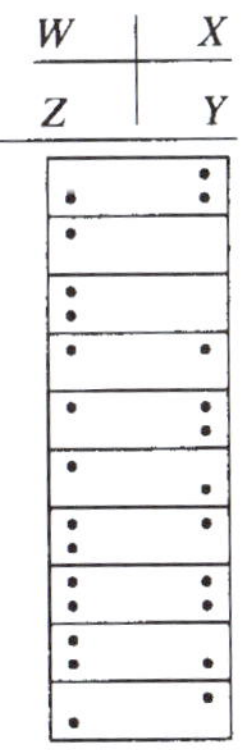

(b) Use a PLA. Give the PLA table.
(c) Specify the connection pattern for the PLA.

9.33 (a) Implement your solution to Problem 7.10 using a PLA. Specify the PLA table and draw the internal connection diagram for the PLA using dots to indicate the presence of switching elements.
(b) Repeat (a) for Problem 7.44.
(c) Repeat (a) for Problem 7.47.

9.34 Show how to make an 8-to-1 MUX using a PAL. Assume that PAL has 14 inputs and six outputs and assume that each output OR gate may have up to four AND terms as inputs, as in Figure 9-33. (*Hint:* Wire some outputs of the PAL around to the inputs, external to the PAL. Some PALs allow this inside the PAL to save inputs.)

9.35 Work Problem 9.34 but make the 8-to-3 priority encoder of Figure 9-20 instead of a MUX.

9.36 The function $F = CD'E + CDE + A'D'E + A'B'DE' + BCD$ is to be implemented in an FPGA which uses 3-variable lookup tables.
(a) Expand F about the variables A and B.
(b) Expand F about the variables B and C.
(c) Expand F about the variables A and C.
(d) Any 5-variable function can be implemented using four 3-variable lookup tables and a 4-to-1 MUX, but this time we are lucky. Use your preceding answers to implement F using only three 3-variable lookup tables and a 4-to-1 MUX. Give the truth tables for the lookup tables.

9.37 Work Problem 9.36 for $F = B'D'E' + AB'C + C'DE' + A'BC'D$.

9.38 Implement a 4-to-1 MUX using a CLB of the type shown in Figure 9-37. Specify the function realized by each function generator.

9.39 Realize the function $f(A, B, C, D) = A'C' + AB'D' + ACD + A'BD$.
(a) Use a single 8-to-1 multiplexer with an active low enable and an active high output. Use A, C, and D as the select inputs where A is the most significant and D is the least significant.
(b) Repeat part (a) assuming the multiplexer enable is active high and output is active low.
(c) Use a single 4-to-1 multiplexer with an active low enable and an active high output and a minimum of additional gates. Show the function expansion both algebraically and on a Karnaugh map.

9.40 Repeat Problem 9.39 for the function
$f(A, B, C, D, E) = A'C'E' + A'B'D'E' + ACDE' + A'BDE'$.

9.41 $F(a, b, c, d) = a' + ac'd' + b'cd' + ad.$

(a) Using Shannon's expansion theorem, expand F about the variable d.

(b) Use the expansion in part (a) to realize the function using two 3-variable LUTs and a 2-to-1 MUX. Specify the LUT inputs.

(c) Give the truth table for each LUT.

9.42 Repeat 9.41 for $F(a, b, c, d) = cd' + ad' + a'b'cd + bc'$.

9.43 Repeat 9.41 for $F(a, b, c, d) = bd + bc' + ac'd + a'd'$.

9.44 The module M below is a demultiplexer (i.e., it routes the input w to one of the four outputs depending on the value of the select lines s and t; thus, an output is 0 or equal to input w depending on the value of s and t). The outputs of module M can be ORed to realize functions of the inputs.

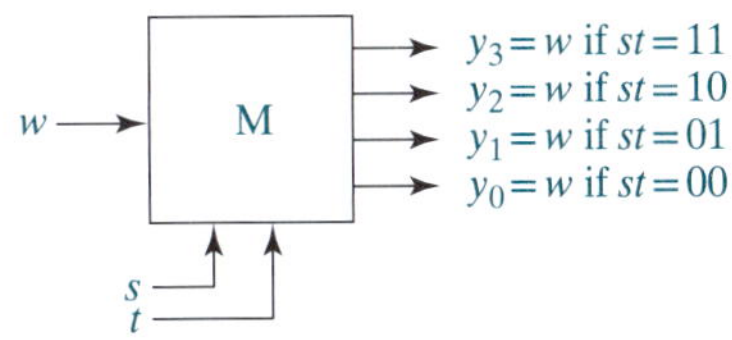

(a) Show how to realize the function $f(a, b, c) = ab' + b'c'$ using one module M and one OR-gate. (Assume that the inputs are available in both true and complement form.)

(b) Using just one module M and one OR gate, is it possible to realize any arbitrary three-variable function? (Again assume inputs are available in both true and complement form.) Justify your answer.

(c) Can the function of part (a) be realized with one module M and one NOR gate? Verify your answer.

(d) If the outputs of module M are active-low, to what type of gate should the outputs connect to realize nontrivial functions? (*Note:* The outputs not selected are logic 1 and the selected output is w'.)

9.45 The circuit below has a 4-input priority encoder connected to a 2-to-4 decoder with enable. The truth table for the priority encoder is given. (The I_3 input is highest priority.) All signals are active high. What functions of A, B, C, and D are realized by Z_3, Z_2, Z_1, and Z_0?

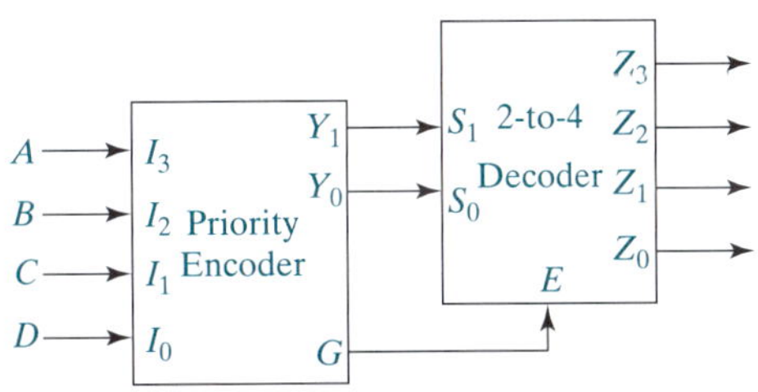

(a)

I_3	I_2	I_1	I_0	G	Y_1	Y_0
0	0	0	0	0	0	0
0	0	0	1	1	0	0
0	0	1	–	1	0	1
0	1	–	–	1	1	0
1	–	–	–	1	1	1

(b)

9.46 The circuit below has a 2-to-4 decoder with active high outputs connected to a 4-to-1 MUX with an active low output.

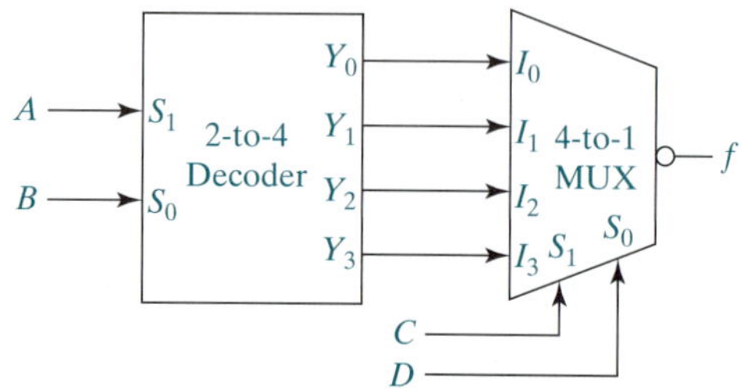

(a) Derive a minimum SOP or a minimum POS expression for the output, $f(A, B, C, D)$.
(b) Repeat part (a) assuming the decoder outputs are active low.

9.47 The Max Selector below has two 4-bit, unsigned inputs, A and B. Its output $Z = A$ if $A \geq B$ and $Z = B$ if $A < B$.

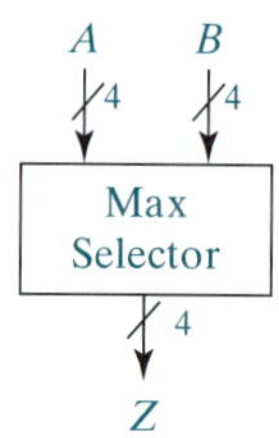

(a) Design the Max Selector in the form shown. The M_i are identical, and a single line connects them with information flowing from right to left. Do one design assuming $c_0 = 0$ and one assuming $c_0 = 1$.

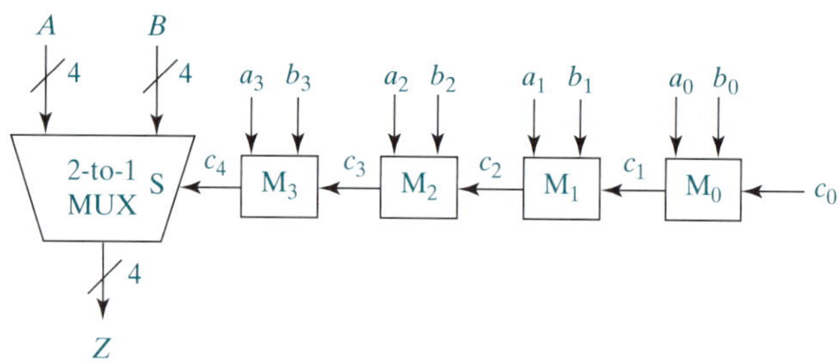

(b) What is the relationship between the design of part (a) and adder or subtractor circuits?

(c) Consider an alternative design of the Max Selector where the information flow is from left to right as shown. Can the Max Selector be designed in this form? If yes, complete the design. If no, explain why not and explain what change

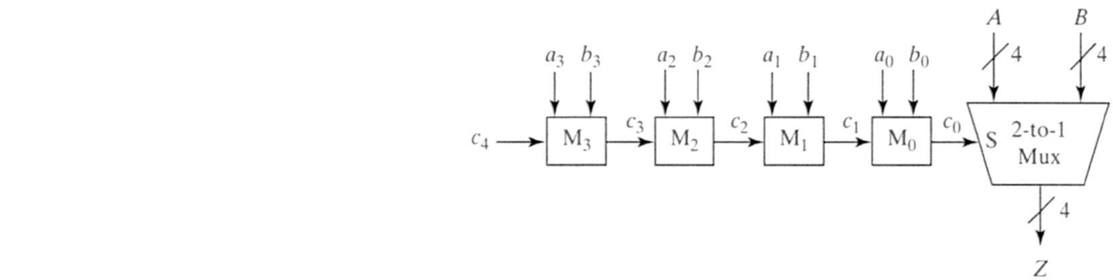

would be required (with information only flowing from left to right between the modules).

9.48 (a) Show that the full adder of Figure 4-5 can be implemented using two 2-input exclusive-OR gates and a 2-to-1 multiplexer. (*Hint:* Rewrite Equation (4-21) in terms of $X \oplus Y$.)

(b) Assume the ripple-carry adder of Figure 4-3 and the full adder of part (a) are implemented using CMOS logic. Which adder would have the smallest maximum addition time? Explain. (You need to be familiar with the material in Appendix A to answer this question.)

9.49 Repeat Problem 9.48 for the full subtractor of Table 4-6.

UNIT 11

Latches and Flip-Flops

Objectives

In this unit you will study one of the basic building blocks used in sequential circuits—the flip-flop. Some of the basic analysis techniques used for sequential circuits are introduced here. In particular, you will learn how to construct timing diagrams which show how each signal in the circuit varies as a function of time. Specific objectives are:

1. Explain in words the operation of S-R and gated D latches.
2. Explain in words the operation of D, D-CE, S-R, J-K, and T flip-flops.
3. Make a table and derive the characteristic (next-state) equation for such latches and flip-flops. State any necessary restrictions on the input signals.
4. Draw a timing diagram relating the input and output of such latches and flip-flops.
5. Show how latches and flip-flops can be constructed using gates. Analyze the operation of a flip-flop that is constructed of gates and latches.

Study Guide

1. Review Section 8.3, *Gate Delays and Timing Diagrams*. Then study Section 11.1, *Introduction*.

   (a) In the circuit shown, suppose that at some instant of time the inputs to both inverters are 0. Is this a stable condition of the circuit?

   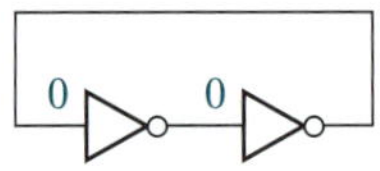

   Assuming that the output of the left inverter changes before the output of the right inverter, what stable state will the circuit reach? (Indicate 0's and 1's on the inverters' inputs and outputs.)

   (b) Work Problem 11.1.

2. Study Section 11.2, *Set-Reset Latch*.

   (a) Build an S-R latch in *SimUaid*, using NOR gates as in Figure 11-3. Place switches on the inputs and probes on the outputs. Experiment with it. Describe in words the behavior of your S-R latch.

   (b) For Figure 11-4(b), what values would P and Q assume if $S = R = 1$?

   (c) What restriction is necessary on S and R so that the two outputs of the S-R latch are complements?

   (d) State in words the meaning of the equation $Q^+ = S + R'Q$.

   (e) Starting with $Q = 0$ and $\overline{S} = \overline{R} = 1$ in Figure 11-10(a), change $\overline{S}$ to 0 and trace signals through the latch until steady-state is reached. Then, change $\overline{S}$ to 1 and $\overline{R}$ to 0 and trace again.

   (f) Work Problems 11.2 and 11.3.

3. Study Section 11.3, *Gated Latches*.

   (a) Build a gated D latch in *SimUaid*. See Figure 11-14. (Construct the S-R latch as in Study Guide Section 2(a).) Place switches on the inputs and probes on the outputs. Experiment with it. Describe in words the behavior of your gated D latch.

   (b) State in words the meaning of the equation $Q^+ = G'Q + GD$.

(c) Given a gated D latch with the following inputs, sketch the waveform for Q.

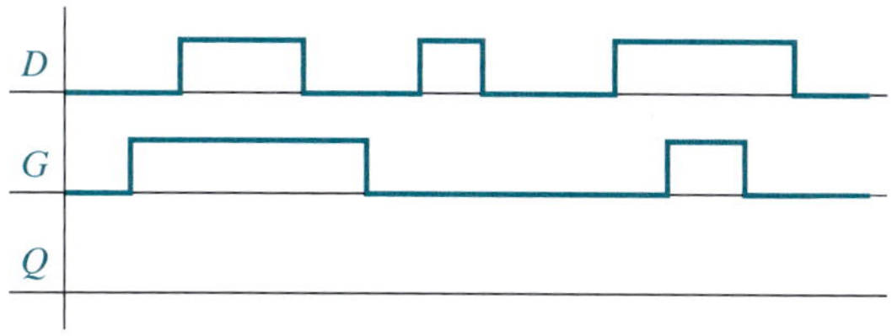

(d) Work Problem 11.4.

4. Study Section 11.4, *Edge-Triggered D Flip-Flop*.

(a) Experiment with a D flip-flop in *SimUaid*. Use the D flip-flop on the parts menu. Place switches on the inputs and probes on the outputs. Describe in words the behavior of your D flip-flop.

(b) Given a rising-edge-triggered D flip-flop with the following inputs, sketch the waveform for Q.

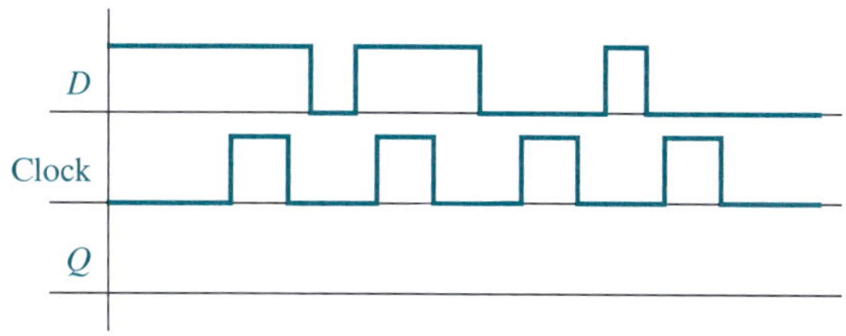

(c) Work Programmed Exercise 11.35.

(d) A D flip-flop with a falling-edge trigger is behaving erratically. It has a setup time of 2 ns and a hold time of 2 ns. The figure shows the inputs to the flip-flop over a typical clock cycle. Why might the flip-flop be behaving erratically?

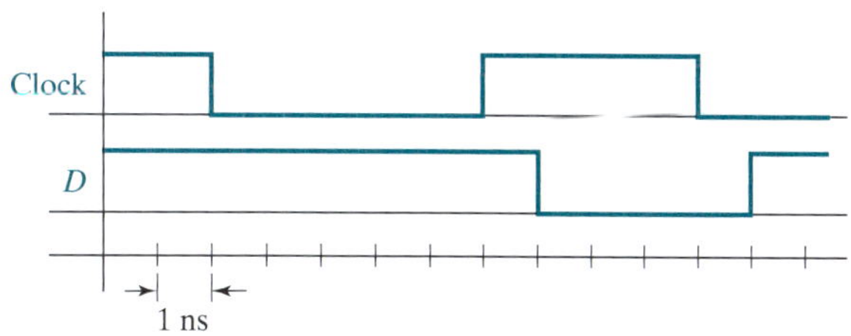

(e) Suppose that for the circuit of Figure 11-21, new semiconductor technology has allowed us to improve the delays and setup times. The propagation delay of the new inverter is 1.5 ns, and the propagation delay and setup times of the new flip-flop are 3.5 ns and 2 ns, respectively. What is the shortest clock period for the circuit of Figure 11-21(a) which will not violate the timing constraints?

(f) Work Problem 11.5.

5. Study Section 11.5, *S-R Flip-Flop*.

(a) Describe in words the behavior of an S-R flip-flop.

(b) Trace signals through the circuit of Figure 11-23(a) and verify the timing diagram of Figure 11-23(b).

(c) What is the difference between a master-slave flip-flop and an edge-triggered flip-flop? Assume that Q changes on the rising clock edge in both cases.

(d) Work Problem 11.6.

6. Study Section 11.6, *J-K Flip-Flop*.

(a) Experiment with a *J-K* flip-flop in SimUaid. Use the *J-K* flip-flop in the parts menu. Place switches on the inputs and probes on the outputs. Describe in words the behavior of your *J-K* flip-flop.

(b) Derive the next-state equation for the J-K flip-flop.

(c) Examine Figures 11-23(a) and 11-25. Construct a J-K flip-flop, using a master-slave S-R flip-flop and two AND gates. (Do not draw the interior of the S-R flip-flop. Just use the symbol in Figure 11-22.)

(d) Work Problem 11.7.

7. Study Section 11.7, *T Flip-Flop*.

(a) Construct a T flip-flop in *SimUaid* from a D flip-flop as in Figure 11-28(b). Place switches on the inputs and probes on the outputs. Experiment with it. Describe in words the behavior of the T flip-flop.

(b) Complete the following timing diagram (assume that $Q = 0$ initially):

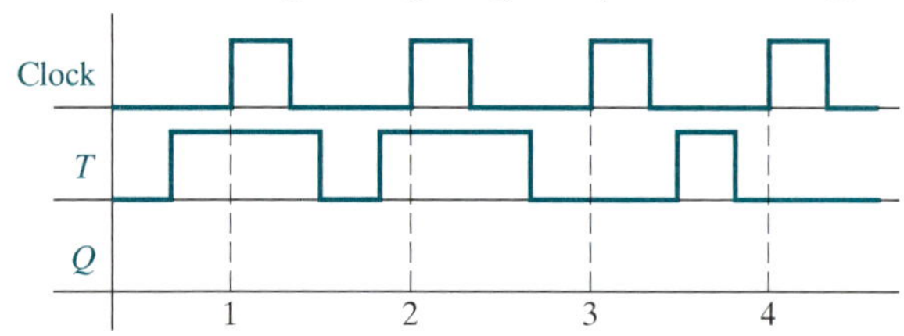

8. Study Section 11.8, *Flip-Flops with Additional Inputs*.

(a) To set the flip-flop of Figure 11-29 to $Q = 1$ without using the clock, the ClrN input should be set to __________ and the PreN input to __________. To reset this flip-flop to $Q = 0$ without using the clock, the

__________ input should be set to __________ and the __________ input to __________.

(b) Complete the following timing diagram for a rising-edge-triggered D flip-flop with ClrN and PreN inputs. Assume Q begins at 0.

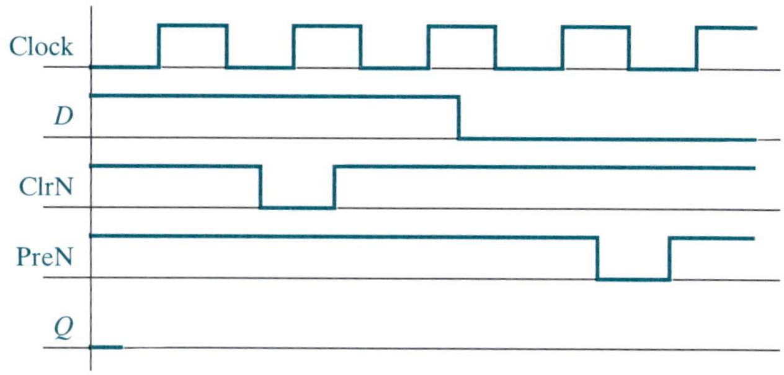

(c) In Figure 11-31(a), what would happen if En changed from 1 to 0 while CLK = 1?
What if En changed when CLK = 0?
In order to have Q change synchronization with the clock, what restriction must be placed on the time at which En can change?

Why does this restriction not apply to Figures 11-31(b) and (c)?

(d) Make a table similar to Figure 11-29(b) that describes the operation of a D flip-flop with a falling-edge clock input, a clock enable input, and an asynchronous active-low clear input (ClrN), but no preset input.

(e) Work Problems 11.8 and 11.9.

9. Study Section 11.10, *Summary*.

(a) Given one of the flip-flops in this chapter or a similar flip-flop, you should be able to derive the characteristic equation which gives the next state of the flip-flop in terms of the present state and inputs. You should understand the meaning of each of the characteristic equations given in Section 11.10.

(b) An S-R flip-flop can be converted to a T flip-flop by adding gates at the S and R inputs. The S and R inputs must be chosen so that the flip-flop will change state whenever $T = 1$ and the clock is pulsed. In order to determine the S and R inputs, ask yourself the question, "Under what conditions must the flip-flop be set to 1, and under what conditions must it be reset?" The flip-flop must be set to 1 if $Q = 0$ and $T = 1$.

Therefore, $S =$ ___________. In a similar manner, determine the equation for R and draw the circuit which converts an S-R flip-flop to a T flip-flop.

(c) Work Problem 11.10.

10. When you are satisfied that you can meet the objectives of this unit, take the readiness test.

Latches and Flip-Flops

11.1 Introduction

Sequential switching circuits have the property that the output depends not only on the present input but also on the past sequence of inputs. In effect, these circuits must be able to "remember" something about the past history of the inputs in order to produce the present output. Latches and flip-flops are commonly used memory devices in sequential circuits. Basically, latches and flip-flops are memory devices which can assume one of two stable output states and which have one or more inputs that can cause the output state to change. Several common types of latches and flip-flops are described in this unit.

In Units 12 through 16, we will discuss the analysis and design of synchronous digital systems. In such systems, it is common practice to synchronize the operation of all flip-flops by a common clock or pulse generator. Each of the flip-flops has a clock input, and the flip-flops can only change state in response to a clock pulse. The use of a clock to synchronize the operation of several flip-flops is illustrated in Units 12 and 13. A memory element that has no clock input is often called a latch, and we will follow this practice. We will then reserve the term flip-flop to describe a memory device that changes its output in response to a clock input and not in response to a data input.

The switching circuits that we have studied so far have not had feedback connections. By feedback we mean that the output of one of the gates is connected back into the input of another gate in the circuit so as to form a closed loop. In order to construct a switching circuit that has memory, such as a latch or flip-flop, we must introduce feedback into the circuit. For example, in the NOR-gate circuit of Figure 11-3(a), the output of the second NOR gate is fed back into the input of the first NOR gate. Sequential circuits must contain feedback, but not all circuits with feedback are sequential. There are a few circuits containing feedback that are combinational.

FIGURE 11-1

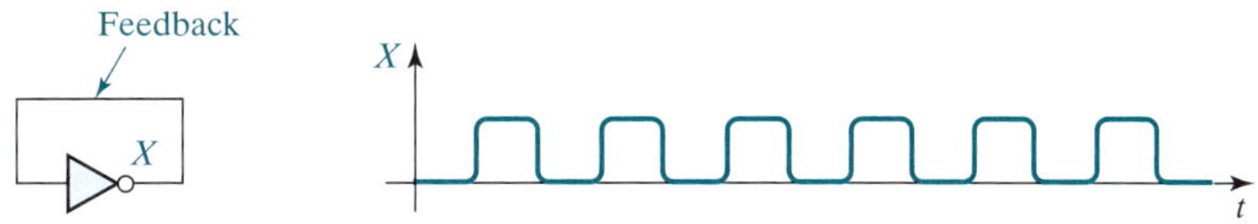

(a) Inverter with feedback (b) Oscillation at inverter output

In simple cases, we can analyze circuits with feedback by tracing signals through the circuit. For example, consider the circuit in Figure 11-1(a). If at some instant of time the inverter input is 0, this 0 will propagate through the inverter and cause the output to become 1 after the inverter delay. This 1 is fed back into the input, so after the propagation delay, the inverter output will become 0. When this 0 feeds back into the input, the output will again switch to 1, and so forth. The inverter output will continue to oscillate back and forth between 0 and 1, as shown in Figure 11-1(b), and it will never reach a stable condition. An oscillator can be created using any odd number of inverters. The oscillator waveform has a high and a low time that is the sum of the propagation times of the inverters. For example, with n inverters and with all having the same delay, the oscillator waveform high time is $(n + 1)/2$ times the high-to-low inverter propagation delay plus $(n - 1)/2$ times the low-to-high inverter propagation delay.

FIGURE 11-2

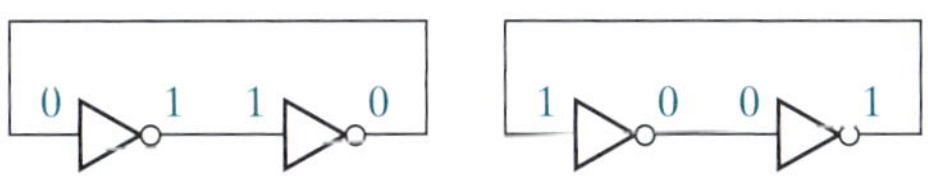

(a) (b)

Next, consider a feedback loop which has two inverters in it, as shown in Figure 11-2(a). In this case, the circuit has two stable conditions, often referred to as stable states. If the input to the first inverter is 0, its output will be 1. Then, the input to the second inverter will be 1, and its output will be 0. This 0 will feed back into the first inverter, but because this input is already 0, no changes will occur. The circuit is then in a stable state. As shown in Figure 11-2(b), a second stable state of the circuit occurs when the input to the first inverter is 1 and the input to the second inverter is 0. The simple loop of two inverters lacks any external means of initializing the state to one of the stable states. The set-reset latches in the next section have inputs for this initialization.

11.2 Set-Reset Latch

We can construct a simple latch by introducing feedback into a NOR-gate circuit, as seen in Figure 11-3(a). As indicated, if the inputs are $S = R = 0$, the circuit can assume a stable state with $Q = 0$ and $P = 1$. Note that this is a stable condition of the circuit because $P = 1$ feeds into the second gate forcing the output to be $Q = 0$, and $Q = 0$ feeds into the first gate allowing its output to be 1. Now if we change S to 1, P will become 0. This is an unstable condition or state of the circuit because both the inputs and output of the second gate are 0; therefore Q will change to 1, leading to the stable state shown in Figure 11-3(b).

FIGURE 11-3

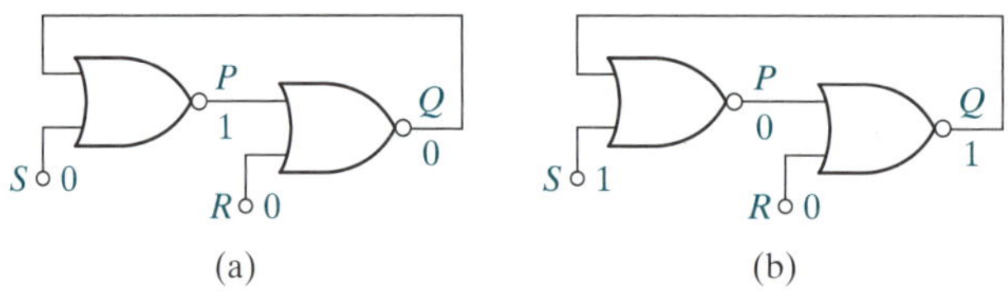

FIGURE 11-4

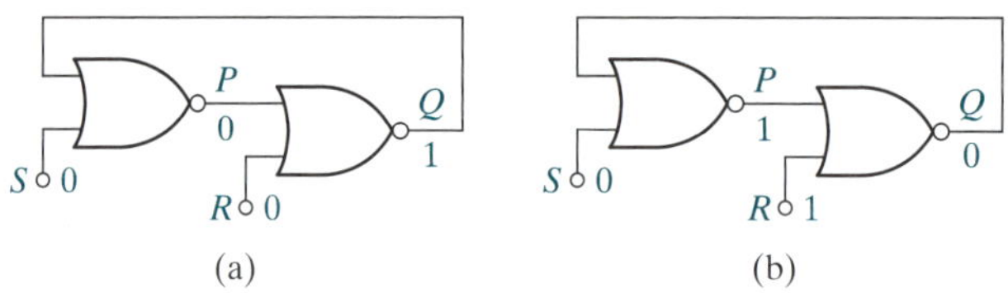

If S is changed back to 0, the circuit will not change state because $Q = 1$ feeds back into the first gate, causing P to remain 0, as shown in Figure 11-4(a). Note that the inputs are again $S = R = 0$, but the outputs are different than those with which we started. Thus, the circuit has two different stable states for a given set of inputs. If we now change R to 1, Q will become 0 and P will then change back to 1, as seen in Figure 11-4(b). If we then change R back to 0, the circuit remains in this state and we are back where we started.

This circuit is said to have memory because its output depends not only on the present inputs, but also on the past sequence of inputs. If we restrict the inputs so that $R = S = 1$ is not allowed, the stable states of the outputs P and Q are always complements, that is, $P = Q'$. To emphasize the symmetry between the operation of the two gates, the circuit is often drawn in *cross-coupled* form (see Figure 11-5(a)). As shown in Figures 11-3(b) and 11-4(b), an input $S = 1$ *sets* the output to $Q = 1$, and an input $R = 1$ *resets* the output to $Q = 0$. When used with the restriction that R and S cannot be 1 simultaneously, the circuit is commonly referred to as a set-reset (S-R) latch and given the symbol shown in Figure 11-5(b). Note that although Q comes out of the NOR gate with the R input, the standard S-R latch symbol has Q directly above the S input.

If $S = R = 1$, the latch will not operate properly, as shown in Figure 11-6. The notation $1 \rightarrow 0$ means that the input is originally 1 and then changes to 0. Note that when S and R are both l, P and Q are both 0. Therefore, P is not equal to Q', and this violates a basic rule of latch operation that requires the latch outputs to be complements. Furthermore, if S and R are simultaneously changed back to 0, P and Q may both change to 1. If $S = R = 0$ and $P = Q = 1$, then after the 1's propagate through the gates, P and Q will become 0 again, and the latch may continue to oscillate if the gate delays are equal.

FIGURE 11-5 S-R Latch

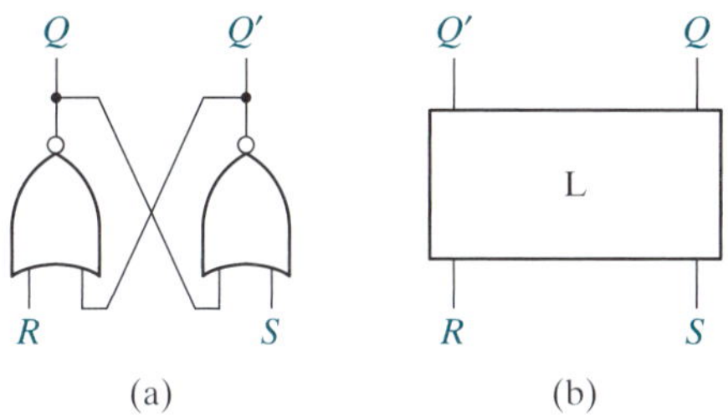

FIGURE 11-6 Improper S-R Latch Operation

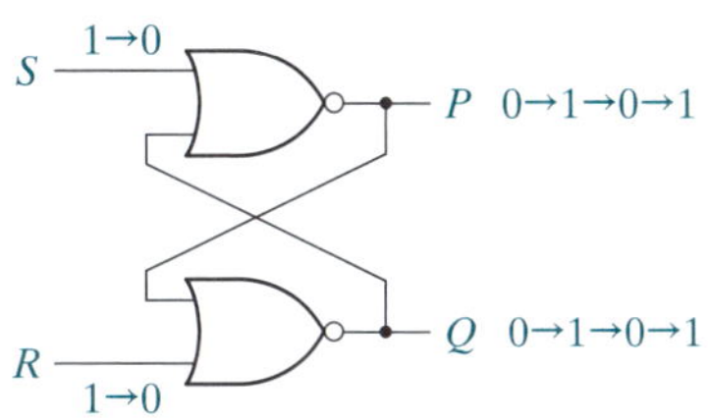

FIGURE 11-7 Timing Diagram for S-R Latch

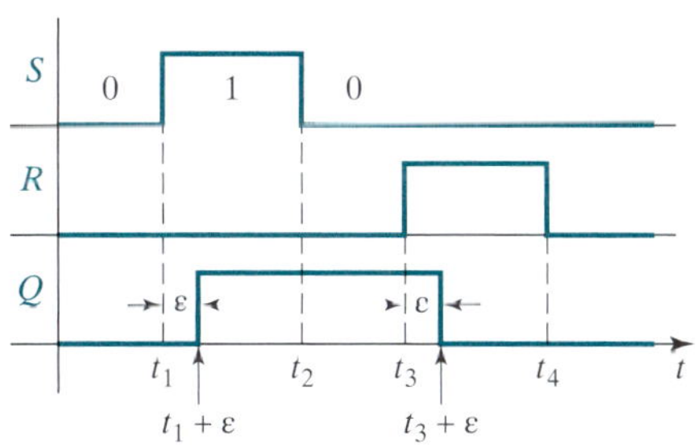

Figure 11-7 shows a timing diagram for the S-R latch. Note that when S changes to 1 at time t_1, Q changes to 1 a short time (ε) later. (ε represents the response time or delay time of the latch.) At time t_2, when S changes back to 0, Q does not change. At time t_3, R changes to 1, and Q changes back to 0 a short time (ε) later. The duration of the S (or R) input pulse must normally be at least as great as ε in order for a change in the state of Q to occur. If $S = 1$ for a time less than ε, the gate output will not change and the latch will not change state.

Theoretically, the two-inverter circuit of Figure 11-2 and the set-reset of Figure 11-3 can exist in a third stable state. This is the situation where the voltage level at the output of the two inverters or gates is approximately halfway between the voltage levels for a logic 0 and a logic 1. This state is referred to as a *metastable* state. It is metastable because any noise existing in the circuit will cause the circuit to transition to one of the truly stable states. However, certain events can cause the latch to enter the metastable state for a short time. The simultaneous change in S and R from 1 to 0 in the set-reset latch can cause the latch to enter the metastable state. Also, starting with $Q = 0$ and applying a pulse on S with a length on the borderline between being too short to cause Q to change and just long enough to cause Q to change may cause the circuit to enter the metastable state.

Even though a circuit in the metastable state will quickly enter a stable state, events causing the circuit to enter the metastable state must be avoided. First, the stable state entered from the metastable state is unpredictable. Second, any gates or latches with an input that is in the metastable state will respond unpredictably; the gate or latch may respond as if the input is a logic 0 or it may respond as if the input is a logic 1. Metastable behavior is discussed further in the next sections.

When discussing latches and flip-flops, we use the term *present state* to denote the state of the Q output of the latch or flip-flop at the time any input signal changes, and the term *next state* to denote the state of the Q output after the latch or flip-flop has reacted to the input change and stabilized. If we let $Q(t)$ represent the present state and $Q(t + \varepsilon)$ represent the next state, an equation for $Q(t + \varepsilon)$ can be obtained from the circuit by conceptually breaking the feedback loop at Q and considering $Q(t)$ as an input and $Q(t + \varepsilon)$ as the output. Then for the S-R latch of Figure 11-3

$$Q(t + \varepsilon) = R(t)'[S(t) + Q(t)] = R(t)'S(t) + R(t)'Q(t) \tag{11-1}$$

and the equation for output P is

$$P(t) = S(t)'Q(t)' \tag{11-2}$$

Normally we write the next-state equation without including time explicitly, using Q to represent the present state of the latch and Q^+ to represent the next state:

$$Q^+ = R'S + R'Q \tag{11-3}$$

$$P = S'Q' \tag{11-4}$$

These equations are mapped in the next-state and output tables of Table 11-1. The stable states of the latch are circled. Note that for all stable states, $P = Q'$ except when $S = R = 1$. As discussed previously, this is one of the reasons why $S = R = 1$ is disallowed as an input combination to the S-R latch. Making $S = R = 1$ a don't-care combination allows simplifying the next-state equation, as shown in Figure 11-8(a). After plotting Equation (11-3) on the map and changing two entries to don't-cares, the next-state equation simplifies to

$$Q^+ = S + R'Q \qquad (SR = 0) \tag{11-5}$$

TABLE 11-1
S-R Latch Next State and Output

Present State	Next State Q^+				Present Output P			
Q	SR 00	SR 01	SR 11	SR 10	SR 00	SR 01	SR 11	SR 10
0	⓪	⓪	⓪	1	1	1	0	0
1	①	0	0	①	0	0	0	0

FIGURE 11-8
Derivation of Q^+ for an S-R Latch

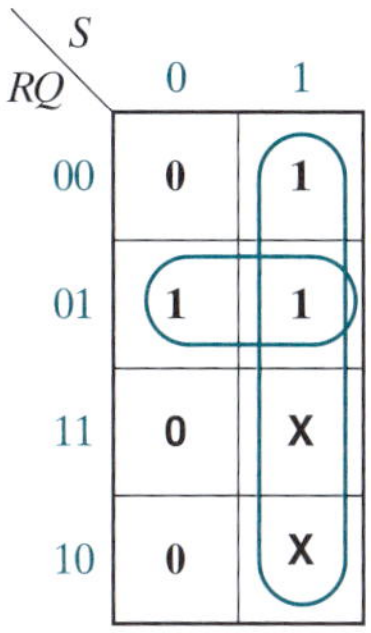

(a) Q^+ map

S	R	Q	Q^+
0	0	0	0
0	0	1	1
0	1	0	0
0	1	1	0
1	0	0	1
1	0	1	1
1	1	0	– } Inputs not
1	1	1	– } allowed

(b) Truth table

In words, this equation tells us that the next state of the latch will be 1 either if it is set to 1 with an S input, or if the present state is 1 and the latch is not reset. The condition $SR = 0$ implies that S and R cannot both be 1 at the same time. An equation that expresses the next state of a latch in terms of its present state and inputs will be referred to as a *next-state equation*, or *characteristic equation*. Note that the characteristic equation is not necessarily the same as the next-state equation derived from the circuit. They both give the functional behavior of the latch (or FF). However, the characteristic equation is a minimal equation and it takes into account any disallowed input combinations. The next-state equation derived from the circuit is not necessarily minimal and it does not take into account any disallowed input combinations. Compare Equations (11-3) and (11-5).

Another approach for deriving the characteristic equation for an S-R latch is based on constructing a truth table for the next state of Q We previously discussed the latch operation by tracing signals through the gates, and the truth table in Figure 11-8(b) is based on this discussion. Plotting Q^+ on a Karnaugh map gives the same result as Figure 11-8(a).

The S-R latch is often used as a component in more complex latches and flip-flops and in asynchronous systems. Another useful application of the S-R latch is for debouncing switches. When a mechanical switch is opened or closed, the switch contacts tend to vibrate or bounce open and closed several times before settling down to their final position. This produces a noisy transition, and this noise can interfere with the proper operation of a logic circuit. The input to the switch in Figure 11-9 is connected to a logic 1 (+V). The pull-down resistors connected to contacts a and b

FIGURE 11-9
Switch Debouncing with an S-R Latch

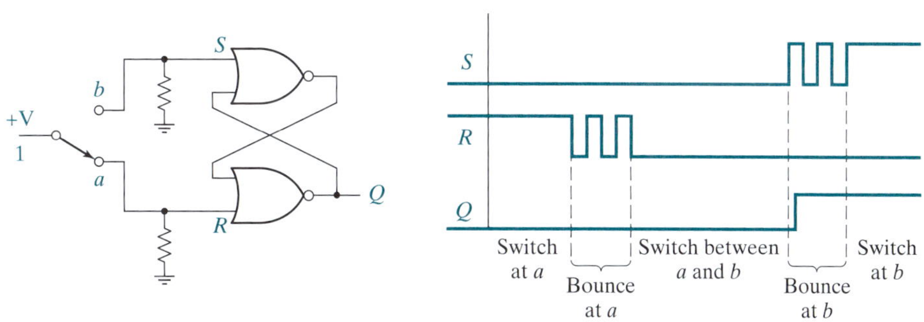

assure that when the switch is between a and b the latch inputs S and R will always be at a logic 0, and the latch output will not change state. The timing diagram shows what happens when the switch is flipped from a to b. As the switch leaves a, bounces occur at the R input; when the switch reaches b, bounces occur at the S input. After the switch reaches b, the first time S becomes 1, after a short delay the latch switches to the $Q = 1$ state and remains there. Thus Q is free of all bounces even though the switch contacts bounce. This debouncing scheme requires a *double throw* switch that switches between two contacts; it will not work with a *single throw* switch that switches between one contact and open.

An alternative form of the S-R latch uses NAND gates, as shown in Figure 11-10. We will refer to this circuit as an $\overline{\text{S}}$-$\overline{\text{R}}$ latch, and the table describes its operation. We have labeled the inputs to this latch $\overline{S}$ and $\overline{R}$ because $\overline{S} = 0$ will set Q to 1 and $\overline{R} = 0$ will reset Q to 0. If $\overline{S}$ and $\overline{R}$ are 0 at the same time, both the Q and Q' outputs are forced to 1. Therefore, for the proper operation of this latch, the condition $\overline{S} = \overline{R} = 0$ is not allowed.

FIGURE 11-10
$\overline{\text{S}}$-$\overline{\text{R}}$ Latch

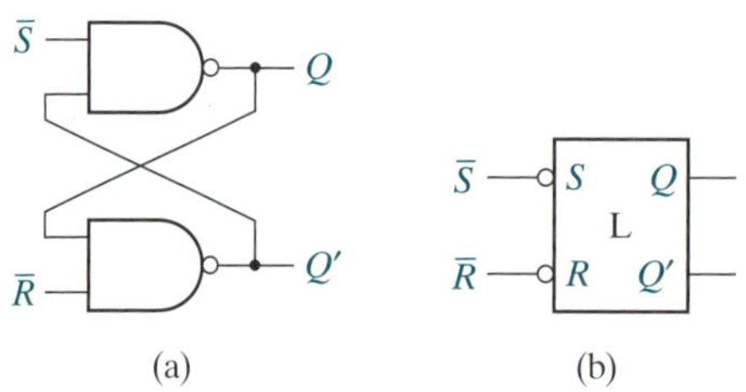

$\overline{S}$	$\overline{R}$	Q	Q^+
1	1	0	0
1	1	1	1
1	0	0	0
1	0	1	0
0	1	0	1
0	1	1	1
0	0	0	– } Inputs not allowed
0	0	1	–

(c)

11.3 Gated Latches

Gated latches have an additional input called the gate or enable input. When the gate input is inactive, which may be the high or low value, the state of the latch cannot change. When the gate input is active, the latch is controlled by the other inputs and operates as indicated in the preceding section. A NAND-gate version of a gated S-R latch is shown in Figure 11-11.

FIGURE 11-11
NAND-Gate Gated S-R Latch

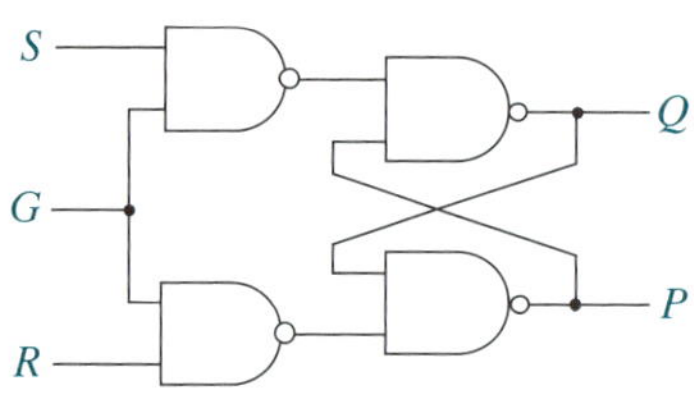

The next-state equation is

$$Q^+ = SG + Q(R' + G')$$

and the equation for the P output is

$$P = Q' + RG$$

The next-state and output tables are shown in Table 11-2. When $G = 0$, the circuit is always in a stable state; when $G = 1$, $S = 1$ sets the latch and $R = 1$ resets the latch. Note that $P = Q'$ whenever the latch is in a stable state except for the input combination $G = S = R = 1$; consequently, as for the basic latch, the $S = R = 1$ input combination is disallowed.

TABLE 11-2 Next-State and Output of Gated S-R Latch

Next State Q^+

Present State Q	G = 0 SR 00	SR 01	SR 11	SR 10	G = 1 SR 00	SR 01	SR 11	SR 10
0	⓪	⓪	⓪	⓪	⓪	⓪	1	1
1	①	①	①	①	①	0	①	①

Present Output P

Present State Q	G = 0 SR 00	SR 01	SR 11	SR 10	G = 1 SR 00	SR 01	SR 11	SR 10
0	1	1	1	1	1	1	1	1
1	0	0	0	0	0	1	1	0

Another reason for disallowing the $S = R = 1$ input combination is illustrated by considering a change in G from 1 to 0 with $S = R = 1$. When G changes, both inputs to the basic S-R latch change from 0 to 1, as shown in Figure 11-12. This causes both gates in the basic S-R latch to attempt to change from 1 to 0; a race condition exists and the propagation delays of the gates determine whether the latch stabilizes with $Q = 0$ or $Q = 1$. This was illustrated in Figure 11-6 for the simple NOR-gate latch.

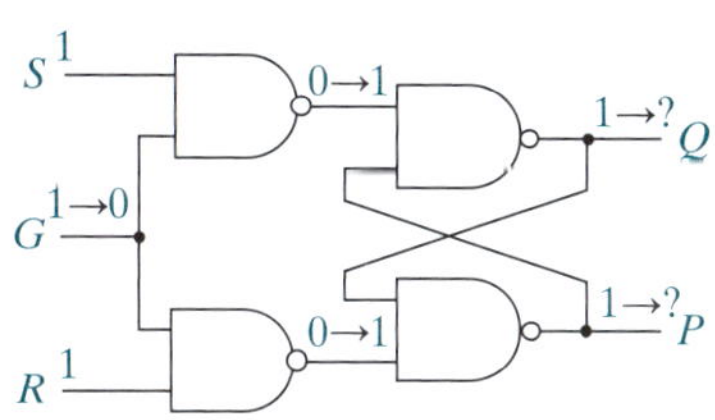

FIGURE 11-12 Race Condition in the Gated S-R Latch

It is also instructive to examine this problem from a different viewpoint. If the equation for Q^+ is plotted on a Karnaugh map (Figure 11-13), it is evident that Q^+ has a static 1-hazard for the input combinations $G = 1, S = 1, R = 1, Q = 1$ and

$G = 0, S = 1, R = 1, Q = 1$. Consequently, when G changes from 1 to 0 between these two input combinations, it is possible for Q to change from 1 to 0 and, because of the feedback, to cause Q to remain at $Q = 0$. This is simply a different interpretation of the race condition described above. Of course, the static 1-hazard exists for G changing from 0 to 1 and, again, Q may change to 0, but in this case Q is forced back to 1 by the $S = R = 1$ input values. So, for this change, the hazard may cause a glitch in Q, but it cannot cause the latch to stabilize with $Q = 0$.

FIGURE 11-13 Karnaugh Map for Q^+

GQ \ SR	00	01	11	10
00	0	0	0	0
01	1	1	1	1
11	1	0	1	1
10	0	0	1	1

Q^+

There is another restriction regarding gated S-R latches. The S and R inputs must not be changing or contain glitches while $G = 1$. For example, assume $Q = 0$ when G changes from 0 to 1 and $S = 0$ and $R = 0$. If S and R remain at 0 until G returns to 0, then Q remains at 0. However, if S contains a 1 glitch, maybe due to a static 1-hazard in its circuit, then Q may be forced to a 1 and will remain there after G becomes 0. A similar problem occurs if S does not change from 1 to 0 until after G changes to 1. This is referred to as the 1's catching problem. A NOR-gate version of the gated S-R latch has a 0's catching problem.

Another gated latch is the gated D latch. It can be obtained from a gated S-R latch by connecting S to D and R to D'. Figure 11-14(a) shows this construction using a basic S-R latch, two AND gates and an inverter; the NAND-gate, gated S-R latch in Figure 11-11 can be converted to a gated D latch with the addition of an inverter. As indicated in Figure 11-14(b), Q of the gated D latch remains unchanged while G is inactive, $G = 0$ in this case, and Q becomes equal to the D after some delay when G is active. The delay is due to the propagation delay of the gates. This latch is also referred to as a *transparent latch* since Q becomes equal to D while G is active. Figure 11-15 shows the state table and characteristic equation for the gated D latch.

FIGURE 11-14 Gated D Latch

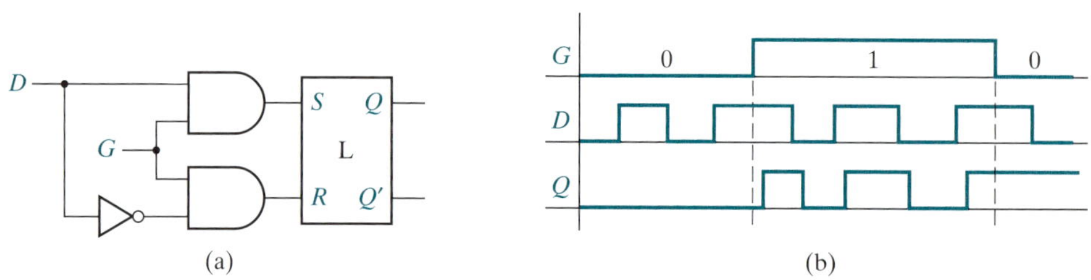

FIGURE 11-15 Symbol and Truth Table for Gated Latch

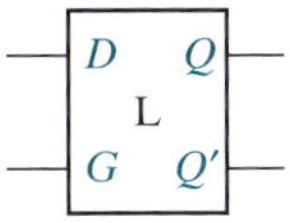

G D Q	Q^+
0 0 0	0
0 0 1	1
0 1 0	0
0 1 1	1
1 0 0	0
1 0 1	0
1 1 0	1
1 1 1	1

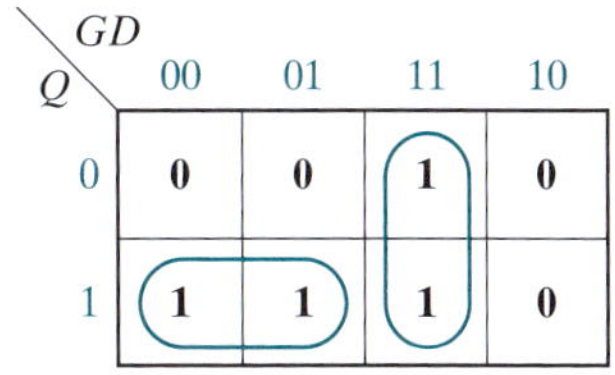

$Q^+ = G'Q + GD$

Most digital systems use a clock signal to synchronize the change in outputs of the system's flip-flops to an edge of the clock signal, either the positive (0 to 1) or the negative (1 to 0) edge of the clock. It is tempting to think that gated latches could be used as flip-flops where the clock signal is connected to the gate inputs of the latches. However, this is not a practical approach. The following example illustrates the difficulty. In the circuit of Figure 11-16, it seems that when the Clk is 1 the next value of Q should be Q' when the input $x = 1$ and should be Q when $x = 0$. However, when Clk $= 1$ and $x = 1$, $D = Q'$ causes Q to change and, if Clk remains 1, the change in Q will feed back and cause Q to change again. If Clk remains at 1, Q will oscillate. Consequently, the circuit will only operate as intended if Clk remains at 1 for a short time; it has to 1 just long enough to allow Q to change but short enough to prevent the change from feeding back and causing a second change. With a single latch, it may be possible to control the clock high time so the latch operates as intended, but in a system with several latches, the variation in gate delays would make it impossible to provide the correct clock width to all latches.

FIGURE 11-16 Unreliable Gated D Latch Circuit

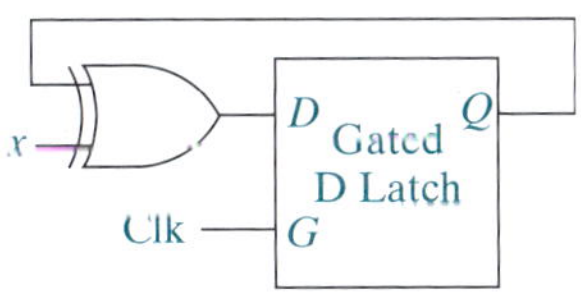

To avoid this timing problem, more complicated flip-flops restrict the flip-flop outputs to only change on an edge of the clock, and the outputs cannot change at other times even if the inputs change. If the inputs to the flip-flop only need to be stable for a short period of time around the clock edge, then we refer to the flip-flop as *edge-triggered*. (See the discussion of setup and hold times in the next section.) The term *master-slave* flip-flop refers to a particular implementation that uses two gated latches in such a way that the flip-flop outputs only change on a clock edge. However, master-slave flip-flops are not necessarily edge-triggered flip-flops because they may require the flip-flop inputs to be stable at times during the clock period other than just around the clock edge. The S-R, J-K, and T master-slave flip-flops in later sections are examples. The master-slave D flip-flop of the next section is an exception; it uses the master-slave implementation and it is also edge triggered.

11.4 Edge-Triggered D Flip-Flop

A D flip-flop (Figure 11-17) has two inputs, D (data) and Ck (clock). The small arrowhead on the flip-flop symbol identifies the clock input. Unlike the D latch, the flip-flop output changes only in response to the clock, not to a change in D. If the output can change in response to a 0 to 1 transition on the clock input, we say that the flip-flop is triggered on the *rising edge* (or positive edge) of the clock. If the output can change in response to a 1 to 0 transition on the clock input, we say that the flip-flop is triggered on the *falling edge* (or negative edge) of the clock. An inversion bubble on the clock input indicates a *falling-edge trigger* (Figure 11-17(b)), and no bubble indicates a *rising-edge trigger* (Figure 11-17(a)). The term *active edge* refers to the clock edge (rising or falling) that triggers the flip-flop state change.

FIGURE 11-17 D Flip-Flops

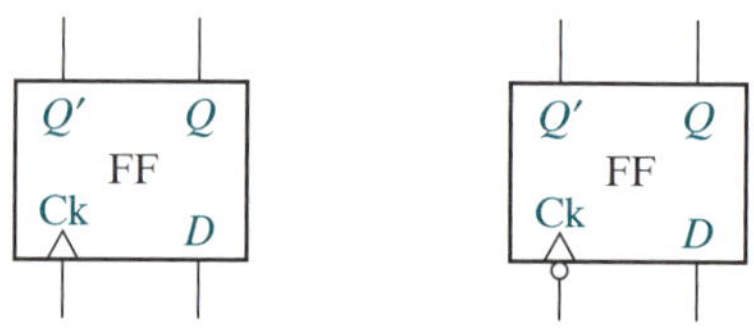

(a) Rising-edge trigger (b) Falling-edge trigger

D Q	Q^+
0 0	0
0 1	0
1 0	1
1 1	1

(c) Truth table

$$\boxed{Q^+ = D}$$

The state of a D flip-flop after the active clock edge (Q^+) is equal to the input (D) before the active edge. For example, if $D = 1$ before the clock pulse, $Q = 1$ after the active edge, regardless of the previous value of Q. Therefore, the characteristic equation is $Q^+ = D$. If D changes at most once following each clock pulse, the output of the flip-flop is the same as the D input, except that the output changes are delayed until after the active edge of the clock pulse, as illustrated in Figure 11-18.

FIGURE 11-18 Timing for D Flip-Flop (Falling-Edge Trigger)

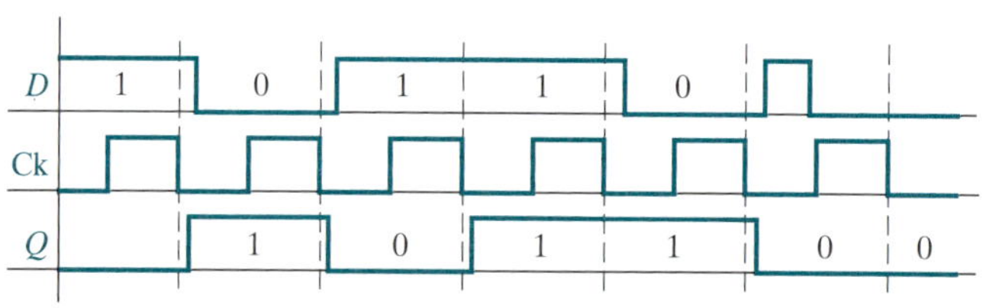

A rising-edge-triggered D flip-flop can be constructed from two gated D latches and an inverter, as shown in Figure 11-19(a). The timing diagram is shown in Figure 11-19(b). When CLK = 0, $G_1 = 1$, and the first latch is *transparent* so that the P output follows the D input. Because $G_2 = 0$, the second latch holds the current value of Q. When CLK changes to 1, G_1 changes to 0, and the current value of D is stored

FIGURE 11-19 D Flip-Flop (Rising-Edge Trigger)

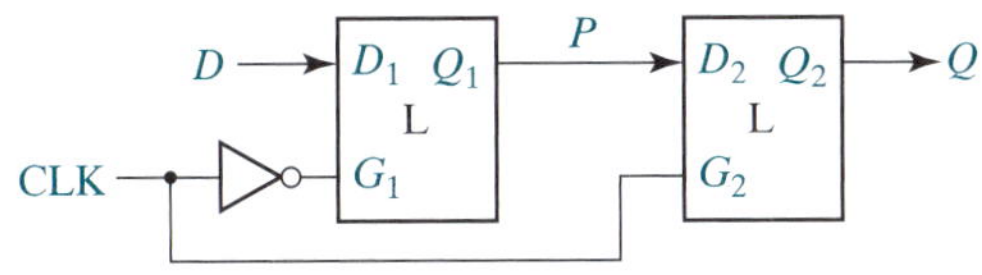

(a) Construction from two gated D latches

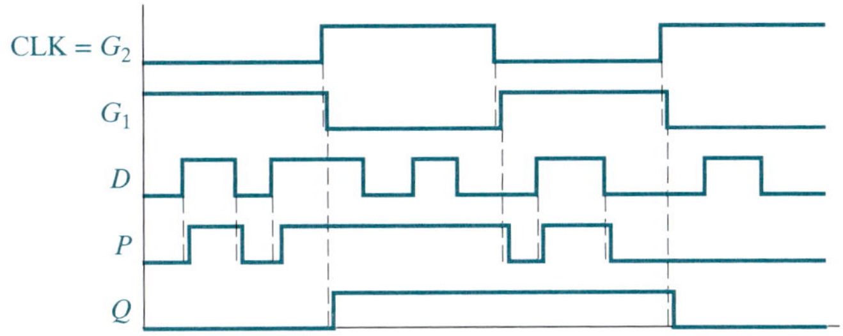

(b) Timing analysis

in the first latch. Because $G_2 = 1$, the value of P flows through the second latch to the Q output. When CLK changes back to 0, the second latch takes on the value of P and holds it and, then, the first latch starts following the D input again. If the first latch starts following the D input before the second latch takes on the value of P, the flip-flop will not function properly. Therefore, the circuit designers must pay careful attention to timing issues when designing edge-triggered flip-flops. With this circuit, output state changes occur only following the rising edge of the clock. The value of D at the time of the rising edge of the clock determines the value of Q, and any extra changes in D that occur between rising clock edges have no effect on Q.

Because a flip-flop changes state only on the active edge of the clock, the propagation delay of a flip-flop is the time between the active edge of the clock and the resulting change in the output. However, there are also timing issues associated with the D input. To function properly, the D input to an edge-triggered flip-flop must be held at a constant value for a period of time before and after the active edge of the clock. If D changes at the same time as the active edge, the behavior is unpredictable. The amount of time that D must be stable before the active edge is called the setup time (t_{su}), and the amount of time that D must hold the same value after the active edge is the hold time (t_h). The times at which D is allowed to change during the clock cycle are shaded in the timing diagram of Figure 11-20. The propagation

FIGURE 11-20 Setup and Hold Times for an Edge-Triggered D Flip-Flop

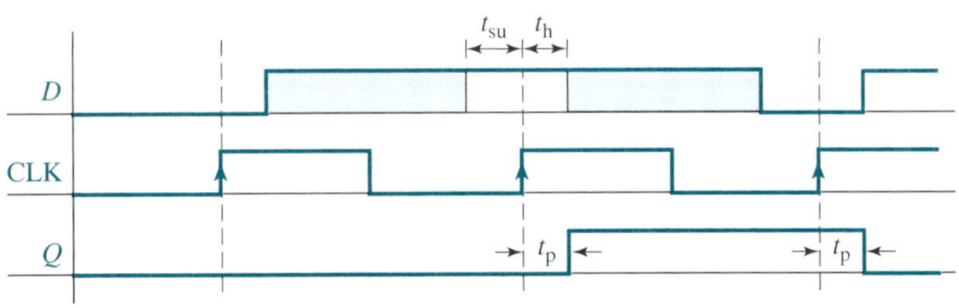

delay (t_p) from the time the clock changes until the Q output changes is also indicated. For Figure 11-19(a), the setup time allows a change in D to propagate through the first latch before the rising edge of Clock. The hold time is required so that D gets stored in the first latch before D changes.

Using these timing parameters, we can determine the minimum clock period for a circuit which will not violate the timing constraints. Consider the circuit of Figure 11-21(a). Suppose the inverter has a propagation delay of 2 ns, and suppose the flip-flop has a propagation delay of 5 ns and a setup time of 3 ns. (The hold time does not affect this calculation.) Suppose, as in Figure 11-21(b), that the clock period is 9 ns, i.e., 9 ns is the time between successive active edges (rising edges for this figure). Then, 5 ns after a clock edge, the flip-flop output will change, and 2 ns after that, the output of the inverter will change. Therefore, the input to the flip-flop will change 7 ns after the rising edge, which is 2 ns before the next rising edge. But the setup time of the flip-flop requires that the input be stable 3 ns before the rising edge; therefore, the flip-flop may not take on the correct value.

Suppose instead that the clock period were 15 ns, as in Figure 11-21(c). Again, the input to the flip-flop will change 7 ns after the rising edge. However, because the clock is slower, this is 8 ns before the next rising edge. Therefore, the flip-flop will work properly. Note in Figure 11-21(c) that there is 5 ns of extra time between the time the D input is correct and the time when it must be correct for the setup time to be satisfied. Therefore, we can use a shorter clock period, and have less extra time, or no extra time. Figure 11-21(d) shows that 10 ns is the minimum clock period which will work for this circuit.

FIGURE 11-21 Determination of Minimum Clock Period

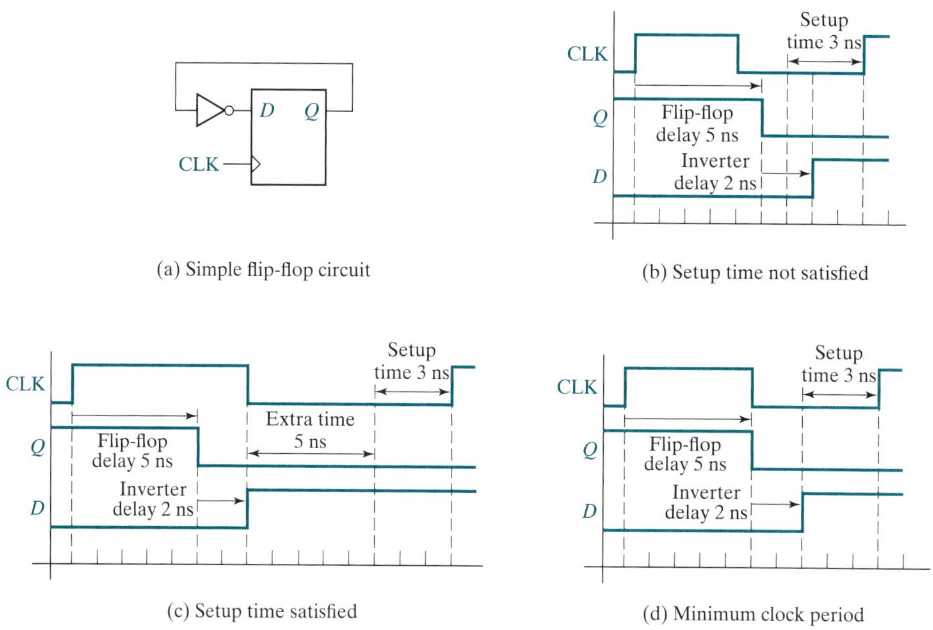

11.5 S-R Flip-Flop

An S-R flip-flop (Figure 11-22) is similar to an S-R latch in that $S = 1$ sets the Q output to 1, and $R = 1$ resets the Q output to 0. The essential difference is that the flip-flop has a clock input, and the Q output can change only after an active clock edge. The truth table and characteristic equation for the flip-flop are the same as for the latch, but the interpretation of Q^+ is different. For the latch, Q^+ is the value of Q after the propagation delay through the latch, while for the flip-flop, Q^+ is the value that Q assumes after the active clock edge.

Figure 11-23(a) shows an S-R flip-flop constructed from two S-R latches and gates. This flip-flop changes state after the rising edge of the clock. The circuit is often referred to as a master-slave flip-flop. When CLK = 0, the S and R inputs set the outputs of the master latch to the appropriate value while the slave latch holds the previous value of Q. When the clock changes from 0 to 1, the value of P is held in the master latch and this value is transferred to the slave latch. The master latch holds the value of P while CLK = 1, and, hence, Q does not change. When the clock changes from 1 to 0, the Q value is latched in the slave, and the master can process

FIGURE 11-22
S-R Flip-Flop

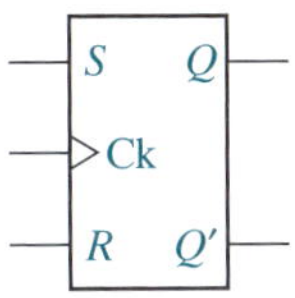

Operation summary:

$S = R = 0$	No state change
$S = 1, R = 0$	Set Q to 1 (after active Ck edge)
$S = 0, R = 1$	Reset Q to 0 (after active Ck edge)
$S = R = 1$	Not allowed

FIGURE 11-23
S-R Flip-Flop Implementation and Timing

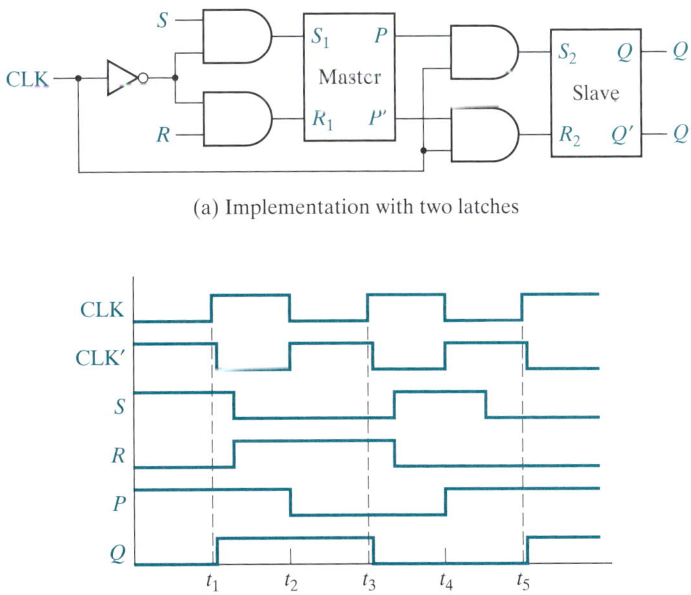

(a) Implementation with two latches

(b) Timing analysis

new inputs. Figure 11-23(b) shows the timing diagram. Initially, $S = 1$ and Q changes to 1 at t_1. Then $R = 1$ and Q changes to 0 at t_3.

At first glance, this flip-flop appears to operate just like an edge-triggered flip-flop, but there is a subtle difference. For a rising-edge-triggered flip-flop the value of the inputs is sensed at the rising edge of the clock, and the inputs can change while the clock is low. For the master-slave flip-flop, if the inputs change while the clock is low, the flip-flop output may be incorrect. For example, in Figure 11-23(b) at t_4, $S = 1$ and $R = 0$, so P changes to 1. Then S changes to 0 at t_5, but P does not change, so at t_5, Q changes to 1 after the rising edge of CLK. However, at t_5, $S = R = 0$, so the state of Q should not change. We can solve this problem if we only allow the S and R inputs to change while the clock is high.

11.6 J-K Flip-Flop

The J-K flip-flop (Figure 11-24) is an extended version of the S-R flip-flop. The J-K flip-flop has three inputs—J, K, and the clock (CK). The J input corresponds to S, and K corresponds to R. That is, if $J = 1$ and $K = 0$, the flip-flop output is set to $Q = 1$ after the active clock edge; and if $K = 1$ and $J = 0$, the flip-flop output is reset to $Q = 0$ after the active edge. Unlike the S-R flip-flop, a 1 input may be applied simultaneously to J and K, in which case the flip-flop changes state after the active clock edge. When $J = K = 1$, the active edge will cause Q to change from 0 to 1, or

FIGURE 11-24 J-K Flip-Flop (Q Changes on the Rising Edge)

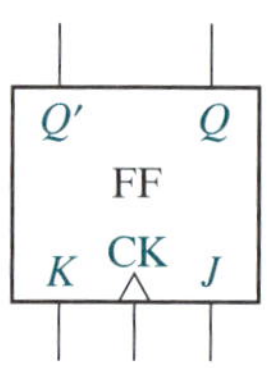

(a) J-K flip-flop

J K Q	Q^+
0 0 0	0
0 0 1	1
0 1 0	0
0 1 1	0
1 0 0	1
1 0 1	1
1 1 0	1
1 1 1	0

$$Q^+ = JQ' + K'Q$$

(b) Truth table and characteristic equation

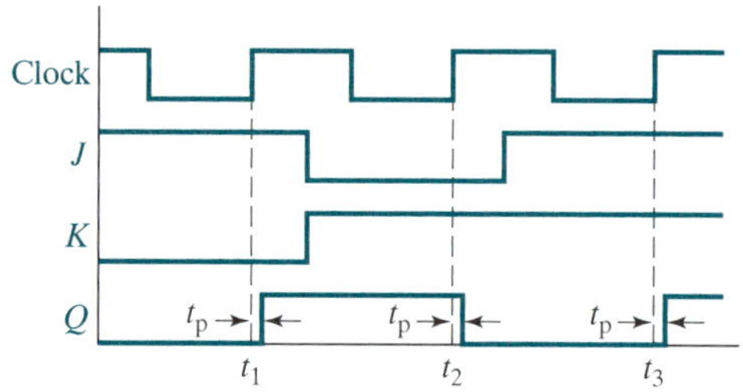

(c) J-K flip-flop timing

from 1 to 0. The next-state table and characteristic equation for the J-K flip-flop are given in Figure 11-24(b).

Figure 11-24(c) shows the timing for a J-K flip-flop. This flip-flop changes state a short time (t_p) after the rising edge of the clock pulse, provided that J and K have appropriate values. If $J = 1$ and $K = 0$ when Clock = 0, Q will be set to 1 following the rising edge. If $K = 1$ and $J = 0$ when Clock = 0, Q will be set to 0 after the rising edge. Similarly, if $J = K = 1$, Q will change state after the rising edge. Referring to Figure 11-24(c), because $Q = 0, J = 1$, and $K = 0$ before the first rising clock edge, Q changes to 1 at t_1. Because $Q = 1, J = 0$, and $K = 1$ before the second rising clock edge, Q changes to 0 at t_2. Because $Q = 0, J = 1$, and $K = 1$ before the third rising clock edge, Q changes to 1 at t_3.

One way to realize the J-K flip-flop is with two S-R latches connected in a master-slave arrangement, as shown in Figure 11-25. This is the same circuit as for the S-R master-slave flip-flop, except S and R have been replaced with J and K, and the Q and Q' outputs are feeding back into the input gates. Because $S = J \cdot Q' \cdot \text{CLK}'$ and $R = K \cdot Q \cdot \text{CLK}'$, only one of S and R inputs to the first latch can be 1 at any given time. If $Q = 0$ and $J = 1$, then $S = 1$ and $R = 0$, regardless of the value of K. If $Q = 1$ and $K = 1$, then $S = 0$ and $R = 1$, regardless of the value of J.

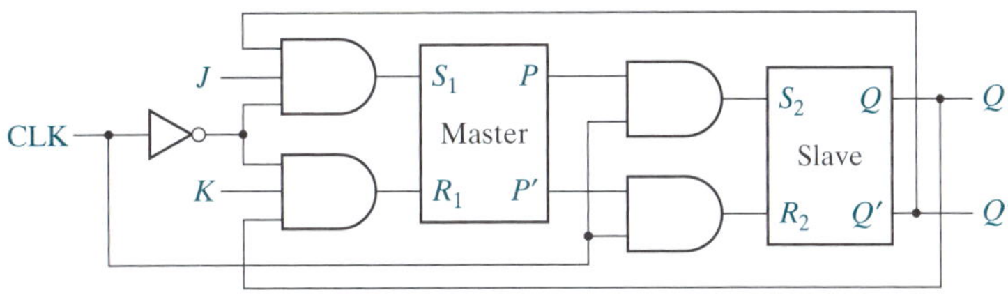

FIGURE 11-25 Master-Slave J-K Flip-Flop (Q Changes on Rising Edge)

11.7 T Flip-Flop

The T flip-flop, also called the toggle flip-flop, is frequently used in building counters. Most CPLDs and FPGAs can be programmed to implement T flip-flops. The T flip-flop in Figure 11-26(a) has a T input and a clock input. When $T = 1$ the flip-flop changes state after the active edge of the clock. When $T = 0$, no state change occurs. The next-state table and characteristic equation for the T flip-flop are given in Figure 11-26(b). The characteristic equation states that the next state of the flip-flop (Q^+) will be 1 iff the present state (Q) is 1 and $T = 0$ or the present state is 0 and $T = 1$.

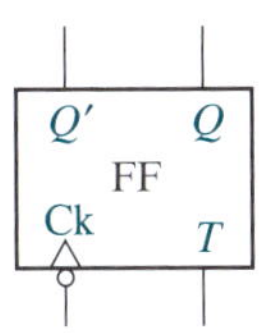

(a)

T Q	Q^+
0 0	0
0 1	1
1 0	1
1 1	0

$$Q^+ = T'Q + TQ' = T \oplus Q$$

(b)

FIGURE 11-26 T Flip-Flop

Figure 11-27 shows a timing diagram for the T flip-flop. At times t_2 and t_4 the T input is 1 and the flip-flop state (Q) changes a short time (t_p) after the falling edge of the clock pulse. At times t_1 and t_3 the T input is 0, and the clock edge does not cause a change of state.

FIGURE 11-27 Timing Diagram for T Flip-Flop (Falling-Edge Trigger)

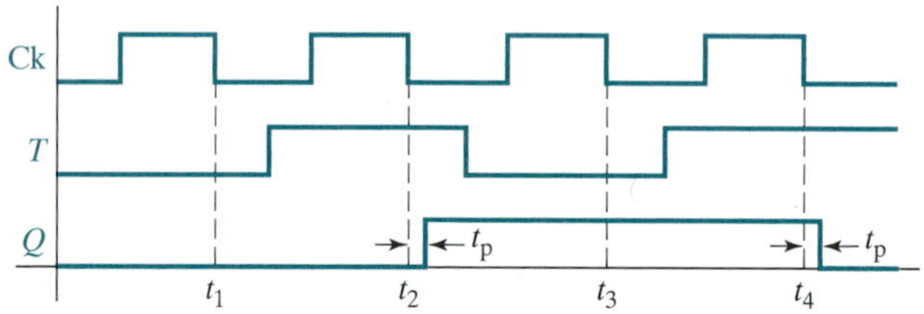

FIGURE 11-28 Implementation of T Flip-Flops

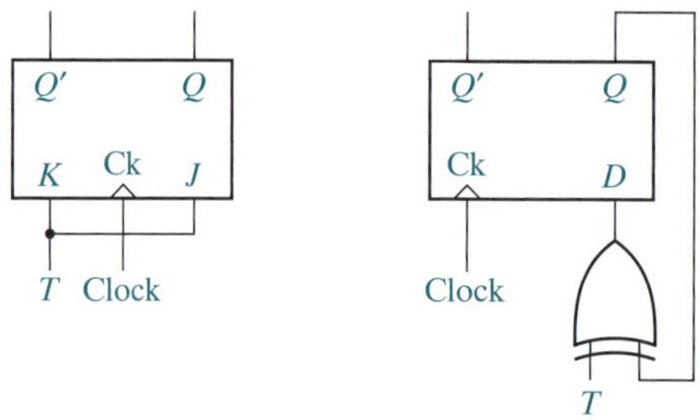

(a) Conversion of J-K to T (b) Conversion of D to T

One way to implement a T flip-flop is to connect the J and K inputs of a J-K flip-flop together, as shown in Figure 11-28(a). Substituting T for J and K in the J-K characteristic equation gives

$$Q^+ = JQ' + K'Q = TQ' + T'Q$$

which is the characteristic equation for the T flip-flop. Another way to realize a T flip-flop is with a D flip-flop and an exclusive-OR gate (Figure 11-28(b)). The D input is $Q \oplus T$, so $Q^+ = Q \oplus T = TQ' + T'Q$, which is the characteristic equation for the T flip-flop.

11.8 Flip-Flops with Additional Inputs

Flip-flops often have additional inputs which can be used to set the flip-flops to an initial state independent of the clock. Figure 11-29 shows a D flip-flop with clear and preset inputs. The small circles (inversion symbols) on these inputs indicate that a logic 0 (rather than a 1) is required to clear or set the flip-flop. This type of input is often referred to as *active-low* because a low voltage or logic 0 will activate the clear or preset function. We will use the notation ClrN or PreN to indicate active-low clear

FIGURE 11-29 D Flip-Flop with Clear and Preset

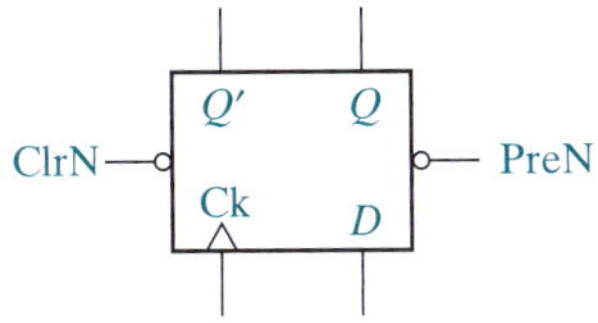

(a)

Ck	D	PreN	ClrN	Q^+
x	x	0	0	(not allowed)
x	x	0	1	1
x	x	1	0	0
↑	0	1	1	0
↑	1	1	1	1
0,1,↓	x	1	1	Q (no change)

(b)

and preset inputs. Thus, a logic 0 applied to ClrN will reset the flip-flop to $Q = 0$, and a 0 applied to PreN will set the flip-flop to $Q = 1$. These inputs override the clock and D inputs. That is, a 0 applied to the ClrN will reset the flip-flop regardless of the values of D and the clock. Under normal operating conditions, a 0 should not be applied simultaneously to ClrN and PreN. When ClrN and PreN are both held at logic 1, the D and clock inputs operate in the normal manner. ClrN and PreN are often referred to as *asynchronous* clear and preset inputs because their operation does not depend on the clock. The table in Figure 11-29(b) summarizes the flip-flop operation. In the table, ↑ indicates a rising clock edge, and X is a don't-care. The last row of the table indicates that if Ck is held at 0, held at 1, or has a falling edge, Q does not change. The clear and preset inputs can also be synchronous, i.e., the clear and set operations occur on the active edge of the clock. Often a synchronous clear or preset will override the other synchronous inputs, e.g., a D flip-flop with a synchronous clear input will clear on the active edge of the clock when the clear input is active independent of the value on D.

Figure 11-30 illustrates the operation of the clear and preset inputs. At t_1, ClrN = 0 holds the Q output at 0, so the rising edge of the clock is ignored. At t_2 and t_3, normal state changes occur because ClrN and PreN are both 1. Then, Q is set to 1 by PreN = 0, but Q is cleared at t_4 by the rising edge of the clock because $D = 0$ at that time.

In synchronous digital systems, the flip-flops are usually driven by a common clock so that all state changes occur at the same time in response to the same clock edge. When designing such systems, we frequently encounter situations where we want some flip-flops to hold existing data even though the data input to the flip-flops may be changing. One way to do this is to gate the clock, as shown in Figure 11-31(a). When En = 0, the clock input to the flip-flop is 0, and Q does not change. This method

FIGURE 11-30 Timing Diagram for D Flip-Flop with Asynchronous Clear and Preset

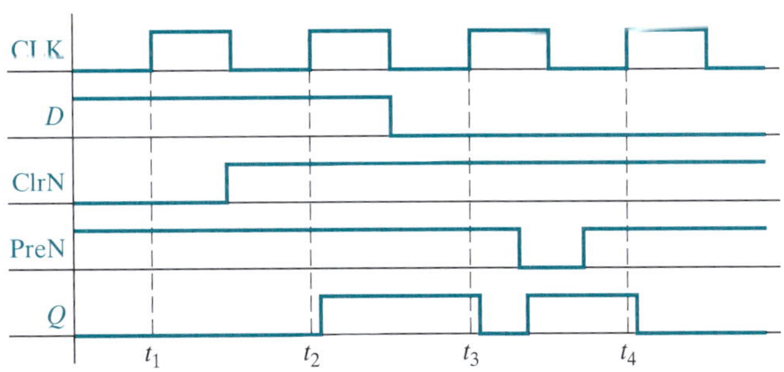

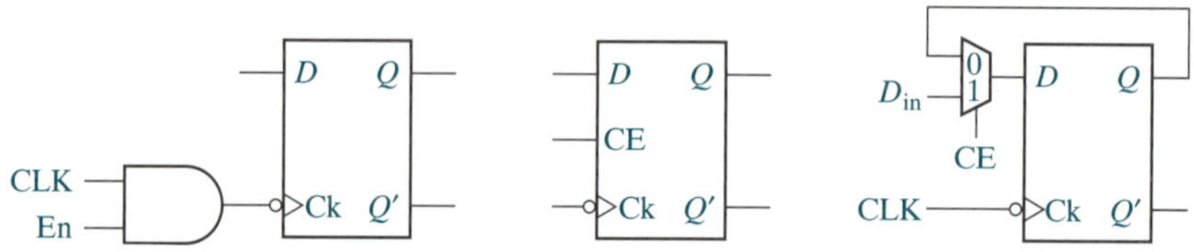

(a) Gating the clock (b) D-CE symbol (c) Implementation

FIGURE 11-31 D Flip-Flop with Clock Enable

has two potential problems. First, gate delays may cause the clock to arrive at some flip-flops at different times than at other flip-flops, resulting in a loss of synchronization. Second, if En changes at the wrong time, the flip-flop may trigger due to the change in En instead of due to the change in the clock, again resulting in loss of synchronization. Rather than gating the clock, a better way is to use a flip-flop with a clock enable (CE). Such flip-flops are commonly used in CPLDs and FPGAs.

Figure 11-31(b) shows a D flip-flop with a clock enable, which we will call a D-CE flip-flop. When CE = 0, the clock is disabled and no state change occurs, so $Q^+ = Q$. When CE = 1, the flip-flop acts like a normal D flip-flop, so $Q^+ = D$. Therefore, the characteristic equation is $Q^+ = Q \cdot CE' + D \cdot CE$. The D-CE flip-flop is easily implemented using a D flip-flop and a multiplexer (Figure 11-31(c)). For this circuit, the MUX output is

$$Q^+ = D = Q \cdot CE' + D_{in} \cdot CE$$

Because there is no gate in the clock line, this cannot cause a synchronization problem.

11.9 Asynchronous Sequential Circuits

The signals on the feedback loops in a sequential circuit define the state of the circuit. In *asynchronous* sequential circuits the state of the circuit can change whenever any input changes. The latches discussed in this chapter are asynchronous sequential circuits. In addition, even though the output of a master-slave or edge-triggered flip-flop only changes on an edge of the clock input, the internal circuitry of the flip-flop is an asynchronous circuit. In order for asynchronous circuits to operate in a well-defined manner, they must satisfy several restrictions. Asynchronous circuits are not discussed in detail in this text,[1] but the examples in this section illustrate some of these restrictions necessary for correct asynchronous circuit operation.

Consider the circuit in Figure 11-32(a). It has two feedback loops and, hence, four states. To analyze the circuit, break the feedback loops at the outputs of gates 4 and 5. Label the gate outputs P^+ and Q^+. The next-state equations are

$$P^+ = x'P + xQ$$
$$Q^+ = x'P' + xQ$$

[1] For more details, see M. Morris Mano and Ciletti, Michael D., *Digital Design (4th Ed)*, (Pearson Prentice Hall, 2007).

Note that both equations contain static 1-hazards. P^+ contains a static 1-hazard for $xPQ = 111 \leftrightarrow 011$.

FIGURE 11-32 Asynchronous Circuit

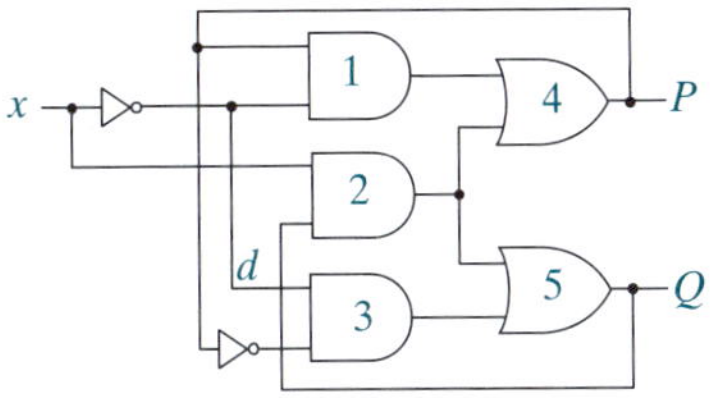

(a)

PQ	x = 0	x = 1
00	01	(00)
01	(01)	11
11	10	(11)
10	(10)	00

(b)

Figure 11-32(b) shows the circuit's next-state table with the stable states circled. Consider starting the circuit in total stable state $xPQ = 111$ and changing x to 0. The table indicates that the next state is $PQ = 10$ (i.e., P does not change and Q changes to 0). However, assume the inverter for x has a delay large compared to the gate delays. Then, when x changes to 0, the output of Gate 2 changes to 0 causing the outputs of both gates 4 and 5 to change to 0. When the x inverter output changes to 1, $P = 0$ prevents the output of Gate 1 from changing to 1 so P remains 0; however, the output of Gate 3 changes to 1 and Q becomes 1. The circuit stabilizes in state $PQ = 01$. This incorrect state transition is due to the static 1-hazard in P^+, which causes a 0 glitch in P if the x inverter delay is large and, because of the feedback, P never changes back to 1. This problem can be eliminated by designing the circuits to be free of hazards.

Even if the circuit is free of hazards, delays in the "wrong" places in the circuit can cause incorrect state transitions. Consider the circuit in Figure 11-32(a) starting the circuit in total stable state $xPQ = 010$ and changing x to 1; the next-state table indicates the next state should be $PQ = 00$. However, if there is a large delay in the line labeled d, the circuit will transition to state $PQ = 11$. The 1 to 0 change propagating through the x inverter, Gate 1 and Gate 4 will change P to 0. When P changes to 0, the lower inverter output changes to 1 and, if line d at Gate 3 is still 1, Gate 3 and Gate 5 change to 1. With $Q = 1$, Gate 2 changes to 1 which causes Gate 4 to change back to 1. Note that this incorrect operation occurs because the x change does not reach a portion of the logic for Q until after the change in P has propagated back to the logic for Q. The potential for this incorrect operation can be detected by examining the next-state table and determining whether the next state after a single change in an input is different from the next state after three changes in that input. In the preceding example, starting in total stable state $xPQ = 010$ and changing x to 1 produces next state $PQ = 00$. Changing x back to 0 produces next state $PQ = 01$. Finally, changing x a third time to 1, produces next state $PQ = 11$. The table is said to contain an *essential hazard* at total state $xPQ = 010$.

Essential hazards are properties of the next-state table; they cannot be eliminated by modifying the circuit's logic. To prevent incorrect operation due to essential hazards, it is necessary to control the delays in the circuit. In general, it may be necessary to insert delays in the feedback loops of the circuit. For the above example, this would require inserting delays at the outputs of Gates 4 and 5. Inserting delays in the feedback loops, assures that input changes propagate through the next state logic before changes in the state variables feedback to the inputs of the logic.

Both of the problems noted above are due to more than one signal in the circuit changing with the circuit's response depending on which signal(s) propagate through the circuit first. The static hazard causes a glitch if an input change propagates over some path in the circuit more quickly than it propagates over another path. For the essential hazard problem the two signals changing are an external input and a state variable. In both cases there is a "race" between two signal changes with the circuit operation depending on which signal wins the race. A similar problem can exist if two (or more) inputs to an asynchronous circuit can change at the same time. Consider the next state table in Figure 11-33 and consider starting in total state $xyPQ = 1011$ and changing both x and y to 0 at the same time. The table indicates the next state should be $PQ = 11$. If the y change propagates through the circuit first, then the circuit enters total state $xyPQ = 1011$. Then after the x change propagates through the circuit, the total state becomes $xyPQ = 0011$. However, if the x change propagates through the circuit first, the total state becomes $xyPQ = 0100$. Then, after the y change propagates through the circuit, the total becomes $xyPQ = 0000$. The final state depends upon which input change propagates fastest and, hence, the final state may not be $PQ = 11$, as indicated by the next-state table.

FIGURE 11-33 Multiple Input Change Example

PQ \ xy	00	01	11	10
00	(00)	(00)	01	(00)
01	11	00	(01)	11
11	(11)	10	(11)	(11)
10	00	(10)	11	00

A similar race problem can occur if more than one state variable changes at (approximately) the same time. Consider the next-state table in Figure 11-34. The total state $xPQR = 0000$ has next state $PQR = 011$ when x changes to 1. If Q changes first, the circuit will transition to state $PQR = 010$ and then to $PQR = 011$. However, if R changes first, the circuit transitions to state $PQR = 001$ and may stabilize there. There is a race between the Q and R changes and, since the circuit can enter two different states depending which signal changes first, the race is called a *critical race*.

FIGURE 11-34 Multiple-State Variable Change Example

PQR \ x	0	1
000	(000)	011
001	101	(001)
011	101	(011)
010	000	011
110	000	(110)
111	101	(111)
101	(101)	110
100	(100)	110

There is a race between the P and Q signals when the circuit starts in total state $xPQR = 1011$ and x changes to 0. In this case, the circuit will enter state $PQR = 101$ no matter which state variable changes first, so the race is not critical.

In general, it is not possible to fix the problem with races between state variables by inserting delays in the circuit. For correct operation an asynchronous circuit must not contain any critical races.

11.10 Summary

In this unit, we have studied several types of latches and flip-flops. Flip-flops have a clock input, and the output changes only in response to a rising or falling edge of the clock. All of these devices have two output states: $Q = 0$ and $Q = 1$. For the S-R latch, $S = 1$ sets Q to 1, and $R = 1$ resets Q to 0. $S = R = 1$ is not allowed. The S-R flip-flop is similar except that Q only changes after the active edge of the clock. The gated D latch transmits D to the Q output when $G = 1$. When G is 0, the current value of D is stored in the latch and Q does not change. For the D flip-flop, Q is set equal to D after the active clock edge. The D-CE flip-flop works the same way, except the clock is only enabled when CE = 1. The J-K flip-flop is similar to the S-R flip-flop in that when $J = 1$ the active clock edge sets Q to 1, and when $K = 1$, the active edge resets Q to 0. When $J = K = 1$, the active clock edge causes Q to change state. The T flip-flop changes state on the active clock edge when $T = 1$; otherwise, Q does not change. Flip-flops can have asynchronous clear and preset inputs that cause Q to be cleared to 0 or preset to 1 independently of the clock.

Flip-flops can be constructed using gate circuits with feedback. Analysis of such circuits can be accomplished by tracing signal changes through the gates. Analysis can also be done using flow tables and asynchronous sequential circuit theory, but that is beyond the scope of this text. Timing diagrams are helpful in understanding the time relationships between the input and output signals for a latch or flip-flops. In general, the inputs must be applied a specified time before the active clock edge (the setup time), and they must be held constant a specified time after the active edge (the hold time). The time after the active clock edge before Q changes is the propagation delay.

The characteristic (next-state) equation for a flip-flop can be derived as follows: First, make a truth table that gives the next state (Q^+) as a function of the present state (Q) and the inputs. Any illegal input combinations should be treated as don't-cares. Then, plot a map for Q^+ and read the characteristic equation from the map.

The characteristic equations for the latches and flip-flops discussed in this chapter are:

$$Q^+ = S + R'Q \ (SR = 0) \qquad \text{(S-R latch or flip-flop)} \qquad (11\text{-}6)$$

$$Q^+ = GD + G'Q \qquad \text{(gated D latch)} \qquad (11\text{-}7)$$

$$Q^+ = D \qquad \text{(D flip-flop)} \qquad (11\text{-}8)$$

$$Q^+ = D \cdot CE + Q \cdot CE' \qquad \text{(D-CE flip-flop)} \qquad (11\text{-}9)$$

$$Q^+ = JQ' + K'Q \qquad \text{(J-K flip-flop)} \qquad (11\text{-}10)$$

$$Q^+ = T \oplus Q = TQ' + T'Q \qquad \text{(T flip-flop)} \qquad (11\text{-}11)$$

In each case, Q represents an initial or present state of the flip-flop, and Q^+ represents the final or next state. These equations are valid only when the appropriate restrictions on the flip-flop inputs are observed. For the S-R flip-flop, $S = R = 1$ is forbidden. For the master-slave S-R flip-flop, S and R should not change during the half of the clock cycle preceding the active edge. Setup and hold time restrictions must also be satisfied.

The characteristic equations given above apply to both latches and flip-flops, but their interpretation is different for the two cases. For example, for the gated D latch, Q^+ represents the state of the flip-flop a short time after one of the inputs changes. However, for the D flip-flop, Q^+ represents the state of the flip-flop a short time after the active clock edge.

Conversion of one type of flip-flop to another is usually possible by adding external gates. Figure 11-28 shows how a J-K flip-flop and a D flip-flop can be converted to a T flip-flop.

Problems

11.1 Assume that the inverter in the given circuit has a propagation delay of 5 ns and the AND gate has a propagation delay of 10 ns. Draw a timing diagram for the circuit showing X, Y, and Z. Assume that X is initially 0, Y is initially 1, after 10 ns X becomes 1 for 80 ns, and then X is 0 again.

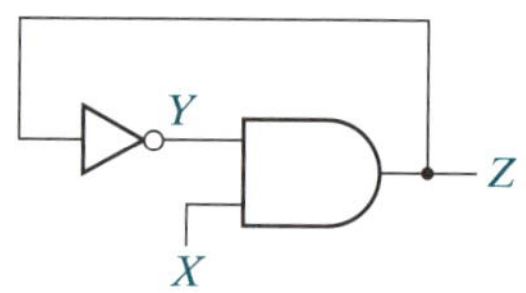

11.2 A latch can be constructed from an OR gate, an AND gate, and an inverter connected as follows:

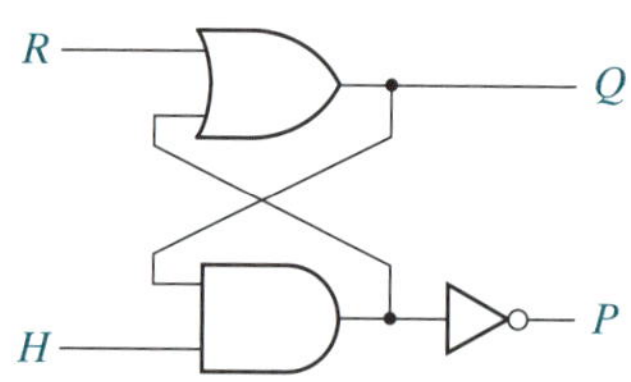

(a) What restriction must be placed on R and H so that P will always equal Q' (under steady-state conditions)?

(b) Construct a next-state table and derive the characteristic (next-state) equation for the latch.

(c) Complete the following timing diagram for the latch.

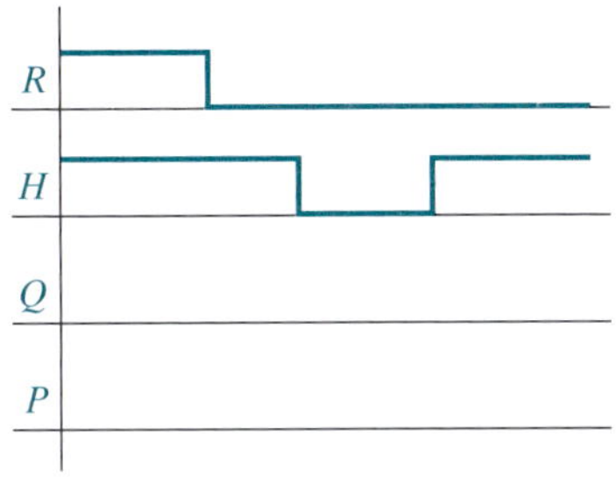

11.3 This problem illustrates the improper operation that can occur if both inputs to an S-R latch are 1 and are then changed back to 0. For Figure 11-6, complete the following timing chart, assuming that each gate has a propagation delay of exactly 10 ns. Assume that initially $P = 1$ and $Q = 0$. Note that when $t = 100$ ns, S and R are both changed to 0. Then, 10 ns later, both P and Q will change to 1. Because these 1's are fed back to the gate inputs, what will happen after another 10 ns?

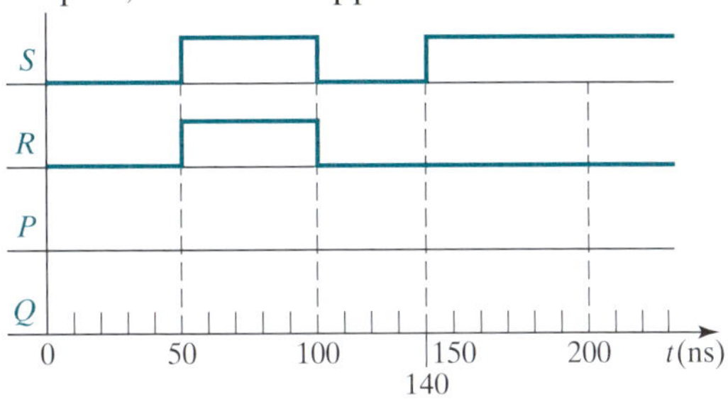

11.4 Design a gated D latch using only NAND gates and one inverter.

11.5 What change must be made to Figure 11-19(a) to implement a falling-edge-triggered D flip-flop? Complete the following timing diagram for the modified flip-flop.

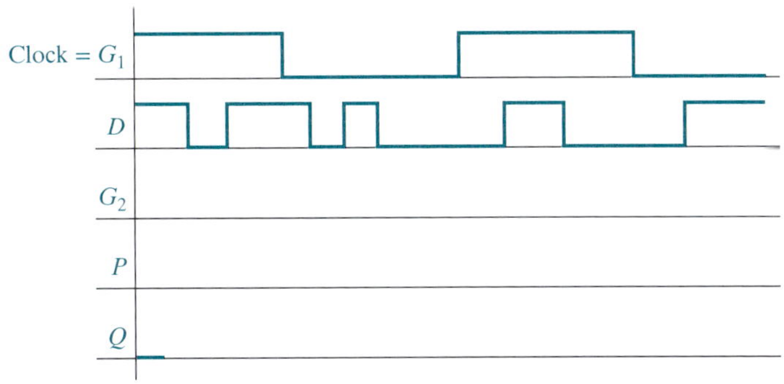

11.6 A reset-dominant flip-flop behaves like an S-R flip-flop, except that the input $S = R = 1$ is allowed, and the flip-flop is reset when $S = R = 1$.

(a) Derive the characteristic equation for a reset-dominant flip-flop.

(b) Show how a reset-dominant flip-flop can be constructed by adding gate(s) to an S-R flip-flop.

11.7 Complete the following timing diagram for the flip-flop of Figure 11-24(a).

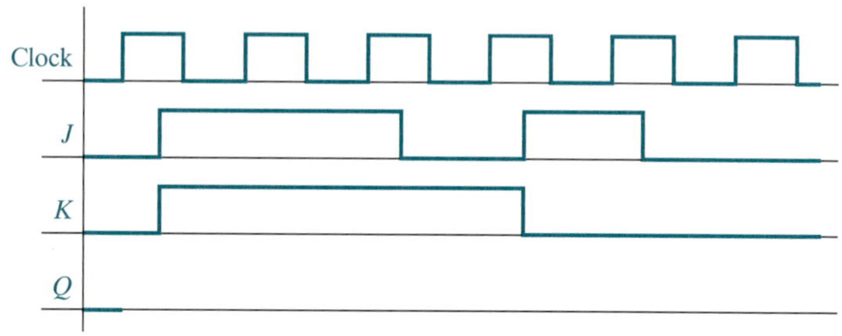

11.8 Complete the following diagrams for the falling-edge-triggered D-CE flip-flop of Figure 11-31(c). Assume Q begins at 1.

(a) First draw Q based on your understanding of the behavior of a D flip-flop with clock enable.

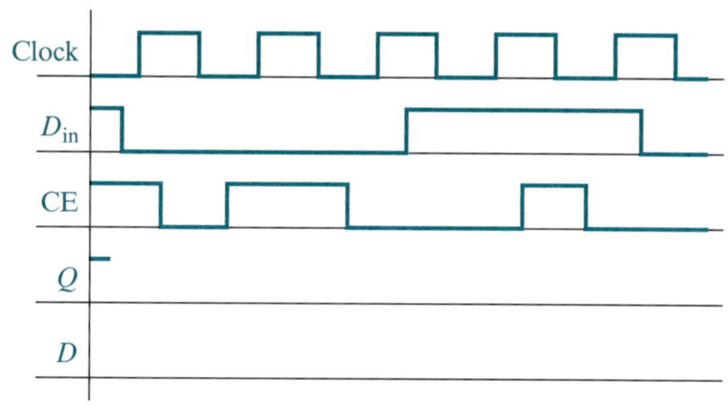

(b) Now draw in the internal signal D from Figure 11-31(c), and confirm that this gives the same Q as in (a).

11.9 (a) Complete the following timing diagram for a J-K flip-flop with a falling-edge trigger and asynchronous ClrN and PreN inputs.

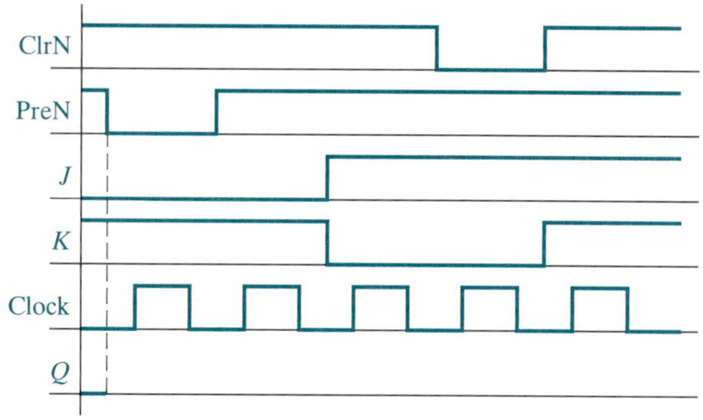

(b) Complete the timing diagram for the following circuit. Note that the Ck inputs on the two flip-flops are different.

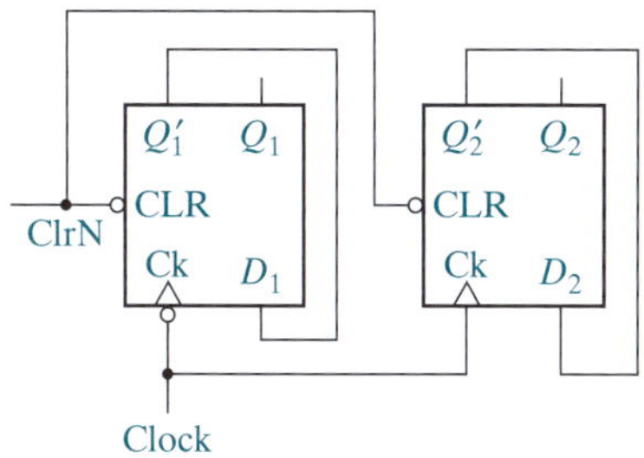

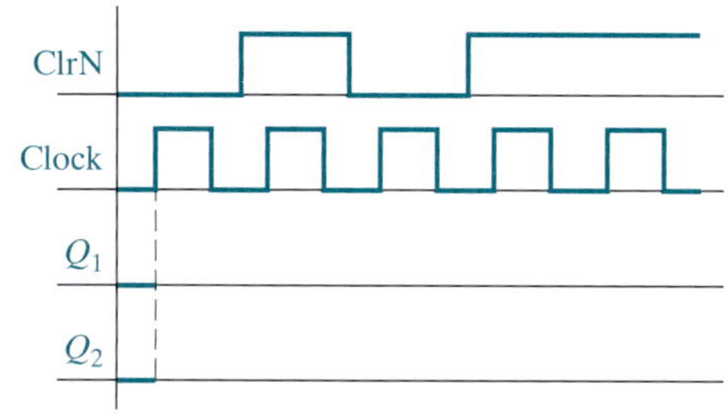

11.10 Convert by adding external gates:
(a) a D flip-flop to a J-K flip-flop.
(b) a T flip-flop to a D flip-flop.
(c) a T flip-flop to a D flip-flop with clock enable.

11.11 Complete the following timing diagram for an S-R latch. Assume Q begins at 1.

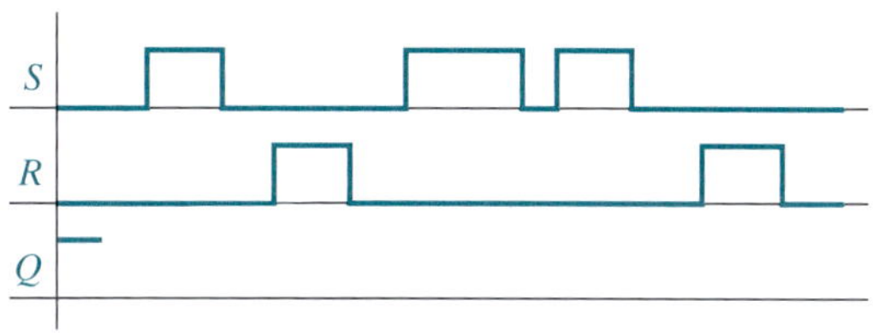

11.12 Using a truth table similar to Figure 11-8(b), confirm that each of these circuits is an S-R latch. What happens when $S = R - 1$ for each circuit?

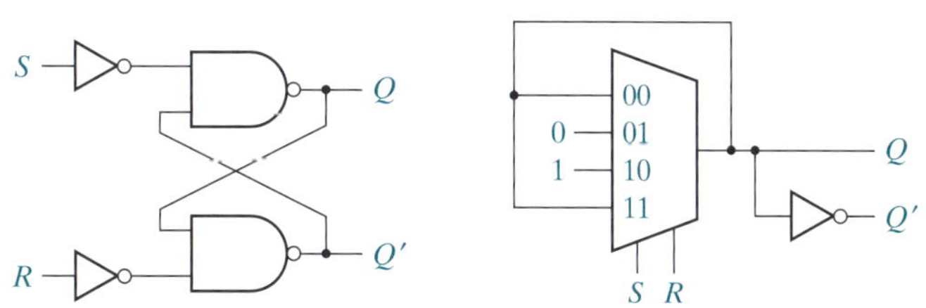

(a) (b)

11.13 An AB latch operates as follows: If $A = 0$ and $B = 0$, the latch state is $Q = 0$; if either $A = 1$ or $B = 1$ (but not both), the latch output does not change; and when both $A = 1$ and $B = 1$, the latch state is $Q = 1$.
(a) Construct the state table and derive the characteristic equation for this AB latch.
(b) Derive a circuit for the AB latch that has four two-input NAND gates and two inverters.
(c) In your circuit of part (b), are there any transitions between input combinations that might cause unreliable operation? Verify your answer.

(d) In your circuit of part (b), is there a gate output that provides the signal Q'? Verify your answer.

(e) Derive a circuit for the AB latch using four two-input NOR gates and two inverters.

(f) Answer parts (c) and (d) for your circuit of part (e).

11.14 (a) Construct a state table for this circuit and identify the stable states of the circuit.

(b) Derive a Boolean algebra equation for the next value of the output Q in terms of Q, A and B.

(c) Analyze the behavior of the circuit. Is it a useful circuit? If not, explain why not; if yes, explain what it does.

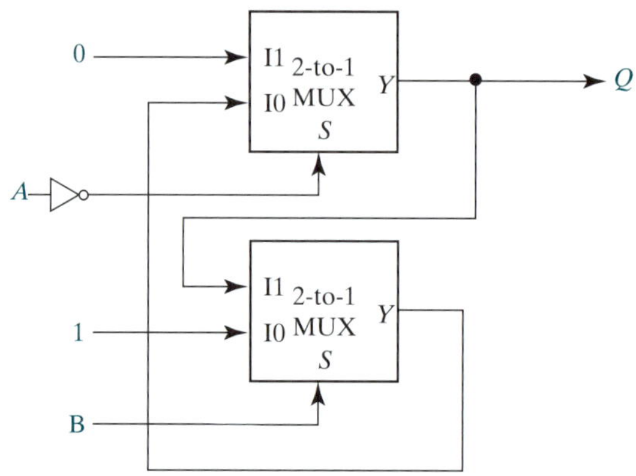

11.15 The following circuit is intended to be a gated latch circuit where the signal G is the gate.

(a) Derive the next-state equation for this circuit using Q as the state variable and P as an output.

(b) Construct the state table and output table for the circuit. Circle the stable states of the circuit.

(c) Are there any restrictions on the allowable input combinations on M and N? Explain your answer.

(d) Is the output P usable as the complement of Q? Verify your answer.

(e) Assume that Gate 1 has a propagation delay of 30 ns and Gates 2, 3, and 4 have propagation delays of 10 ns. Construct a timing diagram for the circuit for the following input change: $M = N = Q = 0$ with G changing from 1 to 0.

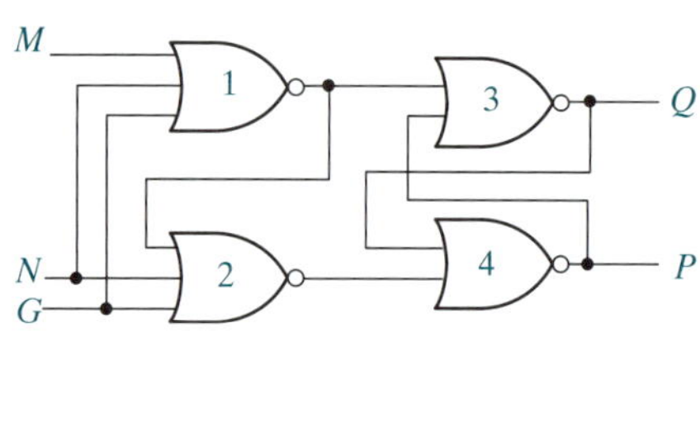

11.16 Analyze the latch circuit shown.

(a) Derive the next-state equation for this circuit using Q as the state variable and P as an output.

(b) Construct the state table and output table for the circuit. Circle the stable states of the circuit.

(c) Are there any restrictions on the allowable input combinations on A and B? Explain your answer.

(d) Is the output P usable as the complement of Q? Verify your answer.

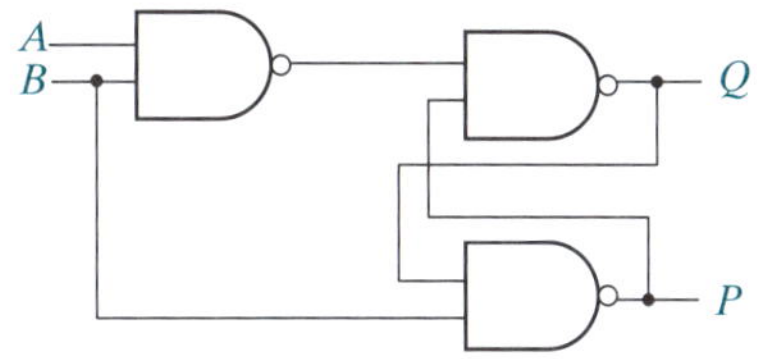

11.17 Derive the characteristic equations for the following latches and flip-flops in product-of-sums form.

(a) S-R latch or flip-flop

(b) Gated D latch

(c) D flip-flop

(d) D-CE flip-flop

(e) J-K flip-flop

(f) T flip-flop

11.18 Complete the following timing diagrams for a gated D latch. Assume Q begins at 0.

(a) First draw Q based on your understanding of the behavior of a gated D latch.

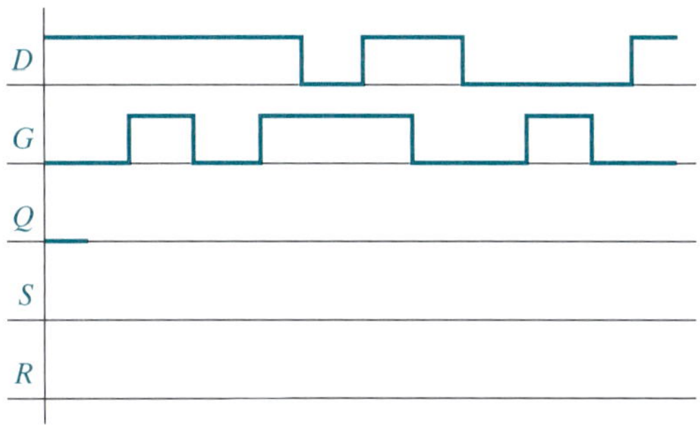

(b) Now draw in the internal signals S and R from Figure 11-14, and confirm that S and R give the same value for Q as in (a).

11.19 Complete the following diagrams for the rising-edge-triggered D flip-flop of Figure 11-19. Assume Q begins at 1.

(a) First draw Q based on your understanding of the behavior of a D flip-flop.

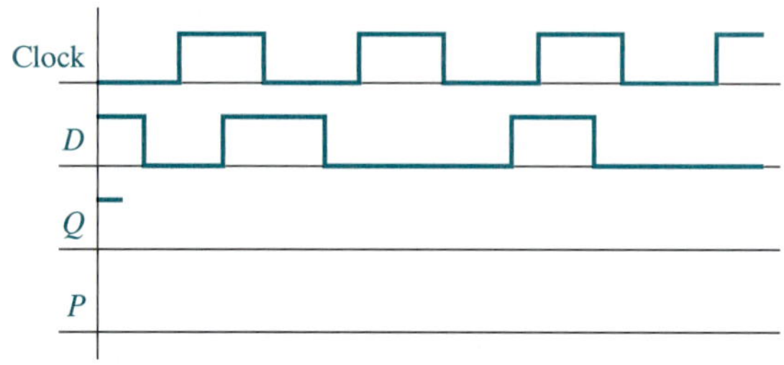

(b) Now draw in the internal signal P from Figure 11-19, and confirm that P gives the same Q as in (a).

11.20 A set-dominant flip-flop is similar to the reset-dominant flip-flop of Problem 11.6 except that the input combination $S = R = 1$ sets the flip-flop. Repeat Problem 11.6 for a set-dominant flip-flop.

11.21 Fill in the timing diagram for a falling-edge-triggered S-R flip-flop. Assume Q begins at 0.

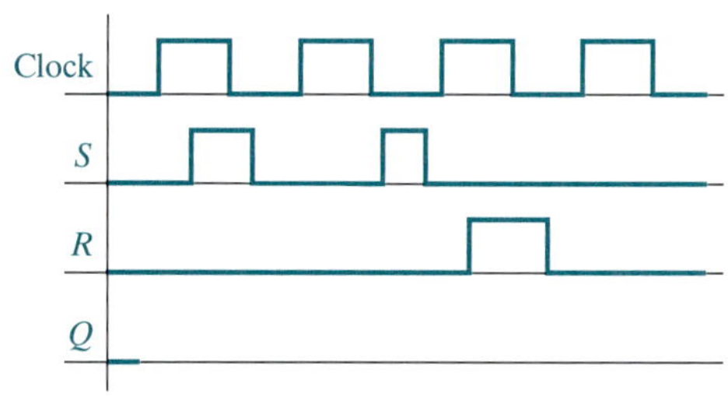

11.22 Fill in the timing diagram for a falling-edge-triggered J-K flip-flop.

(a) Assume Q begins at 0.

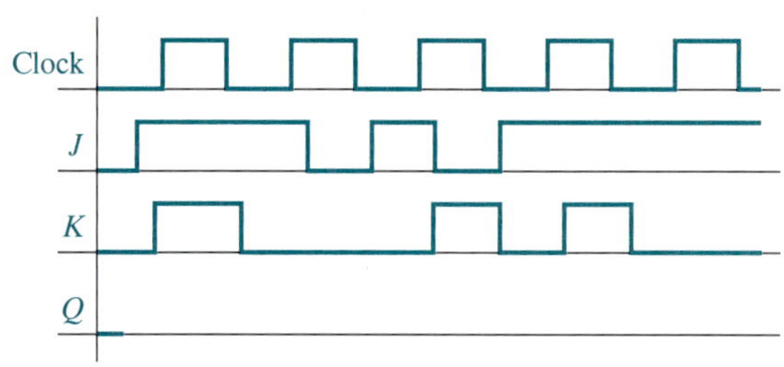

(b) Assume Q begins at 1, but Clock, J, and K are the same.

11.23 (a) Find the input for a rising-edge-triggered D flip-flop that would produce the output Q as shown. Fill in the timing diagram.
(b) Repeat for a rising-edge-triggered T flip-flop.

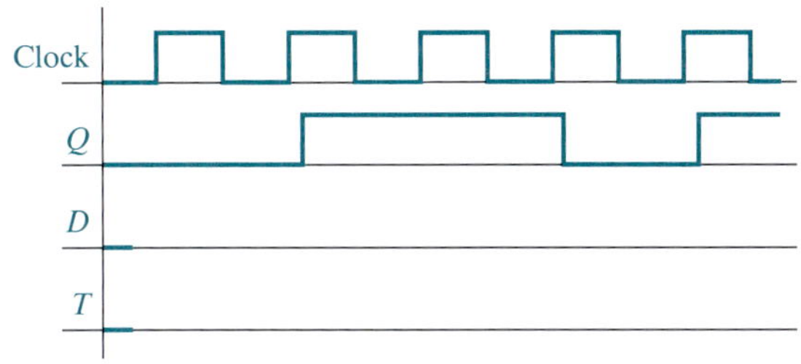

11.24 Here is the diagram of a 3-bit ripple counter. Assume $Q_0 = Q_1 = Q_2 = 0$ at $t = 0$, and assume each flip-flop has a delay of 1 ns from the clock input to the Q output. Fill in Q_0, Q_1, and Q_2 of the timing diagram. Flip-flop Q_1, will be triggered when Q_0 changes from 0 to 1.

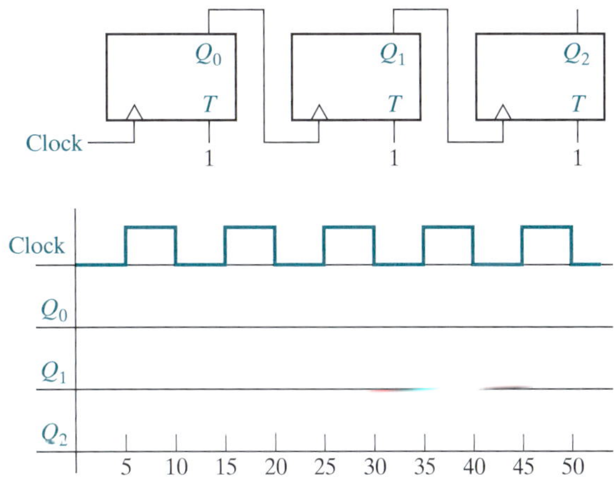

11.25 Fill in the following timing diagram for a rising-edge-triggered T flip-flop with an asychronous active-low PreN input. Assume Q begins at 1.

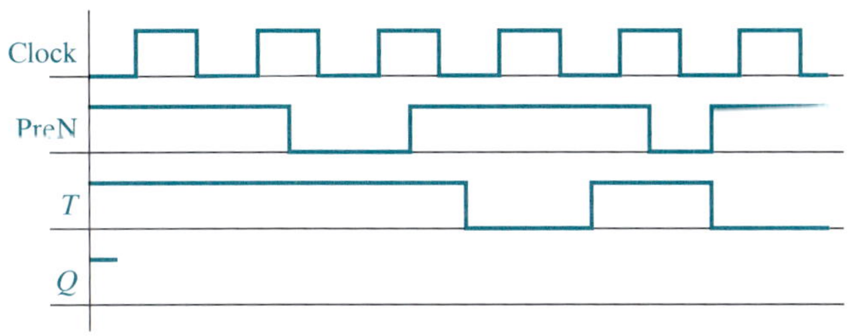

11.26 The ClrN and PreN inputs introduced in Section 11.8 are called asynchronous because they operate independently of the clock (i.e., they are not synchronized with the clock). We can also make flip-flops with synchronous clears or preset inputs.

A D-flip-flop with an active-low synchronous ClrN input may be constructed from a regular D flip-flop as follows.

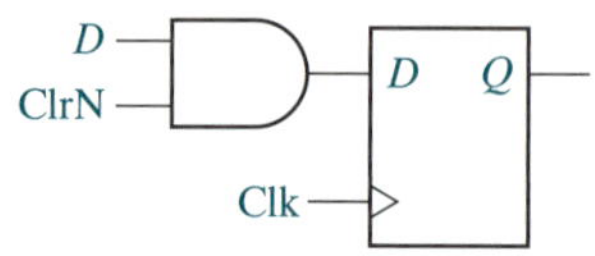

Fill in the timing diagram. For Q_1, assume a synchronous ClrN as above, and for Q_2, assume an asynchronous ClrN as in Section 11.8. Assume $Q_1 = Q_2 = 0$ at the beginning.

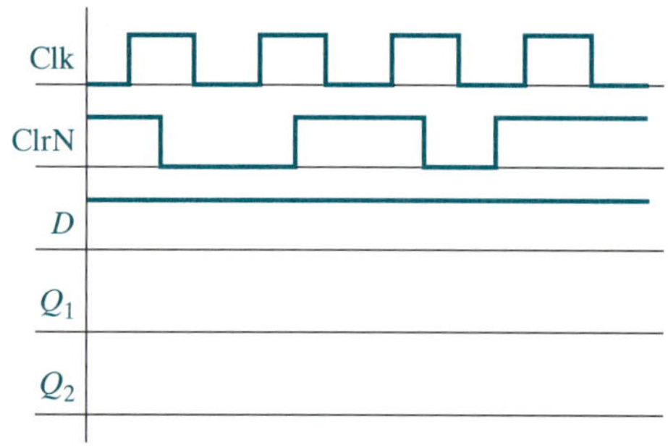

11.27 (a) Construct a D flip-flop using an inverter and an S-R flip-flop.

(b) If the propagation delay and setup time of the S-R flip-flop in (a) are 2.5 ns and 1.5 ns, respectively, and if the inverter has a propagation delay of 1 ns, what are the propagation delay and setup time of the D flip-flop of part (a)?

11.28 Redesign the debouncing circuit of Figure 11-9 using the $\bar{S}$-$\bar{R}$ latch of Figure 11-10.

11.29 (a) Use the characteristic equation for the gated D latch to implement a two-level, NAND-gate circuit for the gated D latch.

(b) What hazards exist in the part (a) circuit for Q^+?

(c) Modify the circuit of part (a) to obtain a two-level, NAND gate circuit for Q^+ that is free of hazards.

(d) Add another output, P, to the circuit of part (a) so that $P = Q'$ whenever the circuit is in a stable state. When the circuit is in an unstable state, the value of P is a don't care. Use the simplest circuit to obtain P. (*Note:* The circuit for P need not be two level.)

11.30 Consider converting the gated S-R latch of Figure 11-11 into a gated D latch by adding an inverter and making $S = D$ and $R = D'$.

(a) Derive the next-state equation for Q^+ from the circuit.

(b) Does the equation from part (a) contain any hazards? Verify your answer.

(c) Show that the equation from part (a) can be simplified to the characteristic equation for the gated D latch.

11.31 (a) Derive a NOR-gate version of the gated S-R latch analogous to Figure 11-11 by replacing each NAND gate with a NOR gate. Re-label the inputs so that Table 11-2 is the next-state table for the circuit; carefully indicate whether the inputs are active low or active high.

(b) Derive the next-state equation for Q^+ from the circuit of part (a).

(c) Determine any hazards contained in the next state equation of part (b). (*Hint:* It may be easiest to write the equation as a product of sums rather than as a sum of products.)

(d) What would be the disadvantage of allowing $S = 1$ and $R = 1$ in this circuit? Consider the effect on the P output, any race conditions that might result, and the effect of any hazards in the Q^+ equation.

11.32 (a) Redesign the circuit of Figure 11-32 so that P^+ and Q^+ are free of hazards. Use 6 two-input NAND gates and inverters as needed.

(b) Add two ouputs, U and V, to your circuit of part (a) so that the values on U and V count the number of times x has changed (0 to 1 and 1 to 0) modulo 4. (U and V will cycle through the values 00, 01, 10, 11, 00, 01, . . . as x changes.) Assume that initially $x = 0$.

11.33 For each of the 8 stable states of Figure 11-33, consider changing both inputs and

(a) determine the correct next state

(b) the sequence of changes if x changes first, and

(c) the sequence of changes if y changes first

For parts (b) and (c), indicate whether the final state is correct or not and whether there might be a glitch in P or Q.

11.34 Label the states in Figure 11-34 from $a = 000$ to $h = 100$. Determine a new assignment of values of P, Q and R to states a through h so that every transition from a stable state is free of races or, else, the race is noncritical. (Retain 000 as the encoding for a.) Identify any noncritical races.

Programmed Exercise 11.35

Cover the bottom part of each page with a sheet of paper and slide it down as you check your answers.

The internal logic diagram of a falling-edge-triggered D flip-flop follows. This flip-flop consists of two basic S-R latches with added gates. When the clock input (CK) is 1, the value of D is stored in the first S-R latch (P). When the clock changes from 1 to 0, the value of P is transferred to the output latch (Q). Thus, the operation is similar to that of the master-slave S-R flip-flop shown in Figure 11-23, except for the edges at which the data is stored.

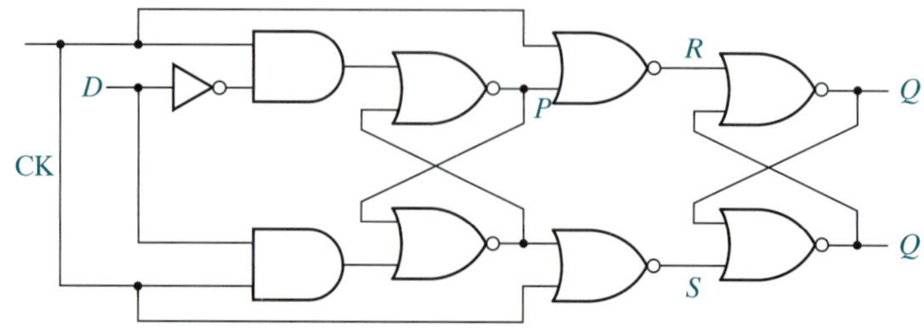

In this exercise you will be asked to analyze the operation of the D flip-flop shown above by filling in a table showing the values of CK, D, P, S, R, and Q after each change of input. It will be helpful if you mark the changes in these values on the circuit diagram as you trace the signals. Initially, assume the following signal values:

CK	*D*	*P*	*S*	*R*	*Q*	
0	0	0	0	1	0	(stable)

Verify by tracing signals through the circuit that this is a stable condition of the circuit; that is, no change will occur in P, S, R, or Q. Now assume that CK is changed to 1:

	CK	*D*	*P*	*S*	*R*	*Q*	
1.	0	0	0	0	1	0	(stable)
2.	1	0	0	0	1	0	?
3.							

Trace the change in CK through the circuit to see if a change in P, S, or R will occur. If a change does occur, mark row 2 of the preceding table "unstable" and enter the new values in row 3.

Answer:

2.	1	0	0	0	1	0	(unstable)
3.	1	0	0	0	0	0	(stable)
4.	1	1	0	0	0	0	(unstable)
5.	1	1					?

Verify that row 3 is stable; that is, by tracing signals show that no further change in P, S, R, or Q will occur. Next D is changed to 1 as shown in row 4. Verify that row 4 is unstable, fill in the new values in row 5, and indicate if row 5 is stable or unstable.

Answer:

	CK	*D*	*P*	*S*	*R*	*Q*	
5.	1	1	1	0	0	0	(stable)
6.	0	1	1	0	0	0	?
7.	0	1					?
8.	0	1					

Then CK is changed to 0 (row 6). If row 6 is unstable, indicate the new value of S in row 7. If row 7 is unstable, indicate the new value of Q in row 8. Then determine whether row 8 is stable or not.

Answer:

	CK	*D*	*P*	*S*	*R*	*Q*	
7.	0	1	1	1	0	0	(unstable)
8.	0	1	1	1	0	1	(stable)
9.	0	0					(stable)
10.	1	0					
11.	1	0					

Next, *D* is changed back to 0 (row 9). Fill in the values in row 9 and verify that it is stable. CK is changed to 1 in row 10. If row 10 is unstable, fill in row 11 and indicate whether it is stable or not.

Answer:

9.	0	0	1	1	0	1	(stable)
10.	1	0	1	1	0	1	(unstable)
11.	1	0	0	0	0	1	(stable)
12.	0	0					
13.	0	0					
14.	0	0					

CK is changed back to 0 in row 12. Complete the rest of the table.

Answer:

12.	0	0	0	0	0	1	(unstable)
13.	0	0	0	0	1	1	(unstable)
14.	0	0	0	0	1	0	(stable)

Using the previous results, plot *P* and *Q* on the following timing diagram. Verify that your answer is consistent with the description of the flip-flop operation given in the first paragraph of this exercise.

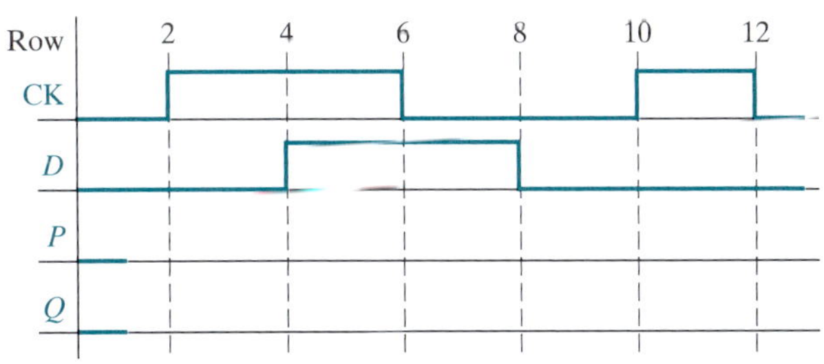

Answer:

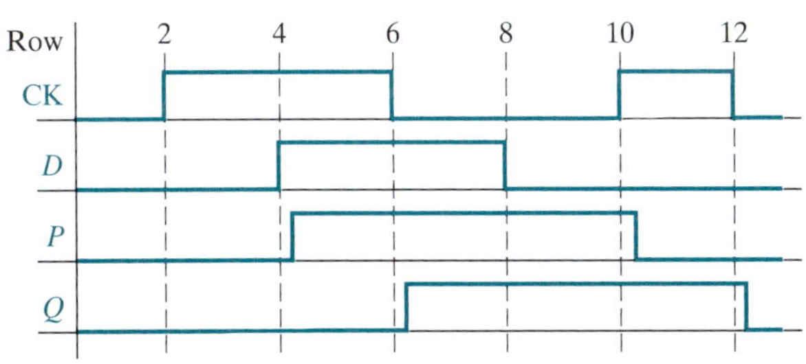

UNIT 12

Registers and Counters

Objectives

1. Explain the operation of registers. Show how to transfer data between registers using a tri-state bus.
2. Explain the operation of shift registers, show how to build them using flip-flops, and analyze their operation. Construct a timing diagram for a shift register.
3. Explain the operation of binary counters, show how to build them using flip-flops and gates, and analyze their operation.
4. Given the present state and desired next state of a flip-flop, determine the required flip-flop inputs.
5. Given the desired counting sequence for a counter, derive the flip-flop input equations.
6. Explain the procedures used for deriving flip-flop input equations.
7. Construct a timing diagram for a counter by tracing signals through the circuit.

Study Guide

1. Study Section 12.1, *Registers and Register Transfers*.
   (a) For the diagram of Figure 12-4, suppose registers A, B, C, and D hold the 8-bit binary numbers representing 91, 70, 249, and 118, respectively. Suppose G and H are both initially 0. What are the contents of G and H (decimal equivalent) after the rising edge of the clock:

   (1) if $EF = 10$, $LdG = 0$, and $LdH = 1$ at the rising edge?
   (2) if $EF = 01$, $LdG = 0$, and $LdH = 1$ at the next rising edge?
   (3) if $EF = 11$, $LdG = 1$, and $LdH = 1$ at the next rising edge?
   (4) if $EF = 00$, $LdG = 1$, and $LdH = 0$ at the next rising edge?
   (5) if $EF = 10$, $LdG = 0$, and $LdH = 0$ at the next rising edge?

   (b) Work Problem 12.1.

2. Study Section 12.2, *Shift Registers*.

   (a) Compare the block diagrams for the shift registers of Figure 12-7 and 12-10. Which one changes state on the rising edge of the clock pulse? The falling edge?

   (b) Complete the following table and timing diagram (see next page) for the shift register of Figure 12-8.

Clock Cycle Number	State of Shift Register When CLK = 1 Q_7	Q_6	Q_5	Q_4	Q_3	Q_2	Q_1	Q_0
1	0	0	0	0	0	0	0	0
2								
3								
4								
5								
6								
7								
8								
9								
10								
11								
12								
13								
14								
15								
16								

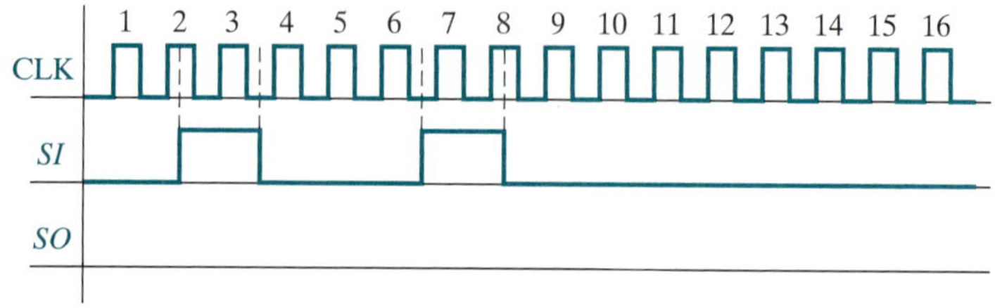

(c) Explain in words the function of the MUX on the D input of flip-flop Q_3 in Figure 12-10(b). Explain in words the meaning of the first of Equation (12-1).

(d) Verify that Equations (12-1) are consistent with Table 12-1.
(e) Work Problem 12.2.

3. Study Section 12.3, *Design of Binary Counters*, and Section 12.4, *Counters for Other Sequences.*

(a) For Figure 12-14, if $CBA = 101$, which of the T inputs is 1?

(b) Complete the following timing diagram for the binary counter of Figure 12-14. The initial value of Clock is 1; this does not count as a rising edge.

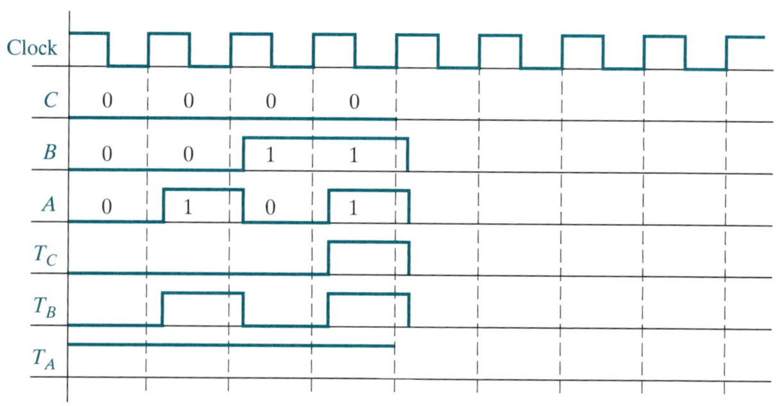

(c) Using the results of (b), draw a transition graph for this binary counter (similar to Figure 12-22).

(d) Complete the following timing diagram for the binary counter of Figure 12-16.

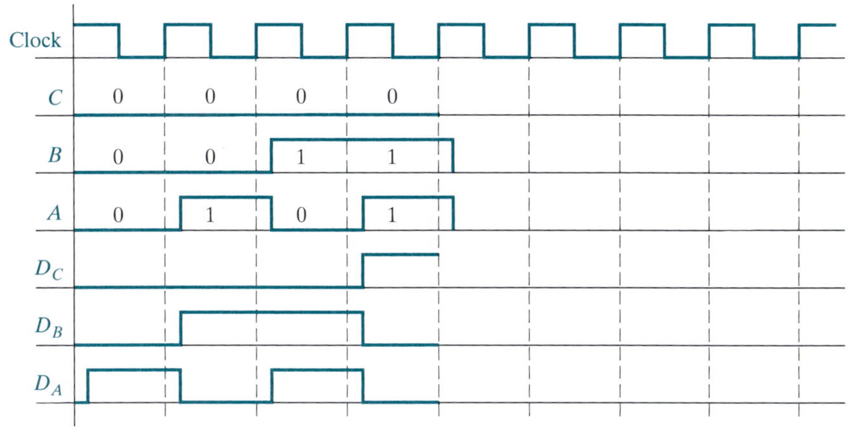

(e) Use Table 12-4 to verify that the values of T_C, T_B, and T_A in Table 12-2 are correct.

(f) What happens if the circuit of Figure 12-24 is started in one of the don't-care states and, then, a clock pulse occurs? In particular, augment the transition graph of Figure 12-26 to indicate the result for starting in states 101 and 110.

(g) What happens if the circuit of Figure 12-27 is started in one of the don't-care states and then a clock pulse occurs? In particular, augment the transition graph of Figure 12-22 to indicate the result for starting in states 001, 101, and 110.

(h) Work Problems 12.3, 12.4, 12.5, 12.6, and 12.7.

4. Study Section 12.5, *Counter Design Using S-R and J-K Flip-Flops.*

(a) Referring to Table 12-5(c):
If $Q = Q^+ = 0$, explain in words why R is a don't-care.

If $Q = Q^+ = 1$, explain in words why S is a don't-care.

If $Q = 0$ and $Q^+ = 1$, what value should S have and why?

If $Q = 1$ and $Q^+ = 0$, what value should R have and why?

(b) For Figure 12-30, verify that the R_B and S_B maps are consistent with the B^+ map, and verify that the R_c and S_c maps are consistent with the C^+ map.

(c) In Figure 12-30, where do the gate inputs (C, B, A, etc.) come from?

(d) For Figure 12-30(c), which flip-flop inputs will be 1 if $CBA = 100$? What will be the state after the rising clock edge?

(e) Complete the following transition graph by tracing signals in Figure 12-30(c). Compare your answer with Figure 12-22. What will happen if the counter is in state 110 and a clock pulse occurs?

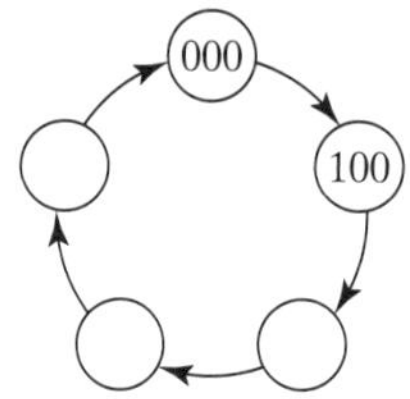

(f) Referring to Table 12-7(c):
If $Q = Q^+ = 0$, explain in words why K is a don't-care.

If $Q = Q^+ = 1$, explain in words why J is a don't-care.

If $Q = 0$ and $Q^+ = 1$, explain why both $JK = 10$ and $JK = 11$ will produce the required state change.

If $Q = 1$ and $Q^+ = 0$, give two sets of values for J and K which will produce the required state change, and explain why your answer is valid.

(g) Verify that the maps of Figure 12-31(b) can be derived from the maps of Figure 12-31(a).

(h) Compare the number of logic gates in Figures 12-30 and 12-31. The J-K realization requires fewer gates than the S-R realization because the J-K maps have more don't-cares than the S-R maps.

(i) Draw in the implied feedback connections on the circuit of Figure 12-31(c).

(j) By tracing signals through the circuit, verify that the state sequence for Figure 12-31(c) is correct.

(k) Find a minimum expression for F_1 and for F_2. (*Hint:* No variables are required.)

F_1:

$BC \backslash A$	0	1
00	X	X
01	1	X
11	X	1
10	X	X

F_2:

$BC \backslash A$	0	1
00	X	X
01	0	X
11	X	X
10	X	X

(l) Work Problems 12.8 and 12.9.

5. Study Section 12.6, *Derivation of Flip-Flop Input Equations—Summary*.

(a) Make sure that you know how to derive input equations for the different types of flip-flops. It is important that you understand the procedures for deriving the equations; merely memorizing the rules is not sufficient.

(b) Table 12-9 is provided mainly for reference. It is not intended that you memorize this table; instead you should understand the reasons for the entries in the table. If you understand the reasons why a given map entry is 0, 1, or X, you should be able to derive the flip-flop input maps without reference to a table.

6. Work the part of Problem 12.10 that you have been assigned. Bring your solution to this problem with you when you come to take the readiness test.

Registers and Counters

A register consists of a group of flip-flops with a common clock input. Registers are commonly used to store and shift binary data. Counters are another simple type of sequential circuit. A counter is usually constructed from two or more flip-flops which change states in a prescribed sequence when input pulses are received. In this unit, you will learn procedures for deriving flip-flop input equations for counters. These procedures will be applied to more general types of sequential circuits in later units.

12.1 Registers and Register Transfers

Several D flip-flops may be grouped together with a common clock to form a register (Figure 12-1(a)). Because each flip-flop can store one bit of information, this register can store four bits of information. This register has a load signal that is ANDed with the clock. When Load = 0, the register is not clocked, and it holds its present value.

FIGURE 12-1
4-Bit D Flip-Flop Registers with Data, Load, Clear, and Clock Inputs

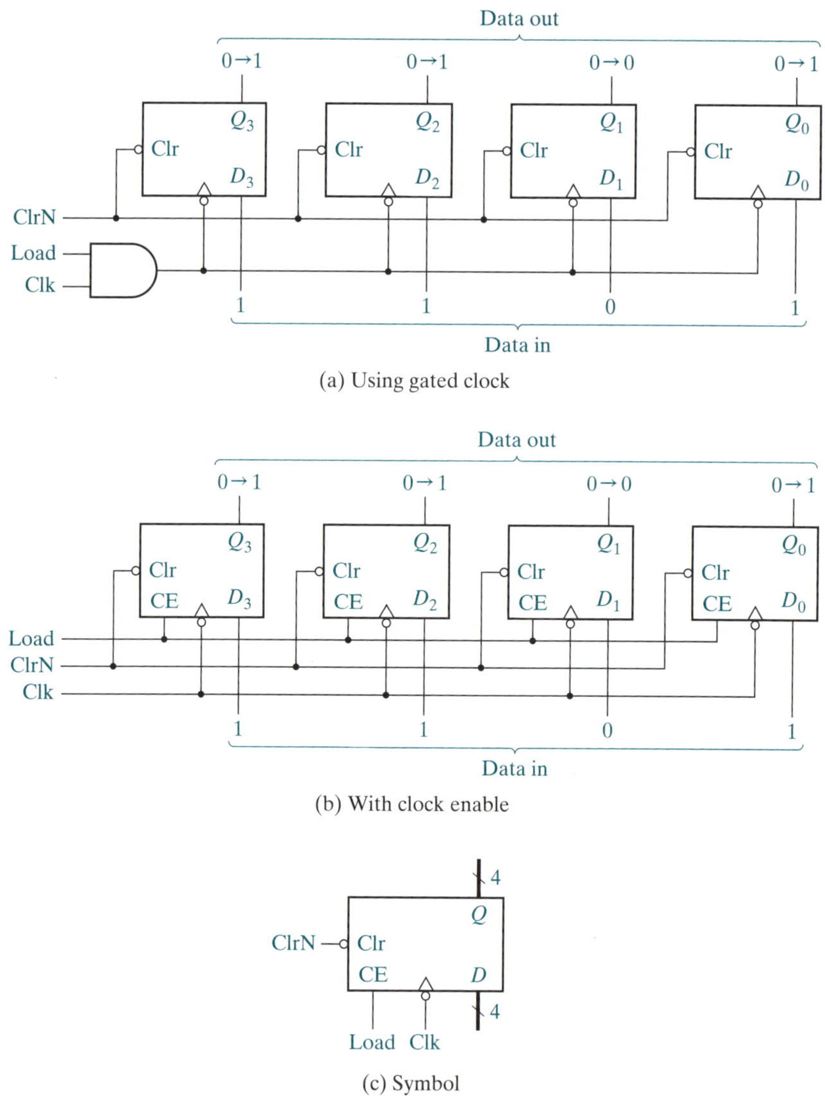

When it is time to load data into the register, Load is set to 1 for one clock period. When Load = 1, the clock signal (Clk) is transmitted to the flip-flop clock inputs and the data applied to the D inputs will be loaded into the flip-flops on the falling edge of the clock. For example, if the Q outputs are 0000 ($Q_3 = Q_2 = Q_1 = Q_0 = 0$) and the data inputs are 1101 ($D_3 = 1$, $D_2 = 1$, $D_1 = 0$ and $D_0 = 1$), after the falling edge of the clock Q will change from 0000 to 1101 as indicated. (The notation $0 \rightarrow 1$ at the flip-flop outputs indicates a change from 0 to 1.)

The flip-flops in the register have asynchronous clear inputs that are connected to a common clear signal, ClrN. The bubble at the clear inputs indicates that a logic 0 is required to clear the flip-flops. ClrN is normally 1, and if it is changed momentarily to 0, the Q outputs of all four flip-flops will become 0.

As discussed in Section 11.8, gating the clock with another signal can cause timing problems. If flip-flops with clock enable are available, the register can be designed as shown in Figure 12-1(b). The load signal is connected to all four CE inputs. When Load = 0, the clock is disabled and the register holds its data. When Load is 1, the clock is enabled, and the data applied to the D inputs will be loaded into the flip-flops, following the falling edge of the clock. Figure 12-1(c) shows a symbol for the 4-bit register using bus notation for the D inputs and Q outputs. A group of wires that perform a common function is often referred to as a bus. A heavy line is used to represent a bus, and a slash with a number beside it indicates the number of bits in the bus.

Transferring data between registers is a common operation in digital systems. Figure 12-2 shows how data can be transferred from the output of one of two registers into a third register using tri-state buffers. If En = 1 and Load = 1, the output of register A is enabled onto the tri-state bus and the data in register A will be stored in Q after the rising edge of the clock. If En = 0 and Load = 1, the output of register B will be enabled onto the tri-state bus and stored in Q after the rising edge of the clock.

FIGURE 12-2
Data Transfer Between Registers

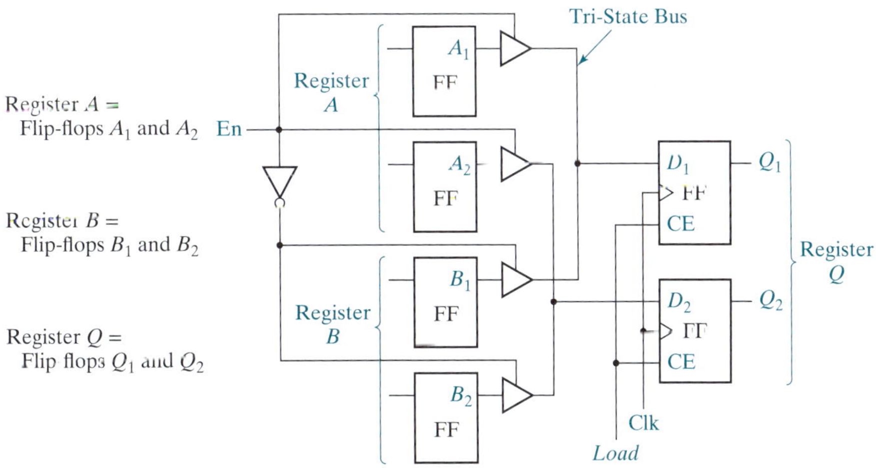

Figure 12-3(a) shows an integrated circuit register that contains eight D flip-flops with tri-state buffers at the flip-flop outputs. These buffers are enabled when En = 0. A symbol for this 8-bit register is shown in Figure 12-3(b).

Figure 12-4 shows how data can be transferred from one of four 8-bit registers into one of two other registers. Registers A, B, C, and D are of the type shown in Figure 12-3.

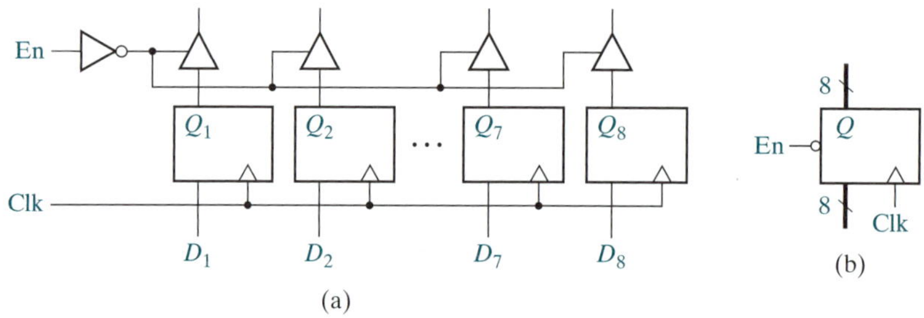

FIGURE 12-3 Logic Diagram for 8-Bit Register with Tri-State Output

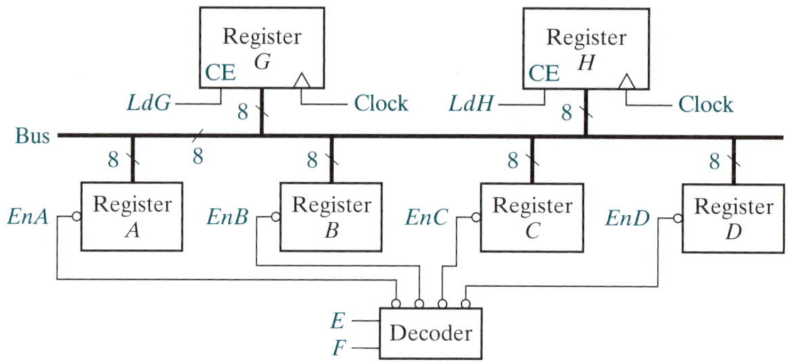

FIGURE 12-4 Data Transfer Using a Tri-State Bus

The outputs from these registers are all connected in parallel to a common tri-state bus. Registers G and H are similar to the register of Figure 12-1 except that they have eight flip-flops instead of four. The flip-flop inputs of registers G and H are also connected to the bus. When $EnA = 0$, the tri-state outputs of register A are enabled onto the bus. If $LdG = 1$, these signals on the bus are loaded into register G after the rising clock edge (or into register H if $LdH = 1$). Similarly, the data in register B, C, or D is transferred to G (or H) when EnB, EnC, or EnD is 0, respectively and $LdG = 1$ (or $LdH = 1$). If $LdG = LdH = 1$, both G and H will be loaded from the bus. The four enable signals may be generated by a decoder. The operation can be summarized as follows:

If $EF = 00$, A is stored in G (or H).
If $EF = 01$, B is stored in G (or H).
If $EF = 10$, C is stored in G (or H).
If $EF = 11$, D is stored in G (or H).

Note that 8 bits of data are transferred in parallel from register A, B, C, or D to register G or H. As an alternative to using a bus with tri-state logic, eight 4-to-1 multiplexers could be used, but this would lead to a more complex circuit.

Parallel Adder with Accumulator

In computer circuits, it is frequently desirable to store one number in a register of flip-flops (called an accumulator) and add a second number to it, leaving the result stored in the accumulator. One way to build a parallel adder with an accumulator is to add a register to the adder of Figure 4-2, resulting in the circuit of Figure 12-5. Suppose that the number $X = x_n \ldots x_2x_1$ is stored in the accumulator. Then, the number $Y =$

FIGURE 12-5 *n*-Bit Parallel Adder with Accumulator

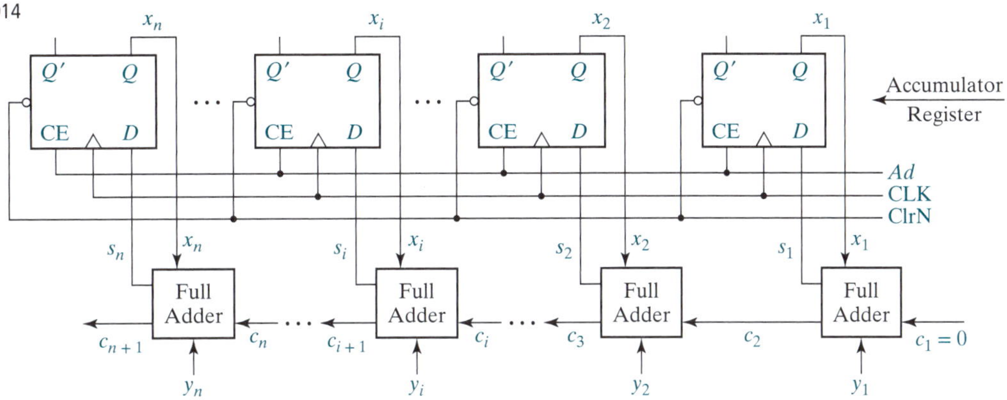

$y_n \cdots y_2y_1$ is applied to the full adder inputs, and after the carry has propagated through the adders, the sum of X and Y appears at the adder outputs. An add signal (Ad) is used to load the adder outputs into the accumulator flip-flops on the rising clock edge. If $s_i = 1$, the next state of flip-flop x_i will be 1. If $s_i = 0$, the next state of flip-flop x_i will be 0. Thus, $x_i^+ = s_i$, and if $Ad = 1$, the number X in the accumulator is replaced with the sum of X and Y, following the rising edge of the clock.

Observe that the adder with accumulator is an iterative structure that consists of a number of identical cells. Each cell contains a full adder and an associated accumulator flip-flop. Cell i, which has inputs c_i and y_i and outputs c_{i+1} and x_i, is referred to as a typical cell.

Before addition can take place, the accumulator must be loaded with X. This can be accomplished in several ways. The easiest way is to first clear the accumulator using the asynchronous clear inputs on the flip-flops, and then put the X data on the Y inputs to the adder and add to the accumulator in the normal way. Alternatively, we could add multiplexers at the accumulator inputs so that we could select either the Y input data or the adder output to load into the accumulator. This would eliminate the extra step of clearing the accumulator but would add to the hardware complexity. Figure 12-6 shows a typical cell of the adder where the accumulator flip-flop

FIGURE 12-6 Adder Cell with Multiplexer

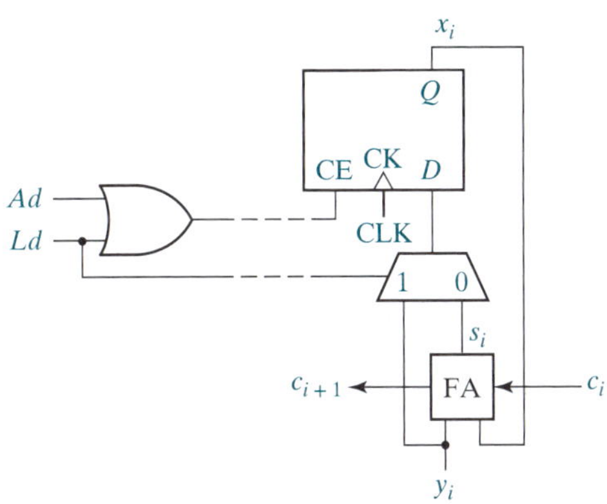

can either be loaded directly from y_i or from the sum output (s_i). When $Ld = 1$ the multiplexer selects y_i, and y_i is loaded into the accumulator flip-flop (x_i) on the rising clock edge. When $Ad = 1$ and $Ld = 0$, the adder output (s_i) is loaded into x_i. The Ad and Ld signals are ORed together to enable the clock when either addition or loading occurs. When $Ad = Ld = 0$, the clock is disabled and the accumulator outputs do not change.

12.2 Shift Registers

A shift register is a register in which binary data can be stored, and this data can be shifted to the left or right when a shift signal is applied. Bits shifted out one end of the register may be lost, or if the shift register is of cyclic type, bits shifted out one end are shifted back in the other end. Figure 12-7(a) illustrates a 4-bit right-shift register with serial input and output constructed from D flip-flops. When Shift = 1, the clock is enabled and shifting occurs on the rising clock edge. When Shift = 0, no shifting occurs and the data in the register is unchanged. The serial input (SI) is loaded into the first flip-flop (Q_3) by the rising edge of the clock. At the same time,

FIGURE 12-7 Right-Shift Register

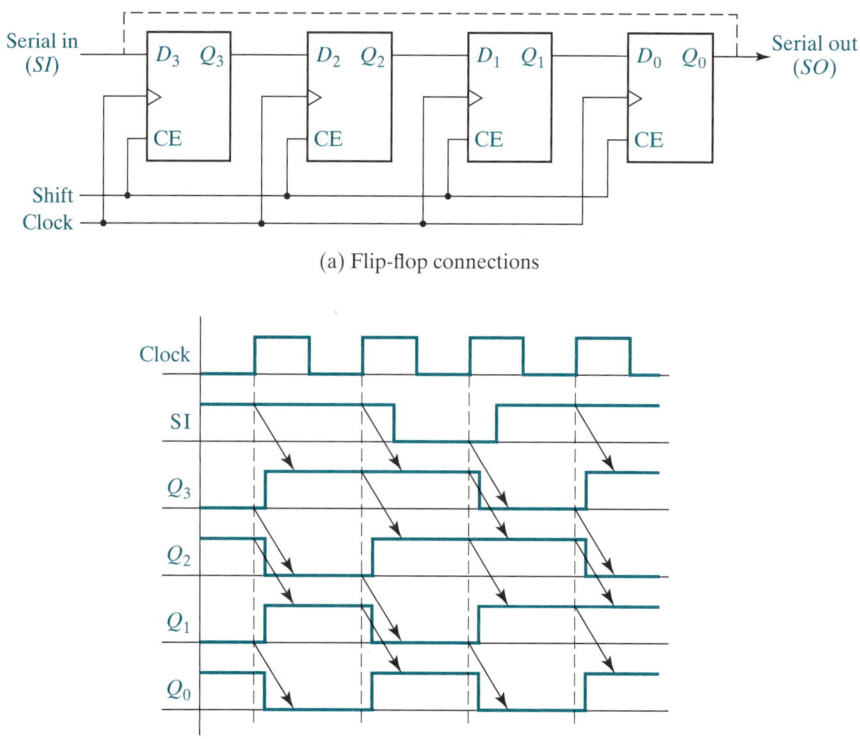

(a) Flip-flop connections

(b) Timing diagram

the output of the first flip-flop is loaded into the second flip-flop, the output of the second flip-flop is loaded into the third flip-flop, and the output of the third flip-flop is loaded into the last flip-flop. Because of the propagation delay of the flip-flops, the output value loaded into each flip-flop is the value before the rising clock edge. Figure 12-7(b) illustrates the timing when the shift register initially contains 0101 and the serial input sequence is 1, 1, 0, 1. The sequence of shift register states is 0101, 1010, 1101, 0110, 1011.

If we connect the serial output to the serial input, as shown by the dashed line, the resulting cyclic shift register performs an end-around shift. If the initial contents of the register is 0111, after one clock cycle the contents is 1011. After a second pulse, the state is 1101, then 1110, and the fourth pulse returns the register to the initial 0111 state.

Shift registers with 4, 8, or more flip-flops are available in integrated circuit form. Figure 12-8 illustrates an 8-bit serial-in, serial-out shift register. *Serial in* means that data is shifted into the first flip-flop one bit at a time, and the flip-flops cannot be loaded in parallel. *Serial out* means that data can only be read out of the last flip-flop and the outputs from the other flip-flops are not connected to terminals of the integrated circuit. The inputs to the first flip-flop are $S = SI$ and $R = SI'$. Thus, if $SI = 1$, a 1 is shifted into the register when it is clocked, and if $SI = 0$, a 0 is shifted in. Figure 12-9 shows a typical timing diagram.

Figure 12-10(a) shows a 4-bit parallel-in, parallel-out shift register. Parallel-in implies that all four bits can be loaded at the same time, and parallel-out implies that all bits can be read out at the same time. The shift register has two control inputs,

FIGURE 12-8 8-Bit Serial-In, Serial-Out Shift Register

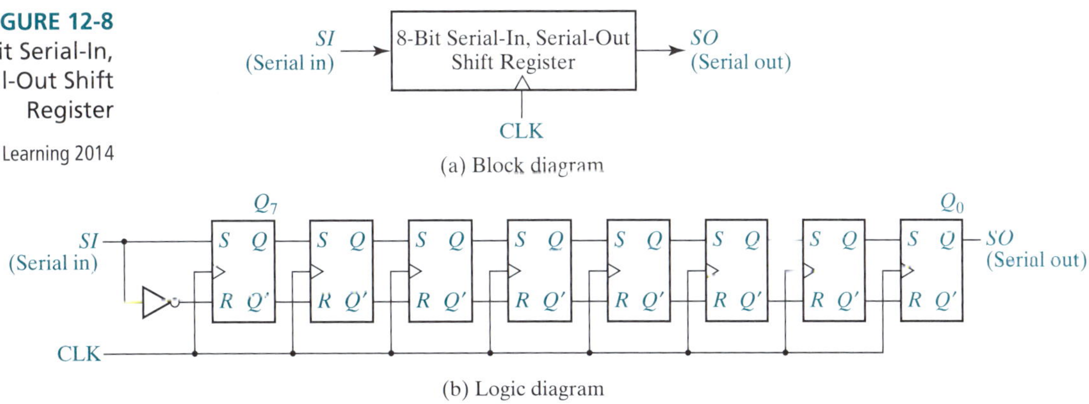

FIGURE 12-9 Typical Timing Diagram for Shift Register of Figure 12-8

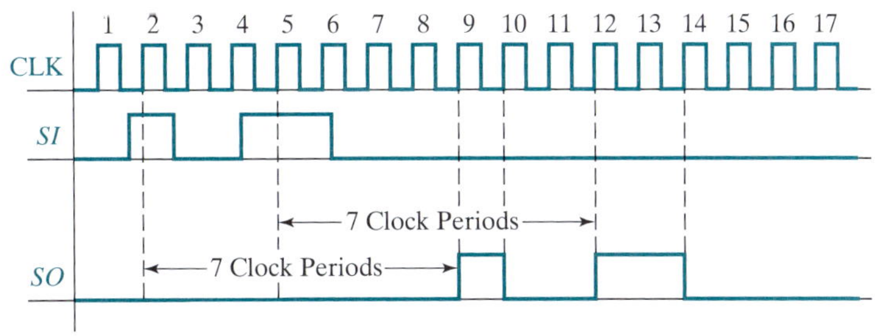

FIGURE 12-10 Parallel-In, Parallel-Out Right-Shift Register

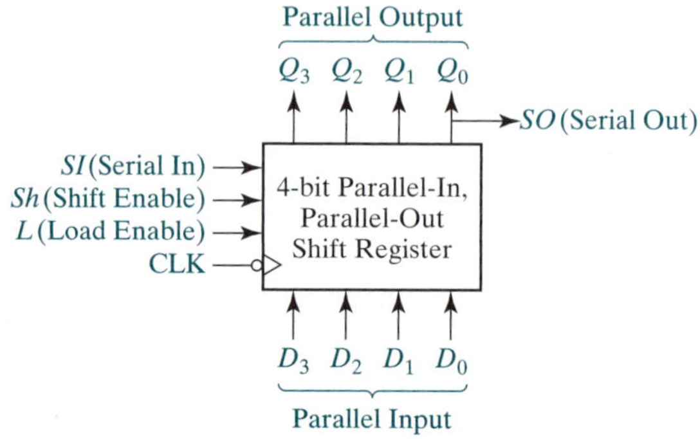

(a) Block diagram

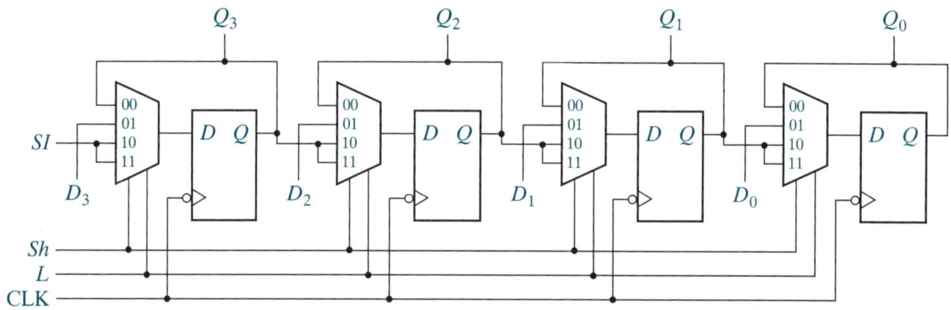

(b) Implementation using flip-flops and MUXes

shift enable (Sh) and load enable (L). If $Sh = 1$ (and $L = 1$ or $L = 0$), clocking the register causes the serial input (SI) to be shifted into the first flip-flop, while the data in flip-flops Q_3, Q_2, and Q_1 are shifted right. If $Sh = 0$ and $L = 1$, clocking the shift register will cause the four data inputs (D_3, D_2, D_1, D_0) to be loaded in parallel into the flip-flops. If $Sh = L = 0$, clocking the register causes no change of state. Table 12-1 summarizes the operation of this shift register. All state changes occur immediately following the falling edge of the clock.

The shift register can be implemented using MUXes and D flip-flops, as shown in Figure 12-10(b). For the first flip-flop, when $Sh = L = 0$, the flip-flop Q_3 output is selected by the MUX, so $Q_3^+ = Q_3$ and no state change occurs. When $Sh = 0$ and $L = 1$, the data input D_3 is selected and loaded into the flip-flop. When $Sh = 1$ and

TABLE 12-1 Shift Register Operation

Inputs		Next State				
Sh (Shift)	L (Load)	Q_3^+	Q_2^+	Q_1^+	Q_0^+	Action
0	0	Q_3	Q_2	Q_1	Q_0	No change
0	1	D_3	D_2	D_1	D_0	Load
1	X	SI	Q_3	Q_2	Q_1	Right shift

$L = 0$ or 1, SI is selected and loaded into the flip-flop. The second MUX selects Q_2, D_2, or Q_3, etc. The next-state equations for the flip-flops are

$$\begin{aligned} Q_3^+ &= Sh' \cdot L' \cdot Q_3 + Sh' \cdot L \cdot D_3 + Sh \cdot SI \\ Q_2^+ &= Sh' \cdot L' \cdot Q_2 + Sh' \cdot L \cdot D_2 + Sh \cdot Q_3 \\ Q_1^+ &= Sh' \cdot L' \cdot Q_1 + Sh' \cdot L \cdot D_1 + Sh \cdot Q_2 \\ Q_0^+ &= Sh' \cdot L' \cdot Q_0 + Sh' \cdot L \cdot D_0 + Sh \cdot Q_1 \end{aligned} \tag{12-1}$$

A typical application of this register is the conversion of parallel data to serial data. The output from the last flip-flop (Q_0) serves as a serial output as well as one of the parallel outputs. Figure 12-11 shows a typical timing diagram. The first clock pulse loads data into the shift register in parallel. During the next four clock pulses, this data is available at the serial output. Assuming that the register is initially clear ($Q_3Q_2Q_1Q_0 = 0000$), that the serial input is $SI = 0$ throughout, and that the data inputs $D_3D_2D_1D_0$ are 1011 during the load time (t_0), the resulting waveforms are as shown. Shifting occurs at the end of t_1, t_2, and t_3, and the serial output can be read during these clock times. During t_4, $Sh = L = 0$, so no state change occurs.

Figure 12-12(a) shows a 3-bit shift register with the Q_1' output from the last flip-flop fed back into the D input of the first flip-flop. If the initial state of the register is 000, the initial value of D_3 is 1, so after the first clock pulse, the register state is 100. Successive states are shown on the transition graph of Figure 12-12(b). When

FIGURE 12-11 Timing Diagram for Shift Register

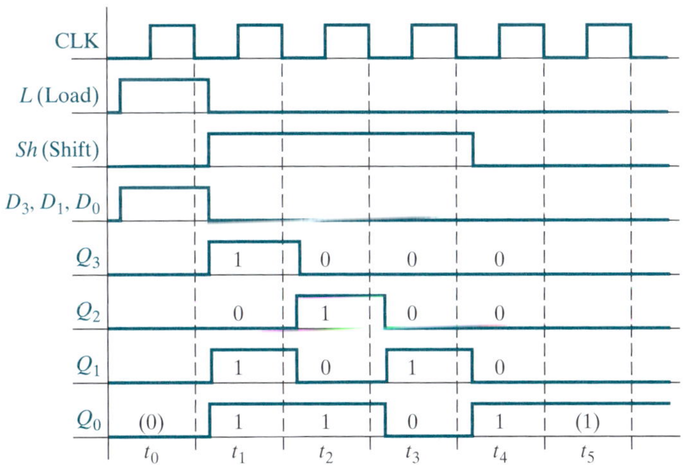

FIGURE 12-12 Shift Register with Inverted Feedback

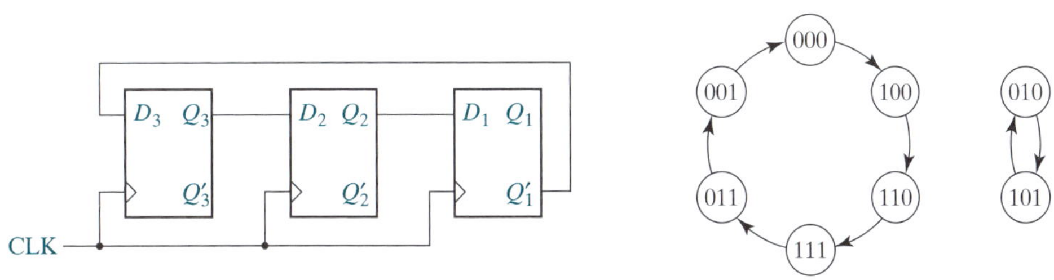

(a) Flip-flop connections (b) Transition graph

the register is in state 001, D_3 is 0, and the next register state is 000. Then, successive clock pulses take the register around the loop again. Note that states 010 and 101 are not in the main loop. If the register is in state 010, then a shift pulse takes it to 101 and vice versa; therefore, we have a secondary loop on the transition graph.

A circuit that cycles through a fixed sequence of states is called a *counter*. A shift register with inverted feedback (Figure 12-12) is called a *Johnson counter* or a *twisted ring counter*. If the feedback is not inverted (e.g., Q_1 connected to D_3 in Figure 12-12), the shift register counter is called a *ring counter*.

Figure 12-13 shows the general form of a shift register counter where the bit being shifted into the leftmost stage can be a general function of the shift register contents. If the gate logic only contains exclusive-OR gates, the counter is called a *linear* (feedback) shift register counter. It can be shown that, for each integer n, there exists a linear n-bit shift register counter that generates a count cycle of length $2^n - 1$; all states are included except for the all 0's state. For many values of n only a small number of exclusive-OR gates is required. Linear shift register counters have many applications, including as random number generators and as encoders and decoders for linear error-correcting codes.

FIGURE 12-13 General Shift Register Counter

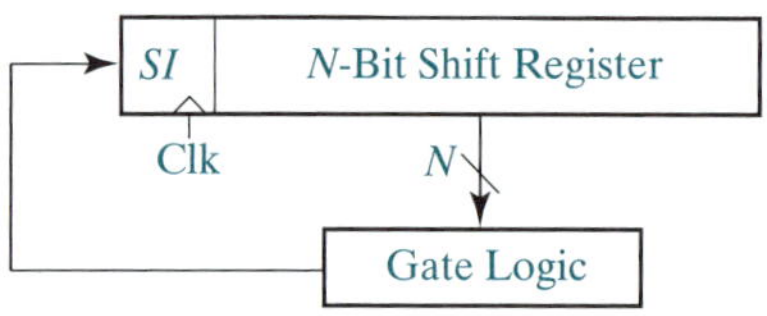

12.3 Design of Binary Counters

The counters discussed in this chapter are all synchronous counters. This means the operation of the flip-flops is synchronized by a common clock pulse so that when several flip-flops must change state, the state changes occur simultaneously. Ripple counters, in which the state change of one flip-flop triggers another flip-flop, are not discussed in detail. See Problems 11.24 and 12.19 for examples.

We will first construct a binary counter using three T flip-flops to count clock pulses (Figure 12-14). We will assume that all the flip-flops change state a short time following the rising edge of the input pulse. The state of the counter is determined by the states of the individual flip-flops; for example, if flip-flop C is in state 0, B in state 1, and A in state 1, the state of the counter is 011. Initially, assume that all flip-flops are set to the 0 state. When a clock pulse is received, the counter will change to state 001; when a second pulse is received, the state will change to 010, etc. The sequence of flip-flop states is CBA = 000, 001, 010, 011, 100, 101, 110, 111, 000, . . . Note that when the counter reaches state 111, the next pulse resets it to the 000 state, and then the sequence repeats.

First, we will design the counter by inspection of the counting sequence; then, we will use a systematic procedure which can be generalized to other types of counters. The problem is to determine the flip-flop inputs—T_C, T_B, and T_A. From the preceding counting sequence, observe that A changes state every time a clock pulse is received.

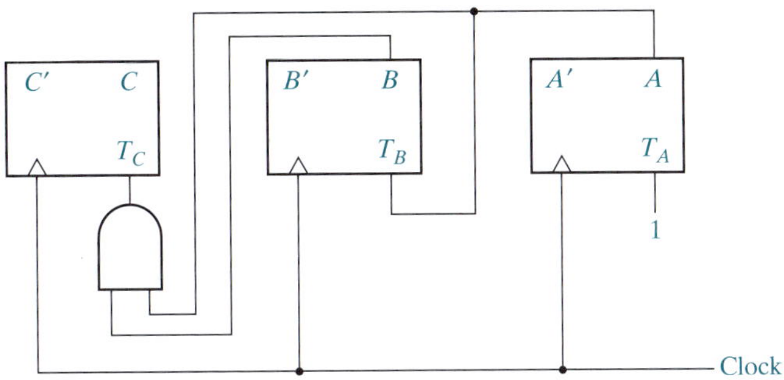

FIGURE 12-14 Synchronous Binary Counter

Because A changes state on every rising clock edge, T_A must equal 1. Next, observe that B changes state only if $A = 1$. Therefore, A is connected to T_B as shown, so that if $A = 1$, B will change state when a rising clock edge occurs. Similarly, C changes state when a rising clock edge occurs only if B and A are both 1. Therefore, an AND gate is connected to T_C so that C will change state if $B = 1$ and $A = 1$ when a rising clock edge occurs.

Now, we will verify that the circuit of Figure 12-14 counts properly by tracing signals through the circuit. Initially, $CBA = 000$, so only T_A is 1 and the state will change to 001 when the first active clock edge arrives. Then, $T_B = T_A = 1$, and the state will change to 010 when the second active clock arrives. This process continues until finally when state 111 is reached, $T_C = T_B = T_A = 1$, and all flip-flops return to the 0 state.

Next, we will redesign the binary counter by using a transition table (Table 12-2). This table shows the present state of flip-flops C, B, and A (before a clock pulse is received) and the corresponding next state (after the clock pulse is received). For example, if the flip-flops are in state $CBA = 011$ and a clock pulse is received, the next state will be $C^+B^+A^+ = 100$. Although the clock is not explicit in the table, it is understood to be the input that causes the counter to go to the next state in sequence. A third column in the table is used to derive the inputs for T_C, T_B, and T_A. Whenever the entries in the A and A^+ columns differ, flip-flop A must change state and T_A must be 1. Similarly, if B and B^+ differ, B must change state so T_B must be 1. For example, if $CBA = 011$, $C^+B^+A^+ = 100$, all three flip-flops must change state, so $T_CT_BT_A = 111$.

TABLE 12-2 Transition Table for Binary Counter

Present State			Next State			Flip-Flop Inputs		
C	B	A	C^+	B^+	A^+	T_C	T_B	T_A
0	0	0	0	0	1	0	0	1
0	0	1	0	1	0	0	1	1
0	1	0	0	1	1	0	0	1
0	1	1	1	0	0	1	1	1
1	0	0	1	0	1	0	0	1
1	0	1	1	1	0	0	1	1
1	1	0	1	1	1	0	0	1
1	1	1	0	0	0	1	1	1

T_C, T_B, and T_A are now derived from the table as functions of C, B, and A. By inspection, $T_A = 1$. Figure 12-15 shows the Karnaugh maps for T_C and T_B, from which $T_C = BA$ and $T_B = A$. These equations yield the same circuit derived previously for Figure 12-14.

Next, we will redesign the binary counter to use D flip-flops instead of T flip-flops. The easiest way to do this is to convert each D flip-flop to a T flip-flop by adding an XOR (exclusive-OR) gate, as shown in Figure 11-28(b). Figure 12-16 shows the resulting counter circuit. The rightmost XOR gate can be replaced with an inverter because $A \oplus 1 = A'$.

FIGURE 12-15
Karnaugh Maps for Binary Counter

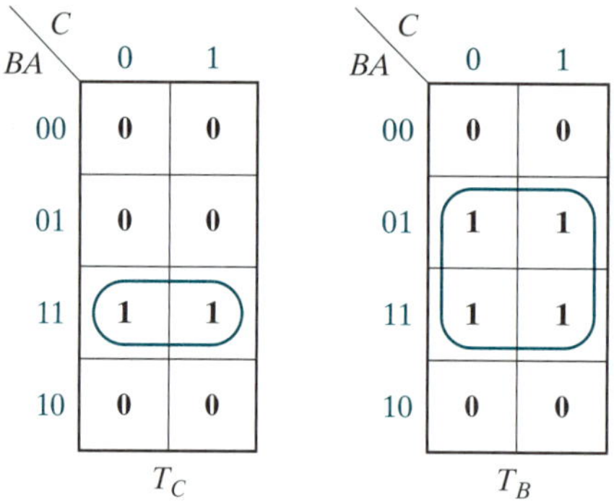

FIGURE 12-16
Binary Counter with D Flip-Flops

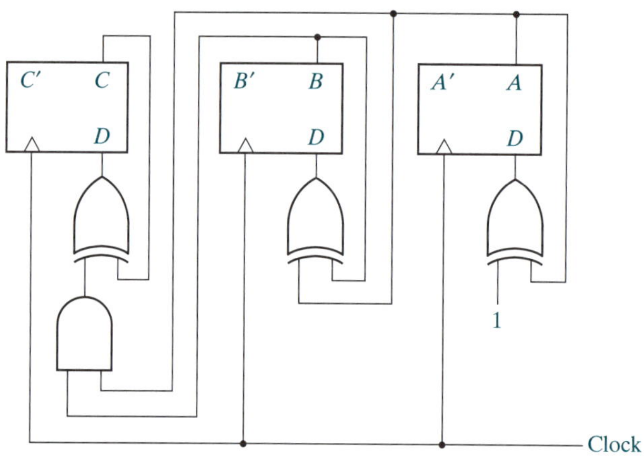

We can also derive the D flip-flop inputs for the binary counter starting with its transition table (Table 12-2). For a D flip-flop, $Q^+ = D$. By inspection of the table, $Q_A^+ = A'$, so $D_A = A'$. The maps for Q_B^+ and Q_C^+ are plotted in Figure 12-17. The D input equations derived from the maps are

$$\begin{aligned} D_A &= A^+ = A' \\ D_B &= B^+ = BA' + B'A = B \oplus A \\ D_C &= C^+ = C'BA + CB' + CA' = C'BA + C(BA)' = C \oplus BA \end{aligned} \tag{12-2}$$

which give the same logic circuit as was obtained by inspection.

FIGURE 12-17 Karnaugh Maps for D Flip-Flops

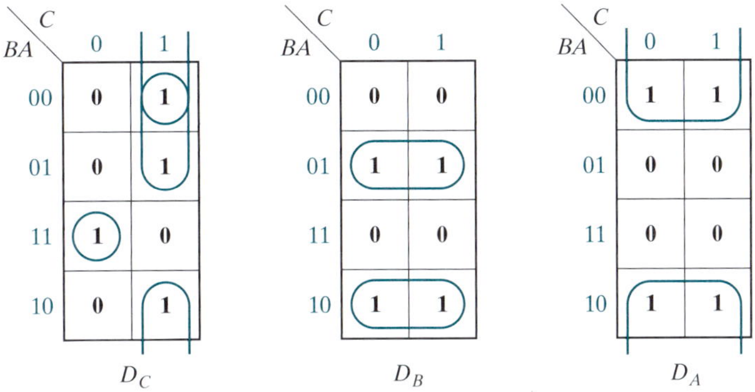

Next, we will analyze an up-down binary counter. The transition graph and table for an up-down counter are shown in Figure 12-18. When $U = 1$, the counter counts up in the sequence 000, 001, 010, 011, 100, 101, 110, 111, 000 . . . When $D = 1$, the counter counts down in the sequence 000, 111, 110, 101, 100, 011, 010, 001, 000 . . . When $U = D = 0$, the counter state does not change, and $U = D = 1$ is not allowed.

FIGURE 12-18 Transition Graph and Table for Up-Down Counter

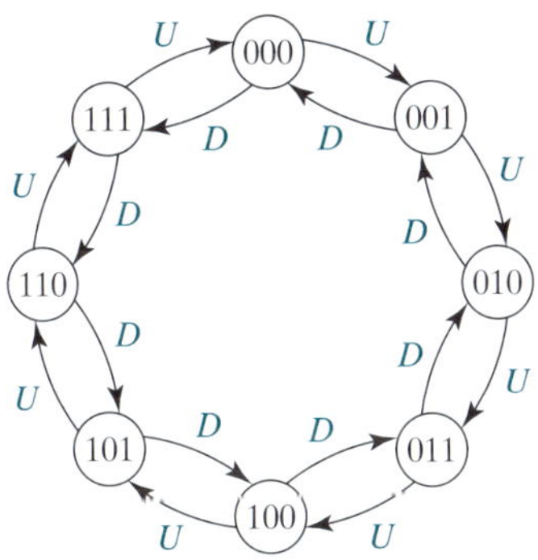

	$C^+B^+A^+$	
CBA	U	D
000	001	111
001	010	000
010	011	001
011	100	010
100	101	011
101	110	100
110	111	101
111	000	110

The up-down counter can be implemented using D flip-flops and gates, as shown in Figure 12-19. The corresponding logic equations are

$$D_A = A^+ = A \oplus (U + D)$$
$$D_B = B^+ = B \oplus (UA + DA')$$
$$D_C = C^+ = C \oplus (UBA + DB'A')$$

When $U = 1$ and $D = 0$, these equations reduce to equations for a binary up counter (Equation (12-2)).

When $U = 0$ and $D = 1$, these equations reduce to

$$D_A = A^+ = A \oplus 1 = A' \quad (A \text{ changes state every clock cycle})$$
$$D_B = B^+ = B \oplus A' \quad (B \text{ changes state when } A = 0)$$
$$D_C = C^+ = C \oplus B'A' \quad (C \text{ changes state when } B = A = 0)$$

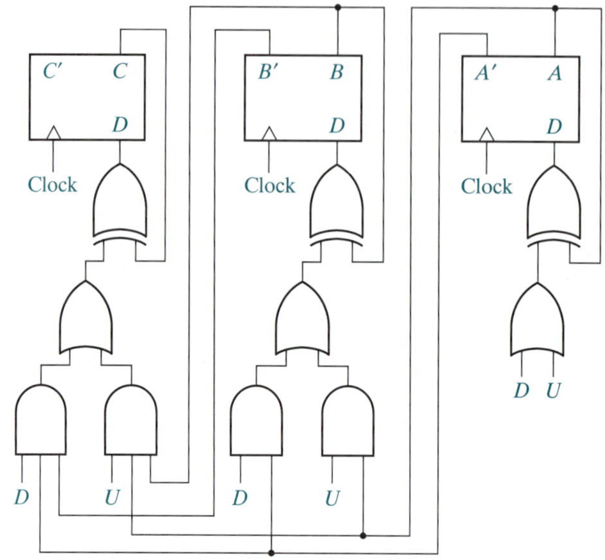

FIGURE 12-19 Binary Up-Down Counter

By inspection of the table in Figure 12-18, we can verify that these are the correct equations for a down counter. For every row of the table, $A^+ = A'$, so A changes state every clock cycle. For those rows where $A = 0$, $B^+ = B'$. For those rows where $B = 0$ and $A = 0$, $C^+ = C'$.

Next, we will design a loadable counter (Figure 12-20(a)). This counter has two control signals Ld (load) and Ct (count). When $Ld = 1$ binary data is loaded into the counter on the rising clock edge, and when $Ct = 1$, the counter is incremented on the rising clock edge. When $Ld = Ct = 0$, the counter holds its present state. When $Ld = Ct = 1$, load overrides count, and data is loaded into the counter. The counter also has an asynchronous clear input that clears the counter when ClrN is 0. Figure 12-20(b) summarizes the counter operation. All state changes occur on the rising edge of the clock (except for the asynchronous clear).

Figure 12-21 shows how the loadable counter can be implemented using flip-flops, MUXes, and gates. When $Ld = 1$, each MUX selects a D_i input, and because the output of each AND gate is 0, the output of each XOR gate is D_i, which gets stored in a flip-flop. When $Ld = 0$ and $Ct = 1$, each MUX selects one of the flip-flop outputs (C, B, or A). The circuit then becomes equivalent to Figure 12-16, and the counter is incremented on the rising clock edge.

FIGURE 12-20 Loadable Counter with Count Enable

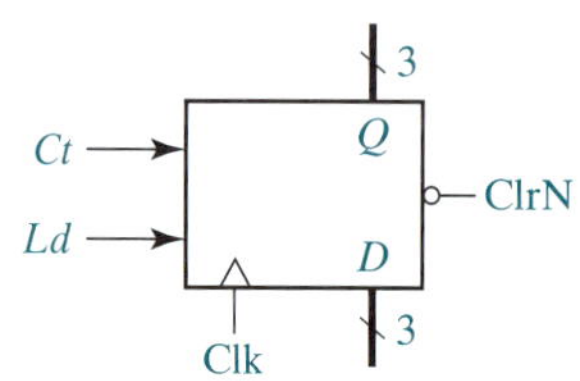

(a)

ClrN	Ld	Ct	C^+	B^+	A^+	
0	X	X	0	0	0	
1	1	X	D_C	D_B	D_A	(load)
1	0	0	C	B	A	(no change)
1	0	1	Present state + 1			

(b)

FIGURE 12-21 Circuit for Figure 12-20

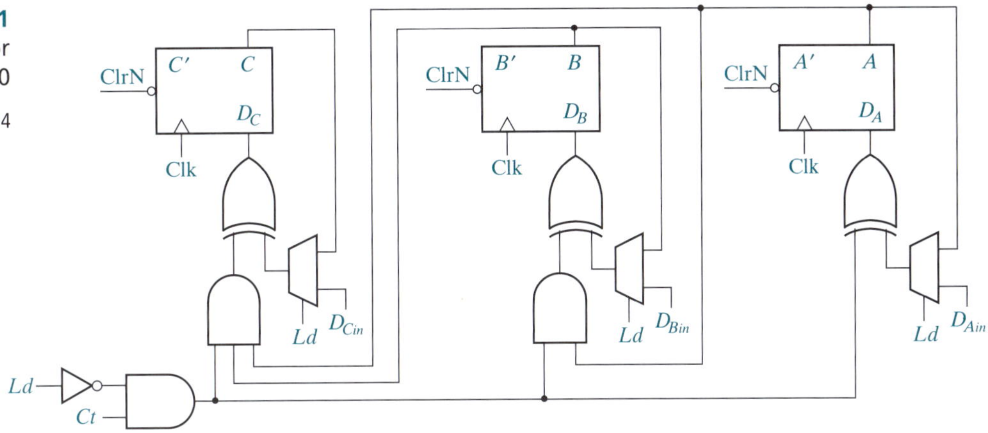

The next-state equations for the counter of Figure 12-21 are

$$A^+ = D_A = (Ld'{\cdot}A + Ld{\cdot}D_{Ain}) \oplus Ld'{\cdot}Ct$$
$$B^+ = D_B = (Ld'{\cdot}B + Ld{\cdot}D_{Bin}) \oplus Ld'{\cdot}Ct{\cdot}A$$
$$C^+ = D_C = (Ld'{\cdot}C + Ld{\cdot}D_{Cin}) \oplus Ld'{\cdot}Ct{\cdot}B{\cdot}A$$

When $Ld = 0$ and $Ct = 1$, these equations reduce to $A^+ = A'$, $B^+ = B \oplus A$, and $C^+ = C \oplus BA$, which are the equations previously derived for a 3-bit counter.

12.4 Counters for Other Sequences

In some applications, the sequence of states of a counter is not in straight binary order. Figure 12-22 shows the transition graph for such a counter. The arrows indicate the state sequence. If this counter is started in state 000, the first clock pulse will take it to state 100, the next pulse to 111, etc. The clock pulse is implicitly understood to be the input to the circuit and not shown on the graph. The corresponding transition table for the counter is Table 12-3. Note that the next state is unspecified for the present states 001, 101, and 110.

We will design the counter specified by Table 12-3 using T flip-flops. We could derive T_C, T_B, and T_A directly from this table, as in the preceding example. However, it is often more convenient to plot next-state maps showing C^+, B^+, and A^+ as functions of C, B, and A, and then derive T_C, T_B, and T_A from these maps. The next-state maps in Figure 12-23(a) are easily plotted from inspection of Table 12-3. From the first row of the table, the $CBA = 000$ squares on the C^+, B^+, and A^+ maps are filled in with 1, 0, and 0, respectively. From the second row, the $CBA = 001$ squares on all three maps are filled in with don't-cares. From the third row, the $CBA = 010$ squares on the C^+, B^+, and A^+ maps are filled in with 0, 1, and 1, respectively. The next-state maps can be quickly completed by continuing in this manner.

FIGURE 12-22 Transition Graph for Counter

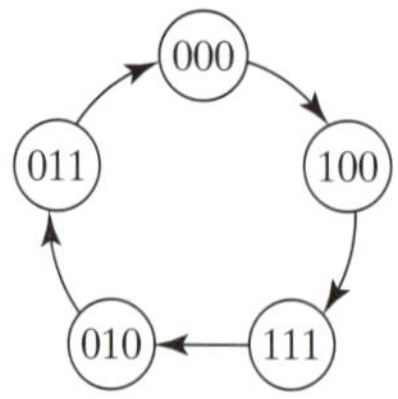

TABLE 12-3 Transition Table for Figure 12-22

C	B	A	C^+	B^+	A^+
0	0	0	1	0	0
0	0	1	–	–	–
0	1	0	0	1	1
0	1	1	0	0	0
1	0	0	1	1	1
1	0	1	–	–	–
1	1	0	–	–	–
1	1	1	0	1	0

FIGURE 12-23

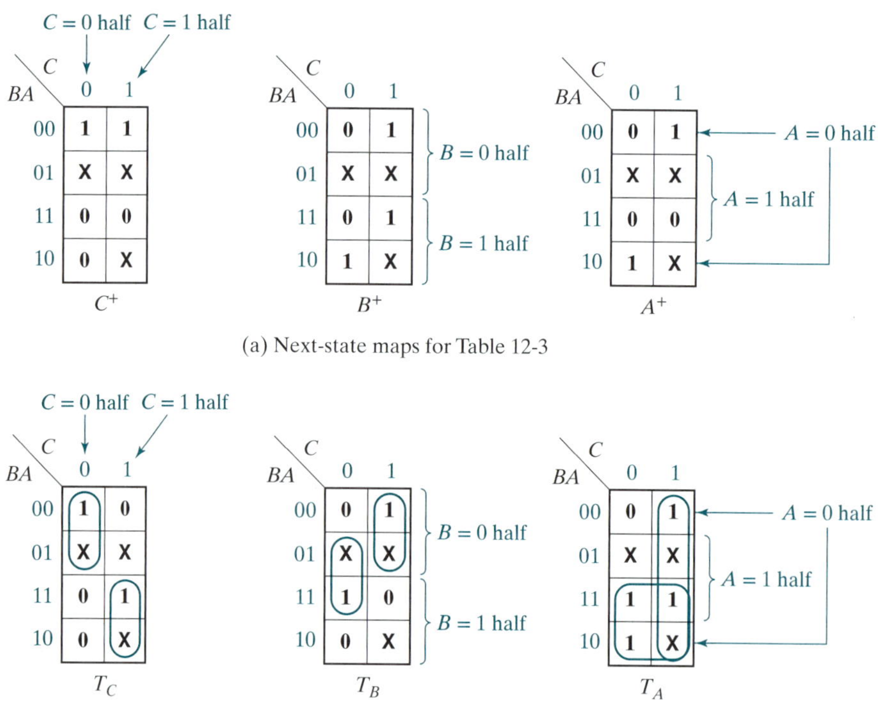

(a) Next-state maps for Table 12-3

$T_C = C'B' + CB$ $T_B = C'A + CB'$ $T_A = C + B$

(b) Derivation of T inputs

Next, we will derive the maps for the T inputs from the next-state maps. In the following discussion, the general symbol Q represents the present state of the flip-flop (C, B, or A) under consideration, and Q^+ represents the next state (C^+, B^+, or A^+) of the same flip-flop. Given the present state of a T flip-flop (Q) and the desired next state (Q^+), the T input must be 1 whenever a change of state is required. Thus, $T = 1$ whenever $Q^+ \neq Q$, as shown in Table 12-4.

TABLE 12-4 Input for T Flip-Flop

Q	Q^+	T
0	0	0
0	1	1
1	0	1
1	1	0

$T = Q^+ \oplus Q$

In general, the next-state map for flip-flop Q gives Q^+ as a function of Q and several other variables. The value written in each square of the map gives the value of Q^+, while the value of Q is determined from the row or column headings. Given the map for Q^+, we can then form the map for T_Q by simply putting a 1 in each square of the T_Q map for which Q^+ is different from Q. Thus, to form the T_C map in Figure 12-23(b) from the C^+ map in Figure 12-23(a), we place a 1 in the $CBA = 000$ square of T_C because $C = 0$ and $C^+ = 1$ for this square. We also place a 1 in the 111 square of T_C because $C = 1$ and $C^+ = 0$ for this square.

If we don't care what the next state of a flip-flop is for some combination of variables, we don't care what the flip-flop input is for that combination of variables. Therefore, if the Q^+ map has a don't-care in some square, the T_Q map will have a don't-care in the corresponding square. Thus, the T_C map has don't-cares for $CBA = 001$, 101, and 110 because C^+ has don't-cares in the corresponding squares.

Instead of transforming the Q^+ map into the T_Q map one square at a time, we can divide the Q^+ map into two halves corresponding to $Q = 0$ and $Q = 1$, and transform each half of the map. From Table 12-4, whenever $Q = 0$, $T = Q^+$, and whenever $Q = 1$, $T = (Q^+)'$. Therefore, to transform the Q^+ map into a T map, we copy the half for which $Q = 0$ and complement the half for which $Q = 1$, leaving the don't-cares unchanged.

We will apply this method to transform the C^+, B^+, and A^+ maps for our counter shown in Figure 12-23(a) into T maps. For the first map, C corresponds to Q (and C^+ to Q^+), so to get the T_C map from the C^+ map, we complement the second column (where $C = 1$) and leave the rest of the map unchanged. Similarly, to get T_B from B^+, we complement the bottom half of the B map, and to get T_A from A^+, we complement the middle two rows. This yields the maps and equations of Figure 12-23(b) and the circuit shown in Figure 12-24. The clock input is connected to the clock (CK) input of each flip-flop so that the flip-flops can change state only in response to a clock pulse. The gate inputs connect directly to the corresponding flip-flop outputs as indicated by the dashed lines. To facilitate reading similar circuit diagrams, such connecting wires will be omitted in the remainder of the book.

The timing diagram of Figure 12-25, derived by tracing signals through the circuit, verifies that the counter functions according to the state diagram of Figure 12-22; for example, starting with $CBA = 000$, $T_C = 1$ and $T_B = T_A = 0$. Therefore, when the clock pulse comes along, only flip-flop C changes state, and the new state is 100.

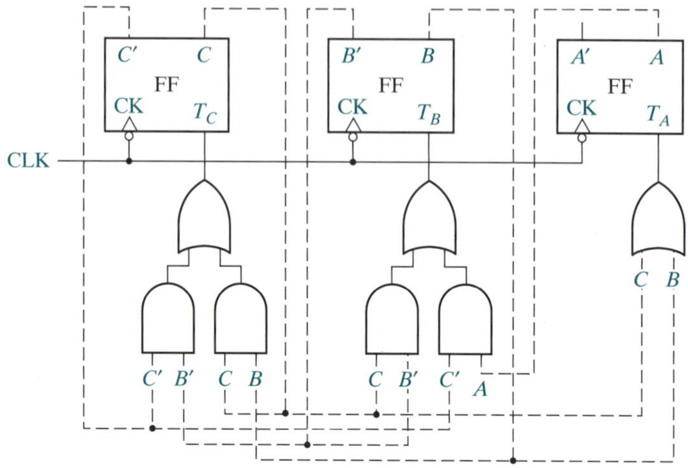

FIGURE 12-24 Counter Using T Flip-Flops

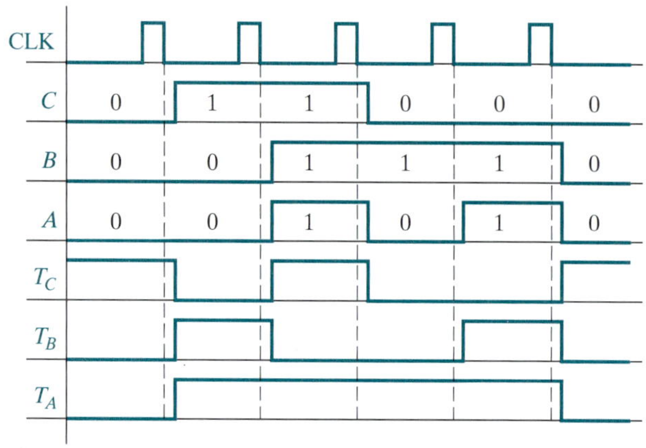

FIGURE 12-25 Timing Diagram for Figure 12-24

Then, $T_C = 0$ and $T_B = T_A = 1$, so flip-flops B and A change state when the next clock pulse occurs, etc. Note that the flip-flops change state following the falling clock edge.

Although the original transition table for the counter (Table 12-3) is not completely specified, the next states of states 001, 101, and 110 have been specified in the process of completing the circuit design. For example, if the flip-flops are initially set to $C = 0$, $B = 0$, and $A = 1$, tracing signals through the circuit shows that $T_C = T_B = 1$ and $T_A = 0$, so that the state will change to 111 when a clock pulse is applied. This behavior is indicated by the dashed line in Figure 12-26. Once state 111 is reached,

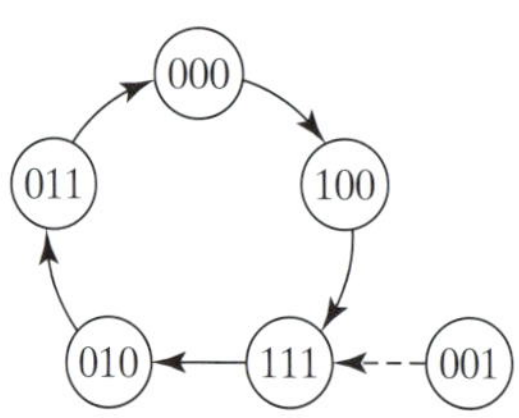

FIGURE 12-26 Transition Graph for Counter

successive clock pulses will cause the counter to continue in the original counting sequence as indicated on the transition graph. When the power in a circuit is first turned on, the initial states of the flip-flops may be unpredictable. For this reason, all of the don't-care states in a counter should be checked to make sure that they eventually lead into the main counting sequence unless a power-up reset is provided. Such a counter is sometimes called *self-starting*.

In summary, the following procedure can be used to design a counter using T flip-flops:

1. Form a transition table which gives the next flip-flop states for each combination of present flip-flop states.
2. Plot the next-state maps from the table.
3. Plot a T input map for each flip-flop. When filling in the T_Q map, T_Q must be 1 whenever $Q^+ \neq Q$. This means that the T_Q map can be formed from the Q^+ map by complementing the $Q = 1$ half of the map and leaving the $Q = 0$ half unchanged.
4. Find the T input equations from the maps and realize the circuit.

Counter Design Using D Flip-Flops

For a D flip-flop, $Q^+ = D$, so the D input map is identical with the next-state map. Therefore, the equation for D can be read directly from the Q^+ map. For the counter of Figure 12-22, the following equations can be read from the next-state maps shown in Figure 12-23(a):

$$D_C = C^+ = B' \qquad D_B = B^+ = C + BA'$$
$$D_A = A^+ = CA' + BA' = A'(C + B)$$

This leads to the circuit shown in Figure 12-27 using D flip-flops. Note that the connecting wires between the flip-flop outputs and the gate inputs have been omitted to facilitate reading the diagram.

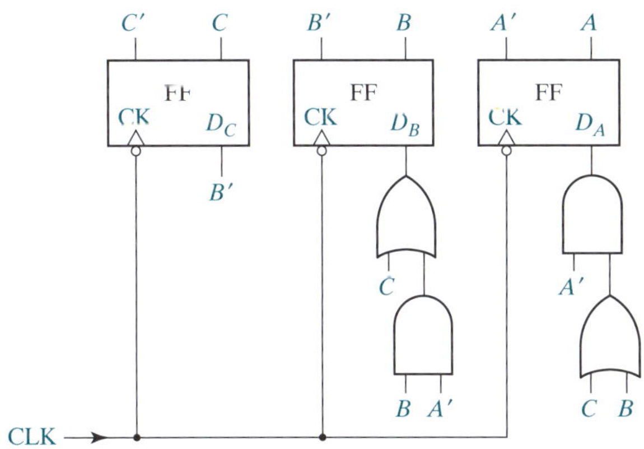

FIGURE 12-27 Counter of Figure 12-22 Using D Flip-Flops

An alternative to directly designing counters for nonbinary sequences is to use a binary counter or some other counter and add decoding logic to generate the desired outputs. This approach may result in a simpler overall circuit. As an example, the count cycle of length 5 of Figure 12-22 can be obtained using a five-stage ring counter with decoding logic. If the stages of the ring counter are $Q_5, Q_4, \ldots, Q_1$, then the outputs are obtained by the decoding equations

$$O_C = Q_4 + Q_3$$
$$O_B = Q_3 + Q_2 + Q_1$$
$$O_A = Q_3 + Q_1$$

Counters and shift registers with clear, preset, or parallel load capability can also be used to generate nonbinary count cycles. Consider a binary counter with a clear input as shown Figure 12-28. The binary counter ($N = 4$) can be converted to a BCD decade (decimal) counter if the gate logic causes the counter to transfer from a count of 9 to a count of 0 (rather than 10). If the clear input is a synchronous clear, then the logic required is

$$\text{Clr} = Q_3Q_0$$

assuming Q_3, Q_2, Q_1, Q_0 are the four counter outputs. Note that the counter never enters states 10 through 15 so these are don't-cares for the Clr function. If the clear input is an asynchronous input, then the logic required is

$$\text{Clr} = Q_3Q_1$$

In this case states 11 through 15 are don't-cares for Clr. On the next active edge of the clock, when the counter is in state 9, the counter will temporarily enter state 10, causing Clr to become active and return the counter to state 0. Such transient states could cause glitches in other logic using the counter outputs.

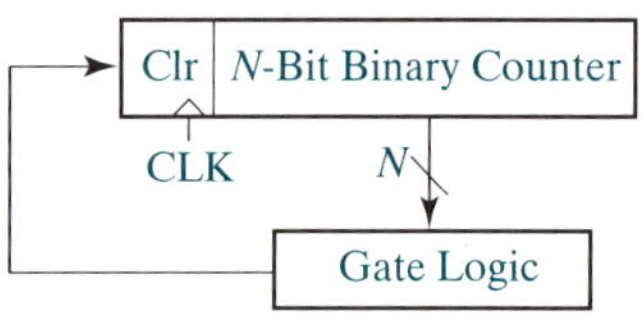

FIGURE 12-28 Binary Counter with Clear

As another example, consider the binary counter of Figure 12-29 ($N = 4$) that has a parallel load capability. It can be used to create a decimal counter that counts using the excess-3 code. The counter must cycle through states 3 through 12. The logic must generate Ld when the counter is in state 12 and the parallel inputs must be 0011.

$$D_3 = 0, D_2 = 0, D_1 = 1, D_0 = 1$$
$$Ld = Q_3Q_2$$

States 0, 1, 2, 13, 14 and 15 are don't-cares.

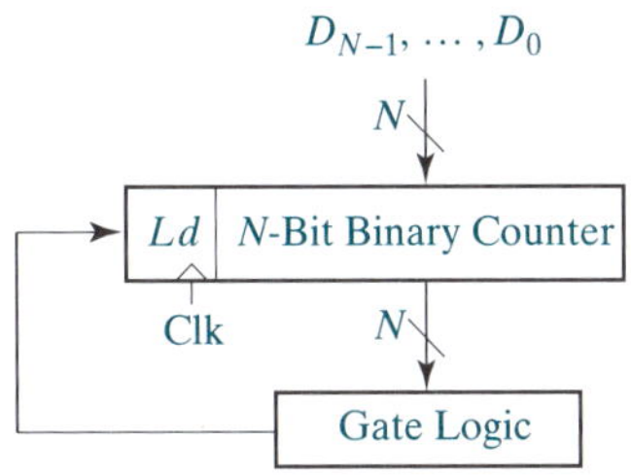

FIGURE 12-29 Binary Counter with Parallel Load

12.5 Counter Design Using S-R and J-K Flip-Flops

The procedures used to design a counter with S-R flip-flops are similar to the procedures discussed in Sections 12.3 and 12.4. However, instead of deriving an input equation for each D or T flip-flop, the S and R input equations must be derived. We will now develop methods for deriving these S and R flip-flop input equations.

Table 12-5(a) describes the behavior of the S-R flip-flop. Given S, R, and Q, we can determine Q^+ from this table. However, the problem we must solve is to determine S and R given the present state Q and the desired next state Q^+. If the present state of the flip-flop is $Q = 0$ and the desired next state is $Q^+ = 1$, a 1 must be applied to the S input to set the flip-flop to 1. If the present state is 1, and the desired next state is 0, a 1 must be applied to the R input to reset the flip-flop to 0. Restrictions on the flip-flop inputs require that $S = 0$ if $R = 1$, and $R = 0$ if $S = 1$. Thus, when forming Table 12-5(b), the rows corresponding to $QQ^+ = 01$ and 10 are filled in with $SR = 10$ and 01, respectively. If the present state and next state are both 0, S must be 0 to prevent setting the flip-flop to 1. However, R may be either 0 or 1 because when $Q = 0$, $R = 1$ has no effect on the flip-flop state. Similarly, if the present state and next state are both 1, R must be 0 to prevent resetting the flip-flop, but S may be either 0 or 1. The required S and R inputs are summarized in Table 12-5(b). Table 12- 5(c) is the same as 12-5(b), except the alternative choices for R and S have been indicated by don't-cares.

TABLE 12-5 S-R Flip-Flop Inputs

(a)

S	R	Q	Q^+
0	0	0	0
0	0	1	1
0	1	0	0
0	1	1	0
1	0	0	1
1	0	1	1
1	1	0	– } inputs not
1	1	1	– } allowed

(b)

Q	Q^+	S	R
0	0	{0	0
		{0	1
0	1	1	0
1	0	0	1
1	1	{0	0
		{1	0

(c)

Q	Q^+	S	R
0	0	0	X
0	1	1	0
1	0	0	1
1	1	X	0

TABLE 12-6

C	B	A	C^+	B^+	A^+	S_C	R_C	S_B	R_B	S_A	R_A
0	0	0	1	0	0	1	0	0	X	0	X
0	0	1	–	–	–	X	X	X	X	X	X
0	1	0	0	1	1	0	X	X	0	1	0
0	1	1	0	0	0	0	X	0	1	0	1
1	0	0	1	1	1	X	0	1	0	1	0
1	0	1	–	–	–	X	X	X	X	X	X
1	1	0	–	–	–	X	X	X	X	X	X
1	1	1	0	1	0	0	1	X	0	0	1

Next, we will redesign the counter of Figure 12-22 using S-R flip-flops. Table 12-3 is repeated in Table 12-6 with columns added for the S and R flip-flop inputs. These columns can be filled in using Table 12-5(c). For $CBA = 000$, $C = 0$ and $C^+ = 1$, so $S_C = 1$, $R_C = 0$. For $CBA = 010$ and 011, $C = 0$ and $C^+ = 0$, so $S_C = 0$ and $R_C = \text{X}$. For $CBA = 100$, $C = 1$ and $C^+ = 1$, so $S_C = \text{X}$ and $R_C = 0$. For row 111, $C = 1$ and $C^+ = 0$, so $S_C = 0$ and $R_C = 1$. For $CBA = 001, 101,$ and 110, $C^+ = \text{X}$, so $S_C = R_C = \text{X}$. Similarly, the values of S_B and R_B are derived from the values of B and B^+, and S_A and R_A are derived from A and A^+. The resulting flip-flop input functions are mapped in Figure 12-30(b).

It is generally faster and easier to derive the S-R flip-flop input maps directly from the next-state maps than to derive them from the transition table as was done in Table 12-6. For each flip-flop, we will derive the S and R input maps from the next-state (Q^+) map using Table 12-5(c) to determine the values for S and R. Just as we did for the T flip-flop, we will use the next-state maps for C^+, B^+, and A^+ in Figure 12-23(a) as a starting point for deriving the S-R flip-flop input equations. For convenience, these maps are repeated in Figure 12-30(a). We will consider one-half of each next-state map at a time when deriving the input maps. We will start with flip-flop $C(Q = C$ and $Q^+ = C^+)$ and consider the $C = 0$ column of the map. From Table 12-5(c), if $C = 0$ and $C^+ = 1$, then $S = 1$ and $R = 0$. Therefore, for every square in the $C = 0$ column where $C^+ = 1$, we plot $S_C = 1$ and $R_C = 0$ (or blank) in the corresponding squares of the input maps. Similarly, for every square in the $C = 0$ column where $C^+ = 0$, we plot $S_C = 0$ and $R_C = \text{X}$ on the input maps. For the $C = 1$ column, if $C^+ = 0$, we plot $S_C = 0$ and $R_C = 1$; if $C^+ = 1$, we plot $S_C = \text{X}$ and $R_C = 0$. Don't-cares on the C^+ map remain don't-cares on the S_C and R_C maps, because if we do not care what the next state is, we do not care what the input is. In a similar manner, we can derive the S_B and R_B maps from the B^+ map by working with the $B = 0$ (top) half of the map and the $B = 1$ (bottom) half of the map. As before, 1's are placed on the S or R map when the flip-flop must be set or reset. S is a don't-care if $Q = 1$ and no state change is required, and $R = \text{X}$ if $Q = 0$ and no state change is required. Finally, S_A and R_A are derived from the A^+ map. Figure 12-30(c) shows the resulting circuit.

The procedure used to design a counter with J-K flip-flops is very similar to that used for S-R flip-flops. The J-K flip-flop is similar to the S-R flip-flop except that J and K can be 1 simultaneously, in which case the flip-flop changes state. Table 12-7(a) gives the next state (Q^+) as a function of J, K, and Q. Using this table, we can derive the required input conditions for J and K when Q and Q^+ are given. Thus if a change from $Q = 0$ to $Q^+ = 1$ is required, either the flip-flop can be set to 1 by using $J = 1$

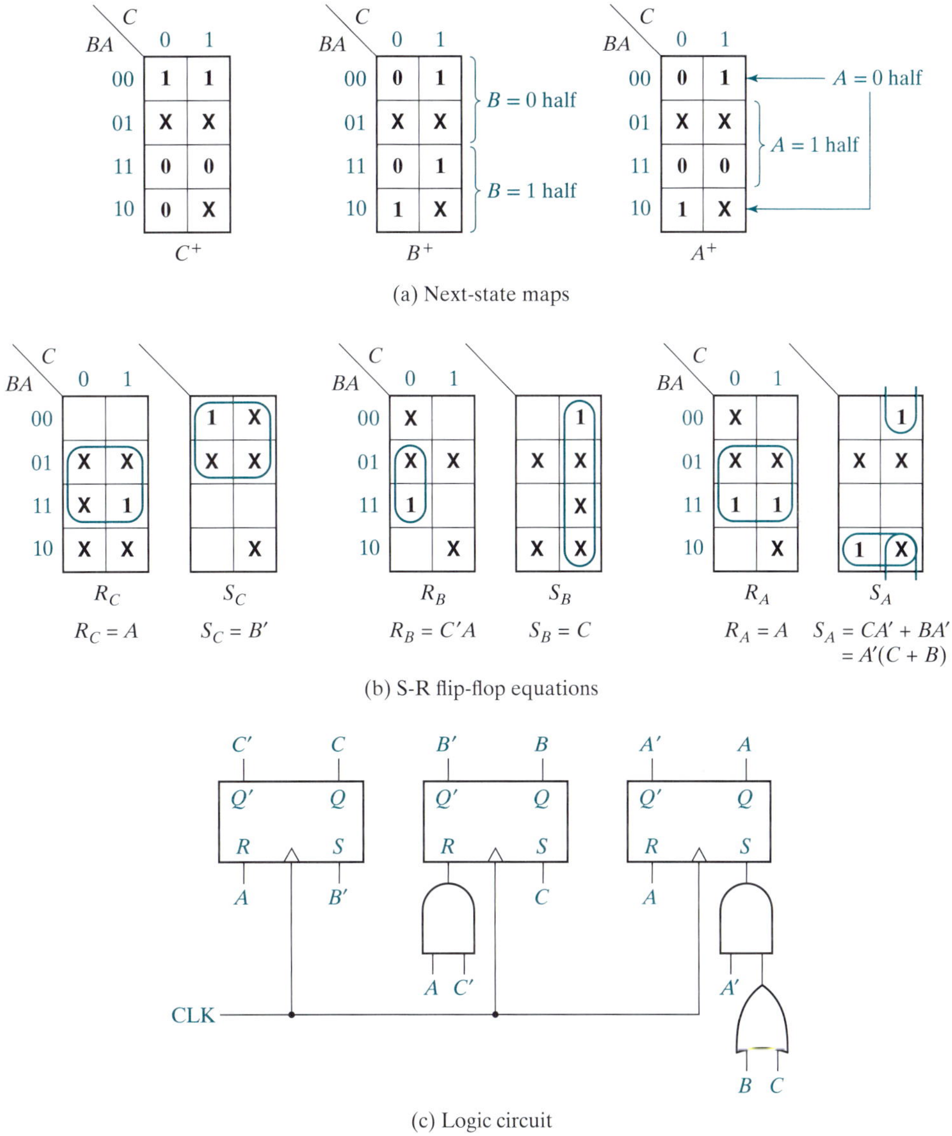

FIGURE 12-30 Counter of Figure 12-22 Using S-R Flip-Flops

(and $K = 0$) or the state can be changed by using $J = K = 1$. In other words, J must be 1, but K is a don't-care. Similarly, a state change from 1 to 0 can be accomplished by resetting the flip-flop with $K = 1$ (and $J = 0$) or by changing the flip-flop state with $J = K = 1$. When no state change is required, the inputs are the same as the corresponding inputs for the S-R flip-flops. The J-K flip-flop input requirements are summarized in Tables 12-7(b) and 12-7(c).

We will now redesign the counter of Figure 12-22 using J-K flip-flops. Table 12-3 is repeated in Table 12-8 with columns added for the J and K flip-flop inputs. We will fill in these columns using Table 12-7(c). For $CBA = 000$, $C = 0$

TABLE 12-7 J-K Flip-Flop Inputs

(a)

J	K	Q	Q^+
0	0	0	0
0	0	1	1
0	1	0	0
0	1	1	0
1	0	0	1
1	0	1	1
1	1	0	1
1	1	1	0

(b)

Q	Q^+	J	K
0	0	{0	0
		{0	1
0	1	{1	0
		{1	1
1	0	{0	1
		{1	1
1	1	{0	0
		{1	0

(c)

Q	Q^+	J	K
0	0	0	X
0	1	1	X
1	0	X	1
1	1	X	0

TABLE 12-8

C	B	A	C^+	B^+	A^+	J_C	K_C	J_B	K_B	J_A	K_A
0	0	0	1	0	0	1	X	0	X	0	X
0	0	1	–	–	–	X	X	X	X	X	X
0	1	0	0	1	1	0	X	X	0	1	X
0	1	1	0	0	0	0	X	X	1	X	1
1	0	0	1	1	1	X	0	1	X	1	X
1	0	1	–	–	–	X	X	X	X	X	X
1	1	0	–	–	–	X	X	X	X	X	X
1	1	1	0	1	0	X	1	X	0	X	1

and $C^+ = 1$, so $J_C = 1$ and $K_C = \text{X}$. For $CBA = 010$ and 011, $C = 0$ and $C^+ = 0$, so $J_C = 0$ and $K_C = \text{X}$. The remaining table entries are filled in similarly. The resulting J-K flip-flop input functions are plotted in Figure 12-31(b). After deriving the flip-flop input equations from the J-K maps, we can draw the logic circuit of Figure 12-31(c).

12.6 Derivation of Flip-Flop Input Equations—Summary

The input equation for the flip-flops in a sequential circuit may be derived from the next-state equations by using truth tables or by using Karnaugh maps. For circuits with three to five variables, it is convenient to first plot maps for the next-state equations, and then transform these maps into maps for the flip-flop inputs.

Given the present state of a flip-flop (Q) and the desired next state (Q^+), Table 12-9 gives the required inputs for various types of flip-flops. For the D flip-flop, the input is the same as the next state. For the T flip-flop, the input is 1 whenever a state change is required. For the S-R flip-flop, S is 1 whenever the flip-flop must be set to 1 and R is 1 when it must be reset to 0. We do not care what S is if the flip-flop state is 1 and must remain 1; we do not care what R is if the flip-flop state is 0 and

FIGURE 12-31 Counter of Figure 12-22 Using J-K Flip-Flops

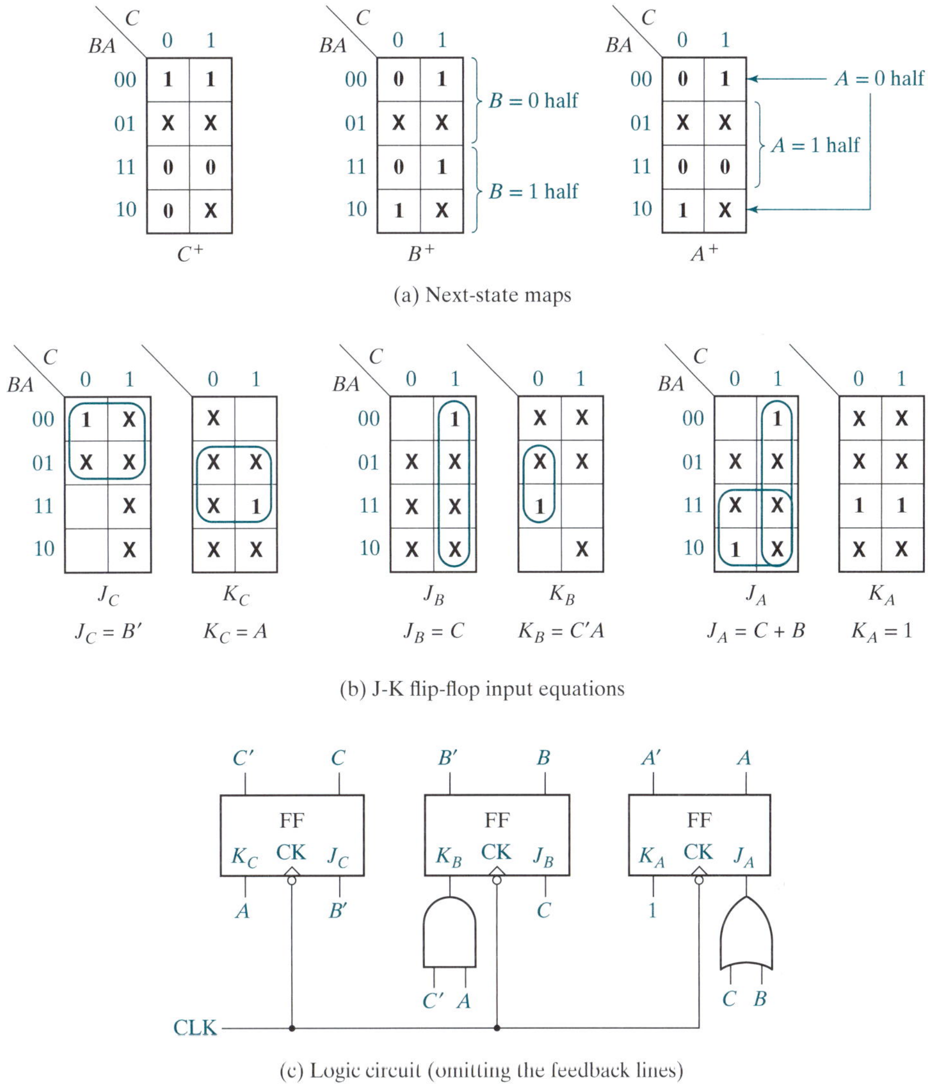

must remain 0. For a J-K flip-flop, the J and K inputs are the same as S and R, respectively, except that when one input is 1 the other input is X. This difference arises because $S = R = 1$ is not allowed, but $J = K = 1$ causes a change of state.

Table 12-9 summarizes the rules for transforming next-state maps into flip-flop input maps. Before applying these rules, we must copy any don't-cares from the next-state maps onto the input maps. Then, we must work with the $Q = 0$ and $Q = 1$ halves of each next-state map separately. The rules given in Table 12-9 are easily derived by comparing the values of Q^+ with the corresponding input values. For example, in the $Q = 0$ column of the table, we see that J is the same as Q^+, so the $Q = 0$ half of the J map is the same as the Q^+ map. In the $Q = 1$ column, $J = \text{X}$ (independent of Q^+), so we fill in the $Q = 1$ half of the J map with X's.

TABLE 12-9 Determination of Flip-Flop Input Equations from Next-State Equations Using Karnaugh Maps

Type of Flip-Flop	Input	$Q = 0$		$Q = 1$		Rules for Forming Input Map From Next-State Map*	
		$Q^+ = 0$	$Q^+ = 1$	$Q^+ = 0$	$Q^+ = 1$	$Q = 0$ Half of Map	$Q = 1$ Half of Map
Delay	D	0	1	0	1	no change	no change
Toggle	T	0	1	1	0	no change	complement
Set-Reset	S	0	1	0	X	no change	replace 1's with X's**
	R	X	0	1	0	replace 0's with X's**	complement
J-K	J	0	1	X	X	no change fill	fill in with X's
	K	X	X	1	0	in with X's	complement

Q^+ means the next state of Q
X is a don't-care
*Always copy X's from the next-state map onto the input maps first.
**Fill in the remaining squares with 0's.

Example (illustrating the use of Table 12-9)

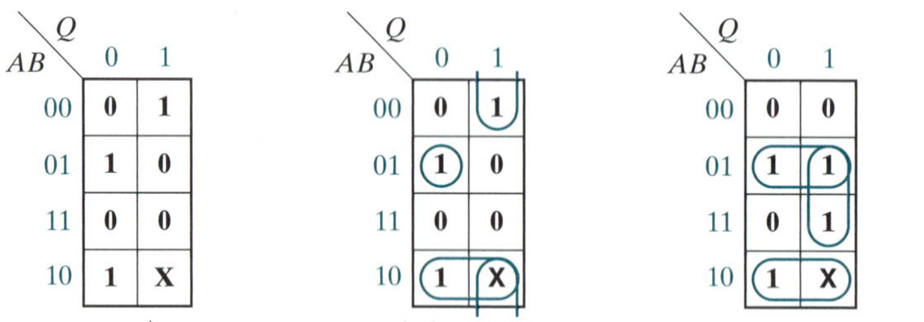

Q^+ — Next-state map

$D = Q'A'B + QB' + AB'$ — D input map

$T = A'B + AB' + QB$ — T input map

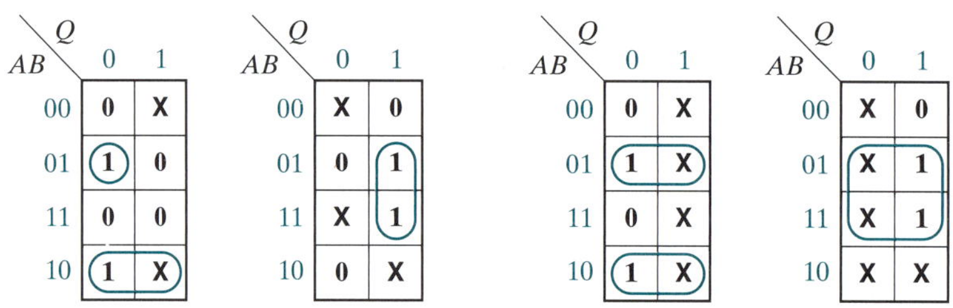

$S = AB' + Q'A'B$ $\quad R = QB$ — S-R input maps

$J = A'B + AB'$ $\quad K = B$ — J-K input maps

For the S-R flip-flop, note that when $Q = 0$, $R = X$ if $Q^+ = 0$; and when $Q = 1$, $R = 1$ if $Q^+ = 0$. Therefore, to form the R map from the Q^+ map, replace 0's with X's on the $Q = 0$ half of the map and replace 0's with 1's on the $Q = 1$ half (and fill in 0's for the remaining entries). Similarly, to form the S map from the Q^+ map, copy the 1's on the $Q = 0$ half of the map, and replace the 1's with X's on the $Q = 1$ half.

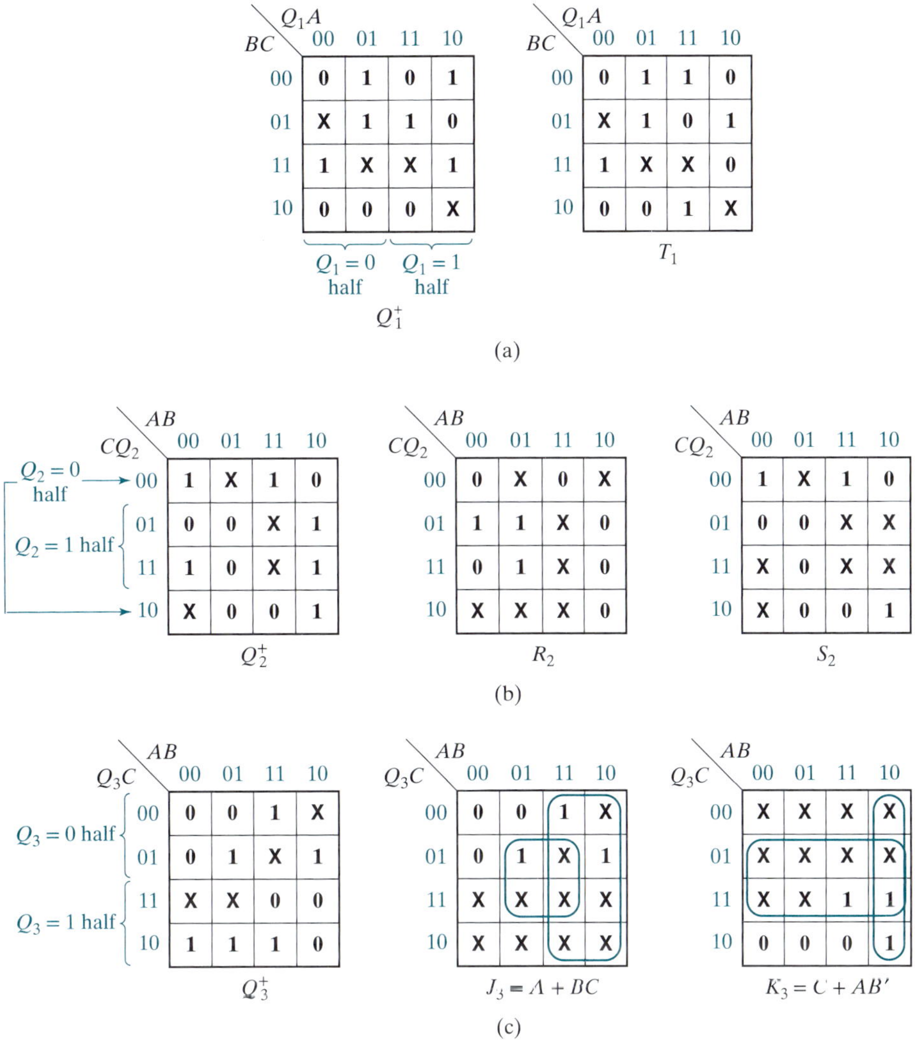

FIGURE 12-32 Derivation of Flip-Flop Input Equations Using Four-Variable Maps

Examples of deriving four-variable input maps are given in Figure 12-32. In each case, Q_i represents the flip-flop for which input equations are being derived A, B, and C represent other variables on which the next state depends. As shown in Figure 12-32(a), a 1 is placed on the T_1 map whenever Q_1 must change state. In Figure 12-32(b), 1's are placed on the $Q_2 = 0$ half of the S^2 map whenever Q_2 must be set to 1, and 1's are placed on the $Q_2 = 1$ half of the R_2 map whenever Q_2 must be reset. Figure 12-32(c) illustrates derivation of J_3 and K_3 by using separate J and K maps. As will be seen in Unit 14, the methods used to derive flip-flop input equations for counters are easily extended to general sequential circuits.

The procedures for deriving flip-flop input equations discussed in this unit can be extended to other types of flip-flops. If we want to derive input equations for a different type of flip-flop, the first step is to construct a table which gives the next state

(Q^+) as a function of the present state (Q) and the flip-flop inputs. From this table, we can construct another table which gives the required flip-flop input combinations for each of the four possible pairs of values of Q and Q^+. Then, using this table, we can plot a Karnaugh map for each input function and derive minimum expressions from the maps.

Problems

12.1 Consider a 6-bit adder with an accumulator, as in Figure 12-5. Suppose the X register contains a number from a previous calculation. We do not want this number. Instead, we want X to equal $3 \times Y$. ($X = x_5x_4x_3x_2x_1x_0$ and $Y = y_5y_4y_3y_2y_1y_0$.) On the timing diagram, give values for Ad and ClrN so that we will have $X = 3 \times Y$ held in the accumulator.

12.2 The shift register of Figure 12-10 can be made to shift to the left by adding external connections between the Q outputs and D inputs. Draw a block diagram like the one in Figure 12-10(a) and indicate the appropriate connections. Which input line would serve as a serial input in this case? With the connections you have made, what should *Sh* and *L* be for a left shift? For a right shift?

12.3 Show how to modify the internal circuitry of the shift register of Figure 12-10 so that it will also shift to the left without external connections as in Problem 12.2. Replace *Sh* and *L* with *A* and *B* and let the register operate according to the following table:

Inputs		Next State				
A	B	Q_3^+	Q_2^+	Q_1^+	Q_0^+	Action
0	0	Q_3	Q_2	Q_1	Q_0	no change
0	1	SI	Q_3	Q_2	Q_1	right shift
1	0	Q_2	Q_1	Q_0	SI	left shift
1	1	D_3	D_2	D_1	D_0	load

12.4 (a) Design a 4-bit synchronous binary counter using T flip-flops. (*Hint:* Add one flip-flop, with necessary gates, to the left side of Figure 12-14. Verify that the gates for the other three flip-flops do not change.)
(b) Repeat (a) using D flip-flops. See Figure 12-16.

12.5 Repeat Problem 12.4(a) using D flip-flops, but implement each D input as a sum of products, without using XOR gates. (*Hint:* Use Equations (12-2). As in Problem 12.4, you will need one more equation.)

12.6 Design a circuit using D flip-flops that will generate the sequence 0, 0, 1, 0, 1, 1 and repeat. Do this by designing a counter for any sequence of states such that the first flip-flop takes on this sequence. There are many correct answers, but do not duplicate states, because each state can have only one next state.

12.7 Design a 3-bit counter which counts in the sequence:
001, 011, 010, 110, 111, 101, 100, (repeat) 001, . . .
(a) Use D flip-flops.
(b) Use T flip-flops.
In each case, what will happen if the counter is started in state 000?

12.8 Design a 3-bit counter which counts in the sequence:
001, 011, 010, 110, 111, 101, 100, (repeat) 001, . . .
(a) Use J-K flip-flops.
(b) Use S-R flip-flops.
In each case, what will happen if the counter is started in state 000?

12.9 An M-N flip-flop works as follows:
If $MN = 00$, the next state of the flip-flop is 0.
If $MN = 01$, the next state of the flip-flop is the same as the present state.
If $MN = 10$, the next state of the flip-flop is the complement of the present state.
If $MN = 11$, the next state of the flip-flop is 1.
(a) Complete the following table (use don't-cares when possible):

Present State Q	Next State Q^+	M N
0	0	
0	1	
1	0	
1	1	

(b) Using this table and Karnaugh maps, derive and minimize the input equations for a counter composed of three M-N flip-flops which counts in the following sequence: CBA = 000, 001, 011, 111, 101, 100, (repeat) 000, . . .

12.10 Design a counter which counts in the sequence that has been assigned to you. Use D flip-flops and NAND gates. Simulate your design using *SimUaid*.
(a) 000, 001, 011, 101, 111, 010, (repeat) 000, . . .
(b) 000, 011, 101, 111, 010, 110, (repeat) 000, . . .
(c) 000, 110, 111, 100, 101, 001, (repeat) 000, . . .
(d) 000, 100, 001, 110, 101, 111, (repeat) 000, . . .
(e) 000, 010, 111, 101, 011, 110, (repeat) 000, . . .

(f) 000, 100, 001, 111, 110, 101, (repeat) 000, . . .
(g) 000, 010, 111, 101, 001, 110, (repeat) 000, . . .
(h) 000, 101, 010, 011, 001, 110, (repeat) 000, . . .
(i) 000, 100, 010, 001, 110, 111, (repeat) 000, . . .
(j) 000, 001, 111, 010, 110, 011, (repeat) 000, . . .
(k) 000, 100, 010, 001, 101, 111, (repeat) 000, . . .
(l) 000, 011, 111, 110, 001, 100, (repeat) 000, . . .
(m) 000, 100, 111, 110, 010, 011, (repeat) 000, . . .
(n) 000, 011, 111, 110, 010, 100, (repeat) 000, . . .

12.11 Redesign the right-shift register circuit of Figure 12-10 using four D flip-flops with clock enable, four 2-to-1 MUXes, and a single OR gate.

12.12 Design a left-shift register similar to that of Figure 12-10. Your register should shift left if $Sh = 1$, load if $Sh = 0$ and $L = 1$, and hold its state if $Sh = L = 0$.
(a) Draw the circuit using four D flip-flops and four 4-to-1 MUXes.
(b) Give the next-state equations for the flip-flops.

12.13 A 74178 shift register is described by the given table. All state changes occur on the 1-0 transition of the clock. The shift register is connected as shown. Complete the timing diagram.

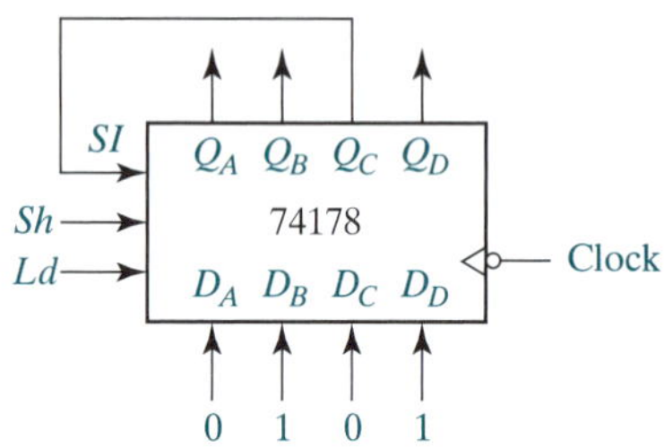

Sh	Ld	Q_A^+	Q_B^+	Q_C^+	Q_D^+
0	0	Q_A	Q_B	Q_C	Q_D
0	1	D_A	D_B	D_C	D_D
1	X	SI	Q_A	Q_B	Q_C

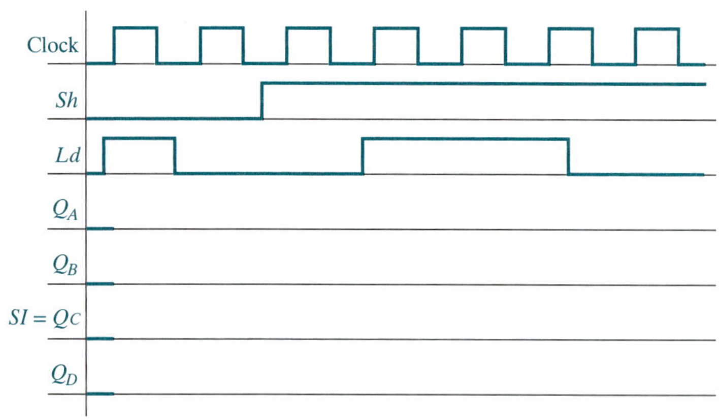

12.14 Design a 5-bit synchronous binary counter. (*Hint:* See Problem 12.4.)
(a) Use T flip-flops.
(b) Use D flip-flops.

12.15 Construct a 4-bit Johnson counter using J-K flip-flops. (See Figure 12-12 for a Johnson counter.) What sequence of states does the counter go through if it is started in state 0000? State 0110?

12.16 Design a 3-bit binary up-down counter which functions the same as the up-down counter of Figure 12-18 and Figure 12-19. Use a 3-bit register of D flip-flops, a 3-bit adder, and one OR gate. (If you are clever enough, you can do it without the OR gate.) (*Hint:* To subtract one, add 111.)

12.17 Design a decade counter which counts in the sequence:
0000, 0001, 0010, 0011, 0100, 0101, 0110, 0111, 1000, 1001, 0000, . . .
(a) Use D flip-flops.
(b) Use J-K flip-flops.
(c) Use S-R flip-flops.
(d) Use T flip-flops.
(e) Draw a complete state diagram for the counter of (b) showing what happens when the counter is started in each of the unused states.

12.18 Repeat Problem 12.17 for the downward decade sequence:
0000, 1001, 1000, 0111, 0110, 0101, 0100, 0011, 0010, 0001, 0000, . . .

12.19 (a) Derive the state graph for the ripple counter below. Show the transient states that occur between stable states.
(b) Modify the ripple counter so that it cycles through the same states but in reverse order. (Use the same type of flip-flop.) (*Hint:* First construct a state graph for the modified counter showing the transient states between stable states.)

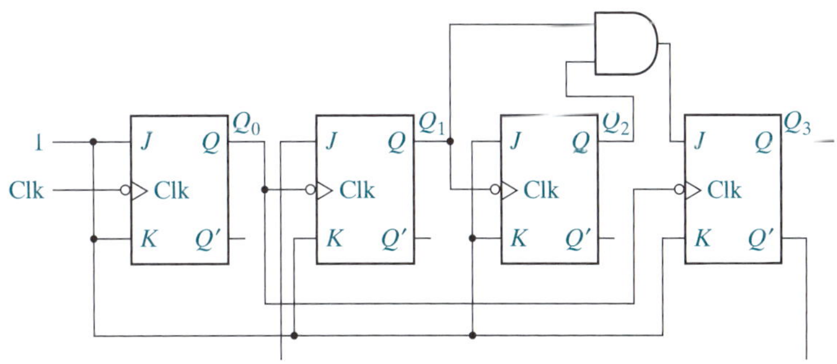

12.20 Design a 3-bit counter which counts in the sequence:
001, 100, 101, 111, 110, 010, 011, 001, . . .
(a) Use D flip-flops.
(b) Use J-K flip-flops.

(c) Use T flip-flops.
(d) Use S-R flip-flops.
(e) What will happen if the counter of (a) is started in state 000?

12.21 Design a decade counter using the following 2-4-2-1 weighted code for decimal digits. Use NAND gates and the indicated flip-flop types.
(a) Use D flip-flops.
(b) Use J-K flip-flops.
(c) Use T flip-flops.
(d) Use S-R flip-flops.

Digit	*ABCD*
0	0000
1	0001
2	0010
3	0011
4	0100
5	1011
6	1100
7	1101
8	1110
9	1111

12.22 Repeat Problem 12.21 using NOR gates instead of NAND gates.

12.23 Design a decade counter using the excess-3 code for decimal digits. Use NAND gates and the indicated flip-flop types.
(a) Use D flip-flops.
(b) Use J-K flip-flops.
(c) Use T flip-flops.
(d) Use S-R flip-flops.

12.24 Repeat Problem 12.23 using NOR gates instead of NAND gates.

12.25 The following binary counter increments on each rising clock edge unless the external clear (ClrN) control input is low.
(a) Implement a modulo 12 counter using this binary counter assuming the Clr control input is a synchronous control input.
(b) Repeat part (a) assuming Clr is an asynchronous control input.

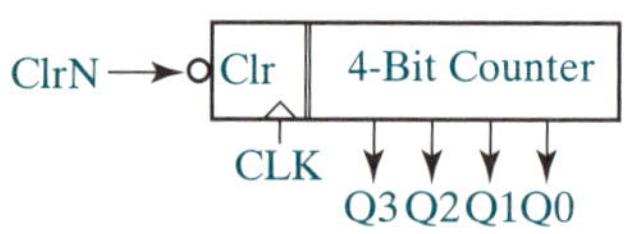

12.26 The following binary counter operates according to the function table given. Using this binary counter, implement a decimal counter that uses the 2-4-2-1 weighted code for representing decimal digits. Minimize the gate logic required by using the parallel load inputs only to change the counting sequence from straight binary to 2-4-2-1 code.

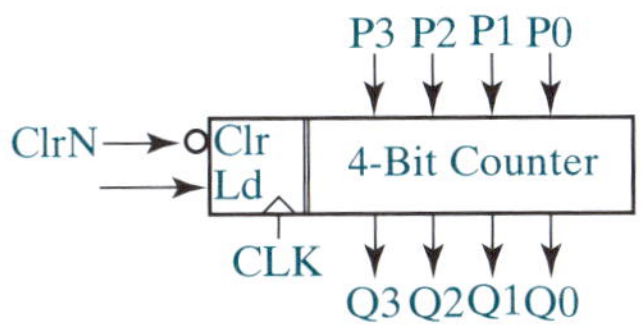

ClrN	Ld	Function
0	—	Clear
1	1	Parallel Load
1	0	Increment

12.27 For each value of N, there exists an exclusive-OR circuit for the linear shift register counter of Figure 12-13 so that the counter cycles through $2^N - 1$ counts.
(a) For $N = 3$, construct the transition graph for the counter if $S_{in} = Q_2 \oplus Q_1$. (The shift register stages are numbered Q_0, Q_1, Q_2 from left to right.)
(b) For $N = 4$, find an exclusive-OR circuit so that the counter cycles through 15 counts.
(c) Make a simple modification to the logic of part (b) so that the counter cycles through 16 counts. (The counter is no longer linear.)

12.28 When started in state 0, an n-stage Johnson (twisted-ring) counter cycles through $2n$ states. Other count cycles will be obtained if the counter is started in other states. Determine all count cycles for the Johnson counter when
(a) $n = 3$,
(b) $n = 4$, and
(c) $n = 5$.

12.29 When started in state 0, an n-stage Johnson (twisted-ring) counter cycles through $2n$ states. A simple modification of the counter will cause it to cycle through $2n - 1$ states. (Assume the counter is implemented with D flip-flops.)
(a) Determine the equation for the shift register input so that the all 1's state is eliminated from the cycle with $2n$ states to obtain a cycle of length $2n - 1$.
(b) Repeat part (a) for the case when the all 0's state is eliminated.

12.30 Repeat Problem 12.29 assuming the counter is implemented with J-K flip-flops.

12.31 Binary up counters can be designed using J-K flip-flops by noting that the least significant stage, Q_0, always toggles and stage Q_i always toggles when stages 0, . . . $(i - 1)$ are 1. This approach can be modified to design counters with a shorter cycle and obtaining nearly minimum equations. Note that an optimum solution may not have the same equations for J and K.

(a) Modify the binary up counter design to obtain a BCD decade up counter using J-K flip-flops.
(b) Modify the binary up counter design to obtain an excess-3 decade up counter using J-K flip-flops.
(c) Modify the design for part (b) so that the counters can be cascaded to obtain excess-3 counters that can count to 99, 999, etc.

12.32 A three-stage binary up-down counter has control input U; when $U = 0$, the counter counts down and when $U = 1$, the counter counts up. Design this counter with a minimum number of NAND gates, using
(a) reset-dominant S-R flip-flops.
(b) D-CE flip-flops.

12.33 A two-stage counter has two input control lines, M and N. The count sequences are as follows:

MN	Sequence
00	0, 1, 2, 3, 0, ...
01	0, 1, 0, 1, 0, 1, . . .
10	2, 0, 2, 0, 2, 0, . . .
11	1, 2, 1, 2, 1, 2, . . .

(a) Design the counter assuming the outputs come directly from J-K flip-flops.
(b) Design the counter assuming a two-stage binary counter is used with the J-K flip-flop outputs decoded.

12.34 A pulse-generating circuit generates eight repetitive pulses as shown in the figure. Implement the pulse-generating circuit using the counter circuits listed and a minimum of gate logic. Use J-K flip-flops for the counters that trigger on the falling edge of a clock that has a frequency eight times the frequency of one of the pulses. The pulses must be free of glitches; explain any restrictions on the propagation delays of gates and flip-flops so that the pulses will be glitch free.
(a) Ring counter (A ring counter is a shift register with end-to-end feedback.)
(b) Johnson counter
(c) Binary counter

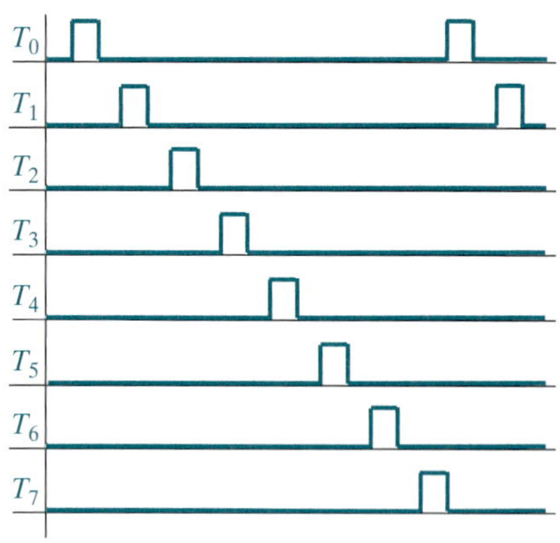

12.35 A U-V flip-flop behaves as follows:
If $UV = 00$, the flip-flop does not change state.
If $UV = 10$, the flip-flop is set to $Q = 0$.
If $UV = 11$, the flip-flop changes state.
The input combination $UV = 01$ is not allowed.
(a) Give the characteristic (next-state) equation for this flip-flop.
(b) Complete the following table, using don't-cares where possible.

Q	Q^+	U	V
0	0		
0	1		
1	0		
1	1		

(c) Realize the following next-state equation for Q using a U-V flip-flop: $Q^+ = A + BQ$. Find equations for U and V.

12.36 A M-F flip-flop behaves as follows:
If $MF = 01$, the flip-flop changes state.
If $MF = 11$, the flip-flop is set to $Q = 0$.
If $MF = 00$, the flip-flop is set to $Q = 1$.
The input combination $MF = 10$ is not allowed.
(a) Give the characteristic (next-state) equation for this flip-flop.
(b) Complete the table, using don't-cares where possible.

Q	Q^+	M	F
0	0		
0	1		
1	0		
1	1		

(c) Realize the following next-state equation for Q using a MF flip-flop: $Q^+ = CQ + DQ'$. Find equations for M and F.

12.37 An L-M flip-flop works as follows:
If $LM = 00$, the next state of the flip-flop is 1.
If $LM = 01$, the next state of the flip-flop is the same as the present state.
If $LM = 10$, the next state of the flip-flop is the complement of the present state.
If $LM = 11$, the next state of the flip-flop is 0.
(a) Complete the following table (use don't-cares when possible):

Present State Q	Next State Q^+	L	M
0	0		
0	1		
1	0		
1	1		

(b) Using this table and Karnaugh maps, derive and minimize the input equations for a counter composed of three L-M flip-flops which counts in the following sequence: $ABC = 000, 100, 101, 111, 011, 001, 000, \ldots$

12.38 A sequential circuit contains a register of four flip-flops. Initially a binary number N $(0000 \leq N \leq 1100)$ is stored in the flip-flops. After a single clock pulse is applied to the circuit, the register should contain $N + 0011$. In other words, the function of the sequential circuit is to add 3 to the contents of a 4-bit register. Design the circuit using J-K flip-flops.

12.39 When an adder is part of a larger digital system, an arrangement like the given figure often works well. For the control signals and the input data in the following table, give the value of the addend, the accumulator, and the bus at the end of each clock cycle (i.e., immediately before the active clock edge). Express the register and bus values in decimal.

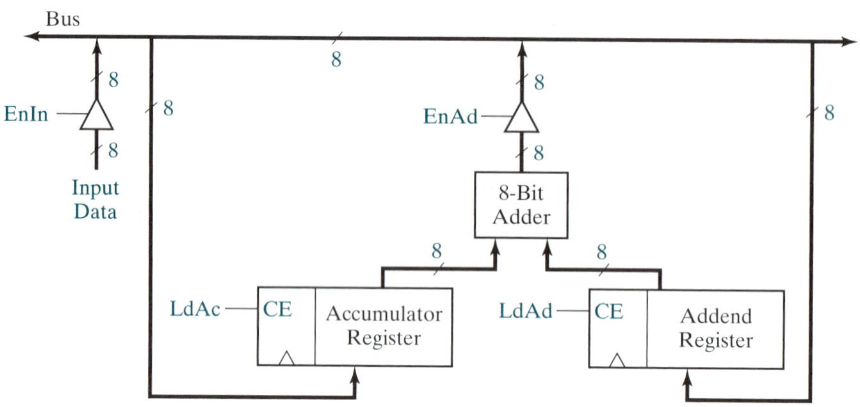

Clock Cycle	Input Data	EnIn	EnAd	LdAc	LdAd	Accumulator Register	Addend Register	Bus
0	18	1	0	1	0	0	0	18
1	13	1	0	0	1			
2	15	0	1	1	0			
3	93	1	0	0	1			
4	47	0	1	1	0			
5	22	1	0	0	1			
6	0	0	1	0	0			

12.40 A digital system can perform any four-variable bitwise logic function, but it may take several clock cycles. (A bitwise logic function performs the same logic function on each bit.) Recall that the NAND operation is functionally complete (i.e., we can do any logic function by a series of NAND operations). On the following 8-bit registers, En is a tri-state buffer enable as in Figure 12-3, and CE is a clock enable as in Figure 12-1.

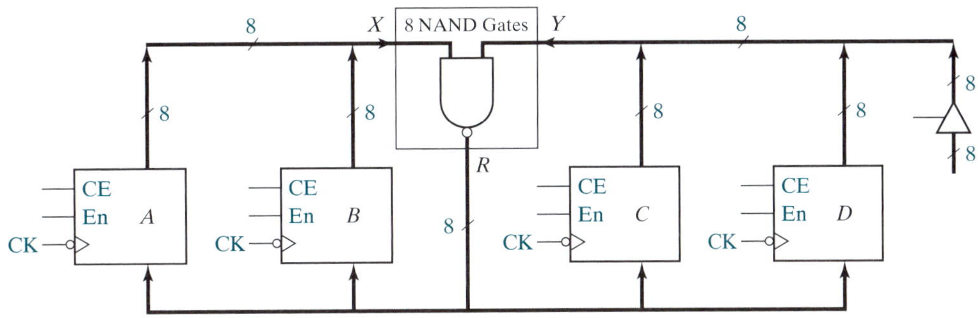

(a) Show how to connect a 2-to-4 decoder (with inverting outputs) so that the next rising edge of the clock will load the result into register A, B, C, or D for control inputs $G_0G_1 = 00, 01, 10$, or 11, respectively.

(b) Show how to connect three control signals, E_0, E_1, and E_2, to the registers so that $E_0 = 0$ places the A register contents on the X bus, $E_0 = 1$ places B onto the X bus, $E_1E_2 = 00$ places C onto the Y bus, $E_1E_2 = 01$ places D onto the Y bus, $E_1E_2 = 10$ places 00000000 on the Y bus, and $E_1E_2 = 11$ places 11111111 on the Y bus. You may use a few additional gates. (*Hint:* Connect E_2 to all 8 data inputs on the tri-state buffer on the right side of the circuit.)

(c) Show how to make the bits in the C register be the OR of the corresponding bits in the A register and in the D register, in four clock cycles. Tell what G_0, G_1, E_0, E_1, and E_2 should be for each cycle. (*Hint:* Use DeMorgan's law and $X' = (X \text{ NAND } 1)$.)

12.41 Show how to make the shift register of Figure 12-10 reverse the order of its bits (i.e., $Q_3^+ = Q_0$, $Q_2^+ = Q_1$, $Q_1^+ = Q_2$, and $Q_0^+ = Q_3$).

(a) Use external connections between the Q outputs and the D inputs. What should the values of Sh and L be for a reversal?

(b) Change the internal circuitry to allow bit reversal, so that the D inputs may be used for other purposes. Replace Sh and L with A and B, and let the register operate according to the following table:

Inputs		Next State				
A	B	Q_3^+	Q_2^+	Q_1^+	Q_0^+	Action
0	0	Q_3	Q_2	Q_1	Q_0	No change
0	1	SI	Q_3	Q_2	Q_1	Right shift
1	0	D_3	D_2	D_1	D_0	Load
1	1	Q_0	Q_1	Q_2	Q_3	Reverse bits

UNIT 13

Analysis of Clocked Sequential Circuits

Objectives

1. Analyze a sequential circuit by signal tracing.
2. Given a sequential circuit, write the next-state equations for the flip-flops and derive the state graph or state table. Using the state graph, determine the state sequence and output sequence for a given input sequence.
3. Explain the difference between a Mealy machine and a Moore machine.
4. Given a state table, construct the corresponding state graph, and conversely.
5. Given a sequential circuit or a state table and an input sequence, draw a timing chart for the circuit. Determine the output sequence from the timing chart, neglecting any false outputs.
6. Draw a general model for a clocked Mealy or Moore sequential circuit. Explain the operation of the circuit in terms of these models. Explain why a clock is needed to ensure proper operation of the circuit.

Study Guide

1. Study Section 13.1, *A Sequential Parity Checker.*
   (a) Explain how parity can be used for error detection.

   (b) Verify that the parity checker (Figure 13-4) will produce the output waveform given in Figure 13-2 when the input waveform is as shown.

2. Study Section 13.2, *Analysis by Signal Tracing and Timing Charts.*
   (a) What is the difference between a Mealy machine and a Moore machine?

   (b) For normal operation of clocked sequential circuits of the types discussed in this section, when should the inputs be changed?

   When do the flip-flops change state?

   At what times can the output change for a Moore circuit?

   At what times can the output change for a Mealy circuit?

   (c) At what time (with respect to the clock) should the output of a Mealy circuit be read?

   (d) Why can false outputs appear in a Mealy circuit and not in a Moore circuit?

   What can be done to eliminate the false outputs?

   If the output of a Mealy circuit is used as an input to another Mealy circuit synchronized by the same clock, will false outputs cause any problem? Explain.

   (e) Examine the timing diagram of Figure 13-8. The value of Z will always be correct just before the falling (active) clock edge that causes the state change. Note there are two types of false outputs. A false 0 output occurs if Z is 1 just before two successive falling clock edges, and Z goes to 0 between the clock edges. A false 1 output occurs if Z is 0 just before two successive falling clock edges and Z goes to 1 between the edges. When the output is 0 (or 1) just before an active clock edge and 1 (or 0) just before the next, the output may be temporarily incorrect after the state changes following the first active edge but before the input has changed to its next value. In this case, we will not say that a false output has occurred because the sequence of outputs is still correct.

3. Study Section 13.3, *State Tables and Graphs.*

(a) In Equations (13-1) through (13-5), at what time (with respect to the clock) is the right-hand side evaluated?

What does Q^+ mean?

(b) Derive the timing chart of Figure 13-6 using Table 13-2(a).

(c) What is the difference between the state graphs for Mealy and Moore machines?

(d) For a state table, Table 13-3(b) for example, what do the terms "*present* state," "*next* state," and "*present* output" mean with respect to the active clock edge?

(e) Why does a Moore state table have only one output column?

(f) For ease in making state tables from Karnaugh maps and vice versa, transition tables with three or four states are often written with states in the order 00, 01, 11, 10. However, this is not necessary. (In fact, for sequential circuits with five or more states, it is impossible.) For example, the following transition table is equivalent to Table 13-2(a); the third and fourth rows have been interchanged. It equally represents the circuit of Figure 13-5.

	A^+B^+		
AB	$X = 0$	$X = 1$	Z
00	10	01	0
01	00	11	1
10	11	01	1
11	01	11	0

4. The following timing chart was derived from the circuit of Figure 13-7.

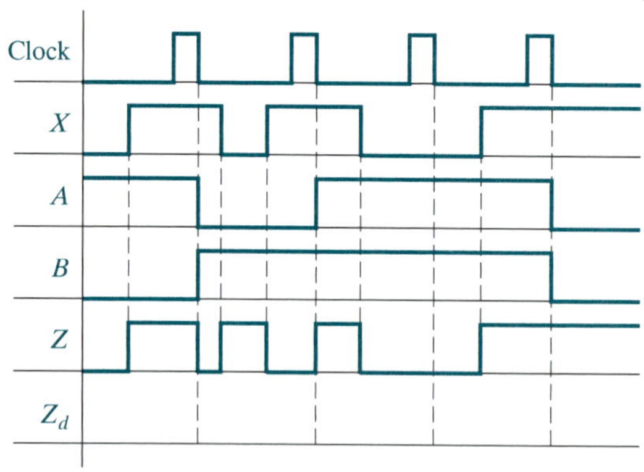

(a) Noting that extra input changes which occur between clock pulses cannot affect the state of the circuit, what is the effective input sequence seen by the flip-flops in the circuit?

(b) Using Table 13-3, verify the waveforms given for A, B, and Z.

(c) Indicate any false outputs. What is the correct output sequence from the circuit?

(d) Using the effective input sequence from (a), determine the output sequence from the state graph (Figure 13-11). This output sequence should be the same as your answer to (c).

(e) The output Z is fed into a clocked D flip-flop, using the same clock (Clock) as the circuit. Sketch the waveform for Z_d. Does Z_d have any false outputs?

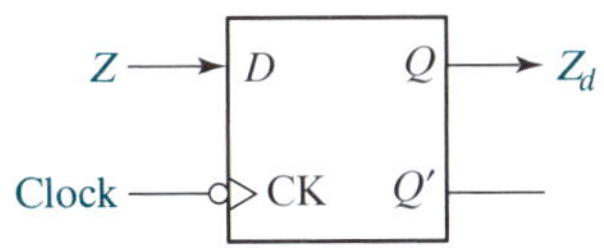

(f) Starting with Figure 13-11, construct the corresponding state table. Verify that your answer is the same as Table 13-3(b). Note that the output label on a given arrow of the graph is associated with the state from which the arrow originates.

(g) Assume that the flip-flops in Figure 13-7 are changed to flip-flops which trigger on the rising edge of the clock; that is, the inversion circles are removed from the clock inputs. Also, the clock waveform in Figure 13-8 is replaced with

The input waveform is left unchanged. What changes, if any, would occur in the remainder of the timing diagram? Explain.

5. Consider the following state tables:

P.S.	Mealy N.S. $X = 0$	1	Z $X = 0$	1
S_0	S_1	S_0	0	0
S_1	S_0	S_2	1	0
S_2	S_0	S_0	1	0

P.S.	Moore N.S. $X = 0$	1	Z
S_0	S_1	S_0	0
S_1	S_3	S_2	0
S_2	S_3	S_0	0
S_3	S_3	S_0	1

(a) Draw the corresponding state graphs.

(b) Show that the same output sequence is obtained from both state graphs when the input sequence is 010 (ignore the initial output for the Moore circuit).

(c) Using the state tables, complete the following timing diagrams for the two circuits. Note that the Mealy circuit has a false output, but the Moore does not. Also note that the output from the Moore circuit is delayed with respect to the Mealy.

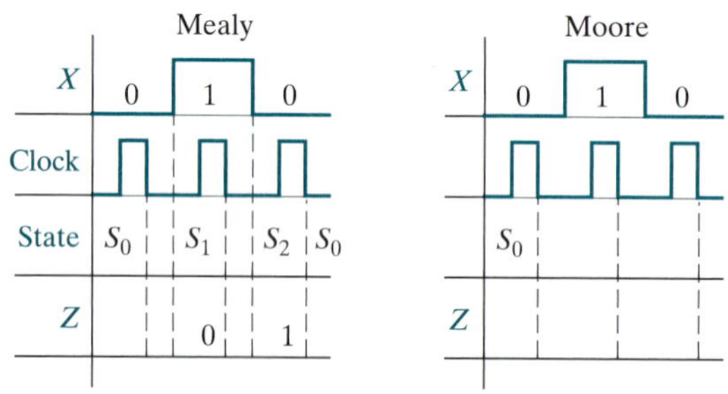

(d) Work Programmed Exercise 13.1.
(e) Work Problems 13.2 and 13.3.

6. Study Section 13.4, *General Models for Sequential Circuits.*

(a) A Mealy sequential circuit has the form shown. The combinational circuit realizes the following equations:

$$Q_1^+ = X_1'Q_1 + X_1Q_1'Q_2' \qquad Z_1 = X_1Q_1$$
$$Q_2^+ = X_1Q_2' + X_2'Q_1 \qquad Z_2 = X_1'Q_1 + X_2Q_2'$$

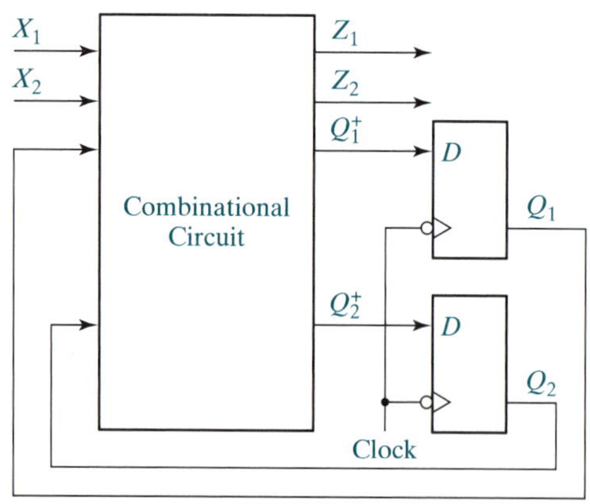

Initially, $X_1 = X_2 = 1$ and $Q_1 = Q_2 = 0$ as shown.

(1) Before the falling clock edge, show the values of the four combinational circuit outputs on the preceding diagram and on the following timing chart.
(2) Show the signal values on the circuit and timing chart immediately after the falling edge.
(3) Show any further changes in signal values which will occur after the new values of Q_1 and Q_2 have propagated through the circuit.
(4) Next change X_1 to 0 and repeat steps (1), (2), and (3). Show the values for each step on the circuit and on the timing chart.
(5) Next change the inputs to $X_1 = 1$ and $X_2 = 0$ and repeat steps (1), (2), and (3).
(6) Change X_2 to 1 and repeat.

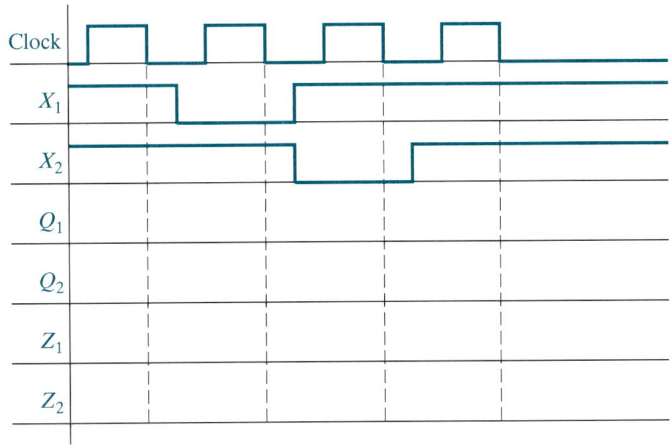

(b) Draw a block diagram for a general model of a Mealy circuit, using J-K flip-flops as memory elements. If the circuit has n output variables and k flip-flops, how many outputs will the combinational subcircuit have?
(c) If the circuit of Figure 13-5 were not synchronized using a clock, but instead, the flip-flops were updated continuously, and if the XOR gate had a longer delay than the OR gate, what problem could appear?

(d) The minimum clock period for a Moore circuit is determined the same way as for a Mealy circuit. Should t_c be determined by the combinational subcircuit for the flip-flop inputs or for the outputs? (See Figure 13-19.)

(e) We can think of the binary counter of Figure 12-15 as a Moore circuit if we say the outputs Z_1, Z_2, and Z_3 are $Z_1 = A$, $Z_2 = B$, and $Z_3 = C$. (The combinational subcircuit for outputs has no gates, but that is okay.) If the XOR gates have a propagation delay of 4 ns and the AND gate has a propagation delay of 2 ns, what is the longest total propagation delay through the combinational subcircuit for flip-flops (i.e., the XOR gates and the AND gate) to the D inputs of the flip-flops? If the flip-flops have $t_{su} = 3$ ns and $t_p = 3$ ns, what is the minimum clock period for the binary counter?

(f) In Equations (13-6) and (13-7), what do the symbols δ and λ mean?

Equation (13-7) is for a Mealy circuit. What is the corresponding equation for a Moore circuit?

(g) For Table 13-5,

$\delta(S_3, 1) =$ ______________ $\lambda(S_3, 1) =$ ______________

$\delta(S_1, 2) =$ ______________ $\lambda(S_1, 2) =$ ______________

7. Work Problems 13.4 through 13.6.

8. When you are satisfied that you can meet the objectives, take the readiness test.

Analysis of Clocked Sequential Circuits

The sequential circuits which we discussed in Chapter 12 perform simple functions such as shifting or counting. The counters we designed go through a fixed sequence of states and have no inputs other than a clock pulse that causes the state to change. We will now consider sequential circuits that have additional inputs. In general, the sequence of outputs and the sequence of flip-flop states for such circuits will depend on the input sequence which is applied to the circuit. Given a sequential circuit and an input sequence, we can analyze the circuit to determine the flip-flop state sequence and the output sequence by tracing the 0 and 1 signals through the circuit. Although signal tracing may be adequate for small circuits, for larger circuits it is better to construct a state graph or state table which represents the behavior of the circuit. Then, we can determine the output and state sequences from the graph or table. Such graphs and tables are also useful for the design of sequential circuits.

In this chapter we will also study the timing relationships between the inputs, the clock, and the outputs for sequential circuits by constructing timing diagrams. These timing relationships are very important when a sequential circuit is used as part of a larger digital system. After analyzing several specific sequential circuits, we will discuss a general model for a sequential circuit which consists of a combinational circuit together with flip-flops that serve as memory.

13.1 A Sequential Parity Checker

When binary data is transmitted or stored, an extra bit (called a parity bit) is frequently added for purposes of error detection. For example, if data is being transmitted in groups of 7 bits, an eighth bit can be added to each group of 7 bits to make the total number of 1's in each block of 8 bits an odd number. When the total number of 1 bits in the block (including the parity bit) is odd, we say that the parity is odd. Alternately, the parity bit could be chosen such that the total number of 1's in the block is even, in which case we would have even parity. Some examples of 8-bit words with odd parity are

7 Data Bits	Parity Bits
0 0 0 0 0 0 0	1
0 0 0 0 0 0 1	0
0 1 1 0 1 1 0	1
1 0 1 0 1 0 1	1
0 1 1 1 0 0 0	0

8-Bit Word

If any single bit in the 8-bit word is changed from 0 to 1 or from 1 to 0, the parity is no longer odd. Thus, if any single bit error occurs in transmission of a word with odd parity, the presence of this error can be detected because the number of 1 bits in the word has been changed from odd to even.

As a simple example of a sequential circuit which has one input in addition to the clock, we will design a parity checker for serial data. (Serial implies that the data enters the circuit sequentially, one bit at a time.) This circuit has the form shown in Figure 13-1. When a sequence of 0's and 1's is applied to the X input, the output of

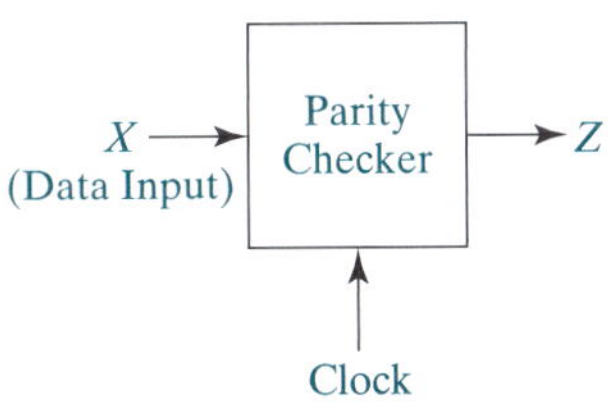

FIGURE 13-1
Block Diagram for Parity Checker

the circuit should be $Z = 1$ if the total number of 1 inputs received is odd; that is, the output should be 1 if the input parity is odd. Thus, if data which originally had odd parity is transmitted to the circuit, a final output of $Z = 0$ indicates that an error in transmission has occurred.

The value of X is read at the time of the active clock edge. The X input must be synchronized with the clock so that it assumes its next value before the next active clock edge. The clock input is necessary in order to distinguish consecutive 0's or consecutive 1's on the X input. Typical input and output waveforms are shown in Figure 13-2.

We will start the design by constructing a state graph (Figure 13-3). The sequential circuit must "remember" whether the total number of 1 inputs received is even or odd; therefore, only two states are required. We will designate these states as S_0 and S_1, corresponding respectively to an even number of 1's received and an odd number of 1's received. We will start the circuit in state S_0 because initially zero 1's have been received, and zero is an even number. As indicated in Figure 13-3, if the circuit is in state S_0 (even number of 1's received) and $X = 0$ is received, the circuit must stay in S_0 because the number of 1's received is still even. However, if $X = 1$ is received, the circuit goes to state S_1 because the number of 1's received is then odd. Similarly, if the circuit is in state S_1 (odd number of 1's received) a 0 input causes no state change, but a 1 causes a change to S_0 because the number of 1's received is then even. The output Z should be 1 whenever the circuit is in state S_1 (odd number of 1's received). The output is listed below the state on the state graph.

Table 13-1(a) gives the same information as the state graph in tabular form. For example, the table shows that if the present state is S_0, the output is $Z = 0$, and if the input is $X = 1$, the next state will be S_1.

Because only two states are required, a single flip-flop (Q) will suffice. We will let $Q = 0$ correspond to state S_0 and $Q = 1$ correspond to S_1. We can then set up a table which shows the next state of flip-flop Q as a function of the present state and X. If we use a T flip-flop, T must be 1 whenever Q and Q^+ differ. From Table 13-1(b), the T input must be 1 whenever $X = 1$. Figure 13-4 shows the resulting circuit.

Figure 13-2 shows the output waveform for the circuit. When $X = 1$, the flip-flop changes state after the falling edge of the clock. Note that the final value of Z is 0 because an even number of 1's was received. If the number of 1's received had been odd, the final value of Z would be l. In this case, it would be necessary to reset the

FIGURE 13-2 Waveforms for Parity Checker

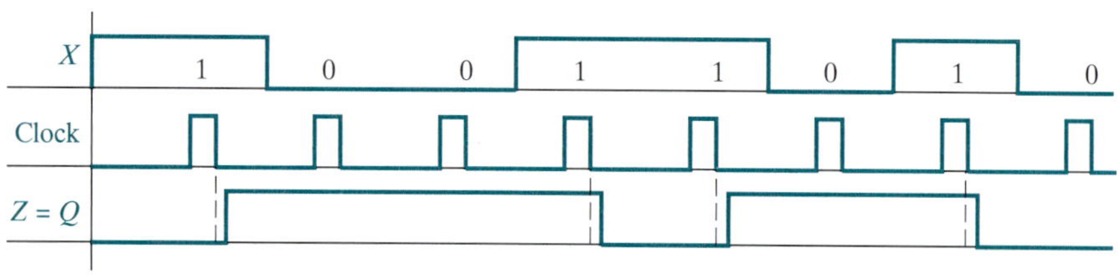

FIGURE 13-3 State Graph for Parity Checker

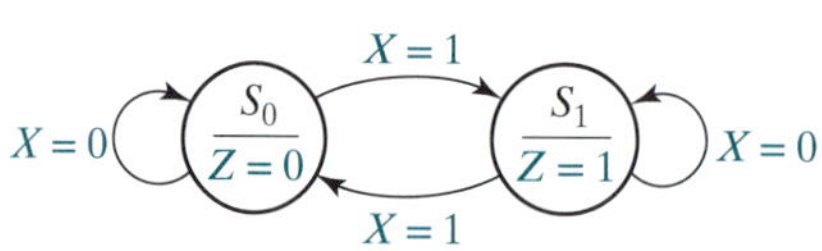

TABLE 13-1 State and Transition Tables for Parity Checker

(a)

Present State	Next State $X = 0$	Next State $X = 1$	Present Output
S_0	S_0	S_1	0
S_1	S_1	S_0	1

(b)

Q	Q^+ $X = 0$	Q^+ $X = 1$	T $X = 0$	T $X = 1$	Z
0	0	1	0	1	0
1	1	0	0	1	1

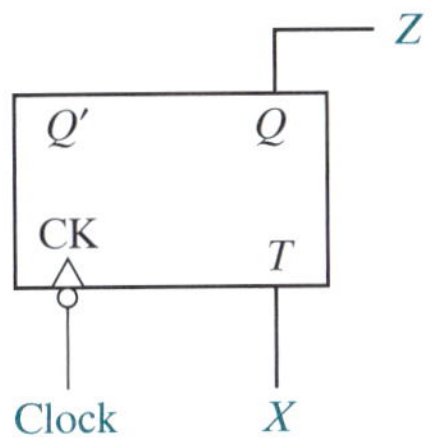

FIGURE 13-4 Parity Checker

flip-flop to the proper initial state ($Q = 0$) before checking the parity of another input sequence.

13.2 Analysis by Signal Tracing and Timing Charts

In this section we will analyze clocked sequential circuits to find the output sequence resulting from a given input sequence by tracing 0 and 1 signals through the circuit. The basic procedure is as follows:

1. Assume an initial state of the flip-flops (all flip-flops reset to 0 unless otherwise specified).
2. For the first input in the given sequence, determine the circuit output(s) and flip-flop inputs.
3. Determine the new set of flip-flop states after the next active clock edge.
4. Determine the output(s) that corresponds to the new states.
5. Repeat 2, 3, and 4 for each input in the given sequence.

As we carry out the analysis, we will construct a timing chart which shows the relationship between the input signal, the clock, the flip-flop states, and the circuit output. We have already seen how to construct timing charts for flip-flops (Unit 11) and counters (Unit 12).

In this unit we will use edge-triggered flip-flops that change state shortly after the active edge (rising or falling edge) of the clock. We will assume that the flip-flop inputs are stable a sufficient time before and after the active clock edge so that setup and hold time requirements are met. When the state of the sequential circuit changes, the change will always occur in response to the active clock edge. The circuit output may change at the time the flip-flops change state or at the time the input changes depending on the type of circuit.

Two types of clocked sequential circuits will be considered—those in which the output depends only on the present state of the flip-flops and those in which the output depends on both the present state of the flip-flops and on the value of the circuit inputs. If the output of a sequential circuit is a function of the present state only (as in Figures 13-4 and 13-5), the circuit is often referred to as a *Moore machine*. The state graph for a Moore machine has the output associated with the state (as in Figures 13-3 and 13-9). If the output is a function of both the present state and the input (as in Figure 13-7), the circuit is referred to as a *Mealy machine*. The state graph for a Mealy machine has the output associated with the arrow going between states (as in Figure 13-11).

As an example of a Moore circuit, we will analyze Figure 13-5 using an input sequence $X = 01101$. In this circuit, the initial state is $A = B = 0$, and all state changes occur after the rising edge of the clock, as shown in Figure 13-6. The X input is synchronized with the clock so that it assumes its next value after each rising edge. Because Z is a function only of the present state (in this case, $Z = A \oplus B$) the output will only change when the state changes. Initially, $X = 0$, so $D_A = 1$ and $D_B = 0$, and the state will change to $A = 1$ and $B = 0$ after the first rising clock edge. Then X is changed to 1, so $D_A = 0, D_B = 1$ and the state changes to $AB = 01$ after the second rising clock edge. After the state change, X remains 1, so $D_A = D_B = 1$, and the next rising edge causes the state to change to 11. When X changes to 0, $D_A = 0$ and $D_B = 1$, and the state changes to $AB = 01$ on the fourth rising edge. Then, with $X = 1, D_A = D_B = 1$, so the

FIGURE 13-5 Moore Sequential Circuit to Be Analyzed

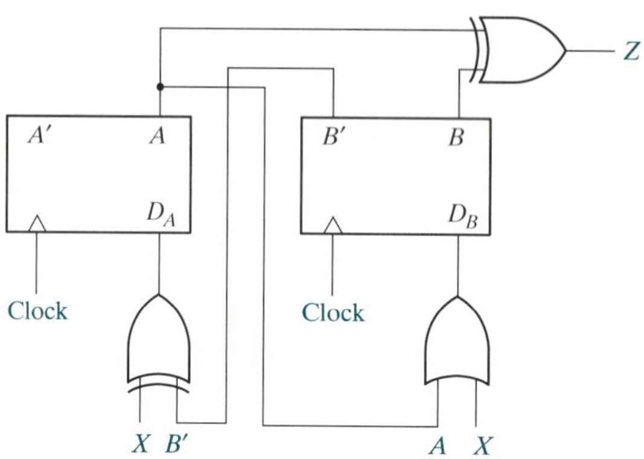

FIGURE 13-6 Timing Chart for Figure 13-5

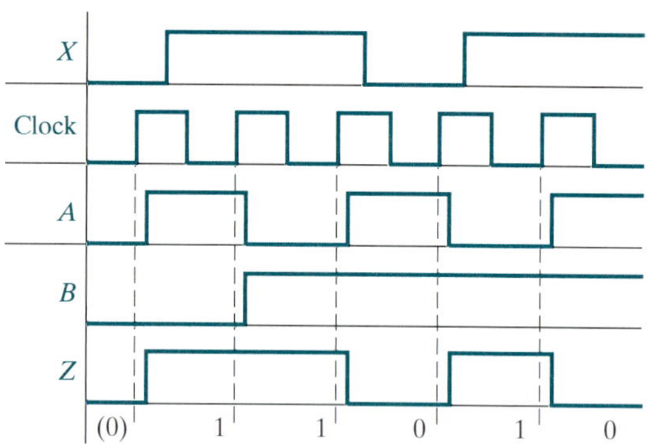

fifth rising clock edge causes the state to change to $AB = 11$. The input, state, and output sequences are plotted on the timing chart of Figure 13-6 and are also listed below.

$X =$	0	1	1	0	1	
$A =$	0	1	0	1	0	1
$B =$	0	0	1	1	1	1
$Z =$	(0)	1	1	0	1	0

When the circuit is reset to its initial state ($A = B = 0$), the initial output is $Z = 0$. Because this initial 0 is not in response to any X input, it should be ignored. The resulting output sequence is $Z = 11010$. Note that for the *Moore* circuit, the output which results from application of a given input does not appear until after the active clock edge; therefore, the output sequence is displaced in time with respect to the input sequence.

As an example of a Mealy circuit, we will analyze Figure 13-7 and construct a timing chart using the input sequence $X = 10101$. The input is synchronized with the clock so that input changes occur after the falling edge, as shown in Figure 13-8. In this example, the output depends on both the input (X) and the flip-flop states (A and B), so Z may change either when the input changes or when the flip-flops change state. Initially, assume that the flip-flop states are $A = 0$, $B = 0$. If $X = 1$, the output is $Z = 1$ and $J_B = K_A = 1$. After the falling edge of the first clock pulse, B changes to 1 so Z changes to 0. If the input is changed to $X = 0$, Z will change back to 1. All flip-flop inputs are then 0, so no state change occurs with the second falling edge. When X is changed to 1, Z becomes 0 and $J_A = K_A = J_B = 1$. A changes to 1 on the third falling clock edge, at which time Z changes to 1. Next, X is changed to 0 so Z becomes 0, and no state change occurs with the fourth clock pulse. Then, X is changed to 1, and Z becomes 1. Because $J_A = K_A = J_B = K_B = 1$, the fifth clock pulse returns the circuit to the initial state. The input, state, and output sequences are plotted on the timing chart of Figure 13-8 and are also listed below.

$X =$	1	0	1	0	1	
$A =$	0	0	0	1	1	0
$B =$	0	1	1	1	1	0
$Z =$	1(0)	1	0(1)	0	1	

(False outputs are indicated in parentheses.)

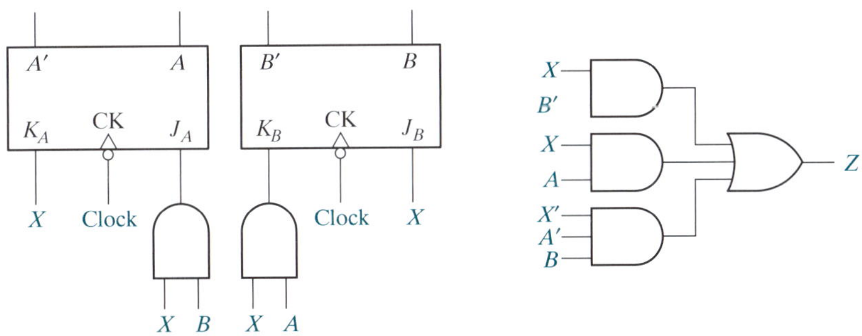

FIGURE 13-7 Mealy Sequential Circuit to Be Analyzed

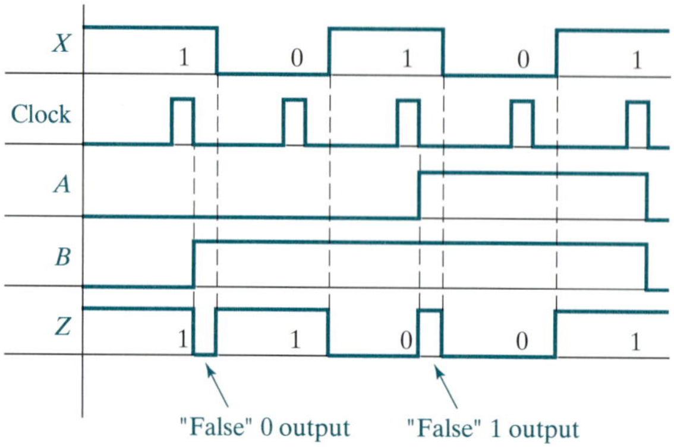

FIGURE 13-8 Timing Chart for Circuit of Figure 13-7

A careful interpretation of the output waveform (Z) of the Mealy circuit is necessary. After the circuit has changed state and before the input is changed, the output may temporarily assume an incorrect value, which we call a *false* output. As indicated on the timing chart, this false value arises when the circuit has assumed a new state but the old input associated with the previous state is still present.

For a clocked sequential circuit, the value of the input immediately preceding the active clock edge determines the next state of the flip-flops. Extra input changes which might occur between active clock edges do not affect the state of the flip-flops. In a similar manner, the output from a Mealy circuit is only of interest immediately preceding the active clock edge, and extra output changes (false outputs) which might occur between active clock edges should be ignored.

Two types of false outputs can occur, as indicated in Figure 13-8. In one case the output Z momentarily goes to 0 and returns to 1 before the active clock edge. In the other case, the output Z momentarily goes to 1 and returns to 0 before the active edge. These false outputs are often referred to as *glitches* and *spikes*. In both cases, two changes of output occur when no change is expected. Ignoring the false outputs by reading the output just before the falling clock edge, the output sequence for the circuit is $Z = 11001$. If circuit delays are negligible, the false outputs could be eliminated if the input X was allowed to change only at the same time as the falling edge of the clock. If the output of the circuit is fed into a second sequential circuit which uses the same clock, the false outputs will not cause any problem because the inputs to the second circuit can cause a change of state only when a falling clock edge occurs. Because the output of a Moore circuit can change state only when the flip-flops change state and not when the input changes, no false outputs can appear in a Moore circuit.

For the *Mealy* circuit, the output which corresponds to a given input appears shortly after the application of that input. Because the correct output appears before the active clock edge, the output sequence is *not* displaced in time with respect to the input sequence as was the case for the Moore circuit. There are a few examples of Moore circuits (i.e., a circuit whose outputs only depend on the present state) whose outputs are not delayed with respect to the input sequence. (See Problem 13.17 for an example.)

13.3 State Tables and Graphs

In the previous section we analyzed clocked sequential circuits by signal tracing and the construction of timing charts. Although this is satisfactory for small circuits and short input sequences, the construction of state tables and graphs provides a more systematic approach which is useful for the analysis of larger circuits and which leads to a general synthesis procedure for sequential circuits.

The transition table and the state table for a sequential circuit specify the next state and output of the circuit in terms of its present state and input. A state in a transition table is specified by the flip-flop output values corresponding to the state. A state in a state table is specified symbolically. When analyzing a sequential circuit, the transition table is derived from the circuit as described below. Given a transition table, a state table for the circuit is derived by assigning symbols to the states. The design process for a sequential circuit reverses this process; usually a state table is determined first; then a transition table is derived by assigning binary combinations to the states.

The following method can be used to construct the transition table:

1. Determine the flip-flop input equations and the output equations from the circuit.
2. Derive the next-state equation for each flip-flop from its input equations, using one of the following relations:

D flip-flop	$Q^+ = D$	(13-1)
D-CE flip-flop	$Q^+ = D{\cdot}CE + Q{\cdot}CE'$	(13-2)
T flip-flop	$Q^+ = T \oplus Q$	(13-3)
S-R flip-flop	$Q^+ = S + R'Q$	(13-4)
J-K flip-flop	$Q^+ = JQ' + K'Q$	(13-5)

3. Plot a next-state map for each flip-flop.
4. Combine these maps to form the transition table. Such a transition table, which gives the next state of the flip-flops as a function of their present state and the circuit inputs.

As an example of this procedure, we will derive the transition table for the circuit of Figure 13-5:

1. The flip-flop input equations and output equation are

$$D_A = X \oplus B' \quad D_B = X + A \quad Z = A \oplus B$$

2. The next-state equations for the flip-flops are

$$A^+ = X \oplus B' \qquad B^+ = X + A$$

3. The corresponding maps are

AB \ X	0	1
00	**1**	**0**
01	**0**	**1**
11	**0**	**1**
10	**1**	**0**

A^+

AB \ X	0	1
00	**0**	**1**
01	**0**	**1**
11	**1**	**1**
10	**1**	**1**

B^+

TABLE 13-2
Moore Transition and State Tables for Figure 13-5

(a)

AB	A^+B^+ $X = 0$	$X = 1$	Z
00	10	01	0
01	00	11	1
11	01	11	0
10	11	01	1

(b)

Present State	Next State $X = 0$	$X = 1$	Present Output (Z)
S_0	S_3	S_1	0
S_1	S_0	S_2	1
S_2	S_1	S_2	0
S_3	S_2	S_1	1

4. Combining these maps yields the transition table in Table 13-2(a), which gives the next state of both flip-flops (A^+B^+) as a function of the present state and input. The output function Z is then added to the table. In this example, the output depends only on the present state of the flip-flops and not on the input, so only a single output column is required.

Using Table 13-2(a), we can construct the timing chart of Figure 13-6 or any other timing chart for some given input sequence and specified initial state. Initially $AB = 00$ and $X = 0$, so $Z = 0$ and $A^+B^+ = 10$. This means that *after* the rising clock edge, the flip-flop state will be $AB = 10$. Then, with $AB = 10$, the output is $Z = 1$. The next input is $X = 1$, so $A^+B^+ = 01$ and the state will change after the next rising clock edge. Continuing in this manner, we can complete the timing chart.

If we are not interested in the flip-flop output values corresponding to the states, we can replace each combination of flip-flop states with a single symbol which represents the state of the circuit. Replacing 00 with S_0, 01 with S_1, 11 with S_2, and 10 with S_3 in Table 13-2(a) yields the state table shown in Table 13-2(b). The Z column is labeled Present Output because it is the output associated with the Present State. The state graph of Figure 13-9 represents Table 13-2(b). Each node of the graph represents a state of the circuit, and the corresponding output is placed in the circle below the state symbol. The arc joining two nodes is labeled with the value of X which will cause a state change between these nodes. Thus, if the circuit is in state S_0 and $X = 1$, a clock edge will cause a transition to state S_1.

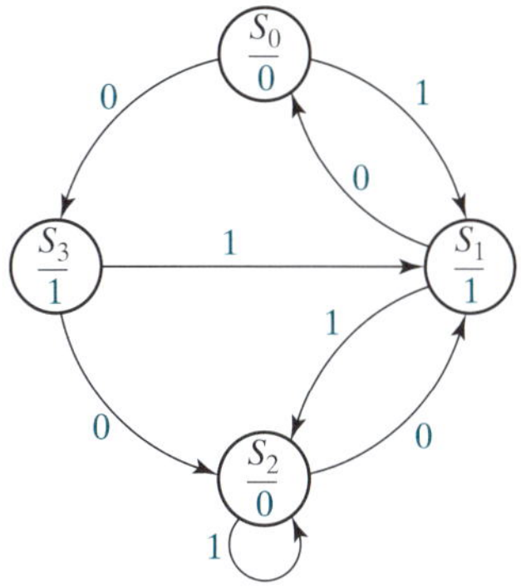

FIGURE 13-9 Moore State Graph for Figure 13-5

Next, we will construct the state table and graph for the Mealy machine of Figure 13-7. The next-state and output equations are

$$A^+ = J_A A' + K_A' A = XBA' + X'A$$
$$B^+ = J_B B' + K_B' B = XB' + (AX)'B = XB' + X'B + A'B$$
$$Z = X'A'B + XB' + XA$$

The next-state and output maps (Figure 13-10) combine to form the transition table in Table 13-3(a). Given values for A, B, and X, the current value of the output is determined from the Z column of this table, and the states of the flip-flops after the active clock edge are determined from the $A^+ \ B^+$ columns.

We can construct the timing chart of Figure 13-8 using Table 13-3(a). Initially with $A = B = 0$ and $X = 1$, the table shows that $Z = 1$ and $A^+B^+ = 01$. Therefore, after the falling clock edge, the state of flip-flop B will change to 1, as indicated in Figure 13-8. Now, from the 01 row of the table, if X is still 1, the output will be 0 until the input is changed to $X = 0$. Then, the output is $Z = 1$, and the next falling clock edge produces no state change. Finish stepping through the state table in this manner and verify that A, B, and Z are as given in Figure 13-8.

If we let $AB = 00$ correspond to circuit state S_0, 01 to S_1, 11 to S_2, and 10 to S_3, we can construct the state table in Table 13-3(b) and the state graph of Figure 13-11. In Table 13-3(b), the Present Output column gives the output associated with the

FIGURE 13-10

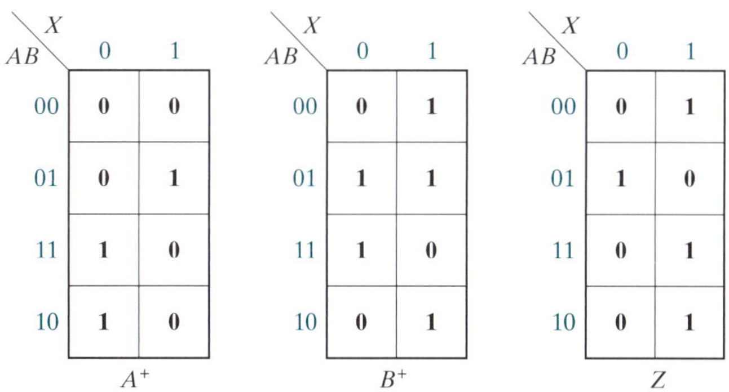

TABLE 13-3 Mealy Transition and State Tables for Figure 13-7

(a)

AB	A^+B^+ $X = 0$	1	Z $X = 0$	1
00	00	01	0	1
01	01	11	1	0
11	11	00	0	1
10	10	01	0	1

(b)

Present State	Next State $X = 0$	1	Present Output $X = 0$	1
S_0	S_0	S_1	0	1
S_1	S_1	S_2	1	0
S_2	S_2	S_0	0	1
S_3	S_3	S_1	0	1

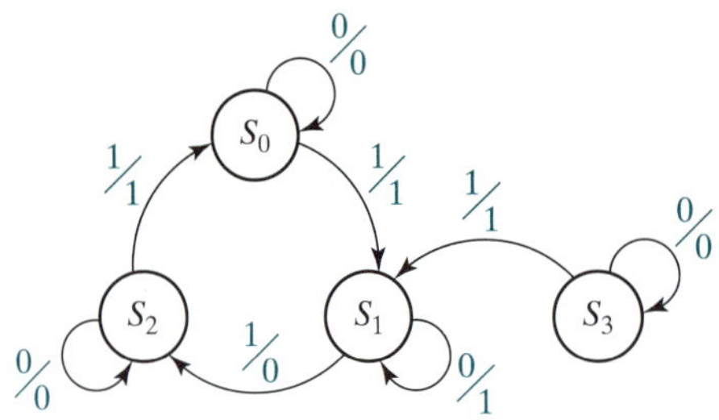

FIGURE 13-11 Mealy State Graph for Figure 13-7

present state and present input. Thus, if the present state is S_0 and the input changes from 0 to 1, the output will immediately change from 0 to 1. However, the state will not change to the next state (S_1) until after the clock pulse. For Figure 13-11, the labels on the arrows between states are of the form X/Z, where the symbol before the slash is the input and the symbol after the slash is the corresponding output. Thus, in state S_0 an input of 0 gives an output of 0, and an input of 1 gives an output of 1. For any given input sequence, we can easily trace out the state and output sequences on the state graph. For the input sequence $X = 10101$, verify that the corresponding output sequence is 11001. This agrees with Figure 13-8 if the false outputs are ignored. Note that the false outputs do not show on the state graph because the inputs are read at the active clock edge, and no provision is made for extra input changes between active edges.

Next, we will analyze the operation of a serial adder (Figure 13-12(a)) that adds two n-bit binary numbers $X = x_{n-1} \ldots x_1x_0$ and $Y = y_{n-1} \ldots y_1y_0$. The operation of the serial adder is similar to the parallel adder of Figure 4-2 except that the binary numbers are fed in serially, one pair of bits at a time, and the sum is read out serially, one bit at a time. First, x_0 and y_0 are fed in; a sum digit s_0 is generated, and the carry c_1 is stored. At the next clock time, x_1 and y_1 are fed in and added to c_1 to give the next sum digit s_1 and the new carry c_2, which is stored. This process continues until all bits have been added. A full adder is used to add the x_i, y_i and c_i bits to form c_{i+1} and s_i. A D flip-flop is used to store the carry (c_{i+1}) on the rising edge of the clock. The x_i and y_i inputs must be synchronized with the clock.

Figure 13-13 shows a timing diagram for the serial adder. In this example we add 10011 + 00110 to give a sum of 11001 and a final carry of 0. Initially the carry flip-flop must be cleared so that $c_0 = 0$. We start by adding the least-significant (rightmost) bits in each word. Adding $1 + 0 + 0$ gives $s_0 = 1$ and $c_1 = 0$, which is stored in the flip-flop at the rising clock edge. Because y_1 is 1, adding $1 + 1 + 0$ gives $s_1 = 0$

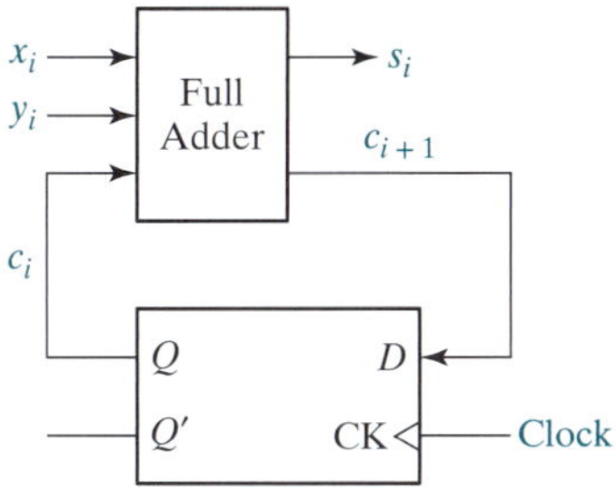

(a) With D flip-flop

x_i	y_i	c_i	c_{i+1}	s_i
0	0	0	0	0
0	0	1	0	1
0	1	0	0	1
0	1	1	1	0
1	0	0	0	1
1	0	1	1	0
1	1	0	1	0
1	1	1	1	1

(b) Truth table

FIGURE 13-12 Serial Adder

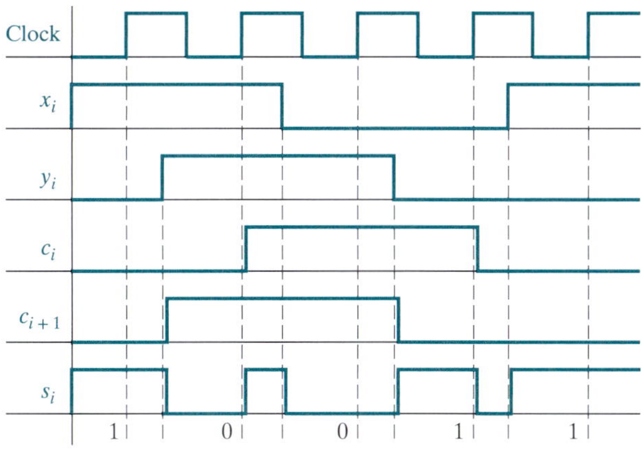

FIGURE 13-13 Timing Diagram for Serial Adder

and $c_2 = 1$, which is stored in the flip-flop on the rising clock edge. This process continues until the addition is completed. Reading the sum output just before the rising edge of the clock gives the correct result.

The truth table for the full adder (Table 4-4) is repeated in Figure 13-12(b) in modified form. Using this table, we can construct a state graph (Figure 13-14) for the serial adder. The serial adder is a Mealy machine with inputs x_i and y_i and output s_i. The two states represent a carry (c_i) of 0 and 1, respectively. From the table, c_i is the present state of the sequential circuit, and c_{i+1} is the next state. If we start in S_0 (no carry), and $x_i\,y_i = 11$, the output is $s_i = 0$ and the next state is S_1. This is indicated by the arrow going from state S_0 to S_1.

Table 13-4 shows a state table for a Mealy sequential circuit with two inputs and two outputs. Figure 13-15 shows the corresponding state graph. The notation 00, 01/00 on the arc from S_3 to S_2 means if $X_1 = X_2 = 0$ or $X_1 = 0$ and $X_2 = 1$, then $Z_1 - 0$ and $Z_2 - 0$.

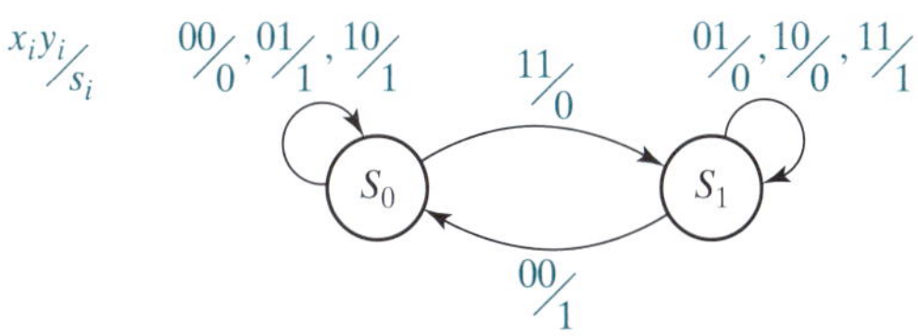

FIGURE 13-14 State Graph for Serial Adder

TABLE 13-4
A State Table with Multiple Inputs and Outputs

Present State	Next State $X_1X_2 = 00$	01	10	11	Present Output (Z_1Z_2) $X_1X_2 = 00$	01	10	11
S_0	S_3	S_2	S_1	S_0	00	10	11	01
S_1	S_0	S_1	S_2	S_3	10	10	11	11
S_2	S_3	S_0	S_1	S_1	00	10	11	01
S_3	S_2	S_2	S_1	S_0	00	00	01	01

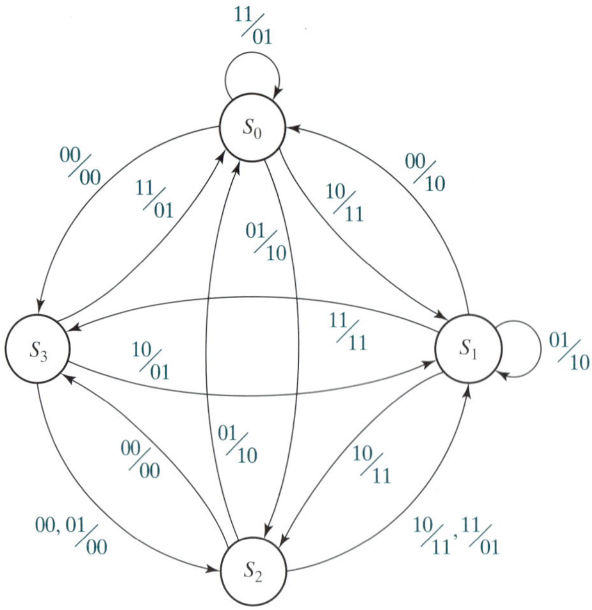

FIGURE 13-15
State Graph for Table 13-4

Construction and Interpretation of Timing Charts

Several important points concerning the construction and interpretation of timing charts are summarized as follows:

1. When constructing timing charts, note that a state change can only occur after the rising (or falling) edge of the clock, depending on the type of flip-flop used.
2. The input will normally be stable immediately before and after the active clock edge.
3. For a Moore circuit, the output can change only when the state changes, but for a Mealy circuit, the output can change when the input changes as well as when the state changes. A false output may occur between the time the state changes and the time the input is changed to its new value. (In other words, if the state has changed to its next value, but the old input is still present, the output may be temporarily incorrect.)
4. False outputs are difficult to determine from the state graph, so use either signal tracing through the circuit or use the state table when constructing timing charts for Mealy circuits.

5. When using a Mealy state table for constructing timing charts, the procedure is as follows:
   (a) For the first input, read the present output and plot it.
   (b) Read the next state and plot it (following the active edge of the clock pulse).
   (c) Go to the row in the table which corresponds to the next state and read the output under the old input column and plot it. (This may be a false output.)
   (d) Change to the next input and repeat steps (a), (b), and (c).
   (Note: If you are just trying to read the correct output sequence from the table, step (c) is naturally omitted.)
6. For Mealy circuits, the best time to read the output is just before the active edge of the clock, because the input(s) must be stable at that time and the output will be correct.

The example in Figure 13-16 shows a state graph, a state table, a circuit that implements the table, and a timing chart. When the state is S_0 and the input is $X = 0$, the output from the state graph, state table, circuit, and timing chart is $Z = 1$ (labeled A on the figure). Note that this output occurs *before* the rising edge of the clock. In a Mealy circuit, the output is a function of the present state and input; therefore, the output should be read just before the clock edge that causes the state to change.

FIGURE 13-16

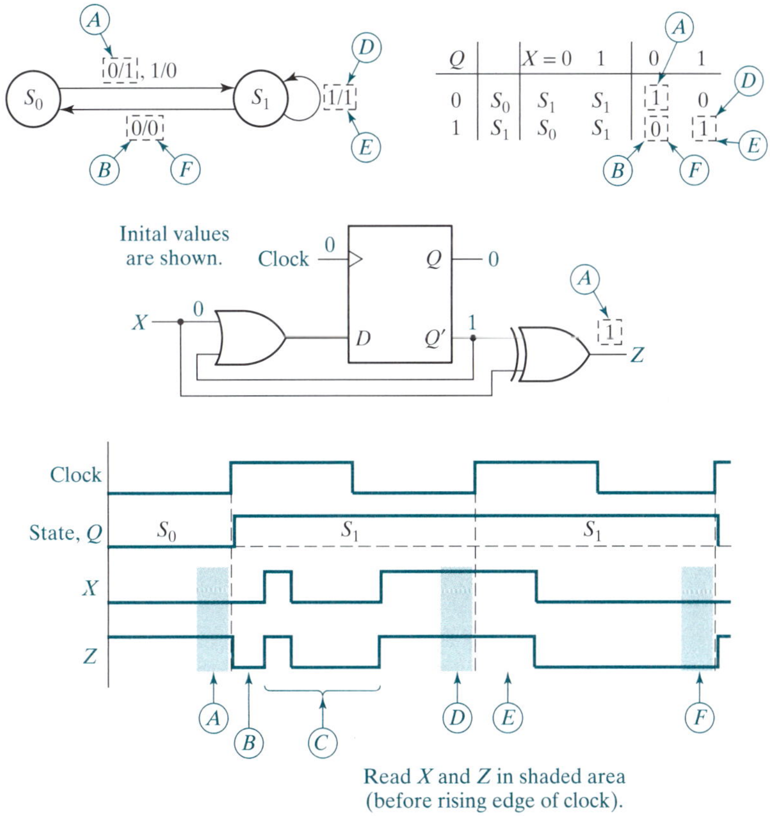

As you continue to study this example, each time the input X changes, trace the changes on the state graph, the state table, the circuit, and the timing chart. Because the input X was 0 before the first rising edge of the clock, the state changes to S_1 after the first rising edge of the clock. Because of the state change, the output also changes (B on the timing chart), but because the input has not yet changed to its new value, the output value may not be correct. We refer to this as a *false output* or *glitch*. If the input changes several times before it assumes its correct value, the output may also change several times (C). The input must assume its correct value before the rising edge of the clock, and the output should be read at this time (D). After the rising clock edge, the state stays the same and the output stays the same for this particular example. In general, the state may change after a rising edge of the clock, and the state change may result in an output change. Again, the output value may be wrong because the input still has the old value (E). When the input is changed to its new value, the output changes to its new value (F), and this value should be read before the next rising clock edge.

If we look at the input and output just before each rising edge of the clock, we find the following sequences:

$$X = 0 \ 1 \ 0$$

$$Z = 1 \ 1 \ 0$$

You should be able to verify the sequence for Z using the state graph, using the state table, and using the circuit diagram.

The synthesis procedure for sequential circuits, discussed in detail in Units 14 through 16, is just the opposite of the procedure used for analysis. Starting with the specifications for the sequential circuit to be synthesized, a state graph is constructed. This graph is then translated to a state table, and the flip-flop output values are assigned for each state. The flip-flop input equations are then derived, and finally, the logic diagram for the circuit is drawn. For example, to synthesize the circuit in Figure 13-7, we would start with the state graph of Figure 13-11. Then, we would derive the state table in Table 13-3(b), the transition table in Table 13-3(a), the next-state and output equations, and, finally, the circuit of Figure 13-7.

13.4 General Models for Sequential Circuits

A sequential circuit can be divided conveniently into two parts—the flip-flops which serve as memory for the circuit and the combinational logic which realizes the input functions for the flip-flops and the output functions. The combinational logic may be implemented with gates, with a ROM, or with a PLA. Figure 13-17 illustrates the general model for a clocked Mealy sequential circuit with m inputs, n outputs, and k clocked D flip-flops used as memory. Drawing the model in this form emphasizes the presence of feedback in the sequential circuit because the flip-flop outputs are fed back as inputs to the combinational subcircuit. The combinational subcircuit realizes

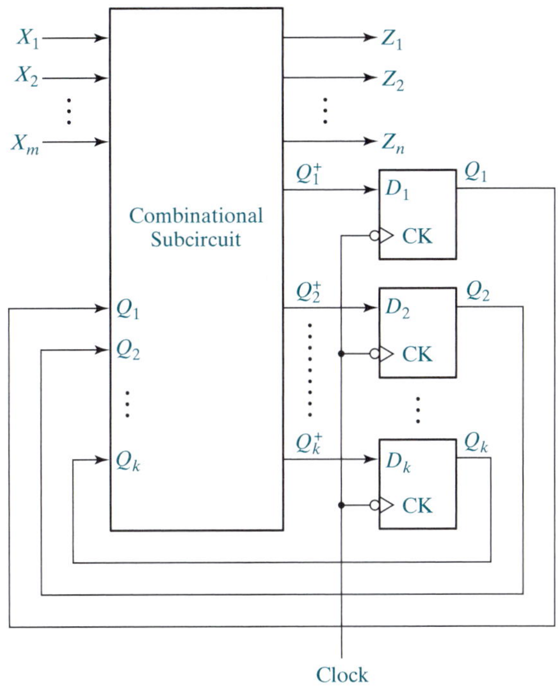

FIGURE 13-17 General Model for Mealy Circuit Using Clocked D Flip-Flops

the n output functions and the k next-state functions, which serve as inputs to the D flip-flops:

$$\left.\begin{aligned} Z_1 &= f_1(X_1, X_2, \ldots, X_m, Q_1, Q_2, \ldots, Q_k) \\ Z_2 &= f_2(X_1, X_2, \ldots, X_m, Q_1, Q_2, \ldots, Q_k) \\ &\vdots \\ Z_n &= f_n(X_1, X_2, \ldots, X_m, Q_1, Q_2, \ldots, Q_k) \end{aligned}\right\} \; n \text{ output functions}$$

$$\left.\begin{aligned} Q_1^+ &= D_1 = g_1(X_1, X_2, \ldots, X_m, Q_1, Q_2, \ldots, Q_k) \\ Q_2^+ &= D_2 = g_2(X_1, X_2, \ldots, X_m, Q_1, Q_2, \ldots, Q_k) \\ &\vdots \\ Q_k^+ &= D_k = g_k(X_1, X_2, \ldots, X_m, Q_1, Q_2, \ldots, Q_k) \end{aligned}\right\} \; k \text{ next-state functions}$$

When a set of inputs is applied to the circuit, the combinational subcircuit generates the outputs $(Z_1, Z_2, \ldots, Z_n)$ and the flip-flop inputs $(D_1, D_2, \ldots, D_k)$. Then, a clock pulse is applied and the flip-flops change to the proper next state. This process is repeated for each set of inputs. Note that at a given point in time, the outputs of the flip-flops represent the present state of the circuit $(Q_1, Q_2, \ldots, Q_k)$. These Q_i's feed back into the combinational circuit, which generates the flip-flop inputs using the Q_i's and the X inputs. When D flip-flops are used, $D_i = Q_i^+$; therefore, the combinational circuit outputs are labeled Q_1^+, Q_2^+, etc. Although the model in Figure 13-17 uses D flip-flops, a similar model may be used for other types of clocked flip-flops, in which case the combinational circuit must generate the appropriate flip-flop inputs instead of the next-state functions.

The clock synchronizes the operation of the flip-flops and prevents timing problems. The gates (or other logic) in the combinational subcircuit have finite propagation delays, so when the inputs to the circuit are changed, a finite time is required before the flip-flop inputs reach their final values. Because the gate delays are not all the same, the flip-flop input signals may contain transients, and they may change at different times. If the next active clock edge does not occur until all flip-flop input signals have reached their final steady-state values, the unequal gate delays will not cause any timing problems. All flip-flops which must change state do so at the same time in response to the active edge of the clock. When the flip-flops change state, the new flip-flop outputs are fed back into the combinational subcircuit. However, no further change in the flip-flop states can occur until the next clock pulse.

We can determine the fastest clock speed (the minimum clock period) from the general model of the Mealy circuit in Figure 13-17. The computation of the minimum clock period is similar to that of Figure 11-21, except that we must also consider the effect of the X inputs. Figure 13-18 shows the sequence of events during one clock period. Following the active edge of the clock the flip-flops change state, and the flip-flop output is stable after the propagation delay (t_p). The new values of Q then propagate through the combinational circuit so that the D values are stable after the combinational circuit delay (t_c). Then, the flip-flop setup time (t_{su}) must elapse before the next active clock edge. Thus, the propagation delay in the flip-flops, the propagation delay in the combinational subcircuit, and the setup time for the flip-flops determine how fast the sequential circuit can operate, and the minimum clock period is

$$t_{\text{clk}}\,(min) = t_p + t_c + t_{su}$$

The preceding discussion assumes that the X inputs are stable no later than $t_c + t_{su}$ before the next active clock edge. If this is not the case, then we must calculate the minimum clock period by

$$t_{\text{clk}}\,(min) = t_x + t_c + t_{su}$$

where t_x is the time after the active clock edge at which the X inputs are stable.

Another property of the circuit affecting the minimum clock period is the clock skew. The clock edge that causes the flip-flops to change state does not necessarily reach all flip-flop clock inputs at the same time; the difference in the arrival time of the clock edges to two flip-flops is the clock skew for those two flip-flops. Clock skew is caused by differing propagation delays over the clock lines and by different clock buffer propagation delays. In Figure 13-18, assume that the clock skew between the two flip-flops is t_{sk}. Furthermore, assume that the clock skew is such that the second flip-flop receives the clock edge before first flip-flop does. Then, the clock skew is added to setup time as part of the clock period. The minimum clock period is

$$T_{\text{clk}}\,(min) = t_p + t_c + t_{su} + t_{sk}$$

A different constraint applies if the clock skew is such that the second flip-flop receives the clock edge after the first flip-flop does. In this case, proper circuit operation requires that the input to the second flip-flop not change until after the clock skew time plus the flip-flop hold time, t_h. It is necessary that

$$t_p + t_c \geq t_{sk} + t_h$$

In the worst case, there may not be any combinational logic between the two flip-flops (e.g., a shift register does not have any logic between the flip-flops). In this case,

$$t_p \geq t_{sk} + t_h$$

In order to satisfy this constraint, it may be necessary to route the clock line so the second flip-flop receives the clock signal before the first clock signal.

The general model for the clocked Moore circuit (Figure 13-19) is similar to the clocked Mealy circuit. The output subcircuit is drawn separately for the Moore circuit because the output is only a function of the present state of the flip-flops and not a function of the circuit inputs. Operation of the Moore circuit is similar to that of the Mealy except when a set of inputs is applied to the Moore circuit, the resulting outputs do not appear until after the clock causes the flip-flops to change state.

To facilitate the study of sequential circuits with multiple inputs and outputs, the assignment of symbols to represent each combination of input values and each combination of output values is convenient. For example, we can replace Table 13-4

FIGURE 13-18 Minimum Clock Period for a Sequential Circuit

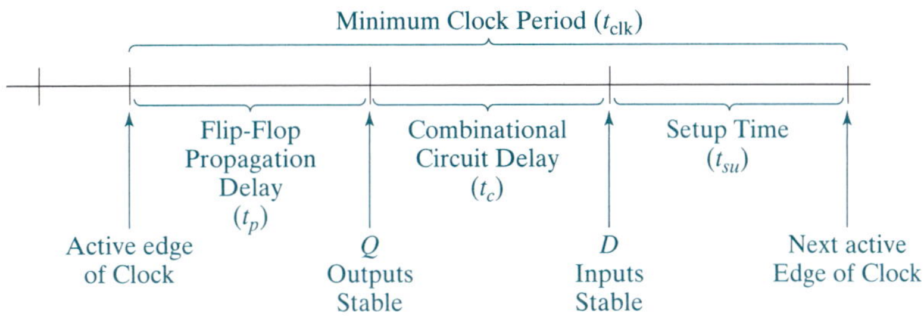

FIGURE 13-19 General Model for Moore Circuit Using Clocked D Flip-Flops

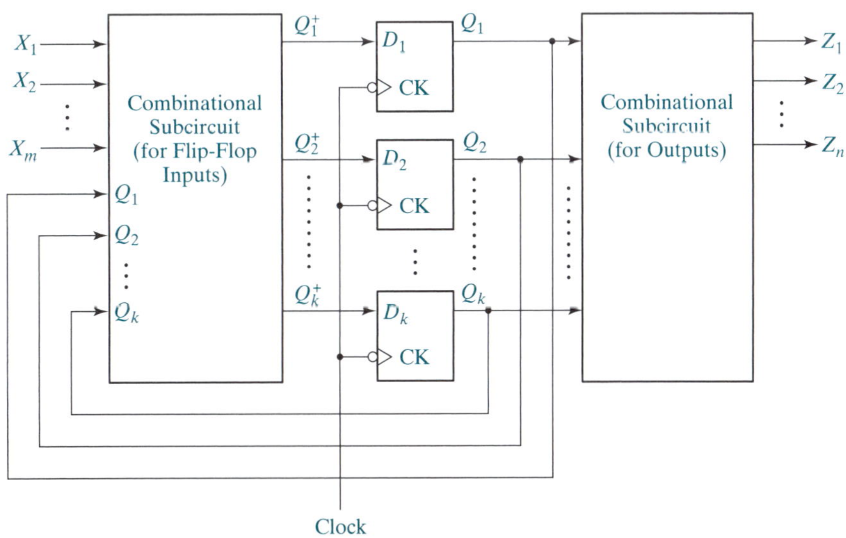

TABLE 13-5 State Table with Multiple Inputs and Outputs

Present State	Next State $X = 0$	1	2	3	Present Output (Z) $X = 0$	1	2	3
S_0	S_3	S_2	S_1	S_0	0	2	3	1
S_1	S_0	S_1	S_2	S_3	2	2	3	3
S_2	S_3	S_0	S_1	S_1	0	2	3	1
S_3	S_2	S_2	S_1	S_0	0	0	1	1

with Table 13-5 if we let $X = 0$ represent the input combination $X_1X_2 = 00$, $X = 1$ represent $X_1X_2 = 01$, etc., and similarly let $Z = 0$ represent the output combination $Z_1Z_2 = 00$, $Z = 1$, represent $Z_1Z_2 = 01$, etc. In this way we can specify the behavior of any sequential circuit in terms of a single input variable X and a single output variable Z.

Table 13-5 specifies two functions, the next-state function and the output function. The next-state function, designated δ, gives the next state of the circuit (i.e., the state after the clock pulse) in terms of the present state (S) and the present input (X):

$$S^+ = \delta(S, X) \tag{13-6}$$

The output function, designated λ, gives the output of the circuit (Z) in terms of the present state (S) and input (X):

$$Z = \lambda(S, X) \tag{13-7}$$

Values of S^+ and Z can be determined from the state table. From Table 13-5, we have

$$\delta(S_0, 1) = S_2 \qquad \delta(S_2, 3) = S_1$$
$$\lambda(S_0, 1) = 2 \qquad \lambda(S_2, 3) = 1$$

We will use the λ and δ notation when we discuss equivalent sequential circuits in Unit 15.

Programmed Exercise 13.1

Cover the answer to each exercise with a sheet of paper and slide it down as you check your answers.

13.1(a) In this exercise you will analyze the following sequential circuit using a state table and a timing chart.
Derive the next-state and output equations.

$A^+ =$ ________________________

$B^+ =$ ________________________

$Z =$ ________________________

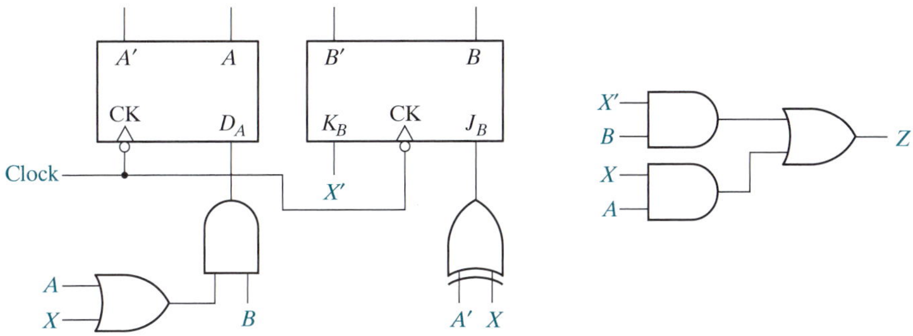

Answer $Z = XA + X'B, \quad B^+ = (A' \oplus X)B' + XB = A'B'X' + AB'X + XB$
$A^+ = B(A + X)$

13.1(b) Plot these equations on maps and complete the transition table.

A^+:

AB \ X	0	1
00		
01		
11		
10		

B^+:

AB \ X	0	1
00		
01		
11		
10		

Z:

AB \ X	0	1
00		
01		
11		
10		

	A^+B^+		Z	
AB	$X = 0$	1	0	1
00				
01				
11				
10				

Answer to 13.1(b)

		A^+B^+		Z	
	AB	$X = 0$	1	0	1
S_0	00	01	00	0	0
S_1	01	00	11	1	0
S_2	11	10	11	1	1
S_3	10	00	01	0	1

13.1(c) Convert your transition table to a state table using the given state numbering.

	Next State		Output	
	$X = 0$	1	0	1
S_0				
S_1				
S_2				
S_3				

Answer to 13.1(c)

	$X = 0$	1	0	1
S_0	S_1	S_0	0	0
S_1	S_0	S_2	1	0
S_2	S_3	S_2	1	1
S_3	S_0	S_1	0	1

13.1(d) Complete the corresponding state graph.

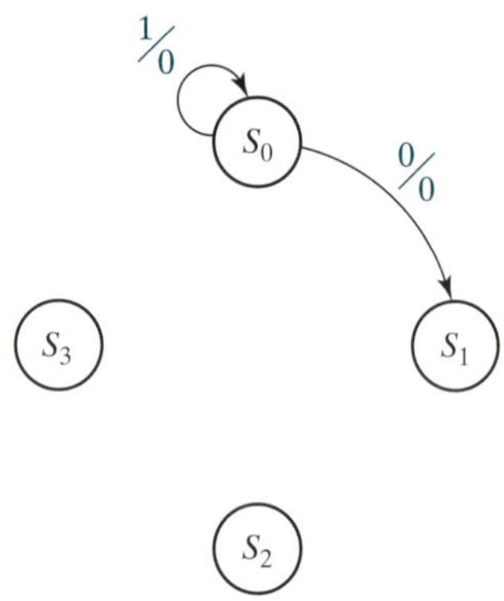

Answer to 13.1(d)

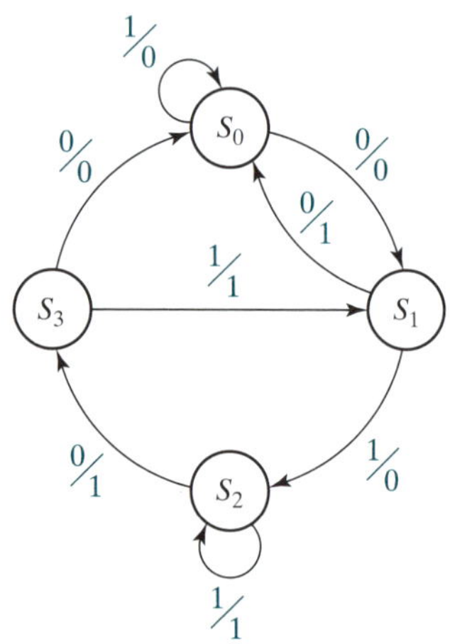

13.1(e) Using this graph, determine the state sequence and output sequence if the initial state is S_0 and the input sequence is $X = 0, 1, 0, 1$.

(1) The initial output with $X = 0$ in state S_0 is $Z =$ __________ and the next state is __________.

(2) The output in this state when the next input ($X = 1$) is applied is $Z =$ __________ and the next state is __________.

(3) When the third input ($X = 0$) is applied, the output is $Z =$ __________ and the next state is __________.

(4) When the last input is applied, $Z =$ __________ and the final state is __________.

In summary, the state sequence is S_0, __________, __________, __________, __________. The output sequence is $Z =$ __________.

Answer to 13.1(e) $S_0, S_1, S_2, S_3, S_1 \quad Z = 0011$

13.1(f) This sequence for Z is the correct output sequence. Next, we will determine the timing chart including any false outputs for Z. Assuming that X changes midway between falling and rising clock edges, draw the waveform for $X (X = 0, 1, 0, 1)$.

Answer

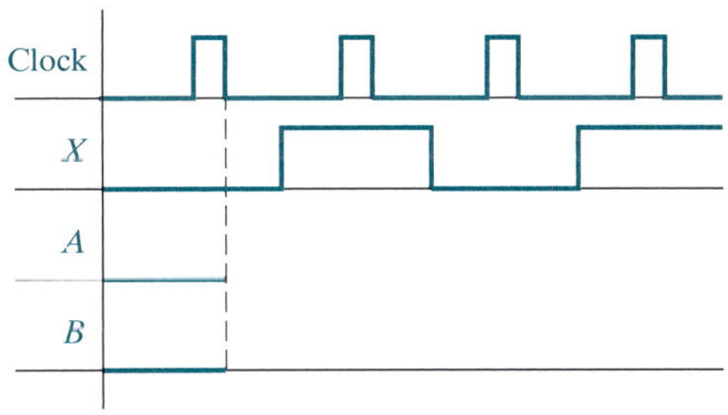

AB	A^+B^+ $X = 0$	1
00	01	00
01	00	11
11	10	11
10	00	01

13.1(g) Referring to the transition table, sketch the waveforms for A and B assuming that initially $A = B = 0$. The state sequence is

$AB = 00$, __________, __________, __________, __________.

Answer (Note that A and B change immediately after the falling clock edge.)

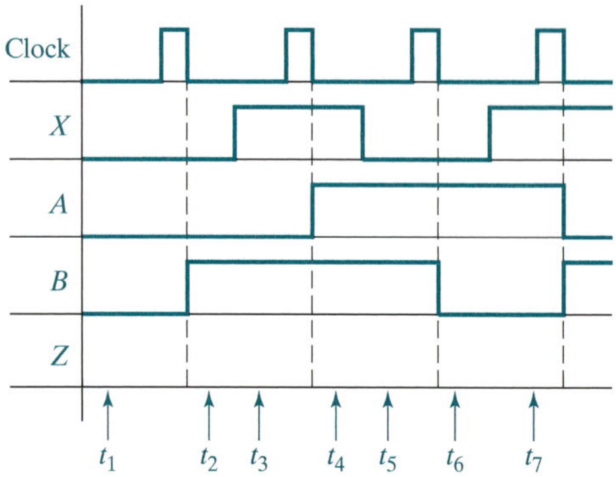

AB	Z $X = 0$	1
00	0	0
01	1	0
11	1	1
10	0	1

Check your state sequence against the answer to 13.1(e), noting that $S_0 = 00$, $S_1 = 01$, $S_2 = 11$ and $S_3 = 10$.

13.1(h) Using the output table, sketch the waveform for Z. At time t_1, $X = A = B = 0$, so $Z =$ __________. At time t_2, $X =$ __________ and $AB =$ __________, so $Z =$ __________. At time t_3, $X =$ __________ and $AB =$ __________, so $Z =$ __________. Complete the waveform for Z, showing the output at t_4, t_5, etc.

Answer (Note that Z can change immediately following the change in X or immediately following the falling clock edge.)

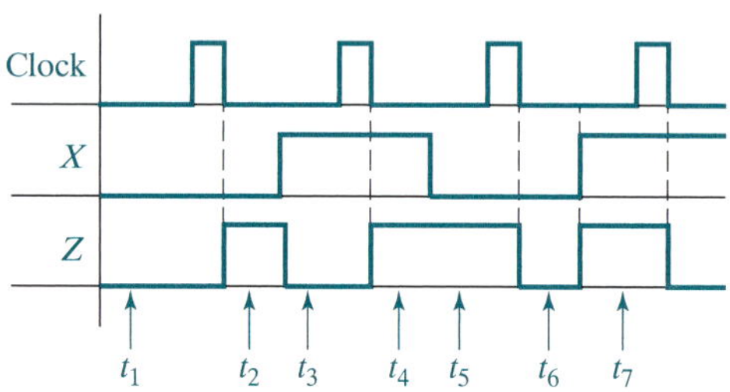

13.1(i) (1) Because this is a Mealy circuit, the correct times to read the output are during intervals t_1, __________, __________, and __________.

(2) The correct output sequence is therefore $Z =$ __________.

(3) False outputs may occur during intervals __________, __________, and __________.

(4) In two of these intervals, false outputs actually occur. These intervals are __________ and __________.

Answer

(1) t_1, t_3, t_5 and t_7
(2) Check your Z sequence against the answer to 13.1(e).
(3) t_2, t_4, and t_6
(4) t_2 and t_6 (output during t_4 is not false because it is the same as t_5).

13.1(j) Finally, we will verify part of the timing chart by signal tracing on the original circuit (see 13.1(a)).

(1) Initially, $A = B = 0$ and $X = 0$, so $D_A =$ ___________, $J_B =$ ___________, $K_B =$ ___________, and $Z =$ ___________.
(2) After the clock pulse $A =$ ___________, $B =$ ___________, and $Z =$ ___________.
(3) After X is changed to 1, $D_A =$ ___________, $J_B =$ ___________, $K_B =$ ___________, and $Z =$ ___________.
(4) After the clock pulse, $A =$ ___________, $B =$ ___________, and $Z =$ ___________.

Check your answers against the timing chart. Answer to (1) corresponds to t_1, (2) to t_2, (3) to t_3, and (4) to t_4.

Problems

13.2 Construct a transition graph for the shift register shown. (X is the input, and Z is the output.) Is this a Mealy or Moore machine?

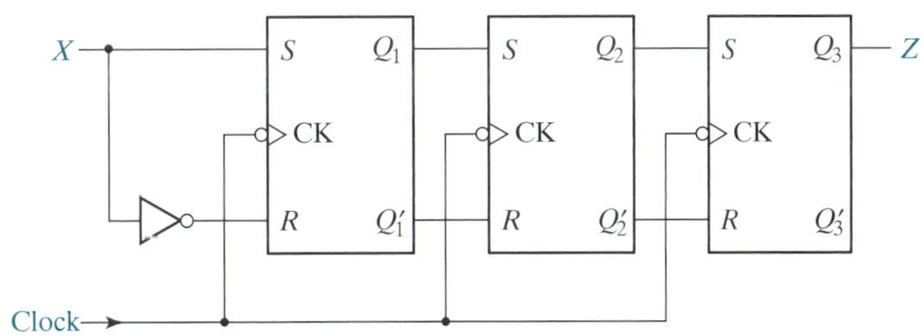

13.3 (a) For the following sequential circuit, find the next-state equation or map for each flip-flop. (Is this a Mealy or Moore machine?) Using these next-state equations or maps, construct a transition table and graph for the circuit.
(b) What is the output sequence when the input sequence is $X = 01100$?
(c) Draw a timing diagram for the input sequence in (b). Show the clock, X, A, B, and Z. Assume that the input changes between falling and rising clock edges.

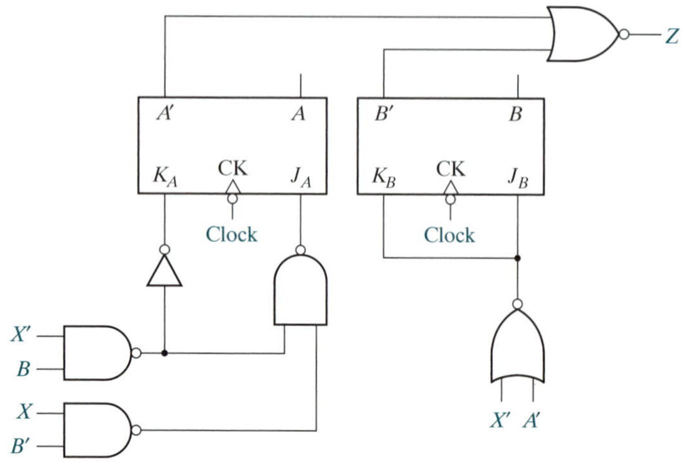

13.4 A sequential circuit has the form shown in Figure 13-17, with

$$D_1 = Q_2Q_3' \qquad D_3 = Q_2' + X$$
$$D_2 = Q_3 \qquad Z = XQ_2' + X'Q_2$$

(a) Construct a transition table and state graph for the circuit. (Is this a Mealy or Moore machine?)
(b) Draw a timing diagram for the circuit showing the clock, X, Q_1, Q_2, Q_3 and Z. Use the input sequence $X = 01011$. Change X between clock edges so that we can see false outputs, and indicate any false outputs on the diagram.
(c) Compare the output sequence obtained from the timing diagram with that from the transition graph.
(d) At what time with respect to the clock should the input be changed in order to eliminate the false output(s)?

13.5 Below is a transition table with the outputs missing. The output should be $Z = X'B' + XB$.

(a) Is this a Mealy machine or Moore machine?
(b) Fill in the outputs on the transition table.
(c) Give the transition graph.
(d) For an input sequence of $X = 10101$, give a timing diagram for the clock, X, A, B, C, and Z. State changes occur on the rising clock edge. What is the correct output sequence for Z? Change X between rising and falling clock edges so that we can see false outputs, and indicate any false outputs on the diagram.

	$A^+B^+C^+$		
ABC	$X = 0$	$X = 1$	
000	011	010	
001	000	100	
010	100	100	
011	010	000	
100	100	001	

13.6 A sequential circuit of the form shown in Figure 13-17 is constructed using a ROM and two rising-edge-triggered D flip-flops. The contents of the ROM are given in the table. Assume the propagation delay of the ROM is 8 ns, the setup time for the flip-flops is 2 ns, and the propagation delay of the flip-flops is 4 ns.

Q_1	Q_2	X	D_1	D_2	Z
0	0	0	1	0	0
0	0	1	1	0	0
0	1	0	0	0	0
0	1	1	1	1	0
1	0	0	1	1	0
1	0	1	0	1	1
1	1	0	0	1	1
1	1	1	1	1	1

(a) What is the minimum clock period for this circuit?

(b) Draw a timing diagram for this circuit, using the given delays and the minimum clock period of part (a). Give the clock, X, D_1, D_2, Q_1, Q_2 and Z. Assume $Q_1Q_2 = 00$ to start with and assume X takes on its new value 4 ns after each rising edge. Use the input sequence $X = 0, 1, 1, 0$. Specify the correct output sequence for Z.

(c) Construct a state table and a state graph for the circuit.

13.7 (a) Construct a transition table and state graph for the circuit shown.

(b) Construct a timing chart for the circuit for an input sequence $X = 10111$. (Assume that initially $Q_1 = Q_2 = 0$ and that X changes midway between the rising and falling clock edges.)

(c) List the output values produced by the input sequence.

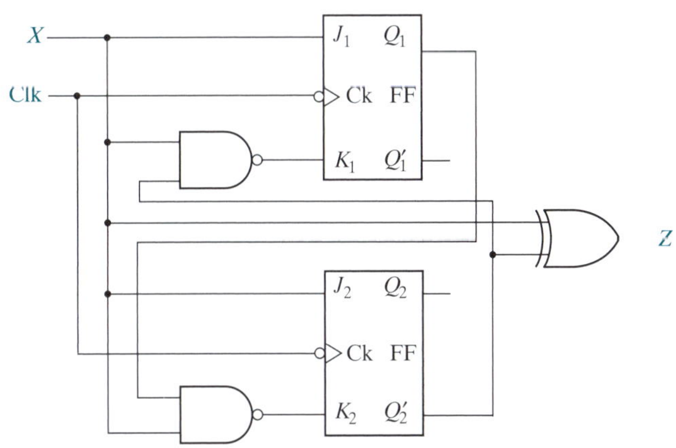

13.8 (a) Construct a transition table and state graph for the circuit shown.
(b) Construct a timing chart for the input sequence $X = 10101$. (Assume that initially $Q_1 = Q_2 = 0$ and that X changes midway between the rising and falling clock edges.) Indicate the times Z has the correct value.
(c) List the output values produced by the input sequence.

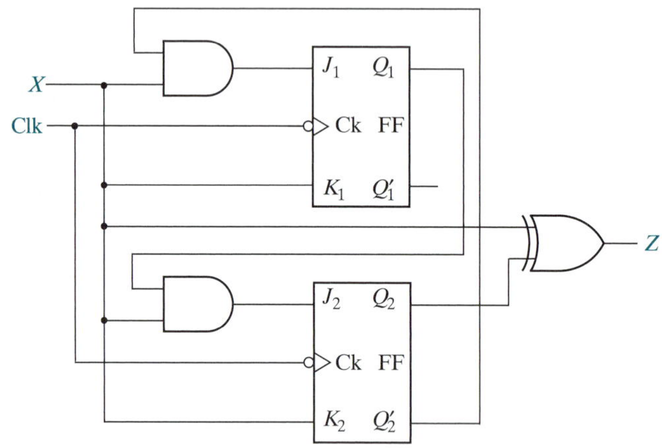

13.9 (a) Construct a transition table and state graph for the circuit shown.
(b) Construct a timing chart for the input sequence $X_1X_2 = 11, 11, 01, 10, 10, 00$. (Assume that initially $Q_1 = Q_2 = 0$ and that X_1 and X_2 change midway between the rising and falling clock edges.)
(c) List the output values produced by the input sequence.

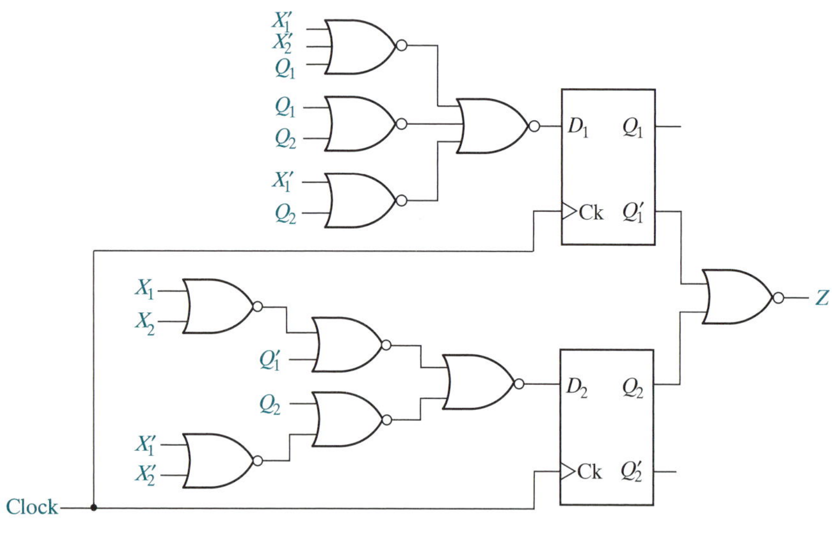

13.10 (a) Construct a transition table and state graph for the circuit shown.

(b) Construct a timing chart for the input sequence $X_1X_2 = 01, 10, 01, 11, 11, 01$. (Assume that initially $Q_1 = Q_2 = 0$ and that X_1 and X_2 change midway between the rising and falling clock edges.)

(c) List the output values produced by the input sequence.

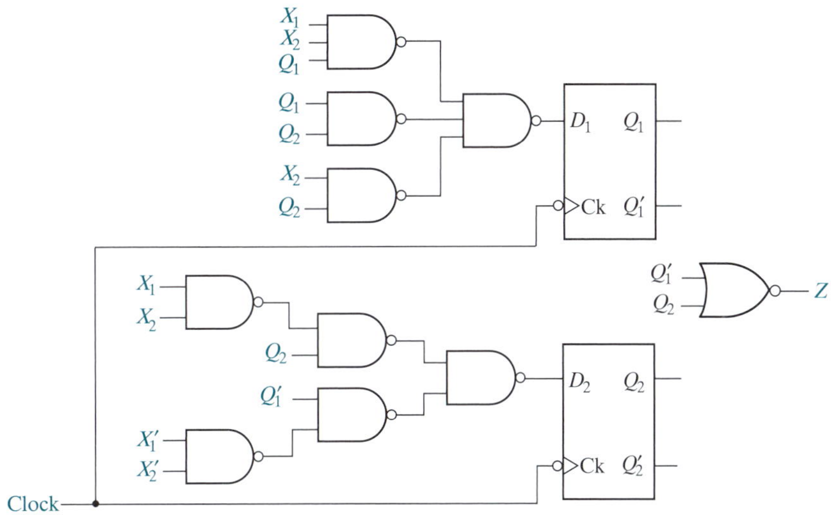

13.11 (a) Construct a transition table and state graph for the given circuit.

(b) Construct a timing chart for the circuit for an input sequence $X = 10011$. Indicate at what times Z has the correct value and specify the correct output sequence. (Assume that X changes midway between falling and rising clock edges.) Initially, $Q_1 = Q_2 = 0$.

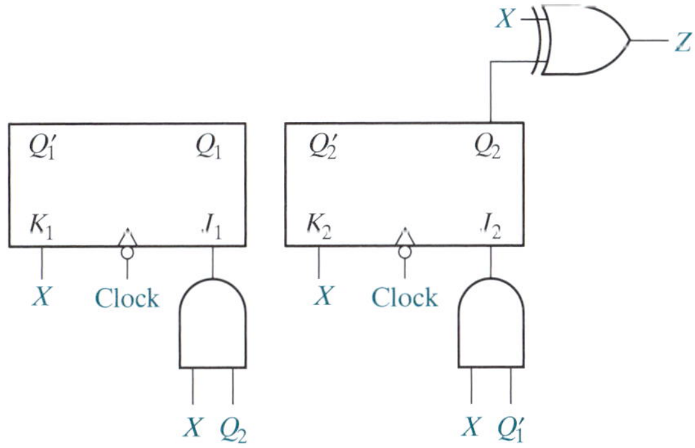

13.12 Repeat Problem 13.11 for the circuit below and $X_1X_2 = 10, 01, 10, 11, 11, 10$.

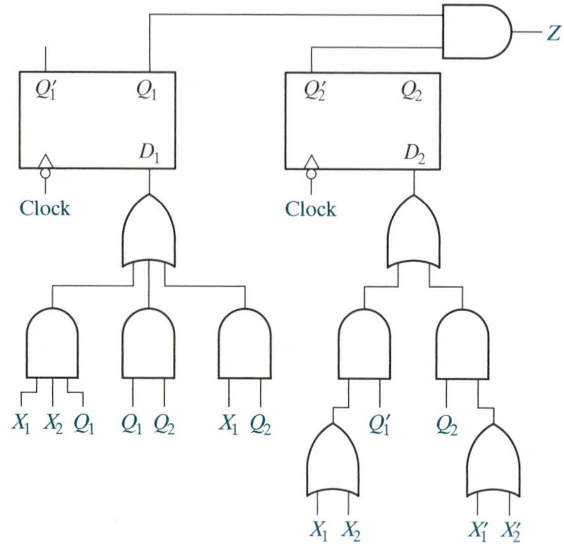

13.13 A sequential circuit has one input X, one output Z, and three flip-flops Q_1, Q_2 and Q_3. The transition and output tables for the circuit follow:

Present State	Next State $X = 0$	Next State $X = 1$	Output (Z) $X = 0$	Output (Z) $X = 1$
000	100	101	1	0
001	100	101	0	1
010	000	000	1	0
011	000	000	0	1
100	111	110	1	0
101	110	110	0	1
110	011	010	1	0
111	011	011	0	1

(a) Construct a timing chart for the input sequence $X = 0101$ and initial state $Q_1Q_2Q_3 = 000$. Identify any false outputs. (Assume that the flip-flops are rising-edge triggered and that the input changes midway between the rising and falling edges of the clock.)

(b) List the output values produced by the input sequence.

13.14 Repeat Problem 13.13 for the input sequence $X = 1001$ and initial state $Q_1Q_2Q_3 = 000$.

13.15 A sequential circuit has the form shown in Figure 13-17 with

$$D_1 = Q_2Q_3' + XQ_1' \qquad D_3 = Q_2' + X$$
$$D_2 = Q_3 + X'Q_2 \qquad Z = XQ_2' + X'Q_2$$

(a) Construct a state table and state graph for the circuit.
(b) Draw a timing diagram for the circuit showing the clock, X, Q_1, Q_2, Q_3 and Z. Use the input sequence $X = 01011$ and assume that X changes midway between falling and rising clock edges. Indicate any false outputs on the diagram.
(c) Compare the output sequence obtained from the timing diagram with that from the state graph.
(d) At what time with respect to the clock should the input be changed in order to eliminate the false output(s)?

13.16 Repeat Problem 13.15 for the given equations and the input sequence $X = 01100$.

$$D_1 = Q_3'X' \qquad D_3 = Q_3'X + Q_1Q_2$$
$$D_2 = Q_3'Q_1 + XQ_2' \qquad Z = XQ_3 + X'Q'$$

13.17 Two sequential circuits have the form shown in Figure 13-17. The initial state of both circuits is $Q_2Q_1Q_0 = 000$. The two circuits perform the same function but with different timing between the input and output.
(a) Construct a state table and state graph for each circuit. (In the state graphs only include states reachable from the initial state.)
(b) Describe in words when each circuit produces $Z = 1$. (*Hint:* Consider three consecutive inputs as a 3-bit binary number with the most significant bit being sent first.)
(c) Considering the dependence of the output on the input, are the circuits Moore or Mealy circuits?
(d) Considering the timing of when $Z = 1$ with respect to the input, do the circuits have Mealy or Moore type timing?

Circuit 1	Circuit 2
$Q_2^+ = Q_2'Q_0$	$Q_2^+ = Q_2'Q_0$
$Q_1^+ = XQ_2'Q_1 + XQ_2'Q_0'$	$Q_1^+ = XQ_1 + XQ_0' + Q_2Q_1$
$Q_0^+ = Q_2'$	$Q_0^+ = Q_2'$
$Z = Q_2Q_1$	$Z = Q_1Q_0'$

13.18 Analyze the following clocked synchronous sequential circuit by performing the following steps:
(a) Write the equations for the flip-flop inputs and the output equation.
(b) Construct the transition and output tables.
(c) Construct the transition graph.
(d) Give a one sentence description of when the circuit produces an output of 0.

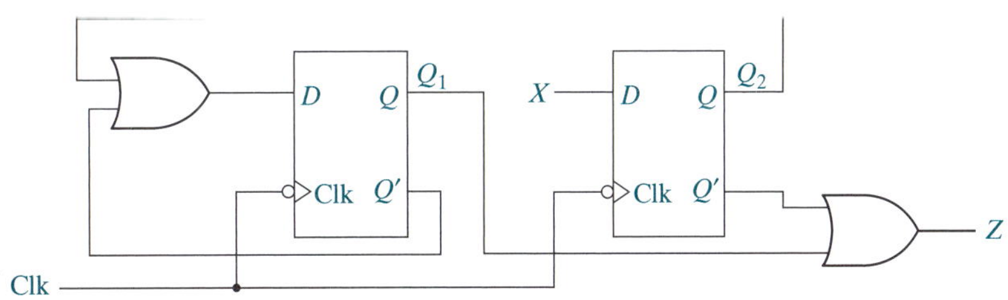

13.19 Repeat Problem 13.18 for the following circuit:

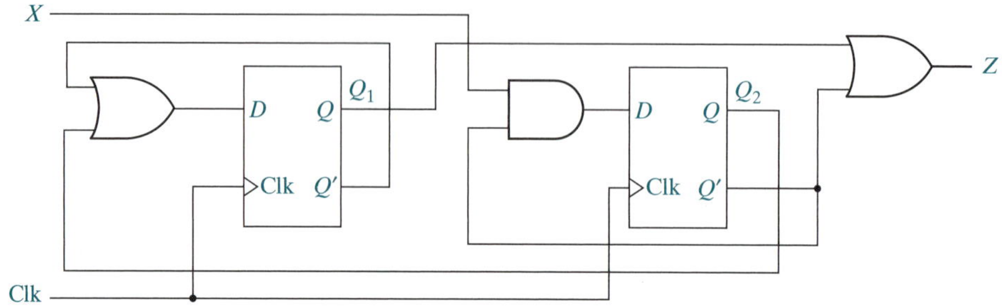

13.20 Consider the circuit shown.

(a) Construct a transition table and state graph for the following circuit. Is the circuit a Mealy or Moore circuit? Does the circuit have any unused states? Assume 00 is the initial state.

(b) Draw a timing diagram for the input sequence $X = 01100$.

(c) What is the output sequence for the input sequence?

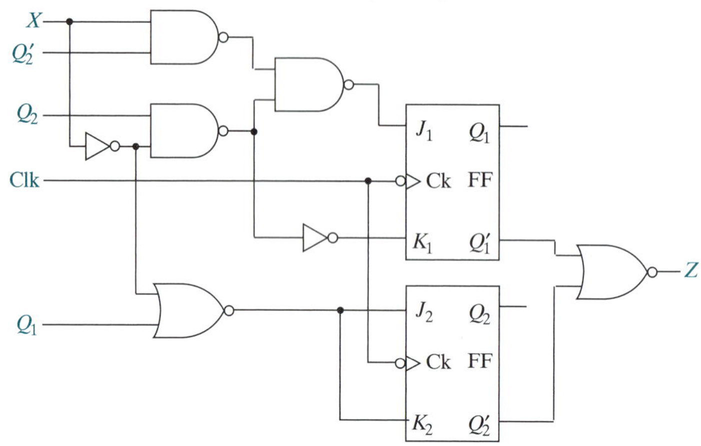

13.21 A Mealy sequential circuit has one input, one output, and two flip-flops. A timing diagram for the circuit follows. Construct a transition table and state graph for the circuit.

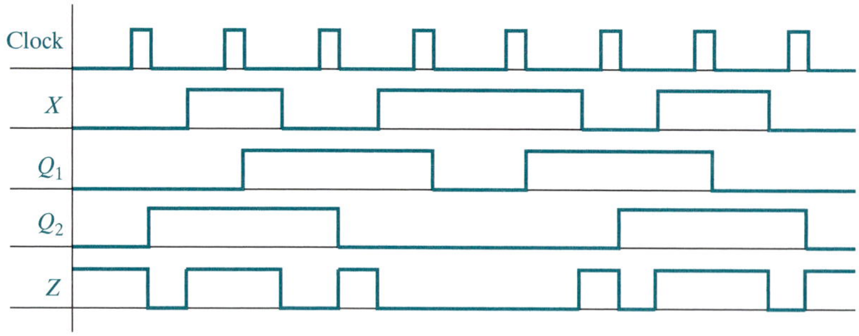

13.22 Repeat Problem 13.21 for the following timing diagram.

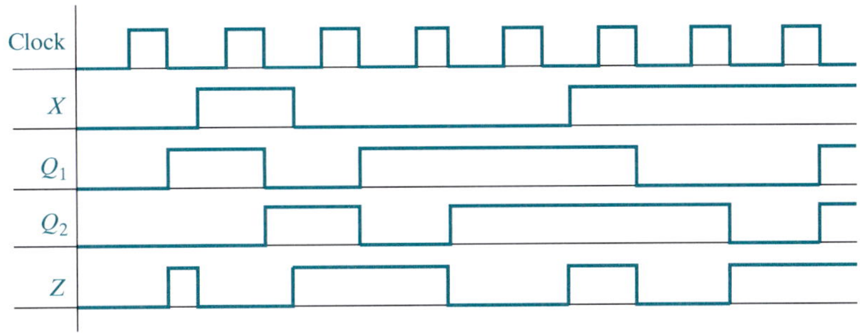

13.23 Given the following timing chart for a sequential circuit, construct as much of the transition table as possible. Is this a Mealy or Moore circuit?

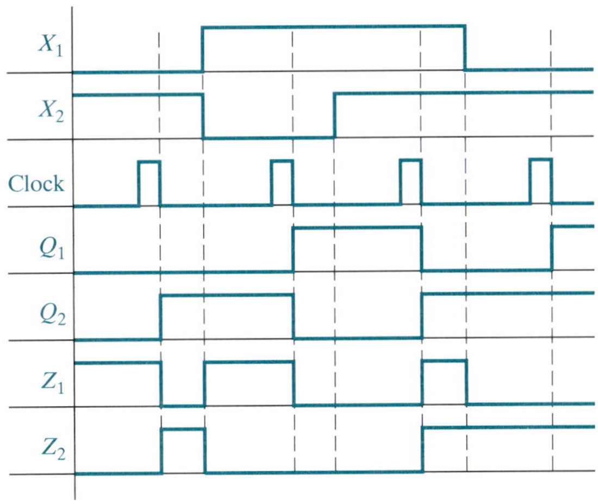

13.24 Given the following timing chart for a sequential circuit, construct as much of the transition table as possible.

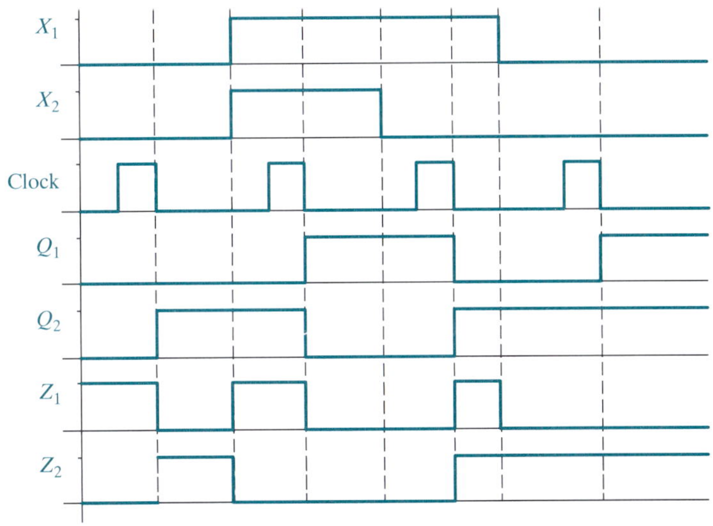

13.25 For the following sequential circuit, the table gives the contents of the PLA. (All PLA outputs are 0 for input combinations not listed in the table.)

(a) Construct a transition table and draw a state graph.

(b) Draw a timing diagram showing the clock, X, Q_1, Q_2, and Z for the input sequence $X = 10011$. Assume that initially $Q_1 = Q_2 = 0$.

(c) Identify any false outputs in the timing diagram. What is the correct output sequence for Z?

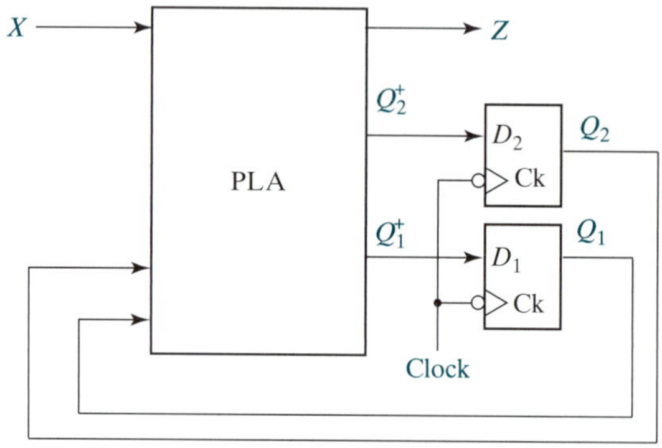

X	Q_1	Q_2	D_1	D_2	Z
0	1	–	1	0	0
1	0	1	1	0	0
0	–	1	0	1	1
1	0	0	0	1	1
1	1	0	0	0	1

13.26 A sequential circuit of the form shown in Figure 13-17 is constructed using a ROM and two D flip-flops. The contents of the ROM are given in the table.

(a) Draw a timing diagram for the circuit for the input sequence $X_1X_2 = 10, 01, 11, 10$. Assume that input changes occur midway between rising and falling clock edges. Indicate any false outputs on the diagram, and specify the correct output sequence for Z_1 and Z_2.

(b) Construct a transition table and state graph for the circuit.

Q_1	Q_2	X_1	X_2	D_1	D_2	Z_1	Z_2
0	0	0	0	0	0	1	0
0	0	0	1	0	0	1	0
0	0	1	0	0	1	1	0
0	0	1	1	0	1	1	0
0	1	0	0	1	1	0	0
0	1	0	1	1	1	1	0
0	1	1	0	1	0	0	1
0	1	1	1	1	0	1	1
1	0	0	0	1	1	0	0
1	0	0	1	0	0	0	0
1	0	1	0	1	1	0	1
1	0	1	1	0	0	0	1
1	1	0	0	1	0	0	0
1	1	0	1	0	1	0	0
1	1	1	0	1	0	0	0
1	1	1	1	0	1	0	0

13.27 For the following state graph, construct a transition table. Then, give the timing diagram for the input sequence $X = 101001$. Assume X changes midway between the falling and rising edges of the clock, and that the flip-flops are falling-edge triggered. What is the correct output sequence?

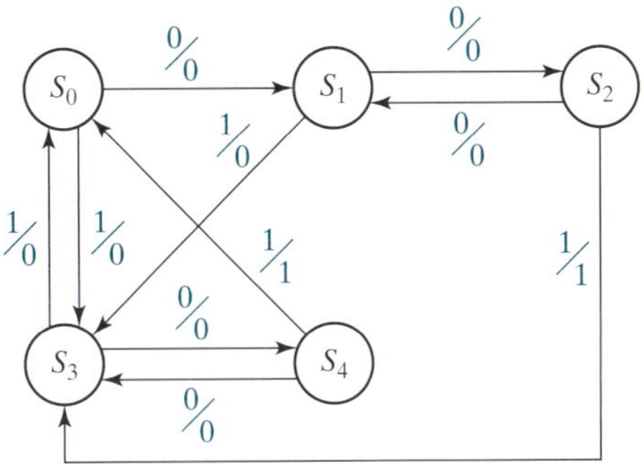

13.28 For the circuit of Problem 13.3, assume the delays of the NAND gates and NOR gates are 3 ns, and assume the delay of the inverter is 2 ns. Assume the propagation delays and setup times for the J-K flip-flops are 4 ns and 2 ns, respectively.

(a) Fill in the given timing diagram. The clock period is 15 ns, and 1-ns increments are marked on the clock signal. Does the circuit operate properly with these timing parameters?

(b) What is the minimum clock period for this circuit, if X is changed early enough? How late may X change with this clock period without causing improper operation of the circuit?

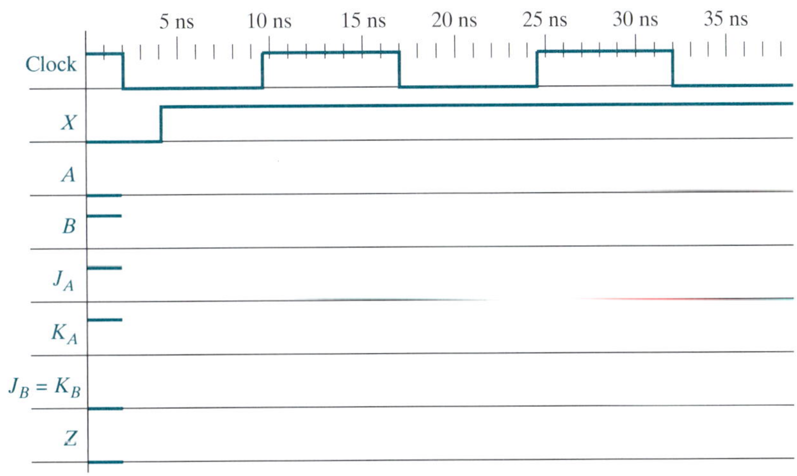

13.29 Draw a timing diagram for the following circuit starting with an initial state $ABC = 000$ and using an input sequence $X = 01010$. Assume that the input changes occur midway between the falling and rising clock edges. Give the output sequence, and indicate false outputs, if any. Verify that your answer is correct by making a state table for the circuit.

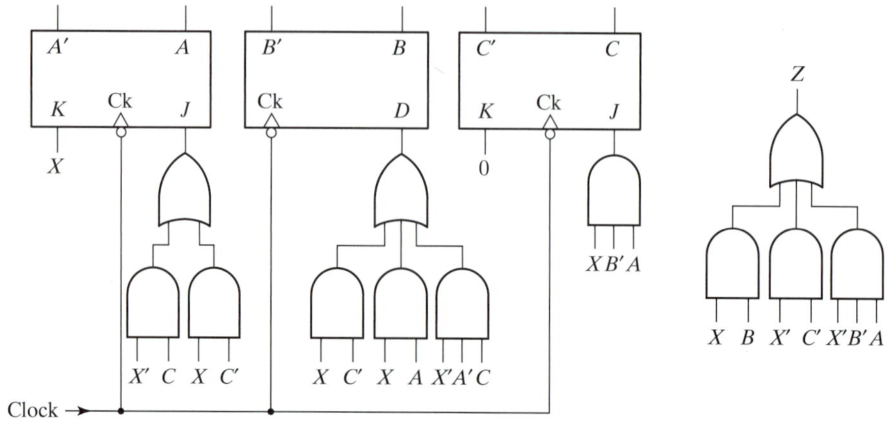

13.30 (a) For the following sequential circuit, write the next-state equations for flip-flops A and B.

(b) Using these equations, find the transition table and draw the state graph.

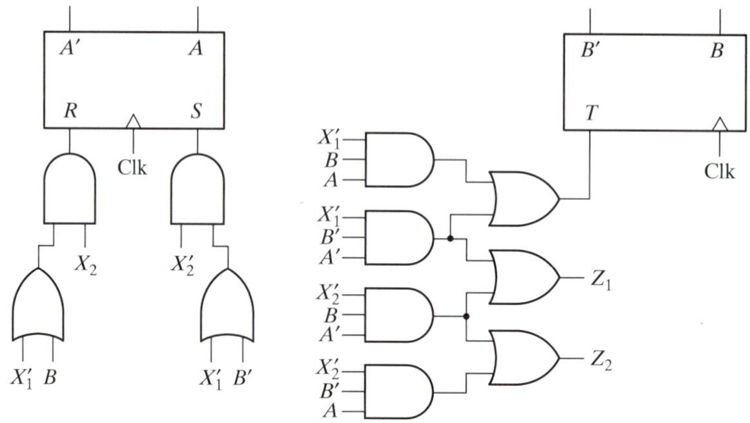

UNIT 14

Derivation of State Graphs and Tables

Objectives

1. Given a problem statement for the design of a Mealy or Moore sequential circuit, find the corresponding state graph and table.
2. Explain the significance of each state in your graph or table in terms of the input sequences required to reach that state.
3. Check your state graph using appropriate input sequences.

Study Guide

1. Study Section 14.1, *Design of a Sequence Detector.*

(a) Verify that the state graph in Figure 14-4 will produce the correct output sequence for Z when the input sequence for X is as given in Equation (14-1).
(b) Using the equations from the Karnaugh maps on p. 460, construct the next-state table for the circuit and verify that it is the same as given in Table 14-2, except that the new table will have four states because the don't-cares were assigned in the process of designing the circuit.
(c) Complete the design of the Moore sequential circuit whose transition table is given by Table 14-4. Use clocked J-K flip-flops for A and B.
(d) Verify that the state graph of Figure 14-6 gives the correct output sequence when the input sequence (14-1) is applied. (Ignore the initial output for the Moore graph.)

2. Study Section 14.2, *More Complex Design Problems.*

3. Study Section 14.3, *Guidelines for Construction of State Graphs.* Study the examples carefully and observe how some of the guidelines were applied.

4. Work through Programmed Exercises 14.1, 14.2, and 14.3.

5. A very important part of deriving state tables or state graphs is knowing how to tell when your answer is right (or wrong!). One way to do this is to make up a suitable list of test sequences, apply them to the state graph, and check the resulting output sequences.

6. To gain proficiency in construction of state tables or graphs requires a fair amount of practice. Work Problems 14.4, 14.5, 14.6, 14.7, and 14.8. The problems on the readiness tests will be about the same order of difficulty as these problems, so make sure that you can work them in a reasonable time.

Note: Do not look at the answers to these problems in the back of the book until you have tried the problems and checked your answers using the following test sequences:

14.4
$X =$ 0 1 1 1 0 1 0 1
$Z =$ (0) 0 0 0 0 1 1 1 1
(Your solution should have five self-loops. A self-loop is an arrow which starts at one state and goes back to the same state.)

14.5
$X =$ 1 0 1 0 1 0 0 1 0 0 0 1 0 0
$Z_1 =$ 0 0 0 1 0 1 0 0 0 0 0 0 0 0
$Z_2 =$ 0 0 0 0 0 0 1 0 0 1 0 0 0 1
(Your solution should have four self-loops.)

14.6
$X_1 =$ 1 0 0 1 0 1 1 0 0 0
$X_2 =$ 1 0 0 0 0 1 0 0 1 0
$Z =$ (0) 0 1 1 1 0 0 0 1 1 0
(Your solution should have at least four self-loops.)

14.7 (a) $X = 0\ 0\ 1\ 1\ 0\ 1\ 0\ 1\ 0\ 1\ 1$
$Z = 1\ 1\ 0\ 0\ 0\ 1\ 1\ 0\ 0\ 0\ 1$
(Your solution should have three self-loops.)
(b) $X = 1\ 1\ 1\ 0\ 0\ 1\ 0\ 1\ 0\ 1$
$Z = 0\ 0\ 0\ 0\ 1\ 0\ 0\ 0\ 0\ 1$

14.8 (a) $X_1 = 0\ 0\ 1\ 1\ 1\ 1\ 0\ 0\ 1\ 0\ 0$
$X_2 = 0\ 1\ 0\ 1\ 1\ 0\ 1\ 0\ 0\ 1\ 1$
$Z_1 = 0\ 1\ 1\ 1\ 0\ 0\ 0\ 0\ 1\ 0\ 0$
$Z_2 = 0\ 0\ 0\ 0\ 0\ 1\ 1\ 1\ 0\ 1\ 0$
(b) You should get the same sequences as in (a) after an initial output of $Z_1Z_2 = 00$.

7. If you have the *LogicAid* program available, use it to check your state tables. This has several advantages over looking at the answers in the back of the book. First, *LogicAid* will determine whether or not your solution is correct even if your states are numbered differently from those in the solution, or even if the number of states is different. Second, if your solution is wrong, *LogicAid* will find a short input sequence for which your state table fails, and you can use this sequence to help locate the error in your solution. If you are having trouble learning to derive state graphs, *LogicAid* has a state graph tutor mode which can be used to check partial state graphs. By using the partial graph checker, you can check your graph after adding each state, and then correct any errors before proceeding to the next state.

8. Read Section 14.4, *Serial Data Code Conversion.*

(a) Complete the following timing diagram, showing waveforms for the NRZ, NRZI, RZ, and Manchester coding schemes:

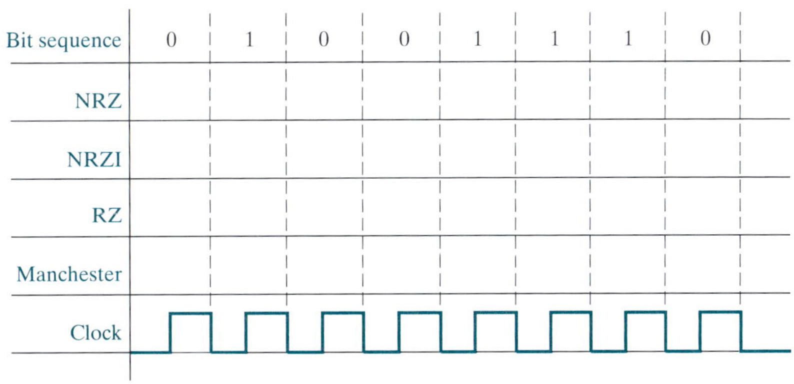

(b) The timing chart of Figure 14-21(b) shows several glitches. By referring to the state table, explain why the second glitch is present.

(c) Consider Figure 14-21. If an error in data transmission occurs, the input sequence $X = 01$ or 10 could occur. Add an Error state to the state diagram. The circuit should go to this error state if such an error occurs.

(d) Work Problem 14.9.

9. Read Section 14.5, *Alphanumeric State Graph Notation*.

(a) Sometimes all outputs are 0 for a given state or arc. We denote this by placing a 0 in the place of the output. For example, a Moore state with all outputs being 0 might be labeled $S_3/0$, and a Mealy arc with all outputs being 0 might be labeled $X'Y/0$.

(b) Try to write the row of a Mealy state table that describes state S_3 in the following partial state graph. You cannot, because there are two contradictory directions when $S = N = 1$. Also, the row is not completely specified. Redraw state S_3 so that S takes priority over N, and so that the circuit stays in state S_3 with no output if no directions are specified by the partial state graph. Then, give the state table row. Show that it has no contradictions and that it is completely specified.

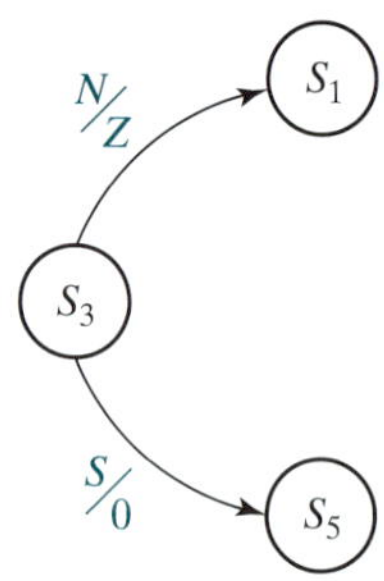

(c) Work Problems 14.10 and 14.11.

10. Read Section 14.6, *Incompletely Specified State Tables*.

Complete the following example:

The sequential circuit shown below has three parts. Assume that circuit A can only generate two possible output sequences, $X = 100$ and $X = 110$. Thus, the sequential circuit subsystem (B) has only two possible input sequences. When the third input in the sequence is received, the output of B is to be $Z = 0$ if 100 was received and $Z = 1$ if 110 was received. Assume that circuit C ignores the value of Z at other times so that we do not care what Z is during the first two inputs in the X sequence.

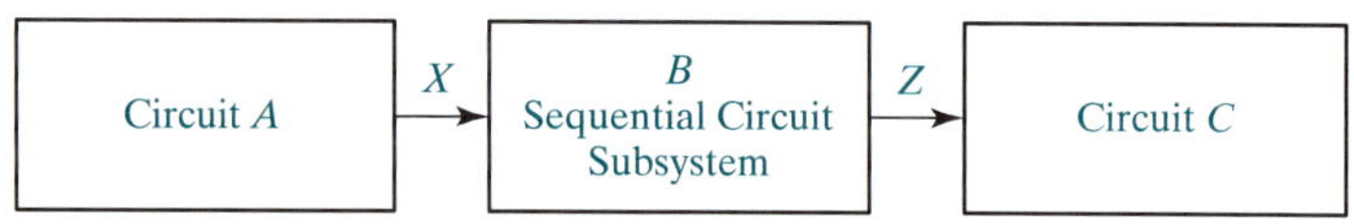

The possible input-output sequences for circuit B are listed in the following table, where t_0, t_1, and t_2 represent successive clock times.

	t_0	t_1	t_2		t_0	t_1	t_2	
$X =$	1	0	0	$Z =$	–	–	0	(– is a don't-care output)
	1	1	0		–	–	1	

Complete the following incompletely specified state table for circuit B:

	$X = 0$	1	0	1
S_0	–	S_1	–	–
S_1				
S_2				
S_3				

Note that the next state entry for S_0 with $X = 0$ is a don't-care because 0 can never occur as the first input in the sequence. When the third input in the sequence is received, why will the next state and output for $X = 1$ be don't-cares?

11. When you are satisfied that you can meet all of the objectives, take the readiness test.

Derivation of State Graphs and Tables

In Unit 13 we analyzed sequential circuits using timing charts and state graphs. Now, we will consider the design of sequential circuits starting from a problem statement which specifies the desired relationship between the input and output sequences. The first step in the design is to construct a state table or graph which specifies the desired behavior of the circuit. Flip-flop input equations and output equations can then be derived from this table. Construction of the state table or graph, one of the most important and challenging parts of sequential circuit design, is discussed in detail in this unit.

14.1 Design of a Sequence Detector

To illustrate the design of a clocked Mealy sequential circuit, we will design a sequence detector. The circuit has the form shown in Figure 14-1.

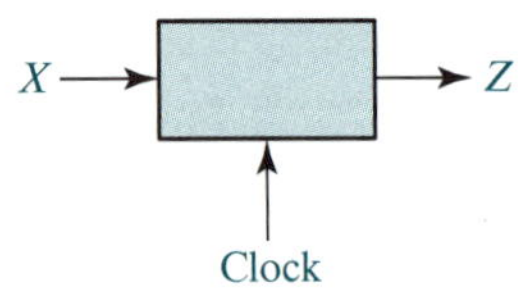

FIGURE 14-1 Sequence Detector to Be Designed

The circuit will examine a string of 0's and 1's applied to the X input and generate an output $Z = 1$ only when a prescribed input sequence occurs. It will be assumed that the input X can only change between clock pulses. Specifically, we will design the circuit so that any input sequence ending in 101 will produce an output $Z = 1$ coincident with the last 1. The circuit does not reset when a 1 output occurs. A typical input sequence and the corresponding output sequence are

$$
\begin{array}{llllllllllllllllll}
X = & 0 & 0 & 1 & 1 & 0 & 1 & 1 & 0 & 0 & 1 & 0 & 1 & 0 & 1 & 0 & 0 \\
Z = & 0 & 0 & 0 & 0 & 0 & 1 & 0 & 0 & 0 & 0 & 0 & 1 & 0 & 1 & 0 & 0 \\
(\text{time:} & 0 & 1 & 2 & 3 & 4 & 5 & 6 & 7 & 8 & 9 & 10 & 11 & 12 & 13 & 14 & 15)
\end{array}
\tag{14-1}
$$

Initially, we do not know how many flip-flops will be required, so we will designate the circuit states as S_0, S_1, etc., and later assign flip-flop states to correspond to the circuit states. We will construct a state graph to show the sequence of states and outputs which occur in response to different inputs. Initially, we will start the circuit in a reset state designated S_0. If a 0 input is received, the circuit can stay in S_0 because the input sequence we are looking for does not start with 0. However, if a 1 is received, the circuit must go to a new state (S_1) to "remember" that the first input in the desired sequence has been received (Figure 14-2). The labels on the graph are of the form X/Z, where the symbol before the slash is the input and the symbol after the slash is the corresponding output.

When in state S_1, if we receive a 0, the circuit must change to a new state (S_2) to remember that the first two inputs of the desired sequence (10) have been received. If a 1 is received in state S_2, the desired input sequence (101) is complete and the output should be l. The question arises whether the circuit should then go to a new state or back to S_0 or S_1. Because the circuit is not supposed to reset when an output occurs, we cannot go back to S_0. However, because the last 1 in a sequence can also be the first 1 in a new sequence, we can return to S_1, as indicated in Figure 14-3.

FIGURE 14-2

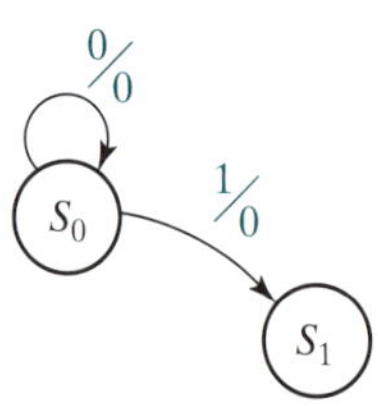

FIGURE 14-3

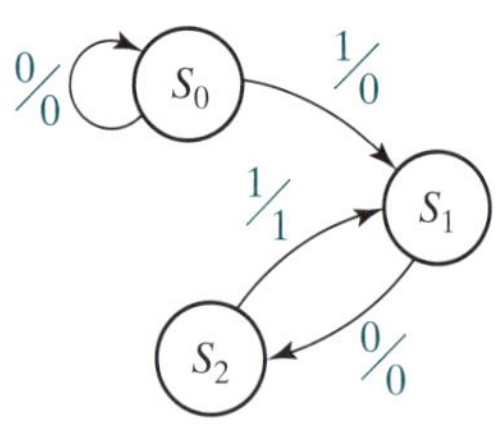

The graph of Figure 14-3 is still incomplete. If a 1 input occurs when in state S_1, we can stay in S_1 because the sequence is simply restarted. If a 0 input occurs in state S_2, we have received two 0's in a row and must reset the circuit to state S_0 because 00 is not part of the desired input sequence, and going to one of the other states could lead to an incorrect output. The final state graph is given in Figure 14-4. Note that for a single input variable each state must have two exit lines (one for each value of the input variable) but may have any number of entry lines, depending on the circuit specifications.

FIGURE 14-4 Mealy State Graph for Sequence Detector

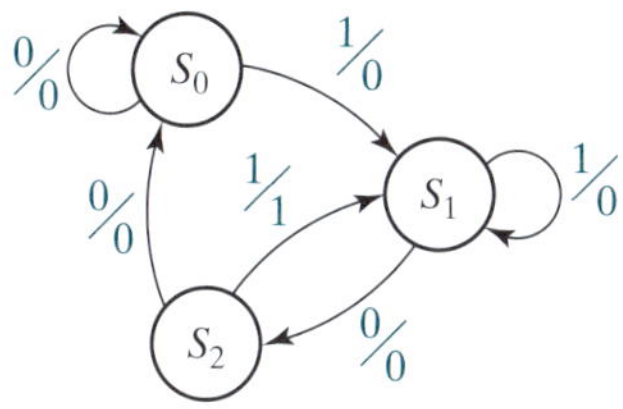

State S_0 is the starting state, state S_1 indicates that a sequence ending in 1 has been received, and state S_2 indicates that a sequence ending in 10 has been received. An alternative way to start the solution would be to first define states in this manner and then construct the state graph. Converting the state graph to a state table yields Table 14-1. For example, the arc from S_2 to S_1 is labeled 1/1. This means that when the present state is S_2 and $X = 1$, the present output is 1. This 1 output is present as soon as X becomes 1, that is, *before* the state change occurs. Therefore, the 1 is placed in the S_2 row of the table.

TABLE 14-1

Present State	Next State		Present Output	
	$X = 0$	$X = 1$	$X = 0$	$X = 1$
S_0	S_0	S_1	0	0
S_1	S_2	S_1	0	0
S_2	S_0	S_1	0	1

At this point, we are ready to design a circuit which has the behavior described by the state table. Because one flip-flop can have only two states, two flip-flops are needed to represent the three states. Designate the two flip-flops as A and B. Let flip-flop states $A = 0$ and $B = 0$ correspond to circuit state S_0; $A = 0$ and $B = 1$ correspond to S_1; and $A = 1$ and $B = 0$ correspond to circuit state S_2. Each circuit state is then represented by a unique combination of flip-flop states. Substituting the flip-flop states for S_0, S_1, and S_2 in the state table yields the transition table (Table 14-2).

TABLE 14-2

	A^+B^+		Z	
AB	$X = 0$	$X = 1$	$X = 0$	$X = 1$
00	00	01	0	0
01	10	01	0	0
10	00	01	0	1

From this table, we can plot the next-state maps for the flip-flops and the map for the output function Z:

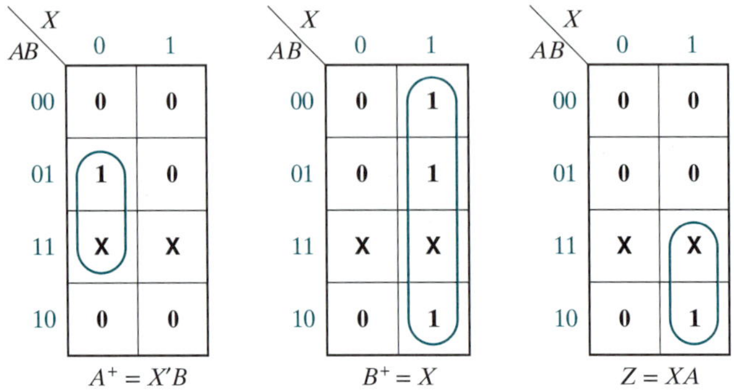

The flip-flop inputs are then derived from the next-state maps using the same method that was used for counters (Section 12.4). If D flip-flops are used, $D_A = A^+ = X'B$ and $D_B = B^+ = X$, which leads to the circuit shown in Figure 14-5. Initially, we will reset both flip-flops to the 0 state. By tracing signals through the circuit, you can verify that an output $Z = 1$ will occur when an input sequence ending in 101 occurs. To avoid reading false outputs, always read the value of Z after the input has changed and before the active clock edge.

FIGURE 14-5

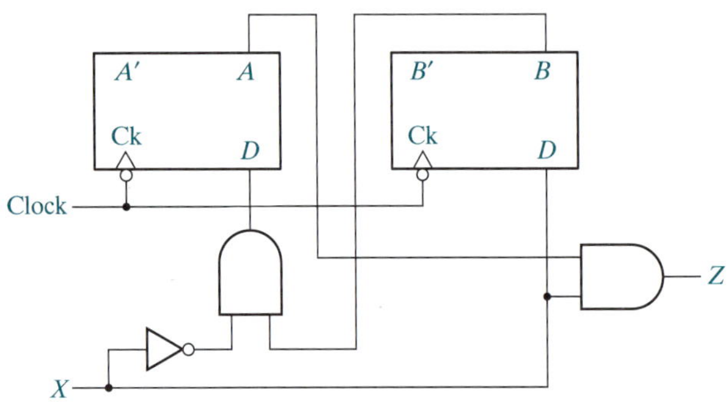

The procedure for finding the state graph for a Moore machine is similar to that used for a Mealy machine, except that the output is written with the state instead of with the transition between states. We will rework the previous example as a Moore machine to illustrate this procedure. The circuit should produce an output of 1 only if an input sequence ending in 101 has occurred. The design is similar to that for the

Mealy machine up until the input sequence 10 has occurred, except that 0 output is associated with states S_0, S_1, and S_2:

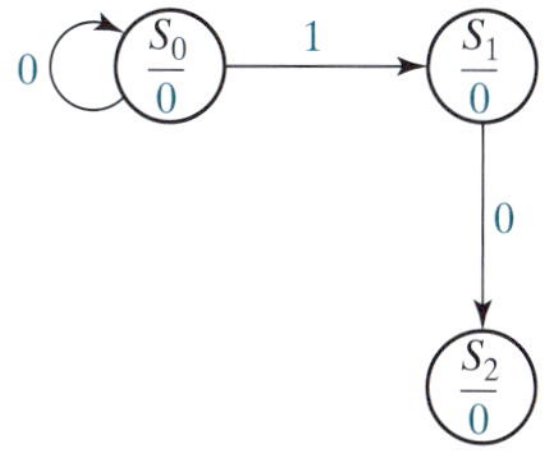

Now, when a 1 input occurs to complete the 101 sequence, the output must become 1; therefore, we cannot go back to state S_1 and must create a new state S_3 with a 1 output:

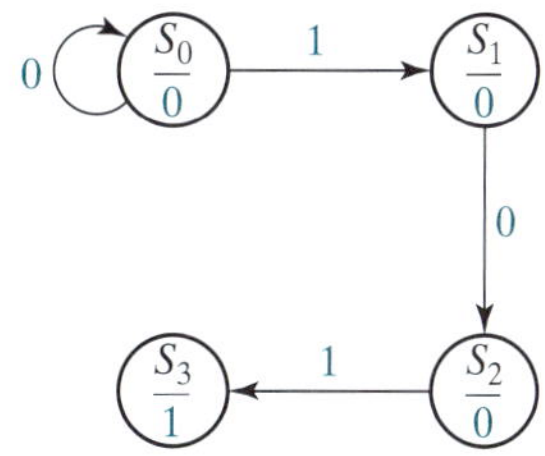

We now complete the graph, as shown in Figure 14-6. Note the sequence 100 resets the circuit to S_0. A sequence 1010 takes the circuit back to S_2 because another 1 input should cause Z to become 1 again.

FIGURE 14-6 Moore State Graph for Sequence Detector

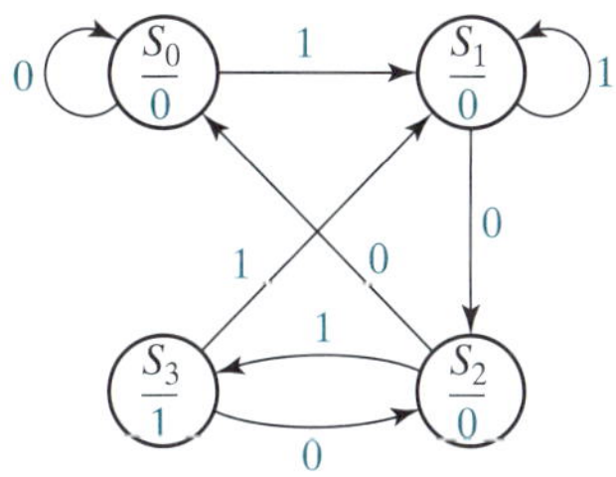

The state table corresponding to the circuit is given by Table 14-3. Note that there is a single column for the output because the output is determined by the present state and does not depend on X. Note that in this example the Moore machine requires one more state than the Mealy machine which detects the same input sequence.

TABLE 14-3

Present State	Next State $X = 0$	Next State $X = 1$	Present Output(Z)
S_0	S_0	S_1	0
S_1	S_2	S_1	0
S_2	S_0	S_3	0
S_3	S_2	S_1	1

Because there are four states, two flip-flops are required to realize the circuit. Using the state assignment $AB = 00$ for S_0, $AB = 01$ for S_1, $AB = 11$ for S_2, and $AB = 10$ for S_3, the following transition table for the flip-flops results (Table 14-4):

TABLE 14-4

AB	A^+B^+ $X = 0$	$X = 1$	Z
00	00	01	0
01	11	01	0
11	00	01	0
10	11	01	1

The output function is $Z = AB'$. Note that Z depends only on the flip-flop states and is independent of Z, while for the corresponding Mealy machine, Z was a function of X. The derivation of the flip-flop input equations is straightforward and will not be given here.

The preceding 101 sequence detectors can be called *sliding window* sequence detectors because the sequence being detected can occur anywhere in the input sequence. The windows can be overlapping or they may be restricted to be nonoverlapping. The first two output sequences in Figure 14-7 illustrate the distinction between overlapping and nonoverlapping windows for an example input sequence and 101 sequence detectors. (The circuits corresponding to Figure 14-4 and Figure 14-6 allow overlapping windows.) In addition, these examples assume that the complete 101 sequence must occur after time 0 to be recognized; this is equivalent to assuming that the circuit responds as if the initial state were preceded by a 0 input. An alternative assumption is that the circuit responds as if the initial state were preceded by a 1 input; the third and fourth output sequences in Figure 14-7 are for this assumption. If state S_1 were made the initial state in Figure 14-4 and Figure 14-6, then the circuits would respond as if the initial state were preceded by a 1 input.

In contrast to a sliding window sequence detector, a *disjoint window* sequence detector divides the input sequence into windows of fixed length (normally the length of the sequence to be detected) and only recognizes the input sequence when confined to one of the windows. The last output sequence in Figure 14-7 is for the 101 sequence detector using disjoint windows of length 3. (Making the window longer than the sequence is equivalent to changing the sequence recognized. For example, detecting 101 anywhere in disjoint windows of length 4 is equivalent to detecting any of the sequences: 0101, 1101, 1010, and 1011.) A sequence detector for disjoint windows will usually be more complex than a similar detector for sliding windows, because the detector must record where bits occur within the window in addition to detecting the sequence.

Disjoint window sequence detectors apply naturally in certain cases. For example, if an input sequence contains successive BCD digits, then the input sequence would be divided into disjoint windows of length 4. A sequence detector might examine each window to determine if it contains a valid BCD digit, or a different detector might determine the parity of each BCD digit.

FIGURE 14-7 Different Types of Sequence Detector

Different 101 Sequence Detectors																			
Different 101 Sequence Detectors	X	0	1	0	1	0	1	0	1	1	1	0	1	0	1	0	1	0	1
Sliding windows, overlapping	Z	0	0	0	1	0	1	0	1	0	0	0	1	0	1	0	1	0	1
Sliding windows, nonoverlapping	Z	0	0	0	1	0	0	0	1	0	0	0	1	0	0	0	1	0	0
Sliding windows, overlapping, initial 1	Z	0	1	0	1	0	1	0	1	0	0	0	1	0	1	0	1	0	1
Sliding windows, nonoverlapping, initial 1	Z	0	1	0	0	0	1	0	0	0	0	0	1	0	0	0	1	0	0
Disjoint windows	Z	0	0	0	0	0	1	0	0	0	0	0	1	0	0	0	0	0	1

14.2 More Complex Design Problems

In this section we will derive a state graph for a sequential circuit of somewhat greater complexity than the previous examples. The circuit to be designed again has the form shown in Figure 14-1. The output Z should be 1 if the input sequence ends in either 010 or 1001, and Z should be 0 otherwise. Before attempting to draw the state graph, we will work out some typical input-output sequences to make sure that we have a clear understanding of the problem statement. We will determine the desired output sequence for the following input sequence:

$$\begin{array}{ccccccccccccccccccc}
X & = & 0 & 0 & 1 & 0 & 1 & 0 & 0 & 1 & 0 & 0 & 0 & 1 & 0 & 0 & 1 & 1 & 0 \\
 & & & & & \uparrow & & \uparrow & & \uparrow & \uparrow & & & & \uparrow & & \uparrow & & \\
 & & & & & a & & b & & c & d & & & & e & & f & & \\
Z & = & 0 & 0 & 0 & 1 & 0 & 1 & 0 & 1 & 1 & 0 & 0 & 0 & 1 & 0 & 1 & 0 & 0
\end{array}$$

At point a, the input sequence ends in 010, one of the sequences for which we are looking, so the output is $Z = 1$. At point b, the input again ends in 010, so $Z = 1$. Note that overlapping sequences are allowed because the problem statement does not say anything about resetting the circuit when a 1 output occurs. At point c, the input sequence ends in 1001, so Z is again 1. Why do we have a 1 output at points d, e, and f? This is just one of many input sequences. A state machine that gives the correct output for this sequence will not necessarily give the correct output for all other sequences.

We will start construction of the state graph by working with the two sequences which lead to a 1 output. Then, we will later add arrows and states as required to make sure that the output is correct for other cases. We start off with a reset state S_0 which corresponds to having received no inputs. Whenever an input is received that corresponds to part of one of the sequences for which we are looking, the circuit should go to a new state to "remember" having received this input. Figure 14-8 shows a partial state graph which gives a 1 output for the sequence 010. In this graph S_1 corresponds to having received a sequence ending in 0, S_2 to a sequence ending in 01, and S_3 to a sequence ending in 010. Now, if a 1 input is received in state S_3, we again have a sequence ending in 01, which is part of the input sequence for which we

FIGURE 14-8

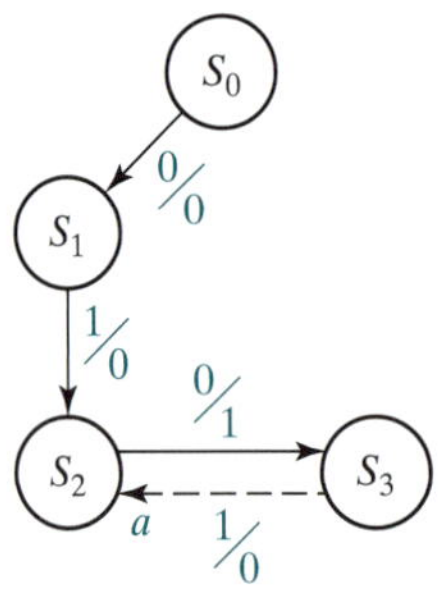

State	Sequence Received
S_0	Reset
S_1	0
S_2	01
S_3	010

are looking. Therefore, we can go back to state S_2 (arrow a) because S_2 corresponds to having received a sequence ending in 01. Then, if we get another 0 in state S_2, we go to S_3 with a 1 output. This is correct because the sequence again ends in 010.

Next, we will construct the part of the graph corresponding to the sequence 1001. Again, we start in the reset state S_0, and when we receive a 1 input, we go to S_4 (Figure 14-9, arrow b) to remember that we have received the first 1 in the sequence 1001. The next input in the sequence is 0, and when this 0 is received, we should ask the question: Should we create a new state to correspond to a sequence ending in 10, or can we go to one of the previous states on the state graph? Because S_3 corresponds to a sequence ending in 10, we can go to S_3 (arrow c). The fact that we did not have an initial 0 this time does not matter because 10 starts off the sequence for which we are looking. If we get a 0 input when in S_3, the input sequence received will end in 100 regardless of the path we took to get to S_3. Because there is so far no state corresponding to the sequence 100, we create a new state S_5 to indicate having received a sequence ending in 100.

If we get a 1 input when in state S_5, this completes the sequence 1001 and gives a 1 output as indicated by arrow e. Again, we ask the question: Can we go back to one of the previous states or do we have to create a new state? Because the end of the sequence 1001 is 01, and S_2 corresponds to a sequence ending in 01, we can go back to S_2 (Figure 14-10). If we get another 001, we have again completed the sequence 1001 and get another 1 output.

FIGURE 14-9

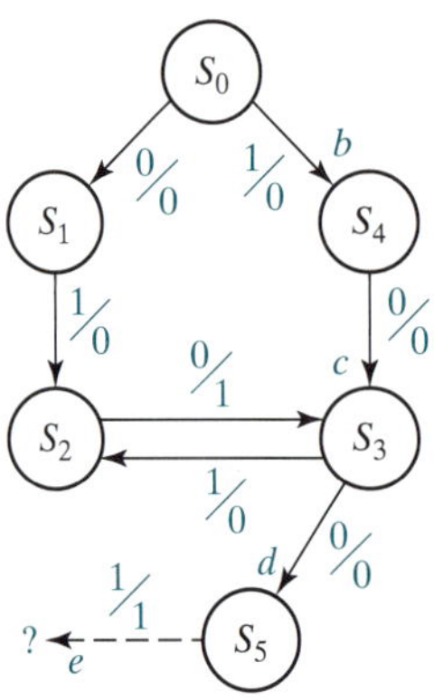

State	Sequence Ends in
S_0	Reset
S_1	0 (but not 10)
S_2	01
S_3	10
S_4	1 (but not 01)
S_5	100

FIGURE 14-10

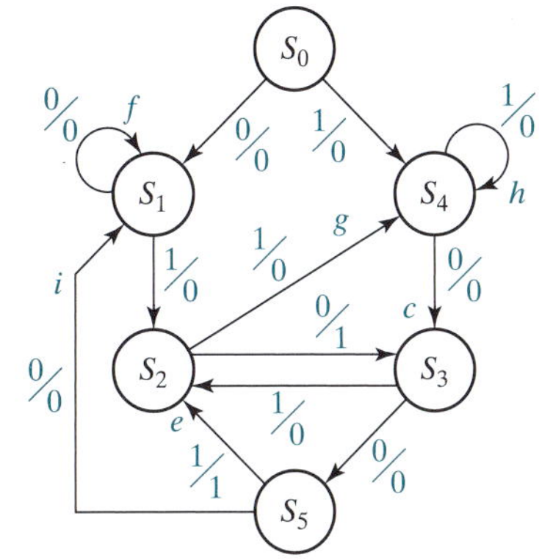

State	Sequence Ends in
S_0	Reset
S_1	0 (but not 10)
S_2	01
S_3	10
S_4	1 (but not 01)
S_5	100

We have now taken care of putting out a 1 when either the sequence 010 or 1001 is completed. Next, we will go back and complete the state graph to take care of the other input sequences, for which we have not already accounted. In state S_1, we have accounted for a 1 input but not a 0 input. If we are in S_1 and get a 0 input, to which state should we go? If a 0 input occurs in S_1, we have a sequence ending in 00. Because 00 is not part of either of the input sequences for which we are looking, we can ignore the extra 0 and stay in S_1 (arrow f). No matter how many extra 0's occur, we still have a sequence ending in 0, and we stay in S_1 until a 1 input occurs. In S_2, we have taken care of the 0 input case but not the 1 input case. If a 1 is received, the input sequence ends in 11. Because 11 is not part of either the sequence 010 or 1001, we do not need a state which corresponds to a sequence ending in 11. We cannot stay in S_2 because S_2 corresponds to a sequence ending in 01. Therefore, we go to S_4, which corresponds to having received a sequence ending in 1 (arrow g). S_3 already has arrows corresponding to 0 and 1 inputs, so we examine S_4 next. If a 1 is received in S_4, the input sequence ends in 11. We can stay in S_4 and ignore the extra 1 (arrow h) because 11 is not part of either sequence for which we are looking. In S_5, if we get a 0 input, the sequence ends in 000. Because 000 is not contained in either 010 or 1001, we can go back to S_1, because S_1 corresponds to having received a sequence ending in one (or more) 0's. This completes the state graph because every state has arrows leaving it which correspond to both 0 and 1 inputs. We should now go back and check the state graph against the original input sequences to make sure that a 1 output is always obtained for a sequence ending in 010 or 1001 and that a 1 output does not occur for any other sequence.

Next, we will derive the state graph for a Moore sequential circuit with one input X and one output Z. The output Z is to be 1 if the total number of 1's received is odd and at least two consecutive 0's have been received. A typical input and output sequence is

$$
\begin{array}{ll}
X = & \,1\ 0\ 1\ 1\ 0\ 0\ 1\ 1 \\
 & \,\uparrow\ \ \ \ \ \ \uparrow\ \ \ \uparrow\ \uparrow\ \uparrow \\
 & \,a\ \ \ \ \ \ \ b\ \ \ c\ d\ e \\
Z = & (0)\,0\ 0\ 0\ 0\ 0\ 1\ 0\ 1
\end{array}
$$

We have shifted the Z sequence to the right to emphasize that for a Moore circuit an input change does not affect Z immediately, but Z can change only after the next active clock edge. The initial 0 in parentheses is the output associated with the reset state. At points a and b in the preceding sequence, an odd number of 1's has been received, but two consecutive 0's have not been received, so the output remains 0. At points c and e, an odd number of 1's and two consecutive 0's have been received, so $Z = 1$. At point d, $Z = 0$ because the number of 1's is even.

We start construction of the Moore state graph (Figure 14-11) with the reset state S_0, and we associate a 0 output with this state. First, we will consider keeping track of whether the number of 1's is even or odd. If we get a 1 input in S_0, we will go to state S_1 to indicate an odd number of 1's received. The output for S_1 is 0 because two consecutive 0's have not been received. When a second 1 is received, should we go to a new state or go back to S_0? For this problem, it is unnecessary to distinguish between an even number of 1's and no 1's received, so we can go back to S_0. A third 1 then takes us to S_1 (odd number of 1's), a fourth 1 to S_0 (even 1's), and so forth.

FIGURE 14-11

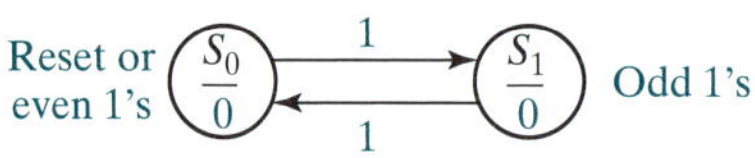

If a 0 is received in S_0, this starts a sequence of two consecutive 0's, so we go to S_2 (0 output) in Figure 14-12. Another 0 then takes us to S_3 to indicate two consecutive 0's received. The output is still 0 in S_3 because the number of 1's received is even. Now if we get a 1 input, we have received an odd number of 1's and go to S_4. (Why can we not go to S_1?) In S_4 we have received two consecutive 0's and an odd number of 1's, so the output is 1.

FIGURE 14-12

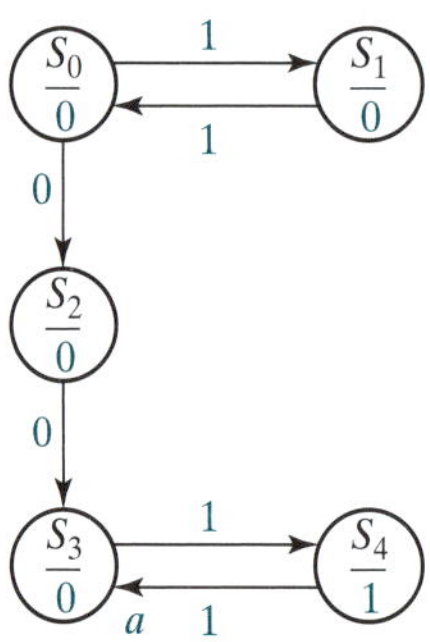

State	Sequence Received
S_0	Reset or even 1's
S_1	Odd 1's
S_2	Even 1's and ends in 0
S_3	Even 1's and 00 has occurred
S_4	00 has occurred and odd 1's

If we receive a 1 in S_4, we have an even number of 1's and two consecutive 0's, so we can return to S_3 (arrow a). The output in S_3 is 0, and when we get another 1 input, the number of 1's is odd, so we again go to S_4 with a 1 output. Now, suppose that we are in S_1 (odd number of 1's received), and we get a 0. We cannot go to S_2 (Why?), so we go to a new state S_5 (Figure 14-13, arrow b) which corresponds to an odd number of 1's followed by a 0. Another 0 results in two consecutive 0's, and we can go to S_4 (arrow c) which gives us a 1 output.

Now, we must go back and complete the state graph by making sure that there are two arrows leaving each state. In S_2, a 1 input means that we have received an odd number of 1's. Because we have not received two consecutive 0's, we must return to S_1 (arrow d) and start counting 0's over again. Similarly, if we receive a 1 in S_5, we return to S_0 (Why?). Now, what should happen if we receive a 0 in S_3? Referring to the original problem statement, we see that once two consecutive 0's have been received, additional 0's can be ignored. Therefore, we can stay in S_3 (arrow f). Similarly, extra 0 inputs can be ignored in S_4 (arrow g). This completes the Moore state diagram, and we should go back and verify that the correct output sequence is obtained for various input sequences.

FIGURE 14-13

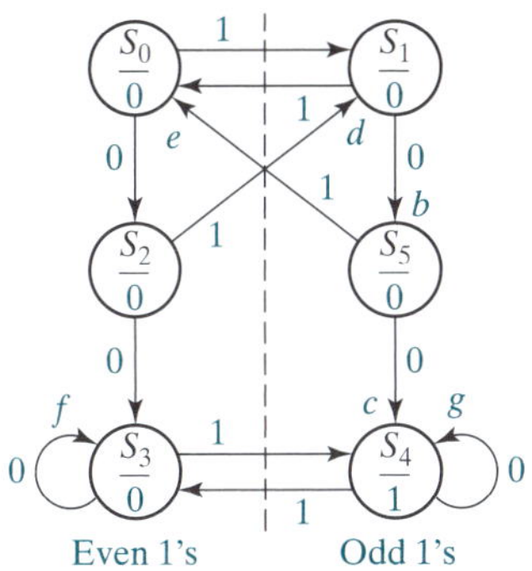

State	Input Sequences
S_0	Reset or even 1's
S_1	Odd 1's
S_2	Even 1's and ends in 0
S_3	Even 1's and 00 has occurred
S_4	Odd 1's and 00 has occurred
S_5	Odd 1's and ends in 0

14.3 Guidelines for Construction of State Graphs

Although there is no one specific procedure which can be used to derive state graphs or tables for every problem, the following guidelines should prove helpful:

1. First, construct some sample input and output sequences to make sure that you understand the problem statement.
2. Determine under what conditions, if any, the circuit should reset to its initial state.
3. If only one or two sequences lead to a nonzero output, a good way to start is to construct a partial state graph for those sequences.
4. Another way to get started is to determine what sequences or groups of sequences must be remembered by the circuit and set up states accordingly.
5. Each time you add an arrow to the state graph, determine whether it can go to one of the previously defined states or whether a new state must be added.
6. Check your graph to make sure there is one and only one path leaving each state for each combination of values of the input variables.
7. When your graph is complete, test it by applying the input sequences formulated in part 1 and making sure the output sequences are correct.

Several examples of deriving state graphs or tables follow.

Example 1

A sequential circuit has one input (X) and one output (Z). The circuit examines groups of four consecutive inputs and produces an output $Z = 1$ if the input sequence 0101 or 1001 occurs. The circuit resets after every four inputs. Find the Mealy state graph.

Solution A typical sequence of inputs and outputs is

$$\begin{array}{rcl|l|l|l} X & = & 0101 & 0010 & 1001 & 0100 \\ Z & = & 0001 & 0000 & 0001 & 0000 \end{array}$$

The vertical bars indicate the points at which the circuit resets to the initial state. Note that an input sequence of either 01 or 10 followed by 01 will produce an output of $Z = 1$. Therefore, the circuit can go to the same state if either 01 or 10 is received. The partial state graph for the two sequences leading to a 1 output is shown in Figure 14-14.

Note that the circuit resets to S_0 when the fourth input is received. Next, we add arrows and labels to the graph to take care of sequences which do not give a 1 output, as shown in Figure 14-15.

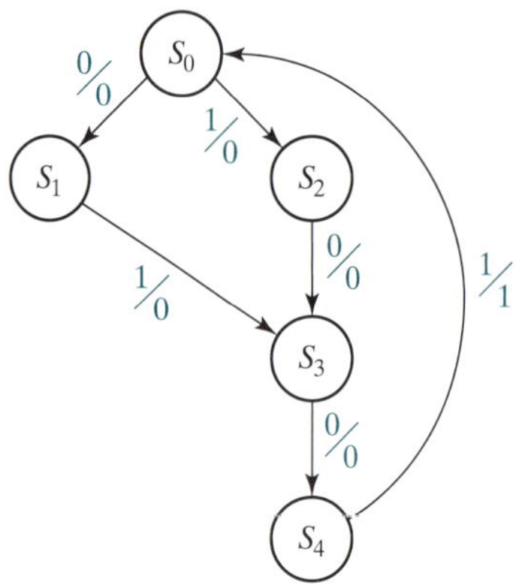

State	Sequence Received
S_0	Reset
S_1	0
S_2	1
S_3	01 or 10
S_4	010 or 100

FIGURE 14-14 Partial State Graph for Example 1

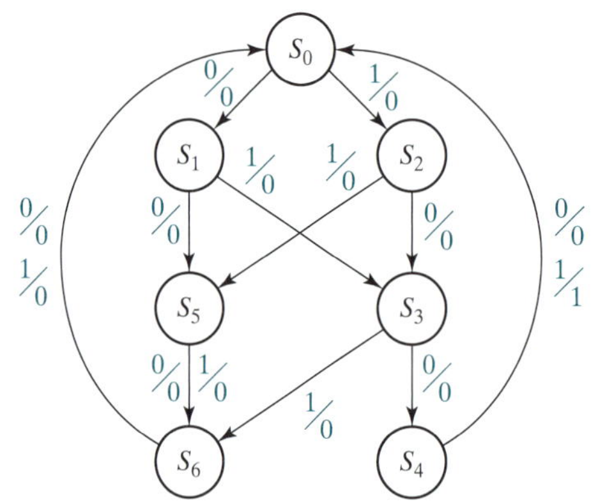

State	Sequence Received
S_0	Reset
S_1	0
S_2	1
S_3	01 or 10
S_4	010 or 100
S_5	Two inputs received, no 1 output is possible
S_6	Three inputs received, no 1 output is possible

FIGURE 14-15 Complete State Graph for Example 1

The addition of states S_5 and S_6 was necessary so that the circuit would not reset to S_0 before four inputs were received. Note that once a 00 or 11 input sequence has been received (state S_5), no output of 1 is possible until the circuit is reset.

Example 2

A sequential circuit has one input (X) and two outputs (Z_1 and Z_2). An output $Z_1 = 1$ occurs every time the input sequence 100 is completed, provided that the sequence 010 has never occurred. An output $Z_2 = 1$ occurs every time the input sequence 010 is completed. Note that once a $Z_2 = 1$ output has occurred, $Z_1 = 1$ can never occur but not vice versa. Find a Mealy state graph and table.

Solution A typical sequence of inputs and outputs is:

$$
\begin{array}{l}
X = 1\ 0\ 0\ 1\ 1\ 0\ 0\ 1\ 0\ 1 \ \vdots\ 0\ 1\ 0\ 0\ 1\ 0\ 1\ 1\ 0\ 1\ 0\ 0 \\
Z_1 = 0\ 0\ 1\ 0\ 0\ 0\ 1\ 0\ 0\ 0 \ \vdots\ 0\ 0\ 0\ 0\ 0\ 0\ 0\ 0\ 0\ 0\ 0\ 0 \\
Z_2 = 0\ 0\ 0\ 0\ 0\ 0\ 0\ 0\ 1\ 0 \ \vdots\ 1\ 0\ 1\ 0\ 0\ 1\ 0\ 0\ 0\ 0\ 1\ 0
\end{array}
$$

Note that the sequence 100 occurs twice before 010 occurs, and $Z_1 = 1$ each time. However, once 010 occurs and $Z_2 = 1, Z_1 = 0$ even when 100 occurs again. $Z_2 = 1$ all five times that 010 occurs. Because we were not told to reset the circuit, 01010 means that 010 occurred twice.

We can begin to solve this problem by constructing the part of the state graph which will give the correct outputs for the sequences 100 and 010. Figure 14-16(a) shows this portion of the state graph.

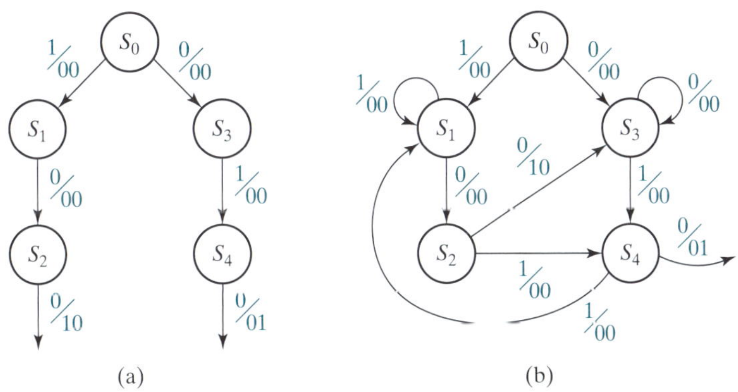

FIGURE 14-16 Partial Graphs for Example 2

An important question to ask at this point is, what does this circuit need to remember to give the correct outputs? The circuit will need to remember how much progress has been made on the sequence 010, so it will know when to output $Z_2 = 1$. The circuit will also need to remember how much progress has been made on the sequence 100 and whether 010 has ever occurred, so it will know when to output $Z_1 = 1$.

Keeping track of what is remembered by each state will help us make the correct state graph. Table 14-5 will help us to do this. State S_0 is the initial state of the

TABLE 14-5
State Descriptions for Example 2

State	Description		
S_0	No progress on 100	No progress on 010	010 has never occurred
S_1	Progress of 1 on 100	No progress on 010	
S_2	Progress of 10 on 100	Progress of 0 on 010	
S_3	No progress on 100	Progress of 0 on 010	
S_4	Progress of 1 on 100	Progress of 01 on 010	
S_5		Progress of 0 on 010	010 has occurred
S_6		Progress of 01 on 010	
S_7		No Progress on 010	

circuit, so there is no progress on either sequence, and 010 has never occurred. State S_1 is the state we go to when a 1 is received from S_0, so in state S_1, we have made progress on the sequence 100 by getting a 1. In state S_1, we have made progress on the sequence 100 by getting 10. Similarly, states S_3 and S_4 represent progress of 0 and 01 toward 010. In S_1, there is no progress toward the sequence 010, and in S_3, there is no progress toward the sequence 100. However, in S_2, we have received 10, so if the next two inputs are 1 and 0, the sequence 010 will be completed. Therefore, in S_2, we have not only made progress of 10 toward 100, but we have also made progress of 0 toward 010. Similarly, in S_4, we have made progress of 1 toward 100, as well as progress of 01 toward 010.

Using this information, we can fill in more of the state graph to get Figure 14-16(b). If the circuit is in state S_1 and a 1 is received, then the last two inputs are 11. The previous 1 is of no use toward the sequence 100. However, the circuit will need to remember the new 1, and there is a progress of 1 toward the sequence 100. There is no progress on the sequence 010, and 010 has never occurred, but this is the same situation as state S_1. Therefore, the circuit should return to state S_1. Similarly, if a 0 is received in state S_3, the last two inputs are 00. There is a progress only of 0 toward the sequence 010, there is no progress toward 100, and 010 has never occurred, so the circuit should return to state S_3. In state S_2, if a 0 is received, the sequence 100 is complete and the circuit should output $Z_1 = 1$. Then, there is no progress on another sequence of 100, and 010 has still not occurred. However, the last input is 0, so there is progress of 0 toward the sequence 010. We can see from Table 14-5 that this is the same situation as S_3, so the circuit should go to state S_3. If, in state S_2, a 1 is received, we have made progress of 01 toward 010 and progress of 1 toward 100, and 010 has still not occurred. We can see from Table 14-5 that the circuit should go to state S_4.

If a 0 is received in state S_4, the sequence 010 is complete, and we should output $Z_2 = 1$. At this point we must go to a new state (S_5) to remember that 010 has been received so that $Z_1 = 1$ can never occur again. When S_5 is reached, we stop looking for 100 and only look for 010. Figure 14-17(a) shows a partial state graph that outputs $Z_2 = 1$ when the input sequence ends in 010. In S_5 we have progress of 0 toward 010 and additional 0's can be ignored by looping back to S_5. In S_6 we have progress of 01 toward 010. If a 0 is received, the sequence is completed, $Z_2 = 1$ and we can go back to S_5 because this 0 starts the 010 sequence again.

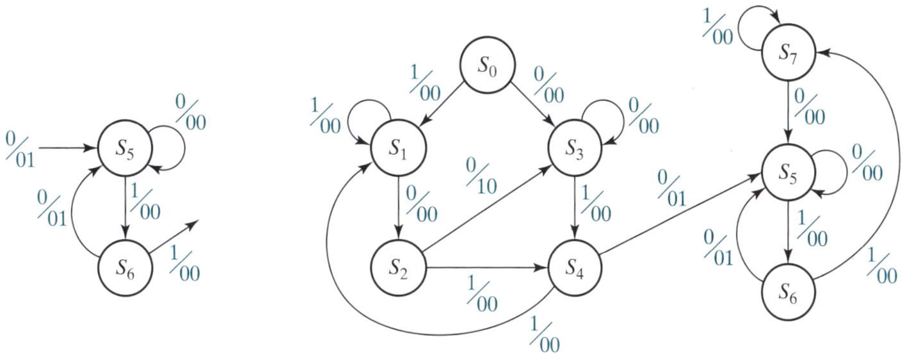

(a) Partial graph for 010 (b) Complete state graph

FIGURE 14-17 State Graphs for Example 2

If we receive a 1 in state S_6, the 010 sequence is broken and we must add a new state (S_7) to start looking for 010 again. In state S_7 we ignore additional 1's, and when a 0 is received, we go back to S_5 because this 0 starts the 010 sequence over again. Figure 14-17(b) shows the complete state graph, and the corresponding table is Table 14-6.

TABLE 14-6

Present State	Next State $X = 0$	$X = 1$	Output (Z_1Z_2) $X = 0$	$X = 1$
S_0	S_3	S_1	00	00
S_1	S_2	S_1	00	00
S_2	S_3	S_4	10	00
S_3	S_3	S_4	00	00
S_4	S_5	S_1	01	00
S_5	S_5	S_6	00	00
S_6	S_5	S_7	01	00
S_7	S_5	S_7	00	00

Example 3

A sequential circuit has two inputs (X_1, X_2) and one output (Z). The output remains a constant value unless one of the following input sequences occurs:

(a) The input sequence $X_1\ X_2 = 01, 11$ causes the output to become 0.
(b) The input sequence $X_1\ X_2 = 10, 11$ causes the output to become 1.
(c) The input sequence $X_1\ X_2 = 10, 01$ causes the output to change value.

(The notation $X_1X_2 = 01,\ 11$ means $X_1 = 0,\ X_2 = 1$ followed by $X_1 = 1,\ X_2 = 1$.) Derive a Moore state graph for the circuit.

Solution The only sequences of input pairs which affect the output are of length two. Therefore, the previous and present inputs will determine the output, and the circuit must remember only the previous input pair. At first, it appears that three states are required, corresponding to the last input received being $X_1X_2 = 01, 10$ and (00 or 11). Note that it is unnecessary to use a separate state for 00 and 11 because neither input starts a sequence which leads to an output change. However, for each of these states the output could be either 0 or 1, so we will initially define six states as follows:

Previous Input (X_1X_2)	Output (Z)	State Designation
00 or 11	0	S_0
00 or 11	1	S_1
01	0	S_2
01	1	S_3
10	0	S_4
10	1	S_5

Using this state designation, we can then set up a state table (Table 14-7). The six-row table given here can be reduced to five rows, using the methods given in Unit 15.

TABLE 14-7

Present State	Z	Next State $X_1X_2 = 00$	01	11	10
S_0	0	S_0	S_2	S_0	S_4
S_1	1	S_1	S_3	S_1	S_5
S_2	0	S_0	S_2	S_0	S_4
S_3	1	S_1	S_3	S_0	S_5
S_4	0	S_0	S_3	S_1	S_4
S_5	1	S_1	S_2	S_1	S_5

The S_4 row of this table was derived as follows. If 00 is received, the input sequence has been 10, 00, so the output does not change, and we go to S_0 to remember that the last input received was 00. If 01 is received, the input sequence has been 10, 01, so the output must change to 1, and we go to S_3 to remember that the last input received was 01. If 11 is received, the input sequence has been 10, 11, so the output should become 1, and we go to S_1. If 10 is received, the input sequence has been 10, 10, so the output does not change, and we remain in S_4. Verify for yourself that the other rows in the table are correct. The state graph is shown in Figure 14-18.

FIGURE 14-18 State Graph for Example 3

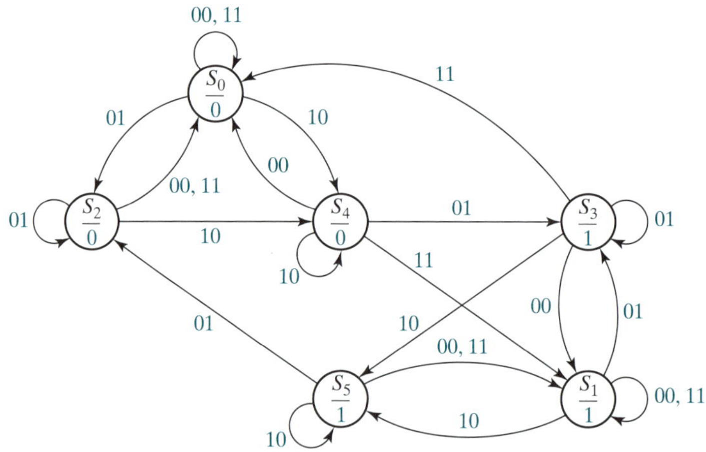

14.4 Serial Data Code Conversion

As a final example of state graph construction, we will design a converter for serial data. Binary data is frequently transmitted between computers as a serial stream of bits. As shown in Figure 14-19(a), a clock signal is often transmitted along with the data, so the receiver can read the data at the proper time. Alternatively (Figure 14-19(b)), only the serial data is transmitted, and a clock recovery circuit (called a digital phase-locked loop) is used to regenerate the clock signal at the receiver.

Figure 14-20 shows four different coding schemes for serial data together with the clock used to synchronize the data transmission. The example shows the transmission of the bit sequence 0, 1, 1, 1, 0, 0, 1, 0. With the NRZ (non-return-to-zero) code, each bit is transmitted for one bit time without any change. With the NRZI (non-return-to-zero-inverted) code, the data is encoded by the presence or absence of transitions in the output signal. For each 0 in the original sequence, the

FIGURE 14-19 Serial Data Transmission

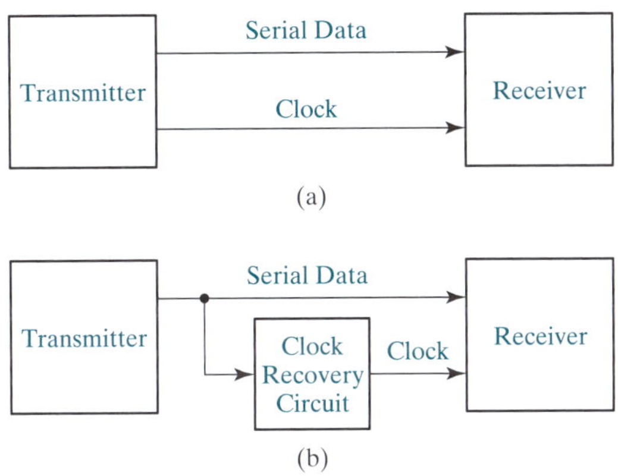

FIGURE 14-20 Coding Schemes for Serial Data Transmission

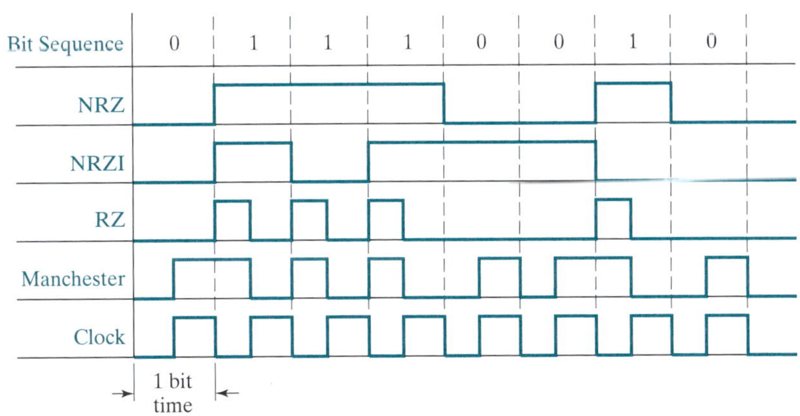

bit transmitted is the same as the previous bit transmitted. For each 1 in the original sequence, the bit transmitted is the complement of the previous bit transmitted. Thus, the preceding sequence is encoded as 0, 1, 0, 1, 1, 1, 0, 0. In other words, a 0 is encoded by no change in the transmitted value, and a 1 is encoded by inverting the previous transmitted value. For the RZ (return-to-zero) code, a 0 is transmitted as a 0 for one full bit time, but a 1 is transmitted as a 1 for the first half of the bit time and, then, the signal returns to 0 for the second half. For the Manchester code, a 0 is transmitted as 0 for the first half of the bit time and 1 for the second half, but a 1 is transmitted as 1 for the first half and 0 for the second half. Thus, the encoded bit always changes in the middle of the bit time. When the original bit sequence has a long string of 1's and 0's, the Manchester code has more transitions. This makes it easier to recover the clock signal.

We will design a sequential circuit which converts an NRZ-coded bit stream to a Manchester-coded bit stream (Figure 14-21(a)). In order to do this, we will use a clock, Clock2, that is twice the frequency of the basic clock (Figure 14-21(b)). In this way, all output changes will occur on the same edge of Clock2, and we can use the standard synchronous design techniques which we have been using in this unit. First, we will design a Mealy circuit to do the code conversion. Note that if the NRZ bit is 0, it will be 0 for two Clock2 periods. Similarly, if the NRZ bit is 1, it will be 1 for two Clock2 periods. Thus, starting in the reset state (S_0 in Figure 14-21(c)), the only two possible input sequences are 00 and 11. For the sequence 00, when the first 0 is received, the output is 0. At the end of the first Clock2 period, the circuit goes to S_1. The input is still 0, so the output becomes 1 and remains 1 for one Clock2 period, and then the circuit resets to S_0. For the sequence 11, when the first 1 is received, the output is 1 for one Clock2 period and, then, the circuit goes to S_2. Then, the output is 0 for one Clock2 period, and the circuit resets to S_0.

When we convert the Mealy graph to a state table (Figure 14-21(d)), the next state of S_1 with an input of 1 is not specified and is represented by a dash. Similarly, the next state of S_2 with a 0 input is not specified. The dashes are like don't-cares, in that we do not care what the next state will be because the corresponding input sequence never occurs. A careful timing analysis for the Mealy circuit shows some possible glitches (false outputs) in the output waveform (Figure 14-21(b)). The input waveform may not be exactly synchronized with the clock, and we have exaggerated this condition in the figure by shifting the input waveform to the right so that the input changes do not line up with the clock edges. For this situation, we will use the state table to analyze the occurrence of glitches in the Z output. The first glitch shown in the timing chart occurs when the circuit is in state S_1, with an input $X = 0$. The state table shows that the output is $Z = 1$, and when the clock goes low, the state changes to S_0. At this time, the input is still $X = 0$, so Z becomes 0. Then X changes to 1, Z becomes 1 again, so a glitch has occurred in the output during the time interval between the clock change and the input change. The next glitch occurs in S_2 with $X = 1$ and $Z = 0$. When the clock goes low, the output momentarily becomes 1 until X is changed to 0.

FIGURE 14-21
Mealy Circuit for NRZ to Manchester Conversion

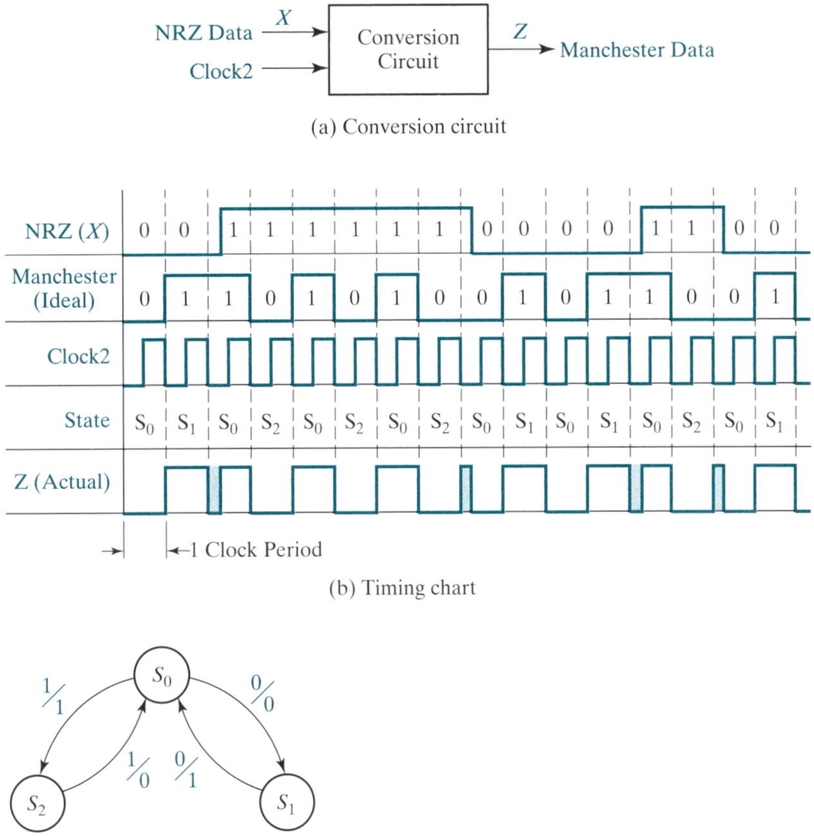

(a) Conversion circuit

(b) Timing chart

(c) State graph

Present State	Next State $X = 0$	Next State $X = 1$	Output (Z) $X = 0$	Output (Z) $X = 1$
S_0	S_1	S_2	0	1
S_1	S_0	–	1	–
S_2	–	S_0	–	0

(d) State table

To overcome the possible glitch problem with the Mealy circuit, we will redesign the circuit in Moore form (Figure 14-22 on the next page). Because the output of a Moore circuit cannot change until after the active edge of the clock, the output will be delayed by one clock period. Starting in S_0, the input sequence 00 takes us to state S_1 with a 0 output and, then, to S_2 with a 1 output. Starting in S_0, 11 takes us to S_3 with a 1 output, and the second 1 can take us back to S_0 which has a 0 output. To complete the graph, we add the two arrows starting in S_2. Note that a 1 input cannot occur in S_1, and a 0 output cannot occur in S_3, so the corresponding state table has two don't-cares.

FIGURE 14-22 Moore Circuit for NRZ to Manchester Conversion

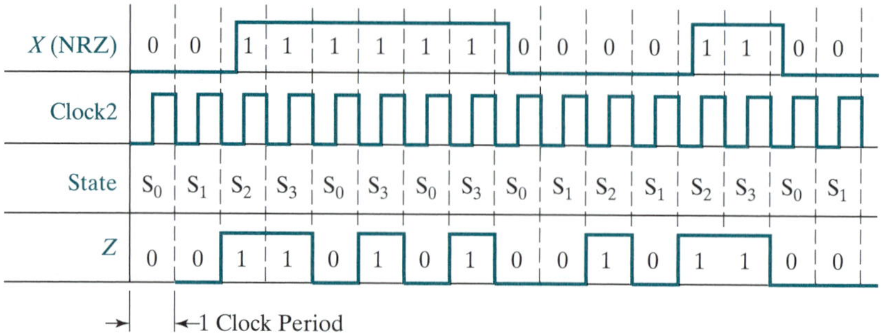

(a) Timing chart

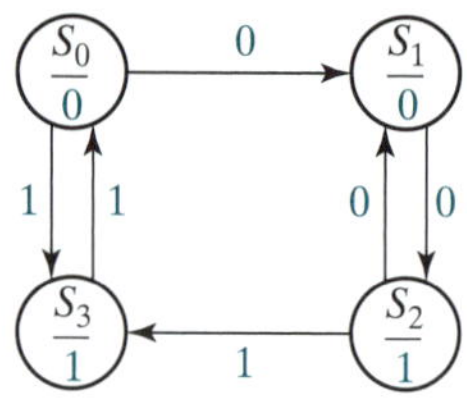

(b) State graph

Present State	Next State $X = 0$	Next State $X = 1$	Present Output (Z)
S_0	S_1	S_3	0
S_1	S_2	–	0
S_2	S_1	S_3	1
S_3	–	S_0	1

(c) State table

14.5 Alphanumeric State Graph Notation

When a state sequential circuit has several inputs, it is often convenient to label the state graph arcs with alphanumeric input variable names instead of 0's and 1's. This makes it easier to understand the state graph and often leads to a simpler state graph. Consider the following example: A sequential circuit has two inputs (F = forward, R = reverse) and three outputs (Z_1, Z_2, and Z_3). If the input sequence is all F's, the output sequence is $Z_1Z_2Z_3Z_1Z_2Z_3$ . . . ; if the input sequence is all R's, the output sequence is $Z_3Z_2Z_1Z_3Z_2Z_1$ . . . . Figure 14-23(a) shows a preliminary Moore state graph that gives the specified output sequences. An arc label F means that the corresponding state transition occurs when $F = 1$. The notation Z_1 within a state means that the output Z_1 is 1, and the other outputs (Z_2 and Z_3) are 0. As long as F is 1, the graph cycles through the states S_0, S_1, S_2, S_0, . . . which gives the output sequence $Z_1Z_2Z_3Z_1$ . . . . When $R = 1$ the state and output sequences occur in reverse order.

At this point the state graph is not completely specified. What happens if both inputs are 0? What happens if both are 1 at the same time? For example, in state S_0 if $F = R = 1$, does the circuit go to state S_1 or to S_2? Because the circuit can only be

FIGURE 14-23 State Graphs with Variable Names on Arc Labels

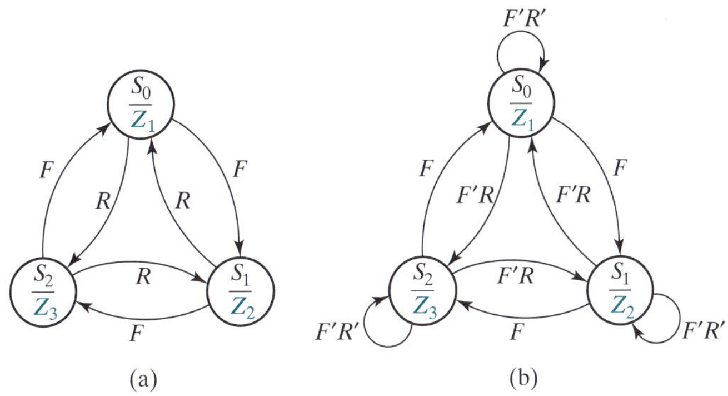

in one state at a time, we must assign a priority. We will assume that input F takes priority over input R. We can then modify the state graph to implement this priority. By replacing R with $F'R$, this means that the corresponding state transition only occurs if $R = 1$ and $F = 0$. When F = R = 0, we will assume that the output should not change. This can be accomplished by adding a self-loop to each state with an arc label $F'R'$. The resulting state graph (Figure 14-23(b)) is completely specified for all combinations of values of F and R, and if both inputs are 1, F takes precedence over R. If we convert the graph to a table, the result is Table 14-8.

TABLE 14-8 State Table for Figure 14-23

PS	NS				Output
	$FR = 00$	01	10	11	$Z_1\,Z_2\,Z_3$
S_0	S_0	S_2	S_1	S_1	1 0 0
S_1	S_1	S_0	S_2	S_2	0 1 0
S_2	S_2	S_1	S_0	S_0	0 0 1

When we construct a state graph using input variable names on the arcs, we should be careful to make sure that the graph is properly specified. To do this, we can check the labels on all the arcs emanating from each state. For state S_0, if we OR together all of the arc labels, we simplify the result to get

$$F + F'R + F'R' = F + F' = 1$$

This result indicates that for any combination of values of the input variables, one of the labels must be 1.

If we AND together every possible pair of arc labels emanating from S_0 we get

$$F \cdot F'R = 0, \quad F \cdot F'R' = 0, \quad F'R \cdot F'R' = 0$$

This result indicates that for any combination of input values, only one arc label can have a value of 1.

In general, a completely specified state graph has the following properties: (1) When we OR together all input labels on arcs emanating from a state, the result reduces to 1. (2) When we AND together any pair of input labels on arcs emanating from a state, the result is 0. Property (1) ensures that for every input combination, at least one next state is defined. Property (2) ensures that for every input combination, no more than one next state is defined. If both properties are true, then exactly one next state is defined, and the graph is properly specified. If we know that certain input combinations cannot occur, then an incompletely specified graph may be acceptable.

We will use the following notation on Mealy state graphs for sequential circuits: X_iX_j/Z_pZ_q means if inputs X_i and X_j are 1 (we don't care what the other input values are), the outputs Z_p and Z_q are 1 (and the other outputs are 0). That is, for a circuit with four inputs (X_1, X_2, X_3, and X_4) and four outputs (Z_1, Z_2, Z_3, and Z_4), X_1X_4'/Z_2Z_3 is equivalent to 1--0/0110. This type of notation is very useful for large sequential circuits where there are many inputs and outputs.

We will use a dash to indicate that all inputs are don't-cares. For example, an arc label $-/Z_1$ means that for any combination of input values, the indicated state transition will occur and the output Z_1 will be 1.

14.6 Incompletely Specified State Tables

In some situations particular input sequences will never occur as inputs to a sequential circuit. In other cases, the output from a sequential circuit is only observed at certain times rather than at every clock time. These situations lead to unspecified next states or outputs in the state table. A state table containing such don't-cares is called an *incompletely specified state table*. Similar to using don't-cares to simplify combinational logic, don't-cares in state tables can be used to simplify the sequential circuit.

As an example of restricted input sequences, assume an input sequence contains successive BCD digits with the most significant bit of each digit occurring first. A Mealy model sequential circuit (a disjoint window sequence detector) examines the input sequence and generates an output of 1 when the last bit of a BCD digit is received if the parity of the digit is even; otherwise, the sequential circuit output is 0. (The parity of a BCD digit is even if it contains an even number of 1's, i.e., zero or two 1's.) A state graph for the sequential circuit is shown in Figure 14-24 and a state table in Table 14-9.

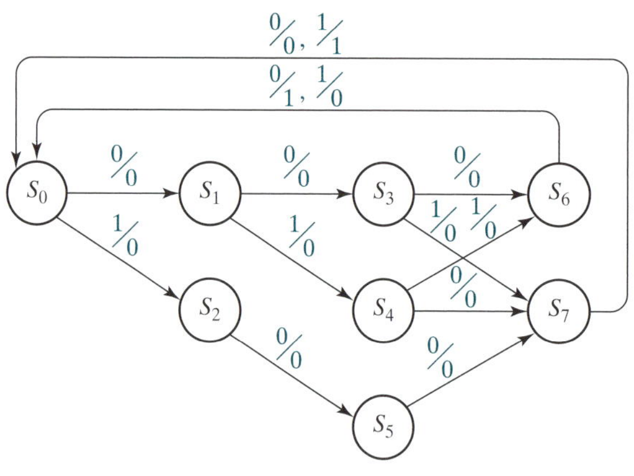

FIGURE 14-24 State Graph for BCD Parity Detector

TABLE 14-9 State Table for Figure 14-24

Present State	Next State $X = 0$	Next State $X = 1$	Output $X = 0$	Output $X = 1$
S_0	S_1	S_2	0	0
S_1	S_3	S_4	0	0
S_2	S_5	–	0	–
S_3	S_6	S_7	0	0
S_4	S_7	S_6	0	0
S_5	S_7	–	0	–
S_6	S_0	S_0	1	0
S_7	S_0	S_0	0	1

State S_0 is the initial state of the circuit and, also, the state at the beginning of each BCD digit. State S_1 (S_2) is the state if the first bit of the BCD digit is 0 (1). After two inputs the possible states are S_3, S_4, and S_5, and after three inputs S_6 and S_7 are the possible states. Note that if the first bit (the most significant bit) is 1, then the next two bits must be 0. (The only BCD digits starting with 1 are 1000 and 1001.) Consequently, the next states and outputs of states S_2 and S_5 for an input of 1 are not specified—they are *don't-cares*. Using the don't-cares to simplify the table (graph) is discussed in Section 15.5.

As an example where the outputs of a sequential circuit are only observed part of the time, consider the 101 sequence detector for disjoint windows and assume that the output from the sequence detector is only used at the end of a window. (A separate counter could be used to determine whether an input is the first, second, or third input of a window.) The outputs from the sequence detector can be left unspecified for the first and second inputs in a window. A state graph for the circuit is shown in Figure 14-25, and Table 14-10 is the corresponding state table. Again the don't-cares allow simplifying the state graph (table); this is discussed in Section 15.5.

FIGURE 14-25 Disjoint Window 101 Detector

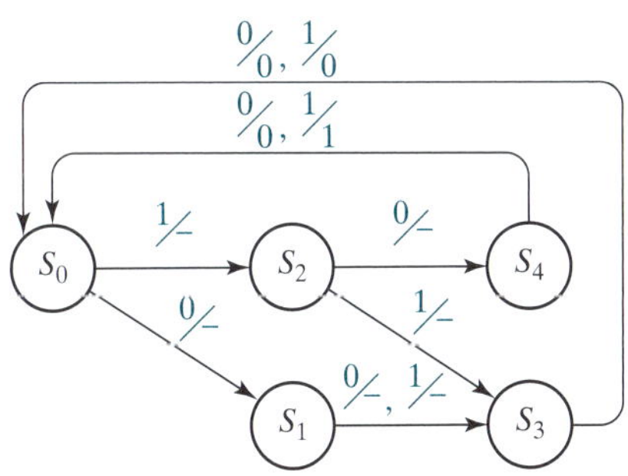

TABLE 14-10 State Table for Figure 14-25

Present State	Next State $X = 0$	Next State $X = 1$	Output $X = 0$	Output $X = 1$
S_0	S_1	S_2	–	–
S_1	S_3	S_3	–	–
S_2	S_4	S_3	–	–
S_3	S_0	S_0	0	0
S_4	S_0	S_0	0	1

Programmed Exercise 14.1

Cover the answer to each problem with a sheet of paper and slide it down as you check your answers. *Write* your answer in the space provided before looking at the correct answers.

Problem: A clocked Mealy sequential circuit with one input (X) and one output (Z) is to be designed. The output is to be 0, unless the input is 0 following a sequence *of exactly* two 0 inputs followed by a 1 input.

To make sure you understand the problem statement, specify the output sequence for each of the following input sequences:

(a) $X = 0010$
$Z =$ ___________

(b) $X = \ldots 10010$ ($\ldots$ means any input sequence not ending in 00)
$Z = \ldots$ ___________

(c) $X = \ldots 00010$ $Z = \ldots$ ___________

(d) $X = 00100100010$
$Z =$ ___________

(e) Does the circuit reset after a 1 output occurs?

Answers

(a) $Z = 0001$ (b) $Z = \ldots 00001$ (c) $Z = \ldots 00000$
(b) $Z = 00010010000$ (e) No
Note that no 1 output occurs in answer (c) because there are three input 0's in a row.

Add arrows to the following graph so that the sequence $X = 0010$ gives the correct output (do not add another state).

Answer

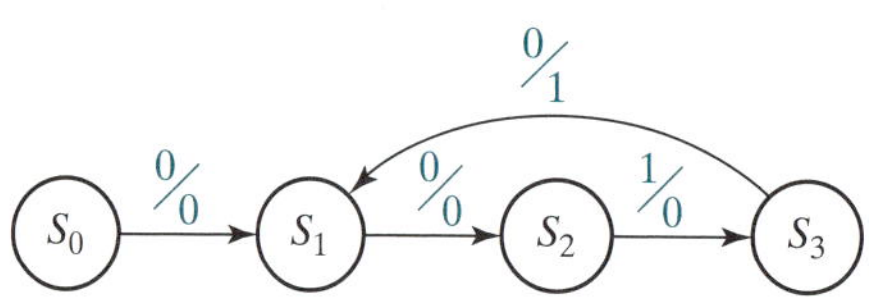

State	Sequence Received
S_0	(Reset)
S_1	0 or 0010
S_2	
S_3	
S_4	

Note that the arrow from S_3 returns to S_1 so that an additional input of 010 will produce another 1 output.

Add a state to the preceding graph which corresponds to "three or more consecutive 0's received." Also complete the preceding table to indicate the sequence received which corresponds to each state.

Answer

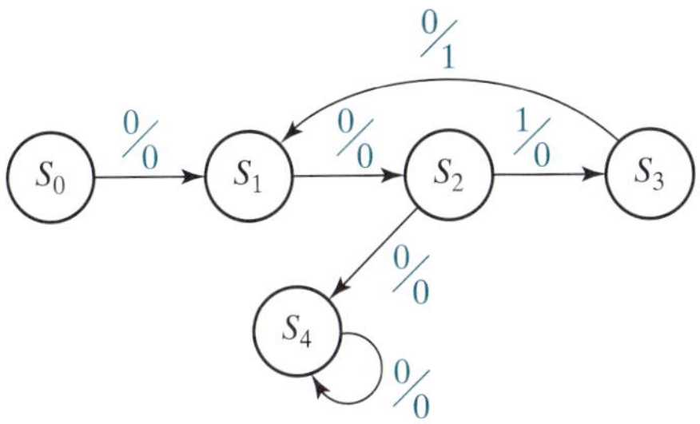

State	Sequence Received
S_0	(Reset)
S_1	0 or 0010
S_2	00
S_3	001
S_4	3 (or more) consecutive 0's

The preceding state graph is not complete because there is only one arrow leaving most states. Complete the graph by adding the necessary arrows. Return to one of the previously used states when possible.

Answer

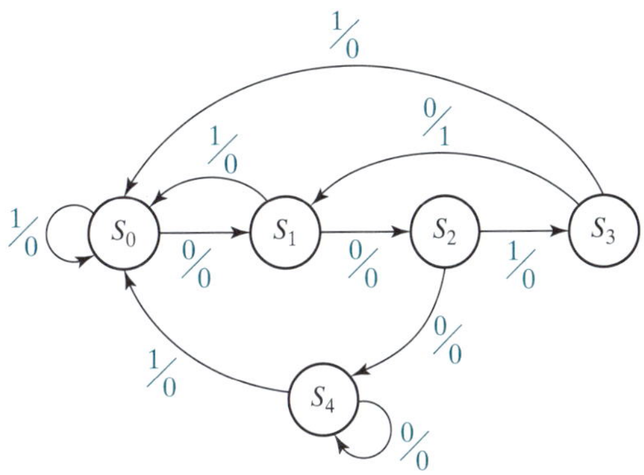

Verify that this state graph gives the proper output sequences for the input sequences listed at the start of this exercise. Write down the Mealy state table which corresponds to the preceding graph.

Answer

Present State	Next State 0	Next State 1	Output 0	Output 1
S_0	S_1	S_0	0	0
S_1	S_2	S_0	0	0
S_2	S_4	S_3	0	0
S_3	S_1	S_0	1	0
S_4	S_4	S_0	0	0

Programmed Exercise 14.2

Problem: A clocked Moore sequential circuit should have an output of $Z = 1$ if the total number of 0's received is an even number greater than zero, provided that two consecutive 1's have never been received.

To make sure that you understand the problem statement, specify the output sequence for the following input sequence:

$X =$ 0 0 0 0 1 0 1 0 1 1 0 0 0 0

$Z =$ (0)_________________

↖ this 0 is the initial output before any inputs have been received

---

Answer $Z =$ (0)01011001100000

Note that once two consecutive 1's have been received, the output can never become 1 again.

To start the state graph, consider only 0 inputs and construct a Moore state graph which gives an ou tput of 1 if the total number of 0's received is an even number greater than zero.

Answer

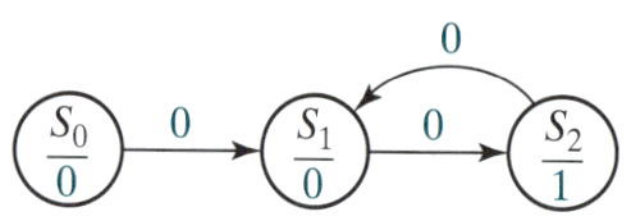

State	Sequence Received
S_0	(Reset)
S_1	Odd number of 0's
S_2	
S_3	
S_4	

Now add states to the above graph so that starting in S_0, if two consecutive 1's are received followed by any other sequence, the output will remain 0. Also, complete the preceding table to indicate the sequence received that corresponds to each state.

Answer

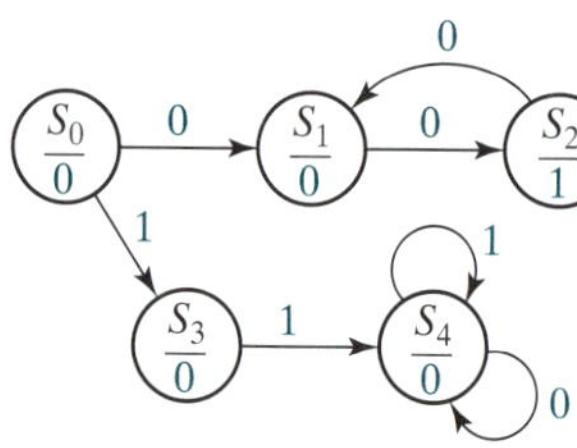

State	Sequence Received
S_0	(Reset)
S_1	Odd number of 0's
S_2	Even number of 0's
S_3	1
S_4	11 (followed by any sequence)
S_5	
S_6	

Now complete the graph so that each state has both a 0 and 1 arrow leading away from it. Add as few extra states to the graph as possible. Also, complete the preceding table.

Answer

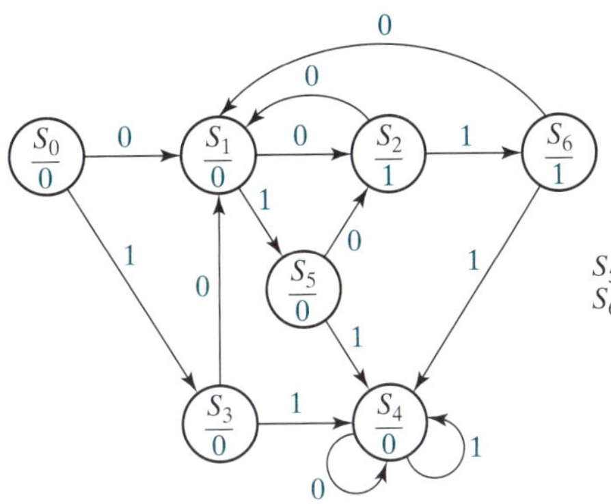

Verify that this state graph gives the proper output sequence for each input sequence at the start of this exercise. Write down the Moore state table which corresponds to the preceding graph. (Note that a Moore table has only one output column.)

Answer

Present State	Next State 0	Next State 1	Output
S_0	S_1	S_3	0
S_1	S_2	S_5	0
S_2	S_1	S_6	1
S_3	S_1	S_4	0
S_4	S_4	S_4	0
S_5	S_2	S_4	0
S_6	S_1	S_4	1

Programmed Exercise 14.3

Derive the state graph and table for a Moore sequential circuit which has an output of 1 iff (1) an even number of 0's have occurred as inputs *and* (2) an odd number of (non overlapping) *pairs* of 1's have occurred. For purposes of this problem, a pair of 1's consists of two consecutive 1's. If three consecutive 1's occur followed by a 0, the third 1 is ignored. If four consecutive 1's occur, this counts as two pairs, etc.

(a) The first step is to analyze the problem and make sure that you understand it. Note that both condition (1) and condition (2) must be satisfied in order to have a 1 output. Consider condition (1) by itself. Would condition (1) be satisfied if zero 0's occurred? __________

If one 0 occurred? __________ Two 0's? __________ Three 0's? __________.

(*Hint*: Is zero an even or odd number? __________)

(b) How many states would it take to determine if condition (1) by itself is satisfied, and what would be the meaning of each state?

____

(c) Now consider condition (2) by itself. For each of the following patterns, determine whether condition (2) is satisfied:

010__________ 0110__________ 01110__________
011110__________ 01010__________ 011010__________
0110110__________

Now check your answers to (a), (b), and (c).

Answers to (a) yes, no, yes, no, even

Answers to (b) two states: even number of 0's, odd number of 0's

Answers to (c) From left to right: no, yes, yes, no, no, yes, no

---

(d) Consider condition (2) by itself and consider an input sequence of consecutive 1's. Draw a Moore state diagram (with only 1 inputs) which will give a 1 output when condition (2) is satisfied. State the meaning of each of the four states in your diagram (for example, odd pairs of 1's).

Answer to (d)

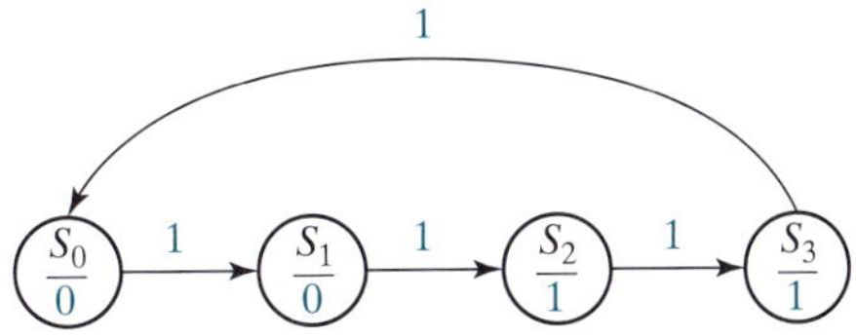

S_0 = even pairs of 1's, S_1 = even pairs of 1's + one 1.
S_2 = odd pairs of 1's, S_3 = odd pairs of 1's + one 1

(e) For the original problem, determine the sequence for Z for the following example:

$X =$ 1 1 0 0 1 1 1 1 1 0 0 0 1 1 1 1 0

$Z = 0$ ______________________________

Answer to (e)

$X =$ 1 1 0 0 1 1 1 1 1 0 0 0 1 1 1 1 0
$Z =$ 0 0 1 0 1 1 0 0 1 1 0 1 0 0 0 0 0 1

(f) Considering that we must keep track of both even or odd 0's, and even or odd pairs of 1's, how many states should the final graph have? ____________

(g) Construct the final Moore state graph. Draw the graph in a symmetric manner with even 0's on the top side and odd 0's on the bottom side. List the meanings of the states such as
S_0 = even 0's and even pairs of 1's.

(h) Check your answer using the test sequence from part (e). Then, check your answers below.

Answer to (f)

Eight states

Answer to (g) S_0 = even 0's and even pairs of 1's, S_1 = even 0's and even pairs of 1's + one 1, S_2 = even 0's and odd pairs of 1's, S_3 = even 0's and odd pairs of 1's + one 1, S_4 = odd 0's and even pairs of 1's, S_5 = odd 0's and even pairs of 1's + one 1, S_6 = odd 0's and odd pairs of 1's, S_7 = odd 0's and odd pairs of 1's + one 1

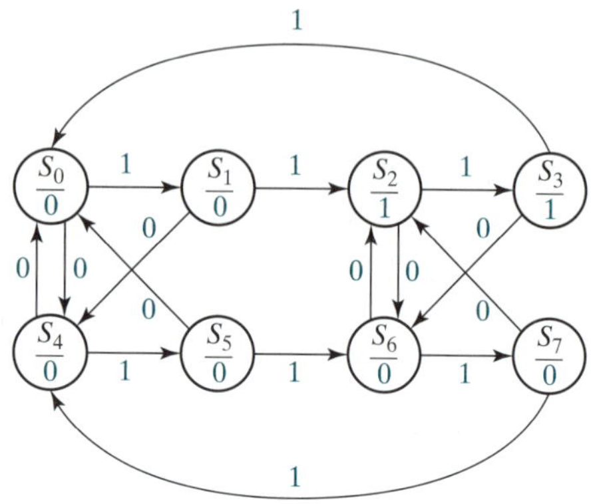

Problems

14.4 A sequential circuit has one input and one output. The output becomes 1 and remains 1 thereafter when at least two 0's *and* at least two 1's have occurred as inputs, regardless of the order of occurrence. Draw a state graph (Moore type) for the circuit (nine states are sufficient). Your final state graph should be neatly drawn with no crossed lines.

14.5 A sequential circuit has one input (X) and two outputs (Z_1 and Z_2). An output $Z_1 = 1$ occurs every time the input sequence 010 is completed, provided that the sequence 100 has never occurred. An output $Z_2 = 1$ occurs every time the input 100 is completed. Note that once a $Z_2 = 1$ output has occurred, $Z_1 = 1$ can never occur but *not* vice versa. Find a Mealy state graph and state table (minimum number of states is eight).

14.6 A sequential circuit has two inputs (X_1 and X_2) and one output (Z). The output begins as 0 and remains a constant value unless one of the following input sequences occurs:

(a) The input sequence $X_1X_2 = 01, 00$ causes the output to become 0.
(b) The input sequence $X_1X_2 = 11, 00$ causes the output to become 1.
(c) The input sequence $X_1X_2 = 10, 00$ causes the output to change value.

Derive a Moore state table.

14.7 A sequential circuit has one input (X) and one output (Z).
Draw a Mealy state graph for each of the following cases:
(a) The output is $Z = 1$ iff the total number of 1's received is divisible by 3. (*Note*: 0, 3, 6, 9,... are divisible by 3.)
(b) The output is $Z = 1$ iff the total number of 1's received is divisible by 3 *and* the total number of 0's received is an even number greater than zero (nine states are sufficient).

14.8 A sequential circuit has two inputs and two outputs. The inputs (X_1 and X_2) represent a 2-bit binary number, N. If the present value of N is greater than the previous value, then Z_1 is 1. If the present value of N is less than the previous value, then Z_2 is 1. Otherwise, Z_1 and Z_2 are 0. When the first pair of inputs is received, there is no previous value of N, so we cannot determine whether the present N is greater than or less than the previous value; therefore, the "otherwise" category applies.
(a) Find a Mealy state table or graph for the circuit (minimum number of states, including starting state, is five).
(b) Find a Moore state table for the circuit (minimum number of states is 11).

14.9 (a) Derive the state graph and table for a Mealy sequential circuit which converts a serial stream of bits from NRZ code to NRZI code. Assume that the clock period is the same as the bit time as in Figure 14-20.
(b) Repeat (a) for a Moore sequential circuit.
(c) Draw a timing diagram for your answer to (a), using the NRZ waveform in Figure 14-20 as the input waveform to your circuit. If the input changes occur slightly after the clock edge, indicate places in the output waveform where glitches (false outputs) can occur.
(d) Draw the timing diagram for your answer to (b), using the same input waveform as in (c).

14.10 For the following state graph, construct the state table, and demonstrate that it is completely specified.

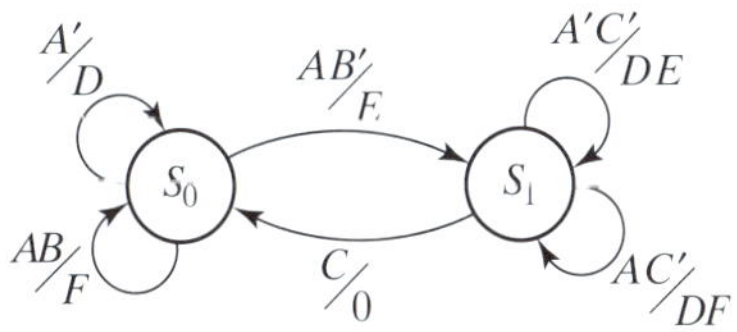

14.11 Design a sequential circuit which will output $Z = 1$ for exactly four clock cycles each time a person pushes a button (which sets $X = 1$). The clock for a digital circuit is usually much faster than a person's finger! The person probably will not have released the button by the time four clock cycles have passed, so X may still be 1 when the four $Z = 1$ outputs have been generated. Therefore, after Z is 1 for four clock cycles, Z should go to 0, until X returns to 0 and then becomes 1 again. Design a Mealy state graph for this circuit, using the alphanumeric state graph notation given in Section 14.5.

14.12 (a) A Moore sequential circuit has one input (x) and one output (z). $z = 1$ if and only if the most recent input *was* 1 and it was preceded by exactly two 0's. Derive a state table for the circuit.

(b) Repeat for a Mealy circuit (i.e., $z = 1$ if and only if the most recent input *is* 1 and it was preceded by exactly two 0's). Derive a state table for the circuit.

14.13 (a) A Mealy sequential circuit has one input (x) and one output (z). z can be 1 when the fourth, eighth, twelfth, etc. inputs are present, and $z = 1$ if and only if the most recent input combined with the preceding three inputs was not a valid BCD encoding for a decimal digit; otherwise, $z = 0$. Assume the BCD digits are received *most* significant bit first. Derive a state table for the circuit. (Eight states are sufficient.)

(b) Repeat for a Moore circuit (i.e., $z = 1$ if and only if, after the fourth, eighth, twelfth, etc. inputs have been received, the previous four inputs were not a valid BCD digit). (Nine states are sufficient.)

(c) Is it possible for a Moore circuit to generate the correct output while the fourth input bit is present rather than after it has been received? Explain your answer.

14.14 (a) A Mealy sequential circuit has one input (x) and one output (z). $z = 1$ if and only if the most recent input, combined with the preceding three inputs, was not a valid BCD encoding of a decimal digit; otherwise, $z = 0$. Assume the BCD digits are received *most* significant bit first. Derive a state table for the circuit. (Seven states are sufficient.)

(b) Repeat for a Moore circuit (i.e., $z = 1$ if and only if the previous four inputs were not a valid BCD digit). (Thirteen states are sufficient.)

(c) Is it possible for a Moore circuit to generate the correct output while the fourth input bit is present rather than after it has been received? Explain your answer.

14.15 (a) A Mealy sequential circuit has one input (x) and one output (z). z can be 1 when the fourth, eighth, twelfth, etc. inputs are present, and $z = 1$ if and only if the most recent input combined with the preceding three inputs was not a valid BCD encoding of a decimal digit; otherwise, $z = 0$. Assume the BCD digits are received *least* significant bit first. Derive a state table for the circuit. (Six states are sufficient.)

(b) Repeat for a Moore circuit (i.e., $z = 1$ if and only if, after the fourth, eighth, twelfth, etc. inputs have been received, the previous four inputs were not a valid BCD digit).

(c) Is it possible for a Moore circuit to generate the correct output while the fourth input bit is present rather than after it has been received? Explain your answer.

14.16 (a) A Mealy sequential circuit has one input (x) and one output (z). $z = 1$ if and only if the most recent input, combined with the preceding three inputs, was not a valid BCD encoding of a decimal digit; otherwise, $z = 0$. Assume the BCD digits are received *least* significant bit first. Derive a state table for the circuit. Assume that in the reset state all previous inputs were 0. (Three states are sufficient.)

(b) Repeat for a Moore circuit (i.e., $z = 1$ if and only if the previous four inputs were not a valid BCD digit). (Four states are sufficient.)

(c) Is it possible for a Moore circuit to generate the correct output while the fourth input bit is present rather than after it has been received? Explain your answer.

14.17 (a) A Mealy sequential circuit has one input (x) and one output (z). z can be 1 when the fourth, eighth, twelfth, etc. inputs are present, and $z = 1$ if and only if the most recent input, combined with the preceding three inputs, was not a valid excess-3 encoding of a decimal digit; otherwise, $z = 0$. Assume the excess-3 digits are received *most* significant bit first. Derive a state table for the circuit. (Ten states are sufficient.)

(b) Repeat for a Moore circuit (i.e., $z = 1$ if and only if, after the fourth, eighth, twelfth, etc. inputs have been received, the previous four inputs were not a valid excess-3 digit). (Eleven states are sufficient.)

(c) Is it possible to for a Moore circuit to generate the correct output while the fourth input bit is present rather than after it has been received? Explain your answer.

14.18 (a) A Mealy sequential circuit has one input (x) and one output (z). $z = 1$ if and only if the most recent input, combined with the preceding three inputs, was not a valid excess-3 encoding of a decimal integer; otherwise, $z = 0$. Assume the excess-3 digits are received *most* significant bit first. Derive a state table for the circuit. Assume that in the reset state all previous inputs were 0. (Eight states are sufficient.)

(b) Repeat for a Moore circuit (i.e., $z = 1$ if and only if the previous four inputs were not a valid excess-3 digit). (Fourteen states are sufficient.)

(c) Is it possible for a Moore circuit to generate the correct output while the fourth input bit is present rather than after it has been received? Explain your answer.

14.19 (a) A Mealy sequential circuit has one input (x) and one output (z). z can be 1 when the fourth, eighth, twelfth, etc. inputs are present, and $z = 1$ if and only if the most recent input, combined with the preceding three inputs, was not a valid excess-3 encoding of a decimal digit; otherwise, $z = 0$. Assume the excess-3 digits are received *least* significant bit first. Derive a state table for the circuit. (Nine states are sufficient.)

(b) Repeat for a Moore circuit (i.e., $z = 1$ if and only if, after the fourth, eighth, twelfth, etc. inputs have been received, the previous four inputs were not a valid excess-3 digit). (Ten states are sufficient.)

(c) Is it possible for a Moore circuit to generate the correct output while the fourth input bit is present rather than after it has been received? Explain your answer.

14.20 (a) A Mealy sequential circuit has one input (x) and one output (z). $z = 1$ if and only if the most recent input combined with the preceding three inputs was not a valid excess-3 encoding of a decimal digit; otherwise, $z = 0$. Assume the excess-3 digits are received *least* significant bit first. Derive a state table for the circuit. Assume that in the reset state all previous inputs were 0. (Six states are sufficient.)

(b) Repeat for a Moore circuit (i.e., $z = 1$ if and only if the previous four inputs were not a valid excess-3 digit). (Eight states are sufficient.)
(c) Is it possible for a Moore circuit to generate the correct output while the fourth input bit is present rather than after it has been received? Explain your answer.

14.21 A sequential circuit has one input and one output. The output becomes 1 and remains 1 thereafter when at least one 1 and three 0's have occurred as inputs, regardless of the order of occurrence. Draw a state graph (Moore type) for the circuit (eight states are sufficient). Your final state graph should be neatly drawn with no crossed lines.

14.22 A sequential circuit has one input (X) and two outputs (Z_1 and Z_2). An output $Z_1 = 1$ occurs every time the input sequence 100 is completed provided that the sequence 011 has never occurred. An output $Z_2 = 1$ occurs every time the input 011 is completed. Note that once a $Z_2 = 1$ output has occurred, $Z_1 = 1$ can never occur but *not* vice versa. Find a Mealy state graph and state table (minimum number of states is eight).

14.23 A sequential circuit has two inputs (X_1 and X_2) and one output (Z). The output begins as 0 and remains a constant value unless one of the following input sequences occurs:
(a) The input sequence $X_1X_2 = 11, 10$ causes the output to become 0.
(b) The input sequence $X_1X_2 = 00, 10$ causes the output to become 1.
(c) The input sequence $X_1X_2 = 01, 10$ causes the output to toggle.
Derive a Moore state table and state graph.

14.24 A sequential circuit has one input (X) and one output (Z).
Draw a Mealy state graph for each of the following cases:
(a) The output is $Z = 1$ iff the total number of 1's received is divisible by 4. (*Note*: 0, 4, 8, 12,... are divisible by 4.)
(b) The output is $Z = 1$ iff the total number of 1's received is divisible by 4 and the total number of 0's received is an odd number (eight states are sufficient).

14.25 A sequential circuit has two inputs and two outputs. The inputs (X_1 and X_2) represent a 2-bit binary number, N. If the present value of N plus the previous value of N is greater than 2, then the Z_1 is 1. If the present value of N times the previous value of N is greater than 2, then Z_2 is 1. Otherwise, Z_1 and Z_2 are 0. When the first pair of inputs is received, use 0 as the previous value of N.
(a) Find a Mealy state table or graph for the circuit (minimum number of states is four).
(b) Find a Moore state table for the circuit (minimum number of states is 10, but any correct answer with 16 or fewer states is acceptable).

14.26 A Moore sequential circuit has one input and one output. When the input sequence 011 occurs, the output becomes 1 and remains 1 until the sequence 011 occurs again in which case the output returns to 0. The output then remains 0 until 011 occurs a third time, etc. For example, the input sequence

$$X = 0\ 1\ 0\ 1\ 1\ 0\ 1\ 0\ 1\ 1\ 0\ 1\ 0\ 0\ 1\ 1\ 1$$

has the output

$$Z = (0)\ 0\ 0\ 0\ 0\ 1\ 1\ 1\ 1\ 1\ 0\ 0\ 0\ 0\ 0\ 0\ 1\ 1$$

Derive the state graph (six states minimum).

14.27 Work Problem 14.26 if the input sequence 101 causes the output to change value. For example, the input sequence

$$X = 0\ 1\ 0\ 1\ 0\ 1\ 0\ 0\ 1\ 0\ 1\ 0\ 1\ 1\ 0\ 1\ 0$$

has the output

$$Z = (0)\ 0\ 0\ 0\ 1\ 1\ 0\ 0\ 0\ 0\ 0\ 1\ 1\ 0\ 0\ 0\ 1\ 1$$

(six states minimum).

14.28 A Mealy sequential circuit has two inputs and one output. If the total number of 0's received is ≥ 4 and at least three pairs of inputs have occurred, then the output should be 1 coincident with the last input pair in the sequence. Whenever a 1 output occurs, the circuit resets. Derive a state graph and state table. Specify the meaning of each state. For example, S_0 means reset, S_1 means one pair of inputs received but no 0's received, etc.

Example:

Input sequence: $X_1 = 1\ 1\ 1\ 0\ 0\ 0\ 1\ 1\ 1\ 0\ 0\ 0\ 1\ 1\ 0\ 0\ 0\ 1\ 0$

$X_2 = 1\ 0\ 0\ 0\ 0\ 0\ 1\ 1\ 1\ 1\ 1\ 1\ 0\ 1\ 0\ 0\ 0\ 1\ 0$

Output sequence: $Z = 0\ 0\ 0\ 1\ 0\ 0\ 1\ 0\ 0\ 0\ 0\ 0\ 1\ 0\ 0\ 1\ 0\ 0\ 1$

14.29 A Moore sequential circuit has one input and one output. The output should be 1 if the total number of 1's received is odd and the total number of 0's received is an even number greater than 0. Derive the state graph and table (six states).

14.30 A Mealy sequential circuit has one input (X) and two outputs (Z_1 and Z_2). The circuit produces an output of $Z_1 = 1$ whenever the sequence 011 is completed, and an output of $Z_2 = 1$ whenever the sequence 0111 is completed. Derive the state graph and table.

14.31 A Moore sequential circuit has two inputs (X_1 and X_2) and one output (Z). Z begins at 0. It becomes 1 when $X_1 = 1$ and $X_2 = 1$ either concurrently, or one after the other (in either order). Z returns to zero when $X_1 = X_2 = 0$. The following input and output sequence should help you understand the problem:

$$X_1 = \quad 0\ 1\ 0\ 0\ 1\ 0\ 0\ 0\ 1\ 1\ 0\ 1\ 1\ 0$$
$$X_2 = \quad 0\ 0\ 1\ 1\ 0\ 0\ 1\ 1\ 0\ 0\ 0\ 1\ 0\ 0$$
$$Z = (0)\ 0\ 0\ 1\ 1\ 1\ 0\ 0\ 0\ 1\ 1\ 0\ 1\ 1\ 0$$

Give the Moore state graph and table.

14.32 A Mealy sequential circuit has one input (X) and one output (Z). The circuit should transmit its input, except that it should prevent the sequence 00110 from occurring. So Z should be the same as X, except that if the input sequence 00110 occurs, Z should be 1 rather than 0 when the last 0 is received, so that the sequence $X = 00110$ is replaced with $Z = 00111$. Derive the state graph and table.

14.33 A Moore sequential circuit has one input and one output. The output is 1 if and only if both of the following conditions are met:
(a) The input sequence contains exactly two groups of 1's, and
(b) Each of these groups contains exactly two 1's.
Each group of 1's must be separated by at least one 0. A single 1 is considered a group of 1's containing one 1. For example, the sequence

$$X = 0\ 1\ 1\ 0\ 0\ 0\ 1\ 1\ 0\ 1\ 1\ 1\ 0$$

satisfies both conditions after the first two pairs of 1's. However, when more 1's appear, condition (a) is no longer satisfied. Therefore, the output sequence should be

$$Z = (0)\ 0\ 0\ 0\ 0\ 0\ 0\ 0\ 1\ 1\ 0\ 0\ 0\ 0$$

On the other hand, the sequence

$$X = 1\ 0\ 1\ 1\ 0\ 1\ 1\ 0$$

never satisfies condition (b), because the first group of 1's contains only one 1. Besides, after the second pair of 1's, (a) is no longer satisfied because the input sequence contains three groups of 1's. Therefore, the output should always be 0.

$$Z = (0)\ 0\ 0\ 0\ 0\ 0\ 0\ 0\ 0$$

Derive a state graph and table.

14.34 A sequential circuit has an input (X) and an output (Z). The output is the same as the input was two clock periods previously. For example,

$$X = 0\ 1\ 0\ 1\ 1\ 0\ 1\ 0\ 1\ 1\ 0\ 1\ 0\ 0\ 0\ 1$$
$$Z = 0\ 0\ 0\ 1\ 0\ 1\ 1\ 0\ 1\ 0\ 1\ 1\ 0\ 1\ 0\ 0$$

The first two values of Z are 0. Find a Mealy state graph and table for the circuit.

14.35 A sequential circuit has an input (X) and an output (Z). The output is the same as the input was three clock periods previously. For example,

$$X = 0\ 1\ 0\ 1\ 1\ 0\ 1\ 0\ 1\ 1\ 0\ 1\ 0\ 0\ 0\ 1$$
$$Z = 0\ 0\ 0\ 0\ 1\ 0\ 1\ 1\ 0\ 1\ 0\ 1\ 1\ 0\ 1\ 0$$

The first three values of Z are 0. Find a Mealy state graph and table for the circuit.

14.36 (a) Construct a Moore state table for the circuit of Problem 14.34. The initial outputs are 0.
(b) How many states are required in a Moore state table for the circuit of Problem 14.35? Explain

14.37 A sequential circuit has an input (X) and two outputs (S and V). X represents a 4-bit binary number N which is input least significant bit first. S represents a 4-bit binary number equal to $N + 2$, which is output least significant bit first. At the time the fourth input occurs, $V = 1$ if $N + 2$ is too large to be represented by four bits; otherwise, $V = 0$. The circuit always resets after the fourth bit of X is received. Find a Mealy state graph and table for the circuit.

Example: $X = 0111$ (binary 14 with the least significant bit first)
$S = 0000$ (because $14 + 2 = 16$, and 16 requires 5 bits)
$V = 0001$

14.38 A sequential circuit has an input (X) and two outputs (D and B). X represents a 4-bit binary number N which is input least significant bit first. D represents a 4-bit binary number equal to $N - 2$, which is output least significant bit first. At the time the fourth input occurs, $B = 1$ if $N - 2$ is less than 0; otherwise $B = 0$. The circuit always resets after the fourth bit of X is received. Find a Mealy state graph and table for the circuit.

Example:

$X = 0001$ 1000 1100
$D = 0110$ 1111 1000
$B = 0000$ 0001 0000

14.39 A sequential circuit has an input (X) and outputs (Y and Z). YZ represents a 2-bit binary number equal to the number of 1's that have been received as inputs. The circuit resets when the total number of 1's received is 3, or when the total number of 0's received is 3. Find a Moore state graph and table for the circuit.

14.40 A sequential circuit has an input X and outputs ($Y$and Z). YZ represents a 2-bit binary number equal to the number of pairs of adjacent 1's that have been received as inputs. For example, the input sequence 0110 contains one pair, the sequence 01110 two pairs, and the sequence 0110111 contains three pairs of adjacent 1's. The circuit *resets* when the total number of pairs of 1's received reaches four. Find a Moore state graph and table for the circuit.

Examples:

Input sequence: $X = 0\,1\,0\,1\,1\,0\,1\,1\,1\,0\,0\,1\,0\,1\,0\,1\,0\,1\,1\,1\,0\,1\,1\,0\,1\,1\,0\,0\,1\,0$
Output sequences: $Y = 0\,0\,0\,0\,0\,0\,0\,1\,1\,1\,1\,1\,1\,1\,1\,1\,1\,1\,0\,0\,0\,0\,0\,0\,0\,1\,1\,1\,1\,1$
$Z = 0\,0\,0\,0\,1\,1\,1\,0\,1\,1\,1\,1\,1\,1\,1\,1\,1\,1\,1\,0\,0\,0\,0\,1\,1\,1\,0\,0\,0\,0\,0$

Input sequence: $X = 1\,1\,1\,1\,1\,1\,1\,1$
Output sequences: $Y = 0\,0\,1\,1\,0\,0\,0\,1$
$Z = 0\,1\,0\,1\,0\,0\,1\,0$

(*Hint*: Be sure that the circuit resets as shown in the examples.)

14.41 A sequential circuit with one input and one output is used to stretch the first two bits of a 4-bit sequence as follows:

Input	Output
00XX	0000
01XX	0011
10XX	1100
11XX	1111

After every 4 bits, the circuit resets. Find a Mealy state graph and table for the circuit. The third and fourth bits of the input sequence can be either 1 or 0, so make sure that the circuit will work for all possible combinations.

14.42 A sequential circuit is to be used to control the operation of a vending machine which dispenses a \$0.25 product. The circuit has three inputs (N, D, and Q) and two outputs (R and C). The coin detector mechanism in the vending machine is synchronized with the same clock as the sequential circuit you are to design. The coin detector outputs a single 1 to the N, D, or Q input for every nickel, dime, or quarter, respectively, that the customer inserts. Only one input will be 1 at a time. When the customer has inserted at least \$0.25 in any combination of nickels, dimes, and quarters, the vending machine must give change and dispense the product. The coin return mechanism gives change by returning nickels to the customer. For every 1 output on C, the coin return mechanism will return one nickel to the customer. The product is dispensed when the circuit outputs a single 1 on output R. The circuit should reset after dispensing the product.

Example:

The customer inserts a nickel, a dime, and a quarter. The circuit inputs and outputs could look like this:

Inputs: $N = 0\,0\,0\,1\,0\,0\,0\,0\,0\,0\,0\,0\,0\,0\,0\,0\,0$
$D = 0\,0\,0\,0\,0\,0\,0\,1\,0\,0\,0\,0\,0\,0\,0\,0\,0$
$Q = 0\,0\,0\,0\,0\,0\,0\,0\,0\,0\,1\,0\,0\,0\,0\,0\,0$

Outputs: $R = 0\,0\,0\,0\,0\,0\,0\,0\,0\,0\,0\,0\,0\,0\,1\,0\,0$
$C = 0\,0\,0\,0\,0\,0\,0\,0\,0\,0\,0\,1\,1\,1\,0\,0\,0$

Note that any number of 0's can occur between 1 inputs.
Derive a Moore state table for the sequential circuit, and for each state indicate how much money the customer has inserted or how much change is due.

14.43 (a) Derive the state graph and table for a Mealy sequential circuit that converts a serial stream of bits from Manchester code to NRZ code. Assume that a double frequency clock (Clock2) is available.
(b) Repeat (a) for a Moore sequential circuit.
(c) Draw a timing diagram similar to Figure 14-21(b) for your answer to (a), using the Manchester waveform in Figure 14-21(b) as the input waveform to your circuit. If the input changes occur slightly after the clock edge, indicate places in the output waveform where glitches (false outputs) can occur. If possible, assign the don't-cares in the output part of your state table to eliminate some of the glitches.
(d) Draw the timing diagram for your answer to (b), using the same input waveform as in (c).

14.44 Design a sequential circuit to control a phone answering machine. The circuit should have three inputs (R, A, and S) and one output (Z). $R = 1$for one clock cycle at the end of each phone ring. $A = 1$ when the phone is answered. S selects whether the machine should answer the phone after two rings ($S = 0$) or four rings ($S = 1$). To cause the tape recorder to answer the phone, the circuit should set the output $Z = 1$ after the end of the second ($S = 0$) or fourth ($S = 1$) ring, and hold $Z = 1$ until the recorder circuit answers the phone (i.e., when A goes to 1). If a person answers the phone at any point, A will become 1, and the circuit should reset. Assume that S is not changed while the phone is counting rings. Give a Moore state graph for this circuit, using the alphanumeric state graph notation given in Section 14.5.

14.45 For the following state graph, derive the state table.

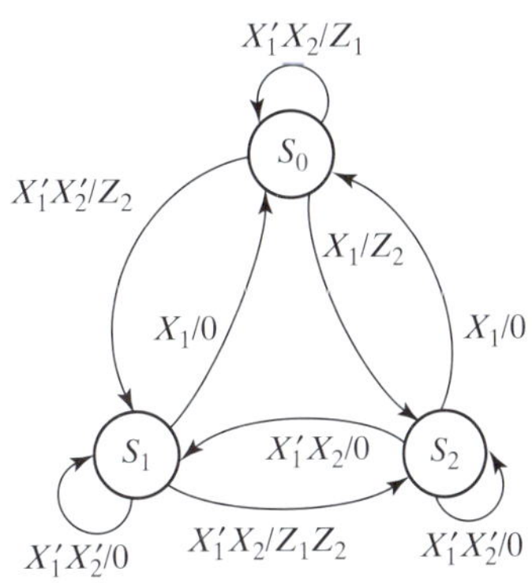

14.46 There are two errors in the state graph shown. One state is not completely specified for one combination of X_1 and X_2. In another state, there is a contradiction for one combination of X_1 and X_2. Correct the state graph by making two minor changes. Demonstrate that the modified state graph is completely specified.

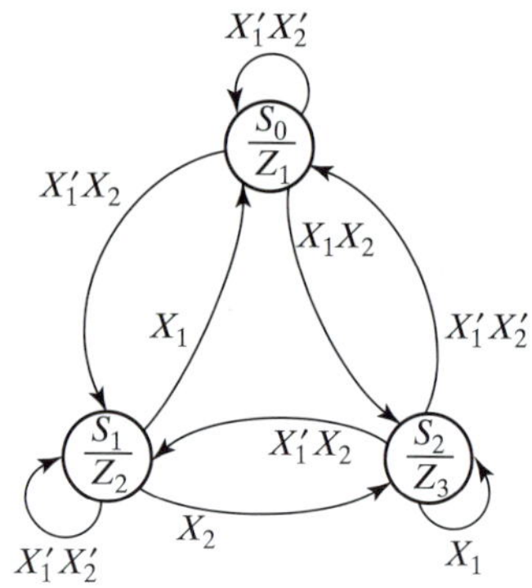

14.47 Modify the BCD parity detector of Table 14-9 for the case where the least significant bit of the BCD digits is received first. Construct the incompletely specified Mealy model state table. (*Hint*: It can be constructed by specifying states to remember the first 3 bits of a BCD digit, but not all sequences of length three have to be distinguished. It requires at most 11 states.)

14.48 In the parity detector of Table 14-9, assume the decimal digits are encoded using the excess-3 code instead of BCD. Construct the incompletely specified state table for this case. (Ten states are required.)

14.49 In the parity detector of Table 14-9, assume the decimal digits are encoded using the Gray code instead of BCD. Construct the incompletely specified state table for this case. (Eleven states are required.)

14.50 The decimal digits 0 through 5 are encoded as 3-bit binary numbers. They are transmitted in disjoint windows of length 3 least significant bit first and are input to a Mealy model circuit. The circuit generates an output of 1 (0) when the third bit is received if the 3 bits have even (odd) parity; the circuit output is a don't-care for the first 2 bits. Construct an incompletely specified table for the circuit.

14.51 Redo Problem 14.50 assuming the bits are sent most significant bit first.

UNIT 15

Reduction of State Tables State Assignment

Objectives

1. Define equivalent states, state several ways of testing for state equivalence, and determine if two states are equivalent.
2. Define equivalent sequential circuits and determine if two circuits are equivalent.
3. Reduce a state table to a minimum number of rows.
4. Specify a suitable set of state assignments for a state table, eliminating those assignments which are equivalent with respect to the cost of realizing the circuit.
5. State three guidelines which are useful in making state assignments, and apply these to making a good state assignment for a given state table.
6. Given a state table and assignment, form the transition table and derive flip-flop input equations.
7. Make a one-hot state assignment for a state graph and write the next-state and output equations by inspection.

Study Guide

1. Study Section 15.1, *Elimination of Redundant States.*
2. Study Section 15.2, *Equivalent States.*
   (a) State in words the meaning of $\lambda_1(p, \underline{X}) = \lambda_2(q, \underline{X})$.

   (b) Assuming that N_1 and N_2 are identical circuits with the following state graph, use Definition 15.1 to show that p is *not* equivalent to q. (Calculate $\lambda(p, \underline{X})$ and $\lambda(q, \underline{X})$ for $\underline{X} = 0$, $\underline{X} = 1$, $\underline{X} = 00$, $\underline{X} = 01$, etc.)

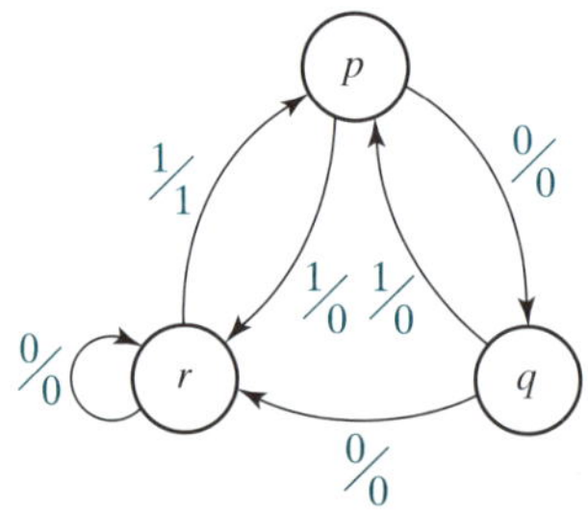

   (c) Suppose you were given two sequential circuits (N_1 and N_2) in black boxes with only input and output terminals available. Each box has a reset button. The button on N_1 resets it to state p and the button on N_2 resets it to state q. Could you experimentally determine if $p \equiv q$ using Definition 15.1? Explain.

   (d) Apply Theorem 15.1 to show that in Table 15-9, $S_2 \not\equiv S_3$.

   (e) Note the difference between the *definition* of state equivalence (Definition 15.1) and the state equivalence *theorem* (Theorem 15.1). The definition requires an examination of output sequences but *not* next states, while the theorem requires looking at *both* the output and next state for each single input. Make sure that you know both the definition and the theorem. Write out the definition of equivalent states:

   Write out the state equivalence theorem:

   When you check your answers, note that the theorem requires *equal* outputs and *equivalent* next states. This distinction between equal and equivalent is very important. For example, in the following state table, no two states have equal next states, but we can still reduce the table to two states, because some next states are equivalent. Note that the state equivalence theorem tells us that $S_3 \equiv S_0$ if $S_3 \equiv S_0$, When this happens, we may say $S_3 \equiv S_0$. What other pair of states are equivalent?

Present State	Next State		Z
S_0	S_1	S_0	0
S_1	S_0	S_2	1
S_2	S_3	S_2	1
S_3	S_1	S_3	0

3. Study Section 15.3, *Determination of State Equivalence Using an Implication Table*.

(a) Fill in the following implication chart to correspond to the given table (first pass only).

	Next State $X = 0$	1	Present Output $X = 0$	1
a	a	b	0	0
b	d	a	0	1
c	a	b	0	1
d	g	f	0	0
f	d	g	0	1
g	d	f	0	1

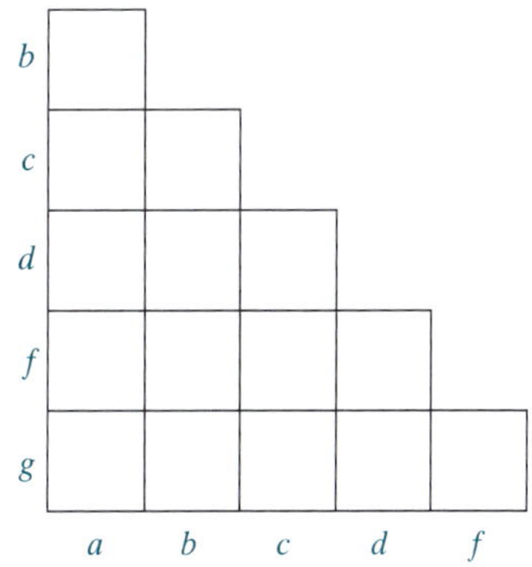

Your answer should have eight squares with X's, two squares with one implied pair, and four squares with two implied pairs. There should be a check in square *f*-*g* because the only nontrivial implication of *f*-*g* is *f*-*g* itself.

(b) Now go through your chart and eliminate all nonequivalent pairs (several passes may be required). What is the only equivalent state pair? According to the state equivalence theorem, why is $b \not\equiv d$? Why is $a \not\equiv b$?

(c) Find all of the equivalent states in the following table using an implication table:

	Next State $X = 0$	1	Present Output $X = 0$	1
a	b	c	0	1
b	d	b	0	0
c	e	a	0	1
d	d	e	0	0
e	e	e	0	0

(You should have found four pairs of equivalent states. If you found only two pairs, reread Section 15.3).
Reduce the table to two rows.

4. Study Section 15.4, *Equivalent Sequential Circuits*. Define equivalent sequential circuits. (Make sure you know the difference between equivalent *states* and equivalent *circuits*.)

5. Work Problems 15.1, 15.2, and 15.3 using the methods of Sections 15.3 and 15.4. When forming the implication charts for state equivalence, follow the convention used in the text. That is, label the bottom of the chart starting with the first state and ending with the next-to-last state. Then, label the left side of the chart starting with the second state at the top and ending with the last state at the bottom.

6. Study Section 15.5, *Reducing Incompletely Specified State Tables*.
   (a) State two reasons why a state table might be incompletely specified.

   (b) For Table 15-5(a), construct the implication chart and find the maximal compatibles. Use the maximal compatibles to construct the reduced table with two states.

   (c) Show that the state table in Study Guide 10 of Unit 14 can be reduced to two states.

7. Read Section 15.6, *Derivation of Flip-Flop Input Equations*.
   (a) Derive J_C and K_C from the C^+ map of Figure 15-10(a).

	00	01	11	10
00				
01				
11				
10				

	00	01	11	10
00				
01				
11				
10				

   (b) Plot the map for the output function (Z) from the transition table of Table 15-9(b) and derive the minimum equation for Z.

	00	01	11	10
00				
01				
11				
10				

(c) Derive the *J-K* input equations for flip-flop *A* from the next-state map of Figure 15-11. Your answers should be

$J_A = X_2B, \qquad K_A = X_2'B'$

(d) Work Problem 15.4.

8. Study Section 15.7, *Equivalent State Assignments*.

(a) Fill in the missing assignments (numbered 8 through 18) in Table 15-11. First, list the remaining assignments with 01 in the first row and then the assignments with 10 in the first row.

(b) Why is it unnecessary to try all possible state assignments to be assured of finding a minimum cost circuit?

(c) For symmetrical flip-flops, why is it always possible to assign all 0's to the starting state and still obtain a minimum circuit?

(d) Complete the following transition table for Table 15-12 using assignment *A*. Then, complete the next-state maps and derive D_1 and D_2.

Q_1Q_2	$X = 0$	1
00	00	10
01		
10		

Q_1Q_2 \ X	0	1
00		
01		
11	X	X
10		

$Q_1^+ = D_1$

Q_1Q_2 \ X	0	1
00		
01		
11	X	X
10		

$Q_2^+ = D_2$

Starting with the equations for assignment *A*, replace all of the 1's with 2's and all 2's with 1's. Verify that the resulting equations are the same as those for assignment *B*.

Starting with the *J* and *K* equations for assignment *A*, replace each *Q* with *Q′* and vice versa. Then, replace the equations for *J* with the corresponding *K* equations and vice versa. (This corresponds to the transformation given in Figure 15-13.) Verify that the resulting equations are the same as for assignment *C*.

Complement the right-hand side of the *D* equations for assignment *A* and, then, replace each *Q* with *Q′* and vice versa. (This corresponds to the transformation given in Figure 15-14.) Verify that the resulting equations are the same as for assignment C.

(e) Show that each of the assignments in Table 15-11 is equivalent to one of the assignments in Table 15-13.

(f) Why are the following two state assignments equivalent in cost?

A	000	011
B	001	111
C	011	101
D	101	110
E	100	010
F	010	001
G	110	000

(g) Show that each of the following assignments can be generated from Table 15-13 by permuting and/or complementing columns:

10	11	01
01	01	11
00	00	00
11	10	10

(h) Why is the trial-and-error method of state assignment of limited usefulness?

(i) Read Problem 15.5, and then answer the following questions regarding state assignments before you work the problem:
(1) Why should a column *not* be assigned all 0's or all 1's?

(2) Why should two columns *not* be given the same assignment?

(3) Does interchanging two columns affect the cost of realizing the circuit?

(4) Does interchanging two rows affect the cost?

(5) Why is an assignment which has two identical rows invalid?

(6) Consider the following two assignments (the number at the top of each column is the decimal equivalent of the binary number in the column):

(1)	(3)	(5)		(3)	(1)	(5)
0	0	0		0	0	0
0	0	1		0	0	1
0	1	0		1	0	0
1	1	1		1	1	1

If we try the column assignment (1) (3) (5), why is it unnecessary to try (3) (1) (5)?

Why is it desirable to assign the column values in increasing numerical order?

9. Study Section 15.8, *Guidelines for State Assignment.*

(a) Why do the guidelines for making state assignments help in making an economical assignment?

(b) What should be done if all the adjacencies specified by the guidelines cannot be satisfied?

(c) The state assignment guidelines for Figure 15-15(a) indicate that the following sets of states should be given adjacent assignments:

$$(1)\ (S_0, S_1, S_3, S_5)\quad (S_3, S_5)\quad (S_4, S_6)\quad (S_0, S_2, S_4, S_6)$$

$$(2)\ (S_1, S_2)\quad (S_2, S_3)\quad (S_1, S_4)\quad (S_2, S_5)2X\quad (S_1, S_6)2X$$

Because the adjacencies from guideline 1 are generally most important, we will start by placing one of the largest groups from guideline 1 in four adjacent squares:

S_0	
S_1	
S_3	
S_5	

Note that (S_3, S_5) is also satisfied by this grouping. Place S_2, S_4, and S_6 in the remaining squares to satisfy as many of the remaining guidelines as possible. Keeping in mind that groups labeled $2X$ should be given preference over groups which are not repeated. Compare your answer with Figure 15-15(b).

(d) Complete the transition table for the state table of Figure 15-17(a), using the assignment of Figure 15-17(b).

	Q_1	Q_2	Q_3	$Q_1^+Q_2^+Q_3^+$ $X = 0$	1	0	1
a	0	0	0	000	100	0	0
b	1	1	1	011	110	0	1
c	1	0	0	100	000	0	0

(e) Complete the next-state and output maps, and verify that the cost of realizing the corresponding equations with an AND-OR gate circuit is 13 gates and 35 gate inputs.

(f) Find J_1 and K_1 from the Q_1^+ map.

$J_1 =$ ____________________

$K_1 =$ ____________________

Q_2Q_3 \ XQ_1	00	01	11	10
00	0	1	0	1
01	X			X
11		0	1	
10				

Q_1^+

Q_2Q_3 \ XQ_1	00	01	11	10
00	0	0	0	0
01	X			X
11		1	1	
10				

Q_2^+

Q_2Q_3 \ XQ_1	00	01	11	10
00	0	0	0	0
01	X			X
11		1	0	
10				

Q_3^+

Q_2Q_3 \ XQ_1	00	01	11	10
00	0	0	0	0
01	X			X
11		0	1	
10				

Z

$D_1 =$ ____________________

$D_2 =$ ____________________

$D_3 =$ ____________________

$Z =$ ____________________

10. Work Problems 15.6, 15.7, and 15.8.

11. Study Section 15.9, *Using a One-Hot State Assignment.*

(a) A one-hot state assignment does not usually give a solution that uses less hardware because of the extra flip-flops required. In what situation is it often advantageous to use it anyway?

(b) It is easy to derive flip-flop input equations directly from a state graph for a one-hot state assignment by inspecting the arcs leading into a given state. Give the next-state equation for Q_5. Only the parts of the state graph which are needed to find Q_5^+ are given.

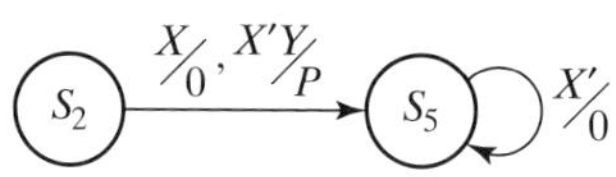

(c) For the state graph in Figure 15-20 and the one-hot state assignment shown, determine the next-state equations for Q_2 and Q_3.

$Q_2^+ =$ ______________ and $Q_3^+ =$ ____________________

(d) For the state graph in Figure 15-20 and the one-hot state assignment shown, determine the output equations for Ad and Done.

$Ad =$ ______________ and Done $=$ ______________

(e) Work Problem 15.9.

12. When you are satisfied that you can meet all of the objectives, take the readiness test.

Reduction of State Tables State Assignment

Given a description of the desired input-output behavior of a sequential circuit, the first step in designing the circuit is to derive a state table using methods similar to the ones discussed in the previous unit. Before we realize this state table using flip-flops and logic gates, reduction of the state table to a minimum number of states is desirable. In general, reducing the number of states in a table will reduce the amount of logic required, and the number of flip-flops may also be reduced. For example, if a table with nine states is reduced to eight states, the number of flip-flops required is reduced from four to three, with a possible corresponding reduction in the amount of input logic for the flip-flops. If the table is further reduced to six states, three flip-flops are still required, but the presence of more don't-cares in the flip-flop input equations will probably further reduce the required logic.

Given the reduced state table, the next step in synthesizing the circuit is to assign binary flip-flop states to correspond to the circuit states. The way in which this assignment is made will determine the amount of logic required for the circuit. The problem of finding a good state assignment which leads to an economical circuit is a difficult one, but some guidelines for achieving this are discussed in Sections 15.7–15.8.

The next step in designing the sequential circuit is to derive the flip-flop input equations. We have already done this for counters in Unit 12, and we will show how to apply these techniques to more general sequential circuits.

15.1 Elimination of Redundant States

In Unit 14, we were careful to avoid introducing unnecessary states when setting up a state graph or table. We will now approach the problem of deriving the state graph somewhat differently. Initially, when first setting up the state table, we will not be overly concerned with inclusion of extra states, but when the table is complete, we will eliminate any redundant states. In previous units, we have used the notation $S_0, S_1, S_2, \ldots$ to represent states in a sequential circuit. In this unit, we will frequently use $A, B, C, \ldots$ (or $a, b, c, \ldots$) to represent these states.

We will rework Example 1 in Section 14.3. Initially, we will set up enough states to remember the first three bits of every possible input sequence. Then, when the fourth bit comes in, we can determine the correct output and reset the circuit to the

TABLE 15-1
State Table for Sequence Detector

Input Sequence	Present State	Next State $X = 0$	Next State $X = 1$	Present Output $X = 0$	Present Output $X = 1$
reset	A	B	C	0	0
0	B	D	E	0	0
1	C	F	G	0	0
00	D	H	I	0	0
01	E	J	K	0	0
10	F	L	M	0	0
11	G	N	P	0	0
000	H	A	A	0	0
001	I	A	A	0	0
010	J	A	A	0	1
011	K	A	A	0	0
100	L	A	A	0	1
101	M	A	A	0	0
110	N	A	A	0	0
111	P	A	A	0	0

initial state. As indicated in Table 15-1, we will designate state A as the reset state. If we receive a 0, we go to state B; if we receive a 1, we go to state C. Similarly, starting in state B, a 0 takes us to state D to indicate that the sequence 00 has been received, and a 1 takes us to state E to indicate that 01 has been received. The remaining states are defined in a similar manner. When the fourth input bit is received, we return to the reset state. The output is 0 unless we are in state J or L and receive a 1, which corresponds to having received 0101 or 1001.

Next, we will attempt to eliminate redundant states from the table. The input sequence information was only used in setting up the table and will now be disregarded. Looking at the table, we see that there is no way of telling states H and I apart. That is, if we start in state H, the next state is A and the output is 0; similarly, if we start in state I, the next state is A and the output is 0. Hence, there is no way of telling states H and I apart, and we can replace I with H where it appears in the next-state portion of the table. Having done this, there is no way to reach state I, so row I can be removed from the table. We say that H is *equivalent* to $I (H \equiv I)$. Similarly, rows K, M, N, and P have the same next state and output as H, so K, M, N, and P can be replaced by H, and these rows can be deleted. Also, the next states and outputs are the same for rows J and L, so $J \equiv L$. Thus, L can be replaced with J and eliminated from the table. The result is shown in Table 15-2.

Having made these changes in the table, rows D and G are identical and so are rows E and F. Therefore, $D \equiv G$, and $E \equiv F$, so states F and G can be eliminated. Figure 15-1 shows a state diagram for the final reduced table. Note that this is identical to the state graph of Figure 14-15, except for the designations for the states. The procedure used to find equivalent states in this example is known as *row matching*. In general, row matching is *not* sufficient to find all equivalent states, except in the special case where the circuit resets to the starting state after receiving a fixed number of inputs.

TABLE 15-2
State Table for Sequence Detector

Present State	Next State $X = 0$	Next State $X = 1$	Present Output $X = 0$	Present Output $X = 1$
A	*B*	*C*	0	0
B	*D*	*E*	0	0
C	~~*F*~~ *E*	~~*G*~~ *D*	0	0
D	*H*	~~*I*~~ *H*	0	0
E	*J*	~~*K*~~ *H*	0	0
~~*F*~~	~~*L* *J*~~	~~*M* *H*~~	~~0~~	~~0~~
~~*G*~~	~~*N* *H*~~	~~*R* *H*~~	~~0~~	~~0~~
H	*A*	*A*	0	0
~~*I*~~	~~*A*~~	~~*A*~~	~~0~~	~~0~~
J	*A*	*A*	0	1
~~*K*~~	~~*A*~~	~~*A*~~	~~0~~	~~0~~
~~*L*~~	~~*A*~~	~~*A*~~	~~0~~	~~1~~
~~*M*~~	~~*A*~~	~~*A*~~	~~0~~	~~0~~
~~*N*~~	~~*A*~~	~~*A*~~	~~0~~	~~0~~
~~*P*~~	~~*A*~~	~~*A*~~	~~0~~	~~0~~

FIGURE 15-1
Reduced State Table and Graph for Sequence Detector

Present State	Next State $X = 0$	Next State $X = 1$	Output $X = 0$	Output $X = 1$
A	*B*	*C*	0	0
B	*D*	*E*	0	0
C	*E*	*D*	0	0
D	*H*	*H*	0	0
E	*J*	*H*	0	0
H	*A*	*A*	0	0
J	*A*	*A*	0	1

(a)

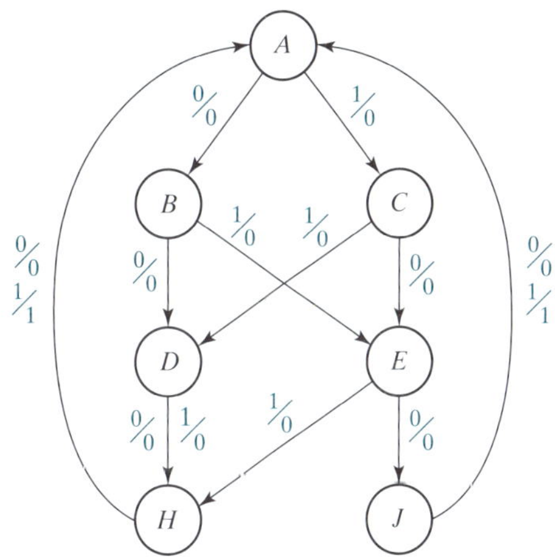

(b)

15.2 Equivalent States

As we have seen in the previous example, state tables can be reduced by eliminating equivalent states. A state table with fewer rows often requires fewer flip-flops and logic gates to realize; therefore, the determination of equivalent states is important in order to obtain economical realizations of sequential circuits.

FIGURE 15-2

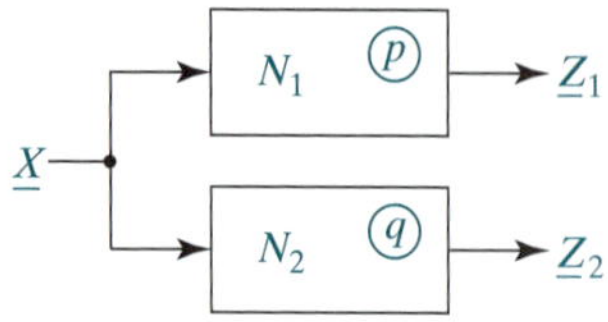

Let us now consider the general problem of state equivalence. Basically, two states are equivalent if there is no way of telling them apart through observation of the circuit inputs and outputs. Consider two sequential circuits (these may be different circuits or two copies of the same circuit), one which is started in state p and one which is started in state q (Figure 15-2): Let $\underline{X}$ represent a sequence of inputs $X_1, X_2, \ldots, X_n$. Feed the same input sequence $\underline{X}$ into both circuits and observe the output sequences $\underline{Z}_1$ and $\underline{Z}_2$. If these output sequences are the same, so far so good. Then, reset the circuits to the states p and q and try a different input sequence for $\underline{X}$ and again compare output sequences. If, for every possible input sequence $\underline{X}$, these output sequences are the same, then there is no way of telling states p and q apart by observing the terminal behavior of the circuits, and we say p is equivalent to $q (p \equiv q)$. On the other hand, if, for some input sequence $\underline{X}$, the output sequences $\underline{Z}_1$ and $\underline{Z}_2$. are different, then we can distinguish between states p and q, and they are not equivalent. Because the output sequence is a function of the initial state and the input sequence, we will write

$$\underline{Z}_1 = \lambda_1(p, \underline{X}) \quad \underline{Z}_2 = \lambda_2(q, \underline{X})$$

We can then state formally the definition of state equivalence as follows:

Definition 15.1

Let N_1 and N_2 be sequential circuits (not necessarily different). Let $\underline{X}$ represent a sequence of inputs of arbitrary length. Then state p in N_1 is equivalent to state q in N_2 iff $\lambda_1(p, \underline{X}) = \lambda_2(q, \underline{X})$ for every possible input sequence $\underline{X}$.

To apply Definition 15.1 directly, we should first test the circuits with $\underline{X} = 0$ and $\underline{X} = 1$. Then, we should test with all input sequences of length 2: $\underline{X} = 00, 01, 10$, and 11. Next, we should test with all input sequences of length 3: $\underline{X} = 000, 001, 010, 011, 100$, 101, 110, and 111. We should then continue this process with all input sequences of length 4, length 5, and so forth. Definition 15.1 is not practical to apply directly in practice because it requires testing the circuit with an infinite number of input sequences in order to prove that two states are equivalent. A more practical way of testing for state equivalence uses the following theorem:

Theorem15.1[1] Two states p and q of a sequential circuit are equivalent iff for every single input X, the outputs are the same and the next states are equivalent, that is,

$$\lambda(p, X) = \lambda(q, X) \qquad \text{and} \qquad \delta(p, X) \equiv \delta(q, X)$$

where $\lambda(p, X)$ is the output given the present state p and input X, and $\delta(p, X)$ is the next state given the present state p and input X. Note that the next states do not have to be equal, just equivalent. For example, in Table 15-1, $D \equiv G$, but the next states (H and N for $X = 0$, and I and P for $X = 1$) are not equal.

[1]See Appendix D for proof.

The row matching procedure previously discussed is a special case of Theorem 15.1 where the next states are actually the same instead of just being equivalent. We will use this theorem to show that Table 13-4 has no equivalent states. By inspection of the output part of the table, the only possible pair of equivalent states is S_0 and S_2. From the table,

$$S_0 \equiv S_2 \quad \text{iff} \quad (S_3 \equiv S_3,\ S_2 \equiv S_0,\ S_1 \equiv S_1, \text{ and } S_0 \equiv S_1)$$

But $S_0 \not\equiv S_1$ (because the outputs differ), so the last condition is not satisfied and $S_0 \not\equiv S_2$.

15.3 Determination of State Equivalence Using an Implication Table

In this section we will discuss a procedure for finding all of the equivalent states in a state table. If the equivalent states found by this procedure are eliminated, then the table can be reduced to a minimum number of states. We will use an implication table (sometimes referred to as a pair chart) to check each pair of states for possible equivalence. The nonequivalent pairs are systematically eliminated until only the equivalent pairs remain.

We will use the example of Table 15-3 to illustrate the implication table method. The first step is to construct a chart of the form shown in Figure 15-3. This chart has a square for every possible pair of states. A square in column i and row j corresponds to state pair i-j. Thus, the squares in the first column correspond to state pairs a-b, a-c, etc. Note that the squares above the diagonal are not included in the chart because if $i \equiv j, j \equiv i$, and only one of the state pairs i-j and j-i is needed. Also, squares corresponding to pairs a-a, b-b, etc., are omitted. To fill in the first column of the chart, we compare row a of Table 15-3 with each of the other rows. Because the output for row a is different than the output for row c, we place an X in a-c square of the chart to indicate that $a \not\equiv c$. Similarly, we place X's in squares a-e, a-f, and a-h to indicate that $a \not\equiv e$, $a \not\equiv f$, and $a \not\equiv h$ because of output differences. States a and b have the same outputs, and thus, by Theorem 15.1,

$$a \equiv b \quad \text{iff} \quad d \equiv f \quad \text{and} \quad c \equiv h$$

To indicate this, we place the *implied pairs*, d-f and c-h, in the a-b square. Similarly, because a and d have the same outputs, we place a-d and c-e in the a-d square to indicate that

$$a \equiv d \quad \text{iff} \quad a \equiv d \quad \text{and} \quad c \equiv e$$

TABLE 15-3

Present State	Next State X = 0	1	Present Output
a	d	c	0
b	f	h	0
c	e	d	1
d	a	e	0
e	c	a	1
f	f	b	1
g	b	h	0
h	c	g	1

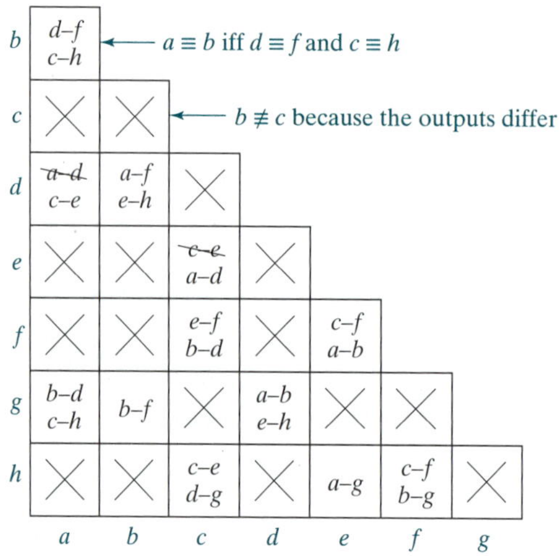

FIGURE 15-3 Implication Chart for Table 15-3

The entries *b-d* and *c-h* in the *a-g* square indicate that

$$a \equiv g \quad \text{iff} \quad b \equiv d \quad \text{and} \quad c \equiv h$$

Next, row *b* of the state table is compared with each of the remaining rows of the table, and column *b* of the implication chart is filled in. Similarly, the remaining columns in the chart are filled in to complete Figure 15-3. Self-implied pairs are redundant, so *a-d* can be eliminated from square *a-d*, and *c-e* from square *c-e*.

At this point, each square in the implication table has either been filled in with an X to indicate that the corresponding state pair is not equivalent (because the outputs are different) or filled in with implied pairs. We now check each implied pair. If one of the implied pairs in square *i-j* is not equivalent, then by Theorem 15.1, $i \not\equiv j$. The *a-b* square of Figure 15-3 contains two implied pairs (*d-f* and *c-h*). Because $d \not\equiv f$ (the *d-f* square has an X in it), $a \not\equiv b$ and we place an X in the *a-b* square, as shown in Figure 15-4. Continuing to check the first column, we note that the *a-d* square contains the implied pair *c-e*. Because square *c-e* does not contain an X, we cannot determine at

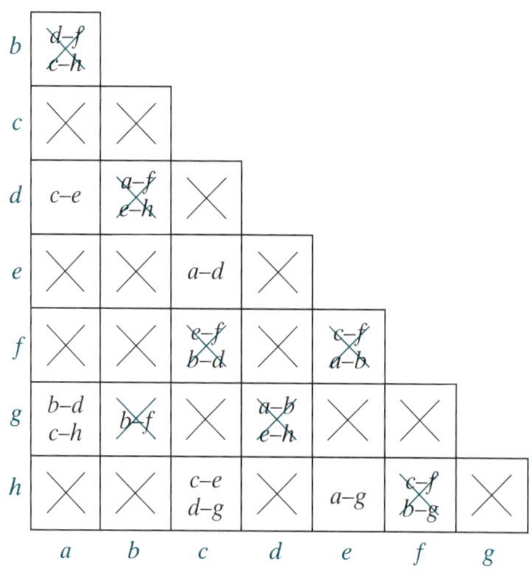

FIGURE 15-4 Implication Chart After First Pass

FIGURE 15-5
Implication Chart After Secong Pass

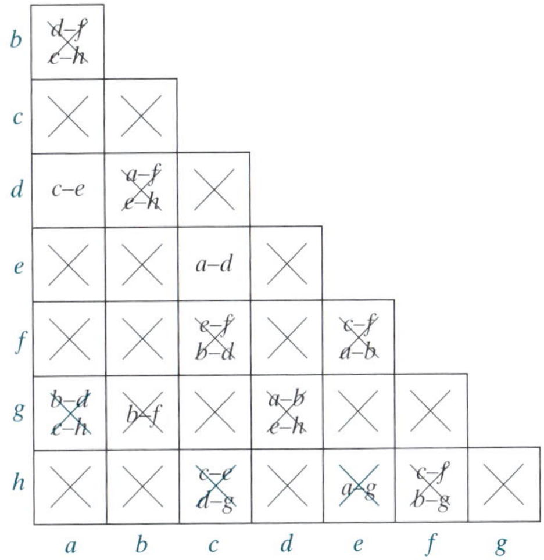

this point whether or not $a \not\equiv d$. Similarly, because neither square *b-d* nor *c-h* contains an X, we cannot determine immediately whether $a \not\equiv g$ or not. Going on to the second column, we place X's in squares *b-d* and *b-g* because we have already shown $a \not\equiv f$ and $b \not\equiv f$. In a similar manner, we check each of the remaining columns and X out squares *c-f*, *d-g*, *e-f*, and *f-h*. Figure 15-4 shows the resulting chart.

In going from Figure 15-3 to Figure 15-4, we found several additional nonequivalent state pairs. Therefore, we must go through the chart again to see if the added X's make any other pairs nonequivalent. Rechecking column *a*, we find that we can place an X in square *a-g* because square *b-d* has an X. Checking the remaining columns, we X out squares *c-h* and *e-h* because *d-g* and *a-g* have X's. This completes the second pass through the implication table, as shown in Figure 15-5. Because we added some X's on the second pass, a third pass is required.

No new X's are added on the third pass through the table, so all squares which correspond to nonequivalent state pairs have been Xed out. The coordinates of the remaining squares must then correspond to equivalent state pairs. Because square *a-d* (in column *a*, row *d*) does not contain an X, we conclude that $a \equiv d$. Similarly, square *c-e* does not contain an X, so $c \equiv e$. All other squares contain X's, so there are no other equivalent state pairs. Note that we determined equivalent states from the column-row coordinates of the squares without X's, *not* by reading the implied pairs contained within the squares.

If we replace *d* with *a* and *e* with *c* in Table 15-3, we can eliminate rows *d* and *e*, and the table reduces to six rows, as shown in Table 15-4.

TABLE 15-4

Present State	Next State $X = 0$	1	Output
a	a	c	0
b	f	h	0
c	c	a	1
f	f	b	1
g	b	h	0
h	c	g	1

The implication table method of determining state equivalence can be summarized as follows:

1. Construct a chart which contains a square for each pair of states.
2. Compare each pair of rows in the state table. If the outputs associated with states i and j are different, place an X in square i-j to indicate that $i \not\equiv j$. If the outputs are the same, place the implied pairs in square i-j. (If the next states of i and j are m and n for some input x, then m-n is an implied pair.) If the outputs and next states are the same (or if i-j only implies itself), place a check ($\surd$) in square i-j to indicate that $i \equiv j$.
3. Go through the table square-by-square. If square i-j contains the implied pair m-n, and square m-n contains an X, then $i \not\equiv j$, and an X should be placed in square i-j.
4. If any X's were added in step 3, repeat step 3 until no more X's are added.
5. For each square i-j which does not contain an X, $i \equiv j$.

If desired, row matching can be used to partially reduce the state table before constructing the implication table. Although we have illustrated this procedure for a Moore table, the same procedure applies to a Mealy table.

15.4 Equivalent Sequential Circuits

In the last section, we found the equivalent states within a single state table so that we could reduce the number of rows in the table. Reducing the number of rows usually leads to a sequential circuit with fewer gates and flip-flops. In this section, we will consider equivalence between sequential circuits. Essentially, two sequential circuits are equivalent if they are capable of doing the same "work." Equivalence between sequential circuits is defined as follows:

Definition 15.2 Sequential circuit N_1 is equivalent to sequential circuit N_2 if for each state p in N_1, there is a state q in N_2 such that $p \equiv q$, and conversely, for each state s in N_2, there is a state t in N_1 such that $s \equiv t$.

Thus, if $N_1 \equiv N_2$, for every starting state p in N_1, we can find a corresponding starting state q such that $\lambda_1(p, \underline{X}) \equiv \lambda_2(q, \underline{X})$ for all input sequences $\underline{X}$ (i.e., the output sequences are the same for the same input sequence). Then, in a given application, N_1 could be replaced with its equivalent circuit N_2.

If N_1 and N_2 have only a few states, one way to show that $N_1 \equiv N_2$ is to match up pairs of equivalent states by inspection and, then, show that Theorem 15.1 is satisfied for each pair of equivalent states. If both N_1 and N_2 have a minimum number of states and $N_1 \equiv N_2$, then N_1 and N_2 must have the same number of states. Otherwise, one circuit would have a state left over which was not equivalent to any state in the other circuit, and Definition 15.2 would not be satisfied.

Figure 15-6 shows two reduced state tables and their corresponding state graphs. By inspecting the state graphs, it appears that if the circuits are equivalent, we must have A

FIGURE 15-6 Tables and Graphs for Equivalent Circuits

	N_1			
	$X = 0$	1	$X = 0$	1
A	B	A	0	0
B	C	D	0	1
C	A	C	0	1
D	C	B	0	0

	N_2			
	$X = 0$	1	$X = 0$	1
S_0	S_3	S_1	0	1
S_1	S_3	S_0	0	0
S_2	S_0	S_2	0	0
S_3	S_2	S_3	0	1

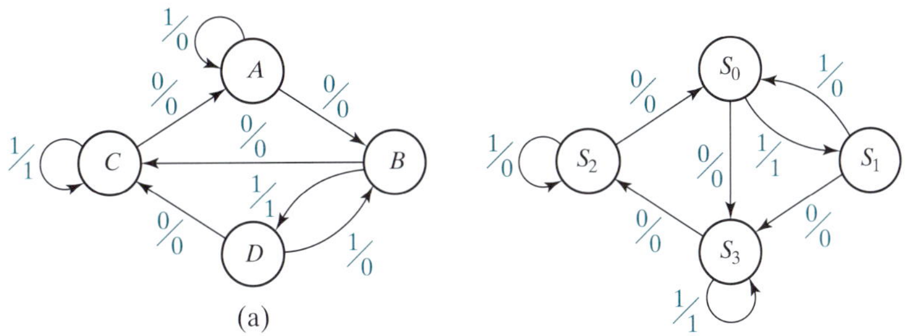

(a) (b)

equivalent to either S_2 or S_3 because these are the only states in N_2 with self-loops. Because the outputs of A and S_2 correspond, the only possibility is $A \equiv S_2$. If we assume that $A \equiv S_2$, this implies that we must have $B \equiv S_0$, which in turn implies that we must have $D \equiv S_1$ and $C \equiv S_3$. Using the state tables, we can verify that these assumptions are correct because for every pair of assumed equivalent states, the next states are equivalent and the outputs are equal when $X \equiv 0$ and also when $X \equiv 1$. This verifies that $N_1 \equiv N_2$.

The implication table can easily be adapted for determining the equivalence of sequential circuits. Because the states of one circuit must be checked for equivalence against states of the other circuit, an implication chart is constructed with rows corresponding to states of one circuit and columns corresponding to states of the other. For example, for the circuits of Figure 15-6 we can set up the implication table of Figure 15-7(a). The first column of Figure 15-7(a) is filled in by comparing row A of the state table in Figure 15-6(a) with each of the rows in Figure 15-6(b). Because states A and S_0 have different outputs, an X is placed in the A-S_0 square. Because states A and S_1 have the same outputs, the implied next-state pairs (B-S_3 and A-S_0) are placed in the A-S_1 square, etc. The remainder of the table is filled in similarly.

In the next step (Figure 15-7(b)), squares corresponding to additional nonequivalent state pairs are crossed out. Thus, square A-S_1 is crossed out because $A \not\equiv S_0$.

FIGURE 15-7 Implication Tables for Determining Circuit Equivalence

	A	B	C	D
S_0	X	C–S_3 D–S_1	A–S_3 C–S_1	X
S_1	B–S_3 A–S_0	X	X	C–S_3 B–S_0
S_2	B–S_0 A–S_2	X	X	C–S_0 B–S_2
S_3	X	C–S_2 D–S_3	A–S_2 C–S_3	X

(a)

	A	B	C	D
S_0	X	C–S_3 D–S_1	~~A–S_3 C–S_1~~	X
S_1	~~B–S_3 A–S_0~~	X	X	C–S_3 B–S_0
S_2	B–S_0 A–S_2	X	X	~~C–S_0 B–S_2~~
S_3	X	~~C–S_2 D–S_3~~	A–S_2 C–S_3	X

(b)

Similarly, square B-S_3 is crossed out because $C \not\equiv S_2$, square C-S_0 because $A \not\equiv S_3$, and square D-S_2 because $B \not\equiv S_2$. Another pass through the table reveals no additional nonequivalent pairs; therefore, the remaining equivalent state pairs are

$$A \equiv S_2 \qquad B \equiv S_0 \qquad C \equiv S_3 \qquad D \equiv S_1$$

Since each state in N_1 has an equivalent state in N_2 and conversely, $N_1 \equiv N_2$.

15.5 Reducing Incompletely Specified State Tables

Reducing incompletely specified state tables is more complicated than reducing completely specified tables. We give only a brief introduction here.[2] At first it might be thought that trying different values for the don't-cares and reducing the resulting completely specified table will produce the minimal state table. We give two examples to show that this would be very complex for tables with many don't-cares and, in fact, often does not produce the minimal table. Table 15-5(a) contains two don't-cares labelled $-^1$ and $-^2$ so we can try reducing it by trying the four combinations of values for the don't-cares. Figure 15-8 shows the implication charts for the four combinations (DC1 = 0and1, DC2 = 0and1); the table only reduces when both don't-cares are 1. In this case the resulting two-state table is the minimal table. However, trying all possibilities if there are many don't-cares would be very complex and, as Table 15-5(b) shows, does not always produce a minimal table.

If the don't-care in Table 15-5(b) is replaced by a 0, the outputs make S_0 and S_2 plus S_1 and S_2 nonequivalent. Then S_0 and S_1 are not equivalent because they have next states S_1 and S_2. Similarly, there are no equivalent states if the don't-care is replaced by a 1. However, this table can be reduced to two states. If S_1 is duplicated with one copy made equivalent to S_0 and the other copy made equivalent to S_2 as shown in Table 15-6, then S_0 is equivalent to S_1^1 and S_2 is equivalent to S_1^2 and the table can be reduced to two states.

TABLE 15-5 Incompletely Specified Examples

Present State	Next State $X=0$	Next State $X=1$	Output $X=0$	Output $X=1$
S_0	S_1	S_3	0	$-^2$
S_1	S_2	S_3	$-^1$	0
S_2	S_1	S_0	1	0
S_3	S_2	S_3	0	1

(a)

Present State	Next State $X=0$	Next State $X=1$	Output $X=0$	Output $X=1$
S_0	S_2	S_1	0	1
S_1	S_1	S_0	–	1
S_2	S_2	S_1	1	1

(b)

[2]See McCluskey, Edward J., *Logic Design Principles* (Upper Saddle River, NJ: Prentice Hall, 1986) for a more complete discussion.

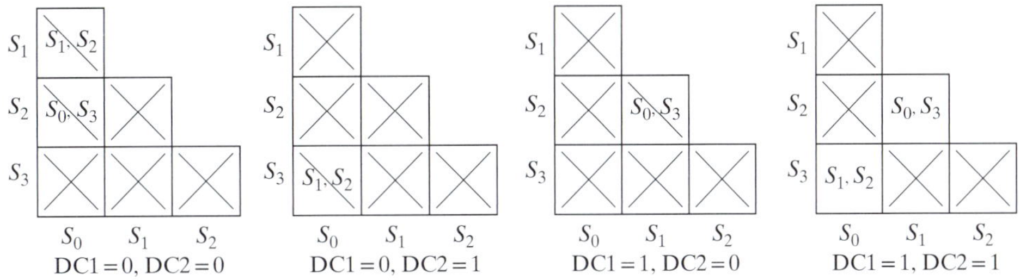

FIGURE 15-8 Implication Charts for Table 15-5(a)

TABLE 15-6 Modified Table 15-5(b)

Present State	Next State $X=0$	Next State $X=1$	Output $X=0$	Output $X=1$
S_0	S_2	S_1^1	0	1
S_1^1	S_1^2	S_0	0	1
S_1^2	S_1^2	S_0	1	1
S_2	S_2	S_1^1	1	1

For a completely specified table, if $S_0 \equiv S_1$ and $S_1 \equiv S_2$, then $S_0 \equiv S_2$. (Equivalence satisfies the transitivity property.) In the preceding example, for any input sequence, S_0 and S_1 will generate the same output sequence values wherever both output sequences have defined values; S_0 and S_1 are said to be *compatible*. (We will use the symbol ~ to denote compatibility.) Similarly, S_1 and S_2 are compatible (i.e., for any input sequence they will generate the same output sequence values wherever both output sequences have defined values). But this is not true for S_0 and S_2; for the single input $x = 0$, they have different outputs. Compatibility does not satisfy transitivity; $S_0 \sim S_1$ and $S_1 \sim S_2$ does not imply that $S_0 \sim S_2$.

To reduce a completely specified table, we find all the maximal sets of states containing pairwise equivalent states; since equivalence is transitive, a state of the table is contained in just one of the maximal equivalence classes. The states of the reduced table correspond to these maximal equivalence classes. The first step in reducing an incompletely specified table is to find all maximal sets of states that are pairwise compatible—the maximal compatibles. However, since compatibility does not satisfy transitivity, a given state may be contained in more than one maximal compatible. To reduce an incompletely specified table, a minimum of the maximal compatibles are selected, say $C_1, C_2, \ldots, C_k$, so that

(1) each state of the table appears in at least one of the C_i, and
(2) for each input combination x and each C_i, the next states of the states in C_i are contained in some C_j. (It may be that $j = i$.)

The second condition allows the construction of a state table with the states corresponding to the selected maximal compatibles; the reduced table will contain k states. (This procedure is valid for most incompletely specified tables. In some cases the states of the reduced table are specified by subsets of the maximal compatibles.)

As an example, consider a circuit that receives binary encodings for decimal digits 0, 1, 2, 3, 4 and 5 in disjoint windows of length 3. The bits are received most significant bit first and the circuit generates an output of 1 (0) for the third input bit if the 3 bits have even (odd) parity. The output is a don't-care for the first 2 bits. Table 15-7 was constructed for the circuit where the states are entered for the input sequences shown. Figure 15-9 is the implication chart for Table 15-7. All pairs of states are compatible except for S_1 and S_2 and S_3 and S_4. $S_0 \sim S_1$, $S_0 \sim S_3$, and $S_1 \sim S_3$ implies that $(S_0\, S_1\, S_3)$ is a compatible set. Similarly, $(S_0\, S_2\, S_3)$, $(S_0\, S_1\, S_4)$, and $(S_0\, S_2\, S_4)$ are compatible sets. $(S_0\, S_1\, S_3)$, and $(S_0\, S_2\, S_3)$ cannot be combined into a larger set since S_1 and S_2 are not compatible. None of these will combine, so $(S_0\, S_1\, S_3)$, $(S_0\, S_2\, S_3)$, $(S_0\, S_2\, S_4)$ and $(S_0\, S_2\, S_4)$ are the maximal compatible sets.

The next states of $(S_0\, S_1\, S_3)$ for $X = 1$ are $(S_0\, S_2\, S_4)$, so if $(S_0\, S_1\, S_3)$ is one of the states in the reduced table, then $(S_0\, S_2\, S_4)$ must be one of the other states. The next states of $(S_0\, S_2\, S_4)$ for $X = 0$ are $(S_0\, S_1\, S_4)$, so if $(S_0\, S_2\, S_4)$ is one of the states in the reduced table, then $(S_0\, S_1\, S_4)$ must be one of the other states. Examining the next states for $X = 0$ and $X = 1$ shows that $(S_0\, S_1\, S_3)$, $(S_0\, S_1\, S_4)$, and $(S_0\, S_2\, S_4)$ can be used to form a state table; Table 15-8 is the resulting table. $(S_0\, S_2\, S_3)$ has next states $(S_0\, S_1\, S_4)$ for $X = 0$, so if $(S_0\, S_2\, S_3)$ were included, the table would have four states; Table 15-8 is the minimal reduced table for Table 15-7.

TABLE 15-7 Even Parity Detector For 0 Through 5

	Present State	Next State $X = 0$	Next State $X = 1$	Output $X = 0$	Output $X = 1$
	S_0	S_1	S_2	–	–
0	S_1	S_3	S_4	–	–
1	S_2	S_4	–	–	–
00	S_3	S_0	S_0	1	0
01, 10	S_4	S_0	S_0	0	1

FIGURE 15-9 Implication Chart for Table 15-7

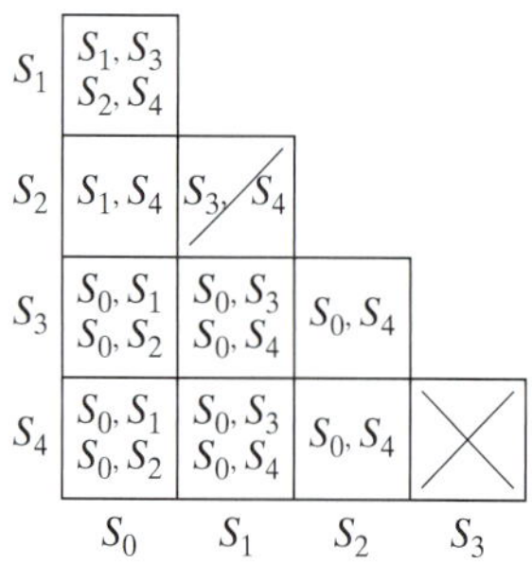

TABLE 15-8 Reduced Table for Table 15-7

	Present State	Next State $X = 0$	Next State $X = 1$	Output $X = 0$	Output $X = 1$
$S_0\, S_1\, S_3$	A	A	B	1	0
$S_0\, S_2\, S_4$	B	C	B	0	1
$S_0\, S_1\, S_4$	C	A	B	0	1

15.6 Derivation of Flip-Flop Input Equations

After the number of states in a state table has been reduced, the following procedure can be used to derive the flip-flop input equations:

1. Assign flip-flop state values to correspond to the states in the reduced table.
2. Construct a transition table which gives the next states of the flip-flops as a function of the present states and inputs.
3. Derive the next-state maps from the transition table.
4. Find flip-flop input maps from the next-state maps using the techniques developed in Unit 12 and find the flip-flop input equations from the maps.

As an example, we will design a sequential circuit to realize Table 15-9(a). Because there are seven states, we will need three flip-flops. We will designate the flip-flop outputs as A, B, and C.

TABLE 15-9

(a) State table

	$X = 0$	1	0	1
S_0	S_1	S_2	0	0
S_1	S_3	S_2	0	0
S_2	S_1	S_4	0	0
S_3	S_5	S_2	0	0
S_4	S_1	S_6	0	0
S_5	S_5	S_2	1	0
S_6	S_1	S_6	0	1

(b) Transition table

	$A^+B^+C^+$		Z	
ABC	$X = 0$	1	0	1
000	110	001	0	0
110	111	001	0	0
001	110	011	0	0
111	101	001	0	0
011	110	010	0	0
101	101	001	1	0
010	110	010	0	1

We could make a straight binary state assignment for which S_0 is represented by flip-flop states $ABC = 000$, S_1 by $ABC = 001$, S_2 by $ABC = 010$, etc. However, because the correspondence between flip-flop states and the state names is arbitrary, we could use many different state assignments. Using a different assignment may lead to simpler or more complex flip-flop input equations. As an example, we will use the following assignment for the states of flip-flops A, B, and C:

$$S_0 = 000, S_1 = 110, S_2 = 001, S_3 = 111, S_4 = 011, S_5 = 101, S_6 = 010 \quad (15\text{-}1)$$

This state assignment is derived in Section 15.8, and the reasons why it leads to an economical solution are given in that section. Starting with Table 15-9(a), we substitute 000 for S_0, 110 for S_1, 001 for S_2, etc. Table 15-9(b) shows the resulting transition table. This table gives the next states of flip-flops A, B, and C in terms of the present states and the input X. We can fill in the next-state maps, Figure 15-10(a), directly from this table. For $XABC = 0000$ the next-state entry is 110, so we fill in $A^+ = 1$, $B^+ = 1$, and $C^+ = 0$; for $XABC = 1000$ the next-state entry is 001, so we fill in $A^+ = 0$, $B^+ = 0$, and $C^+ = 1$; etc. Because the state assignment $ABC = 100$ is not used, the map squares corresponding to $XABC = 0100$ and 1100 are filled with don't-cares.

Once the next-state maps have been plotted from the transition table, the flip-flop input equations can be derived using the techniques developed in Unit 12. As shown in Figure 15-10(a), the D flip-flop input equations can be derived directly from the next-state maps because $D_A = A^+$, $D_B = B^+$, and $D_C = C^+$. If J-K flip-flops are used, the J and K input equations can be derived from the next-state maps as illustrated in Figure 15-10(b). As was shown in Section 12.5, the $A = 0$ half of the J_A map is the same as the A^+ map and the $A = 1$ half is all don't-cares. The $A = 1$ half of the K_A map is the complement of the $A = 1$ half of the A^+ map, and the $A = 0$ half is all don't-cares. We can plot the J_B and K_B in a similar manner by looking at the $B = 0$ and $B = 1$ halves of the B^+ map. Derivation of the J_C and K_C maps from the C^+ map is left as an exercise.

FIGURE 15-10 Next-State Maps for Table 15-9

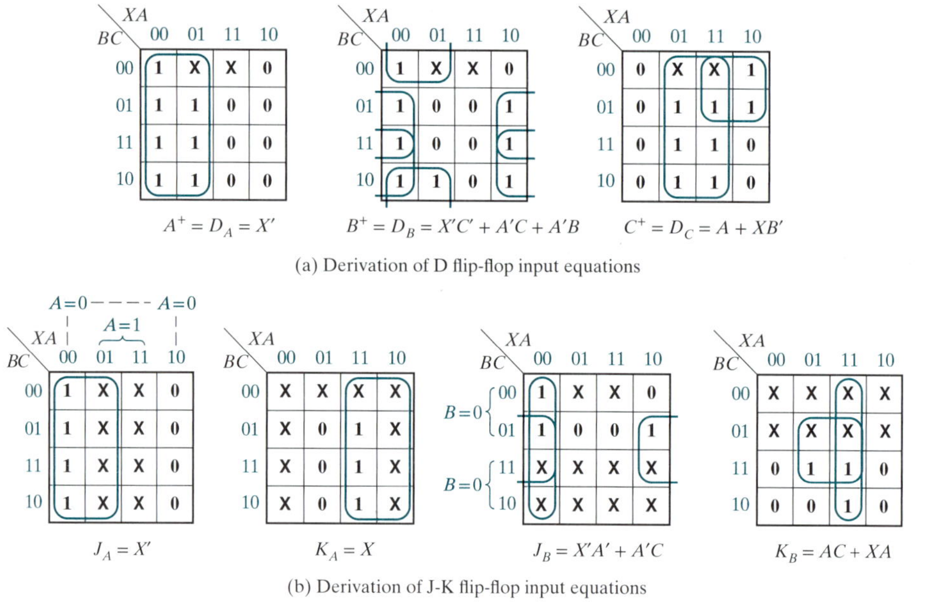

TABLE 15-10

(a) State table

	Next State $X_1X_2 =$				Outputs (Z_1Z_2) $X_1X_2 =$			
P.S.	00	01	11	10	00	01	11	10
S_0	S_0	S_0	S_1	S_1	00	00	01	01
S_1	S_1	S_3	S_2	S_1	00	10	10	00
S_2	S_3	S_3	S_2	S_2	11	11	00	00
S_3	S_0	S_3	S_2	S_0	00	00	00	00

(b) Transition table

	A^+B^+ $X_1X_2 =$				Outputs (Z_1Z_2) $X_1X_2 =$			
AB	00	01	11	10	00	01	11	10
00	00	00	01	01	00	00	01	01
01	01	10	11	01	00	10	10	00
11	10	10	11	11	11	11	00	00
10	00	10	11	00	00	00	00	00

Table 15-10(a) represents a sequential circuit with two inputs (X_1 and X_2) and two outputs (Z_1 and Z_2). Note that the column headings are listed in Karnaugh map order because this will facilitate derivation of the flip-flop input equations. Because the table has four states, two flip-flops (A and B) are required to realize the table. We will use the state assignment $AB = 00$ for S_0, $AB = 01$ for S_1, $AB = 11$ for S_2, and $AB = 10$ for S_3. By substituting the corresponding values of AB for the state names, we obtain the transition table, Table 15-10(b). We can then fill in the next-state and output maps (Figure 15-11) from the transition table. For example, when $X_1X_2AB = 0011$, $A^+B^+ = 10$, and $Z_1Z_2 = 11$; therefore, we fill in the 0011 squares of the A^+, B^+, Z_1, and Z_2 maps with 1, 0, 1, and 1, respectively. We can read the D flip-flop input equations directly from the next-state maps.

FIGURE 15-11 Next-State Maps for Table 15-10

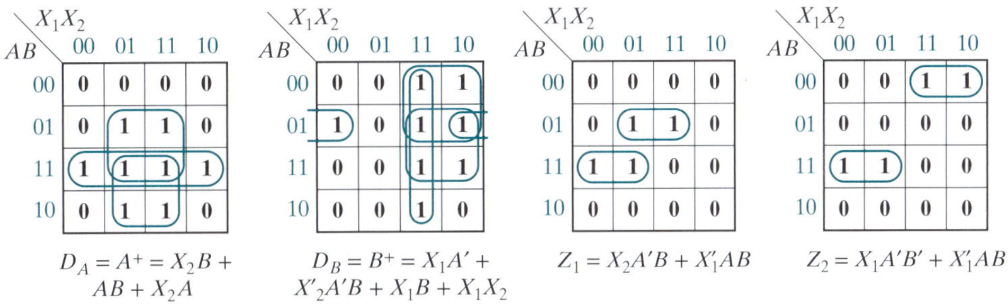

If J-K, T, or S-R flip-flops are used, the flip-flop input maps can be derived from the next-state maps using the techniques given in Section 12.6. As an example, the S-R equations for Table 15-10 are derived in Figure 15-12. The S_A and R_A maps are derived from the A^+ map by applying Table 12-5(c) to the $A = 0$ and $A = 1$ halves of the map. S_B and R_B are derived in a similar manner.

FIGURE 15-12 Derivation of S-R Equations for Table 15-10

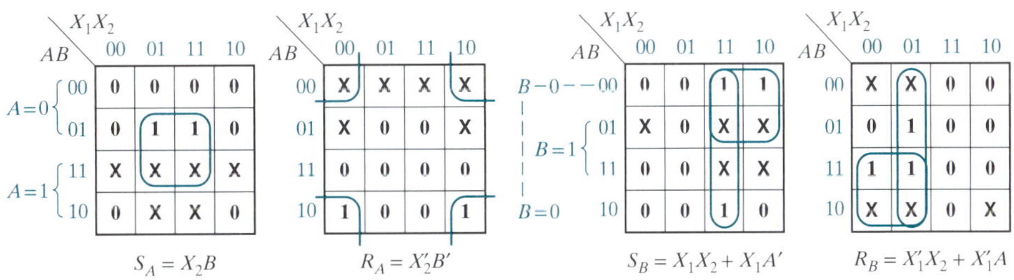

15.7 Equivalent State Assignments

After the number of states in a state table has been reduced, the next step in realizing the table is to assign flip-flop states to correspond to the states in the table. The cost of the logic required to realize a sequential circuit is strongly dependent on the way this state assignment is made. Several methods for choosing state assignments to obtain economical realizations are discussed in this chapter. The trial-and-error method described next is useful for only a small number of states. The guideline method discussed in Section 15.8 produces good solutions for some problems, but it is not entirely satisfactory in other cases.

If the number of states is small, it may be feasible to try all possible state assignments, evaluate the cost of the realization for each assignment, and choose the assignment with the lowest cost. Consider a state table with three states (S_0, S_1, and S_2) as in Table 14-1. Two flip-flops (A and B) are required to realize this table. The four possible assignments for state S_0 are $AB = 00$, $AB = 01$, $AB = 10$, and $AB = 11$. Choosing one of these assignments leaves three possible assignments for state S_1 because each state must have a unique assignment. Then, after state S_1 is assigned, we have two possible assignments for state S_2. Thus, there are $4 \times 3 \times 2 = 24$ possible state assignments for the three states, as shown in Table 15-11. As an example, for assignment 7, the entry 01 in the S_0 row means that flip-flops A and B are assigned values 0 and 1, respectively.

Trying all 24 of these assignments is not really necessary. If we interchange two columns in one of the given assignments, the cost of realization will be unchanged because interchanging columns is equivalent to relabeling the flip-flop variables. For example, consider assignment 1 in Table 15-11. The first column of this assignment shows that flip-flop A is assigned the values 0, 0, and 1 for states S_0, S_1, and S_2, respectively. Similarly, the second column shows that B is assigned the values 0, 1, and 0. If we interchange the two columns, we get assignment 3, for which A has the values 0, 1, and 0 and B has the values 0, 0, and 1. We could have achieved the same result by using assignment 1 and labeling the flip-flop variables BA instead of AB. If we interchange the columns of assignment 2, we get assignment 4, so assignments 2 and 4 have the same cost. Similarly, assignments 5 and 6 have the same cost. Interchanging rows, however, will usually change the cost of realization. Thus, assignments 4 and 6 will have a different cost for many state tables.

TABLE 15-11
State Assignments for 3-Row Tables

	1	2	3	4	5	6	7		19	20	21	22	23	24
S_0	00	00	00	00	00	00	01	. . .	11	11	11	11	11	11
S_1	01	01	10	10	11	11	00		00	00	01	01	10	10
S_2	10	11	01	11	01	10	10		01	10	00	10	00	01

FIGURE 15-13
Equivalent Circuits Obtained by Complementing Q_k

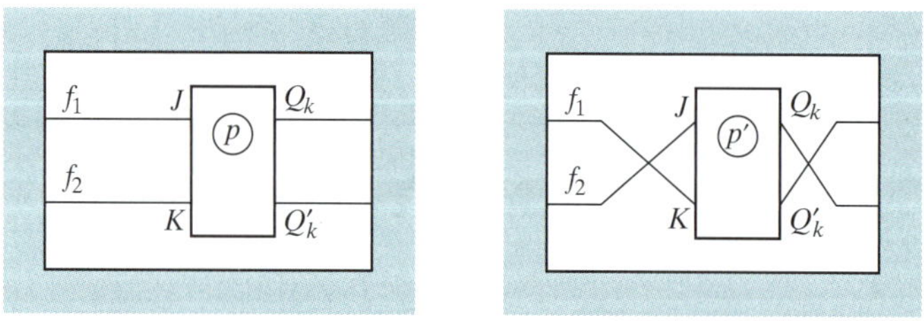

(a) Circuit A

(b) Circuit B (identical to A except leads to flip-flop Q_k are crossed)

If symmetrical flip-flops such as T, J-K, or S-R are used, complementing one or more columns of the state assignment will have no effect on the cost of realization. Consider a J-K flip-flop imbedded in a circuit, Figure 15-13(a). Leave the circuit unchanged and interchange the J and K input connections and the Q_k and Q_k' output connections, Figure 15-13(b). If circuit A is started with $Q_k = p$ and circuit B with $Q_k = p'$, the behavior of the two circuits will be identical, except the value of Q_k will always be complemented in the second circuit because whenever J is 1 in the first circuit, K will be 1 in the second and conversely. The state table for the second circuit is therefore the same as for the first, except the value of Q_k is complemented for the second circuit. This implies that complementing one or more columns in the state assignment will not affect the cost of the realization when J-K flip-flops are used. Similar reasoning applies to T and S-R flip-flops. Thus, in Table 15-11, assignments 2 and 7 have the same cost, and so do assignments 6 and 19.

If unsymmetrical flip-flops are used such as a D flip-flop, it is still true that permuting (i.e., rearranging the order of) columns in the state assignment will not affect the cost; however, complementing a column may require adding an inverter to the circuit, as shown in Figure 15-14. If different types of gates are available, the circuit can generally be redesigned to eliminate the inverter and use the same number of gates as the original. If circuit A in Figure 15-14 is started with $Q_k = p$ and circuit B with $Q_k = p'$, the behavior of the two circuits will be identical, except the value of Q_k will always be complemented in circuit B because f is the same in both circuits and $D = f'$ for circuit B.

Table 15-12 illustrates the effect of interchanging or complementing state assignment columns on the equations for realizing a specific state table.

FIGURE 15-14 Equivalent Circuits Obatined by Complemeting Q_k

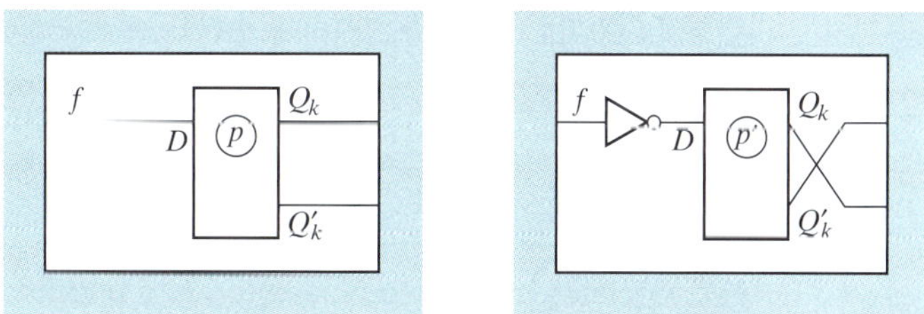

(a) Circuit A

(b) Circuit B (identical to A except for connections to flip-flop Q_k)

TABLE 15-12

Assignments			Present	Next State		Output	
A_3	B_3	C_3	State	$X = 0$	1	0	1
00	00	11	S_1	S_1	S_3	0	0
01	10	10	S_2	S_2	S_1	0	1
10	01	01	S_3	S_3	S_3	1	0

The J-K and D flip-flop input equations for the three assignments can be derived, using Karnaugh maps as explained in Unit 12 and Section 15.6. The resulting J and K input equations are:

Assignment A	Assignment B	Assignment C
$J_1 = XQ_2'$	$J_2 = XQ_1'$	$K_1 = XQ_2$
$K_1 = X'$	$K_2 = X'$	$J_1 = X'$
$J_2 = X'Q_1$	$J_1 = X'Q_2$	$K_2 = X'Q_1'$
$K_2 = X$	$K_1 = X$	$J_2 = X$
$Z = X'Q_1 + XQ_2$	$Z = X'Q_2 + XQ_1$	$Z = X'Q_1' + XQ_2'$
$D_1 = XQ_2'$	$D_2 = XQ_1'$	$D_1 = X' + Q_2'$
$D_2 = X'(Q_1 + Q_2)$	$D_1 = X'(Q_2 + Q_1)$	$D_2 = X + Q_1Q_2$

Note that assignment B in Table 15-12 was obtained by interchanging the columns of A. The corresponding equations for assignment B are the same as for A, except that subscripts 1 and 2 are interchanged. Assignment C was obtained by complementing the columns of A. The Z equation for C is the same as for A, except that Q_1 and Q_2 are complemented. The K and J equations for C are the same, respectively, as the J and K equations for A with the Q's complemented. The D equations for C can be obtained by complementing those for A and, then, complementing the Q's. Thus, the cost of realizing Table 15-12 using J-K flip-flops and any kind of logic gates will be exactly the same for all three assignments. If both AND and OR (or NAND and NOR) gates are available, the cost of realizing the three sets of D equations will be the same. If only NOR gates are available, for example, then realizing D_1 and D_2 for assignment C would require two additional inverters compared with A and B.

By complementing one or more columns, any state assignment can be converted to one in which the first state is assigned all 0's. If we eliminate assignments which can be obtained by permuting or complementing columns of another state assignment, Table 15-11 reduces to three assignments (Table 15-13). Thus, when realizing a three-state sequential circuit with symmetrical flip-flops, it is only necessary to try three different state assignments to be assured of a minimum cost realization. Similarly, only three different assignments must be tried for four states.

TABLE 15-13 Nonequivalent Assignments for Three and Four States

	3-State Assignments			4-State Assignments		
States	1	2	3	1	2	3
a	00	00	00	00	00	00
b	01	01	11	01	01	11
c	10	11	01	10	11	01
d	–	–	–	11	10	10

TABLE 15-14
Number of Distinct (Nonequivalent) State Assignments

Number of States	Minimum Number of State Variables	Number of Distinct Assignments
2	1	1
3	2	3
4	2	3
5	3	140
6	3	420
7	3	840
8	3	840
9	4	10,810,800
⋮	⋮	⋮
16	4	$\approx 5.5 \times 10^{10}$

We will say that two state assignments are *equivalent* if one can be derived from the other by permuting and complementing columns. Two state assignments which are not equivalent are said to be *distinct*. Thus, a four-row table has three distinct state assignments, and any other assignment is equivalent to one of these three. Unfortunately, the number of distinct assignments increases very rapidly with the number of states, as shown in Table 15-14. Hand solution is feasible for two, three, or four states; computer solution is feasible for five through eight states; but for more than nine states it is not practical to try all assignments even if a high-speed computer is used.

15.8 Guidelines for State Assignment

Because trying all nonequivalent state assignments is not practical in most cases, other methods of state assignment are needed. The next method to be discussed involves trying to choose an assignment which will place the 1's on the flip-flop input maps in adjacent squares so that the corresponding terms can be combined. This method does not apply to all problems, and even when applicable, it does not guarantee a minimum solution.

Assignments for two states are said to be adjacent if they differ in only one variable. Thus, 010 and 011 are adjacent, but 010 and 001 are not. The following *guidelines* are useful in making assignments which will place 1's together (or 0's together) on the next-state maps:

1. States which have the same next state for a given input should be given adjacent assignments.
2. States which are the next states of the same state should be given adjacent assignments.

A third guideline is used for simplification of the output function:

3. States which have the same output for a given input should be given adjacent assignments.

The application of Guideline 3 will place 1's together on the output maps.

When using the state assignment guidelines, the first step is to write down all of the sets of states which should be given adjacent assignments according to the guidelines. Then, using a Karnaugh map, try to satisfy as many of these adjacencies as possible. A fair amount of trial and error may be required to fill in the map so that the maximum number of desired state adjacencies is obtained. When filling in the map, keep in mind the following:

(a) Assign the starting state (reset state) to the "0" square on the map. (For an exception to this rule, see the one-hot assignment in Section 15-9.) Nothing is to be gained by trying to put the starting state in different squares on the map because the same number of adjacencies can be found no matter where you put the starting state. Usually, assigning "0" to the starting state simplifies the initialization of the circuit using the clear inputs on the flip-flops.

(b) Adjacency conditions from Guideline 1 and adjacency conditions from Guideline 2 that are required two or more times should be satisfied first.

(c) When guidelines require that three or four states be adjacent, these states should be placed within a group of four adjacent squares on the assignment map.

(d) If the output table is to be considered, then Guideline 3 should also be applied. The priority given to adjacency conditions from Guideline 3 should generally be less than that given to Guidelines 1 and 2 if a single output function is being derived. If there are two or more output functions, a higher priority for Guideline 3 may be appropriate.

The following example should clarify the application of Guidelines 1 and 2. The state table from Table 15-9 is repeated in Figure 15-15(a) so that we can illustrate derivation of the state assignment. According to Guideline 1, S_0, S_2, S_4, and S_6 should be given adjacent assignments because they all have S_1 as a next state (with input 0). Similarly, S_0, S_1, S_3, and S_5 should have adjacent assignments because they have S_2 as a next state (with input 1); also, S_3 and S_5 should have adjacent assignments and so should S_4 and S_6. The application of Guideline 2 indicates that S_1 and S_2 should be

FIGURE 15-15

ABC		X = 0	1	0	1
000	S_0	S_1	S_2	0	0
110	S_1	S_3	S_2	0	0
001	S_2	S_1	S_4	0	0
111	S_3	S_5	S_2	0	0
011	S_4	S_1	S_6	0	0
101	S_5	S_5	S_2	1	0
010	S_6	S_1	S_6	0	1

(a) State table

BC \ A	0	1
00	S_0	
01	S_2	S_5
11	S_4	S_3
10	S_6	S_1

BC \ A	0	1
00	S_0	
01	S_1	S_6
11	S_3	S_4
10	S_5	S_2

(b) Assignment maps

given adjacent assignments because they are both next states of S_0. Similarly, S_2 and S_3 should have adjacent assignments because they are both next states of S_1. Further application of Guideline 2 indicates that S_1 and S_4, S_2 and S_5 (two times), and S_1 and S_6 (two times) should be given adjacent assignments. In summary, the sets of adjacent states specified by Guidelines 1 and 2 are

1. (S_0, S_1, S_3, S_5) (S_3, S_5) (S_4, S_6) (S_0, S_2, S_4, S_6)

2. (S_1, S_2) (S_2, S_3) (S_1, S_4) $(S_2, S_5)2x$ $(S_1, S_6)2x$

We will attempt to fulfill as many of these adjacency conditions as possible. A Karnaugh map will be used to make the assignments so that states with adjacent assignments will appear in adjacent squares on the map. If the guidelines require that three or four states be adjacent, these states should be placed within a group of four adjacent squares on the assignment map. Two possible ways of filling in the assignment maps are shown in Figure 15-15(b). These maps were filled in by trial and error, attempting to fulfill as many of the preceding adjacency conditions as possible. The conditions from Guideline 1 are given preference to conditions from Guideline 2. The conditions which are required two times (such as S_2 adjacent to S_5, and S_1 adjacent to S_6) are given preference over conditions which are required only once (such as S_1 adjacent to S_2, and S_2 adjacent to S_3).

The left assignment map in Figure 15-15(b) implies an assignment for the states of flip-flops A, B, and C which is listed to the left of the state table in Figure 15-15(a). This assignment is the same as the one given in Equations (15-1). We derived the D flip-flop input equations and J and K input equations for this assignment in Section 15.6. The cost of realizing the D flip-flop input equations given in Figure 15-10(a) is six gates and 13 inputs. If a straight binary assignment ($S_0 = 000, S_1 = 001, S_2 = 010$, etc.) were used instead, the cost of realizing the flip-flop input equations would be 10 gates and 39 inputs. Although application of the guidelines gives good results in this example, this is not always the case.

Next, we will explain why the guidelines help to simplify the flip-flop equations when the assignment of Figure 15-15(a) is used. Figure 15-16 shows a next-state map which was constructed using this assignment. Note that if $X = 0$ and $ABC = 000$, the next state is S_1; if $X = 1$ and $ABC = 000$, the next state is S_2. Because Guideline 1 was used in making the state assignment, S_1 appears in four adjacent squares on the next-state map, S_5 appears in two adjacent squares, etc.

The next-state maps for the individual flip-flops, Figure 15-16(b), can be derived in the usual manner from a transition table, or they can be derived directly from Figure 15-16(a). Using the latter approach, wherever S_1 appears in Figure 15-16(a), it is replaced with 110 so that 1, 1 and 0 are plotted on the corresponding squares of the A^+, B^+, and C^+ maps, respectively. The other squares on the next-state maps are filled in similarly.

Because four S_1's are adjacent in Figure 15-16(a), the corresponding squares on the A^+, B^+, and C^+ maps have four adjacent 1's or four adjacent 0's as indicated by the blue shading. This illustrates why Guideline 1 helps to simplify the flip-flop equations. Each time Guideline 2 is applied, two out of the three next-state maps will

FIGURE 15-16
Next-State Maps for Figure 15-15

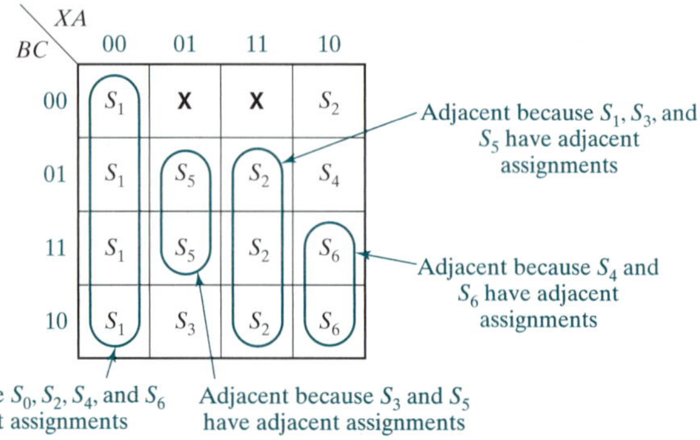

(a) Next-state maps for Figure 15-15

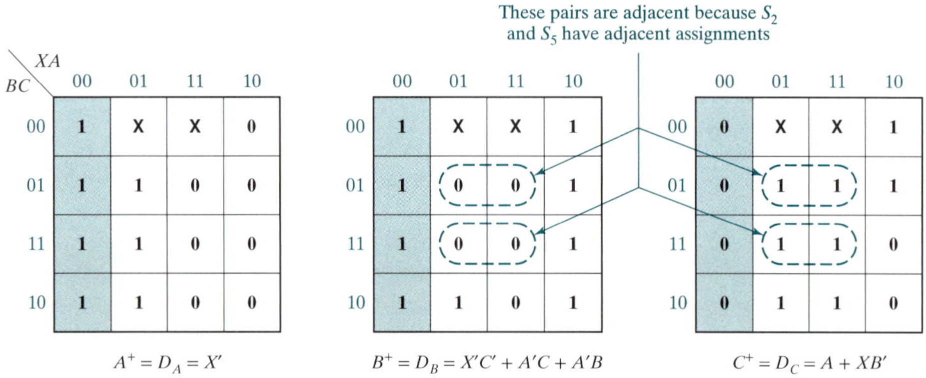

(b) Next-state maps for Figure 15-15 (cont.)

have an additional pair of adjacent 1's or adjacent 0's. This occurs because two of the three state variables are the same for adjacent assignments.

Next, we will apply the state assignment guidelines to Figure 15-17(a). First, we list the sets of adjacent states specified by each Guideline:

1. (b, d) (c, f) (b, e)
2. $(a, c)2x$ (d, f) (b, d) (b, f) (c, e)
3. (a, c) (b, d) (e, f)

Next, we try to arrange the states on a map so as to satisfy as many of these pairs as possible, but giving preference to the duplicated pairs (b, d) and (a, c). Two such arrangements and the corresponding assignments are given in Figure 15-17(b) and (c). For Figure 15-17(c), all adjacencies are satisfied except (b, f), (c, e), and (e, f). We will derive D flip-flop input equations for this assignment. First, we construct the transition table (Table 15-15) from the state table (Figure 15-17(a)) by replacing a with 100, b with 111, c with 000, etc. Then, we plot the next-state and output maps

FIGURE 15-17 State Tables and Assignments

	X = 0	1	X = 0	1
a	a	c	0	0
b	d	f	0	1
c	c	a	0	0
d	d	b	0	1
e	b	f	1	0
f	c	e	1	0

(a)

Q_2Q_3 \ Q_1	0	1
00	a	c
01		e
11	d	b
10		f

$a = 000$, $b = 111$, $c = 100$, $d = 011$, $e = 101$, $f = 110$

(b)

Q_2Q_3 \ Q_1	0	1
00	c	a
01		e
11	d	b
10	f	

$a = 100$, $b = 111$, $c = 000$, $d = 011$, $e = 101$, $f = 010$

(c)

TABLE 15-15 Transition Table for Figure 15-17(a)

	$Q_1^+Q_2^+Q_3^+$			
$Q_1Q_2Q_3$	X = 0	1	X = 0	1
1 0 0	100	000	0	0
1 1 1	011	010	0	1
0 0 0	000	100	0	0
0 1 1	011	111	0	1
1 0 1	111	010	1	0
0 1 0	000	101	1	0

(Figure 15-18) from the transition table. The D flip-flop input equations can be read directly from these maps:

$$D_1 = Q_1^+ = X'Q_1Q_2' + XQ_1'$$
$$D_2 = Q_2^+ = Q_3$$
$$D_3 = Q_3^+ = XQ_1'Q_2 + X'Q_3$$

and the output equation is

$$Z = XQ_2Q_3 + X'Q_2'Q_3 + XQ_2Q_3'$$

The cost of realizing these equations is 10 gates and 26 gate inputs.

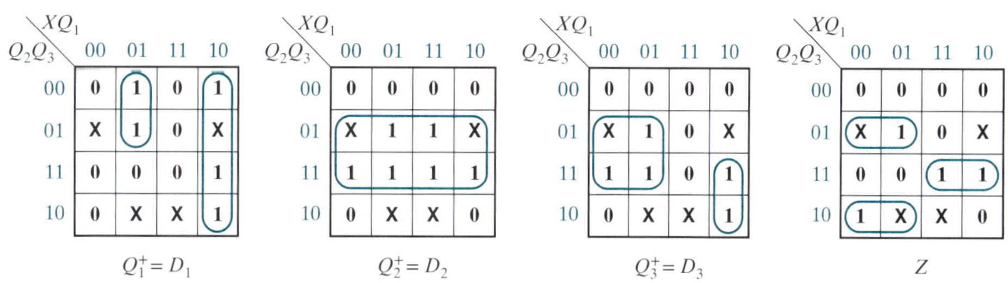

FIGURE 15-18 Next-State and Output Maps for Table 15-15

The assignment of Figure 15-17(b) satisfies all of the guidelines except (d, f) and (e, f). Using this assignment, the cost of realizing the state table with D flip-flops is 13 gates and 35 gate inputs. We would expect that this assignment would produce better results than Figure 15-17(c) because it satisfies one more of the adjacencies given by the guidelines, but just the opposite is true. As illustrated by this example, the assignment which satisfies the most guidelines is not necessarily the best assignment. In general, it is a good idea to try several assignments which satisfy most of the guidelines and choose the one which gives the lowest cost solution.

The guidelines work best for D flip-flops and J-K flip-flops. They do not work as well for T and S-R flip-flops. In general, the best assignment for one type of flip-flop is not the best for another type.

15.9 Using a One-Hot State Assignment

When designing with CPLDs and FPGAs, we should keep in mind that each logic cell contains one or more flip-flops. These flip-flops are there whether we use them or not. This means that it may not be important to minimize the number of flip-flops used in the design. Instead, we should try to reduce the total number of logic cells used and try to reduce the interconnections between cells. When several cells must be cascaded to realize a function as in Figure 9-40(b), the propagation delay is longer, and the logic runs slower. In order to design faster logic, we should try to reduce the number of cells required to realize each equation. Using a *one-hot* state assignment may help to accomplish this.

The one-hot assignment uses one flip-flop for each state, so a state machine with N states requires N flip-flops. Exactly one of the flip-flops is set to one in each state. For example, a system with four states (S_0, S_1, S_2, and S_3) could use four flip-flops (Q_0, Q_1, Q_2, and Q_3) with the following state assignment:

$$S_0\text{: } Q_0 Q_1 Q_2 Q_3 = 1000, \quad S_1\text{: } 0100, \quad S_2\text{: } 0010, \quad S_3\text{: } 0001 \tag{15-2}$$

The other 12 combinations are not used.

We can write next-state and output equations by inspecting the state graph. Consider the partial state graph given in Figure 15-19. Because four arcs lead into S_3, there are four conditions under which the next state is S_3. These conditions are as follows:

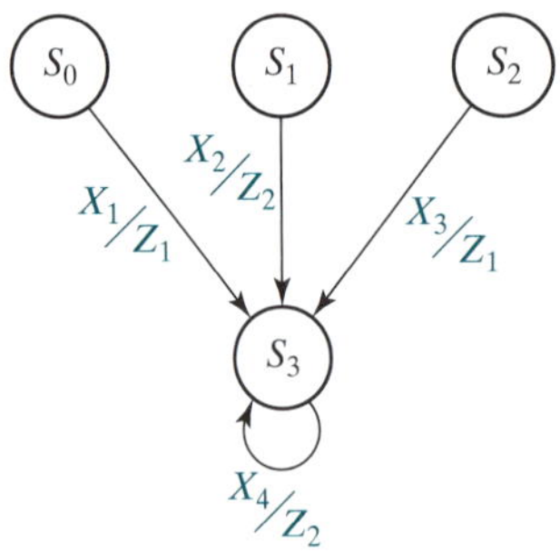

FIGURE 15-19
Partial State Graph

present state (PS) $= S_0$ and $X_1 = 1$, PS $= S_1$ and $X_2 = 1$, PS $= S_2$ and $X_3 = 1$, PS $= S_3$ and $X_4 = 1$. The next state of flip-flop Q_3 is 1 under these four conditions (and 0 otherwise). Therefore, the next-state equation for Q_3 can be written as:

$$Q_3^+ = X_1(Q_0\, Q_1'\, Q_2'\, Q_3') + X_2(Q_0'\, Q_1\, Q_2'\, Q_3') + X_3(Q_0'\, Q_1'\, Q_2\, Q_3') + X_4(Q_0'\, Q_1'\, Q_2'\, Q_3)$$

However, because $Q_0 = 1$ implies $Q_1 = Q_2 = Q_3 = 0$, the $Q_1'\, Q_2'\, Q_3'$ term is redundant and can be eliminated. Similarly, all of the primed state variables can be eliminated from the other terms, so the next-state equation reduces to

$$Q_3^+ = X_1Q_0 + X_2Q_1 + X_3Q_2 + X_4Q_3$$

In general, when a one-hot state assignment is used, each term in the next-state equation for each flip-flop contains exactly one state variable, and the reduced equation can be written by inspecting the state graph.

Similarly, each term in each reduced output equation contains exactly one state variable. Because $Z_1 = 1$ when PS $= S_0$ and $X_1 = 1$, and also when PS $= S_2$ and $X_3 = 1$, we can write $Z_1 = X_1Q_0 + X_3Q_2$. By inspecting the state graph, we can also write $Z_2 = X_2Q_1 + X_4Q_3$.

When a one-hot assignment is used, resetting the system requires that one flip-flop be set to 1 instead of resetting all flip-flops to 0. If the flip-flops used do not have a preset input, then we can modify the one-hot assignment by replacing Q_0 with Q_0' throughout. For the Assignment (15-2), the modification is

$$S_0\text{: } Q_0\, Q_1\, Q_2\, Q_3 = 0000,\quad S_1\text{: } 1100,\quad S_2\text{: } 1010,\quad S_3\text{: } 1001 \qquad (15\text{-}3)$$

and the modified equations are:

$$Q_3^+ = X_1Q_0' + X_2Q_1 + X_3Q_2 + X_4Q_3$$
$$Z_1 = X_1Q_0' + X_3Q_2,\quad Z_2 = X_2Q_1 + X_4Q_3$$

For the Moore machine of Figure 14-23(b), we will make the following one-hot assignment for flip-flops $Q_0\, Q_1\, Q_2$: $S_0 = 100$, $S_1 = 010$, and $S_2 = 001$. When $Q_0 = 1$, the state is S_0; when $Q_1 = 1$, the state is S_1; and when $Q_2 = 1$, the state is S_2. By inspection, because three arcs lead into each state, the next-state equations are

$$Q_0^+ = F'R'Q_0 + FQ_2 + F'RQ_1$$
$$Q_1^+ = F'R'Q_1 + FQ_0 + F'RQ_2$$
$$Q_2^+ = F'R'Q_2 + FQ_1 + F'RQ_0$$

The output equations are trivial because each output occurs in only one state:

$$Z_1 = Q_0,\ Z_2 = Q_1,\ Z_3 = Q_2$$

As another example, consider the state graph in Figure 15-20, which represents a sequential circuit that controls a binary multiplier. The circuit has three inputs (St, M, and K), and four outputs (Load, Ad, Sh, and Done). Starting in state S_0, if $St = 0$, then the circuit stays in S_0. If $St = 1$, the circuit outputs Load $= 1$ (the other outputs are 0), and the next state is S_1. In S_1, if $M = 1$, then $Ad = 1$ and the next state

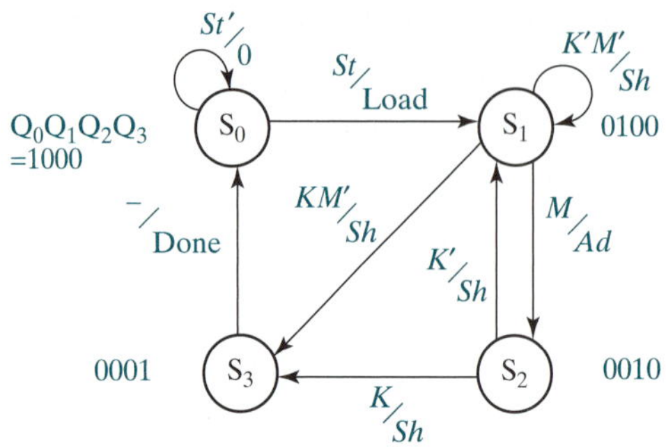

FIGURE 15-20 Multiplier Control State Graph

is S_2. If $M = 0$ and $K = 0$, the output is $Sh = 1$ and the next state is S_1. If $M = 0$ and $K = 1$, the output is $Sh = 1$ and the next state is S_3. In S_2, the output is $Sh = 1$ for both $K = 0$ and $K = 1$. If $K = 0$, the next state is S_1, and if $K = 1$, the next state is S_3. In S_3, the output is Done = 1 and the next state is S_0.

Because there are four states, a one-hot state assignment requires four flip-flops. The one-hot assignments for each state are shown on the state graph. Only Q_0 is 1 in S_0 and the other Q's are 0. Similarly, only Q_1 is 1 in S_1 and the other Q's are 0, etc. To determine the next-state equation for Q_0, note that two arrows lead into state S_0. The loop from state S_0 back to itself indicates $Q_0^+ = 1$ if $Q_0 = 1$ and $St = 0$. The arrow from S_3 to S_0 indicates $Q_0^+ = 1$ if $Q_3 = 1$. Therefore,

$$Q_0^+ = Q_0St' + Q_3$$

Because three arrows lead into S_1, Q_1^+ has three terms:

$$Q_1^+ = Q_0St + Q_1K'M' + Q_2K'$$

We can also determine the output functions by inspection of the state graph. In S_0, Load = 1 when $St = 1$ and Load = 0 for all other states and inputs; therefore, Load = Q_0St. The output Sh appears in four places on the graph. $Sh = 1$ in S_1 if $K'M' = 1$ or if $KM' = 1$; also, $Sh = 1$ in S_2 if $K' = 1$ or $K = 1$. Therefore,

$$Sh = Q_1(K'M' + KM') + Q_2(K' + K) = Q_1M' + Q_2$$

When designing with CPLDs or FPGAs, you should try both an assignment with a minimum number of state variables and a one-hot assignment to see which one leads to a design with the smallest number of logic cells. Alternatively, if the speed of operation is important, the design which leads to the fastest logic should be chosen. When a one-hot assignment is used, more next-state equations are required, but for some state graphs both the next-state and output equations may contain fewer variables. An equation with fewer variables may require fewer logic cells to realize. The more cells which are cascaded, the longer the propagation delay, and the slower the operation.

Problems

15.1 (a) Reduce the following state table to a minimum number of states.

Present State	Next State X = 0	Next State X = 1	Present Output X = 0	Present Output X = 1
A	A	E	1	0
B	C	F	0	0
C	B	H	0	0
D	E	F	0	0
E	D	A	0	0
F	B	F	1	0
G	D	H	0	0
H	H	G	1	0

(b) You are given two identical sequential circuits which realize the preceding state table. One circuit is initially in state *B* and the other circuit is initially in state *G*. Specify an input sequence of length three which could be used to distinguish between the two circuits and give the corresponding output sequence from each circuit.

15.2 Reduce the following state table to a minimum number of states.

Present State	Next State X = 0	Next State X = 1	Present Output (Z)
a	e	e	1
b	c	e	1
c	i	h	0
d	h	a	1
e	i	f	0
f	e	g	0
g	h	b	1
h	c	d	0
i	f	b	1

15.3 Digital engineer B. I. Nary has just completed the design of a sequential circuit which has the following state table:

Present State	Next State X = 0	Next State X = 1	Output X = 0	Output X = 1
S_0	S_5	S_1	0	0
S_1	S_5	S_6	0	0
S_2	S_2	S_6	0	0
S_3	S_0	S_1	1	0
S_4	S_4	S_3	0	0
S_5	S_0	S_1	0	0
S_6	S_5	S_1	1	0

His assistant, F. L. Ipflop, who has just completed this course, claims that his design can be used to replace Mr. Nary's circuit. Mr. Ipflop's design has the following state table:

	Next State $X = 0$	1	Output 0	1
a	a	b	0	0
b	a	c	0	0
c	a	b	1	0

(a) Is Mr. Ipflop correct? (Prove your answer.)
(b) If Mr. Nary's circuit is always started in state S_0, is Mr. Ipflop correct? (Prove your answer by showing equivalent states, etc.)

15.4 Realize the following state table using a minimum number of AND and OR gates together with
(a) a D flip-flop
(b) an S-R flip-flop

	$X_1X_2X_3$ 000	001	010	011	100	101	110	111	Z
A	A	A	B	B	B	B	A	A	0
B	A	B	B	A	A	B	B	A	1

15.5 It is sometimes possible to save logic by using more than the minimum number of flip-flops. For both (a) and (b), fill in each state assignment by columns and, then, check for duplicate rows instead of filling in the assignments by rows and checking for permuted columns. If the columns are assigned in ascending numerical order and the first row is all 0's, then equivalent assignments will not be generated. Do not list degenerate assignments for which two columns are identical or complements of each other, or assignments where one column is all 0's or all 1's.
(a) Consider a state table with three states to be realized using three J-K flip-flops. To be sure of getting the minimum amount of logic, how many different state assignments must be tried? Enumerate these assignments.
(b) For four states and three flip-flops, 29 assignments must be tried. Enumerate 10 of these, always assigning 000 to the first state.

15.6 A sequential circuit with one input and one output has the following state table:

Present State	Next State $X = 0$	$X = 1$	Present Output
S_1	S_5	S_4	0
S_2	S_1	S_6	1
S_3	S_7	S_8	1
S_4	S_7	S_1	0
S_5	S_2	S_3	1
S_6	S_4	S_2	0
S_7	S_6	S_8	0
S_8	S_5	S_3	1

(a) For this part of the problem, do not consider the flip-flop input equations (this means that you can ignore the next-state part of the table). Make a state assignment which might minimize the output equation, and derive the minimum output equation for your assignment.

(b) Forget about your solution to (a). Apply Guidelines 1 and 2 to make a state assignment, assigning 000 to S_1. Derive input equations for D flip-flops using this assignment.

15.7 The following table is to be realized using D flip-flops.

(a) Find a good state assignment using the three guidelines (do not reduce the table first.) Try to satisfy as many of the adjacency conditions as possible.

(b) Using this assignment, derive the D flip-flop input equations and the output equations.

	$X = 0$	1	Z $X = 0$	1
A	F	D	0	0
B	D	B	0	0
C	A	C	0	1
D	F	D	0	0
E	A	C	0	1
F	F	B	0	0

15.8 (a) For the following state table, use the three guidelines to determine which of the three possible nonequivalent state assignments should give the best solution.

	$X_1X_2 = 00$	01	11	10	Z_1Z_2 $X_1X_2 = 00$	01	11	10
A	A	C	B	D	00	00	00	00
B	B	B	D	D	00	00	10	10
C	C	A	C	A	01	01	01	01
D	B	B	C	A	01	01	10	10

(b) Using your answer to (a), derive the T flip-flop input equations and the output equations.

15.9 Implement the given state graph using D flip-flops and gates. Use a one-hot assignment and write down the logic equations by inspecting the state graph.

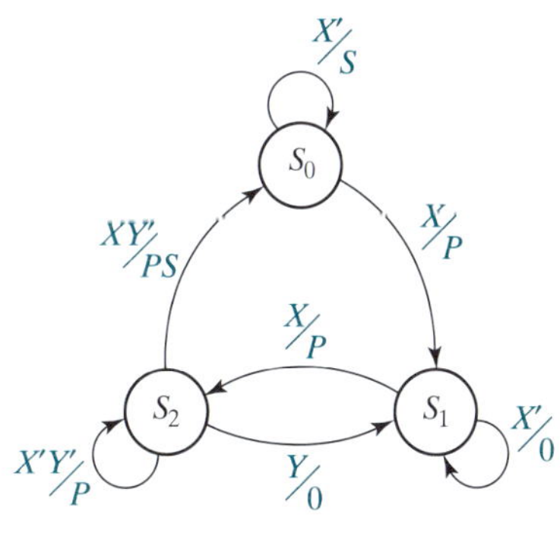

15.10 (a) Reduce the following state table to a minimum number of states.

Present State	Next State X = 0	1	Present Output X = 0	1
a	h	c	1	0
b	c	d	0	1
c	h	b	0	0
d	f	h	0	0
e	c	f	0	1
f	f	g	0	0
g	g	c	1	0
h	a	c	1	0

(b) You are given two identical sequential circuits which realize this state table. One circuit is initially in state *d*, and the other circuit is initially in state *c*. Specify an input sequence of length two which could be used to distinguish between the two circuits, and give the corresponding output sequence from each circuit.

15.11 For the following state table:
(a) Reduce the table to a minimum number of states.
(b) Using the basic definition of state equivalence, show that state *a* is not equivalent to state *b*.

Present State	Next State X = 0	X = 1	Present Output X = 0	X = 1
a	e	g	0	1
b	d	f	0	1
c	e	c	1	0
d	b	f	0	1
e	g	f	0	1
f	b	d	1	0
g	e	c	1	0

15.12 A Moore sequential circuit has a single input (X) and a single output (Z). Z is 1 if the most recent four inputs contained exactly two consecutive 1's or exactly two consecutive 0's (i.e., the input sequences 0011, 0010, 1001, 1100, 0110, 0100, 1011, and 1101). (The initial state S_0 acts as if the preceding inputs were all 0's.) The following state table was constructed using a sufficient number of states to remember the last four inputs and the output for each state assigned according to the sequence remembered by that state.
(a) Reduce the table to a minimum number of states. (*Hint:* First use the simple example of state equivalence used in Section 15.1 to eliminate as many states as possible.)
(b) For each state in the reduced table, give the input pattern remembered by that state.
(c) Convert the reduced table from part (a) into a Mealy state table that produces the same outputs.
(d) Reduce the Mealy state table to a minimum number of states.

Input Pattern	Present State	Next State $X = 0$	Next State $X = 1$	Output Z
0000	S_0	S_0	S_1	0
0001	S_1	S_2	S_3	0
0010	S_2	S_4	S_5	1
0011	S_3	S_6	S_7	1
0100	S_4	S_8	S_9	1
0101	S_5	S_{10}	S_{11}	0
0110	S_6	S_{12}	S_{13}	1
0111	S_7	S_{14}	S_{15}	0
1000	S_8	S_0	S_1	0
1001	S_9	S_2	S_3	1
1010	S_{10}	S_4	S_5	0
1011	S_{11}	S_6	S_7	1
1100	S_{12}	S_8	S_9	1
1101	S_{13}	S_{10}	S_{11}	1
1110	S_{14}	S_{12}	S_{13}	0
1111	S_{15}	S_{14}	S_{15}	0

15.13 A sequential circuit has a single input (X) and a single output (Z). The circuit examines each disjoint block of four inputs and determines whether the block is a valid BCD representation of a decimal digit; if not, $Z = 1$. State S_0 is the initial state, and the circuit enters state S_0 after the fourth input. The BCD digits are received most significant bit first. The following state table was constructed as a Mealy table using a sufficient number of states to remember the last three inputs with the output produced when the fourth input bit of a block is received.

(a) Does the resulting table specify a Mealy or a Moore circuit?

(b) Reduce the state table to a minimum number of states. (*Hint:* Use the simple example of state equivalence used in Section 15.1 to eliminate as many states as possible.)

(c) For each state in the reduced table, give the input pattern remembered by that state.

Input Pattern	Present State	Next State $X = 0$	Next State $X = 1$	Present Output Z $X = 0$	Present Output Z $X = 1$
–	S_1	S_2	S_3	0	0
0	S_2	S_4	S_5	0	0
1	S_3	S_6	S_7	0	0
00	S_4	S_8	S_9	0	0
01	S_5	S_{10}	S_{11}	0	0
10	S_6	S_{12}	S_{13}	0	0
11	S_7	S_{14}	S_{15}	0	0
000	S_8	S_1	S_1	0	0
001	S_9	S_1	S_1	0	0
010	S_{10}	S_1	S_1	0	0
011	S_{11}	S_1	S_1	0	0
100	S_{12}	S_1	S_1	0	0
101	S_{13}	S_1	S_1	1	1
110	S_{14}	S_1	S_1	1	1
111	S_{15}	S_1	S_1	1	1

15.14 A sequential circuit has a single input (X) and a single output (Z). The circuit examines each disjoint block of four inputs and determines whether the block is a valid BCD representation of a decimal digit; if not, $Z = 1$. State S_0 is the initial state, and the circuit enters state S_0 after the fourth input. The BCD digits are received least significant bit first. A Mealy state table can be constructed using a sufficient number of states to remember the last three inputs with the output produced when the fourth input bit of a block is received.

(a) Using the method indicated by Problem 15.13, construct a state table for this circuit.

(b) Reduce the state table to a minimum number of states. (*Hint:* Use the simple example of state equivalence used in Section 15.1 to eliminate as many states as possible.)

(c) For each state in the reduced table, give the input pattern(s) remembered by that state.

15.15 Reduce each of the following state tables to a minimum number of states:

(a)	$XY = 00$	01	11	10	Z
a	a	c	e	d	0
b	d	e	e	a	0
c	e	a	f	b	1
d	b	c	c	b	0
e	c	d	f	a	1
f	f	b	a	d	1

(b)	$X = 0$	1	0	1
a	b	c	1	0
b	e	d	1	0
c	g	d	1	1
d	e	b	1	0
e	f	g	1	0
f	h	b	1	1
g	h	i	0	1
h	g	i	0	1
i	a	a	0	1

15.16 Reduce each of the following tables to a minimum number of states:

(a)	$XY = 00$	01	11	10	Z
a	b	i	c	g	0
b	b	c	f	g	0
c	h	d	d	f	1
d	h	c	e	g	1
e	b	c	i	g	0
f	f	i	i	k	0
g	j	k	g	h	0
h	e	f	c	g	0
i	i	i	i	d	0
j	b	f	c	g	0
k	a	c	e	g	1

(b)	$XY = 00$	01	11	10	$XY = 00$	01	11	10
a	a	a	g	k	1	0	0	0
b	c	f	g	d	0	0	0	0
c	g	c	a	i	1	0	0	0
d	a	d	g	i	1	0	0	0
e	f	h	g	a	0	0	0	0
f	g	c	d	k	1	0	0	0
g	c	j	g	e	0	1	0	0
h	g	h	d	k	1	0	0	0
i	h	h	g	d	0	0	0	0
j	j	j	g	k	1	0	0	0
k	c	c	g	d	0	0	0	0

15.17 Circuits N and M have the state tables that follow.

(a) Without first reducing the tables, determine whether circuits N and M are equivalent.

(b) Reduce each table to a minimum number of states, and then show that N is equivalent to M by inspecting the reduced tables.

M

	$X = 0$	1	
S_0	S_3	S_1	0
S_1	S_0	S_1	0
S_2	S_0	S_2	1
S_3	S_0	S_3	1

N

	$X = 0$	1	
A	E	A	1
B	F	B	1
C	E	D	0
D	E	C	0
E	B	D	0
F	B	C	0

15.18 Below is an incompletely specified state table.

Present State	Next State $X = 0$	1	Present Output (Z)
S_0	S_1	S_0	0
S_1	S_0	S_2	0
S_2	S_3	S_4	1
S_3	S_0	S_3	0
S_4	S_0	–	0

(a) Reduce the state table to four states in two different ways by filling in the don't-care in the state table in different ways.

(b) Show that your two state tables in part (a) are not equivalent, using an implication table similar to Figure 15-7.

(c) Show that your two state tables in (a) are not equivalent by giving a short input sequence which gives different outputs for the two state tables.

15.19 Repeat 15.18 for this state table (four states).

Present State	Next State $X = 0$	1	Present Output (Z) $X = 0$	1
S_0	S_1	S_5	0	0
S_1	S_3	S_2	1	1
S_2	S_2	S_4	0	1
S_3	S_4	S_2	1	1
S_4	S_4	S_2	–	1
S_5	S_5	S_2	0	1

15.20 The following are possible state assignments for a six-state sequential circuit.

	(i)	(ii)	(iii)	(iv)	(v)
S_0	000	010	100	110	001
S_1	001	111	101	010	111
S_2	010	001	000	111	101
S_3	011	110	001	011	011
S_4	100	000	111	000	000
S_5	101	100	110	100	010

(a) Which two state assignments are equivalent?
(b) For each assignment (except (i)), give an equivalent assignment for which state S_0 is assigned to 000.
(c) Give a state assignment which is not equivalent to any of the assignments.

15.21 (a) For an eight-state sequential circuit using three flip-flops, give three state assignments that assign 000 to S_0 and are equivalent to a straight binary assignment.
(b) Give three state assignments that assign 111 to S_0 and are not equivalent to a straight binary assignment or to each other.

15.22 A sequential circuit with one input and one output has the following state table:

Present State	Next State $X = 0$	$X = 1$	Present Output
A	D	G	1
B	E	H	0
C	B	F	1
D	F	G	0
E	C	A	1
F	H	C	0
G	E	A	1
H	D	B	0

(a) For this part of the problem, do not consider the flip-flop input equations (this means that you can ignore the next-state part of the table). Make a state assignment which will minimize the output equation, and derive the minimum output equation for your assignment.

(b) Forget about your solution to (a). Apply Guidelines 1 and 2 to make a state assignment, assigning 000 to A. Derive input equations for D flip-flops using this assignment.

15.23 (a) For the following state table, use the three guidelines to determine which of the three possible nonequivalent state assignments should give the best solution.

	$X_1X_2 = 00$	01	11	10	Z_1Z_2 $X_1X_2 = 00$	01	11	10
A	A	A	C	C	01	01	01	01
B	B	D	B	D	11	11	11	11
C	A	A	B	D	11	11	00	00
D	D	B	A	C	01	01	01	01

(b) Using your answer to (a), derive J-K flip-flop input equations and the output equations.

15.24 Consider the following Moore sequential circuit.
(a) Derive the equations for a one-hot state assignment.
(b) Use Guidelines 1 and 2 to make a "good" state assignment using three state variables. Derive the next-state equations assuming D flip-flops are used.

Present State	Next State $X = 0$	$X = 1$	Present Output Z
A	B	A	0
B	C	A	0
C	E	D	0
D	B	A	1
E	E	A	0

15.25 (a) Reduce the following state table to a minimum number of states using implication charts.
(b) Use the guideline method to determine a suitable state assignment for the reduced table.
(c) Realize the table using D flip-flops.
(d) Realize the table using J-K flip-flops.

	$X = 0$	1	Z
A	A	B	1
B	C	E	0
C	F	G	1
D	C	A	0
E	I	G	1
F	H	I	1
G	C	F	0
H	F	B	1
I	C	E	0

15.26 Repeat Problem 15.25 for the following table:

	$X = 0$	1	Z
A	I	C	1
B	B	I	1
C	C	G	1
D	I	C	0
E	D	E	0
F	I	C	0
G	E	F	0
H	H	A	1
I	A	C	1

15.27 Make a suitable state assignment and realize the state graph of Figure 14-10 using:
(a) D flip-flops (b) S-R flip-flops

15.28 Make a suitable state assignment and realize the state graph of Figure 14-13 using:
(a) J-K flip-flops (b) T flip-flops

15.29 Make a suitable state assignment and realize the state table of Problem 14.22 using D flip-flops and NAND gates.

15.30 Make a suitable state assignment and realize the state table of Problem 14.5 using J-K flip-flops and NAND gates.

15.31 Reduce the state table of Problem 14.6 to a minimum number of rows. Then, make a suitable state assignment and realize the state table using D flip-flops.

15.32 Reduce the state table of Problem 14.23 to a minimum number of rows. Then, make a suitable state assignment and realize the state table using D flip-flops.

15.33 A logic designer who had not taken this course designed a sequential circuit with an input W using three T flip-flops, A, B, and C. The input equations for these flip-flops are

$$T_A = W'A'B + W'BC' + A'BC' + AB'C + WB'C + WAC$$
$$T_B = W'A'C + W'A'B + A'BC + AB'C' + WB'C' + WAC'$$
$$T_C = W'AC + W'B'C' + WBC + WA'C'$$

and the output equation is $Z = W'BC'$. Find an equivalent sequential circuit which uses fewer states. Realize it, trying to minimize the amount of logic required.

15.34 Modify the given state graph so that it is completely specified. Assume that if $X = Y = 1$, X takes precedence. Then implement the state graph using D flip-flops and gates. Use a one-hot assignment and write down the logic equations by inspecting the state graph.

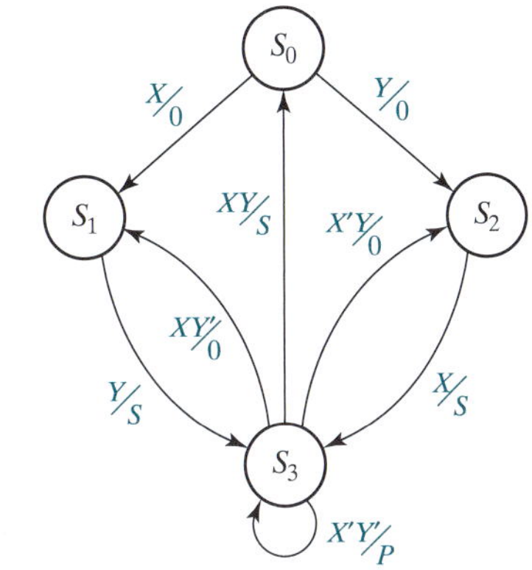

15.35 Implement the following state graph using D flip-flops and gates. Use a one-hot assignment and write down the logic equations by inspecting the state graph.

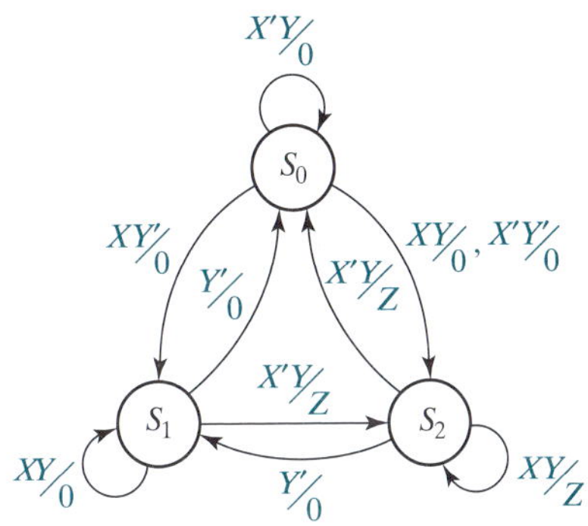

15.36 A state graph for a single-input sequential circuit is given. Implement the circuit using a three-bit parallel loading counter that has the given operation table. Label the counter outputs Q_2, Q_1, Q_0, where Q_0 is the least significant bit and the parallel inputs P_2, P_1, P_0. (*Hint:* Because the Ld signal overrides the Cnt signal, the counting sequence can easily be changed by doing a parallel load at the appropriate times.)

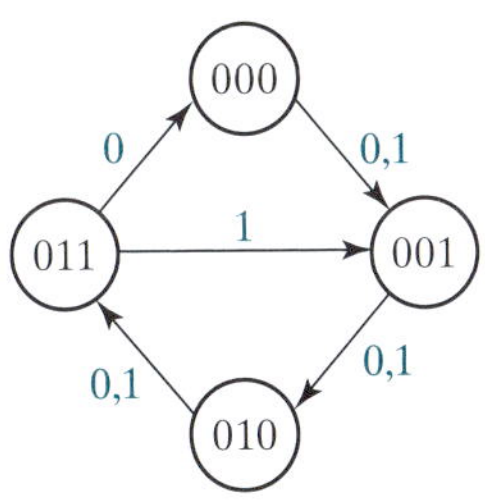

Clr	Ld	Cnt	Function
0	—	—	Clear
1	1	—	Parallel Load
1	0	1	Increment
1	0	0	Hold (No change)

15.37 Consider the following Mealy sequential circuit.

Present State	Next State $X=0$	Next State $X=1$	Present Output $X=0$	Present Output $X=1$
A	B	A	0	0
B	C	A	0	0
C	D	A	0	1
D	D	A	0	0

(a) Use a one-hot state assignment, and implement the circuit using D flip-flops.

(b) Use the state assignment $A = 00$, $B = 01$, $C = 11$, and $D = 10$, and implement the circuit using D flip-flops.

(c) Implement the circuit using a 4-bit parallel loading counter instead of flip-flops for memory. Assume the synchronous counter controls are as follows:

S_1	S_0	Function
0	0	Hold
0	1	Increment
1	0	Parallel load
1	1	Clear

$Q_3Q_2Q_1Q_0$ are the outputs; $P_3P_2P_1P_0$ are the parallel inputs; and Q_0 is the least significant bit of the counter. (With the proper state assignment, this can be done without using the parallel load function of the counter.)

(d) Implement the circuit using a 4-bit parallel loading shift register instead of flip-flops for memory. Assume the synchronous shift register controls are as follows:

S_1	S_0	Function
0	0	Hold
0	1	Shift right
1	0	Parallel load
1	1	Clear

S_{in} is the input for the shift; $Q_3Q_2Q_1Q_0$ are the outputs; $P_3P_2P_1P_0$ are the parallel inputs. When shifting, $S_{in} \rightarrow Q_3$, $Q_3 \rightarrow Q_2$, $Q_2 \rightarrow Q_1$, $Q_1 \rightarrow Q_0$. (With the proper state assignment, this can be done without using the parallel load function of the shift register.)

15.38 A sequential circuit contains two D flip-flops; the excitation equations for the flip-flops are $D_1 = XQ_1 + XQ_2$ and $D_2 = XQ_1 + XQ_2'$.

(a) Convert the circuit into an equivalent one where each D flip-flop is replaced by a T flip-flop. Do this by converting the next-state equations into the form for a T flip-flop. (*Hint:* $MQ + NQ' = (M'Q + NQ')'Q + (M'Q + NQ')Q'$.)

(b) Repeat part (a) by constructing an excitation table for the T flip-flops (i.e., a truth table for T_1 and T_2 as a function of X, Q_1, and Q_2).

(c) Convert the circuit into an equivalent one where each D flip-flop is replaced by a J-K flip-flop. Do this by converting the next-state equations into the form for a J-K flip-flop.

(d) Repeat part (c) by constructing an excitation table for the J-K flip-flops (i.e., a truth table for the J and K flip-flop inputs as a function of X, Q_1, and Q_2).

15.39 A sequential circuit contains two J-K flip-flops; the excitation equations for the flip-flops are $J_1 = Q_2, K_1 = Q_1, J_2 = X + Q_1'$, and $K_2 = 1$.

(a) Convert the circuit into an equivalent one where each J-K flip-flop is replaced by a T flip-flop. Do this by converting the next-state equations into the form for a T flip-flop. (*Hint:* $MQ + NQ' = (M'Q + NQ')'Q + (M'Q + NQ')Q'$.)

(b) Repeat part (a) by constructing an excitation table for the T flip-flops (i.e., a truth table for T_1 and T_2 as a function of X, Q_1, and Q_2).

(c) Convert the circuit into an equivalent one where each D flip-flop is replaced by an S-R flip-flop. Do this by converting the next-state equations into the form for a S-R flip-flop.

(d) Repeat part (c) by constructing an excitation table for the S-R flip-flops (i.e., a truth table for the S and R flip-flop inputs as a function of X, Q_1, and Q_2).

15.40 Show that the following incompletely specified table can be reduced to a table with seven states. (*Hint:* This table can be reduced by applying row matching where a don't-care "matches" anything.)

Present State	Next State $X = 0$	$X = 1$	Output (Z) $X = 0$	$X = 1$
S_0	S_1	S_2	0	0
S_1	S_3	S_4	0	0
S_2	S_5	S_6	0	0
S_3	S_7	S_8	0	0
S_4	S_8	S_9	0	0
S_5	S_{10}	S_9	0	0
S_6	S_9	S_8	0	0
S_7	S_0	S_0	1	0
S_8	S_0	–	0	–
S_9	S_0	–	1	–
S_{10}	S_0	S_0	0	1

15.41 (a) Find the maximal compatibles for Table 14-10.

(b) Reduce the table to a minimum number of states.

(c) Consider a circuit designed from the reduced table. Describe in words the function of the circuit.

15.42 (a) Find the maximal compatibles for Table 14-9. (There are seven.)

(b) Reduce the table to a minimum number of states.

15.43 Modify the BCD Parity Generator of Table 14-9 so that the outputs are don't-cares for the first three inputs. The new state table is given below.
(a) Find the maximal compatibles for this table. (There are eight.)
(b) Reduce this table to a minimum number of states.

Present State	Next State X = 0	Next State X = 1	Output X = 0	Output X = 1
S_0	S_1	S_2	–	–
S_1	S_3	S_4	–	–
S_2	S_5	–	–	–
S_3	S_6	S_7	–	–
S_4	S_7	S_6	–	–
S_5	S_7	–	–	–
S_6	S_0	S_0	1	0
S_7	S_0	S_0	0	1

15.44 (a) Find the maximal compatibles for the incompletely specified table below.
(b) Reduce the table to a minimum number of states.

	Present State	Next State X = 0	Next State X = 1	Output X = 0	Output X = 1
	S_0	S_1	S_2	–	–
0	S_1	S_3	S_4	–	–
1	S_2	S_5	S_6	–	–
00	S_3	S_0	S_0	1	0
01	S_4	S_0	–	0	–
10	S_5	S_0	S_0	0	1
11	S_6	S_0	–	1	–

15.45 (a) Find the maximal compatibles for the table below. (There are eight.)
(b) Try reducing the table using the maximal compatibles. How many states are required?
(c) Show that this table can be reduced to six states using pairs of compatible states. (This is an example of a table that cannot be reduced using only maximal compatibles.)

Present State	Next State X = 0	Next State X = 1	Output (Z) X = 0	Output (Z) X = 1
S_0	S_1	S_2	–	–
S_1	S_3	S_4	–	–
S_2	S_4	S_3	–	–
S_3	S_5	S_6	–	–
S_4	S_6	S_5	–	–
S_5	S_0	S_0	1	0
S_6	S_0	S_0	0	1

UNIT 16

Sequential Circuit Design

Objectives

1. Design a sequential circuit using gates and flip-flops.
2. Test your circuit by simulating it and by implementing it in lab.
3. Design a unilateral iterative circuit. Explain the relationship between iterative and sequential circuits, and convert from one to the other.
4. Show how to implement a sequential circuit using a ROM or PLA and flip-flops.
5. Explain the operation of CPLDs and FPGAs and show how they can be used to implement sequential logic.

Study Guide

1. Study Sections 16.1, *Summary of Design Procedure for Sequential Circuits*, and 16.2, *Design Example—Code Converter*.

   (a) Why are the states in the next-state part of Table 16-2 listed in a different order from the states in Table 15-1?

   (b) Consider the design of a sequential circuit to convert an 8-4-2-1 code to a 6-3-1-1 code (see Table 1-2). If the least significant bit of an 8-4-2-1 coded digit is fed into the circuit at t_0, can the least significant bit of the 6-3-1-1 coded digit be determined immediately? Explain. Why can the technique described in this section not be used to design the 8-4-2-1 to 6-3-1-1 code converter?

2. Study Section 16.3, *Design of Iterative Circuits*

   (a) Draw a state graph for the comparator of Table 16-4. Compare several pairs of binary numbers using the scheme represented by Table 16-4 and make sure you understand why this method works. Draw a circuit similar to Figure 16-6 with five cells. Show the values of all the cell inputs and outputs if $X = 10101$ and $Y = 10011$.

   (b) If the state table for a typical cell of an iterative circuit has n states, what is the minimum number of signals required between each pair of adjacent cells?

   (c) Work Problem 16.17.

3. Study Section 16.4, *Design of Sequential Circuits Using ROMs and PLAs*.

   (a) Review Section 9.5, *Read-Only Memories*, and Section 9.6, *Programmable Logic Devices*.

   (b) What size ROM would be required to realize a state table with 13 states, two input variables, and three output variables?

   (c) In going from Table 16-6(b) to 16-6(c), note that for $X = 0$, $Q_1Q_2Q_3 = 000$, $Z = 1$, and $Q_1^+Q_2^+Q_3^+ = D_1D_2D_3 = 001$; therefore, 1001 is entered in the first row of the truth table. Verify that the other truth table entries are correct.

   (d) Continue the analysis of the PLA realization of the code converter which was started in the paragraph following Table 16-7. In particular, if $Q_1Q_2Q_3 = 100$ and $X = 1$, what will be the PLA outputs? What will the state be after the clock?

(e) Work Problems 16.15 and 16.16.

4. Study Section 16.5, *Sequential Circuit Design Using CPLDs*, and Section 16.6, *Sequential Circuit Design Using FPGAs*.

(a) How many macrocells of a CoolRunner-II are needed to implement a Moore machine with six states, one input, and two outputs? With two inputs?

(b) How many LUT's of a Virtex/Spartan II are needed to implement each of these Moore machines? How many CLB's?

(c) Rewrite the equations for Q_2^+, Q_1^+, and Q_0^+ of Equations 12-1 to fit into one LUT each, as we did for Q_3^+ in Equation (16-2), using CE = *Ld* + *Sh*.

5. Study Section 16.7, *Simulation and Testing of Sequential Circuits*.

(a) Observe the simulator output of Figure 16-23(b), and note the times at which the *Z* output changes. Assuming that each gate and flip-flop in Figure 16-22 has a 10-ns delay, explain the *Z* waveform.

(b) Suppose that you are testing the circuit of Figure 16-4, and that when you set $X = 0$ and $Q_1Q_2Q_3 = 011$ and pulse the clock, the circuit goes to state 100 instead of 000. What would you do to determine the cause of the malfunction?

6. Read Section 16.8, *Overview of Computer-Aided Design*, for general information.

7. Work out your assigned design problem by hand. Then, use *LogicAid* to check your state table using the state table checker, and then verify that your logic equations are correct. Try at least two different state assignments and choose the one which requires the smallest number of logic gates.

8. Answer the following questions before you simulate your circuit or test it in lab. At which of the following times will the output of your circuit be correct? (If you are not absolutely sure that your answer is correct, review Section 13.2, paying particular attention to the timing charts for Mealy circuits.)

(a) Just before the rising clock edge

(b) Just after the rising clock edge (after the state has changed but before the input is changed to the next value)

(c) After the input has been changed to the next value, but before the next rising edge occurs

9. (a) Explain how it is possible to get false outputs from your circuit even though the circuit is correctly designed and working properly.

(b) If the output of your circuit was fed into another sequential circuit using the same clock, would the false outputs cause any problems? Explain.

10. When you get your circuit working properly, determine the output sequences for the given test sequences. Demonstrate the operation of your circuit to a proctor and have him or her check your output sequences. After successful completion of the project, turn in your design and the test results. (No readiness test is required.)

Sequential Circuit Design

We have already studied the various steps in sequential circuit design—derivation of state tables (Unit 14), state table reduction (Unit 15), state assignment (Unit 15), and derivation of flip-flop input equations (Units 12 and 15). This unit contains a summary of the design procedure, a comprehensive design example, and procedures for testing your circuit in lab.

16.1 Summary of Design Procedure for Sequential Circuits

1. Given the problem statement, determine the required relationship between the input and output sequences and derive a state table. For many problems, it is easiest to first construct a state graph.
2. Reduce the table to a minimum number of states. First, eliminate duplicate rows by row matching and, then, form an implication table and follow the procedure in Section 15.3.

3. If the reduced table has m states ($2^{n-1} < m \leq 2^n$), n flip-flops are required. Assign a unique combination of flip-flop states to correspond to each state in the reduced table. The guidelines given in Section 15.8 may prove helpful in finding an assignment which leads to an economical circuit.
4. Form the transition table by substituting the assigned flip-flop states for each state in the reduced state table. The resulting transition table specifies the next states of the flip-flops, and the output in terms of the present states of the flip-flops and the input.
5. Plot next-state maps and input maps for each flip-flop and derive the flip-flop input equations. (Depending on the type of gates to be used, either determine the sum-of-products form from the 1's on the map or the product-of-sums form from the 0's on the map.) Derive the output functions.
6. Realize the flip-flop input equations and the output equations using the available logic gates.
7. Check your design by signal tracing, computer simulation, or laboratory testing.

16.2 Design Example—Code Converter

We will design a sequential circuit to convert BCD to excess-3 code. This circuit adds three to a binary-coded-decimal digit in the range 0 to 9. The input and output will be serial with the least significant bit first. A list of allowed input and output sequences is shown in Table 16-1.

Table 16-1 lists the desired inputs and outputs at times t_0, t_1, t_2, and t_3. After receiving four inputs, the circuit should reset to the initial state, ready to receive another group of four inputs. It is not clear at this point whether a sequential circuit can actually be realized to produce the output sequences as specified in Table 16-1 without delaying the output.

TABLE 16-1

X Input (BCD)				Z Output (excess-3)			
t_3	t_2	t_1	t_0	t_3	t_2	t_1	t_0
0	0	0	0	0	0	1	1
0	0	0	1	0	1	0	0
0	0	1	0	0	1	0	1
0	0	1	1	0	1	1	0
0	1	0	0	0	1	1	1
0	1	0	1	1	0	0	0
0	1	1	0	1	0	0	1
0	1	1	1	1	0	1	0
1	0	0	0	1	0	1	1
1	0	0	1	1	1	0	0

For example, if at t_0 some sequences required an output $Z = 0$ for $X = 0$ and other sequences required $Z = 1$ for $X = 0$, it would be impossible to design the circuit without delaying the output. For Table 16-1 we see that at t_0 if the input is 0 the output is always 1, and if the input is 1 the output is always 0; therefore, there is no conflict at t_0. At time t_1 the circuit will have available only the inputs received at t_1 and t_0. There will be no conflict at t_1 if the output at t_1 can be determined only from the inputs received at t_1 and t_0. If 00 has been received at t_1 and t_0, the output should be 1 at t_1 in all three cases where 00 occurs in the table. If 01 has been received, the output should be 0 at t_1 in all three cases where 01 occurs. For sequences 10 and 11 the outputs at t_1 should be 0 and 1, respectively. Therefore, there is no output conflict at t_1. In a similar manner we can check to see that there is no conflict at t_2, and at t_3 all four inputs are available, so there is no problem.

We will now proceed to set up the state table (Table 16-2), using the same procedure as in Section 15.1. The arrangement of next states in the table is different from that in Table 15-1 because in this example the input sequences are received with least significant bit first, while for Table 15-1 the first input bit received is listed first in the sequence. Dashes (don't-cares) appear in this table because only 10 of the 16 possible 4-bit sequences can occur as inputs to the code converter. The output part of the table is filled in, using the reasoning discussed in the preceding paragraph. For example, if the circuit is in state B at t_1 and a 1 is received, this means that the sequence 10 has been received and the output should be 0.

Next, we will reduce the table using row matching. When matching rows which contain dashes (don't-cares), a dash will match with any state or with any output value. By matching rows in this manner, we have $H \equiv I \equiv J \equiv K \equiv L$ and $M \equiv N \equiv P$. After eliminating I, J, K, L, N, and P, we find $E \equiv F \equiv G$ and the table reduces to seven rows (Table 16-3).

TABLE 16-2
State Table for Code Converter

Time	Input Sequence Received (Least Significant Bit First)	Present State	Next State $X = 0$	1	Present Output (Z) $X = 0$	1
t_0	reset	A	B	C	1	0
t_1	0	B	D	F	1	0
	1	C	E	G	0	1
	00	D	H	L	0	1
t_2	01	E	I	M	1	0
	10	F	J	N	1	0
	11	G	K	P	1	0
	000	H	A	A	0	1
	001	I	A	A	0	1
	010	J	A	–	0	–
	011	K	A	–	0	–
t_3	100	L	A	–	0	–
	101	M	A	–	1	–
	110	N	A	–	1	–
	111	P	A	–	1	–

TABLE 16-3 Reduced State Table for Code Converter

Time	Present State	Next State $X = 0$	1	Present Output (Z) $X = 0$	1
t_0	A	B	C	1	0
t_1	B	D	E	1	0
	C	E	E	0	1
t_2	D	H	H	0	1
	E	H	M	1	0
t_3	H	A	A	0	1
	M	A	–	1	–

An alternate approach to deriving Table 16-2 is to start with a state graph. The state graph (Figure 16-1) has the form of a tree. Each path starting at the reset state represents one of the ten possible input sequences. After the paths for the input sequences have been constructed, the outputs can be filled in by working backward along each path. For example, starting at t_3, the path 0 0 0 0 has outputs 0 0 1 1 and the path 1 0 0 0 has outputs 1 0 1 1. Verify that Table 16-2 corresponds to this state graph.

Three flip-flops are required to realize the reduced table because there are seven states. Each of the states must be assigned a unique combination of flip-flop states. Some assignments will lead to economical circuits with only a few gates, while other assignments will require many more gates. Using the guidelines given in Section 15.8, states B and C, D and E, and H and M should be given adjacent assignments in order to simplify the next-state functions. To simplify the output function, states (A, B, E, and M) and (C, D, and H) should be given adjacent assignments. A good assignment for this example is given on the map and table in Figure 16-2. After the state assignment has been made, the transition table is filled in according to the assignment, and the next-state maps are plotted as shown in Figure 16-3. The D input equations are then read off the Q^+ maps as indicated. Figure 16-4 shows the resulting sequential circuit.

FIGURE 16-1 State Graph for Code Converter

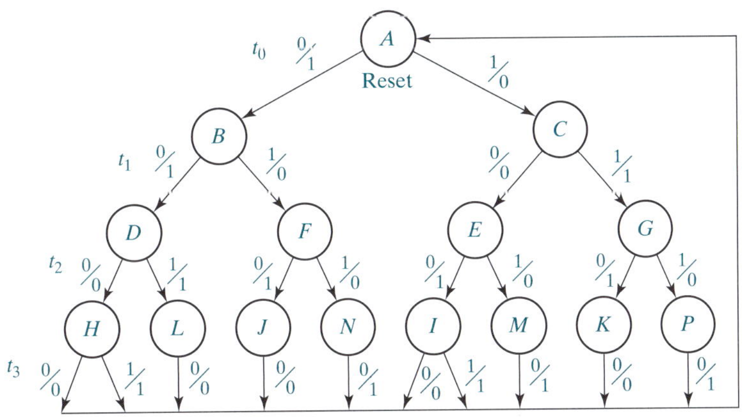

FIGURE 16-2
Assignment Map and Transition Table for Flip-Flops

Q_2Q_3 \ Q_1	0	1
00	*A*	*B*
01		*C*
11	*H*	*D*
10	*M*	*E*

(a) Assignment map

		$Q_1^+Q_2^+Q_3^+$		Z	
	$Q_1Q_2Q_3$	$X = 0$	$X = 1$	$X = 0$	$X = 1$
A	000	100	101	1	0
B	100	111	110	1	0
C	101	110	110	0	1
D	111	011	011	0	1
E	110	011	010	1	0
H	011	000	000	0	1
M	010	000	xxx	1	x
–	001	xxx	xxx	x	x

(b) Transition table

FIGURE 16-3
Karnaugh Maps for Code Converter Design

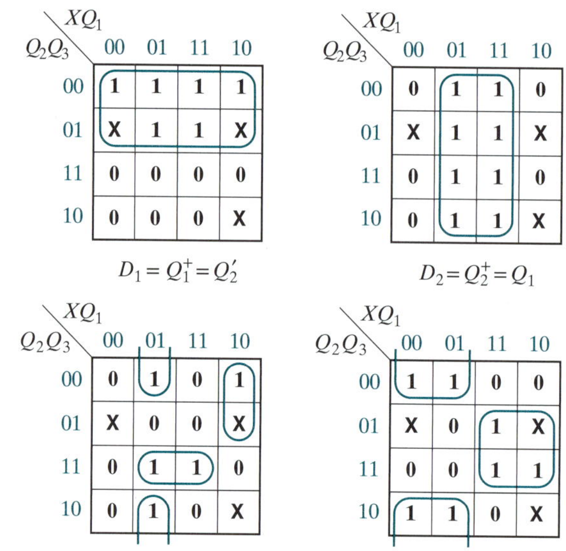

$D_1 = Q_1^+ = Q_2'$

$D_2 = Q_2^+ = Q_1$

$D_3 = Q_3^+ = Q_1Q_2Q_3 + X'Q_1Q_3' + XQ_1'Q_2'$

$Z = X'Q_3' + XQ_3$

FIGURE 16-4
Code Converter Circuit

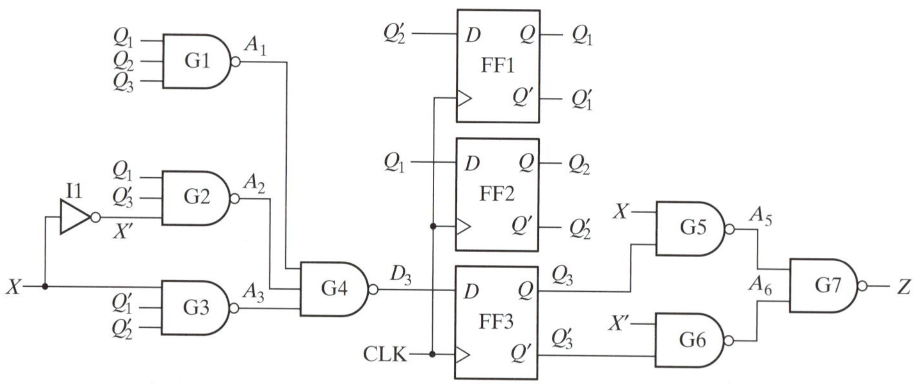

16.3 Design of Iterative Circuits

Many of the design procedures used for sequential circuits can be applied to the design of iterative circuits. An iterative circuit consists of a number of identical cells interconnected in a regular manner. Some operations, such as binary addition, naturally lend themselves to realization with an iterative circuit because the same operation is performed on each pair of input bits. The regular structure of an iterative circuit makes it easier to fabricate in integrated circuit form than circuits with less regular structures.

The simplest form of an iterative circuit consists of a linear array of combinational cells with signals between cells traveling in only one direction (Figure 16-5). Each cell is a combinational circuit with one or more primary inputs (x_i) and possibly one or more primary outputs (z_i). In addition, each cell has one or more secondary inputs (a_i) and one or more secondary outputs (a_{i+1}). The a_i signals carry information about the "state" of one cell to the next cell.

The primary inputs to the cells ($x_1, x_2, \ldots, x_n$) are applied in parallel; that is, they are all applied at the same time. The a_i signals then propagate down the line of cells. Because the circuit is combinational, the time required for the circuit to reach a steady-state condition is determined only by the delay times of the gates in the cells. As soon as steady state is reached, the outputs may be read. Thus, the iterative circuit can function as a parallel-input, parallel-output device, in contrast with the sequential circuit in which the input and output are serial. One can think of the iterative circuit as receiving its inputs as a sequence in space in contrast with the sequential circuit which receives its inputs as a sequence in time. The parallel adder of Figure 4-3 is an example of an iterative circuit that has four identical cells. The serial adder of Figure 13-12 uses the same full adder cell as the parallel adder, but it receives its inputs serially and stores the carry in a flip-flop instead of propagating it from cell to cell.

Design of a Comparator

As an example, we will design a circuit which compares two n-bit binary numbers and determines if they are equal or which one is larger if they are not equal. Direct design as a $2n$-input combinational circuit is not practical for n larger than 4 or 5, so we will try the iterative approach. Designate the two binary numbers to be compared as

$$X = x_1x_2 \ldots x_n \qquad \text{and} \qquad Y = y_1y_2 \ldots y_n$$

We have numbered the bits from left to right, starting with x_1 as the most significant bit because we plan to do the comparison from left to right.

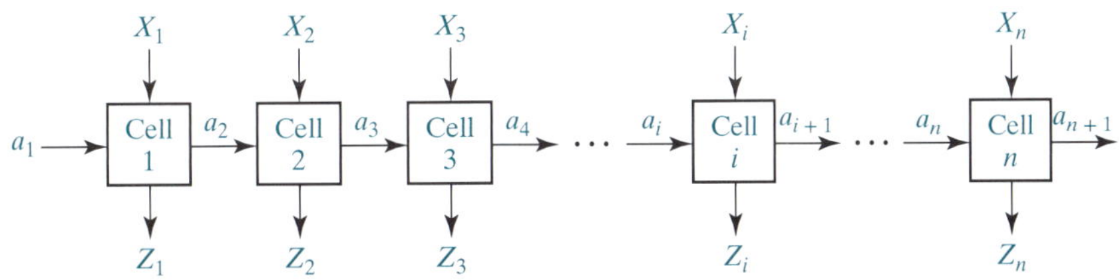

FIGURE 16-5 Unilateral Iterative Circuit

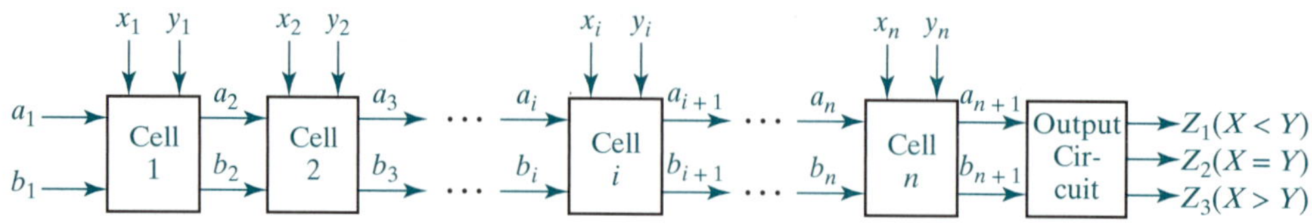

FIGURE 16-6 Form of Iterative Circuit for Comparing Binary Numbers

Figure 16-6 shows the form of the iterative circuit, although the number of leads between each pair of cells is not yet known. Comparison proceeds from left to right. The first cell compares x_1 and y_1 and passes on the result of the comparison to the next cell, the second cell compares x_2 and y_2, etc. Finally, x_n and y_n are compared by the last cell, and the output circuit produces signals to indicate if $X = Y$, $X > Y$, or $X < Y$.

We will now design a typical cell for the comparator. To the left of cell i, three conditions are possible: $X = Y$ so far ($x_1 x_2 \dots x_{i-1} = y_1 y_2 \dots y_{i-1}$), $X > Y$ so far, and $X < Y$ so far. We designate these three input conditions as states S_0, S_1, and S_2, respectively. Table 16-4 shows the output state at the right of the cell (S_{i+1}) in terms of the $x_i y_i$ inputs and the input state at the left of the cell (S_i). If the numbers are equal to the left of cell i and $x_i = y_i$, the numbers are still equal including cell i, so $S_{i+1} = S_0$. However, if $S_i = S_0$ and $x_i y_i = 10$, then $x_1 x_2 \dots x_i > y_1 y_2 \dots y_i$ and $S_{i+1} = S_1$. If $X > Y$ to the left of cell i, then regardless of the values of x_i and y_i, $x_1 x_2 \dots x_i > y_1 y_2 \dots y_i$ and $S_{i+1} = S_1$. Similarly, if $X < Y$ to the left of cell i, then $X < Y$ including the inputs to cell i, and $S_{i+1} = S_2$.

TABLE 16-4 State Table for Comparator

		S_{i+1}				
	S_i	$x_i y_i = 00$	01	11	10	$Z_1 Z_2 Z_3$
$X = Y$	S_0	S_0	S_2	S_0	S_1	0 1 0
$X > Y$	S_1	S_1	S_1	S_1	S_1	0 0 1
$X < Y$	S_2	S_2	S_2	S_2	S_2	1 0 0

The logic for a typical cell is easily derived from the state table. Because there are three states, two intercell signals are required. Using the guidelines from Section 15.8 leads to the state assignment $a_i b_i = 00$ for S_0, 01 for S_1, and 10 for S_2. Substituting this assignment into the state table yields Table 16-5. Figure 16-7 shows the Karnaugh maps, next-state equations, and the realization of a typical cell using NAND gates. Inverters must be included in the cell because only a_i and b_i and not their complements are transmitted between cells.

The $a_1 b_1$ inputs to the left end cell must be 00 because we must assume that the numbers are equal (all 0) to the left of the most significant bit. The equations for the first cell can then be simplified if desired:

$$a_2 = a_1 + x_1' y_1 b_1' = x_1' y_1$$
$$b_2 = b_1 + x_1 y_1' a_1' = x_1 y_1'$$

TABLE 16-5 Transition Table for Comparator

	$a_{i+1} b_{i+1}$				
$a_i b_i$	$x_i y_i = 00$	01	11	10	$Z_1 Z_2 Z_3$
0 0	00	10	00	01	0 1 0
0 1	01	01	01	01	0 0 1
1 0	10	10	10	10	1 0 0

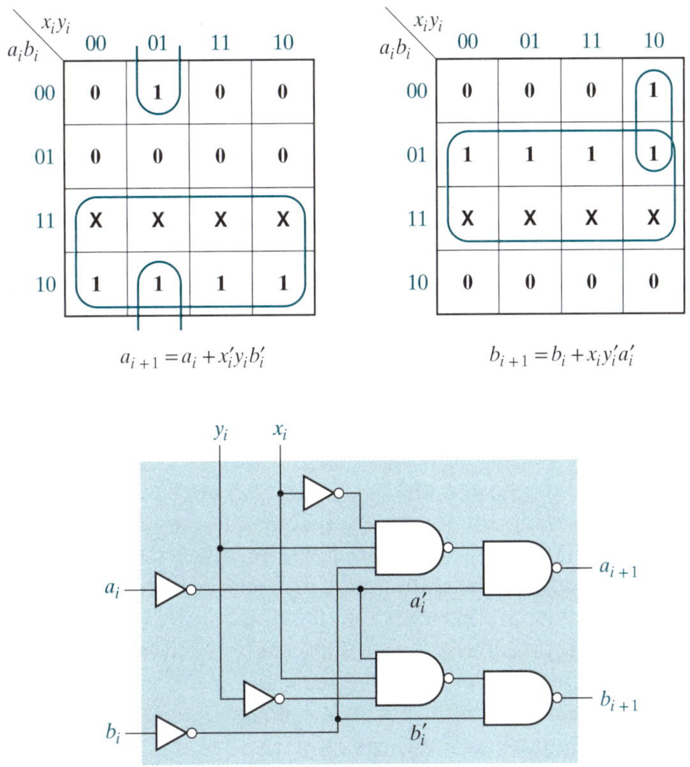

FIGURE 16-7 Typical Cell for Comparator

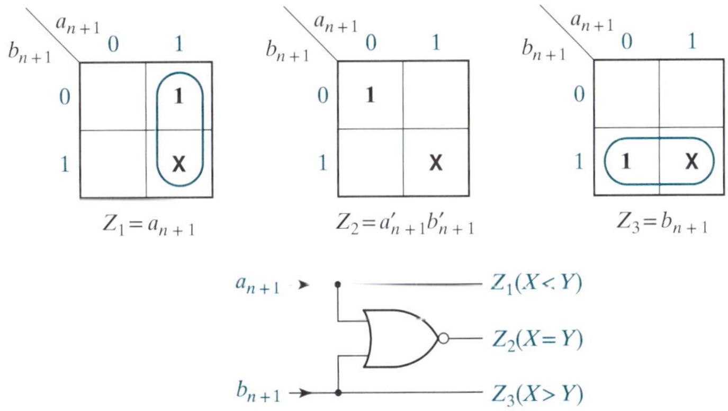

FIGURE 16-8 Output Circuit for Comparator

For the output circuit, let $Z_1 = 1$ if $X < Y$, $Z_2 = 1$ if $X = Y$, $Z_3 = 1$ if $X > Y$. Figure 16-8 shows the output maps, equations, and circuit.

Conversion to a sequential circuit is straightforward. If x_i and y_i inputs are received serially instead of in parallel, Table 16-4 is interpreted as a state table for a sequential circuit, and the next-state equations are the same as in Figure 16-7. If D flip-flops are used, the typical cell of Figure 16-7 can be used as the combinational part of the sequential circuit, and Figure 16-9 shows the resulting circuit. After all of the inputs have been read in, the output is determined from the state of the two flip-flops.

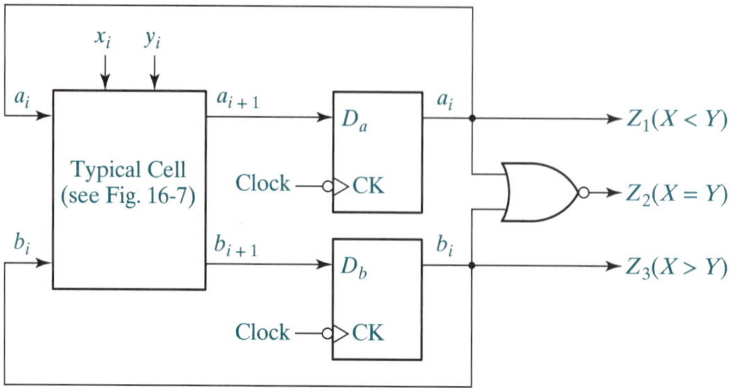

FIGURE 16-9 Sequential Comparator for Binary Numbers

This example indicates that the design of a unilateral iterative circuit is very similar to the design of a sequential circuit. The principal difference is that for the iterative circuit the inputs are received in parallel as a sequence in space, while for the sequential circuit the inputs are received serially as a sequence in time. For the iterative circuit, the state table specifies the output state of a typical cell in terms of its input state and primary inputs, while for the corresponding sequential circuit, the same table specifies the next state (in time) in terms of the present state and inputs. If D flip-flops are used, the typical cell for the iterative circuit can serve as the combinational logic for the corresponding sequential circuit. If other flip-flop types are used, the input equations can be derived in the usual manner.

16.4 Design of Sequential Circuits Using ROMs and PLAs

A sequential circuit can easily be designed using a ROM (read-only memory) and flip-flops. Referring to the general model of a Mealy sequential circuit given in Figure 13-17, the combinational part of the sequential circuit can be realized using a ROM. The ROM can be used to realize the output functions ($Z_1, Z_2, \ldots, Z_n$) and the next-state functions ($Q_1^+, Q_2^+, \ldots, Q_k^+$). The state of the circuit can then be stored in a register of D flip-flops and fed back to the input of the ROM. Thus, a Mealy sequential circuit with m inputs, n outputs, and k state variables can be realized using k D flip-flops and a ROM with $m + k$ inputs (2^{m+k} words) and $n + k$ outputs. The Moore sequential circuit of Figure 13-19 can be realized in a similar manner. The next-state and output combinational subcircuits of the Moore circuit can be realized using two ROMs. Alternatively, a single ROM can be used to realize both the next-state and output functions.

Use of D flip-flops is preferable to J-K flip-flops because use of two-input flip-flops would require increasing the number of outputs from the ROM. The fact that the D flip-flop input equations would generally require more gates than the J-K equations is of no consequence because the size of the ROM depends only on the number of inputs and outputs and not on the complexity of the equations being realized. For this reason, the state assignment which is used is also of little importance, and, generally, a state assignment in straight binary order is as good as any.

In Section 16.2, we realized a code converter using gates and D flip-flops. We will now realize this converter using a ROM and D flip-flops. The state table for the converter is reproduced in Table 16-6(a). Because there are seven states, three D flip-flops are required. Thus, a ROM with four inputs (2^4 words) and four outputs is required, as shown in Figure 16-10. Using a straight binary state assignment, we can construct the transition table, seen in Table 16-6(b), which gives the next state of the flip-flops as a function of the present state and input. Because we are using D flip-flops, $D_1 = Q_1^+$, $D_2 = Q_2^+$, and $D_3 = Q_3^+$. The truth table for the ROM, shown in Table 16-6(c), is easily constructed from the transition table. This table gives the ROM outputs (Z, D_1, D_2, and D_3) as functions of the ROM inputs (X, Q_1, Q_2, and Q_3).

Sequential circuits can also be realized using PLAs (programmable logic arrays) and flip-flops in a manner similar to using ROMs and flip-flops. However, in the case of PLAs, the state assignment may be important because the use of a good state

TABLE 16-6

(a) State table

Present State	Next State $X = 0$	Next State 1	Present Output (Z) $X = 0$	Present Output (Z) 1
A	B	C	1	0
B	D	E	1	0
C	E	E	0	1
D	H	H	0	1
E	H	M	1	0
H	A	A	0	1
M	A	–	1	–

(b) Transition table

	$Q_1Q_2Q_3$	$Q_1^+Q_2^+Q_3^+$ $X = 0$	$Q_1^+Q_2^+Q_3^+$ $X = 1$	Z $X = 0$	Z $X = 1$
A	000	001	010	1	0
B	001	011	100	1	0
C	010	100	100	0	1
D	011	101	101	0	1
E	100	101	110	1	0
H	101	000	000	0	1
M	110	000	–	1	–

(c) Truth Table

X	Q_1	Q_2	Q_3	Z	D_1	D_2	D_3
0	0	0	0	1	0	0	1
0	0	0	1	1	0	1	1
0	0	1	0	0	1	0	0
0	0	1	1	0	1	0	1
0	1	0	0	1	1	0	1
0	1	0	1	0	0	0	0
0	1	1	0	1	0	0	0
0	1	1	1	x	x	x	x
1	0	0	0	0	0	1	0
1	0	0	1	0	1	0	0
1	0	1	0	1	1	0	0
1	0	1	1	1	1	0	1
1	1	0	0	0	1	1	0
1	1	0	1	1	0	0	0
1	1	1	0	x	x	x	x
1	1	1	1	x	x	x	x

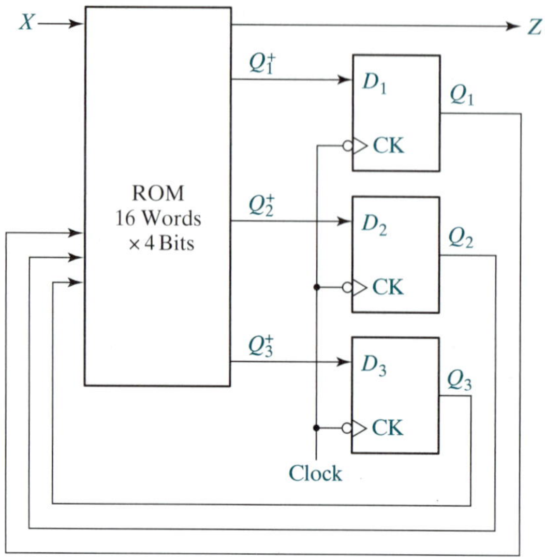

FIGURE 16-10 Realization of Table 16-6(a) Using a ROM

assignment can reduce the required number of product terms and, hence, reduce the required size of the PLA.

As an example, we will consider realizing the state table of Table 16-6(a) using a PLA and three D flip-flops. The circuit configuration is the same as Figure 16-10, except that the ROM is replaced with a PLA of appropriate size. Using a straight binary assignment leads to the truth table given in Table 16-6(c). This table could be stored in a PLA with four inputs, 13 product terms, and four outputs, but this would offer little reduction in size compared with the 16-word ROM solution discussed earlier.

If the state assignment of Figure 16-2 is used, the resulting output equation and D flip-flop input equations, derived from the maps in Figure 16-3, are

$$\begin{aligned} D_1 &= Q_1^+ = Q_2' \\ D_2 &= Q_2^+ = Q_1 \\ D_3 &= Q_3^+ = Q_1Q_2Q_3 + X'Q_1Q_3' + XQ_1'Q_2' \\ Z &= X'Q_3' + XQ_3 \end{aligned} \tag{16-1}$$

The PLA table which corresponds to these equations is in Table 16-7. Realization of this table requires a PLA with four inputs, seven product terms, and four outputs.

TABLE 16-7

X	Q_1	Q_2	Q_3	Z	D_1	D_2	D_3
–	–	0	–	0	1	0	0
–	1	–	–	0	0	1	0
–	1	1	1	0	0	0	1
0	1	–	0	0	0	0	1
1	0	0	–	0	0	0	1
0	–	–	0	1	0	0	0
1	–	–	1	1	0	0	0

Next, we will verify the operation of the circuit of Figure 16-4 using a PLA which corresponds to Table 16-7. Initially, assume that $X = 0$ and $Q_1Q_2Q_3 = 000$. This selects rows --0- and 0--0 in the table, so $Z = 1$ and $D_1D_2D_3 = 100$. After the active clock edge, $Q_1Q_2Q_3 = 100$. If the next input is $X = 1$, then rows --0- and -1-- are selected, so $Z = 0$ and $D_1D_2D_3 = 110$. After the active clock edge, $Q_1Q_2Q_3 = 110$. Continuing in this manner, we can verify the transition table of Figure 16-2.

PALs also provide a convenient way of realizing sequential circuits. PALs are available which contain D flip-flops that have their inputs driven from programmable array logic. Figure 16-11 shows a segment of a sequential PAL. The D flip-flop is driven from an OR gate which is fed by two AND gates. The flip-flop output is fed back to the programmable AND array through a buffer. Thus, the AND gate inputs can be connected to A, A', B, B', Q, or Q'. The X's on the diagram show the connections required to realize the next-state equation

$$Q^+ = D = A'BQ' + AB'Q$$

The flip-flop output is connected to an inverting tri-state buffer, which is enabled when En = 1.

FIGURE 16-11 Segment of a Sequential PAL

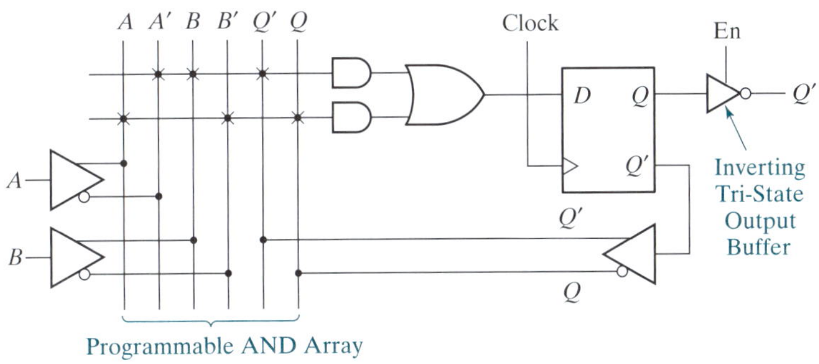

16.5 Sequential Circuit Design Using CPLDs

As discussed in Section 9.7, a typical CPLD contains a number of macrocells that are grouped into function blocks. Connections between the function blocks are made through an interconnection array. Each macrocell contains a flip-flop and an OR gate, which has its inputs connected to an AND gate array. Some CPLDs are based on PALs, in which case each OR gate has a fixed set of AND gates associated with it. Other CPLDs are based on PLAs, in which case any AND gate output within a function block can be connected to any OR gate input in that block.

Figure 16-12 shows the structure of a Xilinx CoolRunner II CPLD, which uses a PLA in each function block. This CPLD family is available in sizes from two to 32 function blocks (32 to 512 macrocells). Each function block has 16 inputs from the AIM (advanced interconnection matrix) and up to 40 outputs to the AIM. Each function block PLA contains the equivalent of 56 AND gates.

FIGURE 16-12 CoolRunner-II Architecture[1] (Figure based on figures and text owned by Xilinx, Inc., Courtesy of Xilinx, Inc. © Xilinx, Inc. 1999–2003. All rights reserved.)

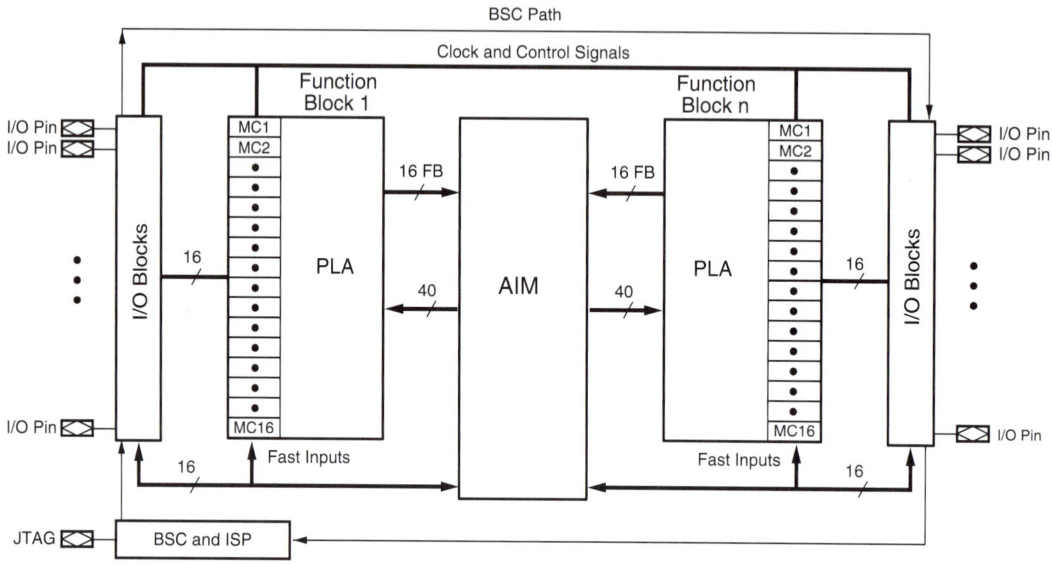

The basic CoolRunner II architecture is similar to that shown in Figure 9-33. Figure 16-13 represents a CoolRunner-II macrocell and the associated AND array. Box (1) represents the AND array which is driven by signals from the AIM. Each of the 56 product terms (P-terms) generated by the AND array (2) can have up to 40 variables. Box (3) represents the OR array which selects the AND gates for each macrocell. The OR gate (4) in a specific macrocell can have any subset of the P-terms as inputs. The MUXes on the diagram do not have control inputs shown because each MUX is programmed to select one of its inputs. For example, MUX (5) can be programmed to select a product term, the complement of a product term, a logic 1, or a logic 0 for the MUX output. If logic 1 is selected, the XOR gate complements the OR gate output; if logic 0 is selected, the XOR gate passes the OR gate output without change. By complementing or not complementing the OR gate output, a function can be implemented as either a product of sums or as a sum of products.

The XOR gate output can be routed directly to an I/O block or to the macrocell flip-flop input. The flip-flop can be programmed as a D-CE flip-flop or as a T flip-flop. The flip-flop can be programmed as an ordinary flip-flop (F/F), a latch, or a dual-edge triggered flip-flop, which can change state on either clock edge. The CK input and the asynchronous *S* and *R* inputs can each be programmed to come from several different sources. MUX (6) can invert the clock input or not, so that the flip-flop can trigger on either clock edge. MUX (7) selects either the flip-flop output or the XOR gate output and passes it to an I/O block.

Figure 16-14 shows how a Mealy sequential machine with two inputs, two outputs, and two flip-flops can be implemented by a CPLD. Four macrocells are required, two

[1]Additional data on Xilinx CPLDs and FPGAs is available from www.Xilinx.com.

FIGURE 16-13 CoolRunner-II Macrocell (Figure based on figures and text owned by Xilinx, Inc., Courtesy of Xilinx, Inc. © Xilinx, Inc. 1999–2003. All rights reserved.)

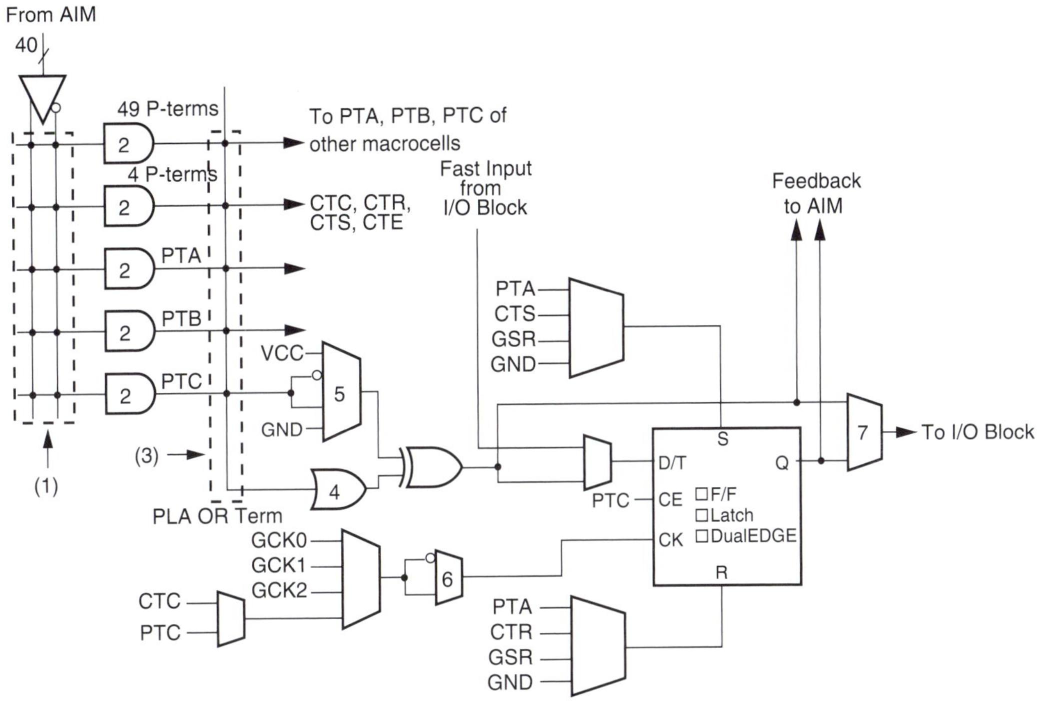

FIGURE 16-14 CPLD Implementation of a Mealy Machine

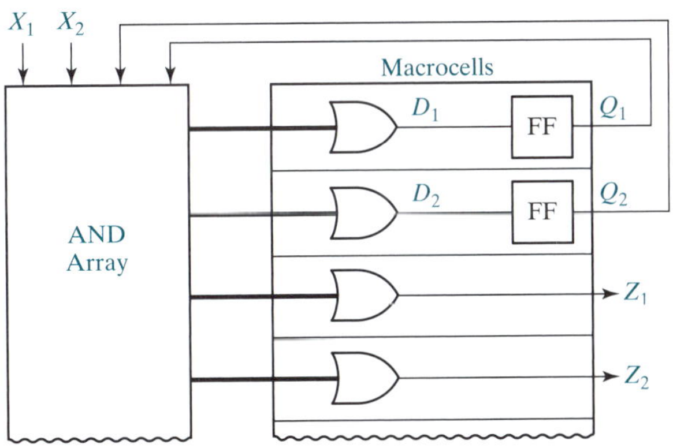

to generate the D inputs to the flip-flops and two to generate the Z outputs. The flip-flop outputs are fed back to the AND array inputs via the interconnection matrix (not shown). The number of product terms required depends on the complexity of the equations for the D's and the Z's.

Figure 16-15 shows how the 4-bit loadable right-shift register of Figure 12-16 can be implemented using four macrocells of a CPLD. The four OR-gate outputs implement the D inputs specified by Equations (12-1). A total of 12 product terms are required. The Q outputs are fed back to the AND array via the interconnection matrix (not shown).

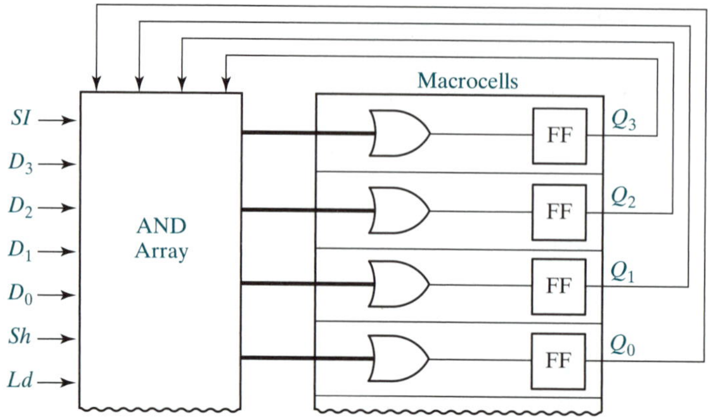

FIGURE 16-15 CPLD Implementation of a Shift Register

Figure 16-16 shows how three bits of the parallel adder with accumulator of Figure 12-5 can be implemented using a CPLD. Each bit of the adder requires two macrocells. One of the macrocells implements the sum function and an accumulator flip-flop. The other macrocell implements the carry, which is fed back into the AND array. The *Ad* signal can be connected to the CE input of each flip-flop via an AND gate (not shown). Each bit of the adder requires eight product terms (four for the sum, three for the carry, and one for CE). If the flip-flops are programmed as T flip-flops, then the logic for the sum can be simplified. For each accumulator flip-flop

$$X_i^+ = X_i \oplus Y_i \oplus C_i$$

Then, the T input is

$$T_i = X_i^+ \oplus X_i = Y_i \oplus C_i$$

which requires only two product terms.

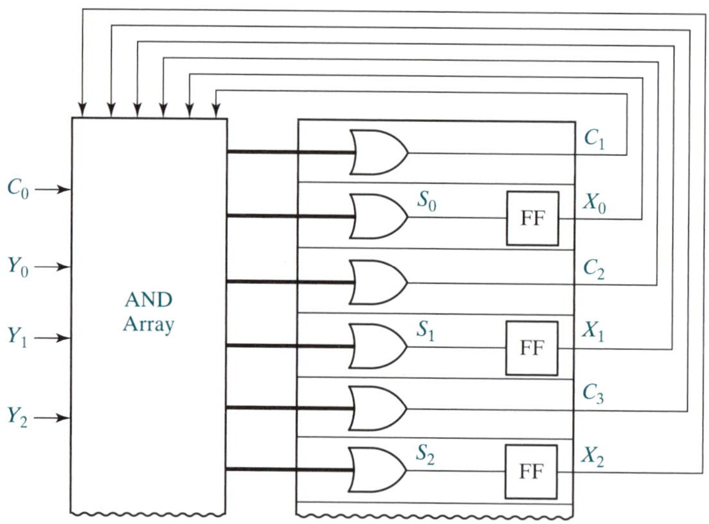

FIGURE 16-16 CPLD Implementation of a Parallel Adder with Accumulator

The add signal can be ANDed with the T_i input so that the flip-flop state can change only when $Ad = 1$:

$$T_i = Ad\,(Y_i \oplus C_i) = Ad\,Y_i C_i' + Ad\,Y_i' C_i$$

16.6 Sequential Circuit Design Using FPGAs

As discussed in Section 9.8, an FPGA usually consists of an array of configurable logic blocks (CLBs) surrounded by a ring of I/O blocks. The FPGA may also contain other components such as memory blocks, clock generators, tri-state buffers, etc. A typical CLB contains two or more function generators, often referred to as look-up tables or LUTs, programmable multiplexers, and D-CE flip-flops (see Figure 9-37). The I/O blocks usually contain additional flip-flops for storing inputs or outputs and tri-state buffers for driving the I/O pins.

Figure 16-17 shows a simplified block diagram for a Xilinx Virtex or Spartan II CLB. This CLB is divided into two nearly identical *slices*. Each slice contains two 4-variable function generators (LUTs), two D-CE flip-flops, and additional logic for carry and control. This additional logic includes MUXes for selecting the flip-flop inputs and for multiplexing the LUT outputs to form functions of five or more variables.

Figure 16-18 shows how a Mealy sequential machine with two inputs, two outputs, and two flip-flops can be implemented by a FPGA. Four LUTs (FGs or function generators) are required, two to generate the D inputs to the flip-flops and two to generate the Z outputs. The flip-flop outputs are fed back to the CLB inputs via interconnections external to the CLB. The entire circuit fits into one Virtex CLB. This implementation works because each D and Z is a function of only four variables

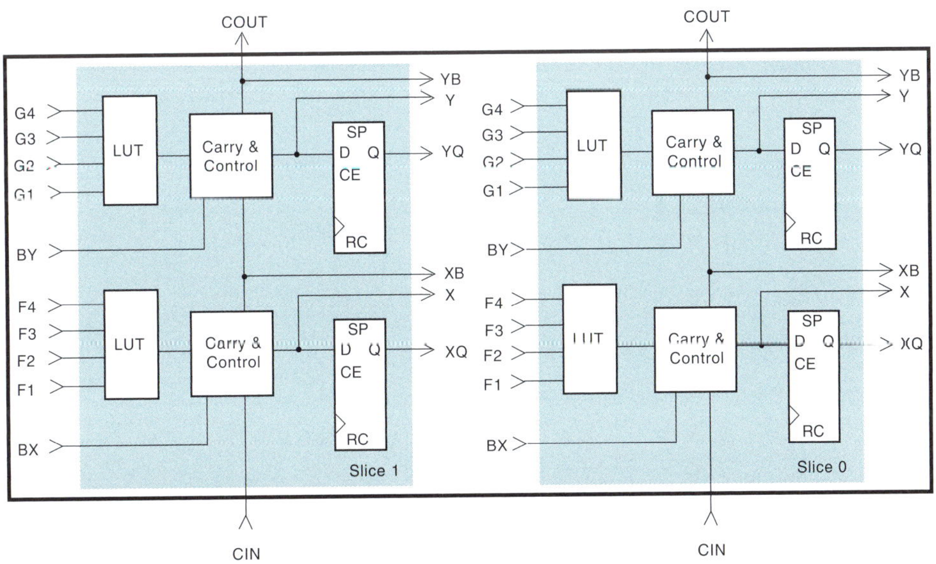

FIGURE 16-17 Xilinx Virtex/Spartan II CLB (Figure based on figures and text owned by Xilinx, Inc., Courtesy of Xilinx, Inc. © Xilinx, Inc. 1999–2003. All rights reserved.)

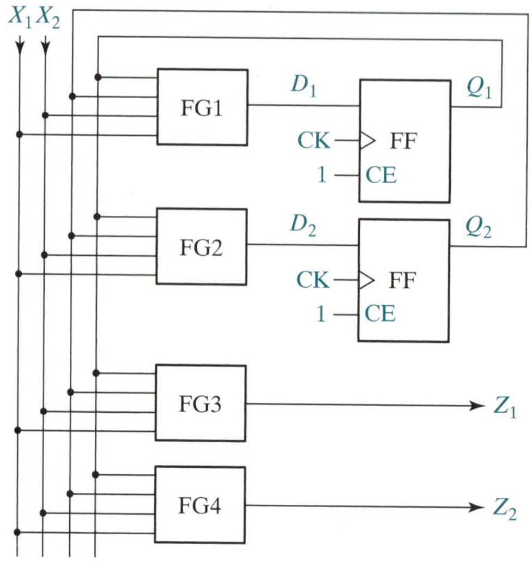

FIGURE 16-18 FPGA Implementation of a Mealy Machine

(X_1, X_2, Q_1, and Q_2). If more flip-flops or inputs are needed, the D or Z functions may have to be decomposed to use additional function generators as in Figure 9-40.

Figure 16-19 shows how the 4-bit loadable right-shift register of Figure 12-16 can be implemented using an FPGA. Four LUTs are used to generate the D inputs to the flip-flops, and a fifth LUT generates the CE input. If we had implemented Equations (12-1) directly without using the CE input, we would need to implement four 5-variable functions. This would require eight LUTs because each 5-variable function requires two 4-variable function generators (see Figure 9-40(a)). However, if we set CE $= Ld + Sh$, then CE $= 0$ when $Ld = Sh = 0$ and the flip-flops hold their current values. Therefore, we do not need the first term in each of Equations (12-1), and the flip-flop D input equations fit into 4-variable function generators. We can rewrite Equation (12-1(a)) in terms of CE as follows:

$$Q_3^+ = \mathrm{CE}'Q_3 + \mathrm{CE}\,D_{3f} = Ld'Sh'Q_3 + (Ld + Sh)(Sh'D_3 + Sh\,SI) \qquad (16\text{-}2)$$

where D_{3f} is the D input to flip-flop 3. The D input to the Q_3 flip-flop is therefore

$$D_{3f} = Sh'D_3 + Sh\,SI$$

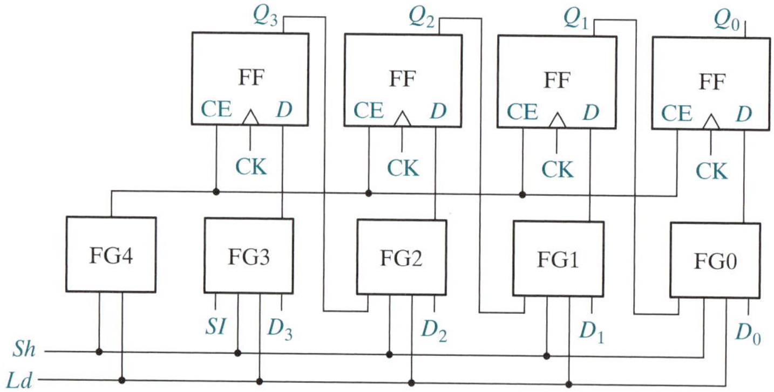

FIGURE 16-19 FPGA Implementation of a Shift Register

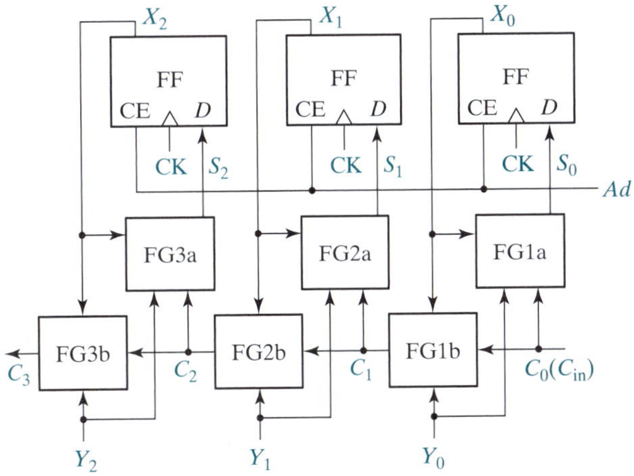

FIGURE 16-20 FPGA Implementation of a Parallel Adder with Accumulator

which is a 3-variable function. We can determine the other three flip-flop D inputs in a similar way.

Figure 16-20 shows how three bits of the parallel adder with accumulator of Figure 12-5 can be implemented using an FPGA. Each bit of the adder can be implemented with two 3-variable function generators, one for the sum and one for the carry. The Ad signal can be connected to the CE input of each flip-flop so that the sum is loaded by the rising clock edge when $Ad = 1$. The arrangement for generating the carries, shown in Figure 16-20, is rather slow because the carry signal must propagate through a function generator and its external interconnections for each bit. Because adders are frequently used in FPGAs, most FPGAs have built-in fast carry logic in addition to the function generators. If the fast carry logic is used, the bottom row of function generators in Figure 16-20 is not needed, and a parallel adder with an accumulator can be implemented using only one function generator for each bit.

16.7 Simulation and Testing of Sequential Circuits

Simulation of a digital system can take place at several levels of detail. At the functional level, system operation is described in terms of a sequence of transfers between registers, adders, memories, and other functional units. Simulation at this level may be used to verify the high-level system design. At the logic level, the system is described in terms of logic elements such as gates and flip-flops and their interconnections. Logic level simulation may be used to verify the correctness of the logic design and to analyze the timing. At the circuit level, each gate is described in terms of its circuit components such as transistors, resistances, and capacitances. Circuit level simulation gives detailed information about voltage levels and switching speeds. In this text, we will consider simulation at the logic level as well as system level simulation using VHDL.

Simulation of sequential circuits is similar to the simulation of combinational circuits described in Section 8.5. However, for sequential circuits, the propagation delays associated with the individual logic elements must be taken into account, and the presence of feedback may cause complications. The simulator output usually includes timing diagrams which show the times at which different signals in the circuit change. The delays in the gates and flip-flops may be modeled in several ways. The simplest method is to assume that each element has one unit of delay. The use of this unit delay model is generally sufficient to verify that the design is logically correct. If a more detailed timing analysis is required, each logic element may be assigned a nominal delay value. The nominal or typical delays for a device are usually provided by the device manufacturer on the specification sheets.

In practice, no two gates of a given type will have *exactly* the same delay, and the value of the delay may change depending on temperature and voltage levels. For these reasons, manufacturers often specify a minimum and maximum delay value for each type of logic element. Some simulators can take the minimum and maximum delay values into account. Instead of showing the exact time at which a signal changes, the simulator output indicates a time interval in which the signal may change. Figure 16-21 shows the output from an inverter which has a nominal delay of 10 ns, a minimum delay of 5 ns, and a maximum delay of 15 ns. The shaded region indicates that the inverter output may change at any time during the interval. Min-max delay simulators can be used to verify that a digital system will operate correctly as long as the delay in each element is within its specified range.

Testing of sequential circuits is generally more difficult than testing combinational circuits. If the flip-flop outputs can be observed, then the state table can be verified directly on a row-by-row basis. The state table can be checked out with a simulator or in lab as follows:

1. Using the direct set and clear inputs, set the flip-flop states to correspond to one of the present states in the table.
2. For a Moore machine, check to see that the output is correct. For a Mealy machine, check to see that the output is correct for each input combination.
3. For each input combination, clock the circuit and check to see that the next state of the flip-flops is correct. (Reset the circuit to the proper state before each input combination is applied.)
4. Repeat steps 1, 2, and 3 for each of the present states in the table.

In many cases when a sequential circuit is implemented as part of an integrated circuit, only the inputs and outputs are available at the IC pins, and observing the state of the internal flip-flops is impossible. In this case, testing must be done by applying input sequences to the circuit and observing the output sequences. Determining a small set of input sequences which will completely test the circuit is generally a

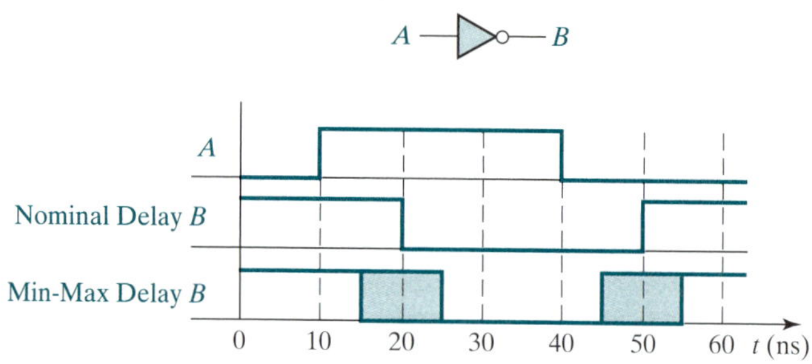

FIGURE 16-21
Simulator Output for an Inverter

FIGURE 16-22 Simulation Screen for Figure 13-7

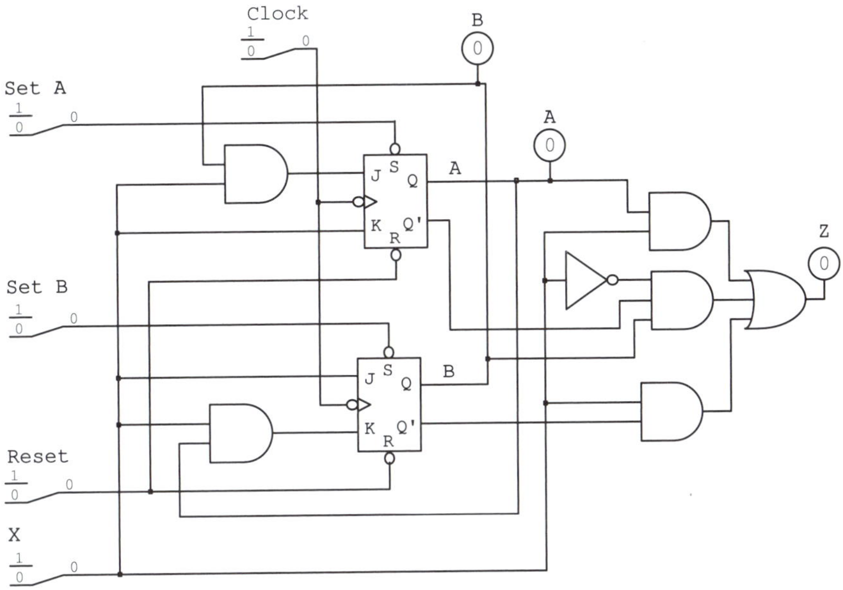

difficult problem that is beyond the scope of this text. The set of test sequences must traverse all arcs on the state graph, but this is generally not a sufficient test.

Figure 16-22 shows a simulator screen for testing the Mealy sequential circuit of Figure 13-7. To step through the circuit one input at a time, switches are used for the Clock and X inputs. Another switch is used to reset both flip-flops, and two switches are used to set flip-flops A and B. Probes are used to observe the Z output and the state of the flip-flops. After X has been set to the desired value, the clock cycle is simulated by flipping the Clock switch to 1 and back to 0. For a Mealy machine, the output should be read just *before* the active edge of the clock.

If an incorrect Z output is found in the process of verifying the state table, the output circuit can be checked using the techniques discussed in Section 8.5. If one of the next states is wrong, this may be due to an incorrect flip-flop input. After determining which flip-flop goes to the wrong state, the circuit should be reset to the proper present state, and the flip-flop inputs should be checked before applying another clock pulse.

Example

Assume that you have built the circuit of Figure 16-4 to implement the state table of Figure 16-2. Suppose that when you set the flip-flop states to 100, set $X = 1$, and pulse the clock, the circuit goes to state 111 instead of 110. This indicates that flip-flop Q_3 went to the wrong state. You should then reset the flip-flops to state 100 and observe the inputs to flip-flop Q_3. Because the flip-flop is supposed to remain in state 0, D_3 should be 0. If $D_3 = 1$, this indicates that either D_3 was derived wrong or that the D_3 circuit has a problem. Check the D_3 map and equation to make sure that $D_3 = 0$ when $X = 1$ and $Q_1Q_2Q_3 = 100$. If the map and equation are correct, then the D_3 circuit should be checked using the procedure in Section 8.5.

After you have verified that the circuit works according to your state table, you must then check the circuit to verify that it works according to the problem statement. To do this, you must apply appropriate input sequences and observe the resulting output sequence. When testing a Mealy circuit, you must be careful to read the outputs at the proper time to avoid reading false outputs (see Section 13.2). The output should be read just before the active edge of the clock. If the output is read immediately following the active clock edge, a false output may be read. See Figure 13-8 for an example.

Instead of manually stepping through the input sequence, simulated input waveforms may be defined for X and Clock. Figure 16-23 shows the simulator input waveform for the example of Figure 16-22, using the test sequence $X = 10101$. When the simulator is run, the timing chart for A, B, and Z will be generated as shown. Note that the simulator output is very similar to the timing chart of Figure 13-8. The simulator output in Figure 16-23(a) assumes the unit delay model, that is, each gate or flip-flop has one unit of delay. Figure 16-23(b) shows the same simulation using a nominal delay of 10 ns for each gate and flip-flop.

So far in our discussion of sequential circuits, we have assumed that the inputs are properly synchronized with the clock. This means that one input in the sequence occurs for each clock cycle, and all input changes satisfy setup and hold time specifications. Synchronization is no problem in the laboratory if we use a manual clock because we can easily change the inputs between active clock edges. However, if we operate our circuits at a high clock rate, then synchronization becomes a problem. We must either generate our input sequences in synchronization with the clock, or we must use a special circuit to synchronize the inputs with the clock. The former can

FIGURE 16-23

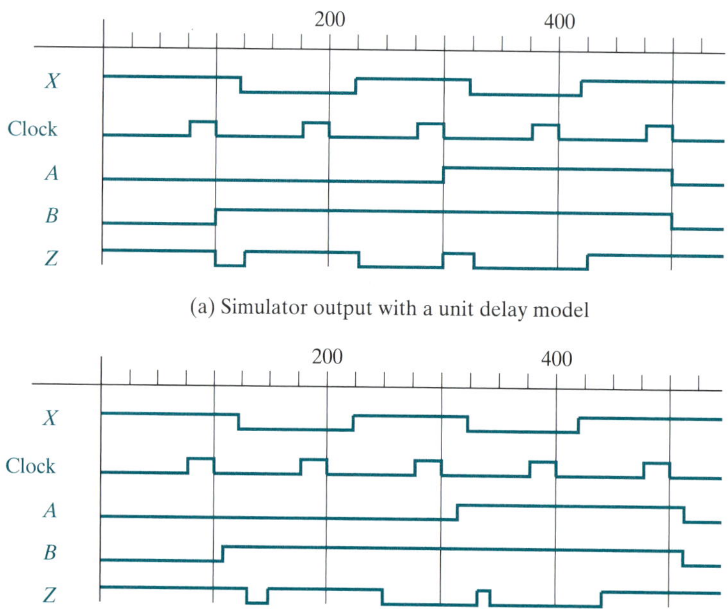

(a) Simulator output with a unit delay model

(b) Simulator output with a nominal delay of 10 ns

FIGURE 16-24 Using a Shift Register to Generate Synchronized Inputs

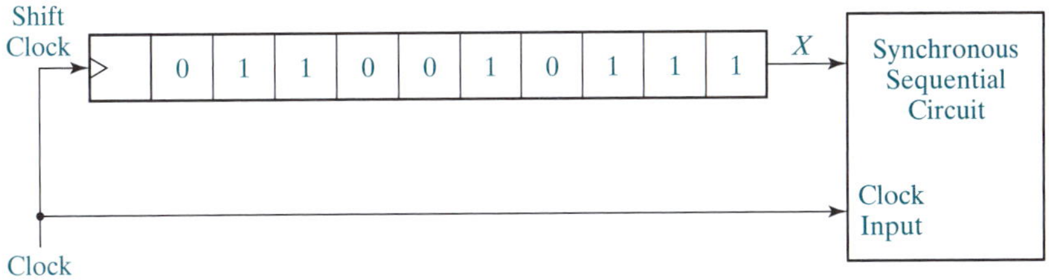

be accomplished by loading the inputs into a shift register, and then using the circuit clock to shift them into the circuit one at a time, as shown in Figure 16-24.

If the input changes are not synchronized with the clock, edge-triggered D flip-flops can be used to synchronize them, as shown in Figure 16-25(a). In this figure, although X_1 and X_2 change at arbitrary times with respect to the clock, X_{1S} and X_{2S} change after the rising clock edge, and the inputs to the sequential circuit should be properly synchronized, as shown in Figure 16-25(b). However, this design has an inherent problem and may occasionally fail to operate properly. If a D input changes very close to the rising clock edge so that setup and hold times are not satisfied (see Figure 11-20), one of the flip-flops may malfunction.

Figure 16-26 shows a more reliable synchronizer[2] that uses two D flip-flops to synchronize a single asynchronous input, X. If X changes from 0 to 1 in the critical region where the setup or hold time is not satisfied, several outcomes could occur: the flip-flop Q_1 output might change to 1; it might remain 0; it might start to change to 1 and then change back or it might oscillate between 0 and 1 for a short time and then settle down to 0 or 1. This region of uncertainty is indicated by the shading on the Q_1 waveform. We will assume that the clock period is chosen so that Q_1 will be settled in either the 0 or 1 state by t_2. If $Q_1 = 1$, Q_2 will change to 1 shortly after t_2. If $Q_1 = 0$, Q_1 will change to 1 shortly after t_2, and Q_2 will change to 1 shortly after t_3. Because X is an asynchronous input, normally it will not matter whether X_{1S} is delayed by one or two clock periods. The important thing is that X_{1S} is a clean signal that is synchronized with the clock.

FIGURE 16-25

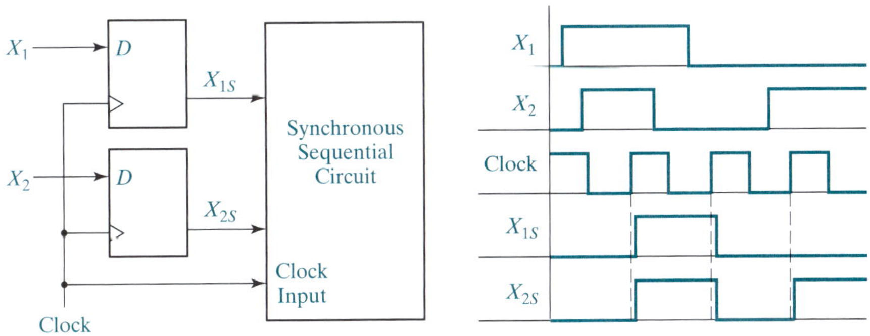

(a) Synchronizer circuit (b) Synchronizer inputs and outputs

[2]For more detailed discussion of synchronizer design, see John F. Wakerly. *Digital Design Design Principles and Practices*, 4th ed. Upper Saddle River, NJ (Prentice Hall, 2006).

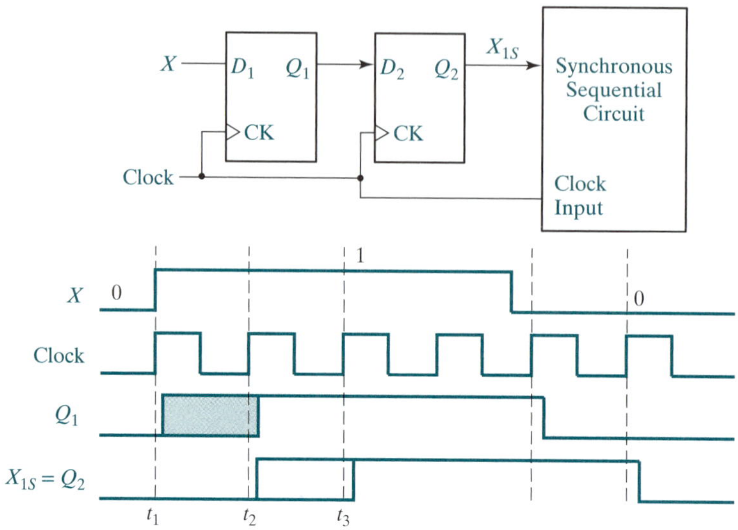

FIGURE 16-26 Synchronizer with Two D Flip-Flops

16.8 Overview of Computer-Aided Design

A wide variety of computer-aided design (CAD) software tools are available to assist in the design of digital systems. Many of these CAD programs will run on a personal computer, but others require a more powerful workstation for execution. Several functions performed by these CAD tools are discussed below.

Generation and minimization of logic equations. Programs of this type accept truth tables, state tables, or state graphs as input and generate minimized logic equations. *LogicAid* is an example of this type of program.

Generation of bit patterns for programming PLDs. These programs generate a file which can be downloaded to a PLD programmer to program PALs and other programmable logic devices.

Schematic capture. This type of program allows the designer to interactively enter and edit a logic diagram for a digital design. The program provides libraries of standard logic components such as gates, flip-flops, registers, adders, counters, etc., which can be selected for inclusion in the diagram. In addition to a plot of the logic diagram, the output from a schematic capture program may include a parts list, a list of interconnections between the ICs, and a circuit description file. This file may be used as input to a simulator, PC board layout program, or other CAD programs.

Simulation. We have already discussed several types of simulators in Sections 10.3 and 16.5. By using such simulators at various points in the design process, designers can correct many errors and resolve critical timing problems before any hardware is actually built. Use of a simulator is essential when an IC is being designed, because the correction of design errors after the IC has been fabricated is very time-consuming and costly.

SimUaid performs the schematic capture and simulation functions for small digital systems. It also automatically generates a structural VHDL description from the schematic.

Synthesis tools. Synthesis software accepts as input a description of a desired digital system written in VHDL, Verilog, or another hardware description language. The HDL code is analyzed and translated into a circuit description that specifies the needed logic components and the connections between these components. The synthesizer output is then fed into software that implements the circuit for a specific target device such as an FPGA, CPLD, or ASIC (application-specific integrated circuit). More details of synthesis and implementation of VHDL code are given in Section 17.5.

IC design and layout. A digital integrated circuit is typically composed of interconnected transistors which are fabricated on a chip of silicon. Such ICs are usually made of several layers of conducting material separated by layers of insulating material with appropriate connections between layers. The patterns for paths on each layer are transferred into the layers during the fabrication process using masks which are similar to photographic negatives. CAD tools for IC design facilitate the process of specifying the geometries of the transistors, placing the transistors on the chip, and routing the interconnections between them. Libraries of standard modules are available for inclusion in the chip designs. Automatic checking of the designs is provided to verify consistency with design rules. The output from the IC design program includes the mask patterns necessary for fabricating the IC.

Test generation. As digital systems become more complex, testing the finished product becomes increasingly difficult. It is not practical to test the system using all possible combinations or sequences of inputs. Automatic test generation programs are available which attempt to generate a relatively small set of input patterns that will adequately test the system in a reasonable length of time.

PC board layout. Most digital systems are built by mounting the integrated circuit components on a printed circuit board. The wiring on such PC boards is made up of thin metallic strips which interconnect the ICs. In order to make all of the required connections, these boards typically have two, three, or more layers of interconnect wiring. PC board layout programs perform two main functions they determine the placement of the ICs on the board, and they route the connections between the ICs. The output of the layout program includes a set of plots which show the wiring on each layer of the PC board.

Many CAD systems integrate several of these CAD tools into a single package so that you can, for example, input a logic diagram, simulate its operation, and then lay out a PC board or IC. The design of large, complex integrated circuits and digital systems would not be feasible without the use of appropriate CAD tools.

One method of designing a small digital system with an FPGA uses the following steps:

1. Draw a block diagram of the digital system. Define the required control signals and construct a state graph that describes the required sequence of operations.
2. Work out a detailed logic design of the system using gates, flip-flops, registers, counters, adders, etc.

3. Construct a logic diagram of the system, using a schematic capture program.
4. Simulate and debug the logic diagram and make any necessary corrections to the design.
5. Run an implementation program that fits the design into the target FPGA. This program carries out the following steps:
   (a) Partition the logic diagram into pieces that will fit into CLBs of the target FPGA.
   (b) Place the CLBs within the logic cell array of the FPGA and route the connections between the logic cells.
   (c) Generate the bit pattern necessary to program the FPGA.
6. Run a timing simulation of the completed design to verify that it meets specifications. Make any necessary corrections and repeat the process as necessary.
7. Download the bit pattern into the internal configuration memory cells in the FPGA and test the operation of the FPGA.

When a hardware description language is used, steps 2 and 3 are replaced with writing HDL code. The HDL code is then simulated and debugged in step 4.

Design Problems

The following problems require the design of a Mealy sequential circuit of the form shown in Figure 16-27. For purposes of testing, the input X will come from a toggle switch, and the clock pulse will be supplied manually from a push button or switch.

FIGURE 16-27

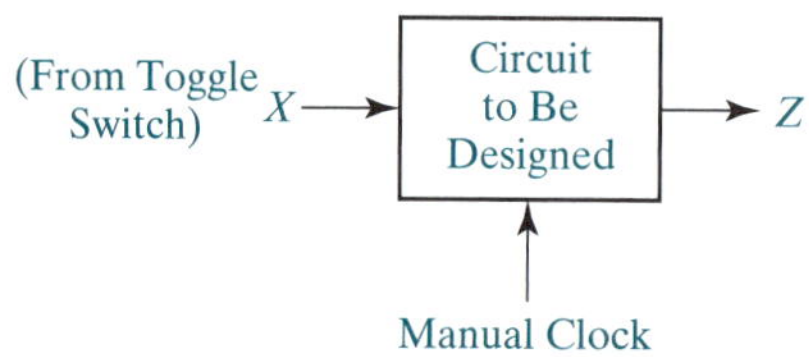

16.1 Design a Mealy sequential circuit (Figure 16-27) which investigates an input sequence X and will produce an output of $Z = 1$ for any input sequence ending in 0010 or 100.

Example:

$$X = 1\ 1\ 0\ 0\ 1\ 0\ 0\ 1\ 0\ 1\ 0\ 0\ 1\ 0\ 1$$
$$Z = 0\ 0\ 0\ 1\ 0\ 1\ 1\ 0\ 1\ 0\ 0\ 1\ 0\ 1\ 0$$

Notice that the circuit does not reset to the start state when an output of $Z = 1$ occurs. However, your circuit should have a start state and should be provided with a method for manually resetting the flip-flops to the start state. A minimum solution requires six states. Design your circuit using NAND gates, NOR gates, and three D flip-flops. Any solution which is minimal for your state assignment and uses 10 or fewer gates and inverters is acceptable. (Assign 000 to the start state.)

Test Procedure: First, check out your state table by starting in each state and making sure that the present output and next state are correct for each input. Then, starting in the proper initial state, determine the output sequence for each of the following input sequences:

(1) 0 0 1 1 0 1 0 0 1 0 1 0 1 0 0 0 1 0 0 1 0 0 1 0
(2) 1 1 0 0 1 1 0 0 1 0 1 0 1 0 0 1 0 1 0 1 0 0 1 0

16.2 Design a Mealy sequential circuit (Figure 16-27) which investigates an input sequence X and will produce an output of $Z = 1$ for any input sequence ending in 1101 or 011.

Example:

$X = 0\ 0\ 1\ 1\ 0\ 1\ 1\ 0\ 1\ 0\ 1\ 1\ 0\ 1\ 0$
$Z = 0\ 0\ 0\ 1\ 0\ 1\ 1\ 0\ 1\ 0\ 0\ 1\ 0\ 1\ 0$

Notice that the circuit does not reset to the start state when an output of $Z = 1$ occurs. However, your circuit should have a start state and should be provided with a method for manually resetting the flip-flops to the start state. A minimum solution requires six states. Design your circuit using NAND gates, NOR gates, and three D flip-flops. Any solution which is minimal for your state assignment and uses nine or fewer gates and inverters is acceptable. (Assign 000 to the start state.)

Test Procedure: First, check out your state table by starting in each state and making sure that the present output and next state are correct for each input. Then, starting in the proper initial state, determine the output sequence for each of the following input sequences:

(1) 1 1 0 0 1 0 1 1 0 1 0 1 0 1 1 1 0 1 1 0 1 1 0 1
(2) 0 0 1 1 0 0 1 1 0 1 0 1 0 1 1 0 1 0 1 0 1 1 0 1

16.3 Design a sequential circuit (Figure 16-27) to convert excess-3 code to BCD code. The input and output should be serial with the least significant bit first. The input X represents an excess-3 coded decimal digit, and the output Z represents the corresponding BCD code. Design your circuit using three D flip-flops, NAND gates, and NOR gates. Any solution which is minimal for your state assignment and uses eight or fewer gates and inverters is acceptable. (Assign 000 to the reset state.)

Test Procedure: First, check out your state table by starting in each state and making sure that the present output and next state are correct for each input. Then, starting in the reset state, determine the output sequence for each of the ten possible input sequences and make a table.

16.4 Design a sequential circuit (Figure 16-27) which adds six to a binary number in the range 0000 through 1001. The input and output should be serial with the least significant bit first. Find a state table with a minimum number of states. Design the circuit using NAND gates, NOR gates, and three D flip-flops. Any solution which is minimal for your state assignment and uses 10 or fewer gates and inverters is acceptable. (Assign 000 to the reset state.)

Test Procedure: First, check out your state table by starting in each state and making sure that the present output and next state are correct for each input. Then, starting in the reset state, determine the output sequence for each of the ten possible input sequences and make a table.

16.5 Design a Mealy sequential circuit (Figure 16-27) which investigates an input sequence X and will produce an output of $Z = 1$ for any input sequence ending in 0110 or 101.

Example:

$$X = 0\ 1\ 0\ 1\ 1\ 0\ 1$$
$$Z = 0\ 0\ 0\ 1\ 0\ 1\ 1$$

Notice that the circuit does not reset to the start state when an output of $Z = 1$ occurs. However, your circuit should have a start state and should be provided with a method for manually resetting the flip-flops to the start state. A minimum solution requires six states. Design your circuit using NAND gates, NOR gates, and three D flip-flops. Any solution which is minimal for your state assignment and uses eight or fewer gates and inverters is acceptable. (Assign 000 to the start state.)

Test Procedure: First, check out your state table by starting in each state and making sure that the present output and next state are correct for each input. Then, starting in the proper initial state, determine the output sequence for each of the following input sequences:

(1) 0 0 1 1 0 1 1 1 1 0 0 1 0 1 0 0
(2) 1 0 1 0 0 0 1 1 1 1 0 1 1 0 0 0

16.6 Design a Mealy sequential circuit which investigates an input sequence X and which will produce an output of $Z = 1$ for any input sequence ending in 0101 provided that the sequence 110 has never occurred.

Example:

$$X = 0\ 1\ 0\ 1\ 0\ 1\ 1\ 0\ 1\ 0\ 1$$
$$Z = 0\ 0\ 0\ 1\ 0\ 1\ 0\ 0\ 0\ 0\ 0$$

Notice that the circuit does not reset to the start state when an output of $Z = 1$ occurs. However, your circuit should have a start state and should be provided with a method for manually resetting the flip-flops to the start state. A minimum solution requires six states. Design your circuit using NAND gates, NOR gates, and three D flip-flops. Any solution which is minimal for your state assignment and uses eight or fewer gates and inverters is acceptable. (Assign 000 to the start state.)

Test Procedure: First, check out your state table by starting in each state and making sure that the present output and next state are correct for each input. Then, starting in the proper initial state, determine the output sequence for the following input sequences:

(1) $X = 0\ 1\ 0\ 1\ 0\ 0\ 0\ 1\ 0\ 1\ 1\ 0$
(2) $X = 1\ 0\ 1\ 0\ 1\ 0\ 1\ 1\ 0\ 1\ 0\ 1$

16.7 Design a Mealy sequential circuit which investigates an input sequence X and which will produce an output of $Z = 1$ if the total number of 1's received is even (consider zero 1's to be an even number of 1's) and the sequence 00 has occurred at least once. *Note*: The *total* number of 1's received includes those received before and after 00.

Example:

$X = 1\ 0\ 1\ 0\ 1\ 0\ 0\ 1\ 1\ 0\ 1$

$Z = 0\ 0\ 0\ 0\ 0\ 0\ 0\ 1\ 0\ 0\ 1$

Notice that the circuit does not reset to the start state when an output of $Z = 1$ occurs. However, your circuit should have a start state and should be provided with a method of manually resetting the flip-flops to the start state. A minimum solution requires six states. Design your circuit using NAND gates, NOR gates, and three D flip-flops. Any solution which is minimal for your state assignment and uses 12 or fewer gates and inverters is acceptable; the best known solution uses seven. (Assign 000 to the start state.)

Test Procedure: First, check out your state table by starting in each state and making sure that the present output and next state are correct for each input. Then, starting in the proper initial state, determine the output sequence for each of the following input sequences:

(1) $X = 0\ 1\ 1\ 0\ 0\ 1\ 0\ 1\ 0\ 0$

(2) $X = 1\ 0\ 1\ 1\ 1\ 1\ 0\ 0\ 1\ 1\ 1\ 0$

16.8 Design a Mealy sequential circuit (Figure 16-27) which investigates an input sequence X and will produce an output of $Z = 1$ for any input sequence ending in 0011 or 110.

Example:

$X = 1\ 0\ 1\ 0\ 0\ 1\ 1\ 0\ 0\ 1\ 1$

$Z = 0\ 0\ 0\ 0\ 0\ 0\ 1\ 1\ 0\ 0\ 1$

Notice that the circuit does not reset to the start state when an output of $Z = 1$ occurs. However, your circuit should have a start state and should be provided with a method for manually resetting the flip-flops to the start state. Design your circuit using NAND gates, NOR gates, and three D flip-flops. Any solution which is minimal for your state assignment and uses 10 or fewer gates and inverters is acceptable; the best known solution uses six. (Assign 000 to the start state.)

Test Procedure: First, check out your state table by starting in each state and making sure that the present output and next state are correct for each input. Then, starting in the reset state, determine the output sequence for each of the following input sequences:

(1) $X = 0\ 0\ 0\ 1\ 0\ 0\ 0\ 1\ 1\ 0\ 1\ 0$

(2) $X = 1\ 1\ 1\ 0\ 0\ 1\ 0\ 0\ 0\ 1\ 1\ 0$

16.9 Design a Mealy sequential circuit which investigates an input sequence X and produces an output Z which is determined by two rules. The initial output from the circuit is $Z = 0$. Thereafter, the output Z will equal the *preceding* value of X (rule 1) until the input sequence 001 occurs. Starting with the next input after 001, the output Z will equal the *complement* of the *present* value of X (rule 2) until the sequence

100 occurs. Starting with the next input after 100, the circuit output is again determined by rule 1, etc. Note that overlapping 001 and 100 sequences may occur.

Example:

$$\begin{aligned}\text{Rule: } & 1\ 1\ 1\ 1\ 2\ 2\ 2\ 2\ 2\ 1\ 1\ 2\\ X &= 1\ 0\ 0\ 1\ 1\ 0\ 1\ 0\ 0\ 0\ 1\ 1\\ Z &= 0\ 1\ 0\ 0\ 0\ 1\ 0\ 1\ 1\ 0\ 0\ 0\end{aligned}$$

Design your circuit using NAND gates, NOR gates, and three D flip-flops. Your circuit should be provided with a method for manually resetting the flip-flops to the start state. A minimum solution requires six states. Any solution which is minimal for your state assignment and uses 12 or fewer gates and inverters is acceptable. (Assign 000 to the start state.)

Test Procedure: First, check out your state table by starting in each state and making sure that the present output and next state are correct for each input. Then, starting in the reset state, determine the output sequence for each of the following input sequences:

(1) $X = 1\ 0\ 0\ 1\ 0\ 0\ 1\ 0\ 0\ 0\ 1\ 1$

(2) $X = 0\ 1\ 1\ 0\ 0\ 0\ 0\ 1\ 1\ 0\ 1\ 1$

16.10 The 8, 4, –2, –1 BCD code is similar to the 8-4-2-1 BCD code, except that the weights are negative for the two least significant bit positions. For example, 0111 in 8, 4, –2, –1 code represents

$$8 \times 0 + 4 \times 1 + (-2) \times 1 + (-1) \times 1 = 1$$

Design a Mealy sequential circuit to convert 8, 4, –2, –1 code to 8-4-2-1 code. The input and output should be serial with the least significant bit first. The input X represents an 8, 4, –2, –1 coded decimal digit and the output Z represents the corresponding 8-4-2-1 BCD code. After four time steps the circuit should reset to the starting state regardless of the input sequence. Design your circuit using three D flip-flops, NAND gates, and NOR gates. Any solution which is minimal for your state assignment and uses eight or fewer gates is acceptable. (Assign 000 to the reset state.)

Test Procedure: First, check out your state table by starting in each state and making sure that the present output and next state are correct for each input. Then, starting in the reset state, determine the output sequence for each of the 10 possible input sequences and make a table.

16.11 Design a Mealy sequential circuit (Figure 16-27) which adds five to a binary number in the range 0000 through 1010. The input and output should be serial with the least significant bit first. Find a state table with a minimum number of states. Design the circuit using NAND gates, NOR gates, and three D flip-flops. Any solution which is minimal for your state assignment and uses nine or fewer gates and inverters is acceptable. (Assign 000 to the reset state.)

Test Procedure: First, check out your state table by starting in each state and making sure that the present output and the next state are correct for each input.

Then, starting in the reset state, determine the output sequence for each of the 11 possible input sequences and make a table.

16.12 Design a Mealy sequential circuit (Figure 16-27) to convert a 4-bit binary number in the range 0000 through 1010 to its 10's complement. (The 10's complement of a number N is defined as $10 - N$.) The input and output should be serial with the least significant bit first. The input X represents the 4-bit binary number, and the output Z represents the corresponding 10's complement. After four time steps, the circuit should reset to the starting state regardless of the input sequence. Find a state table with a minimum number of states. Design the circuit using NAND gates, NOR gates, and three D flip-flops. Any solution which is minimal for your state assignment and uses nine or fewer gates and inverters is acceptable. (Assign 000 to the reset state.)

Test Procedure: First, check out your state table by starting in each state and making sure that the present output and the next state are correct for each input. Then, starting in the reset state, determine the output sequence for each of the 11 possible input sequences and make a table.

16.13 Design a Mealy sequential circuit which investigates an input sequence X and which will produce an output of $Z = 1$ for any input sequence ending in 1010, provided that the sequence 001 has occurred at least once.

Example:

$X = 1\ 0\ 1\ 0\ 0\ 1\ 0\ 1\ 0\ 1\ 0$

$Z = 0\ 0\ 0\ 0\ 0\ 0\ 0\ 0\ 1\ 0\ 1$

Notice that the circuit does not reset to the start state when an output of $Z = 1$ occurs. However, your circuit should have a start state and should be provided with a method of manually resetting the flip-flops to the start state. A minimum solution requires six states. Design your circuit using NAND gates, NOR gates, and three D flip-flops. Any solution which is minimal for your state assignment and uses nine or fewer gates and inverters is acceptable. (Assign 000 to the start state.)

Test Procedure: First, check out your state table by starting in each state and making sure that the present output and the next state are correct for each input. Then, starting in the proper initial state, determine the output sequence for the following input sequences:

(1) $X = 1\ 0\ 0\ 1\ 0\ 0\ 1\ 1\ 0\ 1\ 0\ 1$

(2) $X = 1\ 0\ 1\ 0\ 0\ 0\ 1\ 0\ 1\ 0\ 1\ 0$

16.14 Design a Mealy sequential circuit which investigates an input sequence X and will produce an output of $Z = 1$ whenever the total number of 0's in the sequence is odd, provided that the sequence 01 has occurred at least once.

Example:

$X = 1\ 1\ 0\ 0\ 0\ 1\ 1\ 0\ 1\ 0$

$Z = 0\ 0\ 0\ 0\ 0\ 1\ 1\ 0\ 0\ 1$

A minimum solution requires five states. Design your circuit using NAND gates, NOR gates, and three D flip-flops. Your circuit should have a start state and should be provided with a method of manually resetting the flip-flops to the start state. Any solution which is minimal for your state assignment and which uses 11 or fewer gates and inverters is acceptable. (Assign 000 to the start state.)

Test Procedure: First, check out your state table by starting in each state and making sure that the present output and the next state are correct for each input. Then, starting in the proper initial state, determine the output sequence for the following input sequences:

(1) $X = 1\ 0\ 0\ 0\ 1\ 1\ 0\ 1\ 0\ 0\ 1$

(2) $X = 0\ 0\ 0\ 0\ 1\ 0\ 1\ 0\ 0\ 0\ 1$

Additional Problems

16.15 Draw a block diagram that shows how a ROM and D flip-flops could be connected to realize Table 13-4 (p. 430). Specify the truth table for the ROM using a straight binary state assignment. (Note that a truth table, not a transition table, is to be specified.)

16.16 The state table of Figure 15-15(a) is to be realized using a PLA and D flip-flops.
(a) Draw a block diagram.
(b) Specify the contents of the PLA in tabular form using the state assignment of Figure 15-15(a). (See Figure 15-16(b) for the D equations.)

16.17 An iterative circuit has a form similar to Figure 16-6. The output Z is to be 1 if the total number of X inputs that are 1 is an odd number greater than 2.
(a) Draw a state graph for a typical cell.
(b) Derive the equations and a NAND-gate circuit for a typical cell and for the output circuit.
(c) Specify a_1 and b_1, and simplify the first cell.
(d) Show how a sequential circuit can be constructed using the typical cell and output circuit.

16.18 In the iterative circuit below the input X is a 2's complement number and the output Y is the 2's complement of X. (x_0 and y_0 are the least significant bits.)
(a) Construct a transition table for typical cell. (Note that a_0 is 0.)
(b) Derive expressions for a_{i+1} and y_i.
(c) Construct a sequential circuit that will compute the 2's complement of its input. (Use D flip-flops. Assume the flip-flops are reset at the beginning of an input sequence.)

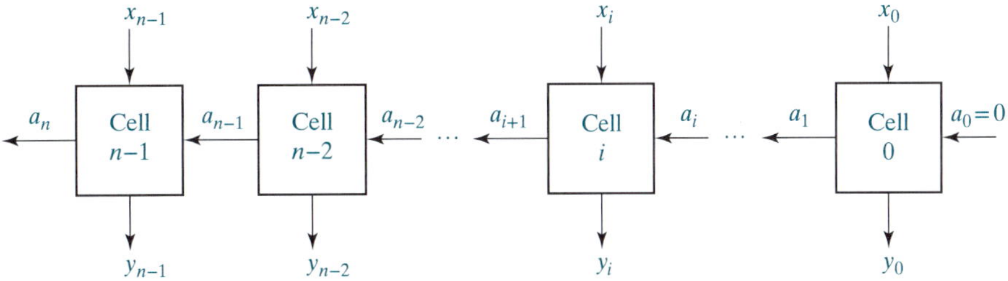

16.19 The iterative circuit below compares two positive, unsigned binary numbers $X = x_{n-1} \ldots x_1x_0$ and $Y = y_{n-1} \ldots y_1y_0$. The numbers are compared from right to left (i.e., least significant bits to most significant bits).

(a) Derive the transition table for a typical cell, cell i, so that $a_nb_n = 00$ if $X = Y$, $a_nb_n = 01$ if $X > Y$, and $a_nb_n = 10$ if $X < Y$.

(b) Derive minimum sum-of-product expressions for a_{i+1} and b_{i+1}.

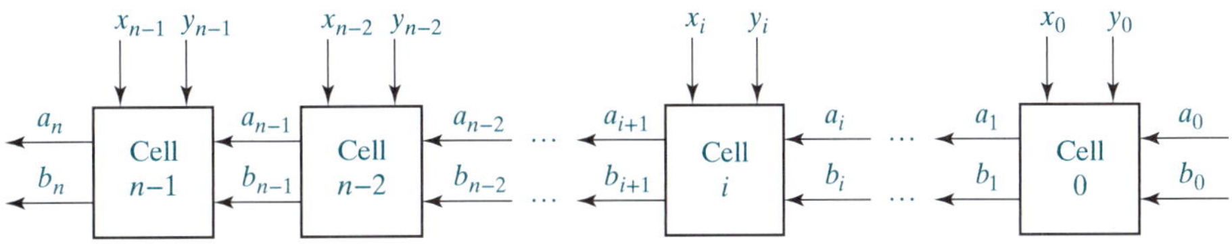

16.20 Modify the binary comparator of Figure 16-6 so it will compare 2's complement numbers. Let $a_1 = x_1$ and $b_1 = y_1$. Note that all four combinations of a_i and b_i are used; 00 indicates that X and Y are positive and equal, 11 indicates that X and Y are negative and equal, 01 indicates $X > Y$, and 10 indicates $X < Y$. (For negative numbers, $-2 > -3$.)

(a) Derive the transition table for this comparator similar to Table 16-5.

(b) Derive minimal sum-of-product expressions for a_{i+1} and b_{i+1}.

(c) Derive a circuit for Z_1, Z_2, and Z_3.

(d) Derive a state table for a sequential circuit that compares 2's complement numbers and produces the outputs Z_1, Z_2, and Z_3 using the transition table of part (a) as a guideline. Note that the circuit will require a distinct initial state since a_1 and b_1 depend upon x_1 and y_1.

16.21 Design a sequential circuit having one input and one output that will produce an output of 1 for every second 0 it receives and for every second 1 it receives.

Example:

$$X(\text{input}) = 0\ 1\ 1\ 0\ 1\ 1\ 1\ 0\ 0\ 0\ 0\ 1\ 0\ 1\ 1\ 0\ 0\ 1\ 0\ 1\ 1\ 0\ 1\ 0$$
$$Z(\text{output}) = 0\ 0\ 1\ 1\ 0\ 1\ 0\ 0\ 1\ 0\ 1\ 1\ 0\ 0\ 1\ 1\ 0\ 0\ 1\ 1\ 0\ 0\ 1\ 1$$

(a) Design a Mealy sequential circuit using D flip-flops, showing a reduced state graph, and equations for the output and D inputs. It should be a reasonably economical design.

(b) Repeat part (a) for J-K flip-flops.

(c) Design a Moore sequential circuit using T flip-flops to do the same task, showing a state graph and input equations for a reasonably economical design.

16.22 Design a sequential circuit to multiply an 8-4-2-1 binary-coded decimal digit by 3 to give a 5-bit binary number. For example, if the input is 0111, the output should be 10101. The input and output to the circuit should be serial with the least significant bit first. Assume that the input will be 0 at the fifth clock time and reset the circuit after the fifth output bit. (*Hint:* As each bit is received, multiply it by 3, giving a product of either 00 or 11. Thus we either output 0 and carry 0 to the next column, or output 1 and carry 1 to the next column. If we carry a 1 to the next column, then the sum of the carry and the next product is either 01 or 100. In this case, we either output 1 and carry 0 or output 0 and carry 10 (2) to the next column. What happens if we carry 10 (2) to the next column?)

(a) Derive a state table with a minimum number of states (3 states).
(b) Design the circuit using J-K flip-flops and NAND and NOR gates.
(c) Design the circuit using a PLA and D flip-flops. Give the PLA table.

16.23 A Moore sequential circuit has three inputs (X_2, X_1, and X_0) that specify a temperature range in a room. The circuit has two outputs (I and D) that control a heater for the room; $I = 1$ causes the heater to increase its heat output, and $D = 1$ causes the heater to decrease its heat output. If the temperature range is 0, 1 or 2, for three successive clock cycles, the circuit generates $I = 1$, and conversely if the temperature range is 5, 6, or 7, for three successive clock cycles, the circuit generates $D = 1$; otherwise, $I = 0$ and $D = 0$.

(a) Construct a state diagram for the circuit.
(b) Encode the states using a one-hot state assignment and derive the D flip-flop input equations and the output equations.
(c) Use a minimum number of D flip-flops and derive the D flip-flop input equations and the output equations.

16.24 Repeat Problem 16.23 using a Mealy circuit.

16.25 A Moore sequential circuit has two inputs (X and Y) and three outputs (Z_2, Z_1, and Z_0). The outputs are a 1's complement number specifying the number of successive times X and Y have been equal or not equal as follows: In decimal, the outputs are 1, 2 and 3, if X and Y have been equal for one time, two successive times, and three or more successive times, and the outputs are, –1, –2, and – 3 if X and Y have been not equal for one time, two successive times, and three or more successive times. Initially, the outputs are all 0.

(a) Construct a state diagram for the circuit.
(b) Encode the states using a one-hot state assignment and derive the D flip-flop input equations and the output equations.
(c) Use a minimum number of D flip-flops and derive the D flip-flop input equations and the output equations.

16.26 Repeat Problem 16.25 using a Mealy sequential circuit.

16.27 A Moore sequential circuit has two inputs (X and Y) and three outputs (Z_2, Z_1, and Z_0). The outputs are a 2's complement number specifying the number of successive times X and Y have been equal or not equal as follows: In decimal, the outputs are 1, 2, and 3 if X and Y have been equal for one time, two successive times, and three or more successive times, and the outputs are –1, –2, –3, and –4 if X and Y have been not equal for one time, two successive times, three successive times, and four or more successive times. Initially, the outputs are all 0.

(a) Construct a state diagram for the circuit.

(b) Encode the states using a one-hot state assignment and derive the D flip-flop input equations and the output equations.

(c) Use a minimum number of D flip-flops and derive the D flip-flop input equations and the output equations.

16.28 Repeat Problem 16.27 using a Mealy sequential circuit.

16.29 The block diagram for an elevator controller for a two-floor elevator follows. The inputs FB_1 and FB_2 are 1 when someone in the elevator presses the first or second floor buttons, respectively. The inputs $CALL_1$ and $CALL_2$ are 1 when someone on the first or second floor presses the elevator call button. The inputs FS_1 and FS_2 are 1 when the elevator is at the first or second floor landing. The output UP turns on the motor to raise the elevator car; $DOWN$ turns on the motor to lower the elevator. If neither UP nor $DOWN$ is 1, then the elevator will not move. R_1 and R_2 reset the latches (described below); and when DO goes to 1, the elevator door opens. After the door opens and remains open for a reasonable length of time (as determined by the door controller mechanism), the door controller mechanism closes the door and sets $DC = 1$. Assume that all input signals are properly synchronized with the system clock.

(a) If we were to realize a control circuit that responded to all of the inputs FB_1, FB_2, $CALL_1$, $CALL_2$, FS_1, FS_2, and DC, we would need to implement logic equations with nine or more variables (seven inputs plus at least two state variables). However, if we combine the signals FB_i and $CALL_i$ into a signal N_i (i – 1 or 2) that indicates that the elevator is needed on the specified floor, we can reduce the number of inputs into the control circuit. In addition, if the

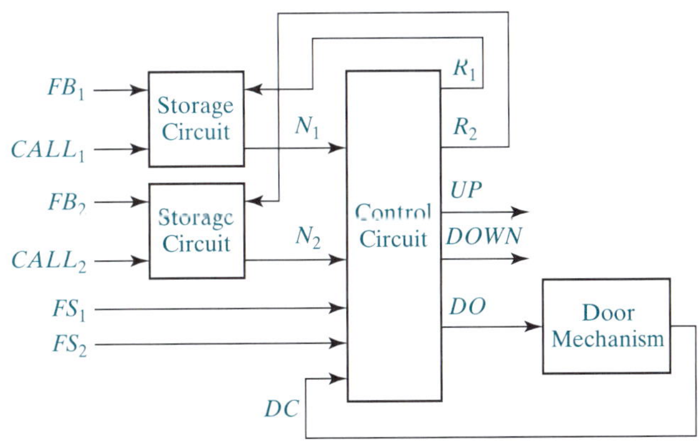

signal N_i is stored so that a single pulse on FB_i or $CALL_i$ will set N_i to 1 until the control circuit clears it, then the control circuit will be simplified further. Using a D flip-flop and a minimum number of added gates, design a storage circuit that will have an output 1 when either input (FB_i or $CALL_i$) becomes 1 and will stay 1 until reset with a signal R_i.

(b) Using the signals N_1 and N_2 that indicate that the elevator is needed on the first or second floor (to deliver a passenger or pick one up or both), derive a state graph for the elevator controller. (Only four states are needed.)

(c) Realize the storage circuits for N_1 and N_2 and the state graph.

16.30 An older model Thunderbird car has three left and three right taillights which flash in unique patterns to indicate left and right turns.

LEFT turn pattern:

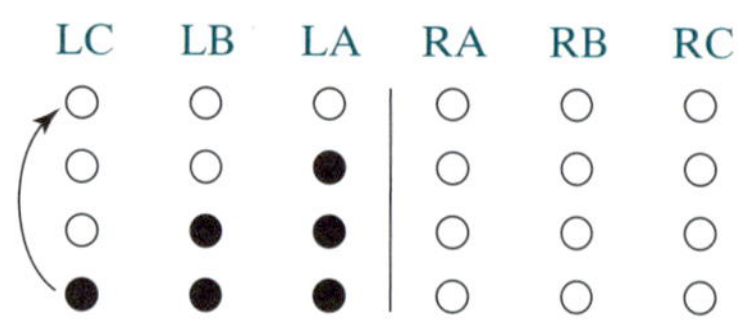

RIGHT turn pattern:

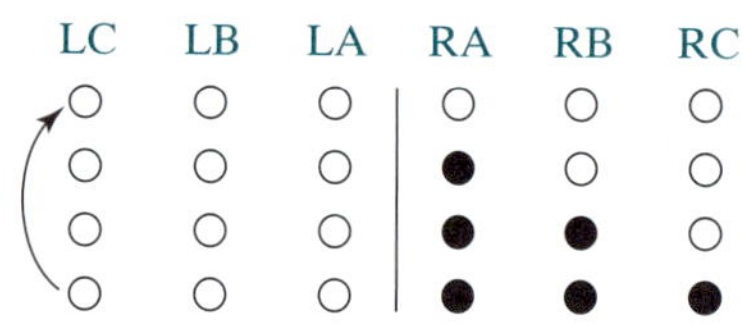

Design a Moore sequential circuit to control these lights. The circuit has three inputs *LEFT*, *RIGHT*, and *HAZ*. *LEFT* and *RIGHT* come from the driver's turn signal switch and cannot be 1 at the same time. As indicated above, when $LEFT = 1$ the lights flash in a pattern *LA* on; *LA* and *LB* on; *LA*, *LB*, and *LC* on; all off; and then the sequence repeats. When $RIGHT = 1$, the light sequence is similar. If a switch from *LEFT* to *RIGHT* (or vice versa) occurs in the middle of a flashing sequence, the circuit should immediately go to the *IDLE* (lights off) state and, then, start the new sequence. *HAZ* comes from the hazard switch, and when $HAZ = 1$, all six lights flash on and off in unison. *HAZ* takes precedence if *LEFT* or *RIGHT* is also on. Assume that a clock signal is available with a frequency equal to the desired flashing rate.

(a) Draw the state graph (eight states).

(b) Realize the circuit using six D flip-flops and make a state assignment such that each flip-flop output drives one of the six lights directly. (Use *LogicAid*.)

(c) Realize the circuit using three D flip-flops, using the guidelines to determine a suitable state assignment. Note the trade-off between more flip-flops and more gates in (b) and (c).

16.31 Design a sequential circuit to control the motor of a tape player. The logic circuit, shown as follows, has five inputs and three outputs. Four of the inputs are the control buttons on the tape player. The input *PL* is 1 if the play button is pressed, the input *RE* is 1 if the rewind button is pressed, the input *FF* is 1 if the fast forward button is pressed, and the input *ST* is 1 if the stop button is pressed. The fifth input to the control circuit is *M*, which is 1 if the special *music sensor* detects music at the current tape position. The three outputs of the control circuit are *P*, *R*, and *F*, which make the tape play, rewind, and fast forward, respectively, when 1. No more than one output should ever be on at a time; all outputs off cause the motor to stop. The buttons control the tape as follows: If the play button is pressed, the tape player will start playing the tape (output $P = 1$). If the play button is held down and the rewind button is pressed and released, the tape player will rewind to the beginning of the current song (output $R = 1$ until $M = 0$) and then start playing. If the play button is held down and the fast forward button is pressed and released, the tape player will fast forward to the end of the current song (output $F = 1$ until $M = 0$) and then start playing. If rewind or fast forward is pressed while play is released, the tape player will rewind or fast forward the tape. Pressing the stop button at any time should stop the tape player motor.

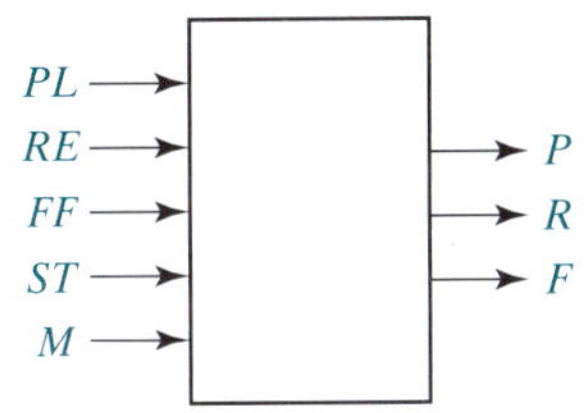

(a) Construct a state graph chart for the tape player control circuit.
(b) Realize the control circuit using a PLA and D flip-flops.

16.32 The circuit below converts a positive, unsigned binary number into a BCD decimal number. The binary number is entered bit serially with the most significant bit first. As each input bit is entered, the BCD number in the registers is multiplied by 2 and the input bit added to it.

(a) Show the contents of the registers after each input bit for the binary number 110110111.
(b) What is the largest binary number that this circuit can correctly convert to BCD?
(c) Derive the minimum sum-of-product equations for the outputs of the BCD multiply by 2 circuit. Let x_3, x_2, x_1, x_0 be the inputs and y_3, y_2, y_1, y_0 be the outputs.

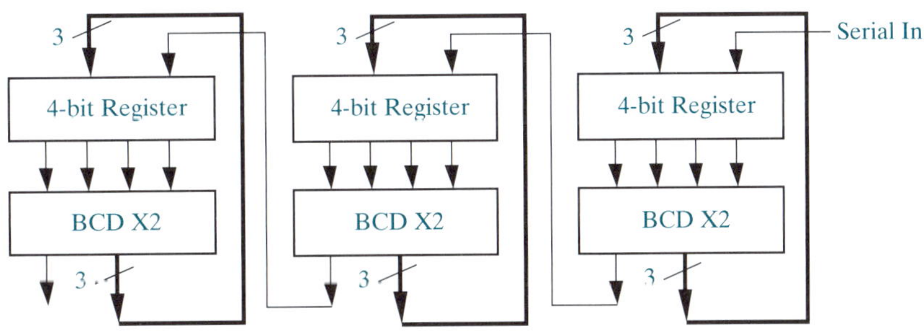

16.33 An iterative circuit has an output of 1 from the last cell if and only if the input pattern 1011 or 1101 has occurred as inputs to any four adjacent cells in the circuit.

(a) Find a Moore state graph or table with a minimum number of states.
(b) Make a suitable state assignment, and derive one of the equations for a typical cell.
(c) Derive the output equation.

16.34 An iterative circuit has a form similar to Figure 16-6. The output Z is to be 1 iff at least one of the X inputs is 1, and no group of two or more consecutive 1 inputs occurs.

Example:

$$0\ 0\ 1\ 0\ 1\ 0\ 0\ 0\ 1\ 0\ 0 \text{ gives an output } Z = 1$$

$$0\ 0\ 1\ 0\ 1\ 1\ 0\ 0\ 0\ 0\ 0 \text{ gives an output } Z = 0$$

(a) Draw a state graph for a typical cell.

(b) Derive the equations and a NOR-gate circuit for a typical cell and for the output circuit.

(c) Specify a_1 and b_1, and simplify the first cell.

(d) Show how a sequential circuit can be constructed using the typical cell and output circuit.

UNIT 18

Circuits for Arithmetic Operations

Objectives

1. Analyze and explain the operation of various circuits for adding, subtracting, multiplying, and dividing binary numbers and for similar operations.
2. Draw a block diagram and design the control circuit for various circuits for adding, subtracting, multiplying, and dividing binary numbers and for similar operations.

Study Guide

1. Study Section 18.1, *Serial Adder with Accumulator.*

(a) Study Figure 18-2 carefully to make sure you understand the operation of this type of adder. Work out a table similar to Table 18-1 starting with $X = 6$ and $Y = 3$:

	X	Y	c_i	s_i	c_i^+
t_0	0110	0011			
t_1					
t_2					
t_3					
t_4					

(b) What changes would be made in this table if the SI input to the addend register (Figure 18-1) was connected to a logic 0 instead of to y_0?

(c) Note in Table 18-1 that when the adding has finished, the full adder still generates a sum and a carry output. The full adder consists of combinational logic, so it will still automatically do the work of calculating its outputs even when they are not needed. What bits are added to generate the last values of s_i and c_i^+? (See Figure 18-2(e).) Are the last values of s_i and c_i^+ useful for anything?

(d) Work Problem 18.3.

2. Study Section 18.2, *Design of a Binary Multiplier.*

(a) For the binary multiplier of Figure 18-7, if the initial contents of the accumulator is 000001101 and the multiplicand is 1111, show the sequence of add and shift signals and the contents of the accumulator at each time step.

(b) For the state graph of Figure 18-8, what is the maximum number of clock cycles required to carry out the multiplication? The minimum number?

(c) For the state graph of Figure 18-9(c), assuming the counter sets $K = 1$ when the counter is in state 3 (11_2), what is the maximum number of clock cycles required to carry out the multiplication? The minimum number?

(d) For Figure 18-7, how many bits would be required for the product register if the multiplier was 6 bits and the multiplicand was 8 bits?

(e) Work Problems 18.4 and 18.5.

(f) Consider the design of a binary multiplier which multiplies 8 bits by 8 bits to give a 16-bit product. What changes would need to be made in Figure 18-7?

If a multiplier control of the type shown in Figure 18-8 were used, how many states would be required?

If a control of the type shown in Figure 18-9 is used, how many bits should the counter have? K should equal 1 in what state of the counter? How many states will the control state graph have?

(g) Work Programmed Exercise 18.1.

3. Study Section 18.3, *Design of a Binary Divider.*

(a) Using the state graph of Figure 18-11 to determine when to shift or subtract, work through the division example given at the start of this section.

(b) What changes would have to be made in Figure 18-12 if the subtraction was done using full adders rather than full subtracters?

(c) For the block diagram of Figure 18-10, under what conditions will an overflow occur and why?

(d) Work Programmed Exercise 18.2.

(e) Derive the control circuit equations, Equations (18-1).

(f) In Figure 18-13, why is one of the inputs to the bus merger at the 0 input of the MUX set to 1?

(g) For a binary multiplier of the type described in Section 18.2, addition is done before shifting. Division requires a series of shift and subtract operations. Since division is the inverse of multiplication, which operation should be done first, subtract or shift?

(h) Work Problems 18.6, 18.7, and 18.8.

4. Optional simulation exercises:

   (a) Simulate the serial adder of Figure 13-12 and test it.

   (b) Connect two 4-bit shift registers to the inputs of the adder that you simulated in (a) to form a serial adder with accumulator (as in Figure 18-1). Supply the shift signal and clock signal from switches so that a control circuit is unnecessary. Test your adder using the following pairs of binary numbers:

   $$0101 + 0110, 1011 + 1101$$

   (c) Input the control circuit from the equations of Figure 18-4, connect it to the circuit which you built in (b), and test it.

5. When you are satisfied that you can meet all of the objectives, take the readiness test.

Circuits for Arithmetic Operations

This unit introduces the concept of using a sequential circuit to control a sequence of operations in a digital system. Such a control circuit outputs a sequence of control signals that cause operations such as addition or shifting to take place at the appropriate times. We will illustrate the use of control circuits by designing a serial adder, a multiplier, and a divider.

18.1 Serial Adder with Accumulator

In this section we will design a control circuit for a serial adder with an accumulator. Figure 18-1 shows a block diagram for the adder. Two shift registers are used to hold the 4-bit numbers to be added, X and Y. The X register serves as an accumulator

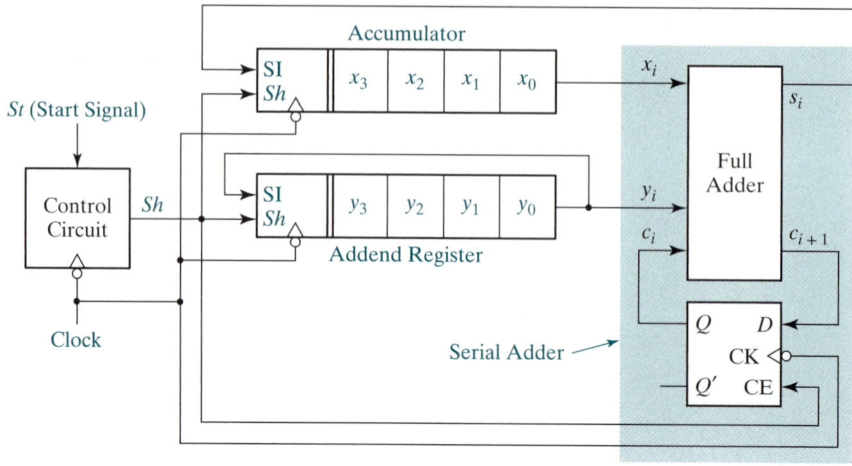

FIGURE 18-1 Block Diagram for Serial Adder with Accumulator

and the Y register serves as an addend register. When the addition is completed, the contents of the X register are replaced with the sum of X and Y. The addend register is connected as a cyclic shift register so that after shifting four times it is back in its original state, and the number Y is not lost. The box at the left end of each shift register shows the inputs: *Sh* (shift signal), *SI* (serial input), and *Clock*. When $Sh = 1$ and an active clock edge occurs, *SI* is entered into x_3 (or y_3) at the same time as the contents of the register are shifted one place to the right. The additional connections required for initially loading the X and Y registers and clearing the carry flip-flop are not shown in the block diagram.

The serial adder, highlighted in blue in the diagram, is the same as the one in Figure 13-12, except the D flip-flop has been replaced with a D flip-flop with clock enable. At each clock time, one pair of bits is added. Because the full adder is a combinational circuit, the sum and carry appear at the full adder output after the propagation delay. When $Sh = 1$, the falling clock edge shifts the sum bit into the accumulator, stores the carry bit in the carry flip-flop, and rotates the addend register one place to the right. Because *Sh* is connected to CE on the flip-flop, the carry is only updated when shifting occurs.

Figure 18-2 illustrates the operation of the adder. Shifting occurs on the falling clock edge when $Sh = 1$. In this figure, t_0 is the time before the first shift, t_1 is the time after the first shift, t_2 is the time after the second shift, etc. Initially, at time t_0, the accumulator contains X and the addend register contains Y. Because the full adder is a combinational circuit, x_0, y_0, and c_0 are added independently of the clock to form the sum s_0 and carry c_1. When the first falling clock edge occurs, s_0 is shifted into the accumulator and the remaining accumulator digits are shifted one position to the right. The same clock edge stores c_1 in the carry flip-flop and rotates the addend register right. The next pair of bits, x_1 and y_1, are now at the full adder input, and the adder generates the sum and carry, s_1 and c_2, as seen in Figure 18-2(b). The second falling edge shifts s_1 into the accumulator, stores c_2 in the carry flip-flop, and

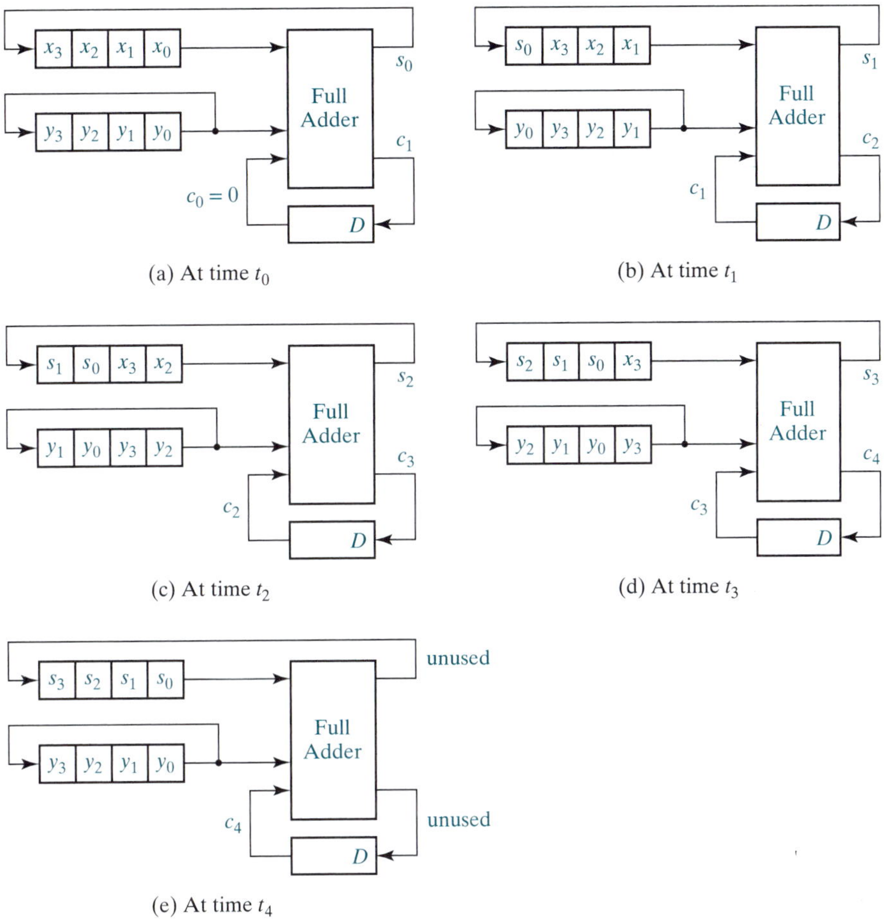

FIGURE 18-2 Operation of Serial Adder

cycles the addend register right. Bits x_2 and y_2 are now at the adder input, as seen in Figure 18-2(c), and the process continues until all bit pairs have been added, as shown in Figure 18-2(e).

Table 18-1 shows a numerical example of the serial adder operation. Initially, the accumulator contains 0101 and the addend register contains 0111. At t_0, the full adder computes $1 + 1 + 0 = 10$, so $s_i = 0$ and $c_i^+ = 1$. After the first falling clock

TABLE 18-1 Operation of Serial Adder

	X	Y	C_i	S_i	C_i^+
t_0	0101	0111	0	0	1
t_1	0010	1011	1	0	1
t_2	0001	1101	1	1	1
t_3	1000	1110	1	1	0
t_4	1100	0111	0	(1)	(0)

edge (time t_1) the first sum bit has been entered into the accumulator, the carry has been stored in the carry flip-flop, and the addend has been cycled right. After four falling clock edges (time t_4), the sum of X and Y is in the accumulator, and the addend register is back to its original state.

The control circuit for the adder must now be designed so that after receiving a start signal, the control circuit will put out four shift signals and then stop. Figure 18-3 shows the state graph and table for the control circuit. The circuit remains in S_0 until a start signal is received, at which time the circuit outputs $Sh = 1$ and goes to S_1. Then, at successive clock times, three more shift signals are put out. It will be assumed that the start signal is terminated before the circuit returns to state S_0 so that no further output occurs until another start signal is received. Dashes appear on the graph because once S_1 is reached, the circuit operation continues regardless of the value of St. Starting with the state table of Figure 18-3 and using a straight binary state assignment, the control circuit equations are derived in Figure 18-4.

A serial processing unit, such as a serial adder with an accumulator, processes data one bit at a time. A typical serial processing unit (Figure 18-5) has two shift registers. The output bits from the shift registers are inputs to a combinational circuit. The combinational circuit generates at least one output bit. This output bit is fed into the input of a shift register. When the active clock edge occurs, this bit is stored in the first bit of the shift register at the same time the registers bits are shifted to the right.

The control for the serial processing unit generates a series of shift signals. When the start signal (St) is 1, the first shift signal (Sh) is generated. If the shift registers

FIGURE 18-3 State Graph for Serial Adder Control

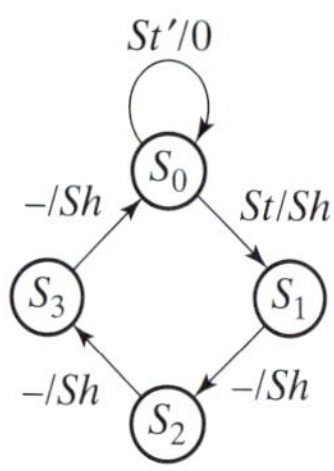

	Next State $St = 0$	1	Sh 0	1
S_0	S_0	S_1	0	1
S_1	S_2	S_2	1	1
S_2	S_3	S_3	1	1
S_3	S_0	S_0	1	1

FIGURE 18-4 Derivation of Control Circuit Equations

	AB	A^+B^+ 0	1
S_0	00	00	01
S_1	01	10	10
S_2	10	11	11
S_3	11	00	00

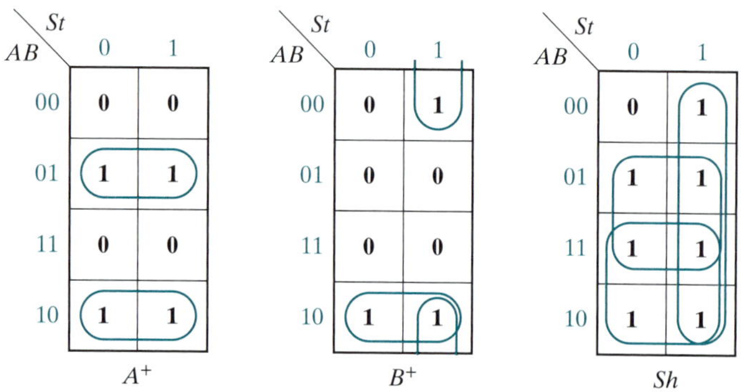

$D_A = A'B + AB'$
$= A \oplus B$

$D_B = St\,B' + AB'$

$Sh = St + A + B$

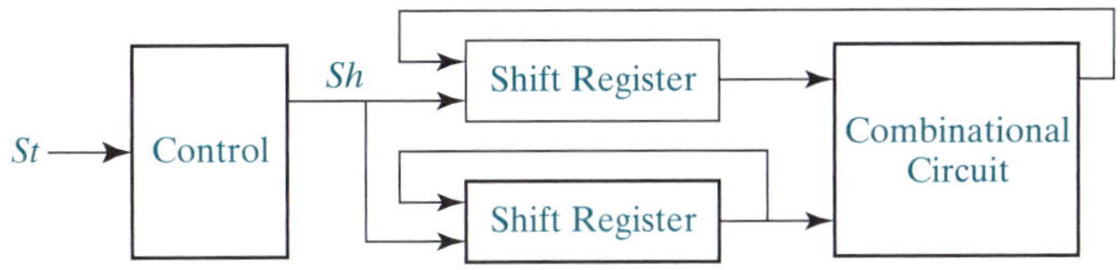

FIGURE 18-5
Typical Serial Processing Unit

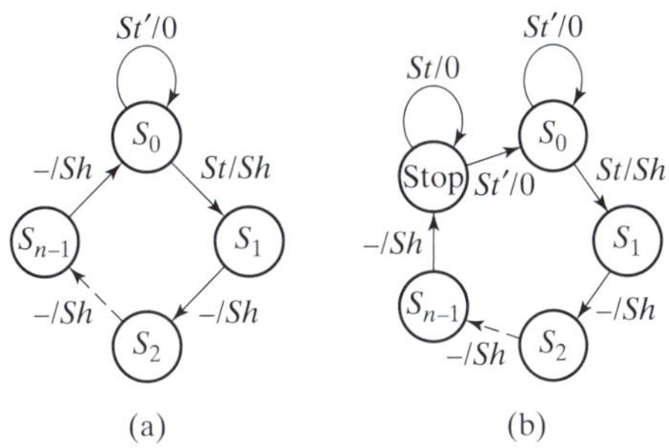

FIGURE 18-6
State Graphs for Serial Processing Unit

have n bits, then a total of n shift signals must be generated. If St is 1 for only one clock time, then the control state graph (Figure 18-6(a)) stops when it returns to state S_0. However, if St can remain 1 until after the shifting is completed, then a separate stop state is required, as shown in Figure 18-6(b). The control remains in the stop state until St returns to 0.

18.2 Design of a Binary Multiplier

Next, we will design a multiplier for positive binary numbers. As illustrated in the example in Section 1.3, binary multiplication requires only shifting and adding. The following example shows how each partial product is added in as soon as it is formed. This eliminates the need for adding more than two binary numbers at a time.

```
Multiplicand ⟶       1101     (13)
Multiplier   ⟶       1011     (11)
                     ----
                     1101
Partial             1101
Products           ------
                   100111
                   0000
                   ------
                   100111
                  1101
                 --------
Product      ⟶   10001111    (143)
```

The multiplication of two 4-bit numbers requires a 4-bit multiplicand register, a 4-bit multiplier register, and an 8-bit register for the product. The product register serves as an accumulator to accumulate the sum of the partial products. Instead of shifting the multiplicand left each time before it is added, as was done in the previous example, it is more convenient to shift the product register to the right each time. Figure 18-7 shows a block diagram for such a multiplier. As indicated by the arrows on the diagram, 4 bits from the accumulator and 4 bits from the multiplicand register are connected to the adder inputs; the 4 sum bits and the carry output from the adder are connected back to the accumulator. (The actual connections are similar to the parallel adder with accumulator shown in Figure 12-5.) The adder calculates the sum of its inputs, and when an add signal (Ad) occurs, the adder outputs are stored in the accumulator by the next rising clock edge, thus causing the multiplicand to be added to the accumulator. An extra bit at the left end of the product register temporarily stores any carry (C_4) which is generated when the multiplicand is added to the accumulator. Note that when ACC is shifted right, the input to bit 8 of ACC must be 0. Then, a following shift will shift the correct value into bit 7 of ACC. The shift input to ACC is not explicitly shown in Figure 18-7.

Because the lower four bits of the product register are initially unused, we will store the multiplier in this location instead of in a separate register. As each multiplier bit is used, it is shifted out the right end of the register to make room for additional product bits.

The Load signal loads the multiplier into the lower four bits of ACC and at the same time clears the upper 5 bits. The shift signal (Sh) causes the contents of the product register (including the multiplier) to be shifted one place to the right when the next rising clock edge occurs. The control circuit puts out the proper sequence of add and shift signals after a start signal ($St = 1$) has been received. If the current multiplier bit (M) is 1, the multiplicand is added to the accumulator followed by a right shift; if the multiplier bit is 0, the addition is skipped and only the right shift occurs. The multiplication example at the beginning of this section (13 × 11) is reworked below showing the location of the bits in the registers at each clock time.

FIGURE 18-7
Block Diagram for Binary Multiplier

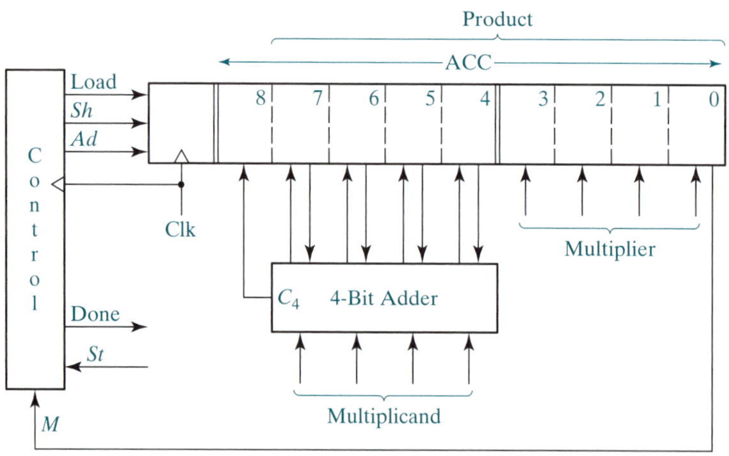

initial contents of product register	0 0 0 0 0 1 0 1 1	←M	(11)
(add multiplicand because $M = 1$)	1 1 0 1		(13)
after addition	0 1 1 0 1 1 0 1 1		
after shift	0 0 1 1 0 1 1 0 1	←M	
(add multiplicand because $M = 1$)	1 1 0 1		
after addition	1 0 0 1 1 1 1 0 1		
after shift	0 1 0 0 1 1 1 1 0	←M	
(skip addition because $M = 0$)			
after shift	0 0 1 0 0 1 1 1 1	←M	
(add multiplicand because $M = 1$)	1 1 0 1		
after addition	1 0 0 0 1 1 1 1 1		
after shift (final answer)	0 1 0 0 0 1 1 1 1		(143)

dividing line between product and multiplier →

The control circuit must be designed to output the proper sequence of add and shift signals. Figure 18-8 shows a state graph for the control circuit. The notation used on this graph is defined in Section 14.5. M/Ad means if $M = 1$, then the output Ad is 1 (and the other outputs are 0). M'/Sh means if $M' = 1$ ($M = 0$), then the output Sh is 1 (and the other outputs are 0). In Figure 18-8, S_0 is the reset state, and the circuit stays in S_0 until a start signal ($St = 1$) is received. This generates a Load signal, which causes the multiplier to be loaded into the lower 4 bits of the accumulator (ACC) and the upper 5 bits of ACC to be cleared on the next rising clock edge. In state S_1, the low order bit of the multiplier (M) is tested. If $M = 1$, an add signal is generated and, then, a shift signal is generated in S_2. If $M = 0$ in S_1, a shift signal is generated because adding 0 can be omitted. Similarly, in states S_3, S_5, and S_7, M is tested to determine whether to generate an add signal followed by shift or just a shift signal. A shift signal is always generated at the next clock time following an add signal (states S_2, S_4, S_6, and S_8). After four shifts have been generated, all four multiplier bits have been processed, and the control circuit goes to a Done state and terminates the multiplication process.

FIGURE 18-8 State Graph for Multiplier Control

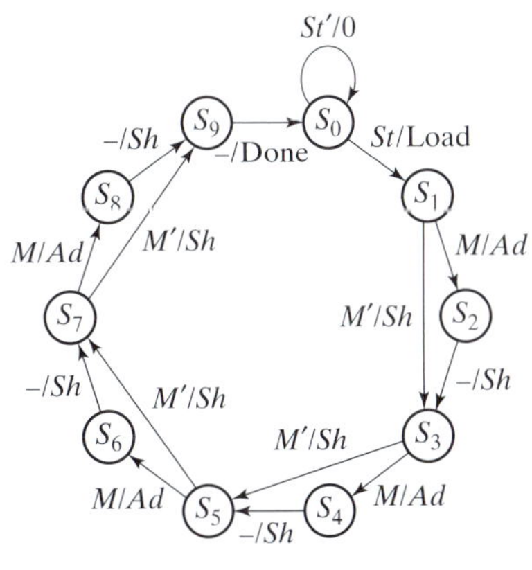

Note that the *Done* signal cannot be turned on in state S_7 or S_8 because the last shift does not occur until the next active clock edge. The last shift occurs at the same time the transition to state S_9 occurs. The control signal *Sh* enables the shifting to occur at the next rising clock edge, but, by itself, it does not cause the shifting to occur. In general, if a control signal is turned on in state S_n, the resulting action does not occur until the next active clock edge takes the circuit to the next state. State S_9 could be eliminated if it was acceptable to turn on the *Done* signal in state S_0 when $St = 0$.

As the state graph indicates, the control performs two functions—generating add or shift signals as needed and counting the number of shifts. If the number of bits is large, it is convenient to divide the control circuit into a counter and an add-shift control, as shown in Figure 18-9(a). First, we will derive a state graph for the add-shift control which tests M and St and outputs the proper sequence of add and shift signals (Figure 18-9(b)). Then, we will add a completion signal (K) from the counter which stops the multiplier after the proper number of shifts have been completed. Starting in S_0 in Figure 18-9(b), when a start signal ($St = 1$) is received, a Load signal is generated. In state S_1, if $M = 0$, a shift signal is generated and the circuit stays in S_1. If $M = 1$, an add signal is generated and the circuit goes to state S_2. In S_2 a shift signal is generated because a shift always follows an add. Back in S_1, the next multiplier bit (M) is tested to determine whether to shift, or add and then shift. The graph of Figure 18-9(b) will generate the proper sequence of add and shift signals, but it has no provision for stopping the multiplier.

In order to determine when the multiplication is completed, the counter is incremented on the active clock edge each time a shift signal is generated. If the multiplier is n bits, a total of n shifts are required. We will design the counter so that a completion signal (K) is generated after $n - 1$ shifts have occurred. When $K = 1$, the circuit should perform one more addition if necessary and then do the final shift. The control operation in Figure 18-9(c) is the same as Figure 18-9(b) as long as $K = 0$. In state S_1, if $K = 1$, we test M as usual. If $M = 0$, we output the final shift signal and stop; however, if $M = 1$, we add before shifting and go to state S_2. In state S_2, if $K = 1$, we output one more shift signal and then go to S_3. The last shift signal will reset the counter to 0 at the same time the add-shift control goes to the Done state.

As an example, consider the multiplier of Figure 18-7, but replace the control circuit with Figure 18-9(a). Because $n = 4$, a 2-bit counter is needed, and $K = 1$ when the counter is in state 3 (11_2). Table 18-2 shows the operation of the multiplier when 1101 is multiplied by 1011. S_0, S_1, and S_2 represent states of the control circuit (Figure 18-9(c)). The contents of the product register at each step is the same as given on p. 635.

FIGURE 18-9

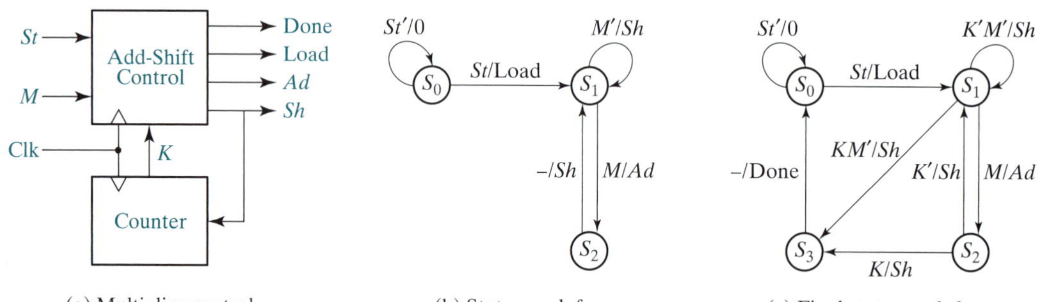

(a) Multiplier control

(b) State graph for add-shift control

(c) Final state graph for add-shift control

TABLE 18-2 Operation of a Multiplier Using a Counter

Time	State	Counter	Product Register	*St*	*M*	*K*	Load	*Ad*	*Sh*	Done
t_0	S_0	00	000000000	0	0	0	0	0	0	0
t_1	S_0	00	000000000	1	0	0	1	0	0	0
t_2	S_1	00	000001011	0	1	0	0	1	0	0
t_3	S_2	00	011011011	0	1	0	0	0	1	0
t_4	S_1	01	001101101	0	1	0	0	1	0	0
t_5	S_2	01	100111101	0	1	0	0	0	1	0
t_6	S_1	10	010011110	0	0	0	0	0	1	0
t_7	S_1	11	001001111	0	1	1	0	1	0	0
t_8	S_2	11	100011111	0	1	1	0	0	1	0
t_9	S_3	00	010001111	0	1	0	0	0	0	1

At time t_0 the control is reset and waiting for a start signal. At time t_1, the start signal $St = 1$, and a Load signal is generated. At time t_2, $M = 1$, so an *Ad* signal is generated. When the next clock occurs, the output of the adder is loaded into the accumulator and the control goes to S_2. At t_3, an *Sh* signal is generated, so, shifting occurs and the counter is incremented at the next clock. At t_4, $M = 1$, so $Ad = 1$, and the adder output is loaded into the accumulator at the next clock. At t_5 and t_6, shifting and counting occurs. At t_7, three shifts have occurred and the counter state is 11, so $K = 1$. Because $M = 1$, addition occurs, and the control goes to S_2. At t_8, $Sh = K = 1$, so at the next clock the final shift occurs, and the counter is incremented back to state 00. At t_9, a Done signal is generated.

The multiplier design given here can easily be expanded to 8, 16, or more bits simply by increasing the register size and the number of bits in the counter. The add-shift control would remain unchanged.

18.3 Design of a Binary Divider

We will consider the design of a divider for positive binary numbers. As an example, we will design a circuit to divide an 8-bit dividend by a 4-bit divisor to obtain a 4-bit quotient. The following example illustrates the division process:

```
                            1010     quotient
divisor              1101 ) 10000111 dividend
                            1101
                            ----
                             0111
                             0000
                             ----
                              1111
(135 ÷ 13 = 10 with           1101
                              ----
a remainder of 5)              0101
                               0000
                               ----
                               0101  remainder
```

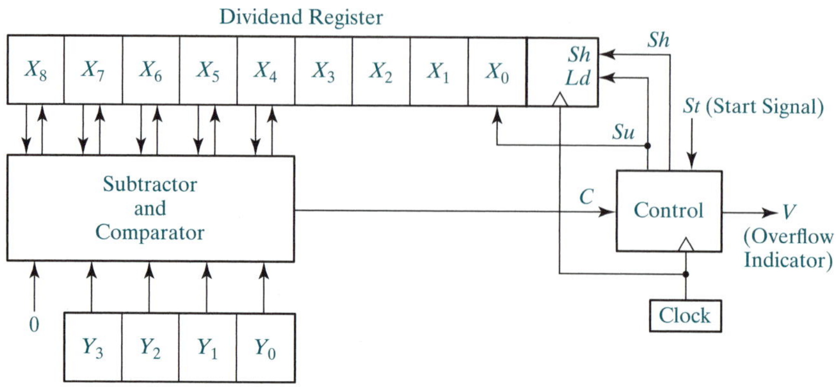

FIGURE 18-10
Block Diagram for Binary Divider

Just as binary multiplication can be carried out as a series of add and shift operations, division can be carried out by a series of subtraction and shift operations. To construct the divider, we will use a 9-bit dividend register and a 4-bit divisor register, as shown in Figure 18-10. During the division process, instead of shifting the divisor to the right before each subtraction as shown in the preceding example, we will shift the dividend to the left. Note that an extra bit is required on the left end of the dividend register so that a bit is not lost when the dividend is shifted left. Instead of using a separate register to store the quotient, we will enter the quotient bit-by-bit into the right end of the dividend register as the dividend is shifted left. Circuits for initially loading the dividend into the register will be added later.

The preceding division example (135 divided by 13) is now reworked, showing the location of the bits in the registers at each clock time. Initially, the dividend and divisor are entered as follows:

0	1	0	0	0	0	1	1	1
	1	1	0	1				

Subtraction cannot be carried out without a negative result, so we will shift before we subtract. Instead of shifting the divisor one place to the right, we will shift the dividend one place to the left:

```
1 0 0 0 0 1 1 1 | 0   ← Dividing line between dividend and quotient
  1 1 0 1
```

← Dividing line between dividend and quotient

↖ Note that after the shift, the rightmost position in the dividend register is "empty".

Subtraction is now carried out, and the first quotient digit of 1 is stored in the unused position of the dividend register:

```
0 0 0 1 1 1 1 1 | 1 ←—— first quotient digit
```

Next, we shift the dividend one place to the left:

```
0 0 1 1 1 1 1 | 1 0
  1 1 0 1
```

Because subtraction would yield a negative result, we shift the dividend to the left again, and the second quotient bit remains 0:

```
0 1 1 1 1 1 ¦ 1 0 0
  1 1 0 1   ¦
```

Subtraction is now carried out, and the third quotient digit of 1 is stored in the unused position of the dividend register:

```
0 0 0 1 0 1 ¦ 1 0 1 ←——— third quotient digit
```

A final shift is carried out and the fourth quotient bit is set to 0:

```
0 0 1 0 1  ¦  1 0 1 0
remainder  ¦  quotient
```

The final result agrees with that obtained in the first example. Note that in the first step the leftmost 1 in the dividend is shifted left into the leftmost position (X_8) in the X register. If we did not have a place for this bit, the division operation would have failed at this step because 0000 < 1101. However, by keeping the leftmost bit in X_8, $10000 \geq 1101$, and subtraction can occur.

If as a result of a division operation, the quotient would contain more bits than are available for storing the quotient, we say that an overflow has occurred. For the divider of Figure 18-10 an overflow would occur if the quotient is greater than 15, because only 4 bits are provided to store the quotient. It is not actually necessary to carry out the division to determine if an overflow condition exists, because an initial comparison of the dividend and divisor will tell if the quotient will be too large. For example, if we attempt to divide 135 by 7, the initial contents of the registers would be:

```
0 1 0 0 0 0 1 1 1
  0 1 1 1
```

Because subtraction can be carried out with a nonnegative result, we should subtract the divisor from the dividend and enter a quotient bit of 1 in the rightmost place in the dividend register. However, we cannot do this because the rightmost place contains the least significant bit of the dividend, and entering a quotient bit here would destroy that dividend bit. Therefore, the quotient would be too large to store in the 4 bits we have allocated for it, and we have detected an overflow condition. In general, for Figure 18-10, if initially $X_8X_7X_6X_5X_4 \geq Y_3Y_2Y_1Y_0$ (i.e., if the left five bits of the dividend register exceed or equal the divisor), the quotient will be greater than 15 and an overflow occurs. Note that if $X_8X_7X_6X_5X_4 \geq Y_3Y_2Y_1Y_0$, the quotient is

$$\frac{X_8X_7X_6X_5X_4X_3X_2X_1X_0}{Y_3Y_2Y_1Y_0} \geq \frac{X_8X_7X_6X_5X_4 0000}{Y_3Y_2Y_1Y_0} = \frac{X_8X_7X_6X_5X_4 \times 16}{Y_3Y_2Y_1Y_0} \geq 16$$

The operation of the divider can be explained in terms of the block diagram of Figure 18-10. A shift signal (Sh) will shift the dividend one place to the left on the next rising clock edge. Because the subtracter is a combinational circuit, it computes

$X_8X_7X_6X_5X_4 - Y_3Y_2Y_1Y_0$, and this difference appears at the subtracter output after a propagation delay. A subtract signal (Su) will load the subtracter output into $X_8X_7X_6X_5X_4$ and set the quotient bit (the rightmost bit in the dividend register) to 1 on the next rising clock edge. To accomplish this, Su is connected to both the Ld input on the shift register and the data input on flip-flop X_0. If the divisor is greater than the five leftmost dividend bits, the comparator output is $C = 0$; otherwise, $C = 1$. The control circuit generates the required sequence of shift and subtract signals. Whenever $C = 0$, subtraction cannot occur without a negative result, so a shift signal is generated. Whenever $C = 1$, a subtract signal is generated, and the quotient bit is set to one.

Figure 18-11 shows the state graph for the control circuit. When a start signal (St) occurs, the 8-bit dividend and 4-bit divisor are loaded into the appropriate registers. Note that this assumes the divisor and dividend are available and stable during the clock cycle when $Load = 1$; they may be available before and after this clock cycle, but they must be available at least during this clock cycle. If C is 1 after the load, the upper half of the dividend is larger than the divisor and the quotient would require five or more bits. Because space is only provided for a 4-bit quotient, this condition constitutes an overflow, so the divider is stopped, and the overflow indicator is set by the V output. Normally, the initial value of C is 0, so a shift will occur first, and the control circuit will go to state S_2. Then, if $C = 1$, subtraction occurs. After the subtraction is completed, C will always be 0, so the next active clock edge will produce a shift. This process continues until four shifts have occurred, and the control is in state S_5. Then, a final subtraction occurs if $C = 1$, and no subtraction occurs if $C = 0$. No further shifting is required, and the control goes to the stop state. For this example, we will assume that when the start signal (St) occurs, it will be 1 for one clock time, and, then, it will remain 0 until the control circuit is back in state S_0. Therefore, St will always be 0 in states S_1 through S_5.

We will now design the control circuit using a one-hot assignment (see Section 15.9) to implement the state graph. One flip-flop is used for each state with $Q_0 = 1$ in S_0, $Q_1 = 1$ in S_1, $Q_2 = 1$ in S_2, etc. By inspection, the next-state and output equations are

$$\begin{aligned} &Q_0^+ = St'Q_0 + CQ_1 + Q_5 && Q_1^+ = StQ_0 \\ &Q_2^+ = C'Q_1 + CQ_2 && Q_3^+ = C'Q_2 + CQ_3 \\ &Q_4^+ = C'Q_3 + CQ_4 && Q_5^+ = C'Q_4 \\ &Load = St\,Q_0 && V = CQ_1 \\ &Sh = C'(Q_1 + Q_2 + Q_3 + Q_4) = C'(Q_0 + Q_5)' \\ &Su = C(Q_2 + Q_3 + Q_4 + Q_5) = C(Q_0 + Q_1)' \end{aligned} \tag{18-1}$$

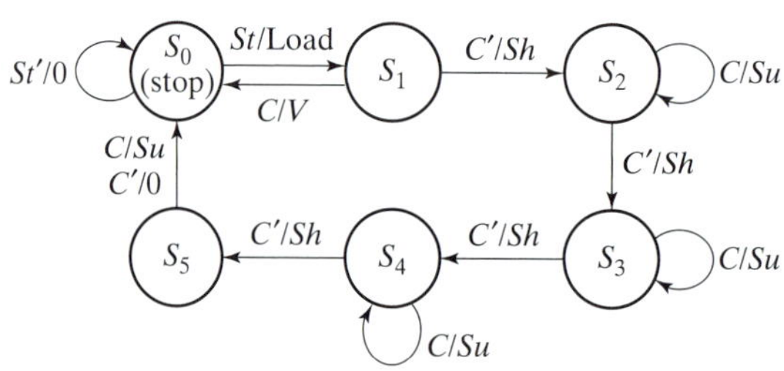

FIGURE 18-11 State Graph for Divider Control Circuit

Because there are three arrows leading into S_0, Q_0^+ has three terms. The equation for *Sh* has been simplified by noting that if the circuit is in state S_1 or S_2 or S_3 or S_4, it is not in state S_0 or S_5.

The subtracter in Figure 18-10 can be constructed using five full subtracters, as shown in Figure 18-12. Because the subtracter is a combinational circuit, whenever the numbers in the divisor and dividend registers change, these changes will propagate to the subtracter outputs. The borrow signal will propagate through the full subtracters before the subtracter output is transferred to the dividend register. If the last borrow signal (b_9) is 1, this means that the result is negative. Hence, if b_9 is 1, the divisor ($Y_3Y_2Y_1Y_0$) is greater than $X_8X_7X_6X_5X_4$, and $C = 0$. Therefore, $C = b_9'$, and a separate comparator circuit is unnecessary. Under normal operating conditions (no overflow) for this divider, we can also show that $C = d_8'$. At any subtraction step, because the divisor is only four bits, $d_8 = 1$ would allow a second subtraction without shifting. However, this can never occur because the quotient digit cannot be greater than 1. Therefore, if subtraction is possible, d_8 will always be 0 after the subtraction, so $d_8 = 0$ implies $X_8X_7X_6X_5X_4$ is greater than $Y_3Y_2Y_1Y_0$ and $C = d_8'$.

The block diagram of Figure 18-10 does not show how the dividend is initially loaded into the X register. This can be accomplished by adding a MUX at the X register inputs, as shown in Figure 18-13. This diagram uses bus notation to avoid drawing multiple wires. When several busses are merged together to form a single bus, a *bus merger* is used. For example, the symbol

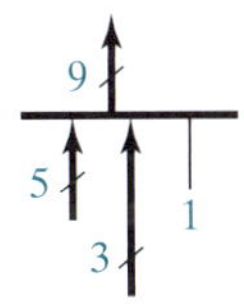

means that the 5-bit subtracter output is merged with bits $X_3X_2X_1$ and a logic 1 to form a 9-bit bus. Thus, the MUX output will be $d_8d_7d_6d_5d_4X_3X_2X_1 1$ when Load = 0.

Similarly, the symbol

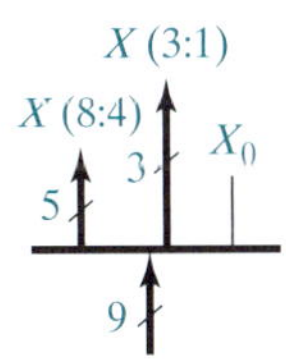

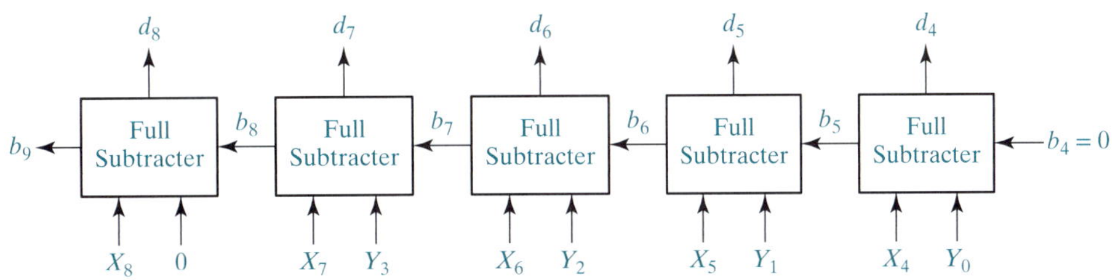

FIGURE 18-12
Logic Diagram for 5-Bit Subtracter

FIGURE 18-13 Block Diagram for Divider Using Bus Notation

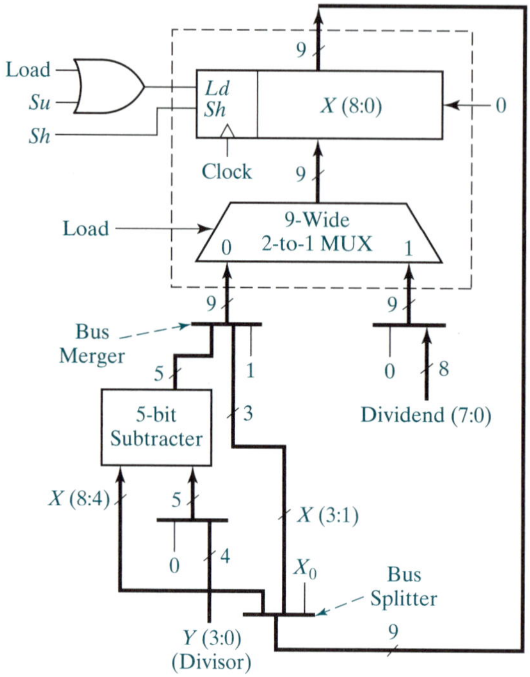

represents a *bus splitter* that splits the 9 bits from the X register into $X_8X_7X_6X_5X_4$ and $X_3X_2X_1$; X_0 is not used. Bus mergers and splitters do not require any actual hardware; they are just a symbolic way of showing bus connections.

The X register is a left-shift register with parallel load capability, similar to the register in Figure 12-10. On the rising clock edge, it is loaded when Ld = 1 and shifted left when Sh = 1. Because the register must be loaded with the dividend when Load = 1 and with the subtracter output when Su = 1, Load and Su are ORed together and connected to the Ld input. The MUX selects the dividend (preceded by a 0) when Load = 1. When Load = 0, it selects the bus merger output which consists of the subtracter output, $X_3X_2X_1$, and a logic 1. When Su = 1 and the clock rises, this MUX output is loaded into X. The net result is that $X_8X_7X_6X_5X_4$ gets the subtracter output, $X_3X_2X_1$ is unchanged, and X_0 is set to 1.

Figure 18-14 shows an alternative version of the divider. The primary difference is the use of a 4-bit subtracter rather than a 5-bit subtracter. The 4-bit subtracter is shown in Figure 18-12 with the leftmost full subtracter deleted. It can be shown that the 4 least significant output bits from the 5-bit subtracter of Figure 18-13 do not depend upon X_8. (See Problem 18.32.) Since the most significant bit of the 5-bit subtracter is discarded by the following shift of the X register, this bit is not needed. However, now the borrow from the 4-bit subtracter, b_8, is not sufficient to determine whether a subtract operation should be done. The state graph of Figure 18-11 still applies, but now C depends on both X_8 and b_8.

FIGURE 18-14 Alternative Divider

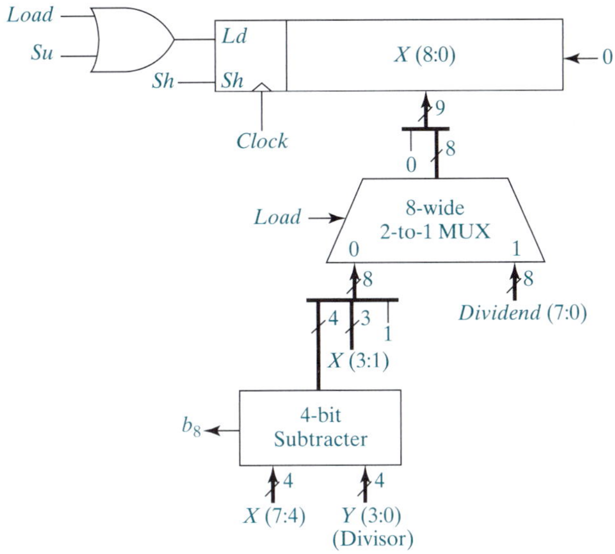

Figure 18-15 shows a second alternative divider. In this version only the upper part of the X register is loaded when a subtract operation is required. Then, on the following shift operation, a quotient bit of 1, $Q = 1$, is shifted into X_0. If no subtract operation is required, a quotient bit of 0, $Q = 0$, is shifted into X_0. This simplifies the circuit since only a 4-wide multiplexer is needed. This circuit is analyzed in Problem 18.33.

FIGURE 18-15 Another Alternative Divider

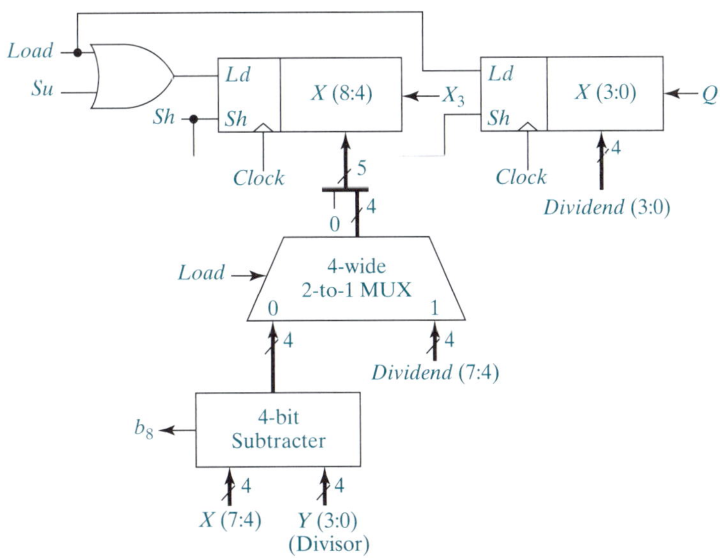

Programmed Exercise 18.1

Cover the answers with a sheet of paper and slide it down as you check your answers. Write your answer in the space provided before looking at the correct answers.

This exercise concerns the design of a circuit which forms the 2's complement of a 16-bit binary number. The circuit consists of three main components—a 16-bit shift register which initially holds the number to be complemented, a control circuit, and a counter which counts the number of shifts. The control circuit processes the number in the shift register one bit at a time and stores the 2's complement back in the shift register. Draw a block diagram of the circuit. Show the necessary inputs and outputs for the control circuit including a start signal (N) which is used to initiate the 2's complement operation.

Answer

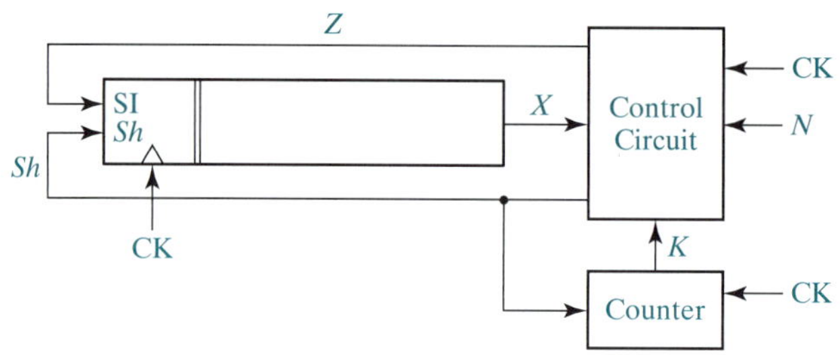

State a rule for forming the 2's complement which is appropriate for use with the preceding block diagram.

Answer Starting with the least significant bit, complement all of the bits to the left of the first 1.

Draw a state graph for the control circuit (three states) which implements the preceding rule. The 2's complement operation should be initiated when $N = 1$. (Assume that N will be 1 for only one clock time.) When drawing your graph, do not include any provision for stopping the circuit. (In the next step you will be asked to add the signal K to your state graph so that the circuit will stop after 16 shifts.) Explain the meaning of each state in your graph.

Answer

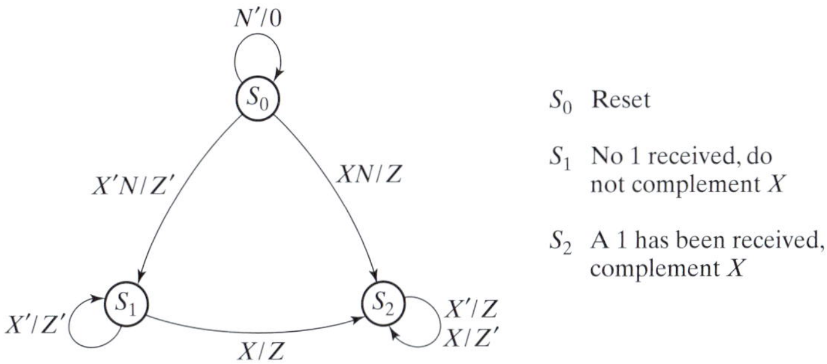

The counter will generate a completion signal (K) when it reaches state 15. Modify your state graph so that when $K = 1$, the circuit will complete the 2's complement operation and return to the initial state. Also, add the Sh output in the appropriate places.

Answer Check the input labels on all arrows leaving each state of your graph. Make sure that two of the labels on arrows leaving a given state cannot have the value 1 at the same time. Make any necessary corrections to your graph, and then check your final answer.

Final Answer

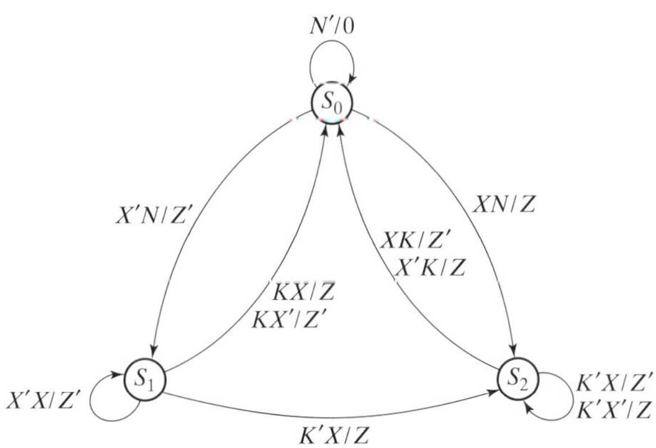

(*Note: Sh* should be added to the graph everywhere Z or Z' appears.)

Programmed Exercise 18.2

This exercise concerns the design of a binary divider to divide a 6-bit number by a 3-bit number to find a 3-bit quotient. The right 3 bits of the dividend register should be used to store the quotient. Draw a block diagram for the divider. Omit the signals required to initially load the dividend register and assume the dividend is already loaded.

Answer

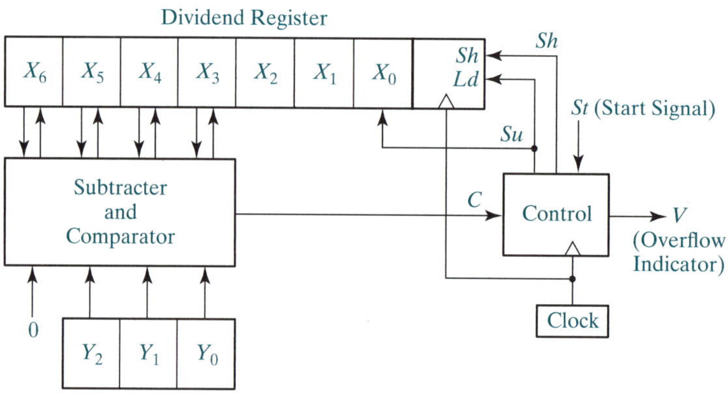

If the contents of the dividend register is initially 0100010 and the divisor is 110, show the contents of the dividend register after each of the first three rising clock edges. Also, indicate whether a shift or a subtraction should occur next.

0	1	0	0	0	1	0	shift

Answer

0	1	0	0	0	1	0	shift
1	0	0	0	1	0	0	subtract
0	0	1	0	1	0	1	shift
0	1	0	1	0	1	0	shift

Now, show the remaining steps in the computation and check your answer by converting to decimal.

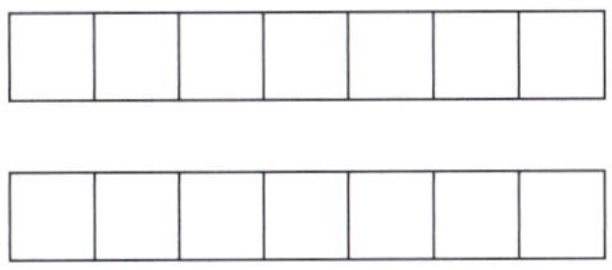

Answer

1	0	1	0	1	0	0	subtract
0	1	0	0	1	0	1	(finished)

If the dividend register initially contained 0011001 and the divisor is 010, can division take place? Explain.

Answer No. Because 011 > 010, subtraction should occur first, but there is no place to store the quotient bit. In other words, the quotient would be greater than three bits, so an overflow would occur.

Draw a state graph for the divider which will produce the necessary sequence of *Su* and *Sh* signals. Assume that the comparator output is $C = 1$ if the upper four bits of the dividend register is greater than the divisor. Include a stop state in your graph which is different than the reset state. Assume that the start signal (*St*) will remain 1 until the division is completed. The circuit should go to the stop state when division is complete or when an overflow is detected. The circuit should then reset when $St = 0$.

Answer

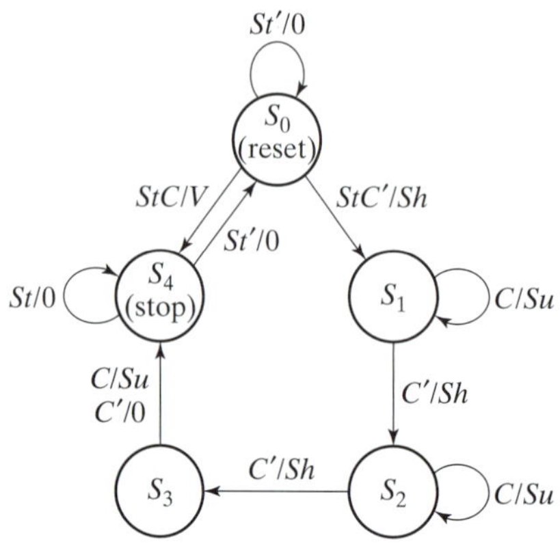

Problems

18.3 Design a serial subtracter with accumulator for 5-bit binary numbers. Assume that negative numbers are represented by 2's complement. Use a circuit of the form of Figure 18-1, except implement a serial subtracter using a D-CE flip-flop and any kind of gates. Give the state graph for the control circuit. Assume that *St* will remain 1 until the subtraction is complete, and the circuit will not reset until *St* returns to 0.

18.4 Design a binary multiplier which multiplies two 3-bit binary numbers to form a 6-bit product. This multiplier is to be a combinational circuit consisting of an array of full adders and AND gates (no flip-flops). Demonstrate that your circuit works by showing all of the signals which are present when 111 is multiplied by 111. (*Hint:* The AND gates can be used to multiply by 0 or 1, and the full adders can be used to add 2 bits plus a carry. Six full adders are required.)

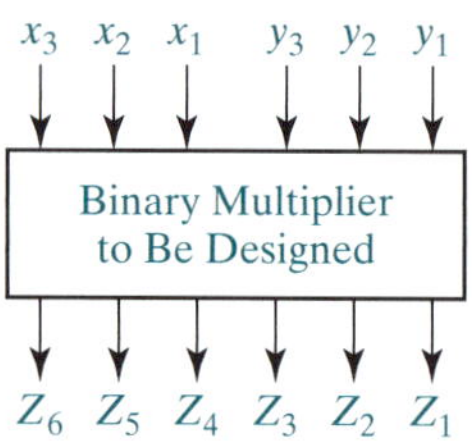

18.5 The binary multiplier of Figure 18-7 has been redesigned so that whenever addition occurs the multiplier bit (M) will be set to 0. Specifically, the *Ad* signal is now connected to a synchronous clear input on only the rightmost flip-flop of the product register of Figure 18-7. Thus, if M is 1 at a given clock time and addition takes place, M will be 0 at the next clock time. Now, we can always add when $M = 1$ and always shift when $M = 0$. This means that the control circuit does not have to change state when $M = 1$, and the number of states can be reduced from ten to six. Draw the resulting state graph for the multiplier control with six states.

18.6 In order to allow for a larger number of bits, the control circuit of the binary divider (Figure 18-10) is to be redesigned so that it uses a separate counter and a subtract-shift control which is analogous to Figure 18-9(a). Draw the state graph for the subtract-shift control.

18.7 Below is the block diagram of a divider which will divide a 5-bit binary number $X_4X_3X_2X_1X_0$ by a 5-bit binary number $Y_4Y_3Y_2Y_1Y_0$. Initially, the 5-bit dividend is loaded into bits X_4 through X_0, and 0's are loaded into bits X_9 through X_5. Because of its design, overflow will only occur if the divisor is 0. This divider operates similarly to the one given in Figures 18-10 and 18-11, except for the starting placement of the dividend.

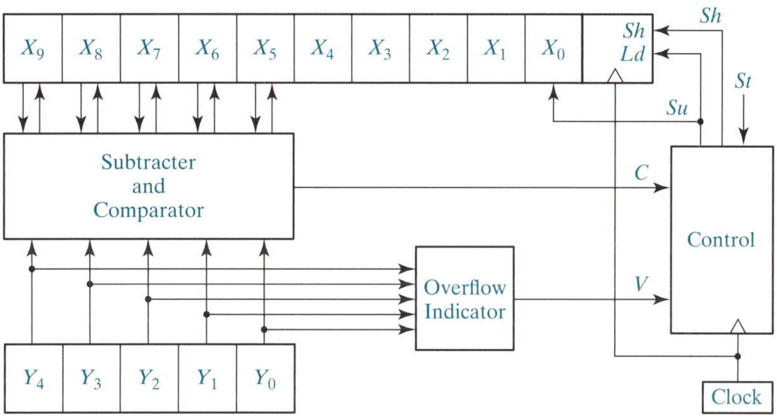

(a) Give the equation for the overflow signal V, generated by the overflow indicator.
(b) Illustrate the operation of the divider when 26 is divided by 5. Specify the sequence of Su and Sh outputs and the contents of the dividend register, and specify the quotient and the remainder.
(c) Draw the state graph for the control circuit. If there is an overflow, the circuit should remain in the starting state. Otherwise, when $St = 1$, the circuit should begin operation. Assume that St will be 1 for only one clock cycle.
(d) In Figure 18-10, the subtracter-comparator and the dividend register have one more bit on the left than the divisor register. Why is that not necessary here?

18.8 A serial logic unit has two 8-bit shift registers, X and Y, shown as follows. Inputs K_1 and K_2 determine the operation to be performed on X and Y. When $St = 1$, X and Y are shifted into the logic circuit one bit at a time and replaced with new values. If $K_1K_2 = 00$, X is complemented and Y is unchanged. If $K_1K_2 = 01$, X and Y

are interchanged. If $K_1K_2 = 10$, Y is set to 0 and each bit of X is replaced with the exclusive-OR of the corresponding bits of X and Y, that is, the new x_i is $x_i \oplus y_i$. If $K_1K_2 = 11$, X is unchanged and Y is set to all 1's.

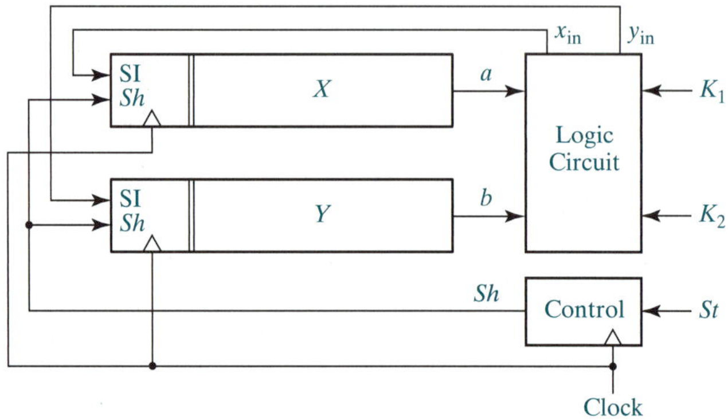

(a) Derive logic equations for x_{in} and y_{in}.
(b) Derive a state graph for the control circuit. Assume that once St is set to 1 it will remain 1 until all 8 bits have been processed. Then, St will be changed back to 0 some time before the start of the next computation cycle.
(c) Realize the logic circuit using two 4-to-1 multiplexers and a minimum number of added gates.

18.9 A circuit for adding one to the contents of a shift register has the following form:

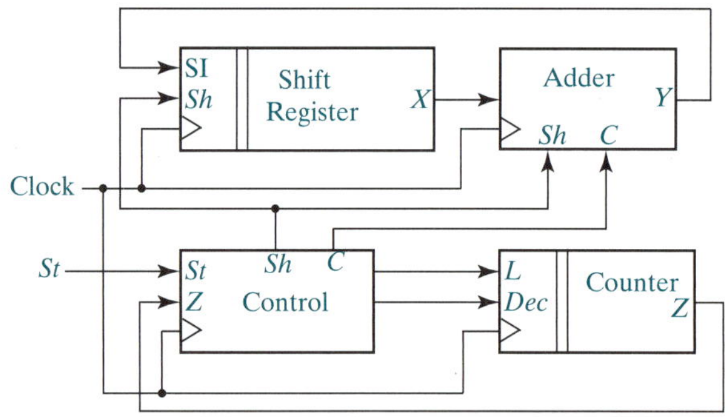

The adder circuit has an internal flip-flop that can be used to store a carry from the adder operation. The control unit has a counter available to determine when the add operation is complete. The counter input L enables a parallel load of the counter with the length of the shift register. The counter input Dec causes the counter to decrement. The counter output Z becomes 1 when the counter value is zero. When St becomes 1, the control unit generates Sh and C the required number of times to cause 1 to be added to the shift register contents. The control unit also generates the signals L and Dec to control the counter.

Design the adder and the control unit, using D flip-flops and NOR gates.

18.10 Repeat Problem 18.9 so that 2 is added to the shift register contents rather than 1.

18.11 Repeat Problem 18.9 so that 3 is added to the shift register contents rather than 1.

18.12 A sequential circuit receives decimal numbers encoded in BCD one digit (4 bits) at a time, starting with the least significant digit. The circuit outputs are the 10's complement of the input number, also encoded in BCD least significant digit first. Input decimal numbers are separated by one or more inputs of all 1's, during which the circuit outputs all 1's. Once valid BCD digits of a new number start, the circuit resumes computing and outputting the 10's complement of the new number.
(a) Construct a state table and output table for the circuit. (Two states are sufficient.)
(b) Realize the circuit using a minimum number of flip-flops.

18.13 Repeat Problem 18.12 assuming the decimal digits are encoded in excess-3 and the separator between decimal numbers is all 0's, which produces all 0's on the outputs.

18.14 A circuit that adds one to the contents of a shift register has the following form:

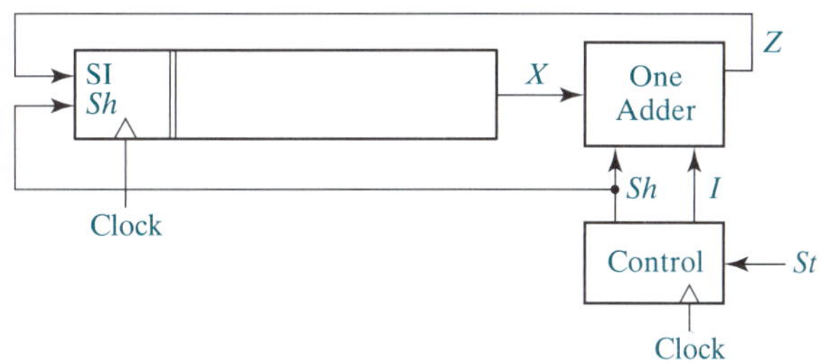

The control circuit outputs I, which should set the ONE ADDER to the proper initial state, and then outputs Sh to the shift register the required number of times. Design the box labeled "ONE ADDER" using NOR gates and a D flip flop with preset and clear inputs.

EXAMPLE:

Contents of shift register before:	000001011
Contents of shift register after I and 9 Sh outputs:	000001100

18.15 (a) Draw a block diagram for a multiplier that can multiply two binary numbers, where the multiplier is 3 bits and the multiplicand is 4 bits. Use an 8-bit shift register along with other necessary blocks.
(b) Draw a state graph for the multiplier control.
(c) Illustrate the operation of the multiplier when 11 is multiplied by 5. Specify the sequence of add and shift outputs generated by the control circuit and specify the contents of the 8-bit register at each clock time.
(d) Draw the logic diagram for the multiplier using an 8-bit shift register of the form of Figure 12-10, a 4-bit adder, three J-K flip-flops, and any necessary gates.

18.16 Work Problem 18.15 if the multiplier is 3 bits and the multiplicand is 5 bits, and show 20 multiplied by 6. Use a 9-bit shift register similar to Figure 12-10, five full adders, three D flip-flops, and a PLA for part (d). Show the PLA table.

18.17 The block diagram for a multiplier for positive binary numbers follows. The counter counts the number of shifts and outputs a signal $K = 1$ after two shifts.

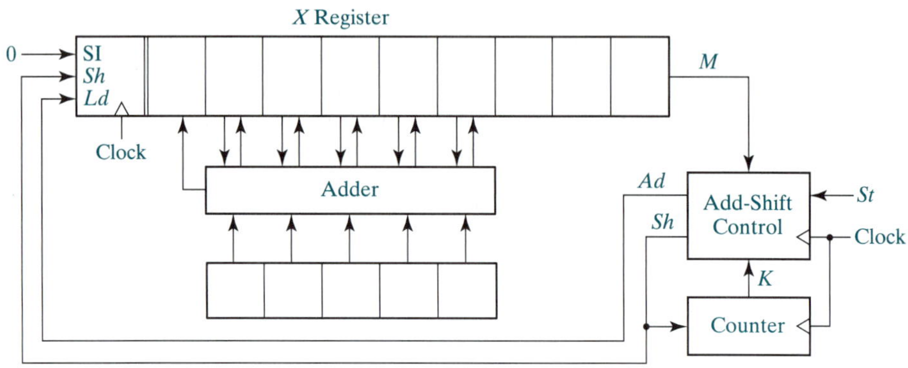

(a) Draw the state graph for the control circuit. Assume that *St* is 1 for one clock period to start the multiplier.
(b) Complete the following table showing the operation of the circuit if the multiplicand is 11001 and the multiplier is 111:

State	Counter	X	St	M	K	Ad	Sh
S_0	00	000000111	1	1	0		

18.18 Design a binary divider which divides a 7-bit dividend by a 2-bit divisor to give a 5-bit quotient. The system has an input *St* that starts the division process.
(a) Draw a block diagram for the subtracter-comparator. You may use full adders or full subtracters.
(b) Draw a block diagram for the rest of the system (do not show the adders or subtracters in the subtracter-comparator block).
(c) Draw the state graph for the control circuit. Assume that the start signal (*St*) is present for one clock period.
(d) Give the contents of the dividend register and the value of C at each time step if initially the dividend is 01010011 and the divisor is 11.

18.19 (a) Draw a block diagram for a divider that is capable of dividing a positive 6-bit binary number by a positive 4-bit binary number to give a 2-bit quotient. Use a dividend register, a divisor register, a subtracter-comparator block, and a control block.

(b) Draw a state graph for the control circuit. Assume that the start signal *St* remains 1 for one or more clock times after the division is complete, and *St* must be set to 0 to reset the circuit.

(c) Show how the subtracter-comparator could be realized using full adders and inverters.

(d) Show the contents of the registers and the value of *C* after each time step if initially the dividend is 101101 and the divisor is 1101.

18.20 Design a controller for an odd-parity generator. The circuit should transmit 7 bits from a shift register onto the output *X*. Then, on the next clock cycle, the eighth value of *X* should be chosen to make the number of 1's be odd. In other words, the last value of *X* should be 1 if there was an even number of 1's in the shift register, so that the 8-bit output word will have odd parity. (Parity was discussed in Section 13.1.) The circuit is shown. *K* will be 1 when the counter reaches 111.

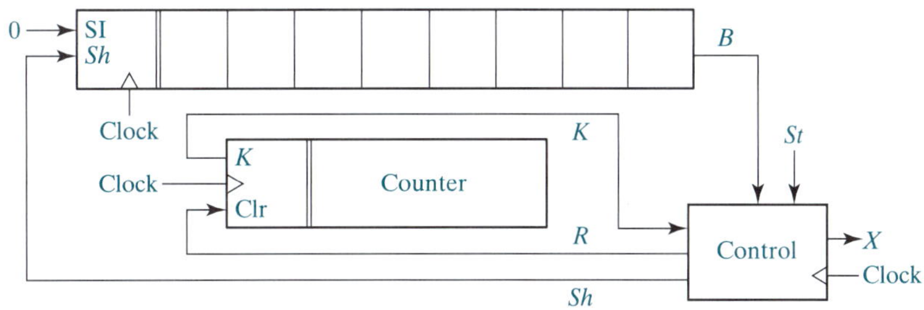

(a) Give the state graph for the control circuit. Assume *St* = 1 for one clock cycle (three states).

(b) Implement the controller using D flip-flops and any necessary gates. Use a one-hot state assignment.

18.21 Design a serial logic unit to multiply a 6-bit number *X* by –1. Assume negative numbers are represented by their 2's complements. Recall that one way to find the 2's complement is to invert all of the bits to the left of the rightmost 1. If the number is –32 = 100000, there is no 6-bit 2's complement representation of +32, so an error signal *Er* should be generated.

(a) Give a block diagram for the circuit, using a control block, a 6-bit right-shift register, and a 3-bit counter. The controller has inputs *St*, *K*, and *SO*, and outputs *Er*, *Clr*, *Sh*, and *SI*. The shift register is like the register of Figure 12-7, but it has 6 bits. The counter has a *Clr* input and an output *K* which is 1 when the counter reaches 6. Assume the shift register contains *X* at the beginning of the operation. The shift register should contain $-X$ when the operation is complete.

(b) Give the state graph for the control circuit. Be sure the circuit will work properly when taking the 2's complement of 0. ($0 \times -1 = 0$.)

(c) Implement the controller using a one-hot state assignment and D flip-flops.

18.22 A serial Boolean logic unit has two 16-bit shift registers, A and B. A control signal (C) is used to select the Boolean operation to be performed. If $C = 0$, the contents of A are serially replaced by the bit-by-bit Boolean AND of A and B. If $C = 1$, the contents of A are serially replaced by the bit-by-bit exclusive-OR of A and B. After the numbers have been placed in A and B, and C is set to 0 or 1, a start signal (St) sets the circuit in operation. A counter is used to count the number of shifts. When the counter reaches state 15, it outputs a signal $K = 1$, which causes the control circuit to stop after one more shift. Assume that St remains 1 and C does not change until the operation is completed. The control then remains in the stop state until St is changed back to 0.

(a) Draw a block diagram of the system, which includes the shift registers, the counter, the control circuit, and a logic circuit that generates the serial input (SI) to the A register.
(b) Draw a state graph for the control circuit (three states).
(c) Design the control circuit using a PLA and D flip-flops.
(d) Design a logic circuit that generates SI.

18.23 Repeat 18.22, but assume that $St = 1$ for only one clock cycle, and that C may change during the operation of the circuit. Therefore, the circuit should operate according to what the value of C was when $St = 1$. Use a one-hot state assignment for (c). (*Hint:* C should be an input to the control circuit, and you will need another output of the control circuit to take the place of C in the logic circuit of part (d) of 18.22.)

18.24 A serial logic unit consists of a 4-bit shift register X and a control unit. The control unit has a start input (St), a shift output (Sh), and an output M which is the serial input to the shift register. In addition, signals C_1 and C_2 are used to select the logic operation performed on the shift register. When $St = 1$, then
If $C_1C_2 = 00$, the contents of register X is serially replaced by all 0's.
If $C_1C_2 = 01$, the contents of register X is serially replaced by all 1's.
If $C_1C_2 = 11$, the contents of register X is serially replaced by its bit-by-bit complement. Assume that C_1C_2 does not change until the selected operation is complete.

(a) Draw a block diagram for the system.
(b) Specify the state graph for the control unit. Assume that St stays 1 for one clock period.
(c) Design the control unit (not the shift register) using J-K flip-flops and any kind of gates. Also, design the logic inside the control unit which generates the serial input M to the shift register. (*Hint: M* depends only on C_1, C_2, and X.)

18.25 Design a circuit which sets a specified number of bits on the right side of a shift register to 0. The number of bits to be set to 0 is in register N before the start of the operation. When $St = 1$, the controller should shift right N times, and then shift left N times. The counter only counts down, and $K = 1$ when the counter reaches 000.

(a) Give the circuit. Use a control block, a 3-bit N register, a 3-bit down counter with load input (Ld) and K output (which is 1 when the counter reaches 000), and an 8-bit right/left shift register which functions according to the table in Problem 12.3 (except that it has 8 bits). Note that the counter does not count up, so you will have to load N into the counter twice. The controller has inputs St and K, and outputs A, B, and Ld.

(b) Give the state graph for the control circuit. Assume $St = 1$ for one clock period.

(c) Implement the controller using two D flip-flops. Use a straight binary assignment.

18.26 Design a controller for the circuit of Problem 12.39 that will add three numbers. Assume each number (including the first one) appears on the 8-bit input data line for two consecutive clock cycles. You may *not* assume that the registers begin with a value of 0. When $St = 1$, the first input appears on the input data line for that clock cycle and the next one. The circuit should halt when the answer goes into the accumulator, and output a signal Done = 1. Done should remain 1 until St returns to 0. You may assume $St = 1$ for enough time for the operation to complete. Give the block diagram and the state graph (seven states), but you do not need to implement the state graph.

18.27 The given multiplier uses only counters to multiply a 4-bit multiplicand by a 4-bit multiplier to obtain an 8-bit product. This Ultra-Slow Multiplier is based on the principle that multiplication is repeated addition and that addition is repeated incrementing. The multiplier works as follows: When the St signal is received, the 8-bit up counter is cleared, N_1 is loaded into 4-bit counter A, and N_2 is loaded into 4-bit counter B. Then, the controller decrements A and increments the up counter until A reaches zero. When A reaches zero, B is decremented and A is reloaded with N_1. Then, the process is repeated until B reaches zero. When B reaches zero, the 8-bit up counter contains the product.

(a) Draw the state graph for the controller. Assume $St = 1$ for only one clock period.

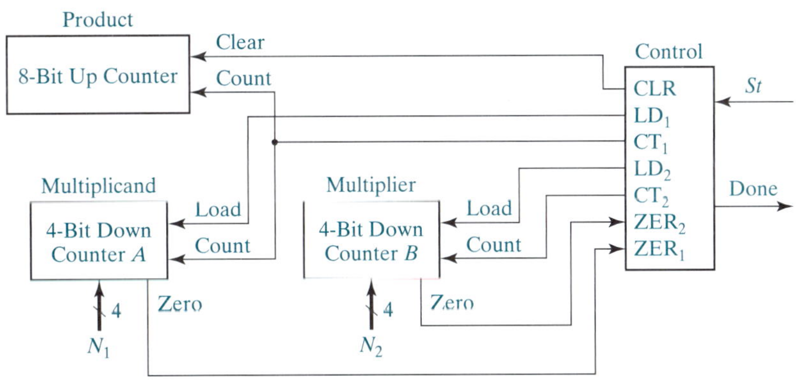

Ultra-Slow Multiplier

(b) Realize the state graph using one or two J-K flip-flops and a minimum number of gates.

(c) If the multiplier is N_1 and the multiplicand is N_2, how many clock periods does it take for the Ultra-Slow Multiplier to calculate the product?

18.28 The following circuit is a multiplier circuit for 4-bit positive numbers. Multiplication is performed by adding the multiplicand to a partial product while decrementing the multiplier. This is continued until the multiplier is decremented to zero. (If the multiplier is initially zero, no additions are done.) When the start input (S) changes to 1, the multiplicand and multiplier are available; the multiplier circuit loads them into A Reg and B Counter, respectively. The partial product register, implemented in two parts (PU and PL), is cleared, as is the carry-out FF for the adder (C FF). To avoid having an adder twice as long as the operands, the addition of the multiplicand to the partial product is done in two steps: First, the multiplicand is added to the lower half of the partial product; second, the carry from the first addition is added to the upper half of the partial product. The multiplier in the B counter is decremented for each addition, and the additions continue until the multiplier has been decremented to zero. Then the done signal (D) is generated, with the product available in the partial product register; D remains asserted and the product available until S returns to 0.

The control signals that the controller must generate are

LB	Load B Counter
DB	Decrement B Counter
CP	Clear PU and PL
LPU	Load PU
LPL	Load PL
LA	Load A Reg
MS	MUX Select Signal
EA	Signal ANDed with A Reg output
CC	Clear C FF
D	Done

The input signals to the controller are start, S, and BZ; BZ = 0 when the B counter is zero.

(a) Determine the contents of the partial product register for each addition step when the multiplicand is 1011 and the multiplier is 0101.

(b) Draw a state graph for the controller. (Four states are sufficient.)

(c) Realize the controller using D FFs and a one-hot state assignment. Give the next-state equations and the controller output equations.

(d) Realize the controller using a minimum number of D FFs.

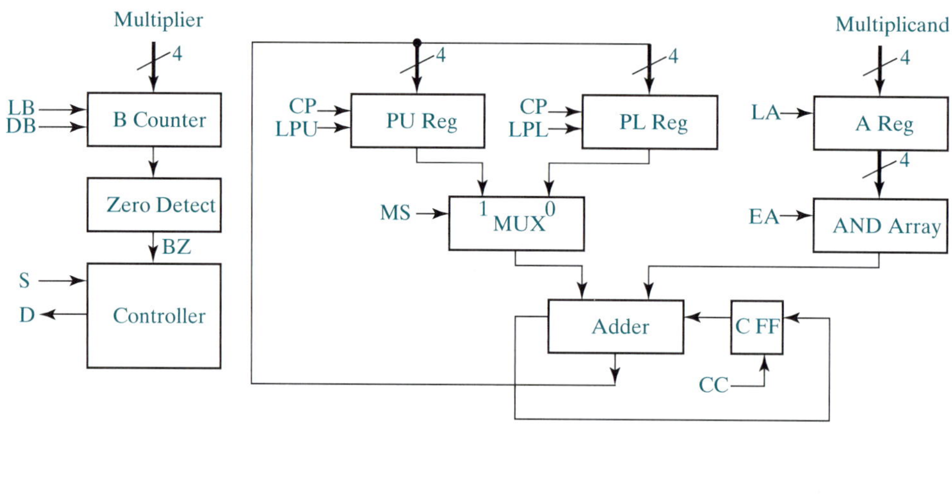

18.29 A few modifications of the circuit of Problem 18.28 are necessary so that it will multiply 2's complement numbers. For example, the Controller must have inputs that are the sign bits of the multiplier and multiplicand; the B Counter must be able to increment a negative multiplier to 0; and the AND Array must be changed so that its outputs can be the multiplicand, all 0's or all 1's.

(a) Redesign the multiplier so that it can multiply 2's complement numbers using these suggested modifications.

(b) Determine the contents of the partial product register for each addition step when the multiplicand is 1011 and the multiplier is 0101. Repeat when the multiplicand is 0101 and the multiplier is 1011.

(c) Draw a state graph for the controller. (At most, five states are required.)

(d) Realize the controller using D FFs and a one-hot state assignment. Give the next-state equations and the controller output equations.

(e) Realize the controller using a minimum number of D FFs.

18.30 The Ultra-Slow Divider, shown in the following block diagram, works on a principle similar to the Ultra-Slow Multiplier in Problem 18.27. When the *St* signal is received, the 8-bit down counter is loaded with the dividend (N_1), the 4-bit down counter is loaded with the divisor (N_2), and the 4-bit quotient up counter is cleared. The dividend counter and the divisor counter are decremented together, and every time the 4-bit divisor counter reaches zero, it is reloaded with the divisor and the quotient up counter is incremented. When the dividend counter reaches zero, the process terminates and the quotient counter contains the result.

(a) Draw the state graph for the controller.

(b) Realize the state graph using one or two D flip-flops and a minimum number of gates.

(c) If the dividend is N_1 and the divisor is N_2, how many clock cycles does it take to calculate the quotient?

(d) How can you tell if an overflow occurs during division?

(e) What will happen in your circuit if the divisor is zero?

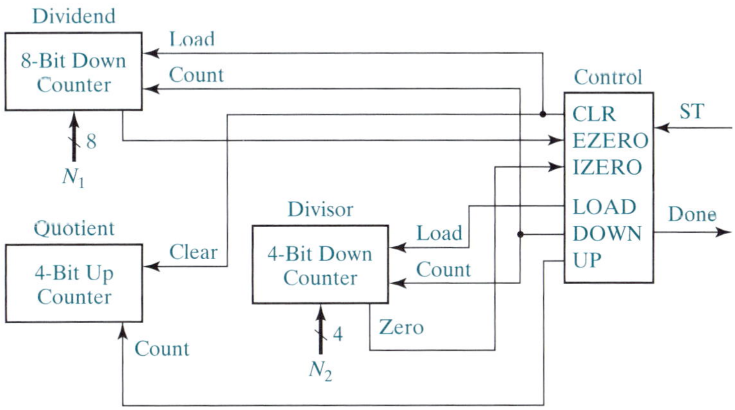

Ultra-Slow Divider

18.31 In Problem 1.42 it was shown that the n-bit 2's complement integer $B = b_{n-1}b_{n-2} \cdots b_1b_0$ has the decimal value $-b_{n-1}2^{n-1} + b_{n-2}2^{n-2} + b_{n-3}2^{n-3} + \cdots + b_12 + b_0$. Based on this expression, multiplication of 2's complement numbers, $B \times A$, can be done by adding $b_i(A2^i)$ to a partial product for $i = 0$ to $n - 2$ and subtracting $b_{n-1}(A2^{n-1})$ from the partial product.

(a) Based on this algorithm, show the contents of ACC in Figure 18-7 for each step of the algorithm for the operands multiplicand $A = 1101(-3)$ and multiplier $B = 1010(-6)$. Repeat for operands $A = 1000(-8)$ and $B = 0110(6)$. Note that the value shifted into bit 7 of ACC will no longer be the carry-out from the adder as shown in Figure 18-7 (i.e., bit 8 of ACC will have to be replaced with a circuit that provides the proper value to shift into bit 7 of ACC).

(b) State in words what value should be shifted into bit 7 of ACC during a shift operation.

(c) Modify the circuit of Figure 18-7 so that 2's complement numbers can be multiplied using this algorithm. Replace bit 8 of ACC with a circuit that provides the value to shift into bit 7 of ACC during a shift operation. Let M_3 be the sign bit of the multiplicand. Let *Su* be the control signal from the controller when a Subtract operation is required.

(d) Using the modified circuit of part (c), modify the state graph of Figure 18-8 to perform 2's complement multiplication.

(e) Using the modified circuit of part (c), modify the state graph of Figure 18-9(c) to perform 2's complement multiplication.

18.32 (a) Consider the 5-bit subtracter in Figure 18-13. Show that the four least significant output bits from the subtracter do not depend upon X_8. Do this by showing examples of the four possible cases of $X_8 = 0$ or 1 and $X(7{:}4)$ is greater than or equal to $Y(3{:}0)$ or less than $Y(3{:}0)$.

(b) Only three of the four cases mentioned in part (a) can occur. Illustrate the case that cannot occur.

(c) If the state graph of Figure 18-11 describes the operation of the circuit of Figure 18-14, what is the expression for C?

(d) Compare the speed and circuit complexity of the dividers in Figures 18-13 and 18-14.

18.33 (a) Construct a state graph for the divider of Figure 18-15.

(b) Compare the speed and circuit complexity of the dividers in Figures 18-13 and 18-15.

18.34 This problem involves the design of a circuit that finds the integer part of the square root of an 8-bit unsigned binary number N using the method of subtracting out odd integers. To find the square root of N, we subtract 1, then 3, then 5, etc., until we can no longer subtract without the result going negative. The number of times we subtract is equal to the integer part of the square root of N. For example, to find $\sqrt{27}$: $27 - 1 = 26$; $26 - 3 = 23$; $23 - 5 = 18$; $18 - 7 = 11$; $11 - 9 = 2$; $2 - 11$ (cannot subtract). Because we subtracted five times, $\sqrt{27}$: $= 5$. Note that the final odd integer is $11_{10} = 1011_2$, and this consists of the square root ($101_2 = 5_{10}$) followed by a 1.

APPENDIX A MOS and CMOS Logic

Most integrated circuits designed today use MOS or CMOS logic. MOS logic is based on the use of MOSFETs (metal-oxide-semiconductor field-effect transistors) as switching elements. Figure A-1 shows the symbols used to represent MOSFETs. The substrate (or body) is a thin slice of silicon. The gate is a thin metallic layer deposited on the substrate and insulated from it by a thin layer of silicon dioxide. A voltage applied to the gate is used to control the flow of current between the drain and source.

In normal operation of an *n*-channel MOSFET, shown in Figure A-1(a), a positive voltage (V_{DS}) is applied between the drain and source. If the gate voltage (V_{GS}) is 0, there is no channel between the drain and source and no current flows. When V_{GS} is positive and exceeds a certain threshold, an *n*-type channel is formed between the drain and source, which allows current to flow from *D* to *S*. Operation of a *p*-channel MOSFET is similar, except V_{DS} and V_{GS} are negative. When V_{GS} assumes a negative value less than the threshold, a *p*-type channel is formed between drain and source, which allows current to flow from *S* to *D*.

The symbol in Figure A-1(c) may be used to represent either a *p*- or *n*-channel MOSFET. When this symbol is used, it is generally understood that the substrate is connected to the most positive circuit voltage for *p*-channel MOSFETs (or the most negative for *n*-channel). If the power supply voltage is V_{DD}, we will use positive logic

FIGURE A-1
MOSFET Symbols

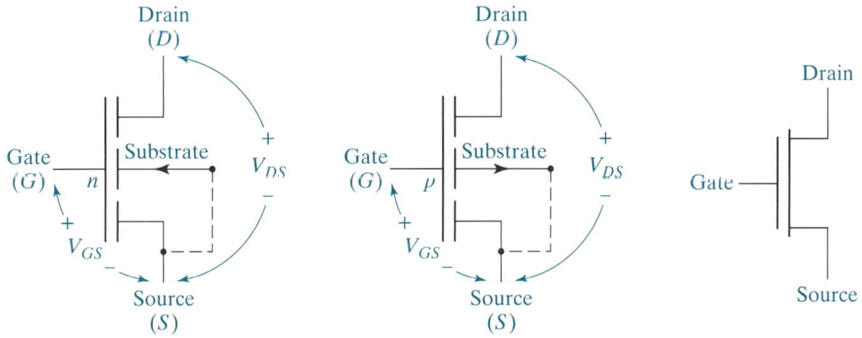

(a) *n*-channel MOSFET (b) *p*-channel MOSFET (c) General MOSFET symbol

(0 volts = logic 0 and V_{DD} volts = logic 1) for *n*-channel MOS circuits and negative logic (V_{DD} volts = logic 0 and 0 volts = logic 1) for *p*-channel MOS circuits. Using this convention, a logic 1 applied to the gate will switch the MOSFET to the ON state (low resistance between drain and source), and a logic 0 will switch it to the OFF state (high resistance between drain and source).

Figure A-2(a) shows a MOS inverter. When a logic 0 is applied to the gate, the MOSFET is in a high-resistance or OFF state, and the output voltage is approximately V_{DD}. When a logic 1 is applied to the gate, the MOSFET switches to a low-resistance or ON state, the output is connected to ground, and the output voltage is approximately 0.

Thus, the operation of the MOSFET is analogous to the operation of a switch in Figure A-2(b) which is open when V_{in} is a logic 0 and closed when V_{in} is a logic 1. In Figure A-2(d), a second MOSFET serves as a load resistor. The geometry of this MOSFET and the gate voltage V_{GG} are chosen so that its resistance is high compared with the ON resistance of the lower MOSFET so that the switching operation of Figure A-2(d) is essentially the same as Figure A-2(a).

As shown in Figure A-3, MOSFETs can be connected in parallel or series to form NOR or NAND gates. In Figure A-3(a), a logic 1 applied to A or B turns on the corresponding transistor and F becomes 0. Thus $F' = A + B$ and $F = (A + B)'$, which is the NOR function. In Figure A-3(c), a logic 1 applied to the A and B inputs turns on both transistors and F becomes 0. In this case $F' = AB$ and $F = (AB)'$, which is the NAND function. More complex functions can be realized by using series-parallel combinations of MOSFETs. For example, the circuit of Figure A-3(e) performs the exclusive-OR function. The output of this circuit has a conducting path to ground, and $F = 0$ if A and B are both 1 or if A' and B' are both 1. Thus, $F' = AB + A'B'$ and $F = A'B + AB' = A \oplus B$. A' and B' are generated by inverters as in Figure A-2(d).

CMOS (complementary MOS) logic performs logic functions using a combination of *p*-channel and *n*-channel MOSFETs. Compared with TTL or other bipolar transistor technologies, CMOS has the advantage of much lower power consumption. Figure A-4(a) shows a CMOS inverter built from a *p*-channel and an *n*-channel MOSFET. When 0 volts (logic 0) is applied to the gate inputs, the *p*-channel transistor (Q_1) is on and the *n*-channel transistor (Q_2) is off, so the output is $+V$(logic 1). When $+V$(logic 1) is applied to the gate inputs, Q_1 is off and Q_2 is on, so the output is 0 volts (logic 0).

FIGURE A-2
MOS Inverter

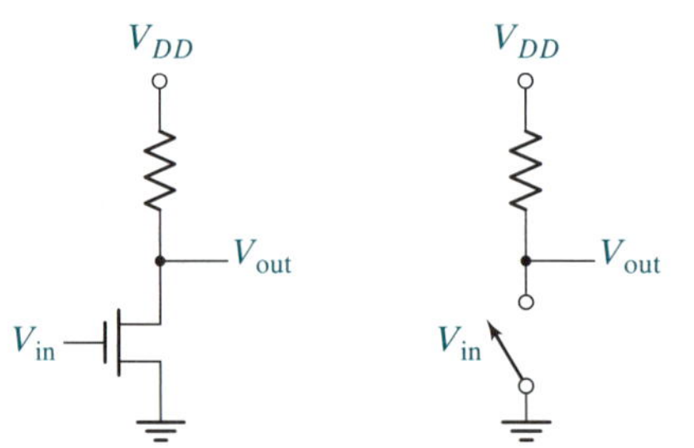

(a) (b)

V_{in}	V_{out}
0	$\approx V_{DD}$
V_{DD}	≈ 0

(c)

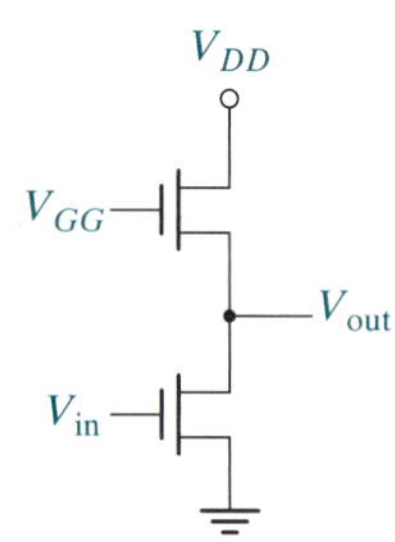

(d)

FIGURE A-3
MOS Gates

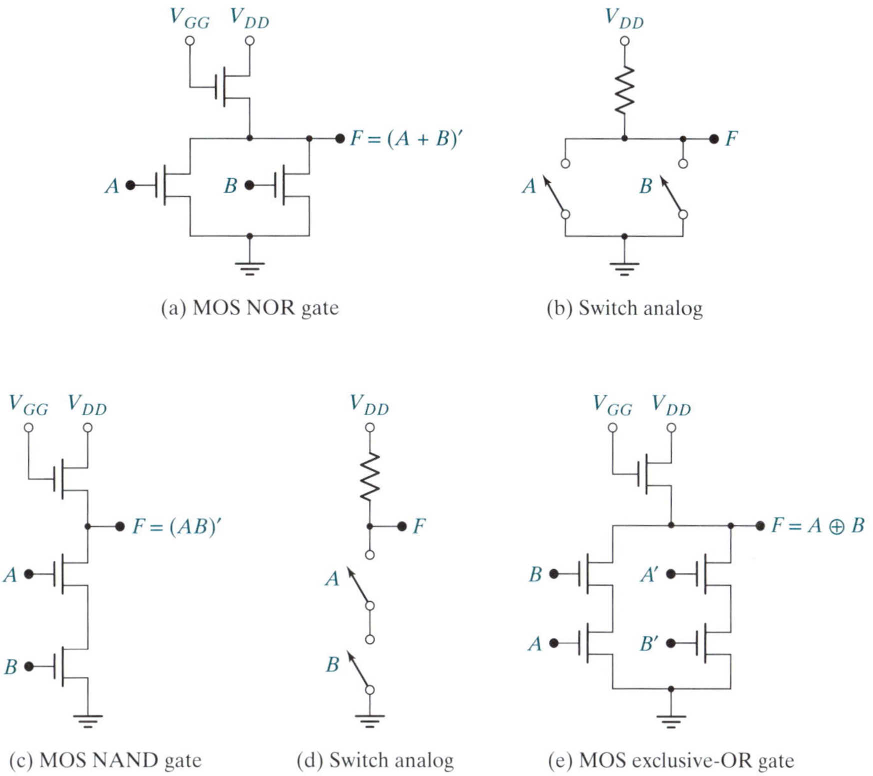

FIGURE A-4
CMOS Inverter

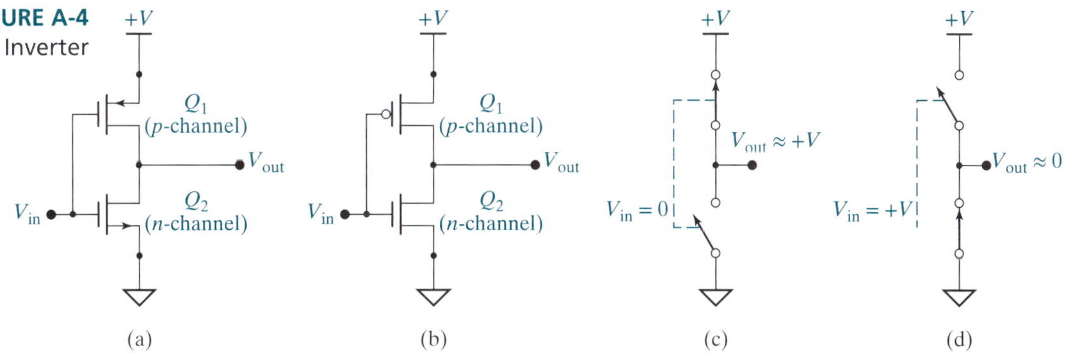

In the remainder of this discussion we will use a bubble at the MOSFET gate input to indicate a *p*-channel transistor, which is turned on by a logic 0. No bubble at the gate input indicates an *n*-channel transistor, which is turned on by a logic 1. Figure A-4(b) shows the CMOS inverter using this bubble notation. The switch analog in Figure A-4(c) illustrates the operation of the inverter when the inverter input is 0. Q_1 is on and Q_2 is off as indicated by the closed and open switches. When the input is $+V$(logic 1), Q_1 is off and Q_2 is on, as indicated by the open and closed switches in Figure A-4(d). The following table summarizes the operation:

V_{in}	V_{out}	Q_1	Q_2
0	$+V$	ON	OFF
$+V$	0	OFF	ON

Figure A-5 shows a CMOS NAND gate. If A or B is 0 volts, then Q_1 or Q_2 is ON while Q_3 or Q_4 is off, and the output is $+V$. If A and B are both $+V$, then Q_3 and Q_4 are both ON while Q_1 and Q_2 are off, and the output is 0 volts. If 0 volts represents a logic 0 and $+V$ represents a logic 1, this gate performs the NAND function, as indicated by the truth table of Figure A-5(b).

FIGURE A-5
CMOS NAND Gate

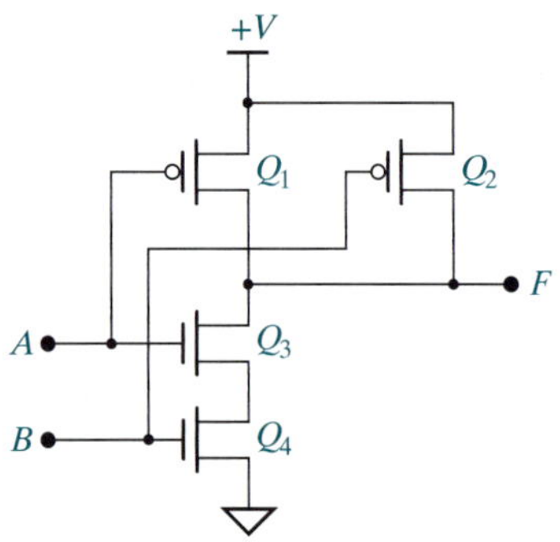

(a) Circuit diagram

A	B	F	Q_1	Q_2	Q_3	Q_4
0	0	$+V$	ON	ON	OFF	OFF
0	$+V$	$+V$	ON	OFF	OFF	ON
$+V$	0	$+V$	OFF	ON	ON	OFF
$+V$	$+V$	0	OFF	OFF	ON	ON

(b) Truth table

Figure A-6 shows a CMOS NOR gate. If $A = 1\ (+V)$, Q_1 is off and Q_4 is on, $F = 0$. Likewise, if $B = 1$, Q_2 is off and Q_3 is on, so $F = 0$. Because $F = 0$ when A or B is 1, $F' = A + B$, and $F = (A + B)'$, which is the NOR function.

FIGURE A-6
CMOS NOR Gate

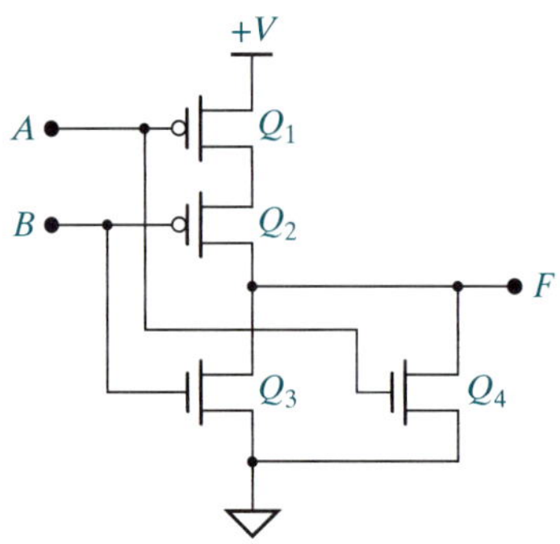

A p-channel and n-channel transistor pair can be connected to form a CMOS transmission gate (TG) as shown in Figure A-7. The two enable inputs are normally complements so that when $En = 1$, both transistors are enabled and a low impedance path connects A and B. When $En = 0$, points A and B are disconnected. In other words, the transmission gate acts like a switch that is closed when $En = 1$ and open

FIGURE A-7 CMOS Transmission Gate and Switch Analog

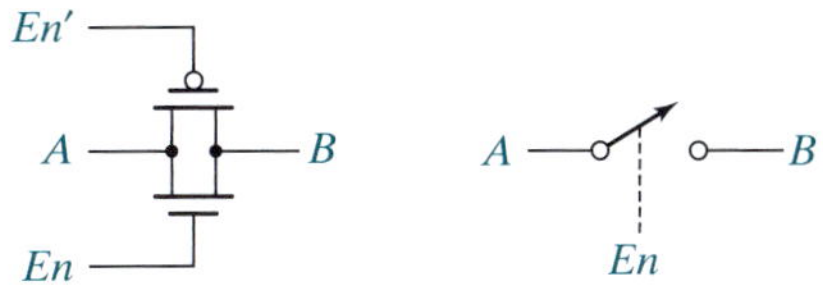

when $En = 0$. Two transistors are used because the p-channel transistor does a good job of transmitting a logic 1 and the n-channel transistor does a good job of transmitting a logic 0.

The 2-to-1 multiplexer of Figure 9-1 can be constructed from two TGs and an inverter, as shown in Figure A-8. When $A = 0$, the upper TG is enabled so that I_0 is connected to F; when $A = 1$, the lower TG is enabled so that I_1 is connected to F.

A CMOS gated D latch, as shown in Figure A-9(a), is easily constructed using two TGs and two inverters. The switch analogs of Figures A-9(b) and (c) represent

FIGURE A-8 CMOS Multiplexer

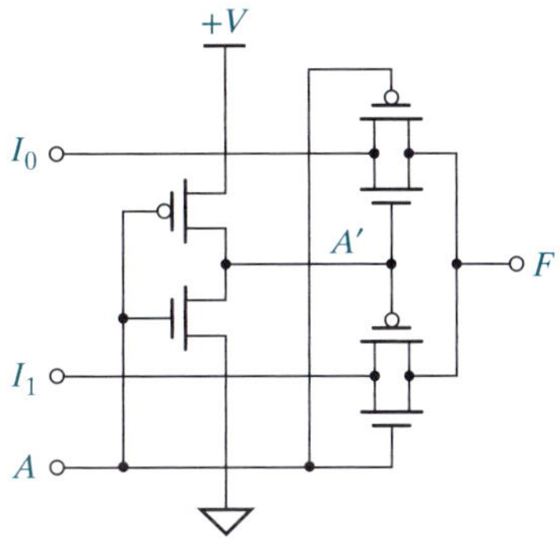

FIGURE A-9 CMOS Latch and Switch Analogs

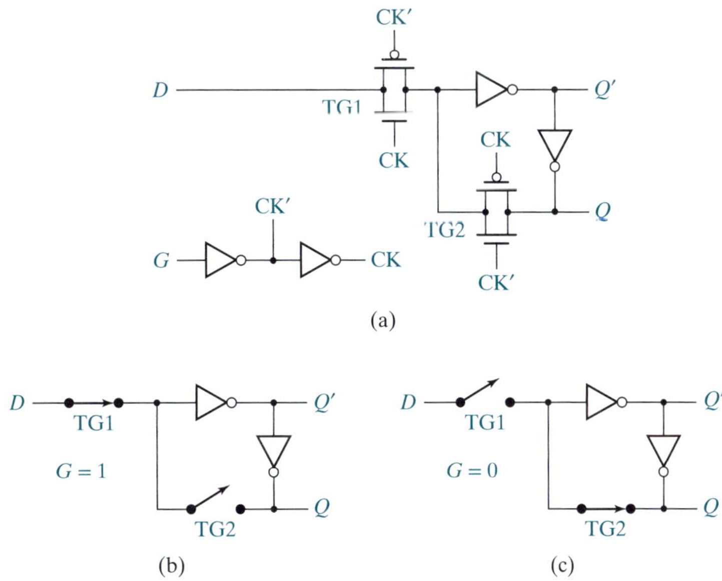

the TGs by switches. When $G = 1$, CK = 1 and TG_1 is closed. Therefore, the latch is transparent, and D is transmitted through the inverters to the Q output. When $G = 0$, TG_2 is closed, and the data in the latch is stored in the closed loop of the two inverters. That is, if $Q = 0$, it is still 0 after going through the two inverters, and if $Q = 1$, it is still 1 after going through the two inverters. Because TG_1 is open, the data does not change when D changes, and the latch holds the stored value of Q.

A CMOS falling-edge-triggered D flip-flop, similar to the type shown in Figure 11-19, can be constructed from two CMOS latches (Figure A-10(a)). The switch analogs of Figures A-10(b) and (c) illustrate the flip-flop operation. When Clock is 1, the input latch is transparent and the output latch holds the current value of Q. When Clock goes to 0, the input latch holds its value, which is transmitted through the output latch to Q. Thus, Q can only change states following the falling edge of Clock.

The technology for implementing a CMOS integrated circuit continues to improve, resulting in smaller transistors, lower voltage levels, faster operation, and very high density logic. When no inputs are changing, the static power dissipation is very low. When the CMOS gates are switching, the power dissipation is proportional to the switching frequency. Thus, the power dissipation at a switching frequency of 100 MHz is ten times that at 10 MHz.

FIGURE A-10 Falling-Edge-Triggered D Flip-Flop

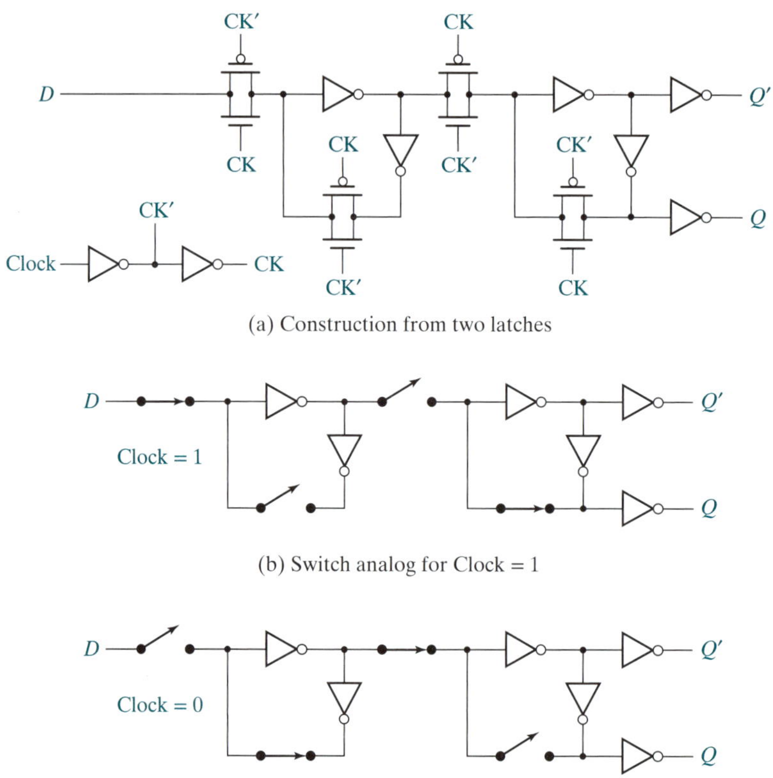

(a) Construction from two latches

(b) Switch analog for Clock = 1

(c) Switch analog for Clock = 0

APPENDIX B

VHDL Language Summary

Reserved words are in boldface type. Square brackets enclose optional items. Curly brackets enclose items which are repeated zero or more times. A vertical bar (|) indicates or.

Disclaimer: This VHDL summary is not complete and contains some special cases. Only VHDL statements used in this text are listed. For a complete description of VHDL syntax, refer to references [1] and [2].

entity declaration

entity entity-name **is**
  port(interface-signal-declaration);
end [entity] [entity-name];

interface-signal declaration

list-of-interface-signals: mode type [:= initial-value]
{; **list**-of-interface-signals: mode type [:= initial-value]}

Note: A signal can be of mode **in, out, inout**, or **buffer**.

architecture declaration

architecture architecture-name **of** entity-name **is**
[declarations]   -- declare internal signals here
begin
  architecture-body
end [**architecture**] [architecture-name];

Note: The architecture body may contain component-instantiation statements, processes, assignment statements, procedure calls, etc.

integer type declaration

type type_name is **range** integer_range;

signal declaration

signal list-of-signal-names : type_name [:= initial_value];

constant declaration

constant constant_name : type_name := constant_value;

alias declaration

alias identifier [:identifier-type] **is** item-name;

Note: Item-name can be a constant, signal, variable, file, type name, etc.

array type and object declaration

type array_type_name **is array** index_range **of** element_type;
signal | **constant** array_name: array_type_name [:= initial_values];

component declaration

```
component component-name
  [generic (list-of-generics-and-their types);]
  port (list-of-interface-signals-and-their-types);
end component;
```

component instantiation (concurrent statement)

```
label: component-name
    [generic map (generic-association-list;)]
    port map (list-of-actual-signals);
```

Note: Use **open** if a component output has no connection.

signal assignment statement (sequential or concurrent statement)

signal <= [**transport**] expression [**after** delay_time];

Note: If concurrent, the signal value is recomputed every time a change occurs on the right-hand side. If [**after** delay-time] is omitted, the signal is updated after Δ time. If [**transport**] is omitted, an inertial delay is assumed.

conditional assignment statement (concurrent statement only)

```
signal <= expression1 when condition1
       else expression2 when condition2
     ...
     [else expression];
```

selected signal assignment statement (concurrent statement only)

```
with expression select
     signal <= expression1 [after delay_time1] when choice1,
     expression2 [after delay_time2] when choice2,
     ...
     [expression [after delay_time] when others];
```

process statement (with sensitivity list)

```
[process-label:] process (sensitivity-list)
      [declarations]           --signal declarations not allowed
begin
      sequential statements
end process [process-label];
```

Note: This form of process is executed initially and thereafter only when an item on the sensitivity list changes value. The sensitivity list is a list of signals. No wait statements are allowed.

process statement (without sensitivity list)

```
[process-label:] process
        [declarations]              --signal declarations not allowed
    begin
        sequential statements
    end process [process-label];
```

Note: This form of process must contain one or more wait statements. It starts execution immediately and continues until a wait statement is encountered.

wait statements

```
wait on sensitivity-list;
wait until Boolean-expression;
wait for time-expression;
```

if statement (sequential statement only)

```
if condition then
        sequential statements
{elsif condition then
        sequential statements }  -- 0 or more elsif clauses may be included
[else sequential statements]
end if;
```

case statement (sequential statement only)

```
case expression is
        when choice1 => sequential statements
        when choice2 => sequential statements
        ...
        [when others => sequential statements]
end case;
```

for loop statement (sequential statement only)

```
[loop-label:] for index in range loop
        sequential statements
end loop [loop-label];
```

Note: You may use **exit** to exit the current loop.

report declaration

```
report string-expression
    [severity severity-level];
```

VHDL Libraries and Packages

VHDL libraries and packages are used to extend the functionality of VHDL by defining types, functions, components, and overloaded operators. The syntax for libraries and packages is as follows:

library declaration

```
library list-of-library names;
```

use statement

```
use library_name.package_name.item;    (.item may be .all)
```

package declaration

```
package package-name is
    package declarations
end [package][package-name];
```

package body

```
package body package-name is
    package body declarations
end [package body][package name];
```

When working with bits and bit_vectors, you may use the following declarations:

```
library BITLIB;
use BITLIB.bit_pack.all;
```

The bit_pack package includes functions and components that work with signals of type bit and bit_vector. For example, the function call vec2int(A) converts a bit_vector A to an integer. The CD contains a complete listing of bit_pack.

When working with std_logic and std_logic_vectors, the following declarations are required:

```
library IEEE;
use IEEE.std_logic_1164.all;
```

The std_logic_1164 package defines the types std_logic and std_logic_vector, a resolution function for these types, conversion functions, and overloaded operators for logic operations. It does not define overloaded operators for arithmetic operations.

In order to perform arithmetic operations on std_logic_vectors, you may add the declaration

```
use IEEE.std_logic_unsigned.all;
```

Although this package is found in the IEEE library, it was written by Synopsys and it is not an IEEE standard. This package treats std_logic_vectors as if they were unsigned numbers and provides overloaded arithmetic operators for +, −, *, =, /=, >, >=, <, and <=. For "+" and "−" if the left operand is a std_logic_vector, the right operand can be the same type, integer type, or std_logic type. For the comparison operators, the right operand can be a std_logic_vector or an integer. The function call CONV_INTEGER(A) converts a std_logic_vector A to an integer.

As an alternative to using std_logic_vectors and the overloaded operators defined in the std_logic_unsigned package, type unsigned may be used. Unsigned type is defined in the Synopsys package std_logic_arith and in the IEEE package numeric_std. To use the former, add the declaration

```
use IEEE.std_logic_arith.all;
```

A vector of type unsigned is similar to a std_logic_vector in that it is an array of std_logic bits, but it has its own overloaded arithmetic operators. Operators for +, −, *, =, /=, >, >=, <, and <= are defined in the std_logic_arith package for various combinations of left and right operands. Unfortunately, logic operators AND, OR, and NOT are not defined for unsigned vectors in this package, so C <= A + B;

works for unsigned vectors, but C <= A **and** B; is not allowed without calling type conversion functions. Some type conversion functions available in this package are as follows:

conv_integer(A) converts an unsigned vector A to an integer
conv_std_logic_vector(A) converts an unsigned vector A to a std_logic_vector
conv_unsigned(B, N) converts an integer B to an unsigned vector of length N

Conversion of a std_logic_vector to unsigned is not defined.

The IEEE numeric_std package, which actually is an IEEE standard, overcomes a number of the deficiencies in the std_logic_arith package. The statement

use IEEE.numeric_std.**all**;

invokes this package. It defines unsigned type and overloaded operators for arithmetic and comparison operations in a way similar to the std_logic_arith package, but in addition it defines overloaded operators for logic operations on unsigned vectors. Useful conversion functions in the package include

TO_INTEGER(A) converts an unsigned vector A to an integer
TO_UNSIGNED(B, N) converts an integer to an unsigned vector of length N

The only significant deficiency is that this package does not define an overloaded operator for adding a std_logic bit to an unsigned type. Thus, a statement of the form

sum <= A + B + carry;

is not allowed when carry is of type std_logic. The carry must be converted to an integer before it can be added to the unsigned vector A + B.

We have used the std_logic_unsigned package in many examples in this book because it is easy to use. For complex VHDL projects, we recommend using the numeric_std package. Most VHDL simulators and synthesizers work well with either package.

APPENDIX C

Tips for Writing Synthesizable VHDL Code

One of our goals throughout this text is to write VHDL code that not only simulates correctly but also synthesizes correctly to implement hardware that works correctly. First and foremost, always remember that when you write VHDL code you are not writing a computer program; you are describing hardware. If you are designing a multiplier for binary numbers, do not simply write a program to multiply binary numbers. Instead think in terms of what registers are required and what sequence of operations on these registers will produce the desired result.

VHDL code that simulates correctly will not always implement correctly in hardware. A frequent cause of problems is the creation of unintended latches. Even though code simulates correctly, the presence of latches may cause timing problems when the code is actually implemented in hardware. After synthesizing your code, check the synthesis report to make sure no latches are present. If latches are present, check your code for the following:

1. Counters, shift registers, flip-flops, and other devices that change state in response to a clock edge must be updated only in a clocked process. The state of these devices should never be changed in a combinational process or in a concurrent statement. All state changes for a given device must be specified within the same process.
   Example: count <= count + 1; should not appear in a combinational process. When this statement, which increments a counter, is placed in a clocked process, any statement that clears the counter must be placed in the same process.
2. If a combinational process sets control signals to '1' at various places in a case statement, all of these signals should be set to '0' *before* the start of the case statement.
3. For every **if** statement in a combinational process, check each signal that is assigned a value in the **then** clause. If such a signal is not assigned a value in step 2, then make sure that it is assigned a value in the **else** clause.
   Example: **if** St = '1' **then** nextstate <=1; load <='1'; **end if;** will create a latch because nextstate is not defined when St = '0'. To eliminate the latch write **if** St = '1' **then** nextstate <= 1; load <= '1'; **else** nextstate <= 0; **end if;** This assumes that load is set to '0' in step (2).

Do not attempt to set the same signal to two different values in two different processes or in a process and in a concurrent statement.

Example

A <='0'; is a concurrent statement, and A <= B; is another concurrent statement or a sequential statement in a process. These statements can attempt to set A to two different values at the same time. If A and B are bit signals, when you try to simulate, you will get an error message that a signal has multiple drivers. That means a conflict exists because A could be driven to '0' and to '1' at the same time. If A and B are std_logic, the conflict still exists, but you will not get the error message. Instead, during simulation A will assume the value 'X' (unknown) if the simulator tries to set A to '0' and '1' at the same time. In both cases, the code will not synthesize properly because it does not correspond to any real hardware.

Also consider the following example:

```
-- Example of what NOT TO DO: output A is assigned values
-- in a concurrent statement and in a process.

entity two_drivers is
  port (B, clk, reset : in bit; A : out bit);
end two_drivers;
architecture arch of two_drivers is
begin
  A <= '0' when reset = '0';
  process (clk)
  begin
    if clk'event and clk = '0' then
      A <= B; end if;
  end process;
end arch;
```

In this example, A is supposed to represent a flip-flop that is reset to '0' when the signal reset is '0' and set equal to B on the falling clock edge. Although this code has correct syntax, it will not simulate properly because the two statements that change A occur as a concurrent statement and also as a sequential statement in a process so that A has two drivers. If the signals are std_logic instead of bits, A will assume a value of 'X' at times during the simulation. The code will not synthesize because all statements that change the output of flip-flop A must be placed in the same process. This also would apply if A were a register or a counter.

Exercise

Change the preceding code so that the reset signal will work properly.

An easy way to write synthesizable VHDL code to perform arithmetic operations is to represent binary numbers as std_logic_vectors so that overloaded operators can be used. This is explained on pages 319–320 of the text.

Overloaded + and – operators cannot be used with bit vectors. If you use overloaded operators with std_logic_vectors in your VHDL code, place the following declarations at the start of your code:

```
library IEEE;                      -- this library contains several useful
                                   --   packages
use IEEE.std_logic_1164.all;       -- this package defines std_logic,
                                   --   std_logic_vectors and logic
                                   --   operations on these types
use IEEE.std_logic_unsigned.all;   -- this package defines overloaded
                                   --   operators for std_logic_vectors
```

Remember that the VHDL operators +, –, and & have the same precedence and will be applied from left to right as they appear in a VHDL statement.

Thus A <= B + C&D; is treated as A <= (B+C)&D;

If you want to do concatenation first, you must use parentheses.

A <= B + (C&D);

APPENDIX D

Proofs of Theorems

Finding Essential Prime Implicants

Section 5.4 presents a method for finding all of the essential prime implicants which is based on finding adjacent 1's on a Karnaugh map. The validity of the method is based on the following theorem:

> If a given minterm m_j of F and all of its adjacent minterms are covered by a single term p_j, then p_j is an essential prime implicant of F.
>
> *Proof:*
>
> **1.** Assume p_j is *not* a *prime implicant.* Then, it can be combined with another term p_k to eliminate some variable x_i and form another term which does not contain x_i. Therefore, $x_i = 0$ in p_j and $x_i = 1$ in p_k, or vice versa. Then, p_k covers a minterm m_k which differs from m_j only in the variable x_i. This means that m_k is adjacent to m_j, but m_k is not covered by p_j. This contradicts the original assumption that all minterms adjacent to m_j are covered by p_j; therefore, p_j is a prime implicant.
>
> **2.** Assume p_j is *not essential.* Then, there is another prime implicant p_h which covers m_j. Because p_h is not contained in p_j, p_h must contain at least one minterm which is adjacent to m_j and not covered by p_j. This is a contradiction, so p_j must be essential.

State Equivalence Theorem

The methods for determining state equivalence presented in Unit 15 are based on Theorem 15.1:

> Two states p and q of a sequential network are equivalent if and only if for every single input x, the outputs are the same and the next states are equivalent.
>
> *Proof:* We must prove both part 1, the "if" part of the theorem, and part 2, the "only if" part.

1. Assume that $\lambda(p, x) = \lambda(q, x)$ and $\delta(p, x) \equiv \delta(q, x)$ for every input x. Then, from Definition 15.1, for every input sequence $\underline{X}$,

$$\lambda[\delta(p, x), \underline{X}] = \lambda[\delta(q, x), \underline{X}]$$

For the input sequence $\underline{Y} = x$ followed by $\underline{X}$, we have

$$\lambda(p, \underline{Y}) = \lambda(p, x) \text{ followed by } \lambda[\delta(p, x), \underline{X})]$$
$$\lambda(q, \underline{Y}) = \lambda(q, x) \text{ followed by } \lambda[\delta(q, x), \underline{X})]$$

Hence, $\lambda(p, \underline{Y}) = \lambda(q, \underline{Y})$ for every input sequence $\underline{Y}$, and $p \equiv q$ by Definition 15.1.

2. Assume that $p \equiv q$. Then, by Definition 15.1, $\lambda(p, \underline{Y}) = \lambda(q, \underline{Y})$ for every input sequence $\underline{Y}$. Let $\underline{Y} = x$ followed by $\underline{X}$. Then,

$$\lambda(p, x) = \lambda(q, x) \text{ and } \lambda[\delta(p, x), \underline{X}] = \lambda[\delta(q, x), \underline{X}]$$

for every sequence $\underline{X}$. Hence, from Definition 15.1, $\delta(p, x) \equiv \delta(q, x)$.

APPENDIX E

Answers to Selected Study Guide Questions and Problems

UNIT 1 Study Guide Answers

2. (e) Two of the rows are:

1110	16	14	*E*
1111	17	15	*F*

3. (b)

$$
\begin{aligned}
1100_2 - 101_2 &= [1 \times 2^3 + 1 \times 2^2 + 0 \times 2^1 + 0 \times 2^0] \\
&\quad - [\qquad 1 \times 2^2 + 0 \times 2^1 + 1 \times 2^0] \quad \text{note borrow from column 1} \\
&= [1 \times 2^3 + 1 \times 2^2 + (0 - 1) \times 2^1 + (10 + 0) \times 2^0] \\
&\quad - [\qquad 1 \times 2^2 + \quad 0 \times 2^1 + \quad 1 \times 2^0] \quad \text{note borrow from column 2} \\
&= [1 \times 2^3 + (1 - 1) \times 2^2 + (10 - 1) \times 2^1 + 10 \times 2^0] \\
&\quad - [\qquad 1 \times 2^2 + \quad 0 \times 2^1 + \quad 1 \times 2^0] \quad \text{note borrow from column 3} \\
&= [(1 - 1) \times 2^3 + (10 - 0) \times 2^2 + 1 \times 2^1 + 10 \times 2^0] \\
&\quad - [\qquad 1 \times 2^2 + 0 \times 2^1 + 1 \times 2^0] \\
&= [\quad 0 \times 2^3 + \quad 1 \times 2^2 + 1 \times 2^1 + 1 \times 2^0] = 111_2
\end{aligned}
$$

5. (f) sign & mag: –0, 2's comp: –32, 1's comp: –31

   (g) Overflow occurs when adding *n*-bit numbers and the result requires $n + 1$ bits for proper representation. You can tell that an overflow has occurred when the sum of two positive numbers is negative or the sum of two negative numbers is positive.

   A carry out of the last bit position does *not* indicate that an overflow has occurred.

6. (a)

BCD:	0001	1000	0111
excess-3:	0100	1011	1010
6-3-1-1:	0001	1011	1001
2-out-of-5:	00101	10100	10010

UNIT 1 Answers to Problems

1.1 (a) $2F5.40_{16} = 001011110101.01000000_2$

(b) $7B.2B_{16} = 01111011.00101011_2$

(c) $164.E3_{16} = 000101100100.11100011_2$

(d) $427.8_{16} = 010000100111.1000_2$

1.2 (a) $7261.3_8 = 3761.4_{10}$, $EB1.6_{16} = 3761.4_{10}$
(b) $2635.6_8 = 1437.8_{10}$, $59D.C_{16} = 1437.8_{10}$

1.3 3252.1002_6

1.4 (a) $5B1.1C_{16}$ (b) $010110110001.00011100_2 = 2661.070_8$
(c) 112301.0130_4 (d) 3564.6_{10}

1.5 (a) Add: 11001. Subtract: 0101. Multiply: 10010110.
(b) Add: 1010011. Subtract: 011001. Multiply: 11000011110.
(c) Add: 111010. Subtract: 001110. Multiply: 1100011000.

1.6 (a) $\begin{array}{r} 1111 \\ 11110100 \\ -\ 1000111 \\ \hline 10101101 \end{array}$ (b) $\begin{array}{r} 111\ \ 1 \\ 1110110 \\ -\ 111101 \\ \hline 0111001 \end{array}$ (c) $\begin{array}{r} 11111\ 1 \\ 10110010 \\ -\ \ 111101 \\ \hline 01110101 \end{array}$

1.7 2's complement:
(a) $\begin{array}{r} 010101 \\ +\ 001011 \\ \hline 100000 \end{array}$ (b) $\begin{array}{r} 110010 \\ +\ 100000 \\ \hline (1)010010 \end{array}$ (c) $\begin{array}{r} 100111 \\ +\ 010010 \\ \hline 111001 \end{array}$

OVERFLOW! (a) **OVERFLOW!** (b)

(d) $\begin{array}{r} 110100 \\ +\ 001101 \\ \hline (1)000001 \end{array}$ (e) $\begin{array}{r} 110101 \\ +\ 101011 \\ \hline (1)100000 \end{array}$

1's complement:
(a) $\begin{array}{r} 010101 \\ +\ 001011 \\ \hline 100000 \end{array}$

OVERFLOW!

(b) not assigned because −32 cannot be represented in 6 bits

(c) $\begin{array}{r} 100110 \\ +\ 010010 \\ \hline 111000 \end{array}$

(d) $\begin{array}{r} 110011 \\ +\ 001101 \\ \hline (1)\ 000000 \\ +\qquad 1 \\ \hline 000001 \end{array}$ (e) $\begin{array}{r} 110100 \\ +\ 101010 \\ \hline (1)\ 011110 \\ +\qquad 1 \\ \hline 011111 \end{array}$

OVERFLOW! (e)

1.8 For a word length of N, the range of 2's complement numbers that can be represented is -2^{N-1} to $2^{N-1} - 1$.
So, for a word length of 8, the range is -2^7 to $2^7 - 1$, or *−128 to 127*. Because 1's complement has a "negative zero" (11111111) in addition to zero (00000000), the values that can be represented range from $-(2^7 - 1)$ to $2^7 - 1$, or *−127 to 127*.

1.9

Dec.	7-3-2-1
0	0000
1	0001
2	0010
3	0011 or 0100
4	0101
5	0110
6	0111

3	6	5	9
0011 or 0100	0111	0110	1010

7 1000
8 1001
9 1010

UNIT 2 Study Guide Answers

2. (d) 1; 0; 1; 1 (e) 1, 1; 0, 0; 0; 1
3. (a) four variables, 10 literals (d) $F = (A'B)'$ (e) $F = (A + B')C$
   (f) *Circuit* should have two OR gates, three AND gates, and three inverters.
4. (b) $A, 0, 0, A; A, 1, A, 1$
6. (c) $Z = ABC$
7. (a) Sum of products
   Neither
   Product of sums (Here, A and B' are each considered to be separate terms in the product.)
   Neither
   (b) Fewer terms are generated.
   (c) $D[A + B'(C + E)] = D(A + B')(A + C + E)$
8. (a) $AE + B'C' + C'D$ (b) $C'DE + AB'CD'E$
10. (a) $a' + b + c$ (b) $ab'c'd$ (c) $a(b' + c')$
    (d) $(a + b)(c' + d')$ (e) $a' + b(c + d')$

UNIT 2 Answers to Problems

2.1 (a) $X(X' + Y) = XX' + XY = 0 + XY = XY$
(b) $X + XY = X(1 + Y) = X(1) = X$
(c) $XY + XY' = X(Y + Y') = X(1) = X$
(d) $(A + B)(A + B') = AA + AB' + AB + BB' = A + AB' + AB + BB'$
$= A(1 + B + B') + 0 = A(1) = A$

2.2

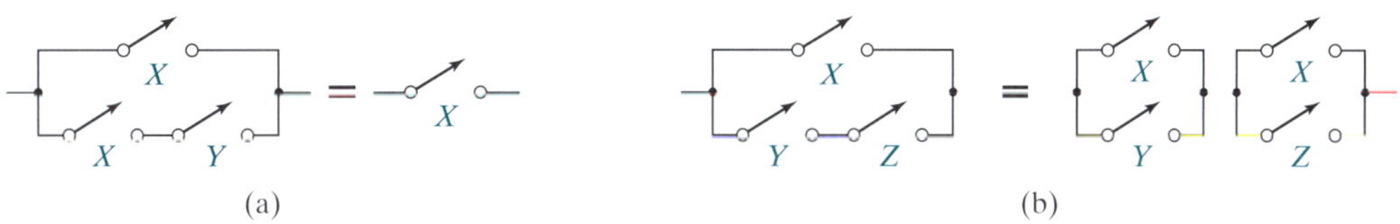

2.3 (a) 1 (Law 5)
(b) $CD + AB'E$ (Law 8D) (technically, we also used Law 3D)
(c) AF (Uniting (1)) (d) $C + D'B + A'$ (Uniting (1))
(e) $A'B + D$ (Absorption (2)) (f) $A + BC + DE + F$ (Elimination (3))

2.4 (a) $F = A + E + BCD$ (one AND gate and one OR gate with three inputs)
(b) $Y = A + B$

2.5 (a) $ACD' + BE$ (b) $A'B' + A'D' + C'B' + C'D'$

2.6 (a) $(A + C')(A + D')(B + C')(B + D')$ (b) $X(W + Z)(W + Y)$
(c) $(A' + E)(B + E)(C + E)(A' + D + F)(B + D + F)(C + D + F)$

(d) $Z(W' + X)(Q' + W' + Y)$ (e) $(A' + D')(C + D')$
(f) $(A + B + D)(A + C + D)(A + B + E)(A + C + E)$

2.7

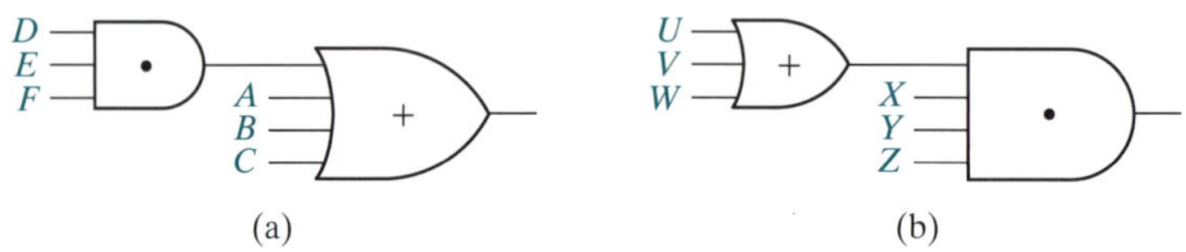

2.8 (a) $ABC + ABD'$ (b) $A'B' + A'CD'$ (c) $A'BC'$

2.9 (a) $F = A'B$ (b) $G = T'$

UNIT 3 Study Guide Answers

1. (b) $(b' + d)(b + a)(b + c)$ $(a + d)(b + d)(a' + b' + c)$
   (c) $w'y' + x'y'z' + xy + wyz$

5. (b) $A'B'C + BC'D' + AB'D' + BCD$
   (c) Add BCD; eliminate $A'BD, ABC$

UNIT 3 Answers to Problems

3.6 (a) $WY'X + WY'Z' + W'X'Y + W'X'Z$ (b) $A'D + AC$

3.7 (a) $(C' + D)(C + D' + B')$
(b) $(D' + A' + B')(D' + C + B')(D + A + C')(D + A' + B)$

3.8 $F = (AB) \oplus [(A \equiv D) + D] = A' + BD' + B'D$

3.9 No. Consider $A = 1, B = 1, C = 0$ or $A = 1, B = 0, C = 1$.

3.10 (a) $W'X + WY'Z + WYZ'$ (b) $BD + A'BC + AB' + AC'$
(c) $(A + C + D)(A' + C' + D')(B + C' + D)$

3.11 $AE' + AC' + B' + CD' + D'E'$

3.12 $A'CD'E + A'B'D' + ABCE + ABD = A'CD'E + BCD'E + A'B'D' + ABCE + ABD = A'B'D' + ABD + BCD'E$

UNIT 4 Study Guide Answers

2. (d) $ab'c'd$ (e) $a + b + c' + d'$
   (g) $(a + b' + c)(a' + b + c')(a' + b' + c)(a' + b' + c')$

3. (c) $m_0 + m_1 + m_3 + m_4 = \Sigma m(0, 1, 3, 4)$ $M_2M_5M_6M_7 = \Pi M(2, 5, 6, 7)$

4. (b) m_{19} (c) $A'BCD'E$
   (e) M_{19} (f) $(A + B' + C' + D + E')$

5. (a) 65536
   (d) $(a_0m_0 + a_1m_1 + a_2m_2 + a_3m_3)(b_0m_0 + b_1m_1 + b_2m_2 + b_3m_3) = \ldots = a_0b_0m_0 + a_1b_1m_1 + a_2b_2m_2 + a_3b_3m_3$
   (f) $f = \Pi M(2, 5, 6)$ $f' = \Sigma m(2, 5, 6) = \Pi M(0, 1, 3, 4, 7)$

6. (b) $\Sigma m(0, 5) + \Sigma d(1, 3, 4)$

UNIT 4 Answers to Problems

4.1 (a) U: Safe unlocked, J: Mr. Jones present, E: Mr. Evans present, B: Normal business hours, S: Security guard present

$$U = (J + E)BS$$

(b) O: Wear overshoes, A: You are outside, R: Raining heavily, S: Wearing suede shoes, M: Mother tells you to

$$O = ARS + M$$

(c) L: Laugh at joke, F: It is funny, G: Good taste, O: Offensive, P: Told by professor

$$L = FGO' + PO'$$

(d) D: Elevator door opens, S: Elevator is stopped, F: Level with floor, T: Timer expired, B: Button pressed

$$D = SFT' + SFB$$

4.2 (a) $Y = A'B'C'D'E' + AB'C'D'E' + ABC'D'E'$ or $Y = C'$
(b) $Z = ABC'D'E' + ABCD'E' + ABCDE'$ or $Z = BE'$

4.3 $F_1 + F_2 = \Sigma\, m(0, 3, 4, 5, 6, 7)$; General rule: $F_1 + F_2$ is the sum of all minterms which are present in either F_1 or F_2, because $F_1 + F_2 = \Sigma\, a_i m_i + \Sigma\, b_i m_i = \Sigma\, (a_i + b_i) m_i$

4.4 (a) 16
(b) $F(x,y) = 0, x'y', x'y, x', xy', y', x'y + xy', x' + y', xy, x'y' + xy, y, x' + y, x, x + y', x + y, 1$

4.5

A	B	C	D	E	F	
0	0	0	1	1	X	
0	0	1	X	X	1	
0	1	0	X	X	X	
0	1	1	X	X	1	or 1 1 X
1	0	0	X	0	0	
1	0	1	X	X	1	
1	1	0	X	X	X	
1	1	1	X	0	0	or 0 X 0

4.6 (a) $F = A'B' + AB (d_1 = 1, d_5 = 0)$ (b) $G = C (d_2 = 0, d_6 = 0)$

4.7 (a) $\Sigma\, m\,(1, 2, 4)$ (b) $\Pi\, M\,(0, 3, 5, 6, 7)$

4.8 (a) $F = A'B'C'D' + A'B'C'D + A'B'CD' + A'B'CD + A'BC'D' + A'BC'D + A'BCD' + AB'C'D' + AB'C'D + ABC'D'$
$F = \Sigma\, m(0, 1, 2, 3, 4, 5, 6, 8, 9, 12)$
(b) $F = (A + B' + C' + D')(A' + B + C' + D)(A' + B + C' + D')(A' + B' + C + D')(A' + B' + C' + D)(A' + B' + C' + D')$
$F = \Pi\, M(7, 10, 11, 13, 14, 15)$

4.9 (a) $F = \Sigma\, m(0, 1, 4, 5, 6)$ (b) $F = \Pi\, M(2, 3, 7)$
(c) $F' = \Sigma\, m(2, 3, 7)$ (d) $F' = \Pi\, M(0, 1, 4, 5, 6)$

4.10 (a) $F = \Sigma\, m(1, 4, 5, 6, 7, 10, 11)$
(b) $F = \Pi\, M(0, 2, 3, 8, 9, 12, 13, 14, 15)$
(c) $F' = \Sigma\, m(0, 2, 3, 8, 9, 12, 13, 14, 15)$
(d) $F' = \Pi\, M(1, 4, 5, 6, 7, 10, 11)$

4.11 (a) $d_i = x_i \oplus y_i \oplus b_i$

$b_{i+1} = b_i x_i' + x_i' y_i + b_i y_i$

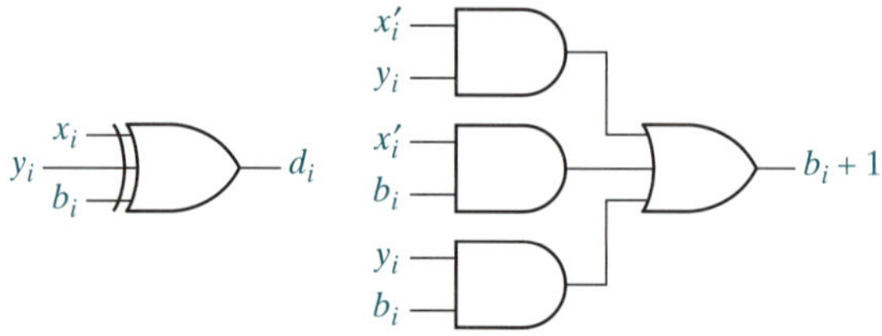

(b) $d_i = s_i$, b_{i+1} is the same as c_{i+1} with x_i replaced by x_i'

4.12

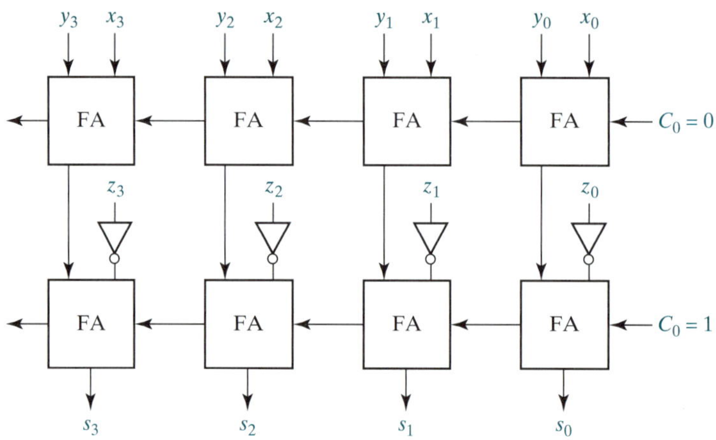

UNIT 5 Study Guide Answers

3. (d) 6, 10, 12, 15; 0, 12, 9, 10
   (g) $f_1 = a'b + bc' + a'cd + ac'd$ $\quad f_2 = b'c + cd + a'bd + ab'd'$
4. (a) $a'b'd'$, $b'c'd'$, $ac'd'$, $ac'd$, also $a'b'cd$, and all the other minterms.
   (b) $AB'C'$ and $AC'D$ are prime implicants.
5. (a) 4 (c) We cannot determine if $B'C'$ is essential.
   (f) Yes (i) $A'D'$ because of m_4, $B'D'$ because of m_{10}
6. (b) $A'D'$ is not essential because all of its minterms are covered by other prime implicants. BC' is essential because of m_{13}. $B'CD$ is essential because of m_{11}. Minimum sum $= B'CD + BC' + BD' + A'B'$.
   (d) $A'C' + ACD + \{AB \text{ or } BC'\}$
8. (a) $F = AB'D' + B'D'E' + A'BDE$ (b) 8,16,25,26,28
   (d) $P_1 + P_2 + P_3 + P_4 + BCDE + AC'E$
   (f) $AC'E' + A'DE + ACE + B'CE + (AB'C \text{ or } ADE' \text{ or } ACD \text{ or } AB'E')$

UNIT 5 Answers to Problems

5.3 (a) $f = bc' + a'c' + ab'c$ (b) $f = e'f' + d'e' + d'f'$
(c) $f = r' + t'$ (d) $f = y + x'z + xz'$

5.4 (a)

CD \ AB	00	01	11	10
00	1 (0)	1 (4)	1 (12)	1 (8)
01	0 (1)	0 (5)	1 (13)	0 (9)
11	1 (3)	0 (7)	1 (15)	1 (11)
10	1 (2)	1 (6)	1 (14)	1 (10)

(b) $F = D' + B'C + AB$
(c) $F = (A + B' + D')(B + C + D')$

5.5 (a)

C_1	C_2	X_1	X_2	Z
0	0	0	0	0
0	0	0	1	1
0	0	1	0	1
0	0	1	1	1
0	1	0	0	0
0	1	0	1	1
0	1	1	0	1
0	1	1	1	0
1	0	0	0	0
1	0	0	1	0
1	0	1	0	0
1	0	1	1	1
1	1	0	0	1
1	1	0	1	0
1	1	1	0	0
1	1	1	1	1

(b) $Z = C_1'X_1'X_2 + C_1'X_1X_1' + C_1X_1X_2 + C_1C_2X_1'X_2' + \{C_1'C_2'X_2$ **or** $C_1'C_2'X_1$ **or** $C_2'X_1X_2\}$

5.6 (a) $f = \underline{a'd} + \underline{a'b'c'} + \underline{b'cd} + \underline{abd'} + \{a'bc$ **or** $bcd'\}$
$a'd \rightarrow m_5; a'b'c' \rightarrow m_0; b'cd \rightarrow m_{11}; abd' \rightarrow m_{12}$
(b) $f = \underline{bd} + \underline{a'c} + \underline{b'd'} + \{a'b$ **or** $a'd'\}$
$bd \rightarrow m_{13}, m_{15}; a'c \rightarrow m_3; b'd' \rightarrow m_8, m_{10}$
(c) $f = \underline{c'd'} + \underline{a'd'} + \underline{b'}$
$c'd' \rightarrow m_{12}; a'd' \rightarrow m_6; b' \rightarrow m_{10}, m_{11}$

5.8 (a) $f = a'bc' + ac'd + b'cd'; f = (b' + c')(c' + d')(a + b + c)(a' + c + d)$
(b) $f = a'b'd + bc'd' + cd; f = (b + d)(b' + d')(a' + c)\{(b' + c')$ **or** $(c' + d)\}$

5.10 (a) $C'D'E' \rightarrow m_{16}, m_{24}; A'CE' \rightarrow m_{14}; ACE \rightarrow m_{31}; A'B'DE \rightarrow m_3$
(b) $A'B'DE, A'D'E', CD'E, A'CE', ACE, A'B'C, B'CE, C'D'E', A'CD'$

5.11 $f = \underline{(a + b + c + d)}(\underline{a + b' + e'})(\underline{a' + d' + e})(\underline{a' + b + c'})(\underline{a + c + e'})$
$(c + d + e')\{(a' + b' + c + d)$ **or** $(a' + b' + c + e)\}$

5.12 (a) $F = \Pi\, M(0, 1, 9, 12, 13, 14)$

$$F = (A + B + C + D)(A + B + C + D')(A' + B' + C + D)$$
$$(A' + B' + C + D')(A' + B' + C' + D)(A' + B + C + D')$$

(b) $F' = A'B'C' + ABD' + AC'D$
(c) $F = (A + B + C)(A' + B' + D)(A' + C + D')$

5.13 $F = A'C' + B'C + ACD' + BC'D$
Minterms $m_0, m_1, m_2, m_3, m_4, m_5, m_7, m_8, m_{10}$, and m_{11} can be made don't-cares individually and will not change the given expression.

UNIT 6 Study Guide Answers

2. (f) (2,6)
3. (a) $m_0 - a'b'c'$ $(m_0,m_1) - a'b'$
$m_1 - a'b'c$ $(m_1,m_5) - b'c$ prime
$m_5 - ab'c$ $(m_5,m_7) - ac$
$m_7 - abc$
(d) $A'B'C'$ and ABC are not prime implicants.
4. (b) $a'c'd'$, bc', $ab'c$
5. (b) $F = bd + a'b, F = bd + bc', F = bc' + a'b, F = a'b + c'd$

UNIT 6 Answers to Problems

6.2 (a)

$a'c'd$	(1,5)
$b'c'd$	(1,9)
$a'bd$	(5,7)
$ab'd$	(9,11)
abd'	(12,14)
bcd	(7,15)
acd	(11,15)
abc	(14,15)

(b)

$a'b'c'$	(0,1)
$b'c'd'$	(0,8)
$ab'd'$	(8,10)
acd'	(10,14)
$a'd$	(1,3,5,7)
bc	(6,7,14,15)

6.3 (a) $f = a'c'd + ab'd + abd' + bcd$ **or** $f = b'c'd + a'bd + abd' + acd$

(b) $$f = a'd + bc + \begin{Bmatrix} a'b'c' + ab'd' \\ \text{OR} \\ b'c'd' + acd' \\ \text{OR} \\ b'c'd' + ab'd' \end{Bmatrix}$$

6.4 $f = b'cd' + bc' + a'd + (a'b \text{ OR } a'c)$ [1 other solution]

6.5 Prime implicants: ab, $c'd$, ac', bc', ad, bd
$F = ab + c'd$ **or** $F = ab + ac'$ **or** $F = ab + ad$ **or** $F = ac' + ad$ **or** $F = ac' + bd$ **or** $F = ad + bc'$

6.6 (a) $F = A'B + A'C'D' + AB'D + A'C'E + BCDE$
(b) $Z = A'B' + ABD + EB'C' + EA'C + FAB + GBD$ [several other solutions]

UNIT 7 Study Guide Answers

1. (b) Z_1: six gates, 13 inputs, four levels Z_2: five gates, 11 inputs, five levels
(d)

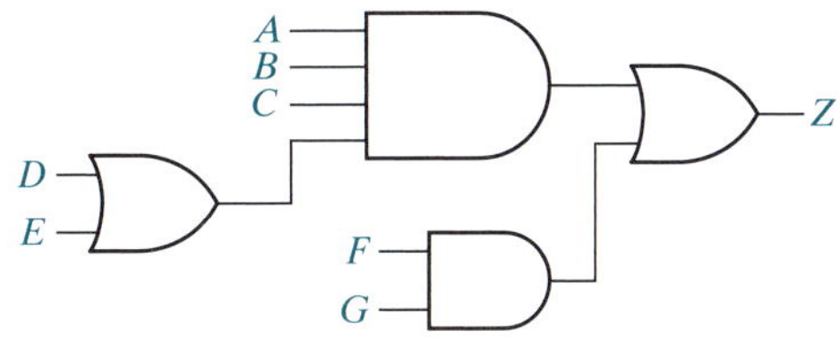

2. (a) 0; 1; 1,1,1; 0,0,0

6. (a) (1) No (2) Yes (3) No (b) because C requires no gate
(c) five gates, 10 inputs; using common gate: four gates, nine inputs
(d) $F_1 = \underline{a'cd} + \underline{acd} + ab'c'$; $F_2 = \underline{a'cd} + \underline{bcd} + a'bc' + acd'$;
$F_3 = \underline{bcd} + \underline{acd} + a'c'd$

UNIT 7 Answers to Problems

7.1 (a) $f = (a + b)(a' + b')(a + c + d')(a' + c' + d')$
OR $f = (a + b)(a' + b')(a + c + d')(b + c' + d')$
OR $f = (a + b)(a' + b')(b' + c + d')(a' + c' + d')$
OR $f = (a + b)(a' + b')(b' + c + d')(b + c' + d')$
(b) $f = a'b(c + d') + ab'(c' + d')$

7.2 (a) $Z = (C' + E')(AD + B) + A'D'E'$ (four levels, 13 inputs)
(b) $Z = (B(C + D) + A)(E + FG)$ (four levels, 12 inputs)

7.3 AND-OR: $F = a'bd + ac'd$; OR-AND: $F = d(a' + c')(a + b)$

NAND-NAND

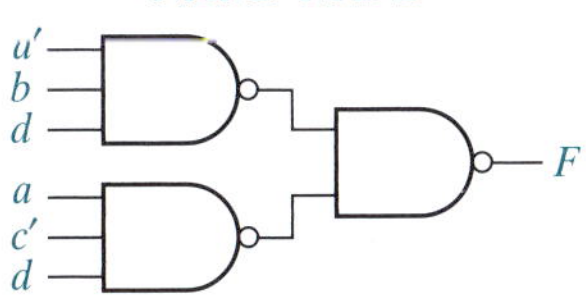

OR-NAND

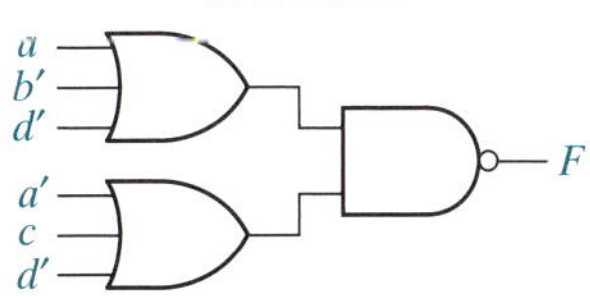

NOR-OR

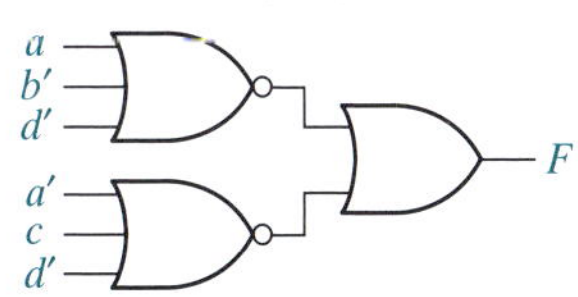

NOR-NOR

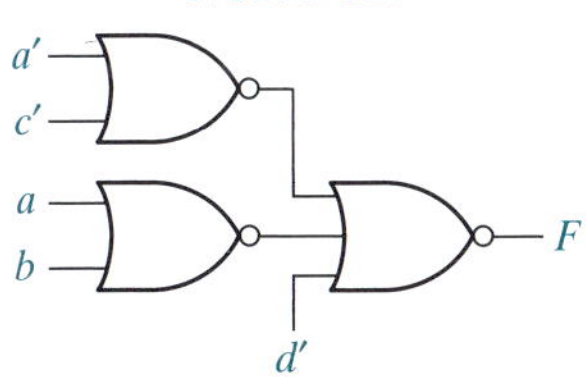

AND-NOR

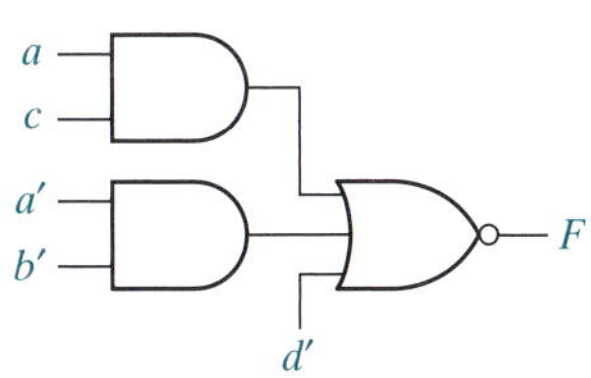

NAND-AND

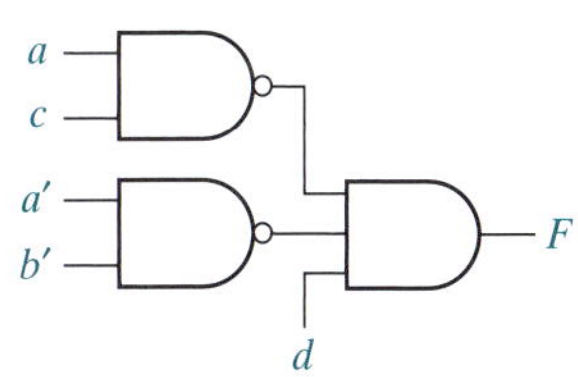

7.4 $F = BC'(A + D) + AB'C$ (three levels, four gates, 10 inputs)

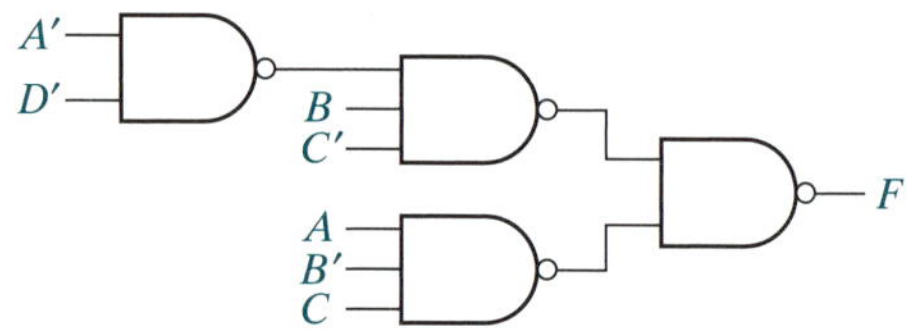

7.5 $Z = (A + C + D)(A' + B'C'D')$ (convert circuit to four NOR gates)

7.6 $Z = A(BC + D) + C'D$ (convert circuit to five NAND gates)

7.7 $Z = E(A + B(D + CF))$ (convert circuit to five NOR gates)

7.8 (a)

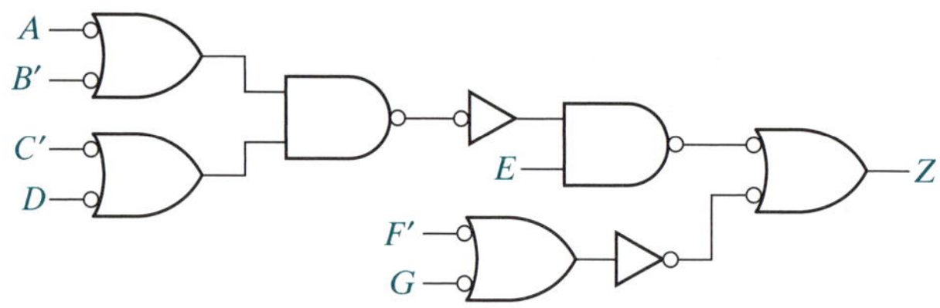

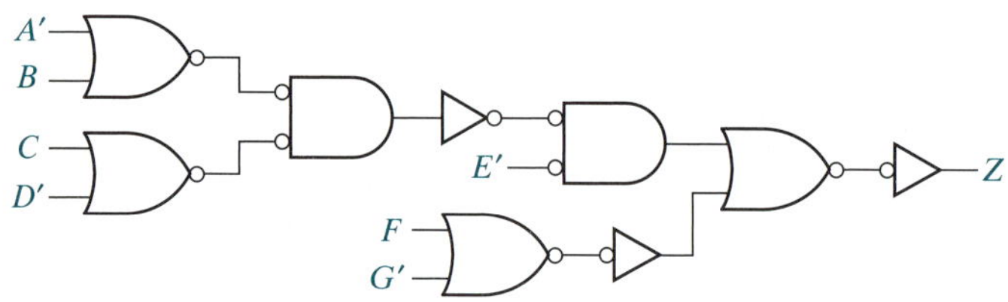

(b)

7.9 $f_1 = \underline{acd'} + ad + \underline{a'b'd}$; $f_2 = a'd' + \underline{a'b'd} + \underline{acd'}$ (six gates, 16 inputs)

7.10 $f_1 = ab'd + \underline{b'cd} + a'bd'$ $f_2 = \underline{ab'c} + b'cd' + bc'd' + \{ac'd'$ **or** $ab'd'\}$
$f_3 = \underline{ab'c} + \underline{b'cd} + a'bc$ (11 gates, 34 inputs)

7.11 $F_1 = (a + c)(a + b')\underline{(a' + b' + c)}\,\underline{(a' + b + c')}$
$F_2 = (a + c')(b' + c + d)\underline{(a' + b' + c)}\,\underline{(a' + b + c')}$ **or**
$(a + c')(a + b' + d)\underline{(a' + b' + c)}\,\underline{(a' + b + c')}$
(eight gates minimum, 23 gate inputs)

7.12 $f_1 = \underline{(a + b + c)}(b' + d)$ $f_2 = \underline{(a + b + c)}\,\underline{(b' + c + d)}(a' + c)$
$f_3 = \underline{(b' + c + d)}(a + c)(b + c')$

7.13 (a) Replace all gates in the AND-OR *circuit* which corresponds to Equations (7-25(b)) with NAND gates. Invert the c input to the f_2 output gate.
(b) Replace all gates in Answer 7.12 with NOR.

UNIT 8 Study Guide Answers

3.

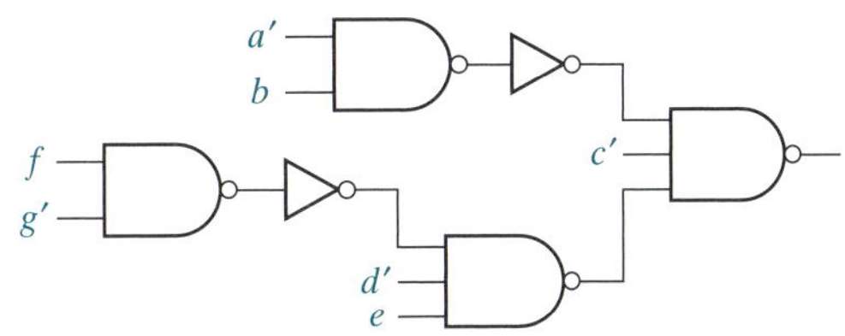

4. (a) Factor out the expression such that the number of inputs on each gate is less than or equal to the maximum allowed. This will result in the addition of more levels of logic.
   (b) Yes.
   (c) Even if the two-level expressions had common terms, most of these common terms would be lost when the expressions are factored.

5. (a) B' goes to 0 at 80 ns. Z goes to 1 at 50 ns and goes to 0 at 110 ns.

6. (a) y_1 goes to 1 at 15 ns. y_2 goes to 0 at 30 ns. Z goes to 1 at 25 ns and, then, goes to 0 at 40 ns.
   (c) A pair of adjacent 1's corresponding to $a'bc$ and abc are not in the same loop in the Karnaugh map, but $a'bc$ and $a'bc'$ are both in $a'b$. Without the map, when $b = c = 1$ and a changes from 0 to 1, $a'b$ may go to 0 before ac becomes 1. But when $a = 0, b = 1$, and c changes from 1 to 0, $a'b$ remains 1.
   (g) The application of DeMorgan's laws to convert a circuit from one form to another will not introduce any hazards.

7. (b) If $G = 0$, gate 4 is faulty. If $G = 1$, gate 1 is faulty.

UNIT 8 Answers to Problems

8.1

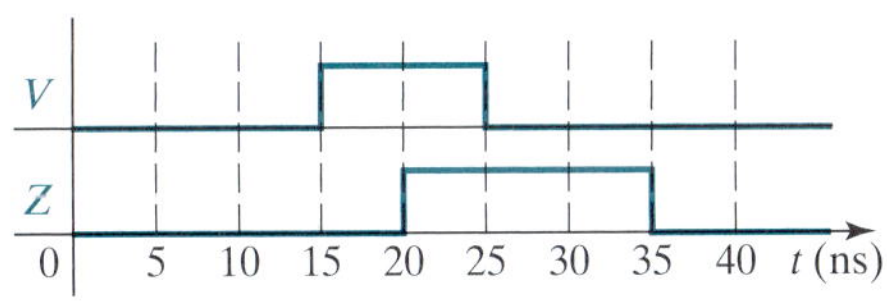

8.2 (a) $F = A'C'D' + BC'D + AC$ (hazards are 1101↔1111 and 0100↔0101 [static 1])
OR $F = (A' + C + D)(B + C + D')(A + C')$ (hazards are 0001↔0011 and 1000 ↔ 1001 [static 0])
(b) $F^t = A'C'D' + BC'D + AC + \underline{A'BC'} + \underline{ABD}$
(c) $F^t = (A' + C + D)(B + C + D')(A + C')\underline{(A' + B + C)}\,\underline{(A + B + D')}$

8.3 (a) Glitch in output of G occurs between 6 ns and 7 ns (static 1-hazard).
(b) Modified equation to avoid hazards: $G = A'C'D + BC + A'BD$

8.4 $A = 1$ $E = \text{X}$
$B = \text{Z}$ $F = 0$
$C = \text{X}$ $G = 0$
$D = 1$ $H = \text{X}$

8.5 Gate 3 is connected incorrectly or is malfunctioning.

UNIT 9 Study Guide Answers

2. (a)

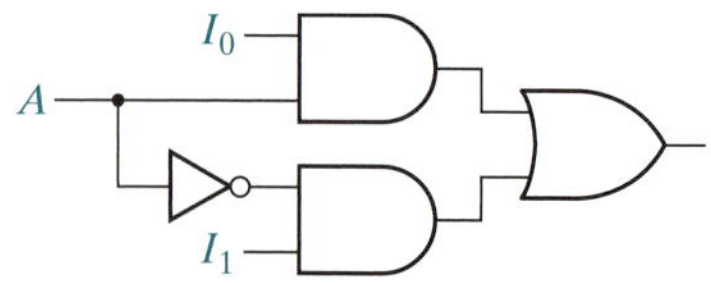

(b) $Z = A'C'I_0 + A'CI_1 + AC'I_2 + ACI_3$
(c) Before C changes, $Z = I_4$, and after C changes, $Z = I_5$.
(d)

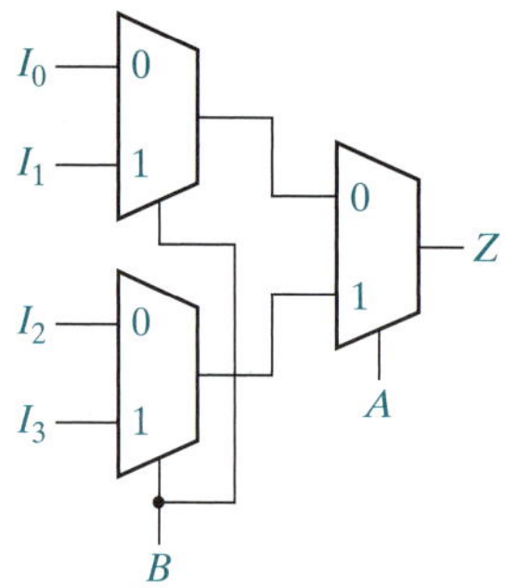

(e) MUX inputs: $I_0 = B$, $I_1 = B'$, control $= A$

3. (f) AND-gate inputs are $A'B'$, $A'B$, AB', and AB

4. (a) Inputs BCD; $A = 0$

5. (b) 32 words $\times$ 4 bits; 1024×8 (c) 16 words $\times$ 5 bits; 16×10

6. (a) Four inputs, seven terms, three outputs
(b) Four inputs, four terms, three outputs
(c)

A	B	C	D	F_1	F_2	F_3
1	1	-	-	1	0	1
1	-	1	1	1	1	0
1	1	0	-	0	1	0
0	-	1	1	0	1	1

(f) When $ABC = 010$, $F_0F_1F_2F_3 = 0111$.

8. (c) $f = c'(d' + a) + c(a'b' + bd)$
(d)

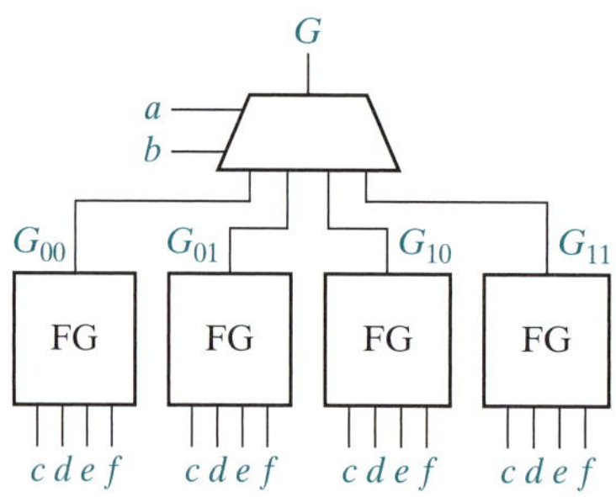

UNIT 9 Answers to Problems

9.1 (a)

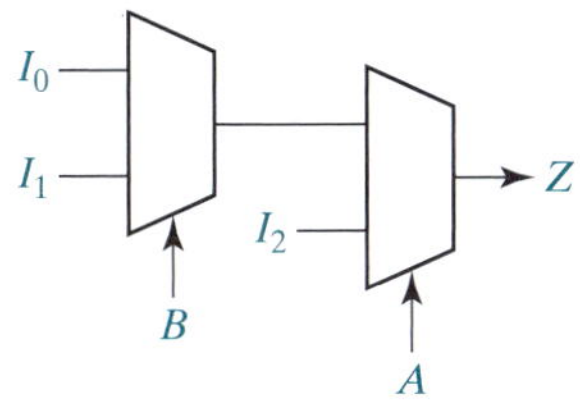

(b)

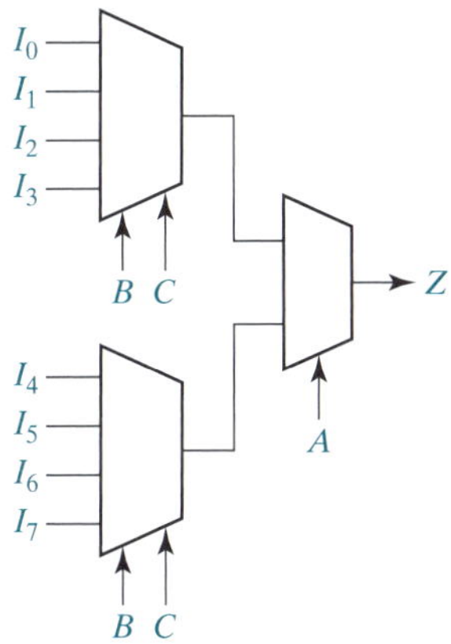

(c)

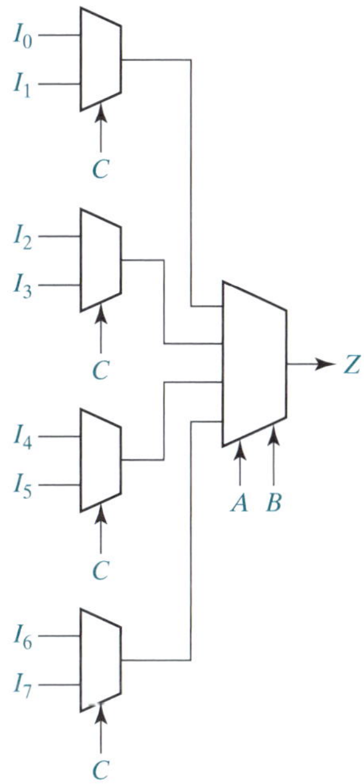

9.2

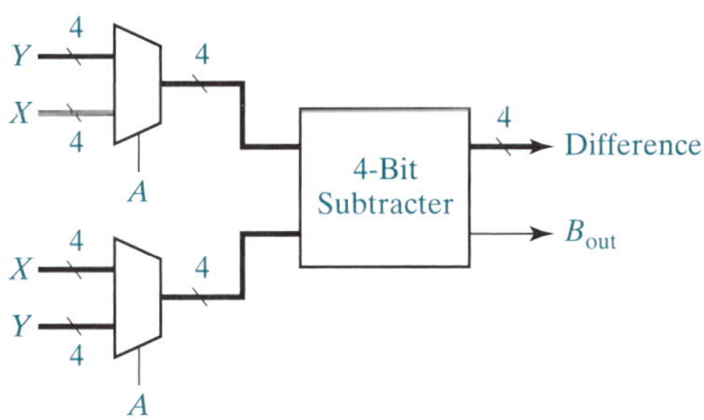

9.3

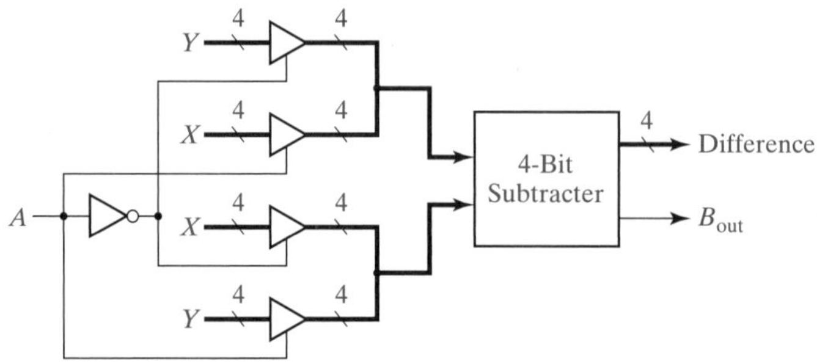

9.4 (a)

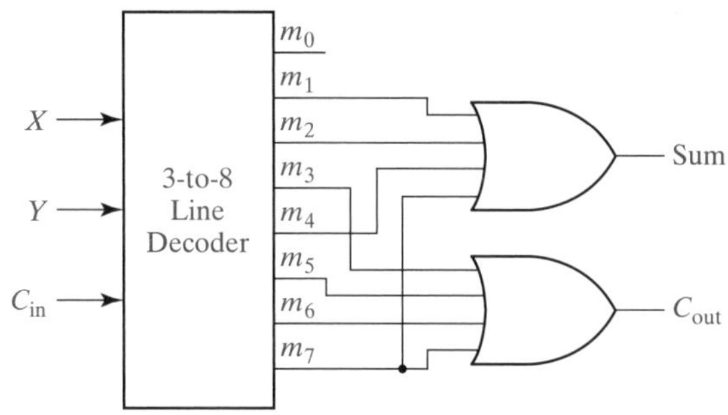

(b)

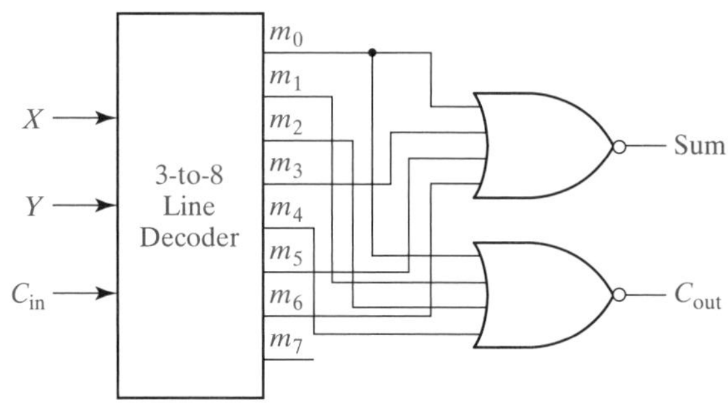

9.5

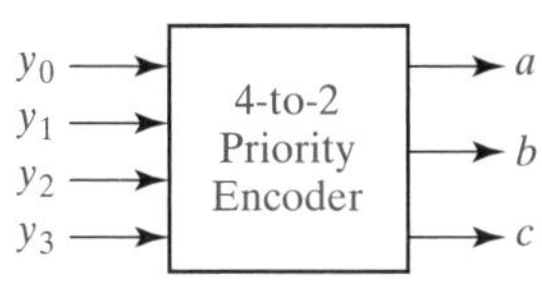

$a = y_2 + y_3$

$b = y_1 y_2' + y_3$

$c = y_0 + y_1 + y_2 + y_3$

9.6

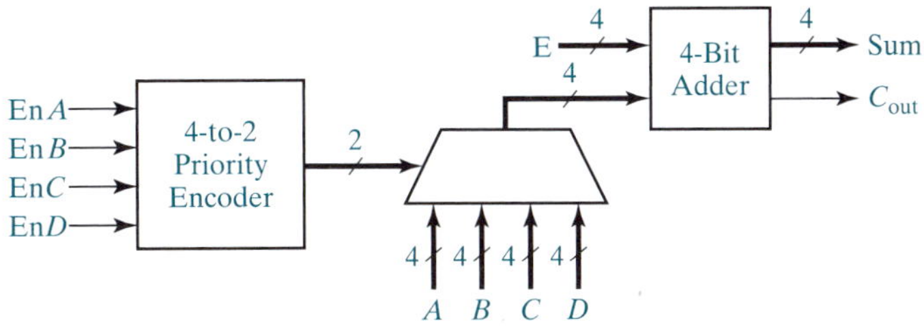

9.7 Block diagram for a Gray code adder:

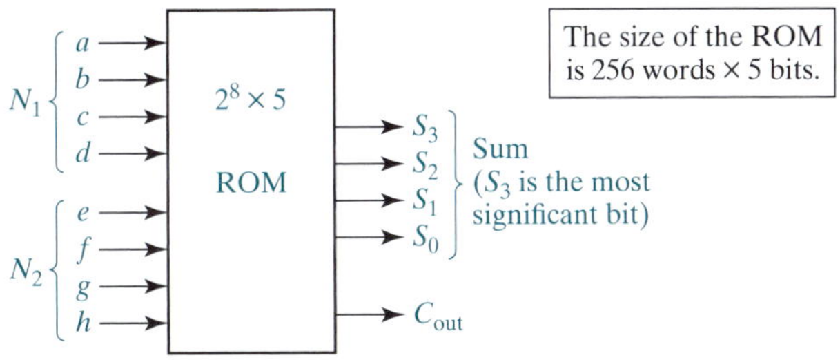

Partial Truth Table

	a	b	c	d	e	f	g	h	S_3	S_2	S_1	S_0	C_{out}
(0 + 0 = 0)	0	0	0	0	0	0	0	0	0	0	0	0	0
(1 + 2 = 3)	0	0	0	1	0	0	1	1	0	0	1	0	0
(5 + 7 = 12)	1	1	1	0	1	0	1	1	0	0	1	1	1
(8 + 9 = 17)	1	0	0	1	1	0	0	0	1	0	1	1	1

9.8 (a)

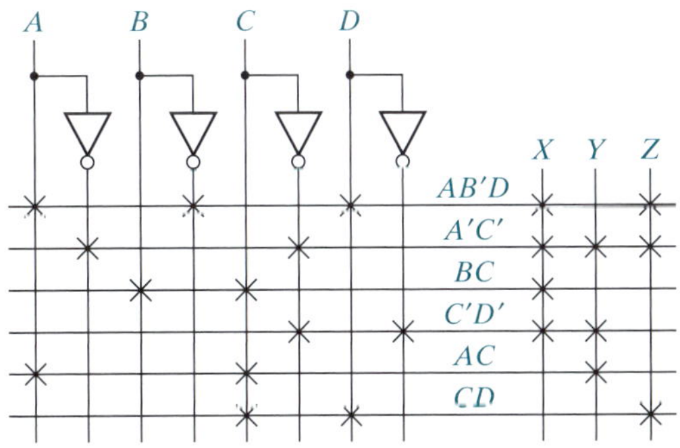

(b) Truth Table for the ROM

A	*B*	*C*	*D*	*X*	*Y*	*Z*
0	0	0	0	1	1	1
0	0	0	1	1	1	1
0	0	1	0	0	0	0
0	0	1	1	0	0	1
0	1	0	0	1	1	1
0	1	0	1	1	1	1
0	1	1	0	1	0	0
0	1	1	1	1	0	1
1	0	0	0	1	1	0
1	0	0	1	1	0	1
1	0	1	0	0	1	0
1	0	1	1	1	1	1
1	1	0	0	1	1	0
1	1	0	1	0	0	0
1	1	1	0	1	1	0
1	1	1	1	1	1	1

9.9

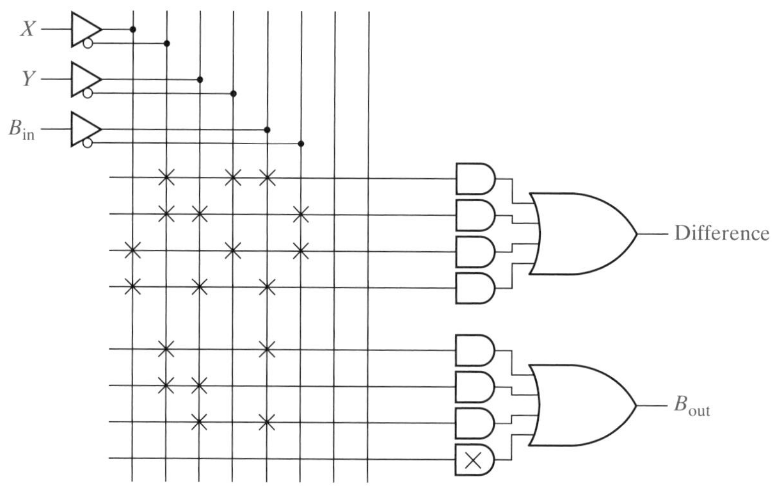

9.10 (a) $A_4 = W' + X'Y' \quad A_3 = WX'Y' \quad A_2 = W'X + XZ + XY$
$A_1 = W'Y + WXY'Z' + YZ \quad A_0 = W'Z + WXZ' + X'Y'Z + WYZ'$

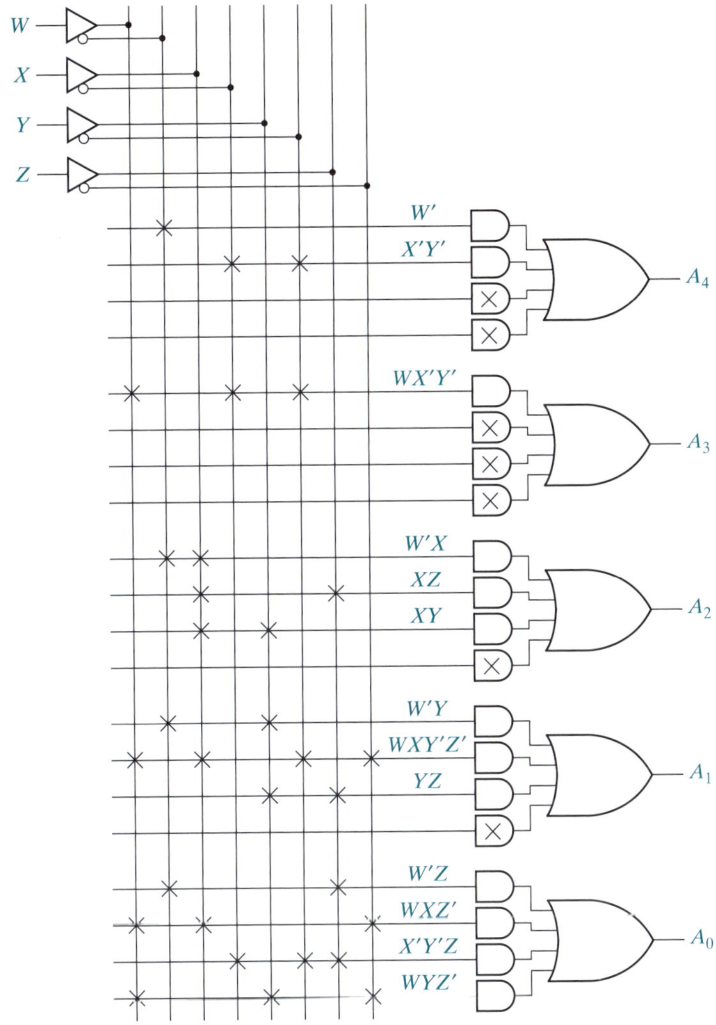

(b)

W	X	Y	Z	A_4	A_3	A_2	A_1	A_0
0	-	-	-	1	0	0	0	0
-	0	0	-	1	0	0	0	0
1	0	0	-	0	1	0	0	0
0	1	-	-	0	0	1	0	0
-	1	-	1	0	0	1	0	0
-	1	1	-	0	0	1	0	0
0	-	1	-	0	0	0	1	0
1	1	0	0	0	0	0	1	0
-	-	1	1	0	0	0	1	0
0	-	-	1	0	0	0	0	1

(continued)

1	1	-	0	0	0	0	0	1
-	0	0	1	0	0	0	0	1
1	-	1	0	0	0	0	0	1

9.11 (a) Not inverting, three AND gates. Inverting, $F = ac + b'c'd$, two AND gates.
(b) Not inverting, two AND gates. Inverting, $F = ad + ac + bd + bc$, four AND gates.

9.12 (b)

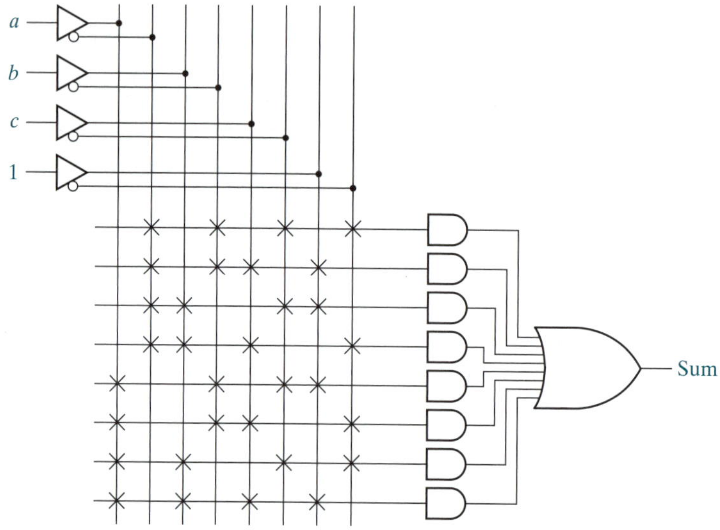

9.13 $F = b'(ade' + a'cd'e) + b((c'd'e + a'cd'e) + ac'de')$

UNIT 10 Study Guide Answers

1. (b) Both statements execute at 5 ns. C and D are updated at $5 + \Delta$ ns.
   (c) M <= **not** M **after** 5 ns;

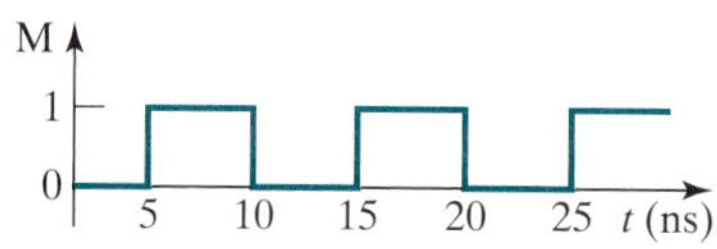

   (d) A <= (**not** B **and** C) or (B **and** **not** C);

2. (a)

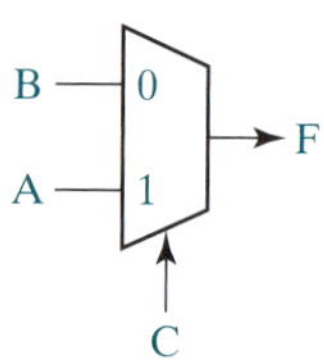

   (b) F <= '1' **when** A&B = "00" **else** '0' **when** A&B = "01" **else** C;

(c) AB <= A&B;
```
with AB select
  F <= '1' when "00", '0' when "01", C when "10", C when "11";
```

3. (c) Change all (3 **downto** 0) to (4 **downto** 0). Change (3 **downto** 1) to (4 **downto** 1). Add another instance of a full adder –
FA4: FullAdder **port map** (A(4), B(4), C(4), Co, S(4));
Change Co in FA3 to C(4).

(f)
```
architecture ckt of fig8_5 is
signal G1: bit;
begin
  G1 <= A and B after 20 ns;
  G2 <= G1 nor C after 20 ns;
end ckt;
```

5. (a) **not** (A&B **xor** "10") **not** (A&B) **xor** "10"
(b) The given statement will keep executing over and over again.

7. (a) A = '1', B = 'X', C = '0', D = '1', E = 'X', F = 'Z'
(b) If *F* is of type bit, compiler will log an error.
If *F* is std_logic, it will be 0 for 2 ns and, then, become X.
(c) Addout 10011, Sum 0011, Cout = 1
(d)
```
Addout <= ('0' & A) + ("000" & B);
Sum <= Addout(5 downto 0);
Cout <= Addout(6);
```
(e)

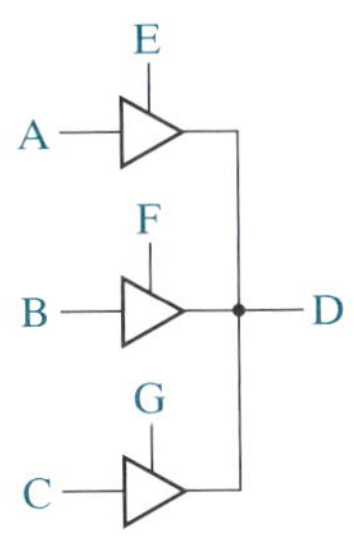

UNIT 10 Answers to Problems

10.1 (a) F <= **not** A **and** B **and** C; G <= D **and not** E; N <= F **xor** G; I <= **not** N;
(b) I <= **not** ((**not** A **and** B **and** C) **xor** (D **and not** E));

10.2

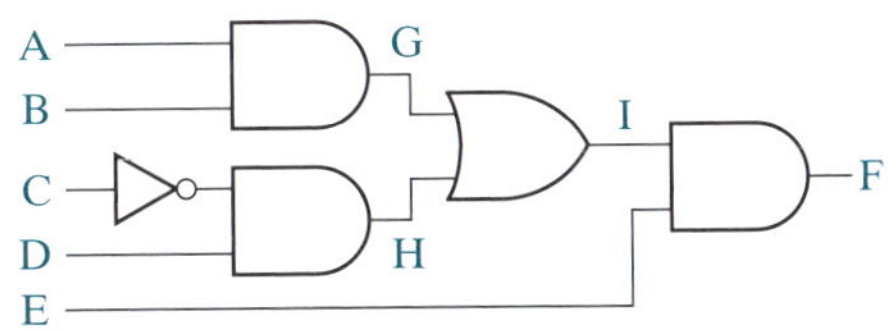

10.3 (a)

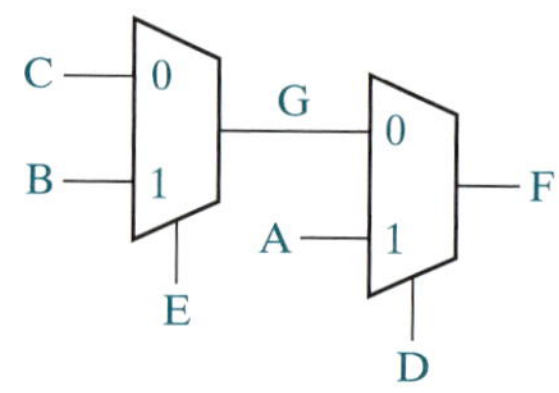

(b)

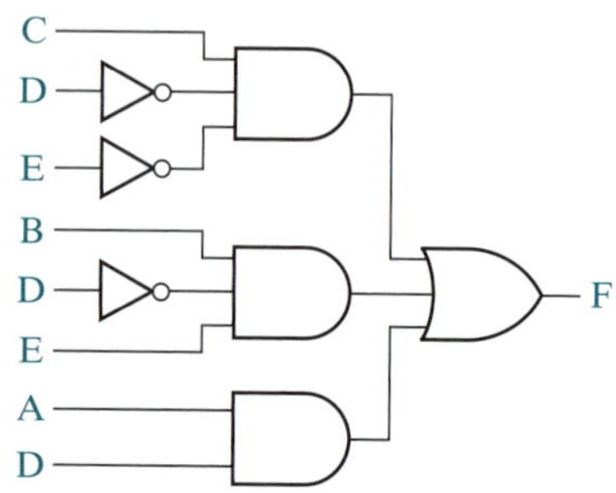

10.4

```
entity quad_mux is
   port (X, Y: in bit_vector(3 downto 0);
      A: in bit;
      Z: out bit_vector(3 downto 0));
end quad_mux;
architecture equations of quad_mux is
begin
   Z <= X when A = '0' else Y;
end equations;
```

10.5

```
entity ROM is
   port (A, B, Cin: in bit;
      Sum, Cout: out bit);
end ROM;
architecture table of ROM is
   type ROM8_2 is array(0 to 7) of bit_vector(1 downto 0);
   constant ROM1: ROM8_2 := ("00", "01", "01", "10", "01", "10", "10","11");
   signal index: integer range 0 to 7;
   signal S: bit_vector(1 downto 0);
   begin
      index <= vec2int(A&B&Cin);
      S <= ROM1(index);
      Sum <= S(0);
      Cout <= S(1);
end table;
```

10.6 (a) $F = 000001101$

(b) The expression evaluates to TRUE.

10.7

```
entity average is
   port (a, b, c, d: in std_logic_vector(15 downto 0);
   f: out std_logic_vector(15 downto 0));
end average;
architecture behavioral of average is
   signal sum: std_logic_vector(17 downto 0);
   begin
   sum <= ("00" & a) + b + c + d;
   f <= sum (17 downto 2) + sum (1);
end behavioral;
```

10.8

```
Bus <= A when EnA = '1' else "ZZZZ";
Bus <= B when EnB = '1' else "ZZZZ";
Bus <= C when EnC = '1' else "ZZZZ";
Bus <= D when EnD = '1' else "ZZZZ";
Addout <= '0' & E + Bus;
Sum <= Addout(3 down to 0);
Cout <= Addout(4);
```

10.9 (a)

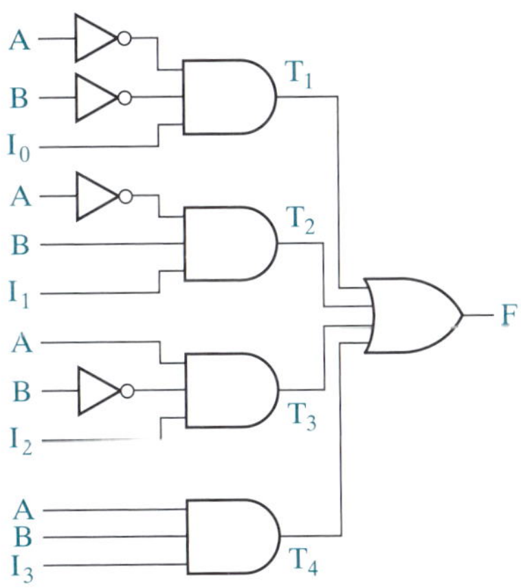

(b)

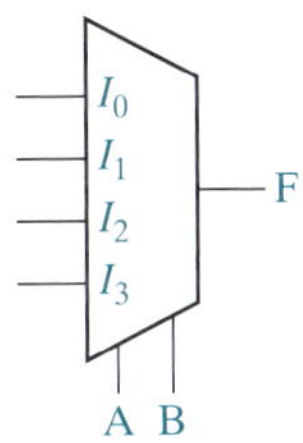


```
sel <= A&B;
with sel select
F <= I0 when "00", I1 when "01",
      I2 when "10", I3 when "11";
```

UNIT 11 Study Guide Answers

1. Left inverter has a 1 output; right inverter has a 0 output.
2. (b) $P = Q = 0$ (c) S and R cannot both be 1 simultaneously.
3. (c)

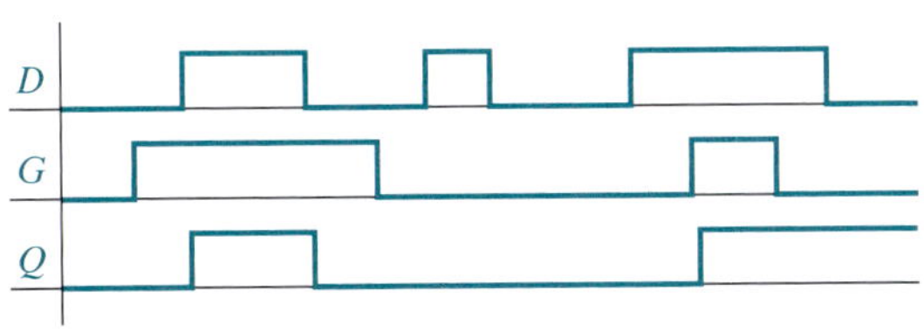

4. (b) Q changes to 1 after first rising clock edge and back to 0 after third rising clock edge.
   (d) Hold time violation (D is not stable for 2 ns after second falling clock edge.)
   (e) = 7 ns.
5. (c) For a rising-edge-triggered flip-flop, the value of the inputs is sensed at the rising edge of the clock, and inputs can change when the clock is low. For a master-slave flip-flop, if the inputs change when the clock is low, the flip-flop outputs may be incorrect.
6. (c)

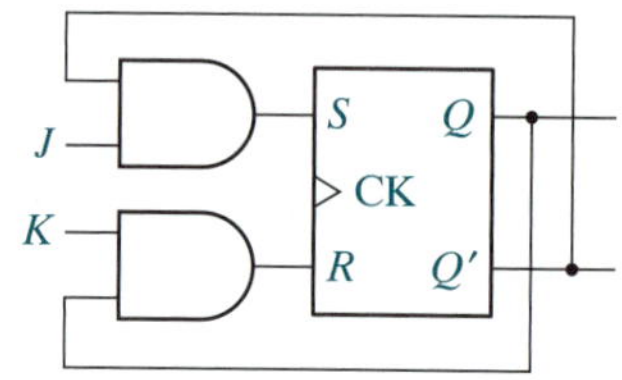

7. (b) Q changes its value at times 1 and 2.
8. (b)

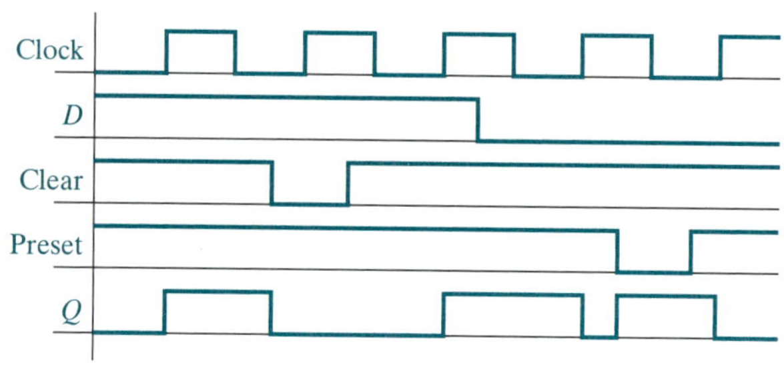

   (c) If CLK = 1, it will produce a falling edge at the clock input of the D flip-flop, causing the output to change. If CLK = 0, only the rising edge is affected, so the output does not change at the wrong time.
   En cannot be changed when the clock is 1.
   The flip-flops in Figures 11-31(b) and (c) can only change on the falling edge of the clock.
   (d)

CK	D	CE	ClrN	Q^+
x	x	x	0	0
x	x	0	1	Q (no change)
↓	0	1	1	0
↓	1	1	1	1
0,1,↑	x	1	1	Q (no change)

9. (b) $S = Q'T$, $R = QT$
   Same as answer to Study Guide 6(c) except connect J and K and label it T.

UNIT 11 Answers to Problems

11.1

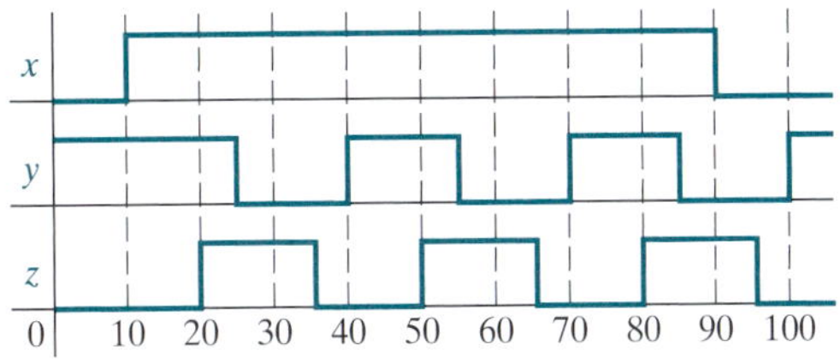

11.2 (a) $R = 1$ and $H = 0$ cannot occur at the same time.

(b)

R	H	Q	Q^+
0	0	0	0
0	0	1	0
0	1	0	0
0	1	1	1
1	0	0	X
1	0	1	X
1	1	0	1
1	1	1	1

$Q^+ = R + HQ$

(c)

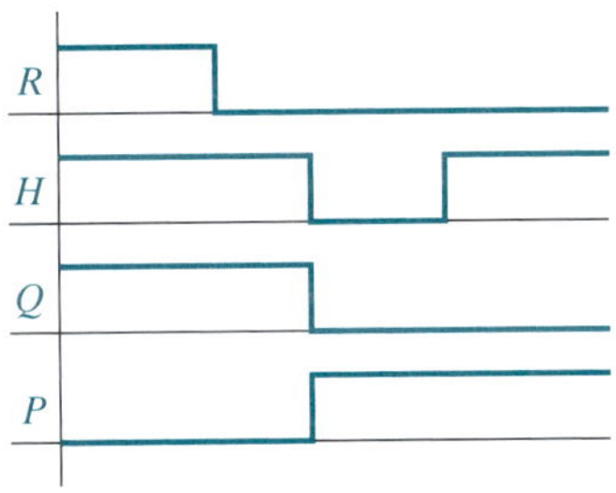

11.3

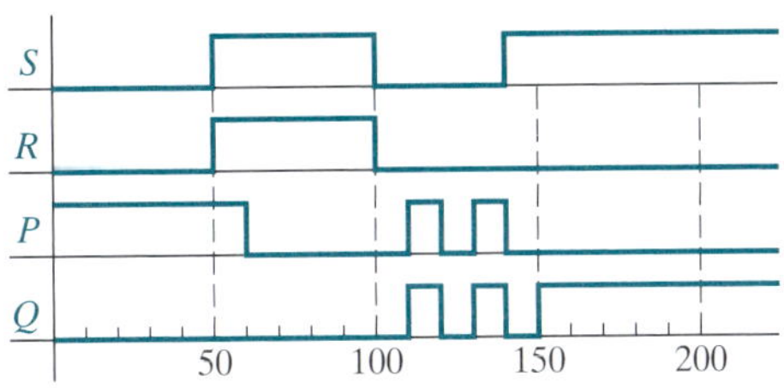

11.4

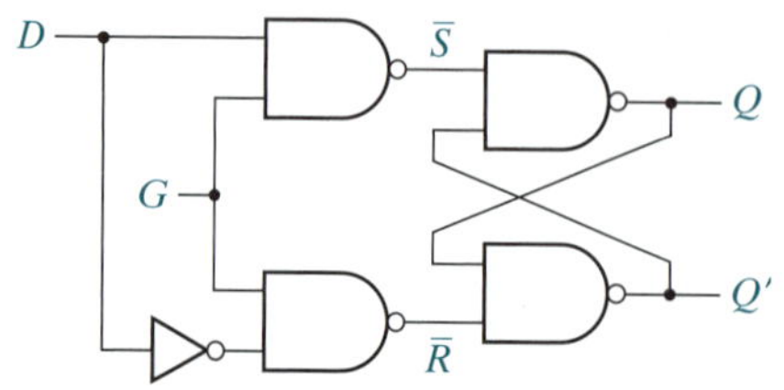

11.5 Connect the clock directly to the input G_1 and connect the clock to G_2 through an inverter.

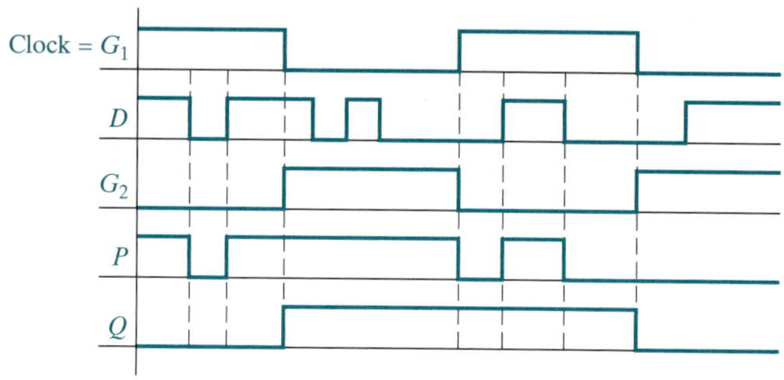

11.6 (a) $Q^+ = SR' + R'Q$ (b)

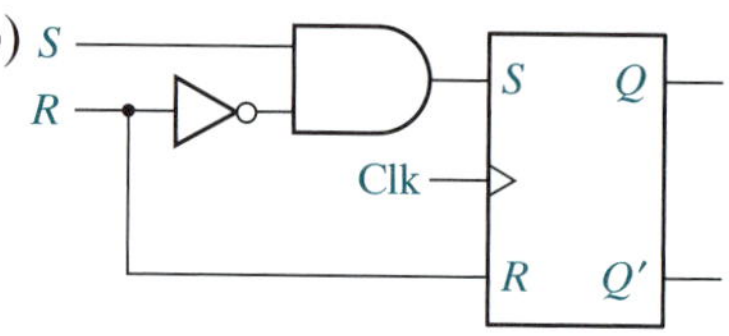

11.7

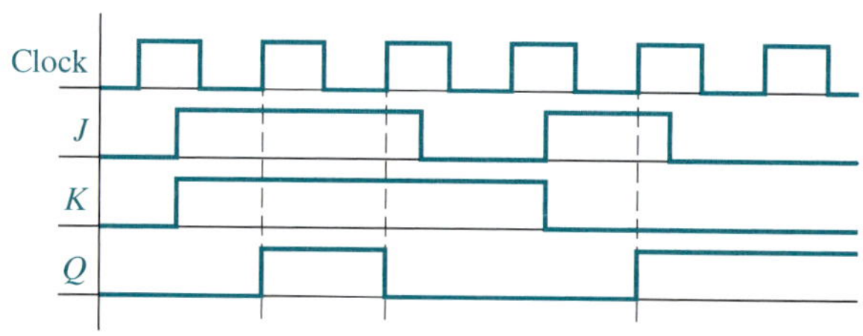

11.8

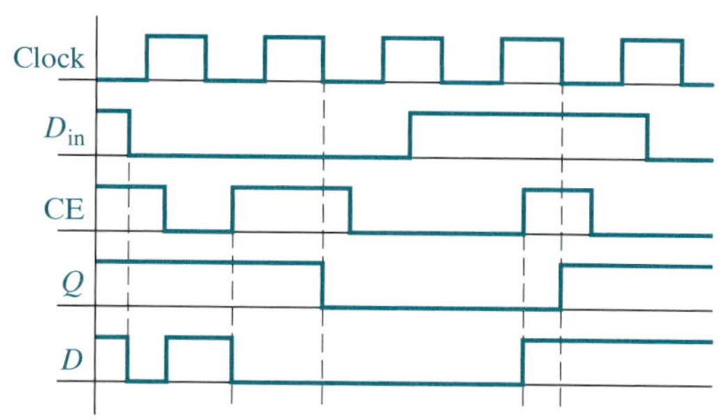

11.9 (a)

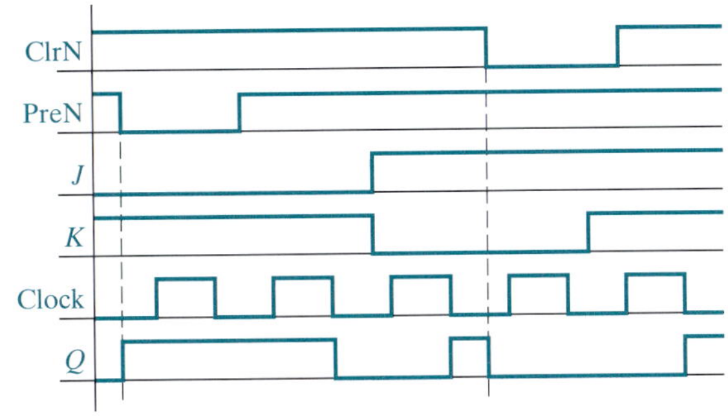

(b)

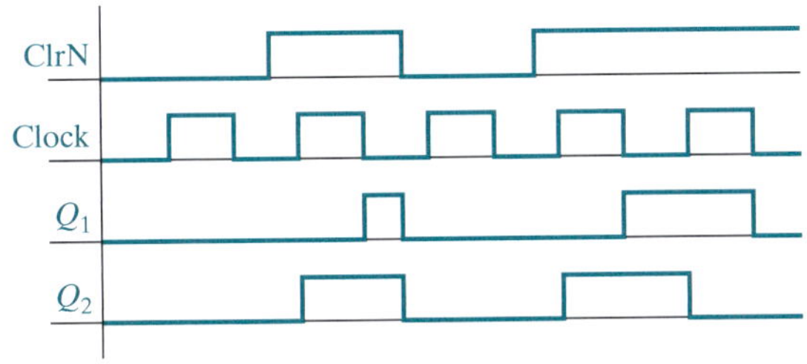

11.10 (a)

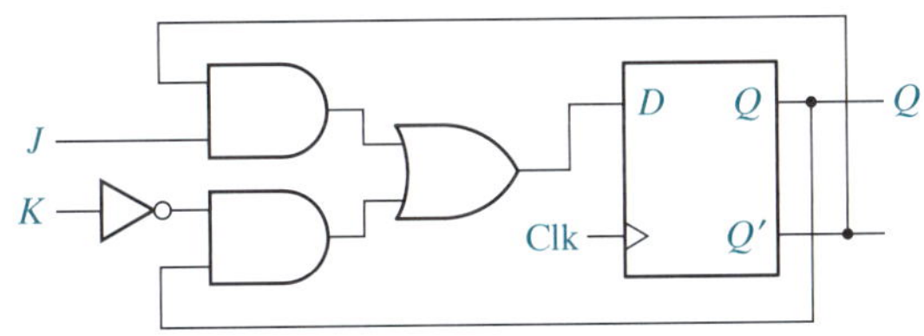

(b)

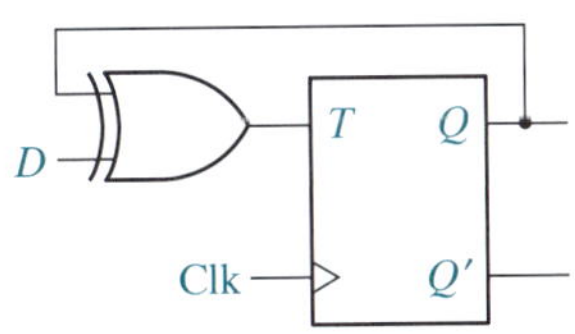

(c)

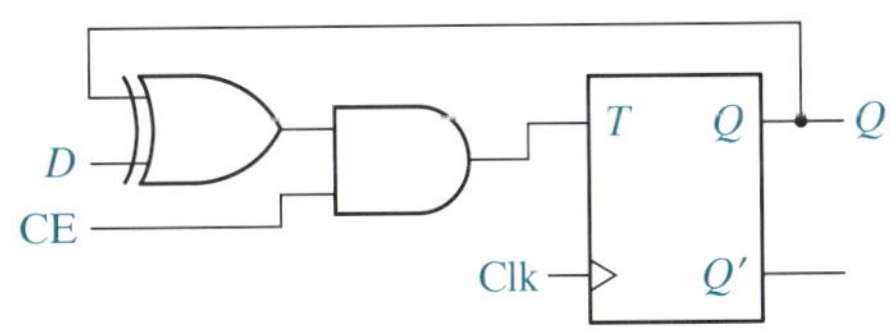

UNIT 12 Study Guide Answers

1. (a) $G = 0, H = 249; G = 0, H = 70; G = 118, H = 118; G = 91, H = 118;$ $G = 91, H = 118$
2. (b) S_0 is 1 between the rising edges of clocks 10 and 11, and also 1 between the rising edges of clocks 14 and 16.

Clock Cycle Number	Q_7	Q_6	Q_5	Q_4	Q_3	Q_2	Q_1	Q_0
	State of Shift Register when Clock = 1							
1	0	0	0	0	0	0	0	0
2	0	0	0	0	0	0	0	0
3	1	0	0	0	0	0	0	0
-	-	-	-	-	-	-	-	-
14	0	0	0	0	0	0	1	1
15	0	0	0	0	0	0	0	1
16	0	0	0	0	0	0	0	0

3. (b)

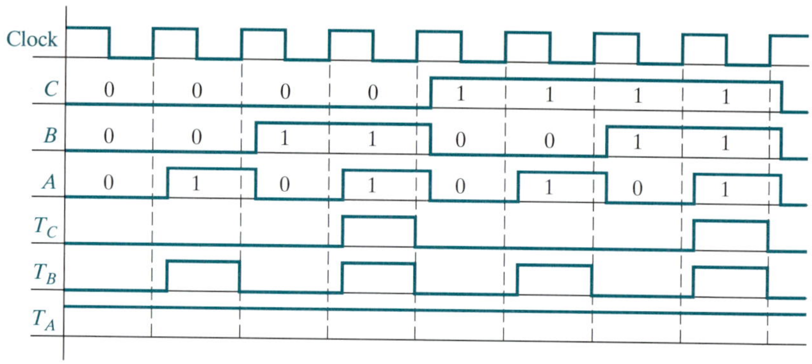

(d)

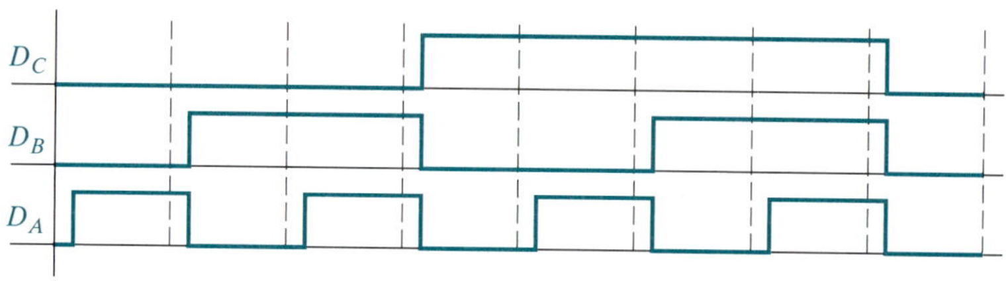

(f) State 101 goes to 110, which goes to 011.
(g) State 001 goes to 100; 101 goes to 110, which goes to 011.

4. (e)

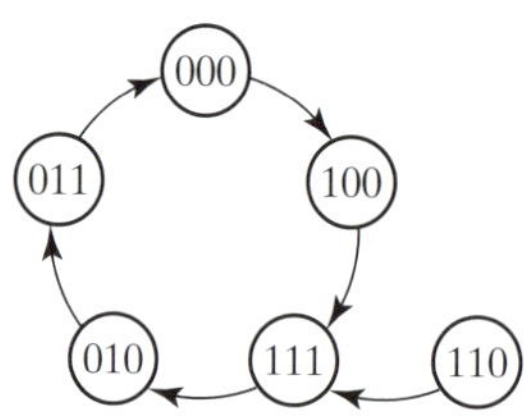

(k) $F_1 = 1 \quad F_2 = 0$

UNIT 12 Answers to Problems

12.1

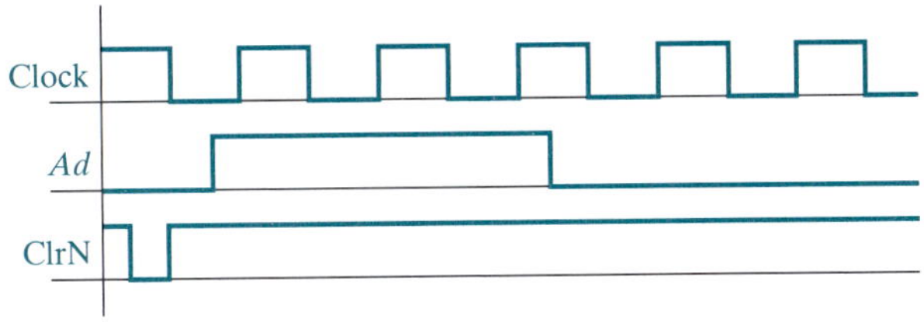

12.2

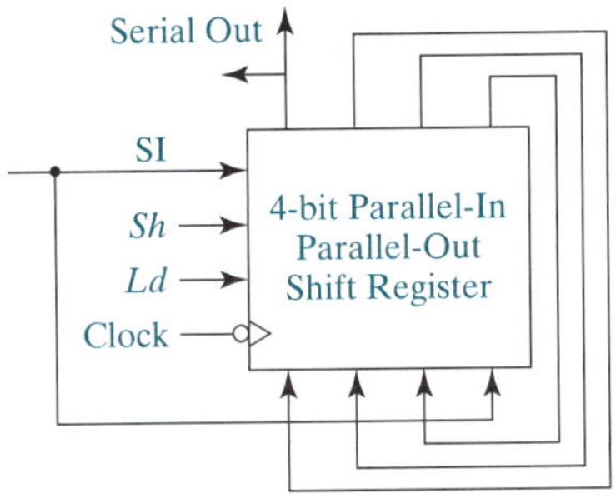

For a right shift, $Sh = 1, Ld = 0$ or 1. For a left shift, $Sh = 0, Ld = 1$.

12.3

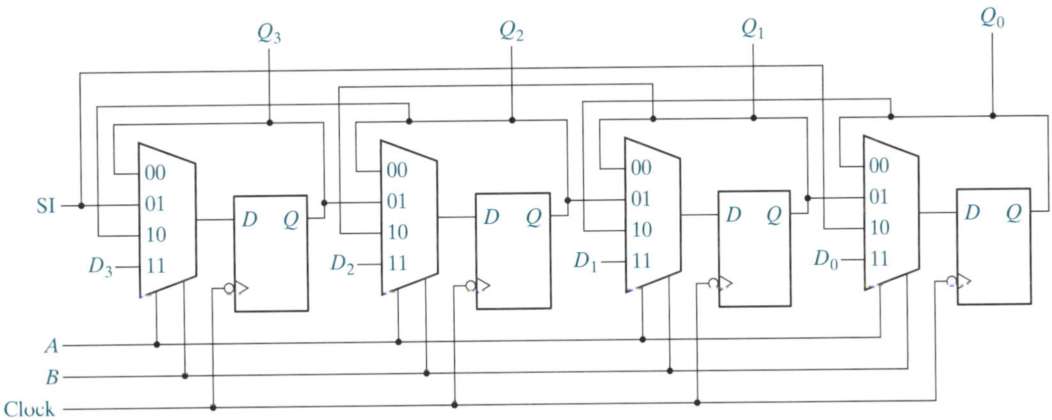

12.4 (a) (b)

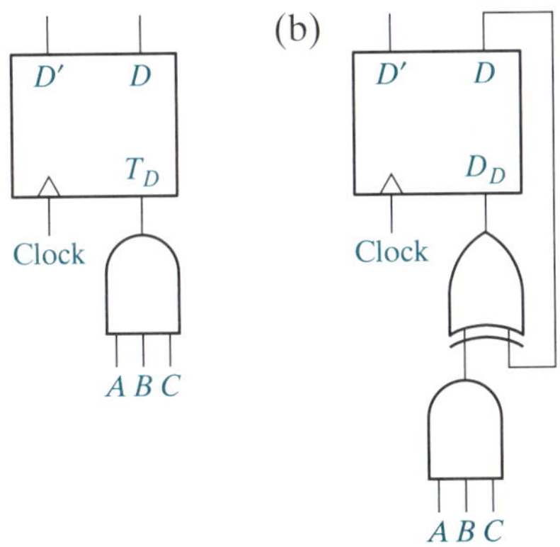

12.5 $D_D = D^+ = D'CBA + DC' + DB' + DA'$ $D_B = B^+ = B'A + BA'$
$D_C = C^+ = C'BA + CB' + CA'$ $D_A = A^+ = A'$

12.6 Uses three flip-flops: $Q_3Q_2Q_1$
Many correct solutions are possible. One is:
$D_3 = Q_1 + Q_2Q_3'$ $D_2 = Q_2'Q_3$
$D_1 = Q_1'Q_3'$

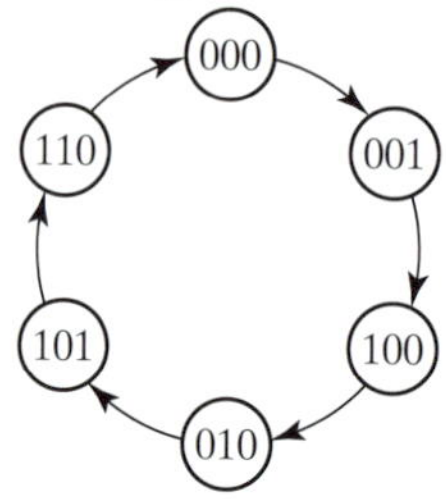

12.7 (a) $D_C = CA + BA'$ $D_B = C' + BA'$
$D_A = B'A' + CB + C'B'$
If $CBA = 000$, next state is 011.

(b) $T_C = B'A' + C'A'$ $T_B = C'B' + CBA$
$T_A = CB' + CA' + C'BA$
If $CBA = 000$, next state is 110.

12.8 (a) $J_C = A'$ $J_B = C'$ $J_A = C$
$K_C = B'A'$ $K_B = CA$ $K_A = CB' + C'B$
If $CBA = 000$, next state is 110.

(b) $S_C = BA'$ $S_B = C'$ $S_A = CA'$
$R_C = B'A'$ $R_B = CA$ $R_A = CB'A + C'B$
If $CBA = 000$, next state is 010.

12.9 (a)

Q	Q^+	M	N
0	0	0	X
0	1	1	X
1	0	X	0
1	1	X	1

(b) $M_C = B$ $M_B = C'A$ $M_A = C'$
$N_C = A$ $N_B = C'$ $N_A = C' + B$

UNIT 13 Study Guide Answers

2. (a) Mealy: output a function of both input and state
Moore: output a function of state only

(b) Before the active clock edge
After the active clock edge
When the flip-flops change state
When the flip-flops change state or when the inputs change

(c) Immediately preceding the active clock edge

(d) Mealy: False outputs can appear when the state has changed to its next value, but the input has not yet changed.
Moore: No false outputs occur because output is not a function of input.
Changing the inputs at the same time the state change occurs will eliminate false outputs.
No, because the output of the first Mealy circuit will still change to its final value before the active clock edge.

3. (a) Before the clock pulse
Q^+ means the state of flip-flop Q after the active clock edge (i.e., the next state of flip-flop Q).

(c) Mealy: output associated with transitions between states
Moore: output associated with state
(d) Present: Before the active clock edge
Next: after the clock pulse
(e) Output depends only on the state and not on the input.

4. (a) 1101 (c) 1001
(e)

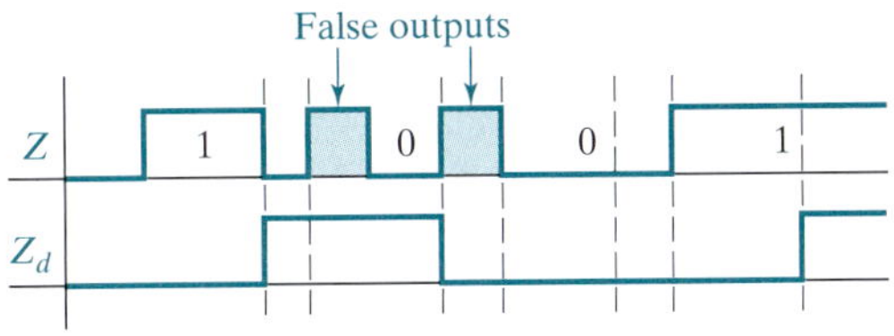

5. (a)

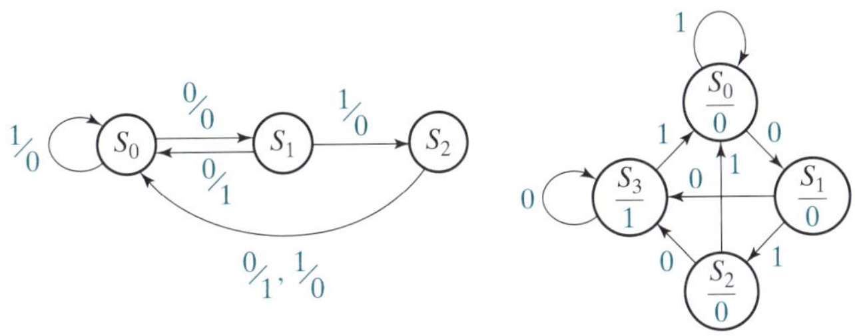

(c)

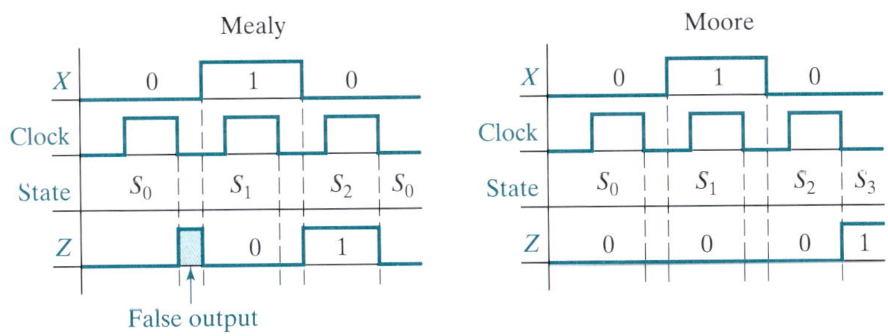

6. (a)

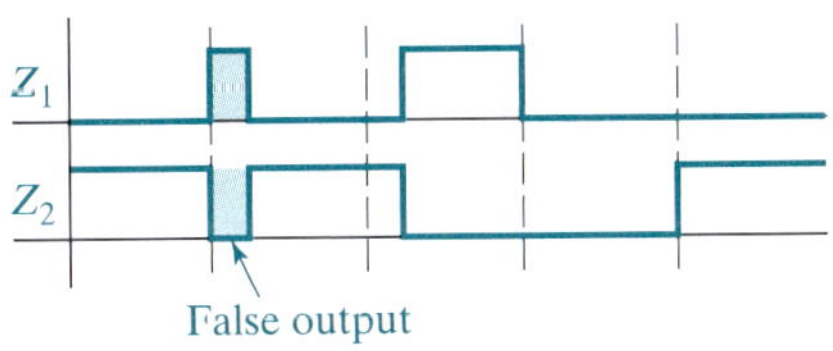

(g) $\delta(S_3, 1) = S_2$, $\lambda(S_3, 1) = 0$, $\delta(S_1, 2) = S_2$, $\lambda(S_1, 2) = 3$

UNIT 13 Answers to Problems

13.2

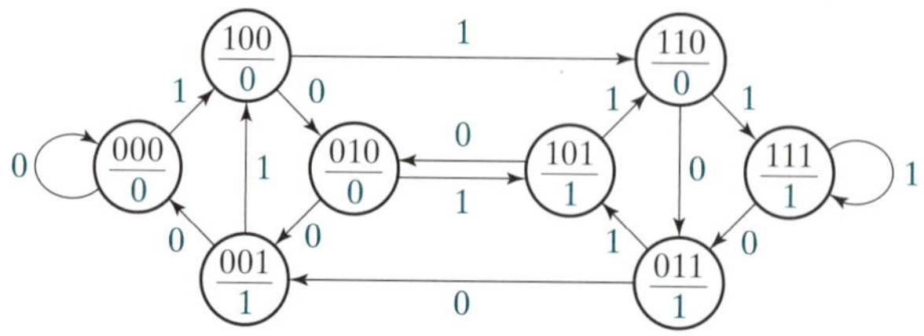

This is a Moore machine.

13.3 (a) $A^+ = A(B' + X) + A'(BX' + B'X) \qquad B^+ = AB'X + B(A' + X')$

Present State	Next State (A^+B^+)		
AB	$X = 0$	$X = 1$	Z
00	00	10	0
01	11	01	0
11	01	10	1
10	10	11	0

This is a Moore machine.

(b) $Z = (0)00101$

(c)

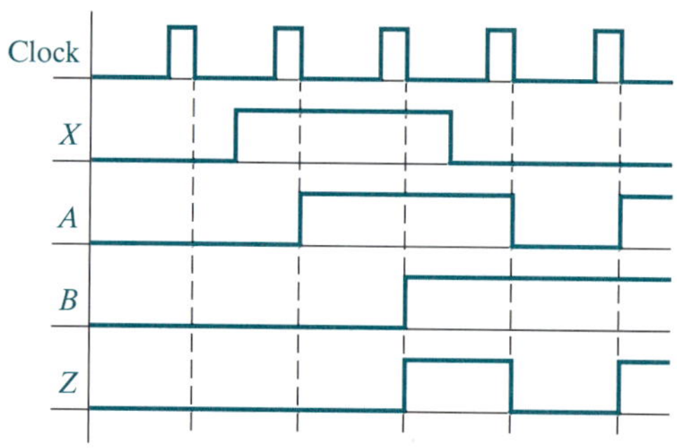

13.4 (a)

	$Q_1Q_2Q_3$	$Q_1^+Q_2^+Q_3^+$ $X = 0$	1	Z 0 1
S_0	0 0 0	001	001	0 1
S_1	0 0 1	011	011	0 1
S_2	0 1 0	100	101	1 0
S_3	0 1 1	010	011	1 0
S_4	1 0 0	001	001	0 1
S_5	1 0 1	011	011	0 1
S_6	1 1 0	100	101	1 0
S_7	1 1 1	010	011	1 0

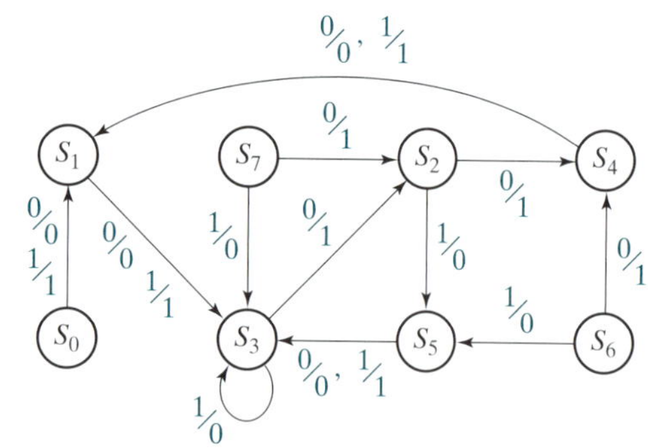

(b)

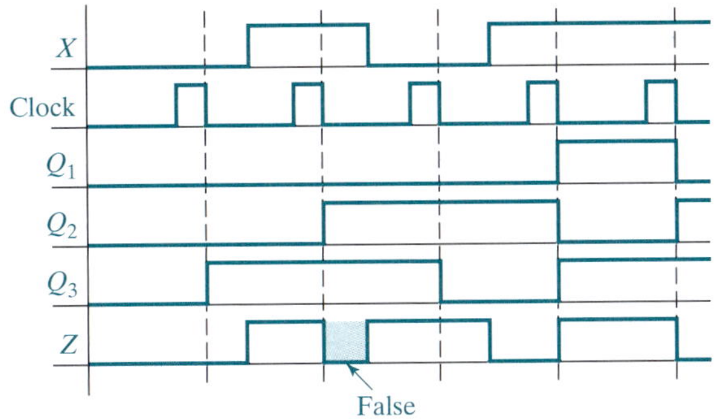

(c) From diagram: 0 1(0) 1 0 1
From graph: 0 1 1 0 1
(same except for false output)

(d) Change input on falling edge of clock

13.5 (a) Mealy Machine

(b)

ABC	$A^+B^+C^+$ X = 0	X = 1	Z X = 0	X = 1
000	011	010	1	0
001	000	100	1	0
010	100	100	0	1
011	010	000	0	1
100	100	001	1	0

(c)

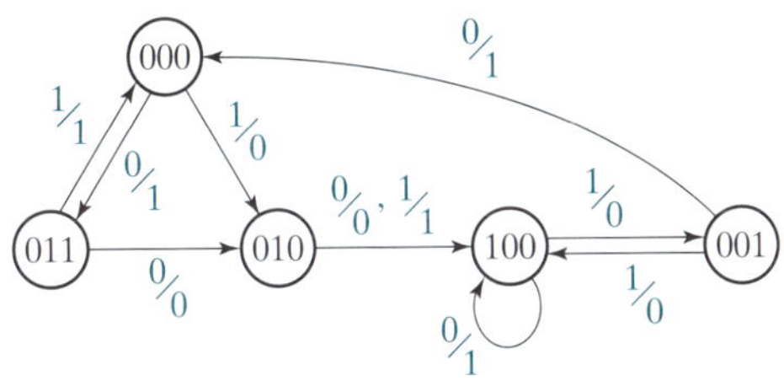

(d)

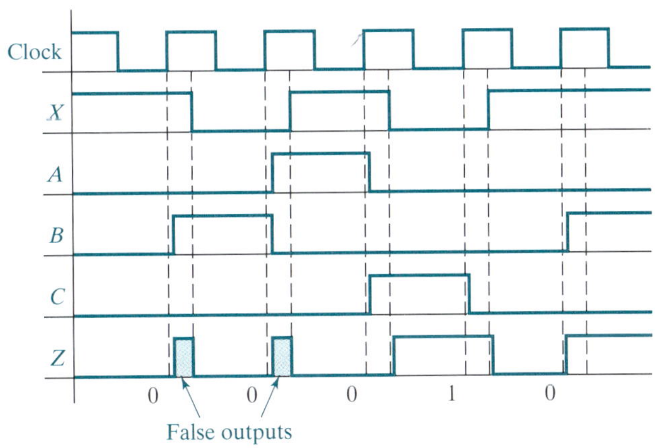

Correct output sequence: 00010

13.6 (a) 14 ns
(b)

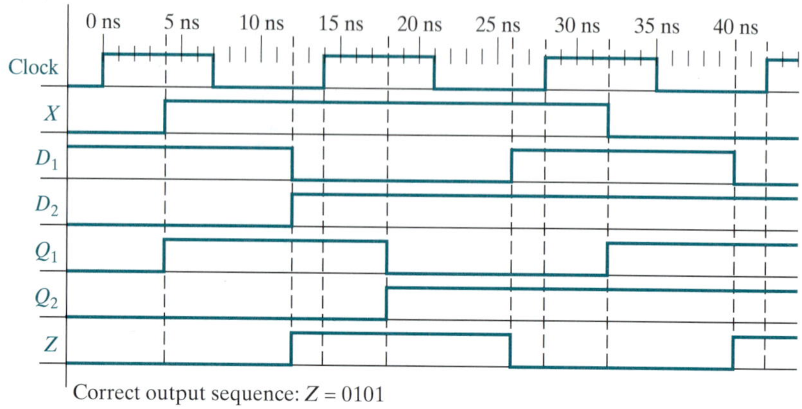

(c)

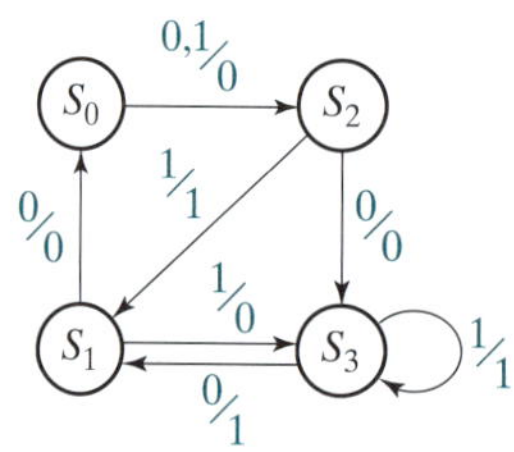

	Next State		Z	
	$X = 0$	$X = 1$	$X = 0$	$X = 1$
S_0	S_2	S_2	0	0
S_1	S_0	S_3	0	0
S_2	S_3	S_1	0	1
S_3	S_1	S_3	1	1

UNIT 14 Study Guide Answers

1. (b) last row: 11 10 01 0 1
   (c) $J_A = BX'$ $K_A = X \oplus B$ $J_B = A + X$ $K_B = A$ $Z = AB'$

8. (a)

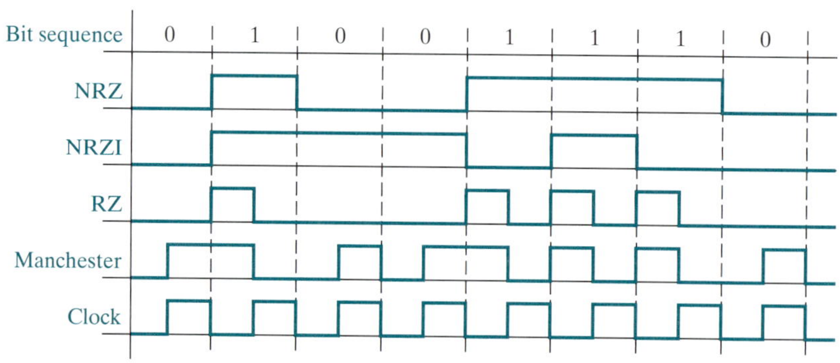

9. (b) Change N to NS'; add loop to S_3: $S'N'/0$

	NS = 00	01	10	11	Z 00	01	10	11
S_3	S_3	S_5	S_1	S_5	0	0	1	0

10. The output for the third input of $X = 1$ is don't care, since the third input can only be 0.

	X = 0	1	0	1
S_0	–	S_1	–	–
S_1	S_2	S_3	–	–
S_2	S_0	–	0	–
S_3	S_0	–	1	–

UNIT 14 Answers to Problems

14.4

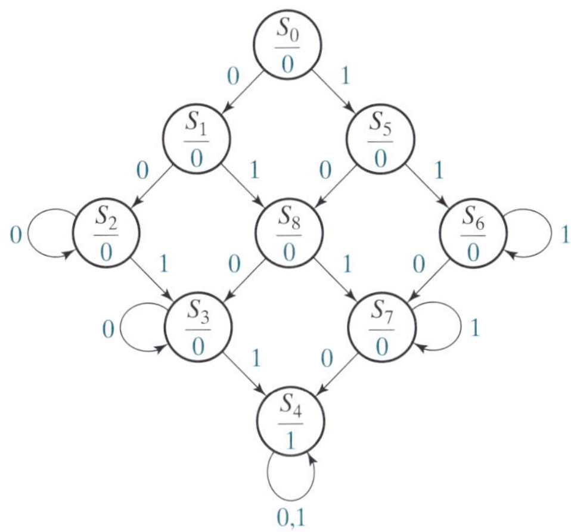

14.5

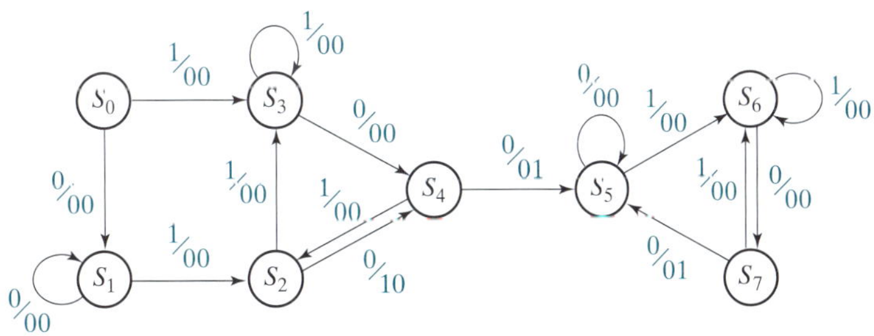

	X = 0	X = 1	Z_1Z_2 X = 0	X = 1
S_0	S_1	S_3	00	00
S_1	S_1	S_2	00	00
S_2	S_4	S_3	10	00
S_3	S_4	S_3	00	00
S_4	S_5	S_2	01	00
S_5	S_5	S_6	00	00
S_6	S_7	S_6	00	00
S_7	S_5	S_6	01	00

14.6

	$X_1X_2 =$ 00	01	10	11	Z
S_0	S_0	S_1	S_3	S_2	0
S_1	S_0	S_1	S_3	S_2	0
S_2	S_4	S_1	S_3	S_2	0
S_3	S_4	S_1	S_3	S_2	0
S_4	S_4	S_5	S_7	S_6	1
S_5	S_0	S_5	S_7	S_6	1
S_6	S_4	S_5	S_7	S_6	1
S_7	S_0	S_5	S_7	S_6	1

(a 4-state solution is also possible)

14.7 (a)

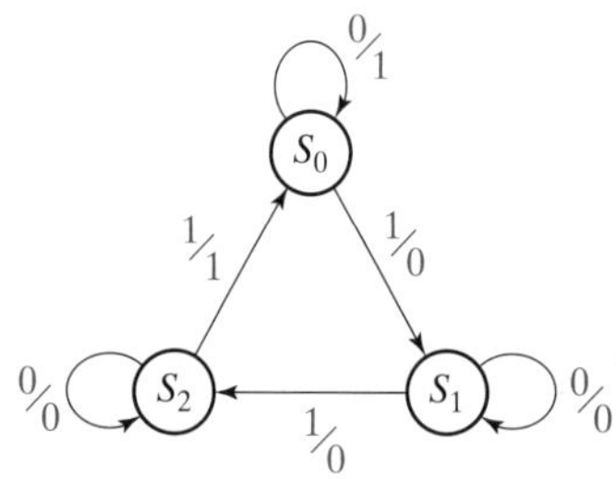

(b)

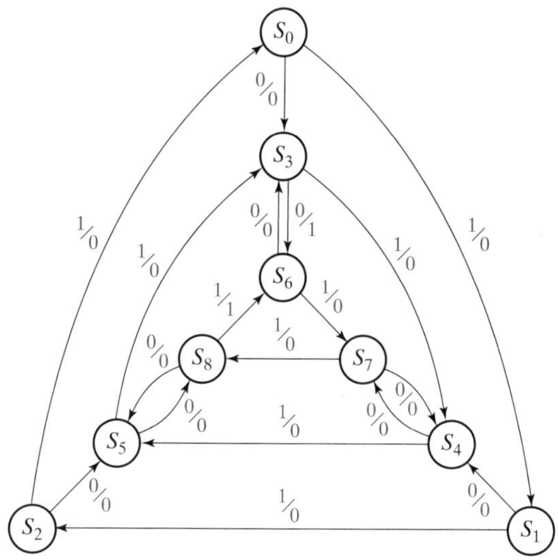

14.8

(a)

	$X_1X_2 =$ 00	01	10	11	Z_1Z_2 00	01	10	11
S_0	S_1	S_2	S_3	S_4	00	00	00	00
S_1	S_1	S_2	S_3	S_4	00	10	10	10
S_2	S_1	S_2	S_3	S_4	01	00	10	10
S_3	S_1	S_2	S_3	S_4	01	01	00	10
S_4	S_1	S_2	S_3	S_4	01	01	01	00

(b)

	$X_1X_2 =$ 00	01	10	11	Z_1Z_2
S_0	S_1	S_4	S_7	S_{10}	00
S_1	S_1	S_3	S_6	S_9	00
S_2	S_1	S_3	S_6	S_9	01
S_3	S_2	S_4	S_6	S_9	10
S_4	S_2	S_4	S_6	S_9	00
S_5	S_2	S_4	S_6	S_9	01
S_6	S_2	S_5	S_7	S_9	10
S_7	S_2	S_5	S_7	S_9	00
S_8	S_2	S_5	S_7	S_9	01
S_9	S_2	S_5	S_8	S_{10}	10
S_{10}	S_2	S_5	S_8	S_{10}	00

14.9

(a)

	$X = 0$	1	$X = 0$	1
S_0	S_0	S_1	0	1
S_1	S_1	S_0	1	0

(b)

	$X = 0$	1	
S_0	S_0	S_1	0
S_1	S_1	S_0	1

(c, d)

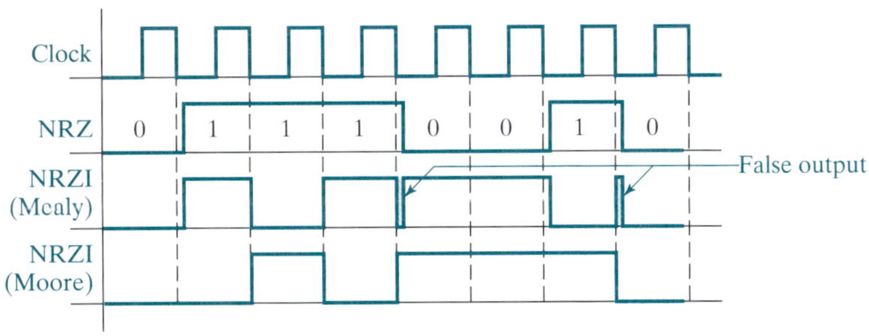

14.10

	Next State 000	001	010	011	100	101	110	111	Output (*DEF*) 000	001	010	011	100	101	110	111
S_0	S_0	S_0	S_0	S_0	S_1	S_1	S_0	S_0	100	100	100	100	010	010	001	001
S_1	S_1	S_0	S_1	S_0	S_1	S_0	S_1	S_0	110	000	110	000	101	000	101	000

For S_0:

$A' + AB + AB' = A' + A = 1$

$A' \cdot AB = 0;\ A' \cdot AB' = 0;$

$AB \cdot AB' = 0$

For S_1:

$A'C' + AC' + C = C' + C = 1$

$A'C' \cdot AC' = 0;$

$A'C' \cdot C = 0;\ AC' \cdot C = 0$

14.11

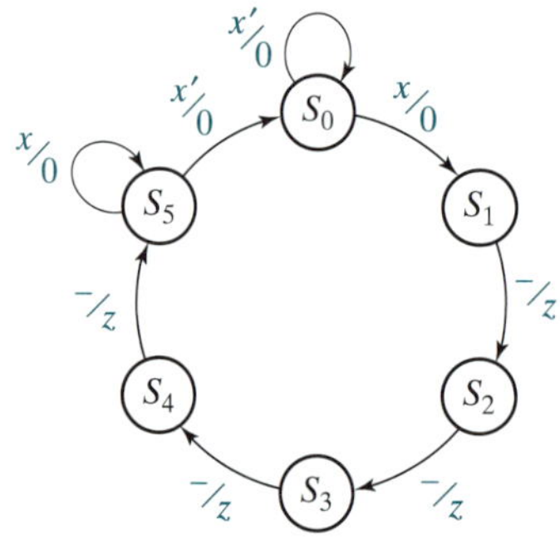

UNIT 15 Study Guide Answers

2. (b) $\lambda(p, 01) = 00$ and $\lambda(q, 01) = 01$; therefore, $p \not\equiv q$
   (c) No. You would have to try an infinite number of test sequences to be sure the *circuits* were equivalent.
   (d) $S_2 \equiv S_3$ iff $S_1 \equiv S_5$ and $S_4 \equiv S_2$. But $S_1 \not\equiv S_5$ because the outputs are different. Therefore, $S_2 \not\equiv S_3$.

3. (a)

b	X				
c	X	a-d a-b			
d	a-g b-f	X	X		
f	X	a-g	a-d b-g	X	
g	X	a-f	a-d b-f	X	✓
	a	*b*	*c*	*d*	*f*

(b) $f \equiv g$
(c) $a \equiv c, b \equiv d, b \equiv e, d \equiv e$

	0	1	0	1
a	b	a	0	1
b	b	b	0	0

6. (a) 1) Some input sequences to the circuit may not occur and 2) the output may be observed at only some time periods.
   (b)

S_1	S_1, S_2		
S_2	X	S_0, S_3	
S_3	S_1, S_2	X	X
	S_0	S_1	S_2

Maximal Compatibles: $(S_0, S_1), (S_1, S_2), (S_0, S_3)$. (S_0, S_3) and (S_1, S_2) produce a two-state reduced table.

	$X = 0$	1	0	1
$(S_0, S_3) = S_0$	S_1	S_0	0	1
$(S_1, S_2) = S_1$	S_1	S_0	1	0

(c)

	$X = 0$	1	0	1
S_0	–	S_1	–	–
S_1	S_2	S_3	–	–
S_2	S_0	–	0	–
S_3	S_0	–	1	–

	$X = 0$	1	0	1
$(S_0, S_1, S_2) = S_0$	S_0	S_1	0	–
$(S_0, S_1, S_3) = S_1$	S_0	S_1	1	–

S_1	S_1, S_3		
S_2	✓	S_0, S_2	
S_3	✓	S_0, S_2	☒
	S_0	S_1	S_2

Maximal Compatibles: (S_0, S_1, S_2), (S_0, S_1, S_3)

7. (b) $Z = X'AB' + XA'BC'$
8. (b) Interchanging columns or complementing columns does not affect circuit cost for symmetric flip-flops.
   (c) Complementing columns (to make the first row all 0's) does not change the cost of the circuit.
   (f) Numbering columns from left to right, column 3 is same as column 4, column 2 is column 5 complemented, column 1 is column 6 complemented.
9. (e) $D_1 = XQ_1' + XQ_3 + Q_2Q_3' + X'Q_1Q_2'$ or $D_1 = XQ_1' + XQ_2 + Q_2'Q_3 + X'Q_1Q_3'$,
   $D_2 = Q_3, D_3 = X'Q_3 + XQ_2Q_3' + (Q_1'Q_3 \text{ or } Q_1'Q_2)$
   $Z = XQ_2Q_3 + X'Q_2'Q_3 + X'Q_2Q_3'$
   (f) $J_1 = X, K_1 = X'Q_2Q_3 + XQ_2'Q_3'$
11. (b) $Q_5^+ = XQ_2 + X'YQ_2 + X'Q_5$
   (c) $Q_2^+ = Q_1M$
   $Q_3^+ = Q_2K + Q_1KM'$
   (d) $Ad = Q_1M$
   $Done = Q_3$

UNIT 15 Answers to Problems

15.1 (a)

	Next State		Output	
Present State	$X = 0$	$X = 1$	$X = 0$	$X - 1$
A	A	C	1	0
B	C	F	0	0
C	B	A	0	0
F	B	F	1	0

(b)

Input	1	0	0
Output (from B)	0	1	0
Output (from G)	0	1	1

15.2

Present State	Next State $X = 0$	$X = 1$	Output
a	*c*	*c*	1
c	*d*	*f*	0
d	*f*	*a*	1
f	*c*	*d*	0

15.3 (a) No, states S_2 and S_4 have no corresponding states in Mr. Ipflop's design.
(b) Because there is no way of reaching S_2 and S_4 by starting from S_0, the two *circuits* would perform the same.

15.4 (a) $D = X_1'X_2Q' + X_1X_2'Q' + X_2'X_3Q + X_2X_3'Q$
(b) $S = X_1'X_2Q' + X_1X_2'Q'$
$R = X_2'X_3'Q + X_2X_3Q$

15.5 (a) Only *one* assignment—000 011 101 OR 001 010 100 etc.

000		001	
011	OR	010	etc.
101		100	

(b)

000	000	000	000	000	000	000	000	000	000	
001	001	001	001	001	001	001	001	001	001	etc.
010	010	010	010	011	011	011	011	110	110	
100	101	110	111	100	101	110	111	010	011	

15.6 (a)

$BC \backslash A$	0	1
00	S_1	S_2
01	S_4	S_3
11	S_6	S_5
10	S_7	S_8

$Z = A$

(b)

$BC \backslash A$	0	1
00	S_1	S_8
01	S_7	S_5
11	S_3	S_4
10	S_2	S_6

$D_A = A^+ = A'B' + XA' + X'AC'$
$D_B = B^+ =$ etc.
$D_C = C^+ =$ etc.

15.7 (a)

$Q_2Q_3 \backslash Q_1$	0	1
00	*A*	*C*
01	*B*	*E*
11	*F*	
10	*D*	

(b) $D_1 = XQ_1$
$D_2 = Q_1'Q_3' + X'Q_1'$
$D_3 = X'Q_1'Q_3' + XQ_1'Q_3 + \{XQ_2' \text{ or } Q_2Q_3\}$
$Z = XQ_1$

15.8 (a) $A = 00,\ B = 01,\ C = 10,\ D = 11$

(b) $T_1 = X_1'X_2Q_2' + X_1'Q_1Q_2 + X_1Q_1'Q_2 + X_1X_2'$

$T_2 = X_1Q_1'Q_2' + X_1Q_1Q_2$

$Z_1 = X_1Q_2,\ Z_2 = X_1'Q_1 + Q_1Q_2'$

15.9

		Q_1	Q_2	Q_3
Assign	S_0	1	0	0
	S_1	0	1	0

$D_1 = X'Q_1 + XY'Q_3$

$D_2 = XQ_1 + YQ_3 + X'Q_2$

$D_3 = XQ_2 + X'Y'Q_3$

$P = XQ_1 + Y'Q_3 + XQ_2$

$S = X'Q_1 + XY'Q_3$

UNIT 16 Study Guide Answers

1. (a) Because the input sequences are listed in reverse order
2. (b) m leads, where $2^{m-1} < n \leq 2^m$
3. (b) 64 words × 7 bits

   (d) $Z = 0,\ D_1 = 1,\ D_2 = 1,\ D_3 = 0;\qquad Q_1Q_2Q_3 = 110$
8. (a) Yes

   (c) Yes
9. (a) After the clock (when the state has just changed) and before the input is set to its new value, the output may be wrong (false output).

   (b) No, because the output is always correct before the active clock edge

UNIT 16 Answers to Problems

16.15

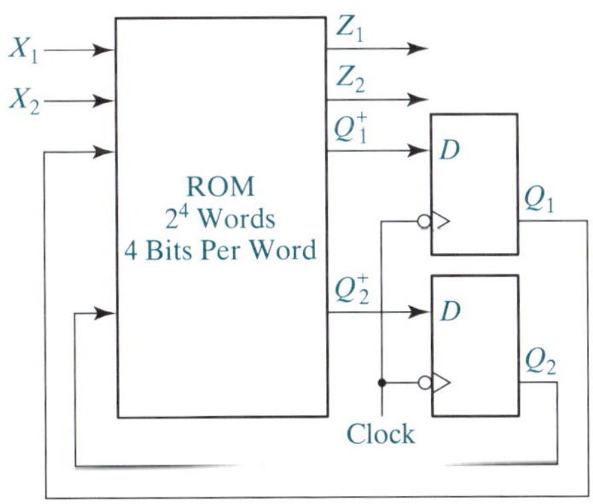

X_1	X_2	Q_1	Q_2	Q_1^+	Q_2^+	Z_1	Z_2
0	0	0	0	1	1	0	0
0	0	0	1	0	0	1	0
0	0	1	0	1	1	0	0
0	0	1	1	1	0	0	0
0	1	0	0	1	0	1	0
0	1	0	1	0	1	1	0
0	1	1	0	0	0	1	0
0	1	1	1	1	0	0	0
1	0	0	0	0	1	1	1
1	0	0	1	1	0	1	1
1	0	1	0	0	1	1	1
1	0	1	1	0	1	0	1
1	1	0	0	0	0	0	1
1	1	0	1	1	1	1	1
1	1	1	0	0	1	0	1
1	1	1	1	0	0	0	1

16.16 (a) Same as Figure 16-10 with ROM replaced by PLA.

(b)

X	A	B	C	Z	D_A	D_B	D_C
0	–	–	–	0	1	0	0
0	–	–	0	0	0	1	0
–	0	–	1	0	0	1	0
–	0	1	–	0	0	1	0
–	1	–	–	0	0	0	1
1	–	0	–	0	0	0	1
0	1	0	1	1	0	0	0
1	0	1	0	1	0	0	0

16.17 (a)

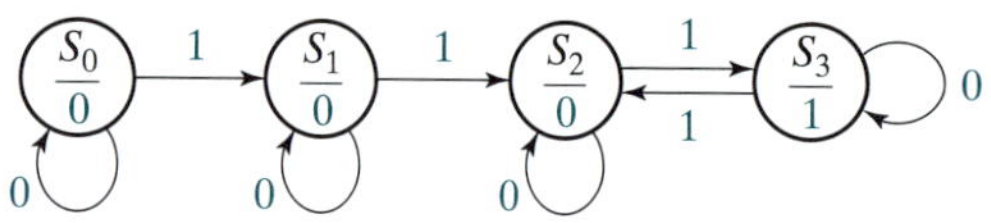

(b) $a_{i+1} = a_i + x_i b_i = [a'_i(x_i b_i)']'$ $\quad b_{i+1} = x_i b'_i + x'_i b_i = [(x_i b'_i)'(x'_i b_i)']'$

$z = a_{n+1} b_{n+1}$

(c) $a_1 = b_1 = 0, a_2 = 0, b_2 = x_1$

(d) Similar to Figure 16-9 with one output

UNIT 17 Study Guide Answers

1. (a) When a falling edge of CLK occurs
   (b) Whenever there is a change in St or Q_1 or V
   (c) Statements S_3 and S_4 will execute
   (d) The code represents a rising-edge-triggered D flip-flop with asynchronous, active-low clear and preset. Because ClrN has higher priority, Q will be set to '0' when both ClrN and SetN are '0'.
   (e) *SN* and *RN* override *J* and *K*, including at a clock edge. Q will be '0'.
   (f) They will get the old value of Q_{int} because of the 8-ns delay.
2. (a)

```
entity register is
port (CLK, CLR, Ld: in bit; D: in bit_vector (3 downto 0);
   Q: out bit_vector (3 downto 0));
end register;


architecture eqn of register is
begin
   -- Given code here
end eqn;
```

   (b) Q will change at time 10 + Δ ns
   (c) On line 6 of the VHDL code, make this change—

```
elsif LS = '1' then Q <= Q (2 downto 0) & Q (3);
```

(d) Change the indicated lines of code as shown—
line 3: **process** (ClrN, CLK)
lines 5 to 7: **if** ClrN = '0' **then** Q <= "0000";
elsif CLK'event **and** CLK = '1' **then**
if En = '1' **then** Q <= Q + 1;

(e) When Carry1 = '1' and Qout2 = "1111"

(f) No overloaded "+" operator is defined for bit_vectors

(g) After the rising edge, Qout1 = "0000" and Qout2 = "1010"

(h)

Control Signals			Next State			
ClrN	LdN	PT	Q_3^+	Q_2^+	Q_1^+	Q_0^+
X	0	X	D_3	D_2	D_1	D_0
X	1	1	Present state + 1			
0	1	0	0	0	0	0
1	1	0	Q_3	Q_2	Q_1	Q_0

3. (a)
```
process (A, B, D)
begin
  E <= (A and B) or D after 5 ns;
end process;
```

4. (a) Nextstate = S4 and Z = 1

(b) Because there are only seven states. Also, specifying the range restricts the number of bits used to represent the integer.
When X changes to 1, Nextstate = 2, and $Z = 0$. Then, when CLK changes to 1, State = 2, Nextstate = 4, and $Z = 1$

(c) The glitch occurs because the change in state and change in the value of X a little while later causes process (State, X) to execute two times, thereby updating the value of Nextstate two times. This glitch does not affect the 'state' because the state will not be updated until the next positive clock edge.

(d) Because Q1, Q2, and Q3 must be updated only on the CLK edge, the other signals must not appear on the sensitivity list. The new values will be computed at 5 ns, and the values are updated at 15 ns.

(e) The statements of lines 13, 14, and 18 will execute.

(f) ROM output = 1100

5. (a) Connect En to CE and A to D

(b) See Figure 11-31(c) (change to rising-edge trigger)

(c) Use four D-CE flip-flops. Connect ASR to every CE input, D3 to Q3, D2 to Q3, D1 to Q2, and D0 to Q1. Label Q3 . . . Q0 as A(3) . . . A(0).

6. (a)
```
process
begin
  A <= B or C;
  wait on B, C;
end process;
```

(b) 2 ns (Both sequential statements execute immediately with no delay.)

UNIT 17 Answers to Problems

17.1 Code to implement a T flip-flop

```
entity tff is
   port (t, clk, clrn: in bit;
               q, qn: out bit);
end tff;
architecture eqn of tff is
signal qint: bit; -- Internal value of q
begin
q <= qint;
qn <= not qint;
process (clk, clrn)
begin
   if clrn = '0' then qint <= '0';
   elsif clk'event and clk = '1' then
      qint <= (t and not qint) or (not t and qint);
   end if;
end process;
end eqn;
```

17.2 Right-shift register with synchronous reset

```
entity rsr is
   Port (clk, clr, ld, rs, lin: in bit;
         d: in bit_vector(3 downto 0);
         q: out bit_vector(3 downto 0));
end rsr;

architecture eqn of rsr is
signal qint: bit_vector(3 downto 0);
begin
q <= qint;

process (clk)
begin
   if clk'event and clk = '1' then
      if clr = '1' then qint <= "0000";
      elsif ld = '1' then qint <= d;
      elsif rs = '1' then qint <= lin & qint(3 downto 1);
      end if;
   end if;
end process;
end eqn;
```

17.3 (a) 4-bit binary up/down counter

```
entity updown is
  Port (clrn, clk, load, ent, enp, up: in std_logic;
        d: in std_logic_vector(3 downto 0);
        q: out std_logic_vector(3 downto 0);
        co: out std_logic);
end updown;

architecture eqn of updown is
signal qint: std_logic_vector(3 downto 0) := "0000";
begin
q <= qint;
co <= (qint(3) and qint(2) and qint(1) and qint(0) and ent and up)
        or (not qint(3) and not qint(2) and not qint(1) and not qint(0)
        and ent and not up);

process (clrn, clk)
begin
  if clrn = '0' then qint <= "0000";
  elsif clk'event and clk = '1' then
  if load = '0' then qint <= d;
    elsif (ent and enp and up) = '1' then qint <= qint + 1;
    elsif (ent and enp and not up) = '1' then qint <= qint - 1;
    end if;
  end if;
end process;
end eqn;
```

17.3 (b) 8-bit binary up/down counter. (For block diagram, connect the carry-out of the first counter to ENT of the second.)

```
entity updown8bit is
Port (clrn, clk, load, ent, enp, up: in std_logic;
          d: in std_logic_vector(7 downto 0);
          q: out std_logic_vector(7 downto 0);
          co: out std_logic);
end updown8bit;
architecture structure of updown8bit is
component updown is
Port (clrn, clk, load, ent, enp, up: in std_logic;
          d: in std_logic_vector(3 downto 0);
          q: out std_logic_vector(3 downto 0);
          co: out std_logic);
end component;
signal co1: std_logic;
signal q1,q2: std_logic_vector(3 downto 0);
```

```
begin
   c1: updown port map (clrn, clk, load, ent, enp, up, d(3 downto 0),q1,co1);
   c2: updown port map (clrn, clk, load, co1, enp, up, d(7 downto 4),q2, co);
   q <= q2 & q1;
end structure;
```

17.4 MUX with a and b as control inputs

```
entity mymux is
Port (a, b, c, d: in bit;
        z: out bit);
end mymux;
architecture eqn of mymux is
signal sel: bit_vector(1 downto 0);
begin
sel <= a & b;
process (a, b, c, d)
begin
   case sel is
      when "00" => z <= not c or d;
      when "01" => z <= c;
      when "10" => z <= not c xor d;
      when "11" => z <= not d;
   end case;
end process;
end eqn;
```

17.5 Implements the state machine of Table 14-1

```
entity sm1 is
Port (x, clk: in bit;
        z: out bit);
end sm1;
architecture table of sm1 is
signal State, Nextstate: integer range 0 to 2 := 0;
begin
process (State, x)
begin
   case State is
   when 0 =>
      if x = '0' then Nextstate <= 0; else Nextstate <= 1; end if;
      z <= '0';
   when 1 =>
      if x = '0' then Nextstate <= 2; else Nextstate <= 1; end if;
      z <= '0';
```

```
      when 2 =>
         if x = '0' then Nextstate <= 0; z <= '0';
         else Nextstate <= 1; z <= '1'; end if;
      end case;
   end process;
   process (clk)
   begin
   if clk'event and clk = '0' then
      State <= Nextstate;
   end if;
   end process;
   end table;
```

17.6 (a) See Figure 13-17, with $m = 2, n = 2$, and $k = 2$.
(b) Implements the state machine of Table 13-4

```
library BITLIB;
use BITLIB.bit_pack.all;
entity sm is
Port (x1, x2, clk: in bit;
   z1,z2: out bit);
end sm;
architecture Behavioral of sm is
type rom16_4 is array (0 to 15) of bit_vector(3 downto 0);
-- Input is in the order X1 X2 Q1 Q2
-- Output in order Q1 Q2 Z1 Z2
constant myrom: rom16_4 := ("1100", "0010", "1100", "1000", "1010", "0110","0010",
            "1000", "0111", "1011", "0111", "0101", "0001", "1111", "0101", "0001");
signal index: integer range 0 to 15;
signal q1,q2: bit;
signal rom_out: bit_vector(3 downto 0);
begin
index <= vec2int(x1&x2&q1&q2);
rom_out <= myrom(index);
z1 <= rom_out(1);
z2 <= rom_out(0);
process(clk)
begin
   if clk'event and clk = '1' then
      q1 <= rom_out(3);
      q2 <= rom_out(2);
   end if;
end process;
end Behavioral;
```

17.7 (a) There are two D-CE flip-flops. For each, CE = LdA + LdB.
D1 = LdA A1 + LdA′LdB B1, D2 = LdA A2 + LdA′LdB B2.

(b) CE does not change. For each D input, replace the gates with a 2-to-1 MUX, with LdA as the control input, and B and A as the data inputs for 0 and 1, respectively. (Alternately, use LdB as the control input, and swap A and B on the data inputs.)

17.8 All statements execute at time = 20 ns
A becomes 1 at 35 ns (not the final value)
B becomes 1 at 20 ns + Δ (not the final value)
C becomes 1 at 30 ns
D becomes 2 at 23 ns
A becomes 5 at 35 ns (overrides the previous value)
B becomes 7 at 20 ns + Δ (overrides the previous value)

UNIT 18 Study Guide Answers

1. (a)

	X	Y	c_i	s_i	c_i^+
t_0	0110	0011	0	1	0
t_1	1011	1001	0	0	1
t_2	0101	1100	1	0	1
t_3	0010	0110	1	1	0
t_4	1001	0011	0	(0)	(1)

(b) Y would fill up with 0's from the left: 0011, 0001, 0000, 0000, 0000.

(c) S_0 and Y_0, no.

2. (a)

add	0	0	0	0	0	1	1	0	1
		1	1	1	1				
shift	0	1	1	1	1	1	1	0	1
shift	0	0	1	1	1	1	1	1	0
add	0	0	0	1	1	1	1	1	1
		1	1	1	1				
shift	1	0	0	1	0	1	1	1	1
add	0	1	0	0	1	0	1	1	1
		1	1	1	1				
shift	1	1	0	0	0	0	1	1	1
	0	1	1	0	0	0	0	1	1

(b) 10, 6. (c) 10, 6. (d) 15 bits

(f) Product register has 17 bits. Adder is 8 bits wide, multiplicand has 8 bits. 18 states. 3-bit counter, $K = 1$ when counter is in state 7 (111_2), control graph unchanged.

3. (b) Change Y to 2's complement by inverting each bit and adding 1 (by setting the carry input of the first full adder to 1). Also change C so that it is equal to the carry out of the last full adder.

(c) An overflow will occur if $X_8X_7X_6X_5X_4 \geq Y_3Y_2Y_1Y_0$, because subtraction is possible but there is no place to store the quotient bit, since there are only 4 bits available to store the quotient.

(f) To set the quotient bit to 1.

UNIT 18 Answers to Problems

18.3

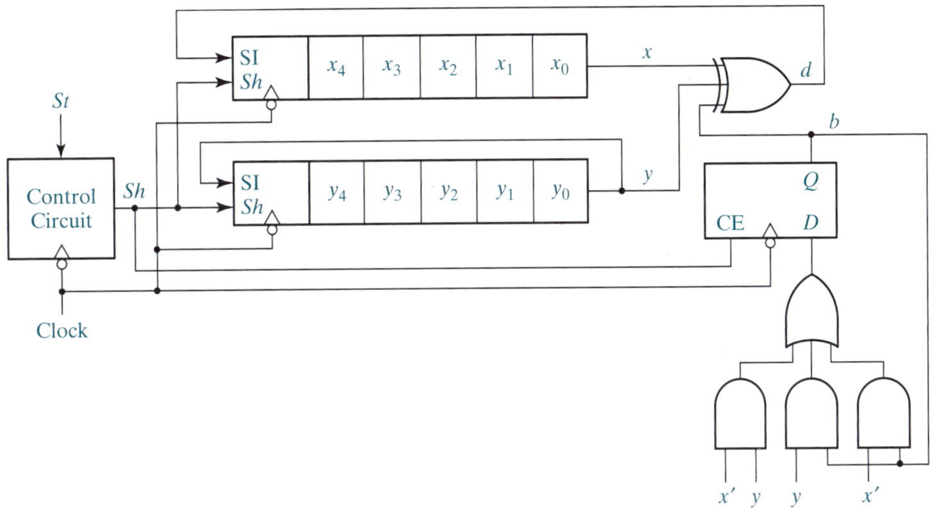

18.4

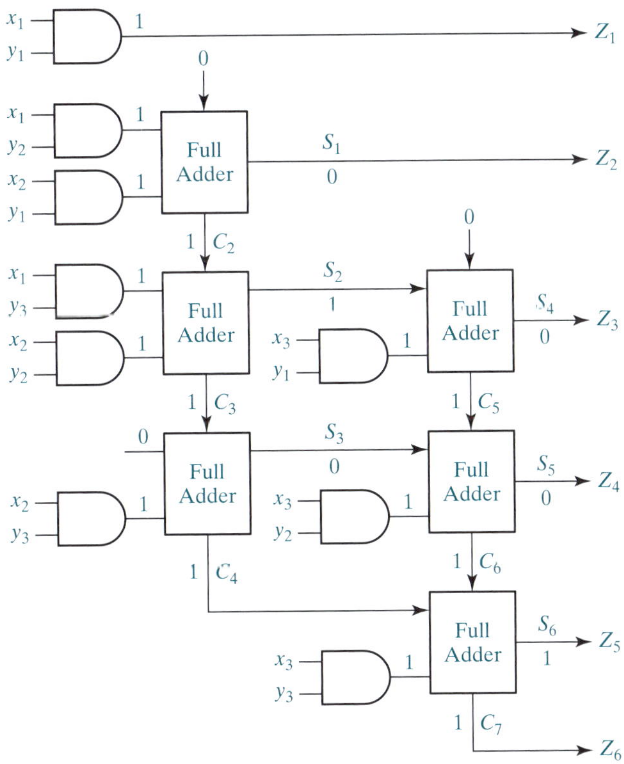

18.5

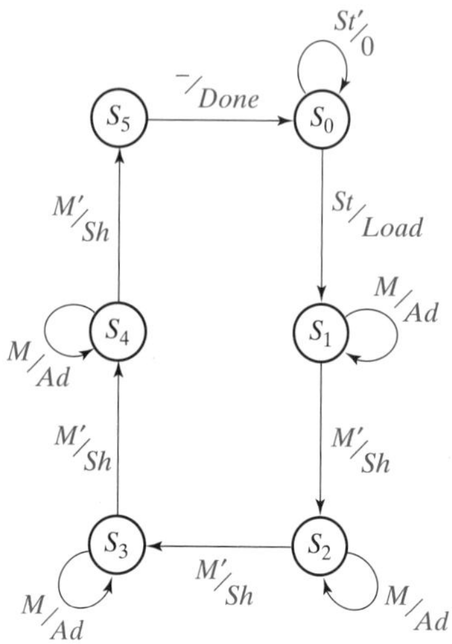

18.6

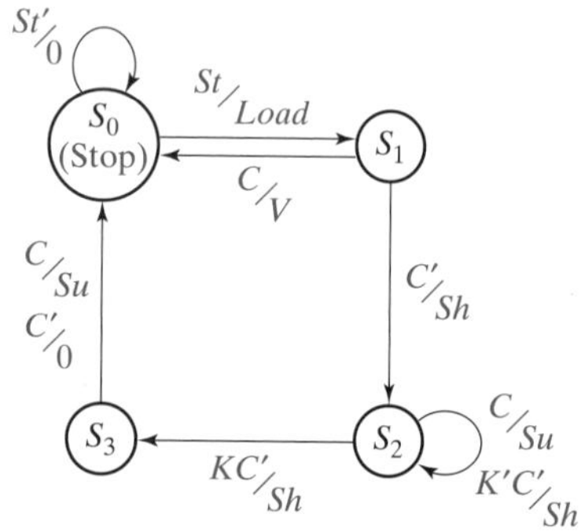

18.7 (a) $V = y_0'y_1'y_2'y_3'y_4' = (y_0 + y_1 + y_2 + y_3 + y_4)'$

(b)

										Sh	*Su*
(0	0	1	0	1)							
0	0	0	0	0	1	1	0	1	0	1	0
0	0	0	0	1	1	0	1	0	0	1	0
0	0	0	1	1	0	1	0	0	0	1	0
0	0	1	1	0	1	0	0	0	0	0	1
0	0	0	0	1	1	0	0	0	1	1	0
0	0	0	1	1	0	0	0	1	0	1	0
0	0	1	1	0	0	0	1	0	0	0	1
0	0	0	0	1	0	0	1	0	1		

remainder = 1 quotient = 5

(c)

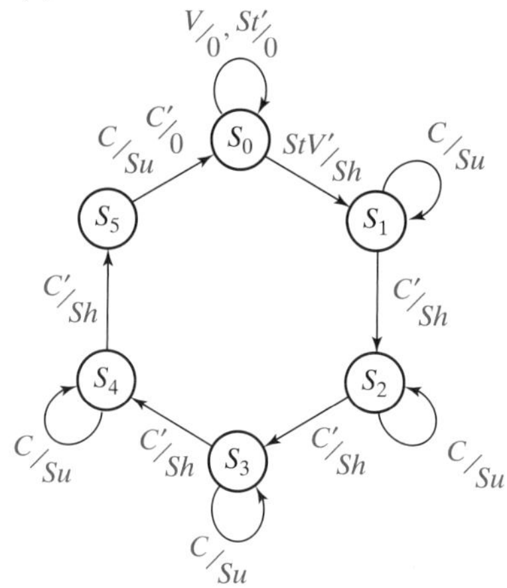

(d) After five shifts, the division is complete.

18.8 (a) $x_{in} = K_1'K_2'a' + K_1'K_2b + K_1K_2'(a \oplus b) + K_1K_2a$

$y_{in} = K_1'K_2'b + K_1'K_2a + K_1K_2' \cdot 0 + K_1K_2 \cdot 1$

(b) Use the state graph of Figure 18-6(b), with nine states total.

(c)

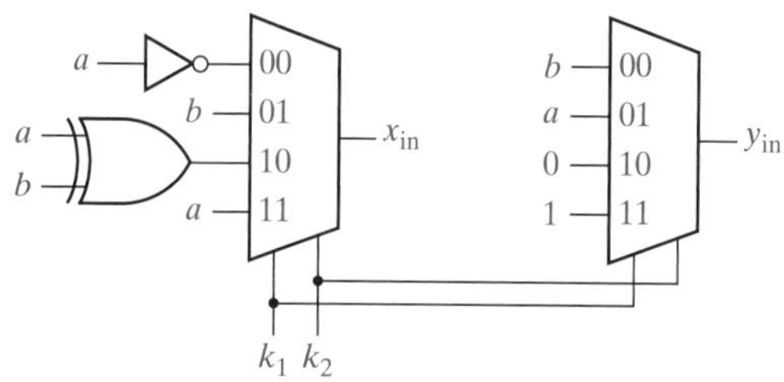

UNIT 19 Study Guide Answers

1. (b) Z_1, Z_2, Z_4 (for both charts) (d)

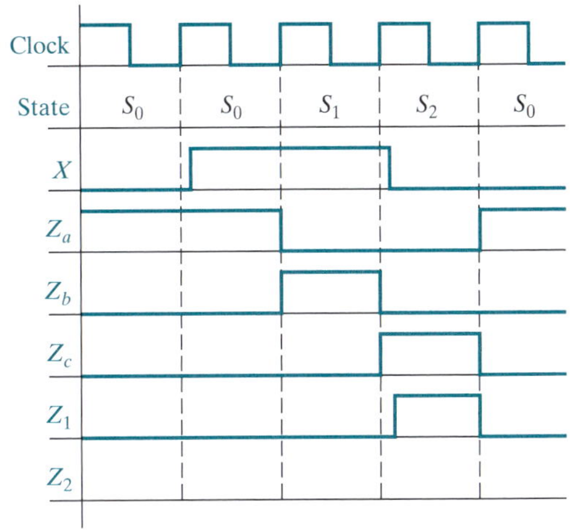

2. (a)

0	0	0	0	1	1	1	0	0	$C = 0$, *Sh*
0	0	0	1	1	1	0	0	0	$C = 0$, *Sh*
0	0	1	1	1	0	0	0	0	$C = 1$, *Su*
0	0	0	1	0	0	0	0	1	$C = 0$, *Sh*
0	0	1	0	0	0	0	1	0	$C = 0$, *Sh*
0	1	0	0	0	0	1	0	0	$C = 1$, *Su*
0	0	0	1	1	0	1	0	1	(result)

3. (a) $A^+ = BX \quad B^+ = A'X + BX$

UNIT 19 Answers to Problems

19.1

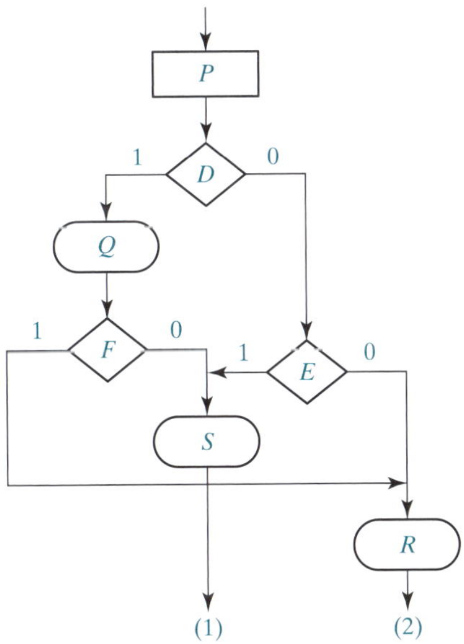

19.2

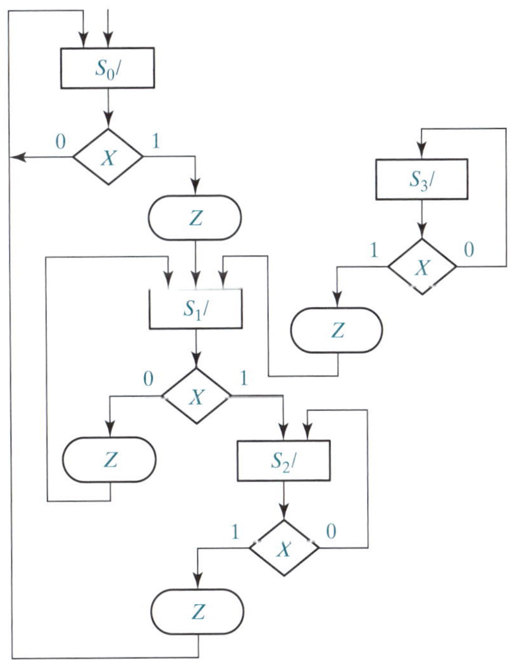

19.3

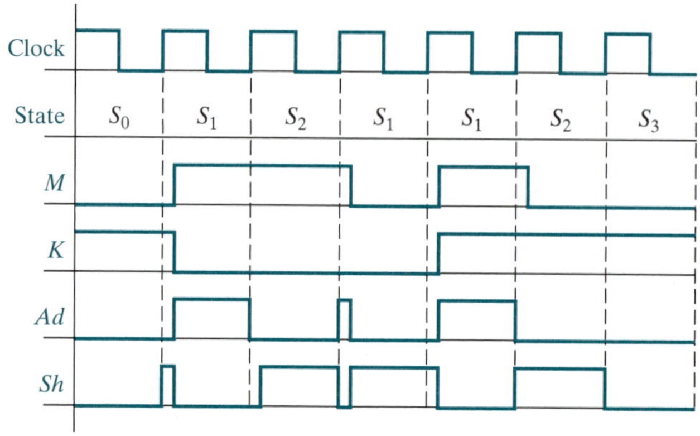
Clock
State
S_0
S_1
S_2
S_1
S_1
S_2
S_3
M
K
Ad
Sh

19.4

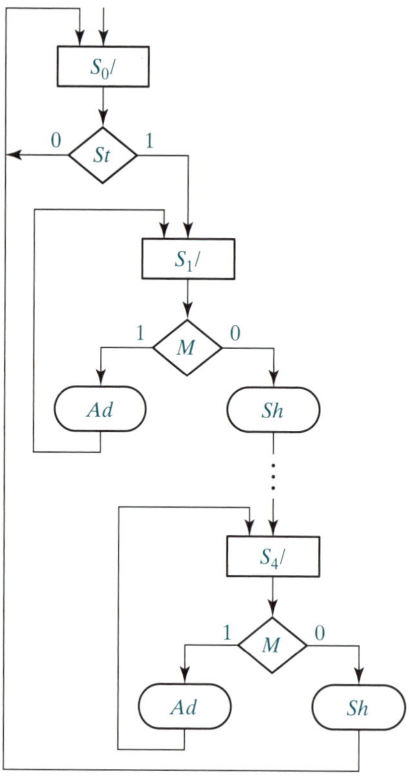
$S_0/$
0
St
1
$S_1/$
1
M
0
Ad
Sh
$S_4/$
1
M
0
Ad
Sh

19.5

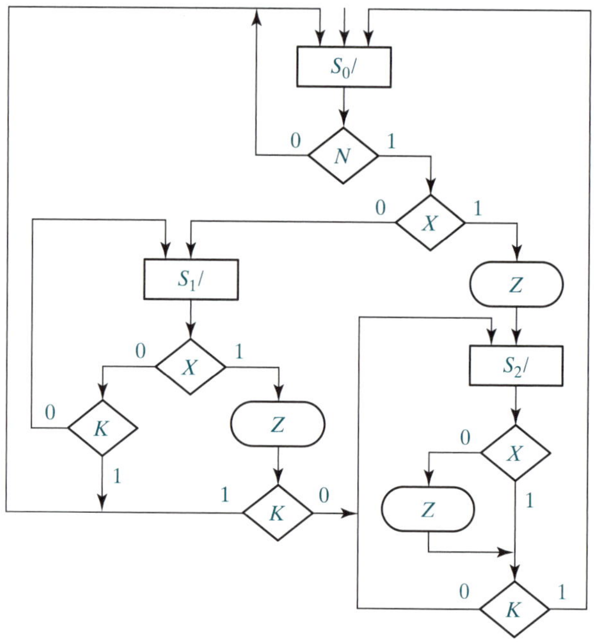
$S_0/$
0
N
1
0
X
1
$S_1/$
Z
0
X
1
$S_2/$
0
K
Z
0
X
1
1
K
0
Z
1
0
K
1

19.6

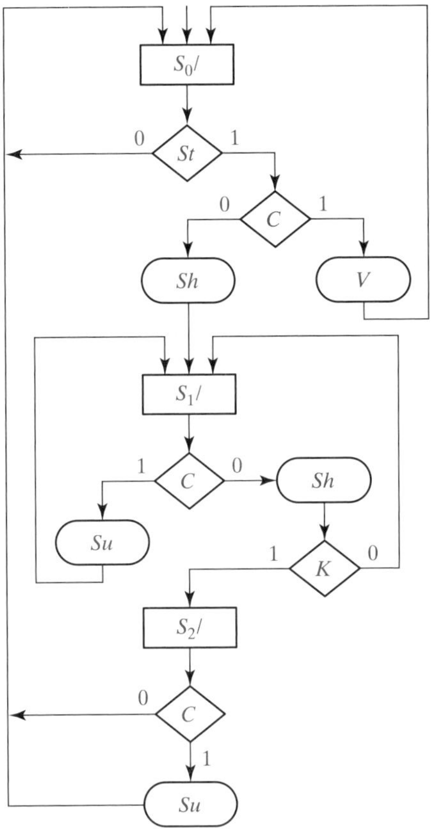

19.7 (a) $Q_0^+ = Q_0'Q_1'Q_2C' + Q_0Q_1' + Q_0Q_2'$

$Q_1^+ = Q_0Q_1'Q_2C' + Q_0Q_1Q_2'$

$Q_2^+ = Q_0'Q_1'Q_2'St + Q_0Q_2'C' + Q_0Q_1'Q_2C$

$Load = Q_0'Q_1'Q_2'St,\ Su = Q_0C,\ Sh = Q_1'Q_2C' + Q_0Q_2'C',\ V = Q_0'Q_1'Q_2C$

(These equations could be further simplified using don't-cares.)

(b) $Q_0^+ = Q_0'Q_1M + Q_0'Q_1M'K + Q_0Q_1K$

$Q_1^+ = Q_0'Q_1'St + Q_0'Q_1M'K' + Q_0'Q_1M + Q_0Q_1K'$

$Load = Q_0'Q_1'St,\ Sh = Q_0Q_1 + Q_0'Q_1M',\ Ad = Q_0'Q_1M,\ \text{Done} = Q_0Q_1'$

19.8 (a) $A^+ = A'B'C \cdot Rb'D_{711}'D_{2312}' + AB'C' + AB' \cdot Rb + AB'Eq'D_7'$

$B^+ = A'B'C \cdot D_{711} + A'B \cdot Reset' + AB'C \cdot Rb'Eq'D_7'$

$C^+ = A'B'Rb + A'BC \cdot Reset' + B'C'Rb + AB'C \cdot Rb'Eq'D_7$

$Roll = B'C \cdot Rb\ Sp = A'B'C \cdot Rb'D_{711}'D_{2312}'$

$Win = A'BC'\ \ Lose = A'BC$

(These equations could be further simplified using don't-cares.)

(b) If the input from the adder is $S_3S_2S_1S_0$, then the equations realized by the test logic block are

$D_7 = S_2S_1S_0 \quad D_{711} = S_1S_0(S_2 + S_3) \quad D_{2312} = S_3'S_2' + S_3S_2$

19.9 (a) $A^+ = BX \qquad Z_a = A'B' \qquad Z_1 = ABX'$

$B^+ = A'X + BX \qquad Z_b = A'B \qquad Z_2 = ABX$

$Z_c = AB$

X	A	B	A^+	B^+	Z_a	Z_b	Z_c	Z_1	Z_2
1	–	1	1	1	0	0	0	0	0
1	0	–	0	1	0	0	0	0	0
–	0	0	0	0	1	0	0	0	0
–	0	1	0	0	0	1	0	0	0
–	1	1	0	0	0	0	1	0	0
0	1	1	0	0	0	0	0	1	0
1	1	1	0	0	0	0	0	0	1

19.10 (a)

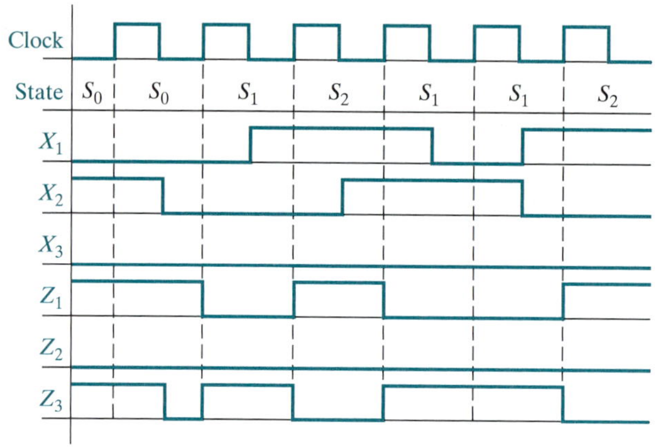

(b) $D_1 = Q_1'X_1'X_2'X_3 + Q_2X_2'$ $D_2 = Q_2'X_1 + Q_1'Q_2'X_2'X_3' + Q_2X_2$

$Z_1 = Q_2'X_1' + Q_1$ $Z_2 = Q_1'Q_2'X_1$ $Z_3 = Q_1'X_1'X_2 + Q_2$

(c)

Q_1	Q_2	X_1	X_2	X_3	D_1	D_2	Z_1	Z_2	Z_3
0	0	1	–	–	0	1	0	1	0
0	0	0	1	–	0	0	1	0	1
0	0	0	0	1	1	0	1	0	0
0	0	0	0	0	0	1	1	0	0
0	1	–	0	–	1	0	0	0	1
0	1	–	1	–	0	1	0	0	1
1	0	0	–	–	0	0	1	0	0
1	0	1	–	–	0	1	1	0	0

(d) $2^5 \times 5$ ROM

Q_1	Q_2	X_1	X_2	X_3	D_1	D_2	Z_1	Z_2	Z_3
0	0	0	0	0	0	1	1	0	0
0	0	0	0	1	1	0	1	0	0
0	0	0	1	0	0	0	1	0	1
0	0	0	1	1	0	0	1	0	1
0	0	1	0	0	0	1	0	1	0

UNIT 20 Study Guide Answers

1. (a) lines 15 and 16
   The full adder is combinational logic.
   Lines 34 and 35, which are in clocked process because it is a clocked register
   (b) In line 31, change clk = ‘0’ to clk = ‘1’.
2. (a) So we can use the overloaded “+” operator
   The change from state 9 to state 0
   To make the result be 5 bits
   Lines 35 and 36 will execute when State is 2, 4, 6, or 8.
   (b) ACC is uninitialized and is not loaded until St = ‘1’ at a rising clock edge.
   When Done = 1, i.e., in state 9, 160–180 ns.
   (c) $X = 101111001$, 60 ns
   (d) Mcand = 1101, Mplier = 1011, and product is 10001111 = 143
   (e) Line 19
   To avoid having to set them to 0 in each case where they are not 1. When they are set to 1, it overrides line 22 because these are sequential statements.
   ACC <= “00000” & Mplier;
   Because it is a clocked register that is updated on the rising clock edge.
   The process executes on the rising clock edge, and when state is 9 at the rising clock edge, it is too late; the state is about to change to 0.
   The process of lines 20–34 is not clocked; it executes when State changes to 9.
   (f) Whenever the value of count changes.
   Lines 51 and 52.
   10 ns + Δ.
   Sequential statements execute in 0 time, so A and B update simultaneously.
   (g) At time 60 ns, we are in state 2 when K = 0, so Sh = 1. So A = $00B_{16} = 000001011_2$ shifts to the right to become $005_{16} = 000000101_2$. At time 140 ns, we are in state 1 and M = 1, so Ad = 1. So we add the multipicand, 000001011_2, to A = $006_{16} = 000000110_2$ to get $011_{16} = 000010001_2$.
3. (a) 0; 1
   C should be 1 iff we can subtract, i.e., Dividend(8 **downto** 4) > Divisor.

UNIT 20 Answers to Problems

20.1 First process executes at t = 2 ns. Lines 22–25 execute.
Second process executes at t = 10 ns. Lines 38–40 and 43 execute.
Because the state changes, first process executes again at 10 + Δ ns. Lines 22–23 and lines 27–30 execute.

20.2

```
entity complementer is
  Port (clk, n: in std_logic;
  Regout: out std_logic_vector(15 downto 0));
end complementer;
architecture Behavioral of complementer is
signal State, NextState: integer range 0 to 2 := 0;
signal count: std_logic_vector(3 downto 0) := "0000"; --4-bit counter
```

```
signal X, Z, Sh: std_logic;
signal K: std_logic := '0';
signal Reg: std_logic_vector(15 downto 0);
begin
    Regout <= Reg;
    X <= Reg(0);
    K <= '1' when count = "1111" else '0';
    process (State, X, N, K)
    begin
        case State is
        when 0 =>
            if N = '0' then NextState <= 0; Sh <= '0'; Z <= '0';
            elsif X = '1' then NextState <= 2; Sh <= '1'; Z <= '1';
        else NextState <= 1; Sh <= '1'; Z <= '0'; end if;
      when 1 => Sh <= '1';
        if K = '1' then NextState <= 0;
            if X = '1' then Z <= '1';
            else Z <= '0'; end if;
        elsif X = '0' then NextState <= 1; Z <= '0';
        else NextState <= 2; Z <= '1'; end if;
      when 2 => Sh <= '1';
         if K = '1' then NextState <= 0;
            if X = '1' then Z <= '0';
            else Z <= '1'; end if;
         elsif X = '0' then NextState <= 2; Z <= '1';
         else NextState <= 2; Z <= '0'; end if;
      end case;
    end process;
    process (clk)
    begin
        if clk'event and clk = '1' then
            if Sh = '1' then Reg <= Z & Reg(15 downto 1);
            count <= count + 1; end if;
            State <= NextState;
        end if;
    end process;
end Behavioral;
```

20.4

```
entity test is
end test;
architecture Behavioral of test is
component sm17_2 is
  Port (x,clk: in std_logic;
        z: out std_logic);
end component;
constant N: integer:= 40;
```

```
signal flag: std_logic:= '0';
signal clk: std_logic:= '1';
signal x,z: std_logic;
constant x_seq: std_logic_vector(1 to 40) :=
                         ("0000100001001100001010100110111000011001");
constant z_seq: std_logic_vector(1 to 40) :=
                         ("1100001010100110111000011001010111010011");
begin
  sm1: sm17_2 port map(x,clk,z);
  clk <= not clk after 10 ns;   -- clock has 20 ns period
  process
  begin
    for i in 1 to N loop
      x <= x_seq(i);
      wait for 5 ns;   -- wait for z to become stable
      if z = z_seq(i) then flag <= '0'; else flag <= '1'; end if;
      wait until clk'event and clk = '1';
      wait for 5 ns;
    end loop;
  end process;
end Behavioral;
```

20.5

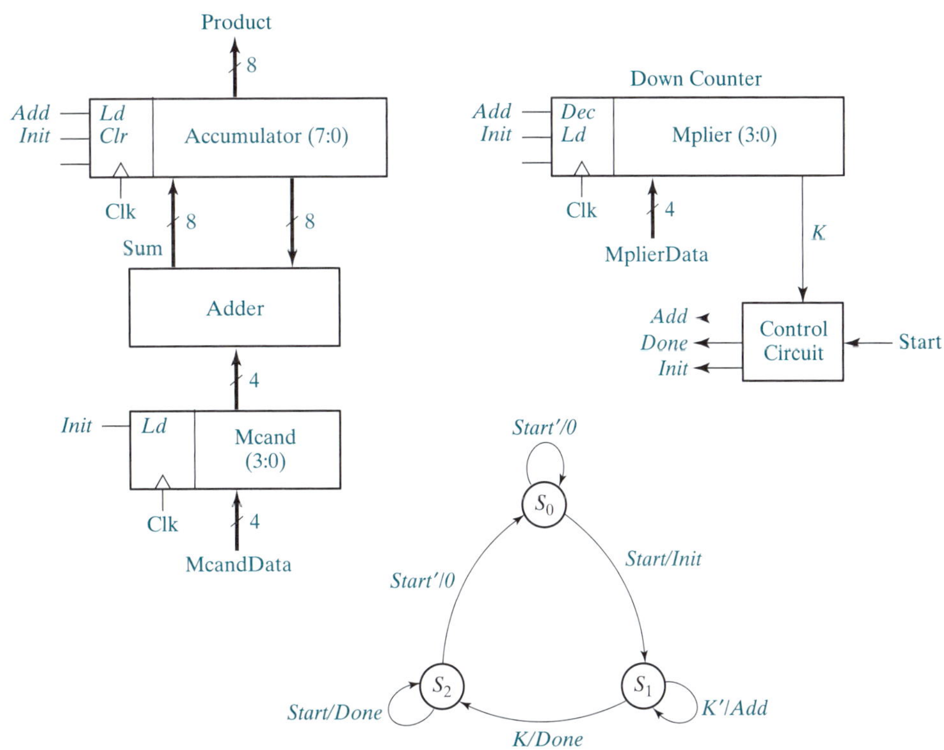

20.6

```
entity prob20_6 is
   Port (st, clk: in std_logic;
         X: in std_logic_vector(7 downto 0);
         Z: out std_logic_vector(9 downto 0));
end prob20_6;
architecture Behavioral of prob20_6 is
   signal State, NextState: integer range 0 to 3 := 0;
   signal lda, ldb, ldc, ad: std_logic;
   signal B, C: std_logic_vector(7 downto 0);
   signal A: std_logic_vector(9 downto 0);
   signal sumAB: std_logic_vector(8 downto 0);
   signal sumABC: std_logic_vector(9 downto 0);
begin
   sumAB <= ("0"&A(7 downto 0)) + B;
   sumABC <= ("0"&sumAB) + C;
   Z <= A;
process (st, State)
begin
   lda <= '0'; ldb <= '0'; ldc <= '0'; ad <= '0';
   case State is
      when 0 =>
         if st = '1' then lda <= '1'; NextState <= 1;
         else NextState <= 0; end if;
      when 1 =>
         ldb <= '1'; NextState <= 2;
      when 2 =>
         ldc <= '1'; NextState <= 3;
      when 3 =>
         ad <= '1'; lda <= '1'; NextState <= 0;
   end case;
end process;
process(clk)
begin
   if clk'event and clk = '1' then
      if lda = '1' then
         if ad = '1' then A <= sumABC;
         else A <= ("00" & X); end if;
      elsif ldb = '1' then B <= X;
      elsif ldc = '1' then C <= X;
      end if;
      State <= NextState;
   end if;
end process;
end Behavioral;
```

References

1. Ashenden Peter J. *The Designer's Guide to VHDL*, 3rd ed. San Francisco: Morgan Kaufmann Publishers, 2008.
2. Bhasker, J. *A Guide to VHDL Syntax.* Upper Saddle River, NJ: Prentice-Hall, 1995.
3. Bhasker, J. *VHDL Primer*, 3rd ed. Upper Saddle River, NJ: Prentice-Hall, 1999.
4. Brayton, Robert, et al. *Logic Minimization Algorithms for VLSI Synthesis*. Secaucus, NJ: Springer, 1984.
5. Givone, Donald D. *Digital Principles and Design*. New York: McGraw-Hill, 2003.
6. Katz, Randy H. and Gaetano Borriello. *Contemporary Logic Design*, 2nd ed. Upper Saddle River, NJ: Prentice Hall, 2004.
7. Mano, M. Morris and Michael D. Ciletti, *Digital Design*, 5th ed. Upper Saddle River, NJ: Prentice Hall, 2012.
8. Mano, M. Morris and Charles R. Kime. *Logic and Computer Design Fundamentals,* 4th ed. Old Tappan, NJ: Pearson Prentice Hall, 2008.
9. Marcovitz, Alan B. *Introduction to Logic Design*, 3rd ed. New York: McGraw-Hill, 2009.
10. McCluskey, Edward J. *Logic Design Principles*. Upper Saddle River, NJ: Prentice Hall, 1986.
11. Miczo, Alexander. *Digital Logic Testing and Simulation*, 2nd ed. New York: John Wiley & Sons, Ltd West Sussex, England, 2003.
12. Patt, Yale N. and Sanjay J. Patel. *Introduction to Computing Systems: From Bits and Gates to C and Beyond*, 3rd ed. New York: McGraw-Hill, 2013.
13. Roth, Charles H. Jr. and Lizy Kurian John. *Digital Systems Design Using VHDL*, 2nd ed. Toronto, Ontario: Cengage Learning, 2008.
14. Rushton, Andrew. *VHDL for Logic Synthesis*, 3rd ed. West Sussex, England: John Wiley & Sons, Ltd, 2011.
15. Wakerly, John F. *Digital Design Principles & Practices*, 4th ed. Upper Saddle River, NJ: Prentice Hall, 2006.
16. Weste, Neil and David Harris. *CMOS VLSI Design: A Circuits and Systems Perspective*, 4th ed. Boston: Addison-Wesley, 2010.

Index

Description of the CD

The CD that accompanies this text contains three programs that are useful in the computer-aided design and simulation of digital logic—LogicAid, SimUaid, and DirectVHDL-PE. Principal features of these programs are listed below. User manuals for LogicAid and SimUaid are provided on the CD in PDF format. The user manuals for DirectVHDL, which are provided in the form of HTML help files, will be installed when you run setup from the DirectVHDL directory. The CD also contains copies of the VHDL code found in Units 10, 17, and 20 of the text. The software on the CD is compatible with Windows XP, Vista, Windows 7, and Windows 8.

The software may be installed in computer labs on any campus where the *Fundamentals of Logic Design* text is assigned as a required textbook.

LogicAid Features:

- Logic functions may be input in the following forms: sum-of-products, product-of-sums, truth table, PLA table, Karnaugh map, minterm or maxterm expansion
- Choice of logic simplification algorithms provides for finding a fast solution or all minimum solutions
- Sequential logic may be input as Mealy or Moore state tables, state graphs (with either binary or alphanumeric input/output), or SM charts
- Reduces state tables to a minimum number of rows and derives flip-flop input equations for D, T, J-K, and S-R flip-flops
- Creates JEDEC files for programming 22V10 PALs
- Tutorial aids include Karnaugh map tutor, state table checker, and partial graph checker

SimUaid Features:

- Friendly user interface allows easy placement and wiring of components
- Available devices include basic gates, flip-flops, switches, probes, registers, counters, adders, multiplexers, decoders, 7-segment indicators, clocks, tri-state buffers, and state machines
- Four-valued logic simulation (0, 1, X, Z)
- Displays all device inputs and outputs for ease of signal tracing and debugging
- Probe placement automatically sets up waveform display
- Live simulation mode allows immediate observation of response to input switch changes
- Synchronous simulation allows stepping one clock period at a time
- Asynchronous simulation allows stepping until a signal changes
- Converts a circuit diagram to synthesizable VHDL code

DirectVHDL Features:

- Edits, compiles, and simulates VHDL code
- Easy to learn user interface
- VHDL editor highlights syntax errors as you type
- Simulator displays waveforms and listing output
- Command interface allows forcing input values interactively or from a command file
- Compatible with IEEE Standard 1076-1993 VHDL